Instructor's Manual for
APPLIED CALCULUS
Third Edition

and

CALCULUS
FOR MANAGEMENT, SOCIAL AND LIFE SCIENCES
Third Edition

by Berkey

FRED WRIGHT
Iowa State University

SAUNDERS COLLEGE PUBLISHING
Harcourt Brace College Publishers
Fort Worth Philadelphia San Diego New York
Orlando Austin San Antonio Toronto
Montreal London Sydney Tokyo

Printed in the United States of America.

Fred Wright: Instructor's Manual to accompany APPLIED CALCULUS, 3/E and CALCULUS FOR MANAGEMENT, SOCIAL AND LIFE SCIENCES, 3/E

ISBN 0-03-076174-3

456 017 987654321

Contents

(Numbers in parentheses refer to the order of chapters in *Calculus for Management, Social and Life Science, 3/e.*)

Preface

This *Instructor's Manual* is a supplement for the third editions of Berkey's *Calculus for Management, Social and Life Sciences* and *Applied Calculus*. It contains detailed solutions to all the exercises in the text (both section exercises and review exercises) to assist you in the classroom and in grading assignments. The answers to all the odd-numbered section and review exercises also appear at the back of the text. Below is an overview of the entire supplements package to adopters of these texts summarizing the features and uses for each ancillary item.

Supplements

The set of supplements for *Calculus for Management, Social, and Life Sciences, Third Edition,* and *Applied Calculus, Third Edition,* consists of the following items: Instructor's Manual, Student Solutions Manual, Test Bank, Computerized Test Banks, Graph 2D/3D Software, and Visualization of Calculus Concepts for the TI-85, TI-81, SHARP EL9300, and CASIO fx-7700G. Each item is described below:

Instructor's Manual

* Solutions to all exercises in the text

Student Solutions Manual

* Review material for each chapter
* Solutions for every odd-numbered exercise in the text (both section and review exercises)
* 13-20 practice problems per chapter with solutions at the end of the chapter

Test Bank

* 5 tests of 20 questions each for each chapter
* 3 tests per chapter are open-ended, 2 are multiple-choice
* 2 final examinations (open-ended and multiple-choice)
* Answers to all questions are included

Computerized Test Banks

* Available for IBM® and Macintosh® computers
* Mixture of open-ended and multiple-choice questions for each chapter
* Can preview, select, edit, add and delete test items to tailor test to course or select items at random
* Can add or edit graphics (IBM® version)

* Option to arrange tests by chapter or from several chapters
* Option to print separate answer key and student answer sheet
* Can print up to 99 different versions of the same test and answer sheet
* ESAGradebook™ software is available with the IBM® version

Graph 2D/3D Software

* Available for IBM® and Macintosh® computers
* Full-featured graphing package for functions of 1 and 2 variables
* Function Graphing module allows user to graph 1-4 functions simultaneously, zoom in and out, display tables of function values, trace values, and save and retrieve set-ups via data files
* Surface Graphing module allows user to control view via rotating or raising/lowering surfaces, compute function values for any x, y values, save and retrieve data files

Visualization of Calculus Concepts for the TI-85, TI-81, SHARP EL9300, and CASIO fx-7700G

* Helps students become proficient in the use of graphics calculators and enhances their understanding of calculus concepts
* Includes programs for each graphics calculator on key calculus concepts, such as limits, derivatives, and integrals
* Over 300 exercises similar to ones in the text

If you have any comments or suggestions about this *Student Solutions Manual and Study Guide*, please address your correspondence to: Mathematics Editor, Saunders College Publishing, The Public Ledger Building, Suite 560, 620 Chestnut Street, Philadelphia, PA 19106.

CHAPTER 1

Solutions to Exercise Set 1.1

1. Any real number
2. Any real number
3. Any real number
4. Integers
5. Integers
6. Any real number
7. Integers
8. Any real number
9. Any real number
10. Any real number
11. d
12. h
13. a
14. i
15. b
16. c
17. g
18. j
19. f
20. e
21. True
22. False. There is no largest real number.
23. Two, $m = -1$ or $m = 1$
24. $2x > 10, \quad x > 5, \quad (5, \infty)$
25. $6x - 2 \le 16, \quad 6x \le 18, \quad x \le 3, \quad (-\infty, 3]$

26. $6 - 5x \geq 16, \quad -5x \geq 10, \quad x \leq -2, \quad (-\infty, -2]$
27. $4 - 2x \leq 12, \quad -2x \leq 8, \quad x \geq -4, \quad [-4, \infty)$
28. $4 - 3x > 5, \quad -3x > 1, \quad x < -\frac{1}{3}, \quad (-\infty, -\frac{1}{3})$
29. $6 + x \leq 2x - 5, \quad -x \leq -11, \quad x \geq 11, \quad [11, \infty)$
30. $x - 7 \geq 3 - x, \quad 2x \geq 10, \quad x \geq 5, \quad [5, \infty)$
31. $5 - 3x < 7 - 2x, \quad -x < 2, \quad x > -2, \quad (-2, \infty)$
32. $(x-4)^2 \geq x^2 - x + 12, \quad x^2 - 8x + 16 \geq x^2 - x + 12, \quad -7x \geq -4, \quad x \leq \frac{4}{7}, \quad (-\infty, \frac{4}{7}]$
33. $x(6+x) \geq 3 + x^2, \quad 6x + x^2 \geq 3 + x^2, \quad 6x \geq 3, \quad x \geq \frac{1}{2}, \quad [\frac{1}{2}, \infty)$
34. $28 + g \leq 108, \quad g \leq 80$
35. $200P + 250B \leq 10,000$
36. $250B \leq 10,000 - 200P, \quad 250B \leq 10,000 - 200(15), \quad 250B \leq 10,000 - 3000, \quad 250B \leq 7000, \quad B \leq 28$
37. Let x = number of guests. $12x + 50 \leq 300, \quad 12x \leq 250, \quad x \leq 20$
38. $2 \cdot 10 \cdot \pi r^2 + 30 \cdot 2\pi r \cdot h \leq 1000, \quad 20\pi r^2 + 60\pi rh \leq 1000$
39. $1296 < 54w < 1620, \quad 24 < w < 30$
40. x = number of \$10,000 cars he sells per month $\quad 900 + .02(10,000)(x) \geq 2000, \quad 200x \geq 1100, \quad x \geq 6$
41. a. $1100 + 0.02(20,000)(x) > 0.03(20,000)(x)$, $1100 + 400x > 600x$, $-200x > -1100$, $x < 6$

    b. $.03(20,000)(x) > 1100 + .02(20,000)(x), \quad 600x > 1100 + 400x, \quad 200x > 1100, \quad x \geq 6$
42. $0.04(20,000)(x) > 1100 + 0.02(20,000)(x)$, $800x > 1100 + 400x$, $\quad 400x > 1100, \quad x \geq 3$
43. Suppose 11 oranges weight 3 ounces each. Then the twelth orange could weigh at most 15 ounces.

Solutions to Exercise Set 1.2

1.

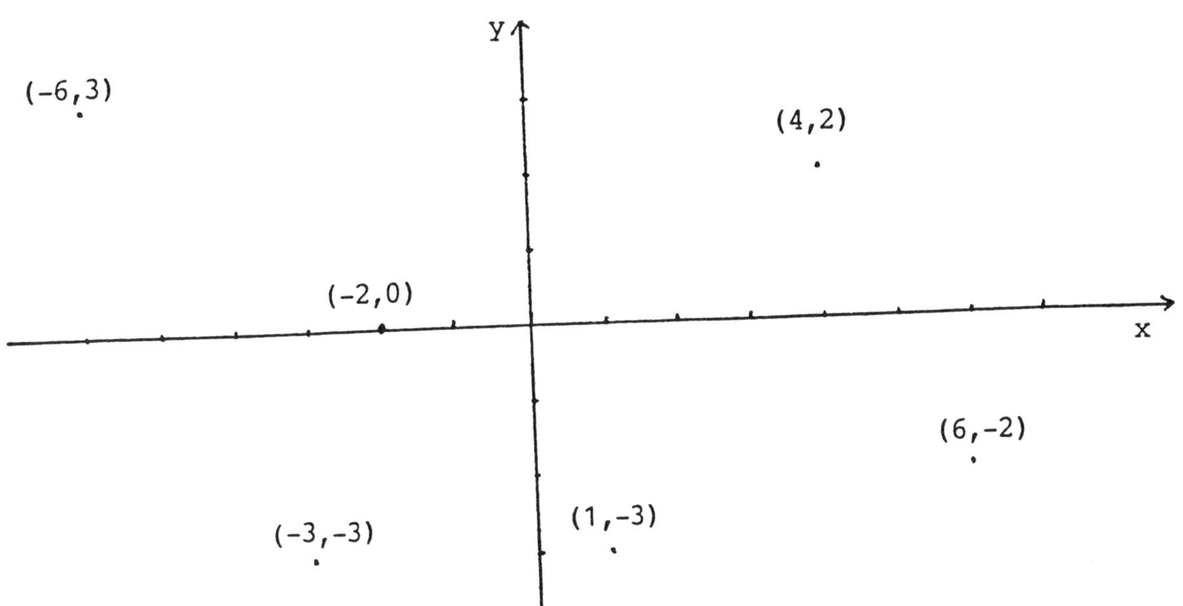

2.

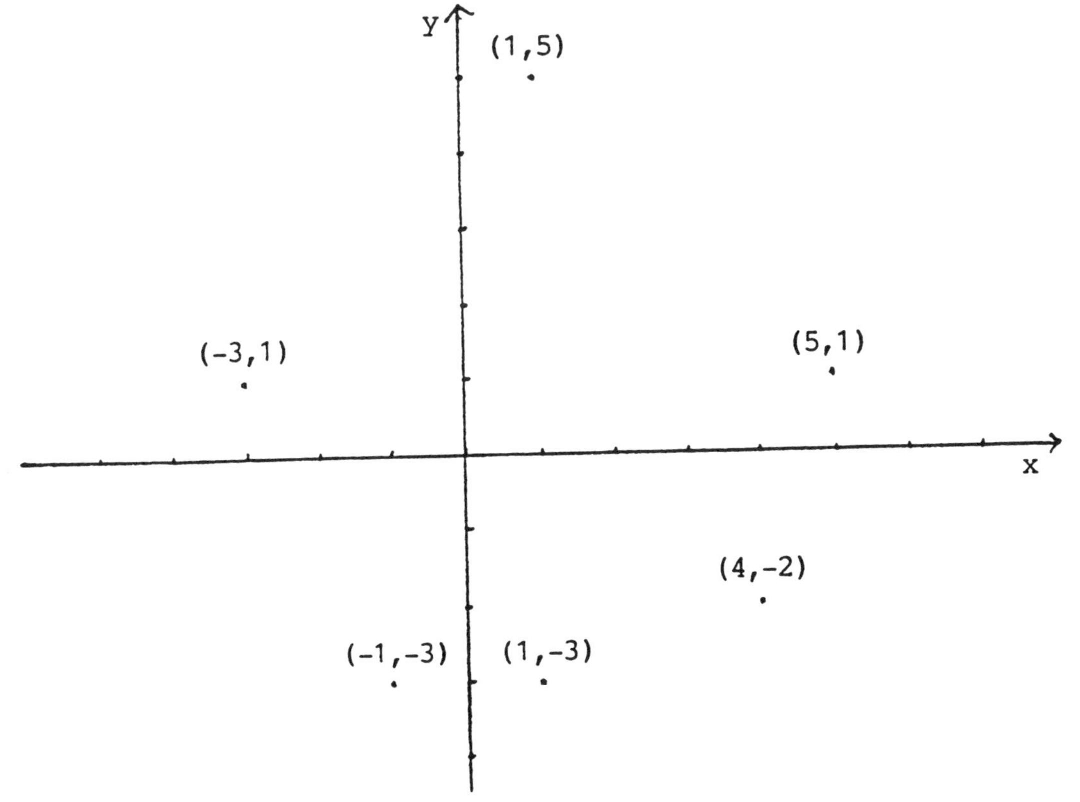

3. $3x - y = 5, \quad -y = 5 - 3x,\ y = 3x - 5$

x	-3	-2	-1	0	1	2	3
y	-14	-11	-8	-5	-2	1	4

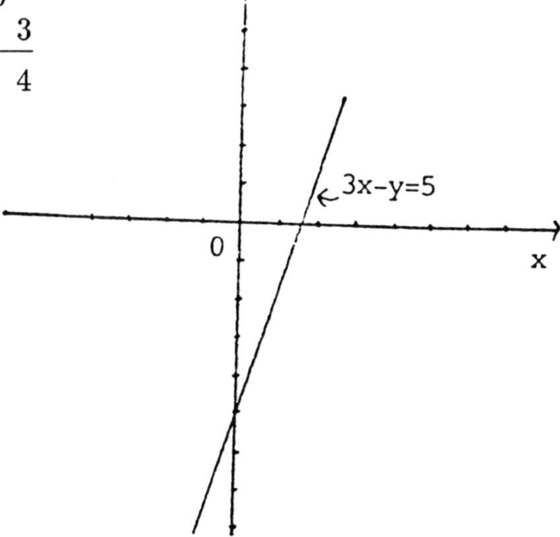

4. $y = x^2 - 3$

x	-3	-2	-1	0	1	2	3
y	6	1	-2	-3	-2	1	6

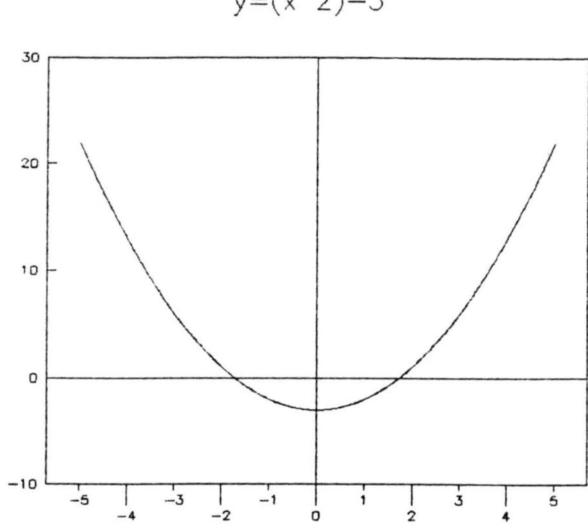

5. $y = \sqrt{x} + 2$

x	9	4	1	0
y	5	4	3	2

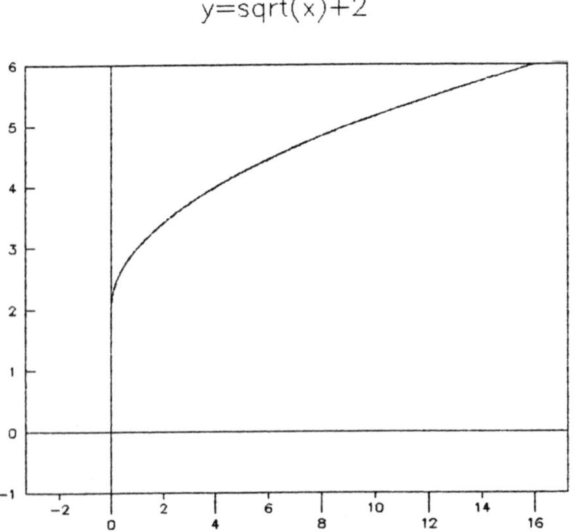

6. $y - x^2 = 4, \quad y = 4 + x^2$

x	-2	-1	0	1	2
y	8	5	4	5	8

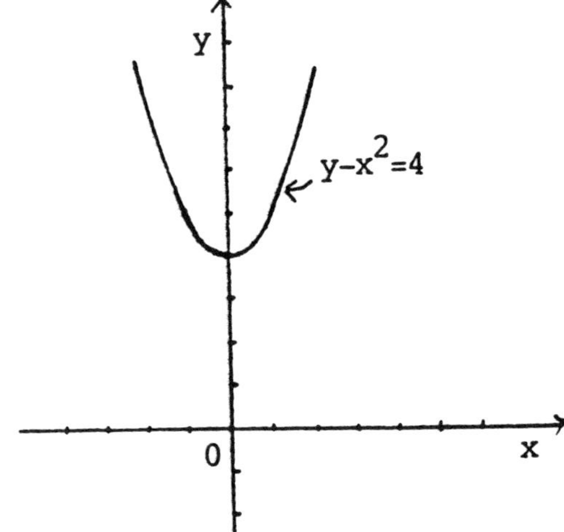

7. $x - y = 1, \quad -y = 1 - x, \quad y = x - 1$

x	-2	-1	0	1	2	3
y	-3	-2	-1	0	1	2

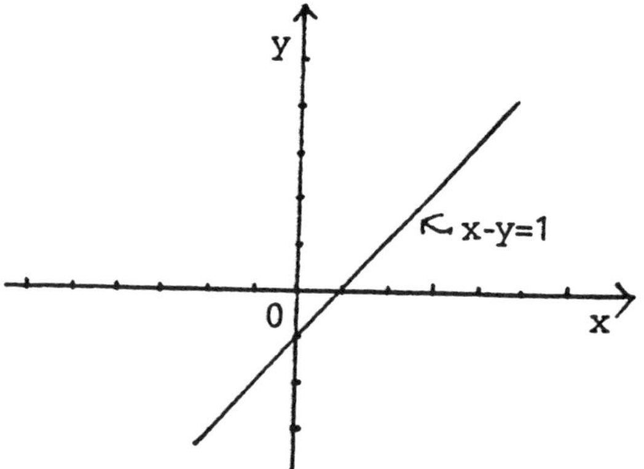

8. $x = y^2$, $y = \pm\sqrt{x}$

x	4	1	0	1	4
y	-2	-1	0	1	2

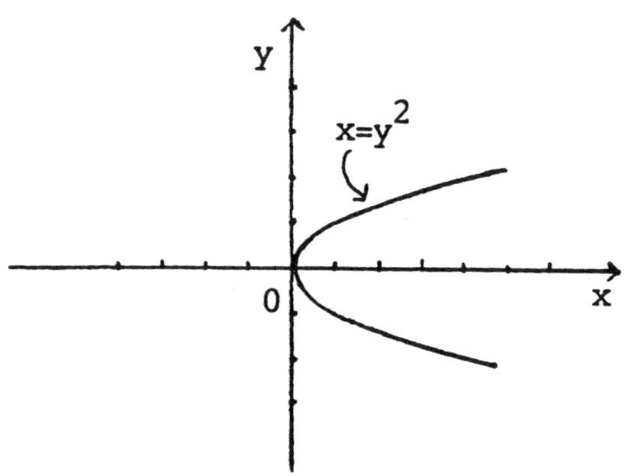

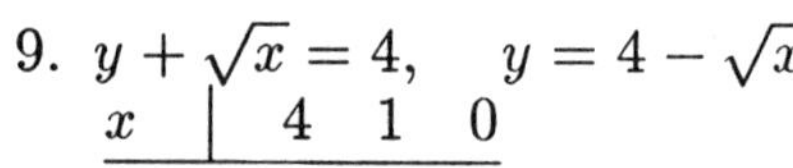

9. $y + \sqrt{x} = 4, \quad y = 4 - \sqrt{x}$

x	4	1	0
y	2	3	4

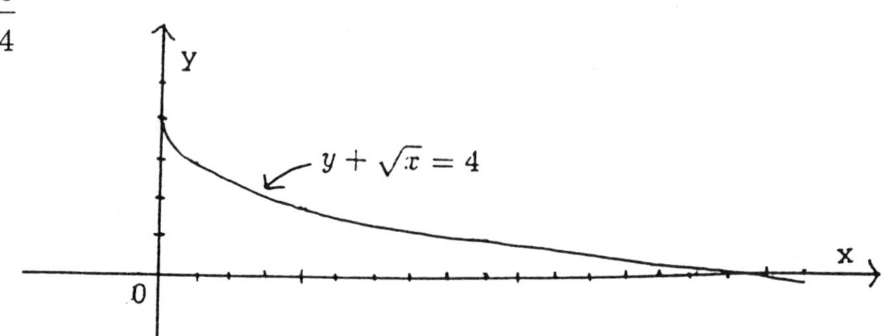

10. $x = \frac{4}{y} \implies y = \frac{4}{x}$

x	-2	-1	1	2
y	-2	-4	4	2

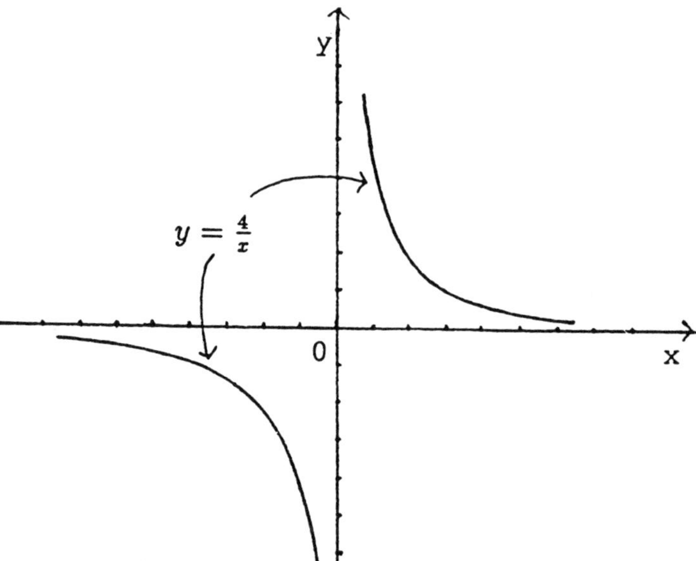

11. $y = \frac{1}{x-2}$

x	3	5	0	-5	1
y	1	$\frac{1}{3}$	$-\frac{1}{2}$	$-\frac{1}{7}$	-1

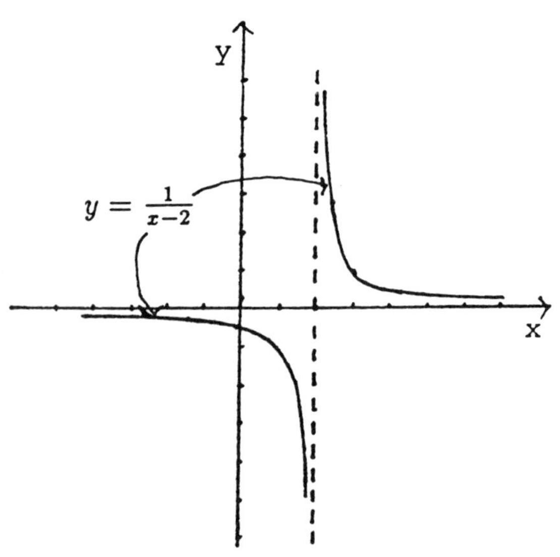

12. $2x^2 + y = 3$, $y = 3 - 2x^2$

x	-2	-1	0	1	2
y	-5	1	3	1	-5

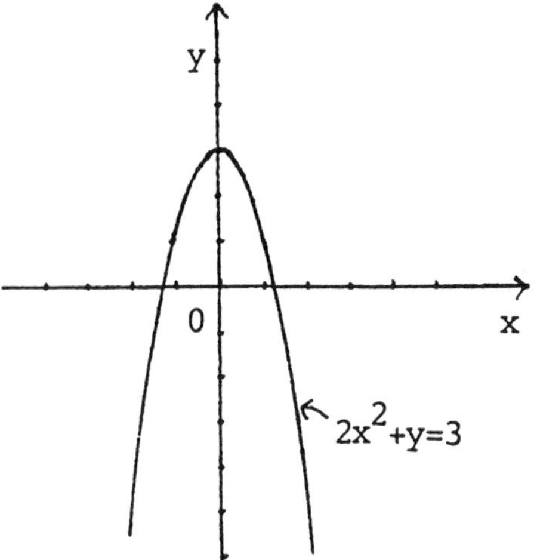

13. False, a vertical line does not have slope.

14. True

15. $\frac{a-4}{-2-2} = 3$, $\frac{a-4}{-4} = 3$, $a - 4 = -12$, $a = -8$

16. $\frac{-5-1}{1-b} = 2$, $\frac{-6}{1-b} = 2$, $-6 = 2 - 2b$, $-8 = -2b$, $b = 4$

17. $y = 4x - 2$

18. $y = -2x - 5$

19. $y = -5$

20. $m = \frac{2-6}{3-(-1)} = \frac{-4}{4} = -1$

    $y - 6 = (-1) \cdot (x - (-1))$, $y - 6 = -x - 1$, $y = -x + 5$

21. $y - 1 = (-4) \cdot (x - 4)$, $y - 1 = -4x + 16$, $y = -4x + 17$

22. $y - 3 = -3(x - 1)$, $y - 3 = -3x + 3$, $y = -3x + 6$

23. $m = \frac{6-0}{0-(-3)} = \frac{6}{3} = 2$

    $y = 2x + 6$

24. $x = -3$

25. $y = -3$

26. $m = \frac{8-4}{-6-(-2)} = \frac{4}{-4} = -1$

    $y - 4 = -1\,(x - (-2))$, $y - 4 = -x - 2$, $y = -x + 2$

27. $m = \frac{-4-2}{-1-0} = \frac{-6}{-1} = 6$

    $y - 2 = 6(x - 0)$, $y - 2 = 6x$, $y = 6x + 2$

28. Find the slope of the line $2x - 6y + 5 = 0$. $-6y = -2x - 5$, $y = \frac{1}{3}x + \frac{5}{6}$, slope $m_1 = \frac{1}{3}$.

Equation for requested line: $y - 4 = \frac{1}{3}(x - 1)$, $y - 4 = \frac{1}{3}x - \frac{1}{3}$, $y = \frac{1}{3}x + \frac{11}{3}$

29. Find the slope of the line $3x - 5y = 15$. $-5y = -3x + 15$, $y = \frac{3}{5}x - 3$.

The slope of this line is $m_1 = \frac{3}{5}$. The slope of the requested line is $m_2 = m_1 = \frac{3}{5}$.

Equation for requested line: $y - (-2) = \frac{3}{5}(x - 5)$, $5y + 10 = 3x - 15$, $3x - 5y = 25$.

30. Find the slope of the line $3x + y = 7$. $y = -3x + 7$, slope $m_1 = -3$.

Let m_2 = the slope of the requested line.

$m_1 m_2 = -1$, $-3 \cdot m_2 = -1$, $m_2 = \frac{1}{3}$.

Equation for the requested line: $y - 3 = \frac{1}{3}(x - 1)$, $y - 3 = \frac{1}{3}x - \frac{1}{3}$, $y = \frac{1}{3}x + \frac{8}{3}$

31. Slope $m_1 = \frac{9-5}{-1-(-2)} = \frac{4}{1} = 4$.

$m_1 m_2 = -1$, $4 \cdot m_2 = -1$, $m_2 = -\frac{1}{4}$

Equation for the requested line: $y - (-1) = -\frac{1}{4}(x - 4)$, $y + 1 = -\frac{1}{4}x + 1$, $y = -\frac{1}{4}x$

32. $x = 7 - y$, $-y = x - 7$, $y = -x + 7$,

$m = -1$, $b = 7$

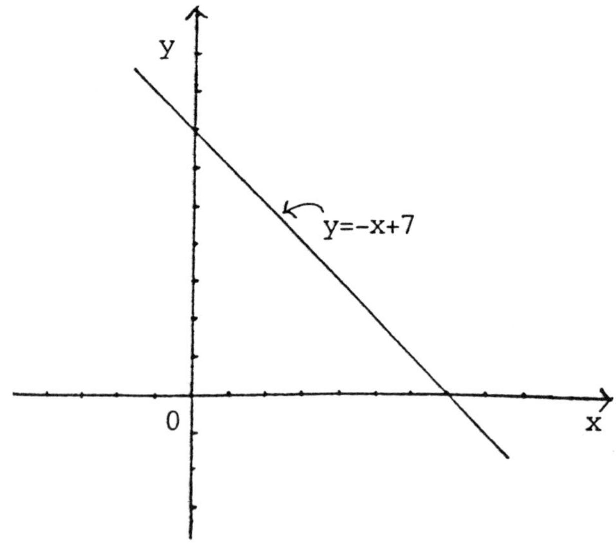

33. $3x - 5y = 6, \quad -5y = 6 - 3x, \quad y = \frac{3}{5}x - \frac{6}{5},$
$m = \frac{3}{5}, \quad b = -\frac{6}{5}$

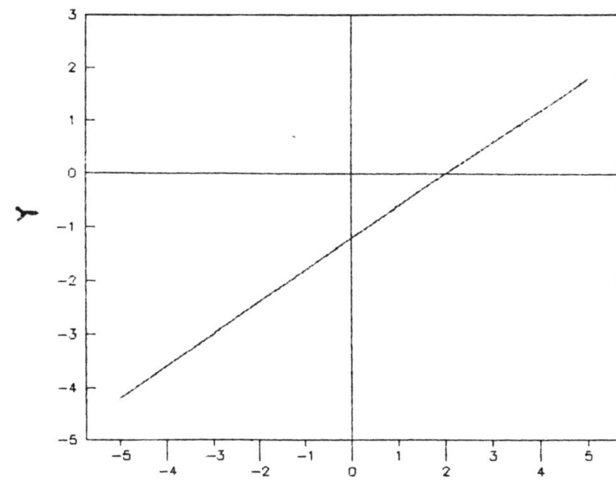

34. $x + y + 3 = 0, \quad y = -x - 3,$
$m = -1, \quad b = -3$

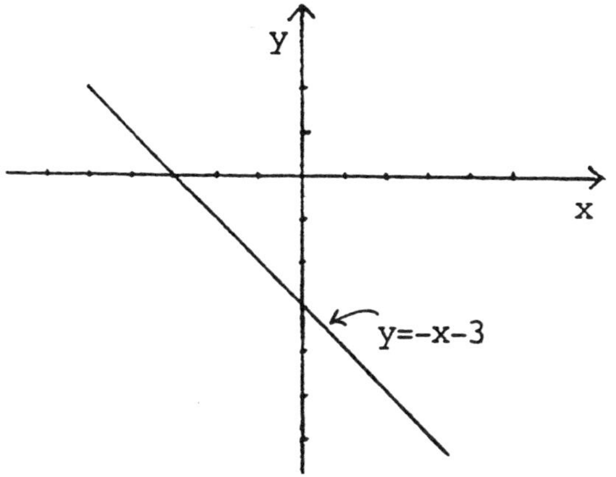

35. $2x = 10 - 3y, -3y = 2x - 10, \quad y = -\frac{2}{3}x + \frac{10}{3},$
$m = -\frac{2}{3}, \quad b = \frac{10}{3}$

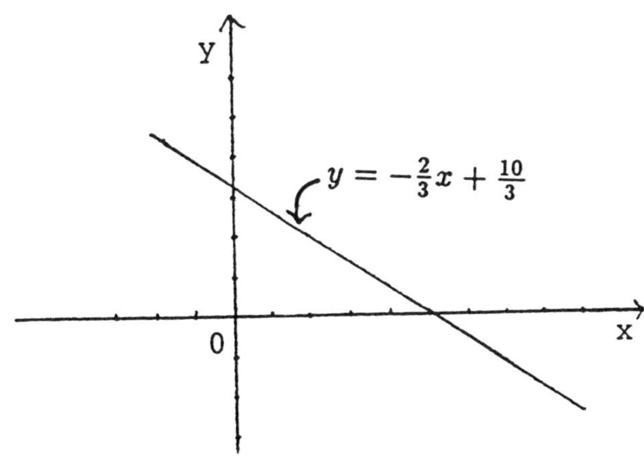

36. $y = 5$

$m = 0, \quad b = 5$

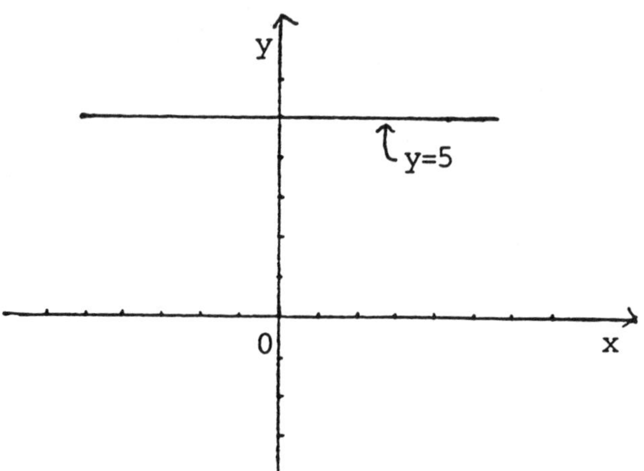

37. $y - 2x = 9, \quad y = 2x + 9$

$m = 2, \quad b = 9$

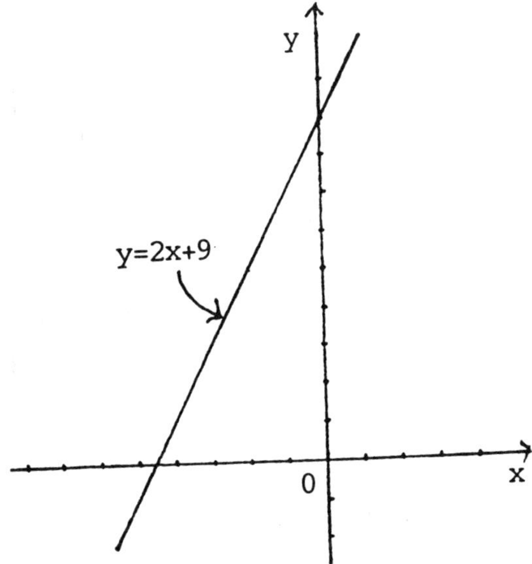

38. $x = -3$

There is no slope and y-intercept.

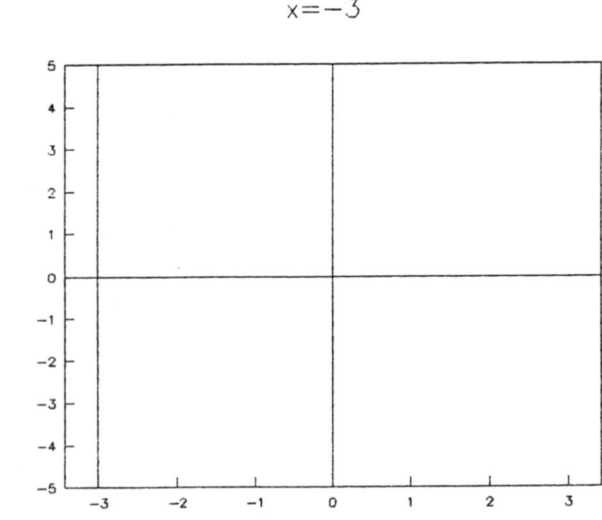

39. $y = x$

$m = 1, \quad b = 0$

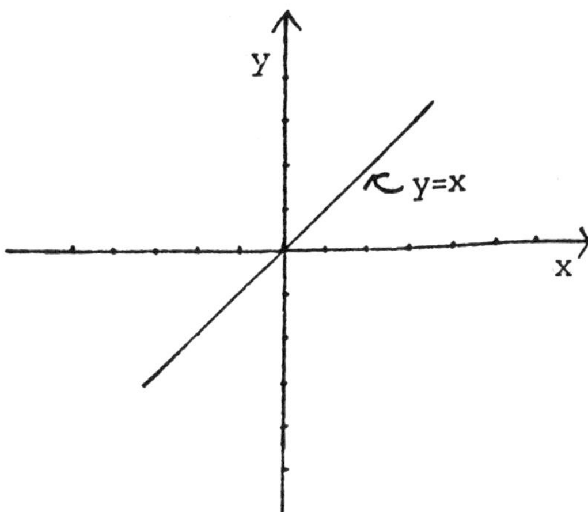

40. $3x - y = 9, \quad -y = 9 - 3x, \quad y = 3x - 9$

$m = 3, \quad b = -9$

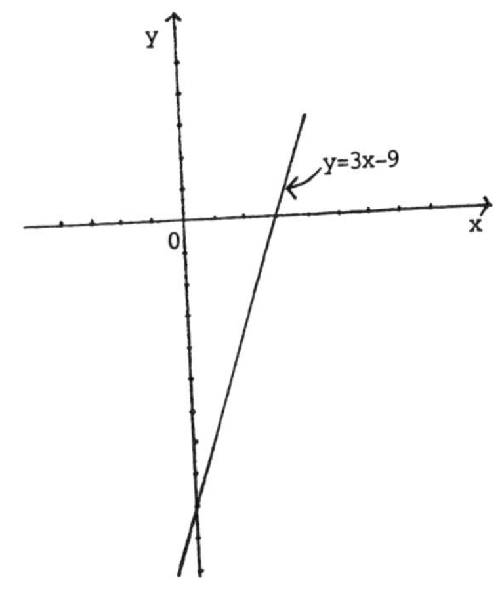

41. $9x - 5y = 45, \quad -5y = 45 - 9x, \quad y = \frac{9}{5}x - 9.$

$m = \frac{9}{5}, \quad b = -9.$

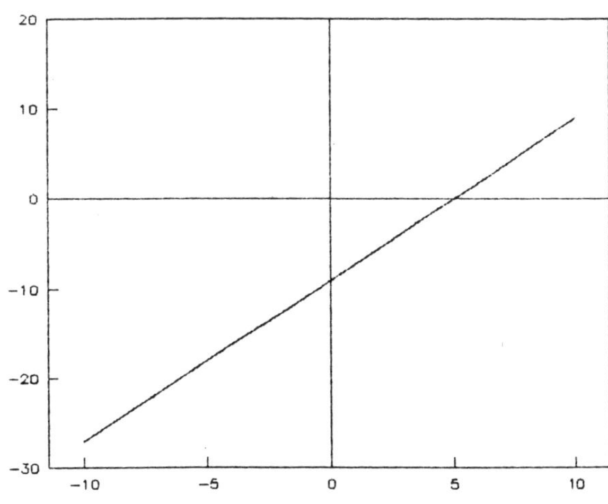

42. a. $T = 0.05(i - 5000)$

b. $T = 800, \quad 800 = 0.05(i - 5000), \quad 800 = 0.05i - 250, \quad 0.05i = 1050,$

$i = \$21,000$

43. a. $m = \frac{600-200}{100-20} = \frac{400}{80} = 5$

$200 = 5(20) + b, \quad 200 = 100 + b, \quad b = 100$

b. $C = 5(150) + 100, \quad C = \$850$

c. \$100/week

44. Let t = the number of years after 1986 and let y denote the corresponding world production of crude steel in billion metric tons.

Slope $m = \frac{770-700}{4} = 17.5$

$y - 700 = 17.5(t - 0) \implies y = 17.5t + 700.$

a. $t = 8 \implies y = 17.5(8) + 700 = 840.$

b. $t = 12 \implies y = 17.5(12) + 700 = 910.$

45. t = the number of years after 1967

Slope $m = \frac{293-100}{15} = \frac{193}{15} = 12.87.$

When $t = 0, \quad C = 100.$

a. $C = 12.87t + 100$

b. $C = 12.87(22) + 100 = 383.14$

46. Let t = the number of years after 1981 and let y denote the number of billion shares of stock traded on the New York Stock Exchange.

Slope $m = \frac{45-12}{10} = \frac{33}{10} = 3.3.$

$y - 12 = 3.3(t - 0) \implies y = 3.3t + 12$

a. $t = 15 \implies y = 3.3(15) + 12 = 61.5.$

b. $t = 20 \implies y = 3.3(20) + 12 = 78.$

47. Let t = the number of years after 1988, let y denote the corresponding industrial production for Germany.

Slope $m = \frac{120-105}{3} = 5$

$y - 105 = 5(t - 0) \implies y = 5t + 105.$

a. $t = 6 \implies y = 5(6) + 105 = 135.$

b. $t = 9 \implies y = 5(9) + 105 = 150.$

48. Let C denote the acquisition cost of the assembly plant, let N denote the useful lifetime of the plant, and let S denote the salvage value of the plant. Let D denote the total depreciation after t years, let V denote the depreciated value of the plant after t years.

Here $C = 100$ million dollars, $N = 20$, $S = 0$.

$D = t\left(\frac{C-S}{N}\right) = t\left(\frac{100-0}{20}\right) = 5t.$

$V = C - D = C - t\left(\frac{C-S}{N}\right).$

a. $V = 100 - 5t.$

b. $t = 12 \implies V = 100 - 5(12) = 40$ million dollars.

c. The annual rate of depreciation is 5 million dollars per year.

49. In this problem, $C = 20,000 dollars$, N $= 4$ years, and $S = 2,000$ dollars.

a. $V = C - t\left(\frac{C-S}{N}\right) = 20,000 - t\left(\frac{20,000-2,000}{4}\right) = 20,000 - 4,500t.$

b. $t = 1.5 \implies V = 20,000 - 4,500(1.5) = 13,250$ dollars.

c. When $V = 10,000$,

$$10,000 = 20,000 - 4,500t \implies 4,500t = 10,000$$
$$\implies t = \tfrac{20}{9} \text{ years.}$$

d. The annual rate of depreciation is 4,500 dollars per year.

50. a. $\frac{C-2,500}{3} = 4,600 \implies C - 2,500 = 3(4,600) \implies C = 16,300$ dollars.

b. When $V = 9,400$,

$$9,400 = 16,300 - 4,600t \implies 4,600t = 6,900 \implies t = 1.5 \text{ years.}$$

51. a. $V = 12,000 - 200t.$

b. $A = 6,000 + 400t.$

c. $V = A \implies 12,000 - 200t = 6,000 + 400t \implies 600t = 6,000 \implies t = 10$ months.

d. The selling price will be $(12,000 - 200(10)) = 10,000$ dollars.

52. Let V denote the value of the ice cream truck in dollars t years after Paul's purchase.

$V = 10,000 - t\left(\frac{10,000-1,000}{2}\right) = 10,000 - 4,500t.$

53. Let D denote the demand in reservations per month for the tour, let p denote the price in dollars.

a. The slope $m = \frac{40-200}{3,600-2,000} = -.1$

$D - 200 = -.1(p - 2000) \implies D = -0.1p + 400.$

b. If $p = 3000$, then $D = -0.1(3,000) + 400 = 100$ reservations per month.

c. When $D = 0$

$$0 = -0.1p + 400 \implies p = 4,000 \text{ dollars.}$$

d. The vertical intercept for this linear model is the value of D when $p = 0$. This value of D is 400 reservations per month, which is an upper bound for the number of reservations possible.

Solutions to Exercise Set 1.3

1. $D = \sqrt{(0-2)^2 + (2-(-1))^2}, \quad D = \sqrt{4+9}, \quad D = \sqrt{13}$
2. $D = \sqrt{(1-3)^2 + (3-1)^2}, \quad D = \sqrt{4+4}, \quad D = \sqrt{8} = 2\sqrt{2}$
3. $D = \sqrt{(1-0)^2 + (-9-2)^2}, \quad D = \sqrt{1+121}, \quad D = \sqrt{122}$
4. $D = \sqrt{(6-1)^2 + (6-(-3))^2}, \quad D = \sqrt{25+81}, \quad D = \sqrt{106}$
5. $D = \sqrt{(-1-1)^2 + (-1-1)^2}, \quad D = \sqrt{4+4}, \quad D = \sqrt{8} = 2\sqrt{2}$
6. $D = \sqrt{(1-(-2))^2 + (2-(-2))^2}, \quad D = \sqrt{9+16}, \quad D = \sqrt{25} = 5$
7. $D = \sqrt{(-3-(-4))^2 + (-5-2)^2}, \quad D = \sqrt{1+49}, \quad D = \sqrt{50} = 5\sqrt{2}$
8. $D = \sqrt{(-6-0)^2 + (0-(-9))^2}, \quad D = \sqrt{36+81}, \quad D = \sqrt{117} = 3\sqrt{13}$

9. First we verify that the distance d_1 between (x_1, y_1) and $\left(\frac{x_1+x_2}{2}, \frac{y_1+y_2}{2}\right)$ equals the distance d_2 between (x_2, y_2) and $\left(\frac{x_1+x_2}{2}, \frac{y_1+y_2}{2}\right)$.

$$d_1 = \sqrt{\left(\frac{x_1+x_2}{2} - x_1\right)^2 + \left(\frac{y_1+y_2}{2} - y_1\right)^2} = \sqrt{\left(\frac{x_2-x_1}{2}\right)^2 + \left(\frac{y_2-y_1}{2}\right)^2}$$

$$d_2 = \sqrt{\left(\frac{x_1+x_2}{2} - x_2\right)^2 + \left(\frac{y_1+y_2}{2} - y_2\right)^2} = \sqrt{\left(\frac{x_1-x_2}{2}\right)^2 + \left(\frac{y_1-y_2}{2}\right)^2}$$

Thus $d_1 = d_2$.

Now we show that the point $\left(\frac{x_1+x_2}{2}, \frac{y_1+y_2}{2}\right)$ is on the line joining (x_1, y_1) and (x_2, y_2).

Case 1: $x_2 \neq x_1$. The line joining (x_1, y_1) and (x_2, y_2) has the equation

$$y - y_1 = \left(\frac{y_2-y_1}{x_2-x_1}\right)(x - x_1).$$

$$\frac{y_1+y_2}{2} - y_1 \stackrel{?}{=} \left(\frac{y_2-y_1}{x_2-x_1}\right)\left(\frac{x_1+x_2}{2} - x_1\right)$$

$$\frac{y_2-y_1}{2} \stackrel{?}{=} \left(\frac{y_2-y_1}{x_2-x_1}\right)\frac{(x_2-x_1)}{2} \qquad \text{true.}$$

Case 2: $x_2 = x_1$. The line joining (x_1, y_1) and (x_2, y_2) has the equation

$$x = x_1.$$

$$\frac{x_1+x_2}{2} \stackrel{?}{=} x_1$$

$$\frac{x_1+x_1}{2} \stackrel{?}{=} x_1 \qquad \text{true.}$$

Therefore the point $\left(\frac{x_1+x_2}{2}, \frac{y_1+y_2}{2}\right)$ is the midpoint of the line segment joining the points (x_1, y_2) and (x_2, y_2).

10. a. midpoint $= \left(\frac{0+0}{2}, \frac{0+6}{2}\right) = (0, 3)$

    b. midpoint $= \left(\frac{1+4}{2}, \frac{3+7}{2}\right) = \left(\frac{5}{2}, 5\right)$

    c. midpoint $= \left(\frac{(-1)+7}{2}, \frac{2+4}{2}\right) = (3, 3)$

11. $x^2 + y^2 = 3^2 = 9.$

12. $(x-2)^2+(y-4)^2=6^2, \quad x^2-4x+4+y^2-8y+16=36, \quad x^2-4x+y^2-8y-16=0$

13. $(x-4)^2+(y-3)^2=2^2, \quad x^2-8x+16+y^2-6y+9=4, \quad x^2-8x+y^2-6y+21=0$

14. $(x-(-2))^2+(y-4)^2=3^2, \quad x^2+4x+4+y^2-8y+16=9,$

 $x^2+4x+y^2-8y+11=0$

15. $(x-(-6))^2+(y-(-4))^2=10^2, \quad x^2+12x+36+y^2+8y+16=100,$

 $x^2+12x+y^2+8y-48=0$

16. $x^2+6x=(x^2+6x+3^2)-3^2=(x+3)^2-9$

17. $x^2-10x=\left(x^2-10x+(-5)^2\right)-(-5)^2=(x-5)^2-25$

18. $t-t^2=-(t^2-t)=-\left[\left(t^2-t+\left(-\frac{1}{2}\right)^2\right)-\left(-\frac{1}{2}\right)^2\right]=-\left(t-\frac{1}{2}\right)^2+\frac{1}{4}$

19. $8x-x^2=-(x^2-8x)=-\left[(x^2-8x+(-4)^2)-(-4)^2\right]=-(x-4)^2+16$

20. $x^2-6x+5=\left(x^2-6x+(-3)^2\right)-(-3)^2+5=(x-3)^2-4$

21. $3-2t-2t^2=-2(t^2+t)+3=-2\left[\left(t^2+t+\left(\frac{1}{2}\right)^2\right)-\left(\frac{1}{2}\right)^2\right]+3=-2\left(t+\frac{1}{2}\right)^2+\frac{7}{2}$

22. $(x^2-2x)+y^2-8=0$

 $(x^2-2x)=(x^2-2x+1)-1=(x-1)^2-1$

 $y^2=y^2$

 $\left[(x-1)^2-1\right]+y^2-8=0,$

 $(x-1)^2+y^2=9$

 The circle has the center $(1,0)$ and radius $r=\sqrt{9}=3$.

23. $(x^2-2x)+(y^2+6y)-12=0$

 $(x^2-2x)=(x^2-2x+1)-1=(x-1)^2-1$

 $(y^2+6y)=(y^2+6y+9)-9=(y+3)^2-9$

 $\left[(x-1)^2-1\right]+\left[(y+3)^2-9\right]-12=0, \quad (x-1)^2+(y+3)^2=22$

 The circle has the center $(1,-3)$ and radius $r=\sqrt{22}$.

24. $(x^2+4x)+(y^2+2y)-11=0$

 $(x^2+4x)=(x^2+4x+4)-4=(x+2)^2-4$

 $(y^2+2y)=(y^2+2y+1)-1=(y+1)^2-1$

 $\left[(x+2)^2-4\right]+\left[(y+1)^2-1\right]-11=0, \quad (x+2)^2+(y+1)^2=16$

 The circle has the center $(-2,-1)$ and radius $r=\sqrt{16}=4$.

25. $(x^2 + 14x) + (y^2 - 10y) + 70 = 0$

$(x^2 + 14x) = (x^2 + 14x + 49) - 49 = (x + 7)^2 - 49$

$(y^2 - 10y) = (y^2 - 10y + 25) - 25 = (y - 5)^2 - 25$

$\left[(x+7)^2 - 49\right] + \left[(y-5)^2 - 25\right] + 70 = 0, \quad (x+7)^2 + (y-5)^2 = 4$

The circle has the center $(-7, 5)$ and radius $r = \sqrt{4} = 2$.

26. $(x^2 - 6x) + (y^2 - 4y) + 8 = 0$

$(x^2 - 6x) = (x^2 - 6x + 9) - 9 = (x - 3)^2 - 9$

$(y^2 - 4y) = (y^2 - 4y + 4) - 4 = (y - 2)^2 - 4$

$\left[(x-3)^2 - 9\right] + \left[(y-2)^2 - 4\right] + 8 = 0, \quad (x-3)^2 + (y-2)^2 = 5$

The circle has the center $(3, 2)$ and radius $r = \sqrt{5}$.

27. $(x^2 - 2x) + (y^2 - 6y) + 3 = 0$

$(x^2 - 2x) = (x^2 - 2x + 1) - 1 = (x - 1)^2 - 1$

$(y^2 - 6y) = (y^2 - 6y + 9) - 9 = (y - 3)^2 - 9$

$\left[(x-1)^2 - 1\right] + \left[(y-3)^2 - 9\right] + 3 = 0, \quad (x-1)^2 + (y-3)^2 = 7$

The circle has the center $(1, 3)$ and radius $r = \sqrt{7}$.

28. $(x^2 - 4x) + (y^2 + 6y) - 3 = 0$

$(x^2 - 4x) = (x^2 - 4x + 4) - 4 = (x - 2)^2 - 4$

$(y^2 + 6y) = (y^2 + 6y + 9) - 9 = (y + 3)^2 - 9$

$\left[(x-2)^2 - 4\right] + \left[(y+3)^2 - 9\right] - 3 = 0, \quad (x-2)^2 + (y+3)^2 = 16$

The circle has the center $(2, -3)$ and radius $r = \sqrt{16} = 4$

The distance between $(2, -3)$ and $(1, 1)$ is $\sqrt{(1-2)^2 + (1-(-3))^2} = \sqrt{1+16} = \sqrt{17}$ which is greater than the radius $r = \sqrt{16}$. So point $(1, 1)$ lies outside the circle.

29. $(x^2 - 2x) + (y^2 - 4y) + 1 = 0$

$(x^2 - 2x) = (x^2 - 2x + 1) - 1 = (x - 1)^2 - 1$

$(y^2 - 4y) = (y^2 - 4y + 4) - 4 = (y - 2)^2 - 4$

$\left[(x-1)^2 - 1\right] + \left[(y-2)^2 - 4\right] + 1 = 0, \quad (x-1)^2 + (y-2)^2 = 4$

The circle has the center $(1, 2)$ and radius $r = \sqrt{4} = 2$.

The distance between $(1,2)$ and $(2,1)$ is $\sqrt{(2-1)^2+(1-2)^2} = \sqrt{1+1} = \sqrt{2}$ which is smaller than the radius $r = \sqrt{4}$. So point $(2,1)$ lies inside the circle.

30. $(x^2-6x)+(y^2-8y)=0$

$(x^2-6x) = (x^2-6x+9)-9 = (x-3)^2-9$

$(y^2-8y) = (y^2-8y+16)-16 = (y-4)^2-16$

$\big[(x-3)^2-9\big]+\big[(y-4)^2-16\big]=0, \quad (x-3)^2+(y-4)^2=25$

The circle has the center $(3,4)$ and radius $r=\sqrt{25}=5$.

The slope for the radius of the circle at the origin is $m_1 = \frac{4-0}{3-0} = \frac{4}{3}$

The slope m_2 for the tangent line at the origin is $m_2 = -\frac{1}{m_1} = -\frac{3}{4}$.

Equation for this tangent line: $y-0 = -\frac{3}{4}(x-0), \quad y=-\frac{3}{4}x$.

31. We assume that the points (x,y) located within 5 units of the tree are the solutions of the inequality.

$$(x-2)^2+(y-4)^2<5^2.$$

The point $(0,0)$ satisfies this inequality since

$$(0-2)^2+(0-4)^2=4+16<25.$$

32. a. Consider first the region inside the circle defined by the equation

$$x^2-4x+y^2-6y+4=0 \implies (x^2-4x+4)+(y^2-6y+9)+4= 4+9 \implies (x-2)^2+(y-3)^2=3^2.$$

Therefore the acorns from tree A remain within the region inside the circle with center at the point $(2,3)$ and with radius 3.

Consider next the region inside the circle defined by the equation

$$x^2+2x+y^2+2y-7=0 \implies (x^2+2x+1)+(y^2+2y+1)-7=1+1 \implies (x+1)^2+(y+1)^2=3^2.$$

Therefore the acorns from tree B remain within the region inside the circle with center at the point $(-1,-1)$ and with radius 3.

b. Tree A is located at the point $(2,3)$, and tree B is located at the point $(-1,-1)$.

c. We can see from the picture that the regions overlap.

For instance, the point $(1,1)$ is in both regions.

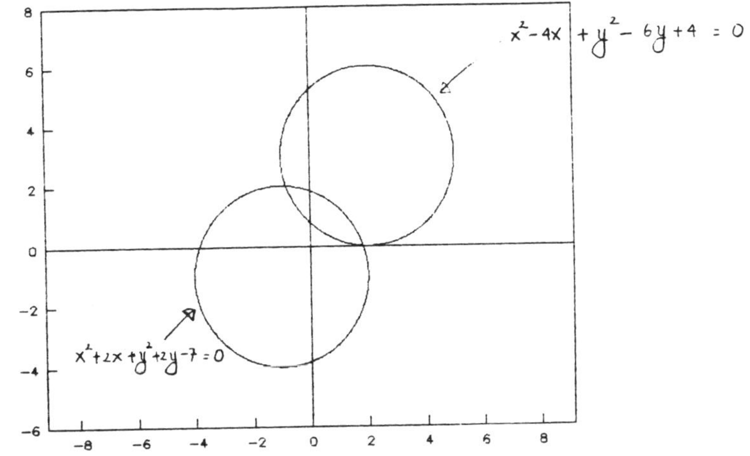

Solutions to Exercise Set 1.4

1. Function, because each package has a unique postage charge.
2. Function, because each taxpayer has a unique federal income tax.
3. Not necessarily a function, because a professor might have more than one child.
4. Not necessarily a function, because a company might have more than one product.
5. Function, because each day has a unique average temperature.
6. Function, because each person has a unique daily weight.
7. Not necessarily a function, because a course might have more than one textbook.
8. Function, because each home has a unique electric bill (we assume each person has a unique home.)
9. a. $f(0) = 7 - 3(0) = 7$

   b. $f(3) = 7 - 3(3) = -2$

c. $f(a) = 7 - 3a$

d. $f(x+h) = 7 - 3(x+h)$

e. The domain of f is the set of all real numbers, and the range of f is also the set of all real numbers.

10. a. $f(8) = \frac{1}{8-7} = \frac{1}{1} = 1$

b. $f(b) = \frac{1}{b-7}$

c. $f(-1) = \frac{1}{-1-7} = \frac{1}{-8} = -\frac{1}{8}$

d. $f(x+h) = \frac{1}{(x+h)-7}$

e. The domain of f is the set of all real numbers x except for $x = 7$.

f. The range of f is the set of all real numbers y except for $y = 0$.

11. a. $f(0) = \frac{0+3}{0-0+2} = \frac{3}{2}$

b. $f(-3) = \frac{-3+3}{9+3+2} = \frac{0}{14} = 0$

c. $f(-1) = \frac{(-1)+3}{(-1)^2-1+2} = \frac{2}{4} = \frac{1}{2}$

d. $f(x+h) = \frac{(x+h)+3}{(x+h)^2-(x+h)+2}$

12. a. $f(2) = \frac{\sqrt{6-2}}{1+4} = \frac{2}{5}$

b. $f(6) = \frac{\sqrt{6-6}}{1+36} = \frac{0}{37} = 0$

c. $f(-3) = \frac{\sqrt{6+3}}{1+9} = \frac{3}{10}$

d. $f(x+h) = \frac{\sqrt{6-(x+h)}}{1+(x+h)^2}$

13. Not a function

14. Function

15. Not a function

16. Not a function

17. Function

18. Function

19. Function

20. Function

21. $f(x) = \frac{1}{x^2-7}$. The domain is the set of all real numbers x except $x = \sqrt{7}$ and $x = -\sqrt{7}$, because division by zero is not allowed.

22. $f(x) = \frac{x-\sqrt{x}}{x-3}$. The domain is all nonnegative real numbers x except for $x = 3$, because the square roots of negative numbers are not defined and division by zero is not allowed.

23. $f(s) = \frac{s^2-1}{s+1} = s - 1$. The domain is the set of all real numbers s except $s = -1$, because division by zero is not allowed in the original formula for $f(s)$.

24. $f(t) = \frac{t}{1+\sqrt{t}}$. $\sqrt{t}$ is not defined for negative numbers. So the domain is $[0, \infty)$.

25. $g(x) = \frac{1}{\sqrt{1-x^2}}$. Because square roots of negative numbers are not allowed and division by zero also is not allowed, the domain is the set of all real numbers x such that $(1 - x^2) > 0$, $x^2 < 1$, $-1 < x < 1$. The domain is the open interval $(-1, 1)$.

26. $f(x) = \frac{1}{x^2-3x-4} = \frac{1}{(x-4)(x+1)}$. The domain is the set of all real numbers except 4 and -1 because division by zero is not allowed.

27. $f(x) = \frac{\sqrt{x}}{x^3-x}$. The domain is the set of all positive real numbers x except $x = 1$, because division by zero is not allowed, and square roots of negative numbers are not allowed.

28. $f(x) = \sqrt{(x-1)(x+1)}$. The domain is the set of all real numbers x such that $(x - 1)(x + 1) \geq 0$. This is the set of all real numbers x such that either both $(x+1)$ and $(x-1)$ are non-negative, or both $(x+1)$ and $(x-1)$ are non-positive. This is the set of all real numbers x such that either $x \geq 1$ or $x \leq -1$. The domain consists of the intervals $[1, \infty)$ and $(-\infty, -1]$.

29. $f(x) = x^2 + 6x + 7 = \left(x^2 + 6x + (3)^2\right) - 9 + 7 = (x + 3)^2 - 2$.
The range is the interval $[-2, \infty)$.

30. $f(x) = x^3 - x^5$. This function can assume any real numbers, so the range is $(-\infty, \infty)$.

31. $C(x) = 2x + 0.5$ dollars

32. $y = C(x) = 0.2x + 10$ dollars

33. $C(x) = 70x + 1000$ dollars

34. a) $R = 150x$ dollars

b) $P = R - C = 150x - (70x + 1000) = 80x - 1000$ dollars

c) $P > 0$, $80x - 1000 > 0$, $80x > 1000$, $x > 12$. At least 13 lawnmowers must be sold per week to show a profit.

35. a) $D(5) = 800 - 40(5) = 800 - 200 = 600$

b) $D(0)$ gives the greatest demand. (At $p = 0$, $D(p)$ is greatest.)

c) $D(p) = 0$ for $800 - 40p = 0$, $40p = 800$, $p = 20$.

36. Let $p =$ daily rental price in dollars of a jackhammer.

Let $y =$ the corresponding number of jackhammers rented per year.

We assume that each of these jackhammers is rented for the entire year.

Let $R =$ the corresponding yearly revenue in dollars from jackhammer rentals.

y is linearly related to p. So $y = mp + b$. $m = -20$.

When $p = 30$, $y = 500$. So $500 = -20 \cdot 30 + b$, $b = 1100$.

$y = -20p + 1100$.

a) $R = 365p \cdot y = 365p \cdot (-20p + 1100) = -7300p^2 + 401500p$.

b) $R = 0$, $-7300p^2 + 401500p = 0$, $p = 0$ or $p = \frac{401500}{7300} = 55$.

37. Let $p =$ daily room rent in dollars, and let $y =$ the corresponding number of rooms rented per day.

Let $R(p) =$ daily revenue in dollars from room rentals.

If $p < 50$, then $y = 400$, $R(p) = p \cdot y = 400p$ dollars.

Suppose $p \geq 50$. Note that y is linearly related to p, so $y = mp + b$. $m = -10$, because for each dollar increase in the daily rate above 50 dollars, 10 rooms will remain vacant. Because $y = 400$ when $p = 50$, $400 = (-10)(50) + b \implies b = 900$. Therefore $y = -10p + 900$. Thus $R(p) = p \cdot y = p(-10p + 900) = -10p^2 + 900p$.

38. $N(x) = 32 - \frac{24}{\sqrt{x+1}}$, $x \geq 0$

a. $N(0) = 32 - \frac{24}{\sqrt{0+1}} = 32 - 24 = 8$ customers/hr.

b. The bigger x is, the smaller $\frac{24}{\sqrt{x+1}}$ is, so the larger $N(x)$ is. Thus, according to this model, the employee's speed never stops improving.

c. The "upper limit" of the speed is 32 customers per hour, but this cannot be achieved.

39. a.

n	0	5	10	15	20
$W(n)$	20	28.28	34.64	40	44.72

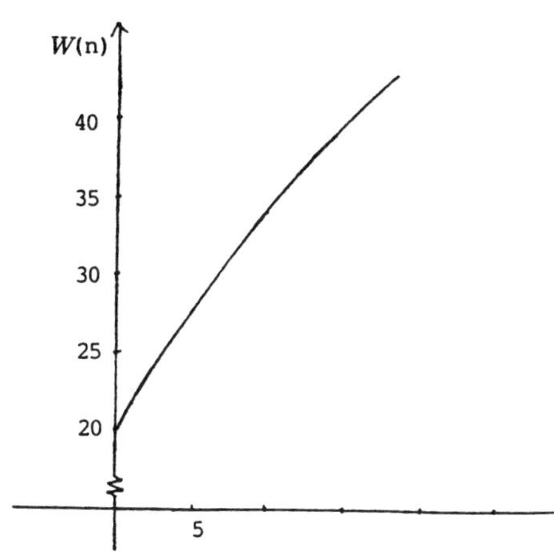

b. $W(0) = \sqrt{400} = 20$ words.

c. There is no maximum speed since $W(n)$ continually increases as n increases.

40. Let $t =$ the number of years the funds have been on deposit.

When $t = 1$, $P(t) = 1000 + 0.08(1000) = 1000(1 + 0.08)$.

When $t = 2$, $P(t) = 1000(1 + 0.08) + 0.08\left(1000(1 + 0.08)\right)$

$$= 1000(1 + 0.08) \cdot (1 + 0.08) = 1000 \cdot (1 + 0.08)^2.$$

When $t = 3$, $P(t) = 1000(1 + 0.08)^2 + 0.08\left(1000(1 + 0.08)^2\right)$

$$= 1000(1 + 0.08)^2 \cdot (1 + 0.08) = 1000(1 + 0.08)^3.$$

In general, $P(t) = 1000 \cdot (1 + 0.08)^t$.

41. a. $P(x) = 0.005x + 100$.

b. When $x = 150,000$, $P(x) = 0.005(150,000) + 100 = \$850$.

42. Let $x =$ face amount of policy. Let $P_2(x) =$ the annual premium charged by the Bjax Insurance Company.

$P_2(x) = 0.003x + 200$.

a. When $x = 150,000$, $P_2(x) = 0.003(150,000) + 200 = \$650$. Therefore the Bjax Company has a lower premium for a \$150,000 policy.

b. When the two companies' premiums are equal,

$$0.005x + 100 = 0.003x + 200 \implies 0.002x = 100 \implies x = \$50,000.$$

43. $t=$ number of years since 1958.

$L=$ carbon dioxide level at the South Pole in parts per million.

$L(t)=mt+b$.

a. When $t=0$, $L=315$;

when $t=8$, $L=322$;

when $t=16$, $L=329$.

$m=\frac{322-315}{8}=\frac{7}{8}$

$b=315$

$L(t)=\frac{7}{8}t+315$

b. When $t=40$, $L=\frac{7}{8}(40)+315=350$ parts per million.

44. a. $Y(x)=B(x)-C(x)=(8x-x^2)-(x^2/4)=8x-\frac{5}{4}x^2$.

$Y(4)=8(4)-\frac{5}{4}(4)^2=12$.

b. $B(x)=0 \implies 8x-x^2=0 \implies x(8-x)=0 \implies x=0$ or $x=8$.

c. $C(x)=B(x) \implies 8x-\frac{5}{4}x^2=0 \implies x\left(8-\frac{5}{4}x\right)=0 \implies x=0$ or $x=\frac{32}{5}=6.4$.

Solutions to Exercise Set 1.5

1. $27^{2/3}=\left(27^{1/3}\right)^2=3^2=9$
2. $81^{5/4}=\left(81^{1/4}\right)^5=3^5=243$
3. $4^{-3/2}=\frac{1}{4^{3/2}}=\frac{1}{(4^{1/2})^3}=\frac{1}{2^3}=\frac{1}{8}$
4. $9^{-3/2}=\frac{1}{9^{3/2}}=\frac{1}{(9^{1/2})^3}=\frac{1}{3^3}=\frac{1}{27}$
5. $\frac{2(3x)^4}{6x^2}=\frac{2(81x^4)}{6x^2}=27x^2$
6. $\frac{4x^3\sqrt{y}}{(x^2y)^{2/3}}=\frac{4x^3y^{1/2}}{x^{4/3}y^{2/3}}=\frac{4x^{3-\frac{4}{3}}}{y^{\frac{2}{3}-\frac{1}{2}}}=\frac{4x^{5/3}}{y^{1/6}}$
7. $\frac{(3x^2y^{-1})^3}{\sqrt{xy}}=\frac{27x^6y^{-3}}{x^{1/2}\cdot y^{1/2}}=\frac{27x^{6-\frac{1}{2}}}{y^{\frac{1}{2}+3}}=\frac{27x^{11/2}}{y^{7/2}}$.
8. $\frac{\sqrt{4x^4y^3}}{\sqrt{y}}=\frac{2x^2y^{3/2}}{y^{1/2}}=2x^2y$.
9. $\frac{6(xy)^3}{x^{2/3}y^{-3/2}}=\frac{6x^3y^3}{x^{2/3}y^{-3/2}}=6x^{3-\frac{2}{3}}y^{3-\left(-\frac{3}{2}\right)}=6x^{7/3}y^{9/2}$
10. $\frac{(27x^6y^2)^{-1/3}}{3x^{-2}y^{4/3}}=\frac{1}{(27x^6y^2)^{1/3}}\cdot\frac{1}{3x^{-2}y^{4/3}}=\frac{1}{9x^2y^{2/3}x^{-2}y^{4/3}}=\frac{1}{9y^2}$

11. True

12. True

13. $f(x) = \frac{1}{2}x^2 - 3$

x	$f(x)$
0	−3.0
±1	−2.5
±2	−1.0
±3	1.5
±4	5.0

The function is symmetric about the y-axis.

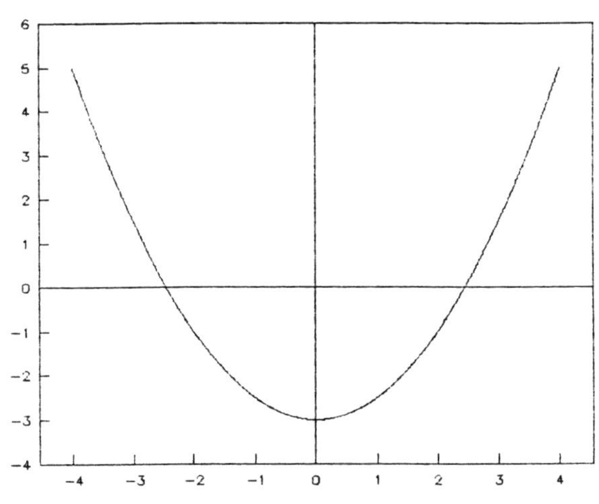

14. $f(x) = -x^3$

x	$f(x)$
0	0
1	−1
2	−8
3	−27
4	−64
−1	1
−2	8
−3	27
−4	64

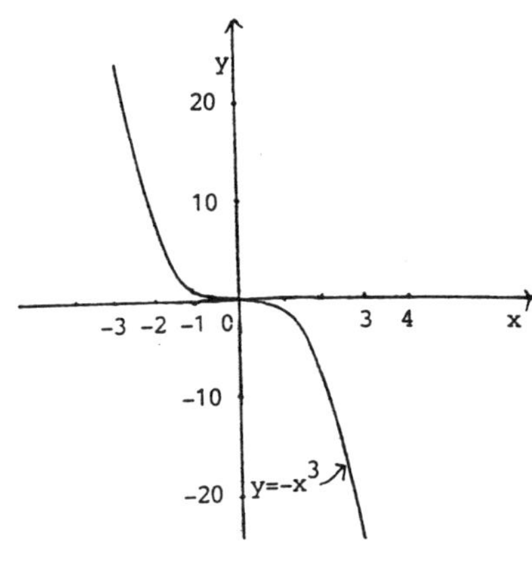

15. $f(x) = 3x^{2/3}$

x	$f(x)$
0	0
±1	3
±4	7.56
±8	12

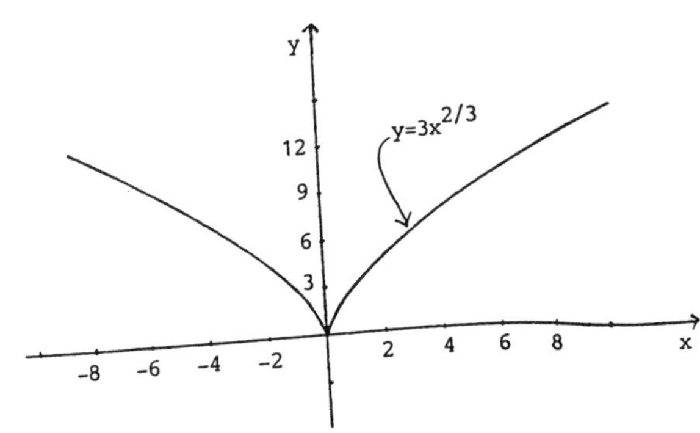

16. $f(x) = 2x^{3/2}$

x	$f(x)$
0	0
1	2
4	16
9	54

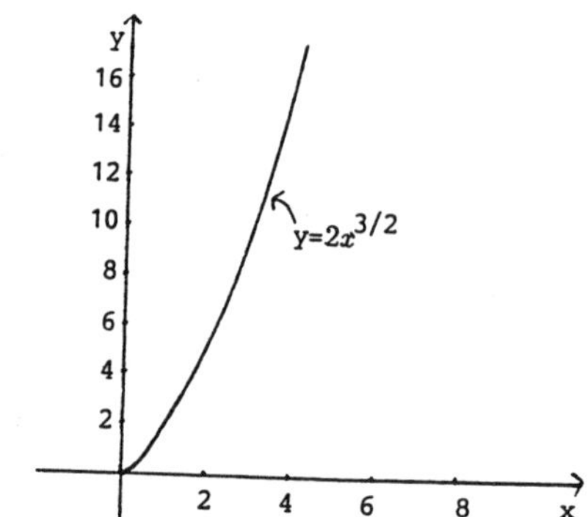

17. $f(x) = -4x^{1/3}$

x	$f(x)$
0	0
± 1	∓ 4
± 8	∓ 8

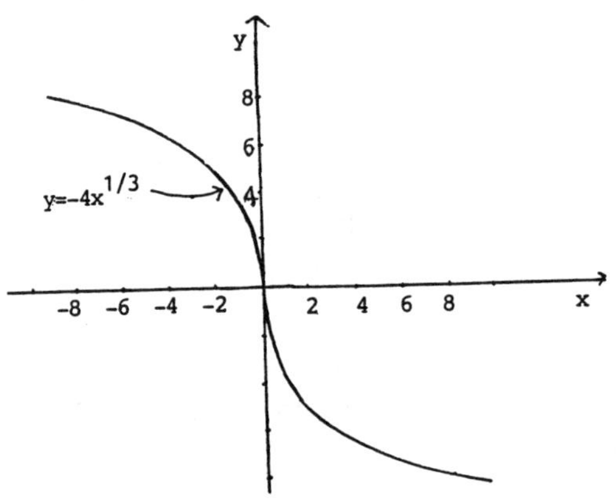

18. $y = (x-4)^{-1/2}$

x	y
4.1	3.16
4.5	1.41
5.0	1.00
8.0	0.50

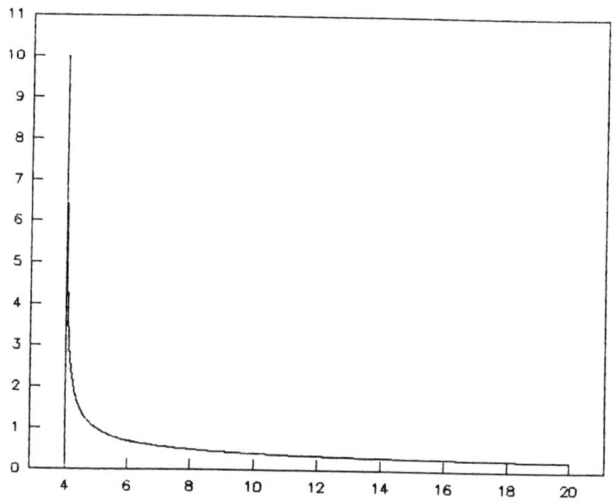

19. $y = -x^{-2/3}$

x	y
$\pm.1$	-4.64
$\pm.5$	-1.59
±1	-1
±4	$-.4$
±8	$-.25$

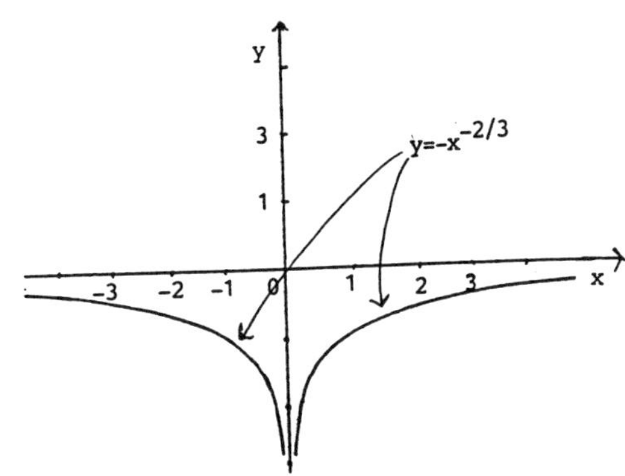

20. $f(x) = 2x^{-2}$

x	$f(x)$
$\pm.5$	8
±1	2
±2	.5
±3	.22

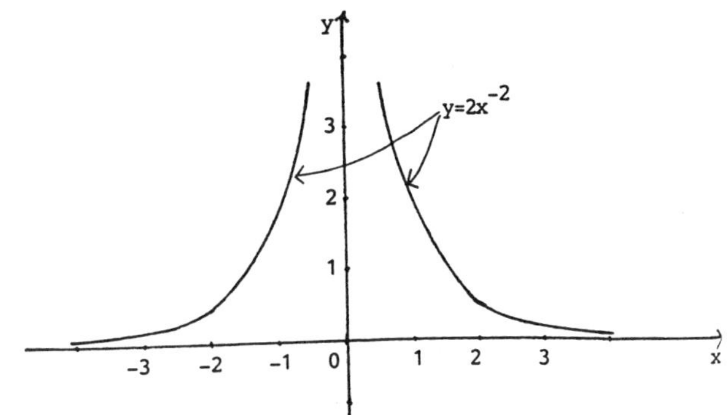

21. $f(x) = (x-3)^2 - 4$

x	$f(x)$
-1	12
0	5
1	0
2	-3
3	-4
4	-3
5	0
6	5

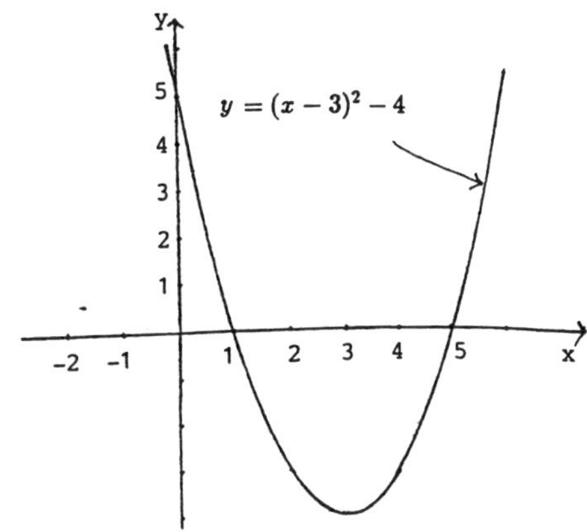

22. $f(x) = \frac{1}{2}(x+1)^2 - 2$

x	$f(x)$
-4	$5/2$
-3	0
-2	$-3/2$
-1	-2
0	$-3/2$
1	0
2	$5/2$

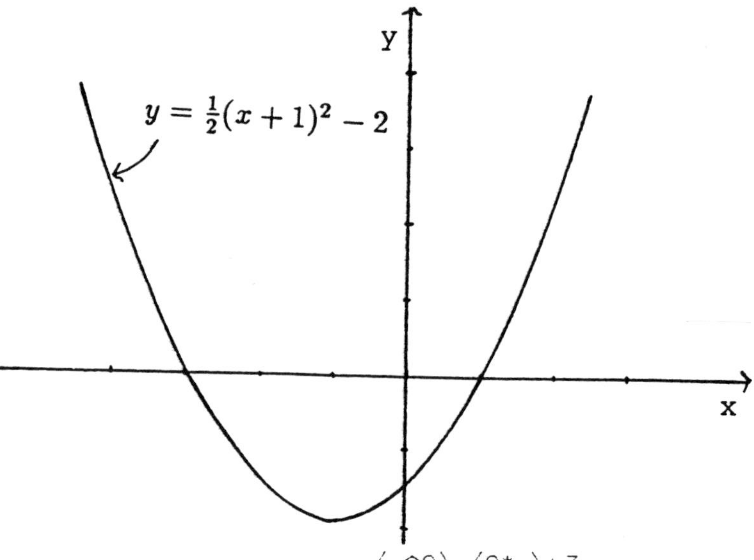

23. $f(x) = x^2 - 2x + 3$

x	$f(x)$
-1	6
0	3
1	2
2	3
3	6

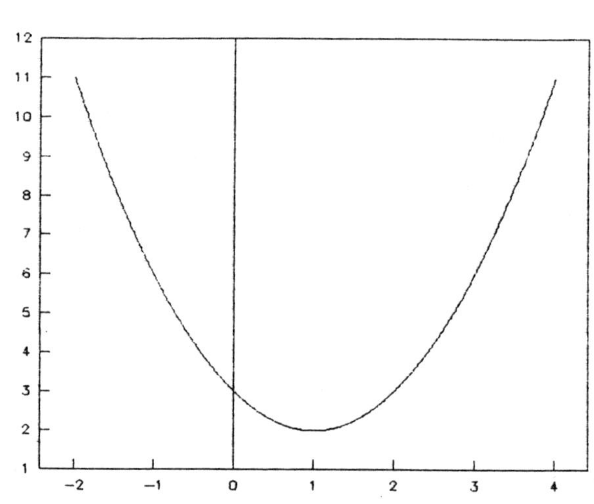

24. $f(x) = x^2 + 4x + 6$

x	$f(x)$
-4	6
-3	3
-2	2
-1	3
0	6

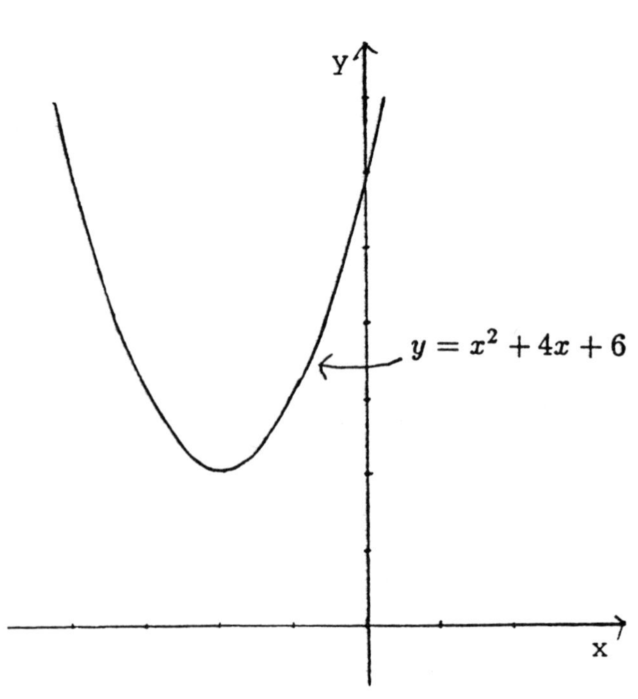

25. $f(x) = -2x^2 - 4x + 1$

x	$f(x)$
-3	-5
-2	1
-1	3
0	1
1	-5

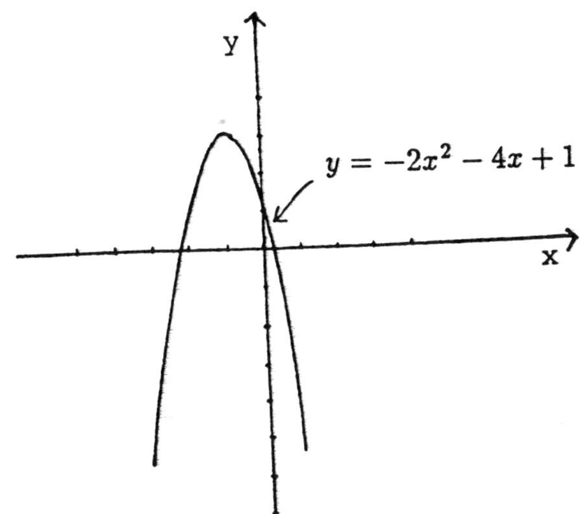

26. $f(x) = -\frac{3}{2}x^2 - 6x - 3$

x	$f(x)$
-4	-3
-3	1.5
-2	3
-1	1.5
0	-3

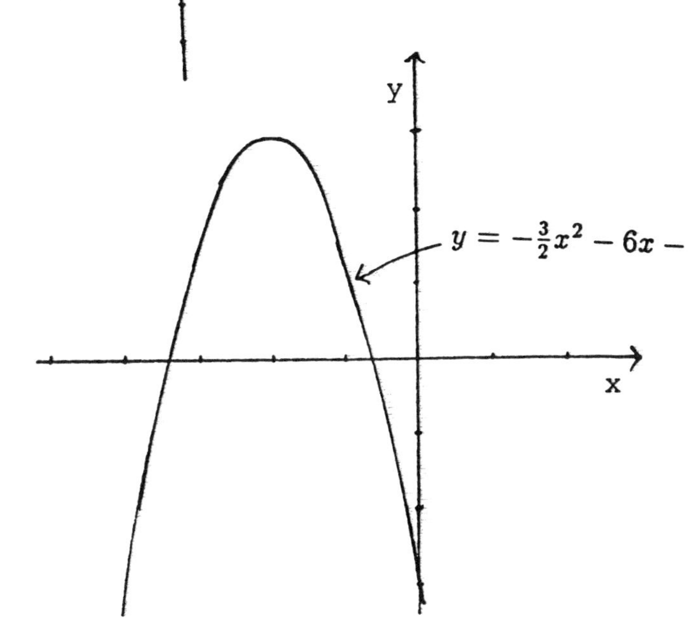

27. $f(x) = x^3 + 4$

x	$f(x)$
-2	-4
-1	3
0	4
1	5
2	12

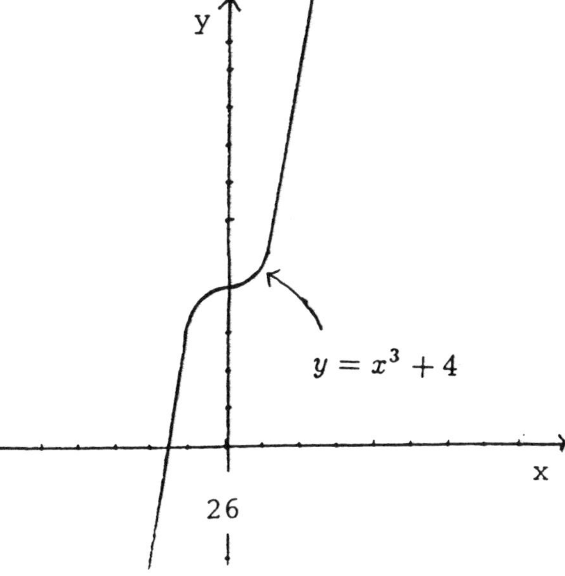

28. $f(x) = x^3 + x$

x	$f(x)$
-2	-10
-1	-2
0	0
1	2
2	10

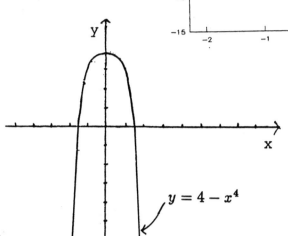

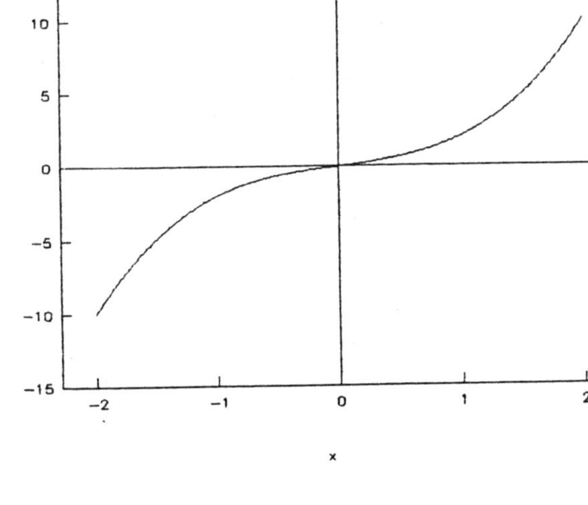

29. $f(x) = 4 - x^4$

x	$f(x)$
-2	-12
-1	3
0	4
1	3
2	-12

30. $f(x) = x^3 - 4x + 2$

x	$f(x)$
-3	-13
-2	2
-1	5
0	2
1	-1
2	2

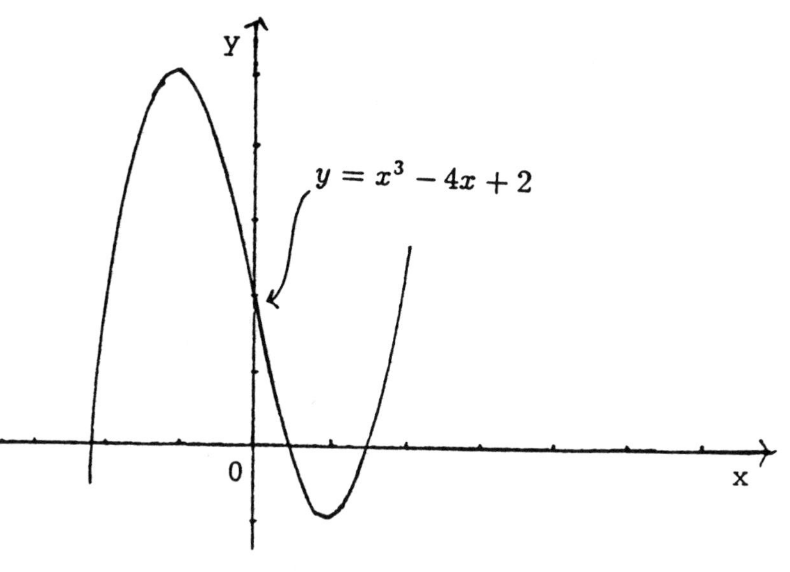

31. $f(x) = x^3 - x^2 - 2x + 2$

x	$f(x)$
-2	-6
-1	2
$-\frac{1}{2}$	$\frac{21}{8}$
0	2
1	0
$\frac{3}{2}$	$\frac{1}{8}$
2	2

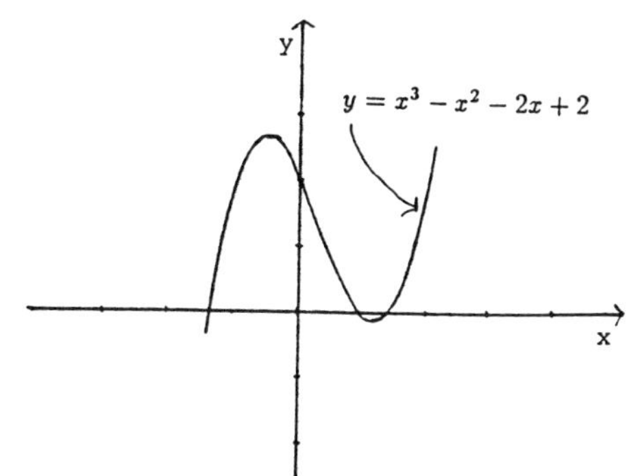

32. $f(x) = 1 - 3x + x^3$

x	$f(x)$
-2	-1
-1	3
0	1
1	-1
2	3

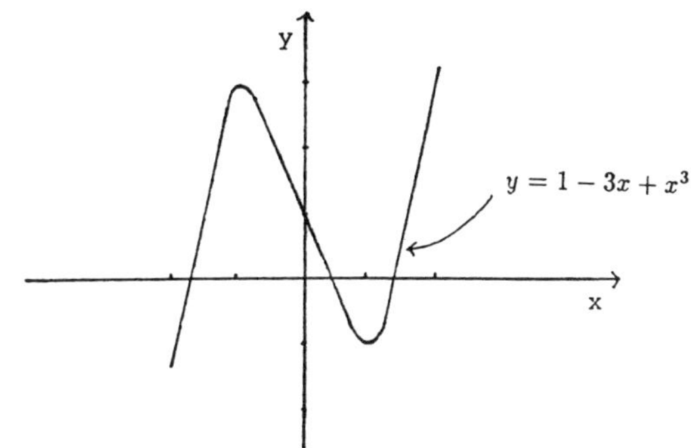

33. $f(x) = \frac{1}{x-3}$

x	$f(x)$
0	$-\frac{1}{3}$
1	$-\frac{1}{2}$
2	-1
4	1
5	$\frac{1}{2}$
6	$\frac{1}{3}$

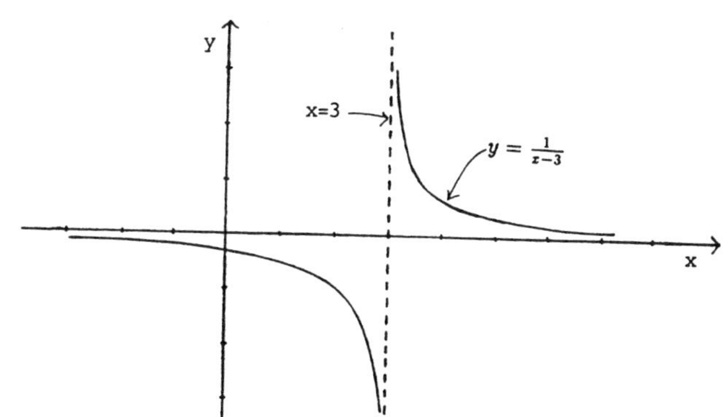

34. $f(x) = \frac{3}{2+x}$

x	$f(x)$
-5	-1
-4	$-\frac{3}{2}$
-3	-3
-1	3
0	$\frac{3}{2}$
1	1
2	$\frac{3}{4}$

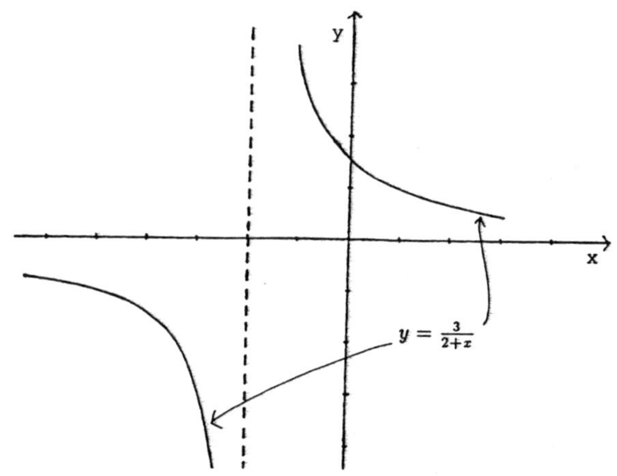

35. $f(x) = \frac{1-x}{1+x}$

x	$f(x)$
-10.0	$-\frac{11}{9} = -1.22$
$-\ 4.0$	$-\frac{5}{3}$
$-\ 3.0$	-2
$-\ 2.0$	-3
$-\ 1.5$	-5

x	$f(x)$
-0.5	3
0	1
1.0	0
2.0	$-\frac{1}{3}$
3.0	$-\frac{1}{2}$
10.0	$-\frac{9}{11} = -0.82$

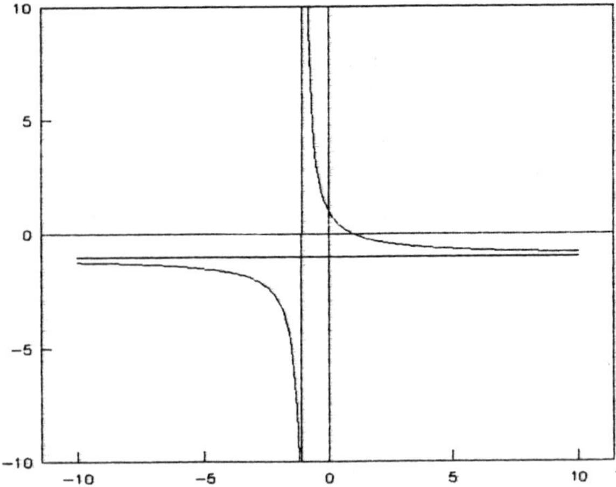

36. $f(x) = \frac{x}{1+x^2}$

x	$f(x)$
-3	$-\frac{3}{10}$
-2	$-\frac{2}{5}$
-1	$-\frac{1}{2}$
0	0
1	$\frac{1}{2}$
2	$\frac{2}{5}$
3	$\frac{3}{10}$

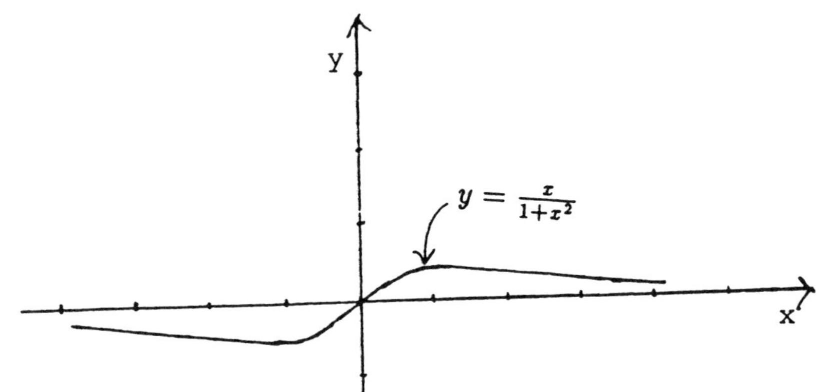

37. $f(x) = \frac{x^3}{1+x}$

x	$f(x)$
-4	$\frac{64}{3}$
-3	$\frac{27}{2}$
-2	8
$-\frac{3}{2}$	$\frac{27}{4}$
$-\frac{5}{4}$	$\frac{125}{16}$
$-\frac{1}{2}$	$-\frac{1}{4}$
0	0
1	$\frac{1}{2}$
2	$\frac{8}{3}$

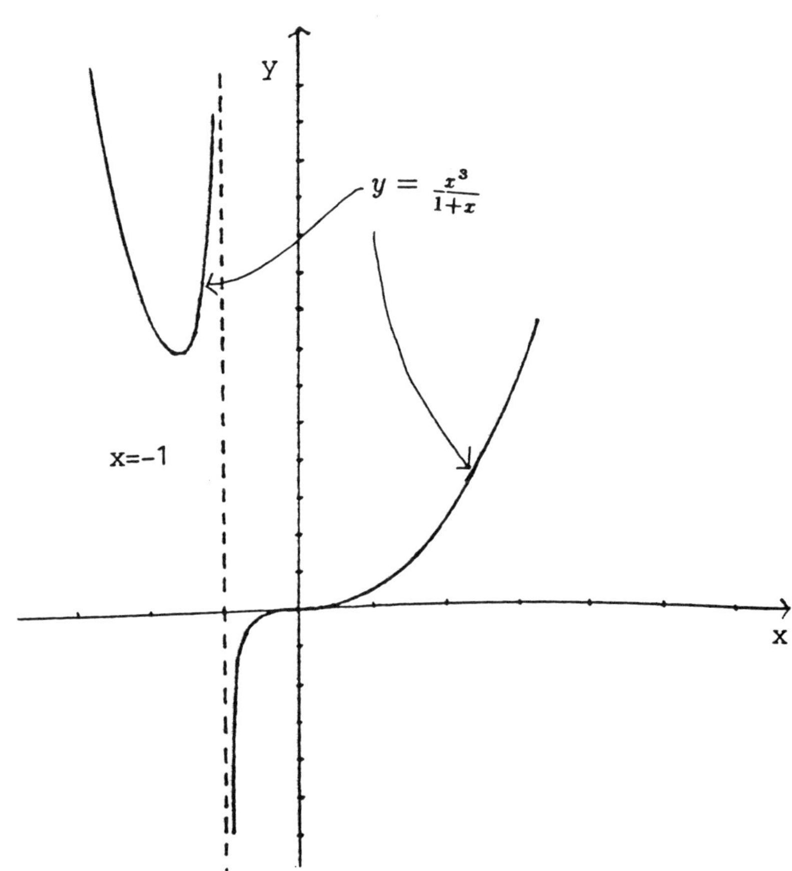

38. $f(x) = \frac{x+1}{1-x^2} = \frac{1}{1-x}$ for $x \neq \pm 1$

x	$f(x)$
-4	$\frac{1}{5}$
-3	$\frac{1}{4}$
-2	$\frac{1}{3}$
0	1
$\frac{1}{2}$	2
$\frac{3}{2}$	-2
2	-1
3	$-\frac{1}{2}$
4	$-\frac{1}{3}$

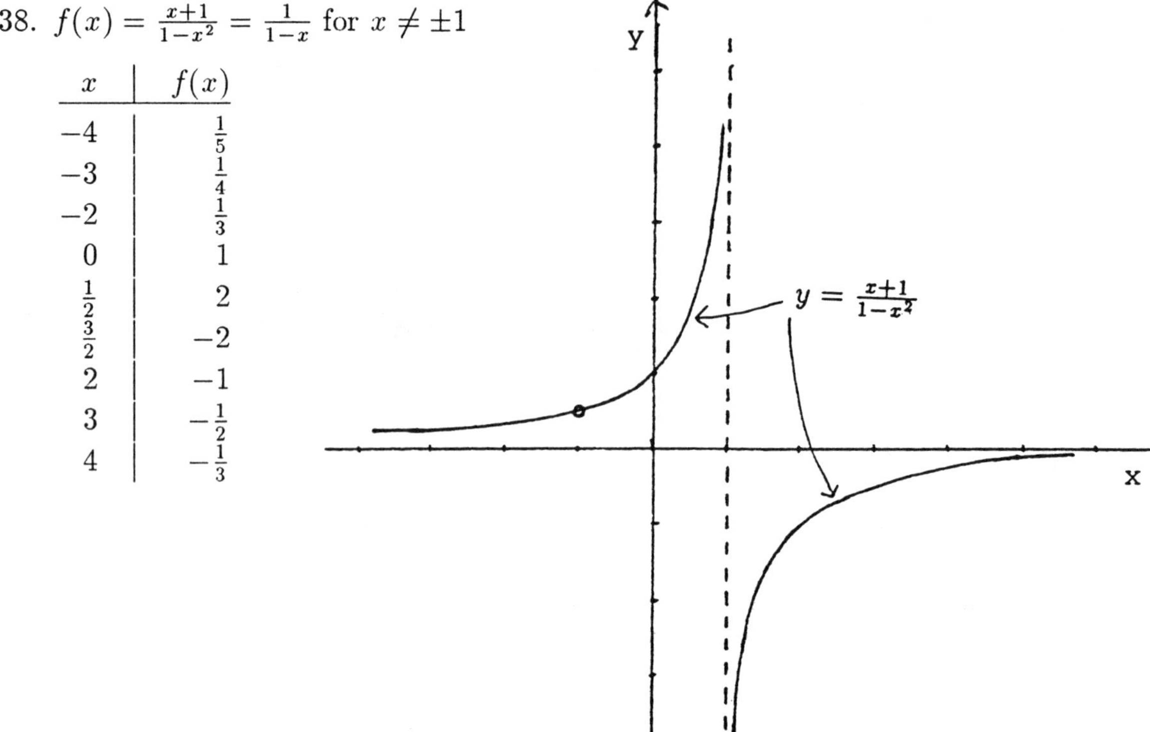

39. Let $w =$ width in cm, $\ell =$ length in cm.

$w \cdot \ell = 80.$ $w = \frac{80}{\ell}.$

The domain of this function is the set of all positive real numbers ℓ.

40. $R = k \cdot v^2$ for some positive constant k. The graph is similar to the graph of the function in problem 13.

41. a.

x	0	1	2	3	4	14
$U_1(x)$	1.4	1.7	2	2.2	2.4	4

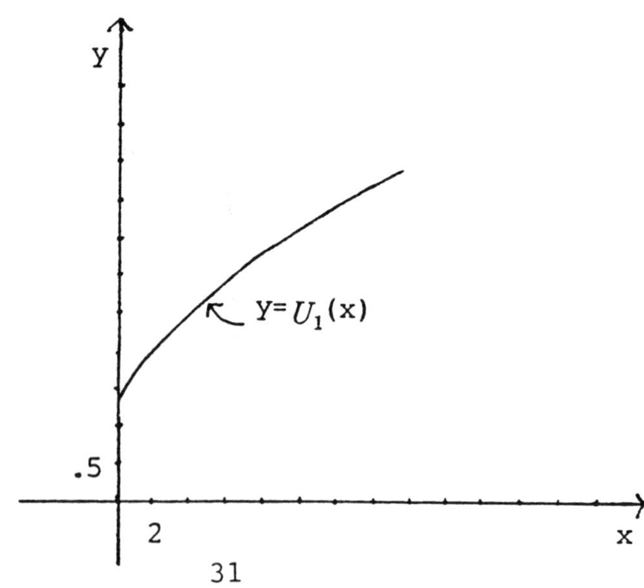

b.

x	0	1	2	3	4	5	6
$U_2(x)$	0	5	8	9	8	5	0

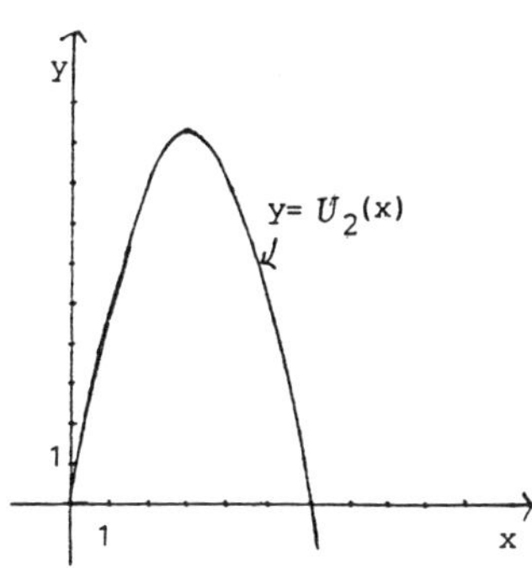

c. $U_2(x)$ would be more appropriate because after eating a certain number of slices of pizza, your satisfaction goes down, not up.

d. $U_1(x)$ because the more money you win the more satisfaction you get.

42. $D(p) = \sqrt{400 - p^2}$.

a. The domain of D consists of all p such that $p \geq 0$ and $400 - p^2 \geq 0 \Rightarrow p^2 \leq 400 \Rightarrow p \leq 20$. Thus, the domain of D is the interval $[0, 20]$.

b. $D(12) = \sqrt{400 - 12^2} = \sqrt{400 - 144} = \sqrt{256} = 16$ cars/week.

c.

p	0	4	8	12	16	20
$D(p)$	20	19.5	18.3	16	12	0

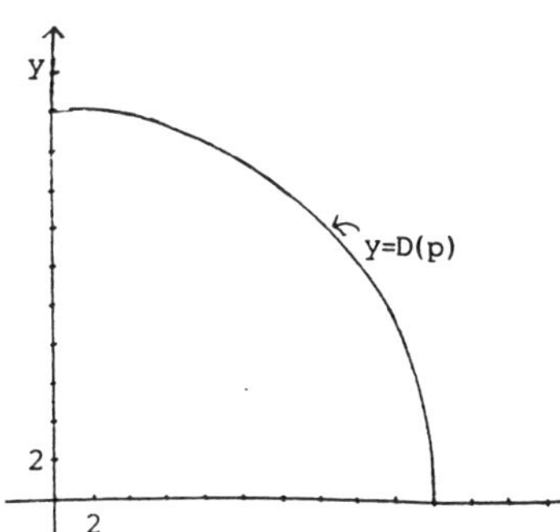

43. $T(x) = 30\left(1 + \frac{2}{\sqrt{x+1}}\right)$

a. $T(4) = 30\left(1 + \frac{2}{\sqrt{5}}\right) = 56.83$ seconds.

b.

x	0	1	2	3	4	5	10	15
$T(x)$	90	72.43	64.64	60	56.83	54.49	48.09	45

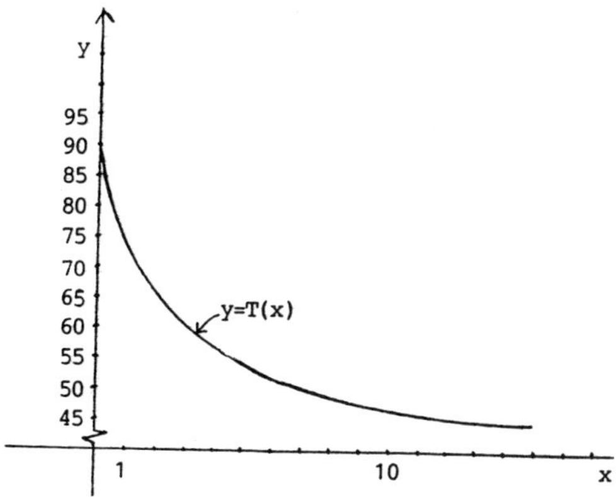

44. $N(t) = 10t - 0.5t^2$

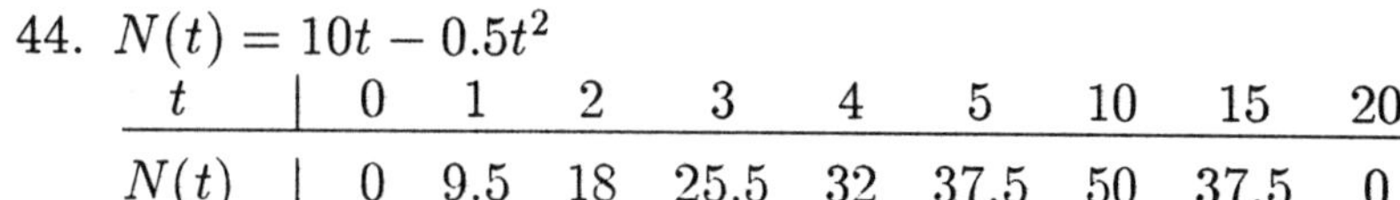

t	0	1	2	3	4	5	10	15	20
$N(t)$	0	9.5	18	25.5	32	37.5	50	37.5	0

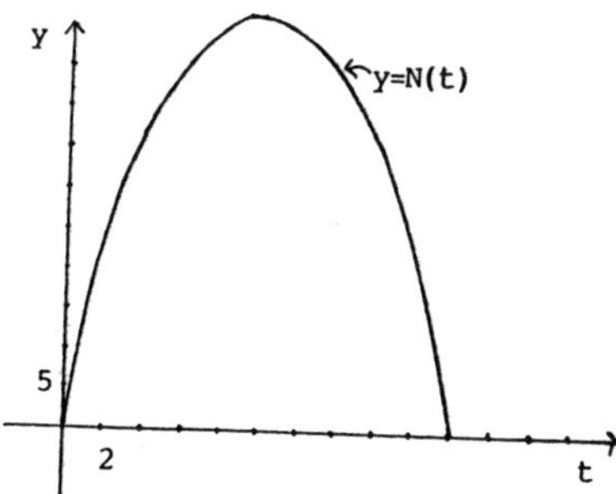

45. $C(x) = 400 + 40x + 0.2x^2$

x	0	40	80	120	160	200	240	280	320
$C(x)$	400	2320	4880	8080	11920	16400	21520	27280	33680

Fixed costs $= C(0) = 400$.

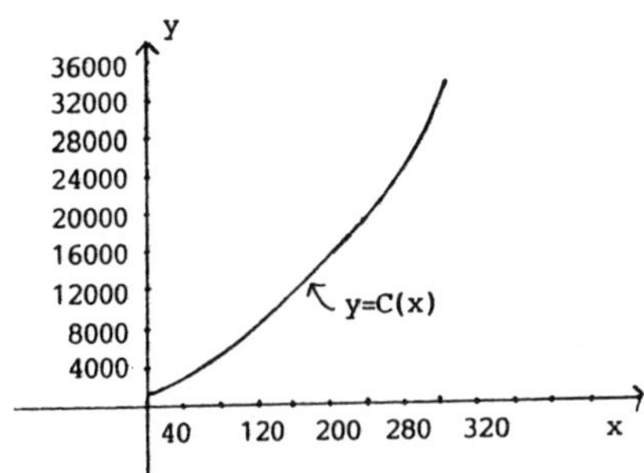

46. a. $R(x) = 100x$

b. $P(x) = R(x) - C(x) = 100x - (400 + 40x + 0.2x^2)$
$= -0.2x^2 + 60x - 400$ dollars.

x	0	40	80
$R(x)$	0	4000	8000

x	0	40	80	120	160	200	240	280	320
$P(x)$	−400	1680	3120	3920	4080	3600	2480	720	−1680

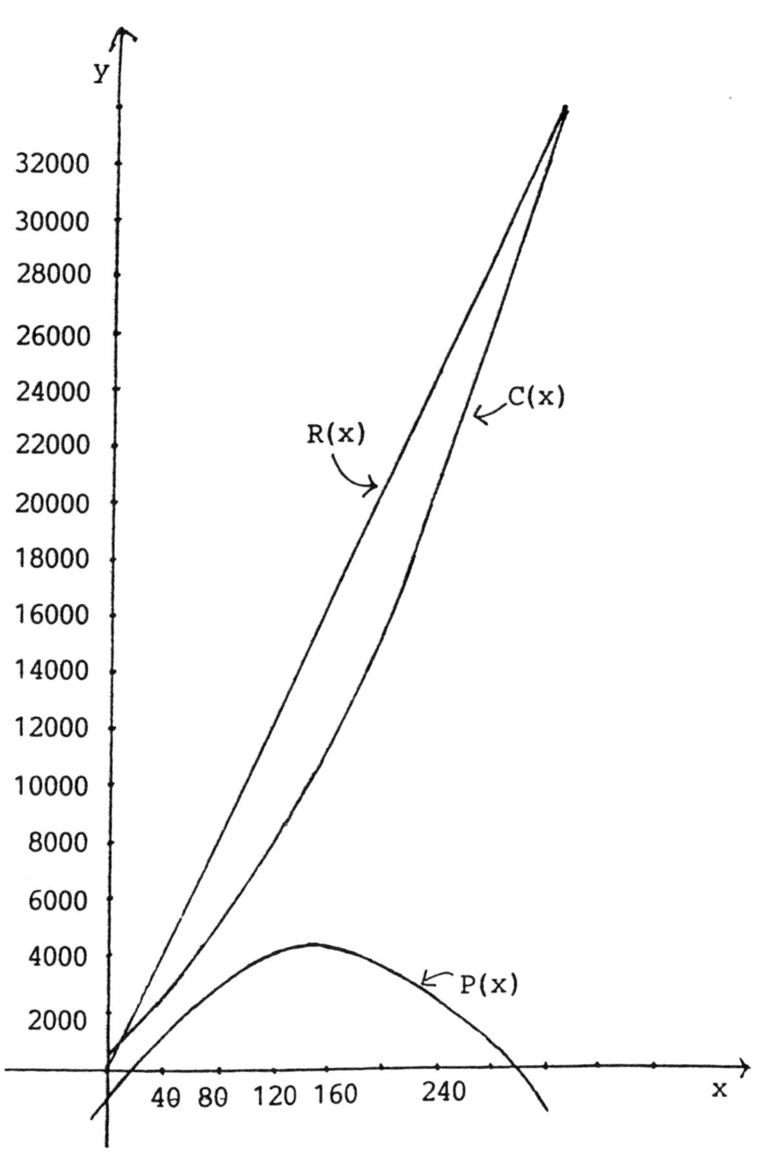

47. a. $C(5) = 500 + 10\left(\frac{5-5}{5}\right)^3 = 500$
$C(30) = 500 + 10\left(\frac{30-5}{5}\right)^3 = 1,750$
$C(40) = 500 + 10\left(\frac{40-5}{5}\right)^3 = 3,930$
$C(50) = 500 + 10\left(\frac{50-5}{5}\right)^3 = 7,790$
b. $R(x) = px = 100x$
c. $P(x) = R(x) - C(x) = 100x - 500 - 10\left(\frac{x-5}{5}\right)^3$
d. $P(t) = 100t - 500 - 10\left(\frac{t-5}{5}\right)^3$
$P(30) = 100(30) - 1750 = 1,250.$
$P(40) = 100(40) - 3,930 = 70.$
$P(50) = 100(50) - 7,790 = -2790.$
e. The profits are not largest at the largest production level because the cost $C(x)$ keeps increasing as x increases.

48. a. $C(10) = 600 + 5(10 - 10)^{1/3} + 15 \cdot (10) = 750.$
$C(37) = 600 + 5(37 - 10)^{1/3} + 15 \cdot (37) = 1,170.$
$C(135) = 600 + 5(135 - 10)^{1/3} + 15 \cdot (135) = 2,650.$
b. $R(x) = px = 30x$
c. $P(x) = R(x) - C(x) = 30x - \left[600 + 5(x - 10)^{1/3} + 15x\right]$
$= 15x - 600 - 5(x - 10)^{1/3}.$
d. $P(10) = 30 \cdot (10) - 750 = -450.$
$P(37) = 30 \cdot (37) - 1,170 = -60.$
$P(135) = 30 \cdot (135) - 2,650 = 1,400.$

Solutions to Exercise Set 1.6

1. $f(x) = x^2 - 4 = (x + 2)(x - 2)$
$f(x) = 0$ for $x = -2$, $x = 2$.
2. $f(x) = x - x^3 = x(1 - x^2) = x(1 - x)(1 + x)$
$f(x) = 0$ for $x = 0$, $x = 1$, $x = -1$.
3. $f(x) = x^2 + x - 6 = (x + 3)(x - 2)$
$f(x) = 0$ for $x = -3$, $x = 2$.
4. $f(x) = x^2 - 9x + 20 = (x - 4)(x - 5)$
$f(x) = 0$ for $x = 4$, $x = 5$.
5. $f(x) = x^2 - 2x - 35 = (x - 7)(x + 5)$
$f(x) = 0$ for $x = 7$, $x = -5$.
6. $f(x) = x^3 - 5x^2 - 36x = x(x^2 - 5x - 36) = x(x - 9)(x + 4)$
$f(x) = 0$ for $x = 0$, $x = 9$, $x = -4$.
7. $f(x) = \frac{\sqrt{x-2}}{x+1}$
$f(x) = 0$ for $x - 2 = 0 \implies x = 2$.
8. $f(x) = \frac{x^{4/3}-16}{x^2+1}$
$f(x) = 0$ when $x^{4/3} = 16 \implies x = -8$ or $x = 8$.

9. $f(x) = x^{3/2} - 3x^{1/2} = x^{1/2}(x-3) = 0$ for $x = 0$, $x = 3$.
10. $f(x) = x - 2\sqrt{x} + 1 = (\sqrt{x} - 1)^2 = 0$ for $\sqrt{x} = 1 \Rightarrow x = 1$.
11. $f(x) = x^2 + x - 1 = 0$ for $x = \frac{-1 \pm \sqrt{1^2 - 4(1)(-1)}}{2(1)} = \frac{-1 \pm \sqrt{5}}{2}$.
12. $f(x) = x^2 + 3x + 1 = 0$ for $x = \frac{-3 \pm \sqrt{3^2 - 4(1)(1)}}{2(1)} = \frac{-3 \pm \sqrt{5}}{2}$.
13. $f(x) = 2x^2 + 3x - 1 = 0$ for $x = \frac{-3 \pm \sqrt{3^2 - 4(2)(-1)}}{2(2)} = \frac{-3 \pm \sqrt{17}}{4}$
14. $f(x) = 1 + 4x - x^2 = 0$ for $x = \frac{-4 \pm \sqrt{4^2 - 4(-1)(1)}}{2(-1)} = \frac{-4 \pm \sqrt{20}}{-2}$
$= \frac{4 \pm 2\sqrt{5}}{2} = 2 \pm \sqrt{5}$.
15. $f(x) = 3 + 2x - x^2 = 0$ for $x = \frac{-2 \pm \sqrt{2^2 - 4(-1)(3)}}{2(-1)} = \frac{-2 \pm 4}{-2} = 3$ or -1.

16. $f(x) = 2x^2 - 2 - 2x = 0$ for $x = \frac{2 \pm \sqrt{(-2)^2 - 4(2)(-2)}}{2(2)} = \frac{2 \pm \sqrt{20}}{4}$
$= \frac{2 \pm 2\sqrt{5}}{4} = \frac{1}{2}\left(1 \pm \sqrt{5}\right)$.
17. $f(x) = x - 3x^2 + 4 = 0$ for $x = \frac{-1 \pm \sqrt{(1)^2 - 4(-3)(4)}}{2(-3)}$
$= \frac{-1 \pm \sqrt{49}}{-6} = \frac{-1 \pm 7}{-6} = \frac{4}{3}$ or -1.
18. $f(x) = 3 - 3x - 2x^2 = 0$ for $x = \frac{3 \pm \sqrt{3^2 - 4(-2)(3)}}{2(-2)} = \frac{3 \pm \sqrt{33}}{-4}$.
19. $f(x) = g(x)$ for $2x + 1 = 7 - x$, $3x = 6$, $x = 2$.
20. $f(x) = g(x)$ for $1 - \frac{x}{3} = 8 + 2x$, $3 - x = 24 + 6x$, $-7x = 21$, $x = -3$.
21. $f(x) = g(x)$ for $x^2 - 2 = 2x + 1 \implies x^2 - 2x - 3 = 0 \implies (x-3)(x+1) = 0$
$\implies x = 3$ or $x = -1$.
22. $f(x) = g(x)$ for $4 - x^2 = 2 - x$, $-x^2 + x + 2 = 0$, $(-x+2)(x+1) = 0$
$\Rightarrow x = 2, x = -1$.
23. $f(x) = g(x)$ for $\sqrt{x} = 4x - 3 \implies x = 16x^2 - 24x + 9 \implies 16x^2 - 25x + 9 = 0$
$\implies x = \frac{25 \pm \sqrt{(25)^2 - 4 \cdot (16) \cdot (9)}}{2 \cdot (16)} = \frac{25 \pm 7}{32} \implies x = 1$ or $x = \frac{9}{16}$. $x = \frac{9}{16}$ does not work in the original equation.
24. $f(x) = g(x)$ for $7 - x^2 = x^2 - 1$, $-2x^2 = -8$, $x^2 = 4 \Rightarrow x = \pm 2$.
25. $f(x) = g(x)$ for $x^3 + 2 = x + 2$, $x^3 - x = 0$, $x(x+1)(x-1) = 0$
$\Rightarrow x = 0$, $x = -1$, $x = 1$.
26. $f(x) = g(x)$ for $\sqrt{x+2} = \frac{1}{2}x + 1 \implies x + 2 = \frac{1}{4}x^2 + x + 1 \implies x^2 = 4 \implies$
$x = 2$ or $x = -2$.
27. $f(x) = g(x)$ for $3x^2 - 12x = 5x - 20$, $3x^2 - 17x + 20 = 0$,
$(3x-5)(x-4) = 0 \Rightarrow x = \frac{5}{3}$, $x = 4$.
28. $f(x) = g(x)$ for $\sqrt{x} + 1 = \frac{1}{3}x + 1$, $\sqrt{x} = \frac{1}{3}x$, $x = \frac{1}{9}x^2$,
$\frac{1}{9}x^2 - x = 0$, $\frac{1}{9}x(x-9) = 0 \Rightarrow x = 0$, $x = 9$.
29. The demand is greater than the supply.
30. The supply is greater than the demand.
31. $L(x) = S(x)$ for $100\sqrt{x} = \frac{5}{3}x$, $10,000x = \frac{25}{9}x^2$,

$25x^2 - 90,000x = 0$, $25x(x-3600) = 0$, $x = \$3600$.

32. $C(x) = 400 + 15x$, $R(x) = 25x$.
Break-even production level: $400 + 15x = 25x$, $10x = 400$, $x = 40$ pairs.

33. a. $C(x) = 500 + 20x + x^2$, $R(x) = 80x$,
$P(x) = R(x) - C(x) = 80x - (500 + 20x + x^2) = -x^2 + 60x - 500$
Break-even production level: $-x^2 + 60x - 500 = 0$,
$(-x+50)(x-10) = 0 \Rightarrow x = 10$, $x = 50$.
b. $P(x) > 0$ for $10 < x < 50$.

34. $D(p) = S(p)$ for $20 - p = p^2$, $p^2 + p - 20 = 0$
$(p-4)(p+5) = 0$, $p = 4$.

35. $f(t) = g(t)$ for $\frac{25}{t} = \frac{50}{1+(t-2)^2}$, $50t = 25\left(1 + (t-2)^2\right)$,
$50t = 25\left(1 + (t^2 - 4t + 4)\right)$, $50t = 25t^2 - 100t + 125$,
$25t^2 - 150t + 125 = 0$, $25(t^2 - 6t + 5) = 0$,
$25(t-1)(t-5) = 0 \Rightarrow t = 1$, $t = 5$ hours.

36. t = number of years since 1830.
L = % of U.S. labor force involved in farming. $L = mt + b$.
$m = \frac{20-70}{100} = \frac{-50}{100} = -0.5$, $b = 70$, $L = -0.5t + 70$.
When $L = 40$, $40 = -0.5t + 70$; $0.5t = 30$, $t = 60$.
It was 40% in 1890.

37. Selling price = average cost when $70 = \frac{500+10x+x^2}{x}$
$500 + 10x + x^2 = 70x \implies x^2 - 60x + 500 = 0 \implies (x-50)(x-10) = 0$
$x = 50$, $x = 10$.

38. The total weekly cost function for producing x jackets is given by $C(x) = 400 + 10x + \frac{1}{2}x^2$. The average cost per jacket is the ratio $c(x) = \frac{C(x)}{x} = \frac{400}{x} + 10 + \frac{1}{2}x$.

a. $c(x) = 40 \implies \frac{400}{x} + 10 + \frac{1}{2}x = 40 \implies \frac{1}{2}x - 30 + \frac{400}{x} = 0 \implies$
$x^2 - 60x + 800 = 0 \implies (x-20)\cdot(x-40) = 0 \implies x = 20$ or $x = 40$.

b. From a graph of $y = c(x)$ for $0 < x \le 100$ obtained by plotting points one can discern the following information. $c(x)$ becomes very large for x very close to 0, and $c(x)$ also becomes very large as x becomes very large. First $c(x)$ decreases as x increases, and then $c(x)$ increases as x increases. $c(x)$ attains a minimum value for x approximately 28.3, and this minimum value is approximately 38.28. Therefore $c(x) < 40$ when $20 < x < 40$.

c. From the above information it follows that $c(x) > 40$ when either $x < 20$ or $x > 40$.

39. $C(x) = 2,000 + 80\sqrt{x}$.
The average cost $c(x) = \frac{C(x)}{x} = \frac{2000}{x} + \frac{80}{\sqrt{x}}$.
$c(x) = 1 \implies 1 = \frac{2000}{x} + \frac{80}{\sqrt{x}} \implies x - 80\left(\sqrt{x}\right) - 2000 = 0$
$\implies \left(\sqrt{x}\right)^2 - 80\sqrt{x} - 2000 = 0 \implies \left(\sqrt{x} - 100\right)\cdot\left(\sqrt{x} + 20\right) = 0 \implies \sqrt{x} = 100 \implies x = 10,000$.

40. The mistake in using the given procedure is that we lose the zero $x = 0$ of the function $f(x)$ by cancelling the common factor x in step c.

Solutions to Exercise Set 1.7

1. $h_1(x) = (f+g)(x) = f(x) + g(x) = (x-7) + (x^2+2) = x^2 + x - 5.$
2. $h_2(x) = \left(\frac{f}{g}\right)(x) = \frac{f(x)}{g(x)} = \frac{x-7}{x^2+2}.$
3. $h_1(5) = 5^2 + 5 - 5 = 25.$
4. $h_2(-1) = \frac{-1-7}{(-1)^2+2} = -\frac{8}{3}.$
5. $h_2(-4) = \frac{-4-7}{(-4)^2+2} = -\frac{11}{18}.$
6. $h_1(-3) = (-3)^2 + (-3) - 5 = 1.$
7. $h_1(x) = (f - 2g)(x) = f(x) - 2g(x) = \frac{1}{x+2} - 2\sqrt{3+x}.$
8. $h_2(x) = \left(\frac{g}{f}\right)(x) = \frac{g(x)}{f(x)} = \frac{\sqrt{3+x}}{\frac{1}{x+2}} = (x+2)\sqrt{3+x}.$
9. $h_2(-3) = ((-3)+2)\sqrt{3+(-3)} = 0.$
10. $h_1(-3) = \frac{1}{(-3)+2} - 2\sqrt{3+(-3)} = -1.$
11. $h_1(1) = \frac{1}{1+2} - 2\sqrt{3+1} = \frac{1}{3} - 4 = -\frac{11}{3}.$
12. $h_2(6) = (6+2)\sqrt{3+6} = 24.$
13. $(f+g)(x) = 4 - x.$
    Since $(f+g)(x) = f(x) + g(x) = (x^2 - x) + g(x)$, $(x^2 - x) + g(x) = 4 - x$, $g(x) = (4-x) - (x^2 - x)$, $g(x) = -x^2 + 4.$
14. $(fg)(x) = x - 4.$
    Since $(fg)(x) = f(x)g(x) = (\sqrt{x} - 2)\,g(x)$, $(\sqrt{x} - 2)\,g(x) = x - 4$, $g(x) = \frac{x-4}{\sqrt{x}-2}.$
15. $(f \circ u)(x) = f(u(x)) = f(\sqrt{x}) = 3\sqrt{x} + 1.$
16. $(u \circ h)(x) = u(h(x)) = u\left(\frac{1}{x+1}\right) = \sqrt{\frac{1}{x+1}} = \frac{1}{\sqrt{x+1}}.$
17. $(g \circ f)(x) = g(f(x)) = g(3x+1) = (3x+1)^3.$
18. $(u \circ 2g)(x) = u(2g(x)) = u(2x^3) = \sqrt{2x^3} = \sqrt{2} \cdot x^{3/2}.$
19. $(h^2 \circ u)(x) = [h(u(x))]^2 = \left[\frac{1}{\sqrt{x}+1}\right]^2 = \frac{1}{x + 2\sqrt{x} + 1}.$
20. $(h \circ f \circ u)(x) = h((f \circ u)(x)) = h(f(u(x))) = h(f(\sqrt{x})) = h(3\sqrt{x} + 1)$
    $= \frac{1}{(3\sqrt{x}+1)+1} = \frac{1}{3\sqrt{x}+2}.$
21. $(f \circ g \circ u)(x) = f(g(u(x))) = f(g(\sqrt{x})) = f\left((\sqrt{x})^3\right) = 3(\sqrt{x})^3 + 1 = 3x^{3/2} + 1.$
22. $(u \circ g \circ f)(x) = u(g(f(x))) = u(g(3x+1)) = u\left((3x+1)^3\right) = (3x+1)^{3/2}.$
23. Since $u(x) = x - 3$, $f(u(x)) = f(x-3) = x^3 + 5.$
    So let $f(x) = (x+3)^3 + 5.$
24. Since $u(x) = x^2$, $f(u(x)) = f(x^2) = \sqrt{x^2+1}.$
    So, let $f(x) = \sqrt{x+1}.$
25. $f(g(u(x))) = 4x^2 - 8x + 8.$
    Since $f(g(u(x))) = f\left((u(x)-2)^2\right) = (u(x)-2)^2 + 4,$
    then $(u(x)-2)^2 + 4 = 4x^2 - 8x + 8,$
    $(u(x)-2)^2 = 4x^2 - 8x + 4 = 4(x-1)^2.$

Take $u(x) - 2 = 2(x-1) \Rightarrow u(x) = 2(x-1) + 2 = 2x$.

26. c.
27. a.
28. f.
29. b.
30. e.
31. d.
32. b or c.
33. a.
34. b or c.
35. d.
36. $f(x) = \begin{cases} x^2 - 2, & x \le 2 \\ 4 - x, & x > 2 \end{cases}$

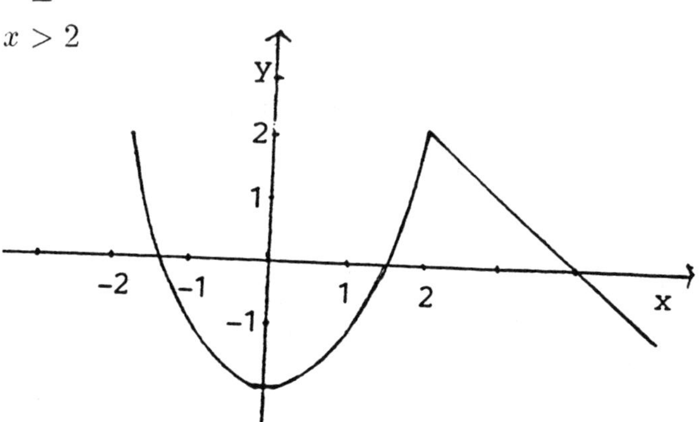

37. $f(x) = \begin{cases} \sqrt{4-x}, & x \le 0 \\ 2 - x^2, & x > 0. \end{cases}$

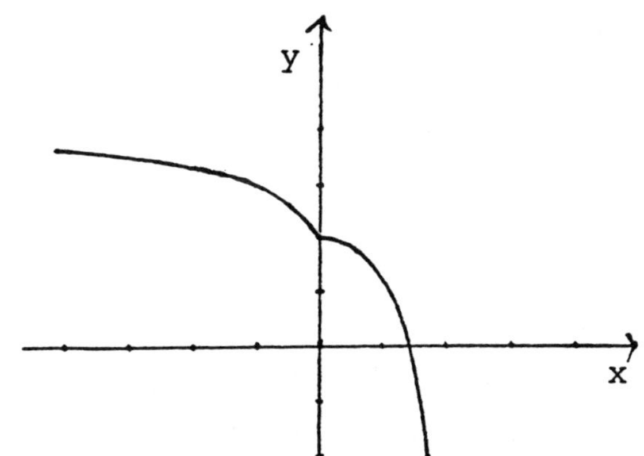

38. $f(x) = |x^2 - x - 6| = |(x-3)(x+2)|$. $x^2 - x - 6 = 0$ for $x = -2$, $x = 3$.
$x < -2$ or $x > 3 \implies (x^2 - x - 6) > 0$.
$-2 < x < 3 \implies (x^2 - x - 6) < 0$.

$$f(x) = \begin{cases} x^2 - x - 6, & x < -2 \\ -x^2 + x + 6, & -2 < x < 3 \\ x^2 - x - 6, & x > 3 \end{cases}$$

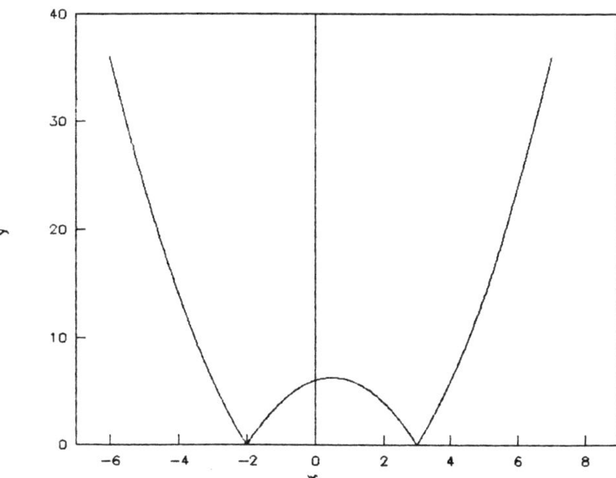

39. $f(x) = x - |x|$.

$$f(x) = \begin{cases} 2x, & x < 0 \\ 0, & x \geq 0 \end{cases}$$

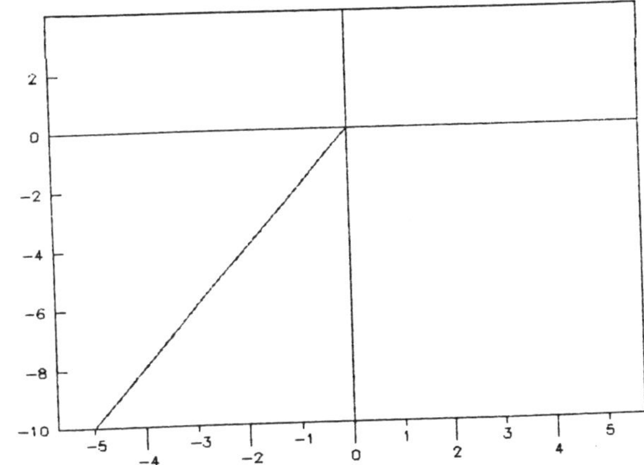

40. $f(x) = |x^2 - 4| = \begin{cases} x^2 - 4, & x < -2 \text{ or } x > 2 \\ 4 - x^2, & -2 \leq x \leq 2 \end{cases}$

41. $f(x) = |x^2 - x - 2|$

$$f(x) = \begin{cases} x^2 - x - 2, & x < -1 \text{ or } x > 2 \\ -x^2 + x + 2, & -1 \le x \le 2 \end{cases}$$

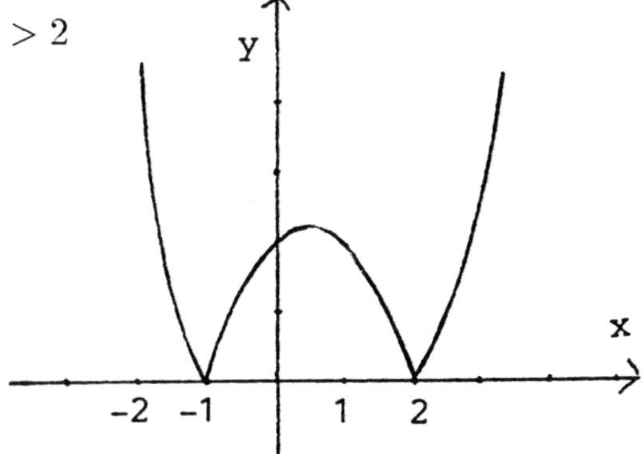

42. Since $x = 2t$, the total variable costs are
$C(2t) = 30(2t) + 0.02(2t)^2 = 60t + 0.08t^2$.

43. a. $p(x) = \begin{cases} 500, & 0 < x \le 5 \\ 500 - 10(x-5) = 550 - 10x, & 5 < x \le 25 \\ 300, & 25 < x \end{cases}$

b.

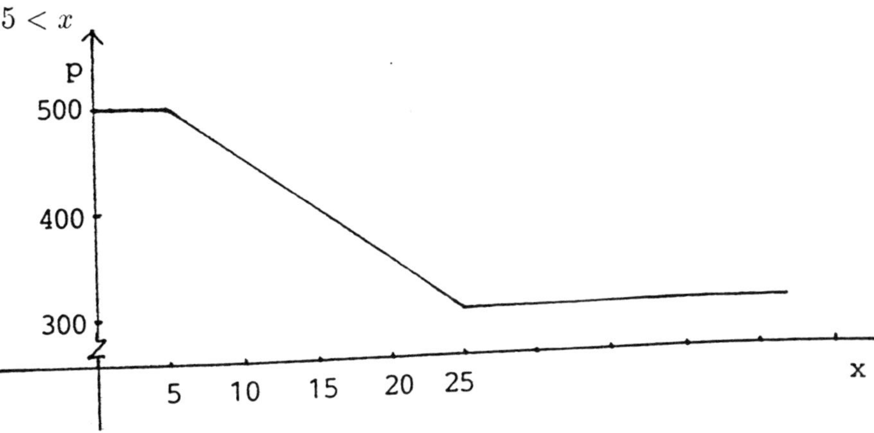

44. When $x \le 5$, $R(x) = 500x$.
When $5 < x \le 25$, $R(x) = 550x - 10x^2$.
When $x > 25$, $R(x) = 300x$.

$$R(x) = \begin{cases} 500x, & 0 < x \le 5 \\ 550x - 10x^2, & 5 < x \le 25 \\ 300x, & x > 25. \end{cases}$$

45. $(f \circ r)(c) = f(r(c)) = f(10 + 0.2c + 0.01c^2) = \sqrt{10 + 0.2c + 0.01c^2 + 5}$.

46. Let $V(t)$ = the value of a truck after t years.
Then $V(t) = 20000 \cdot (1 - 0.3)^t = 20000(0.7)^t$.
Since $C(V) = 0.8V$, $C(V(t)) = .08 \cdot (20000(0.7)^t) = 1600(.7)^t$ dollars.

47. a. $C(x) = 200x + 75$ dollars.
b. $R(x) = (1 + .4)\cdot(200x + 75) = 280x + 105$ dollars.
c. The retail price per lawnmower $= \frac{R(x)}{x} = p(x) = \frac{280x+105}{x} = 280 + \frac{105}{x}$ dollars.

48. a. The profit $P(x)$ from the production and sales of x computers per week is given by

$$P(x) = R(x) - C(x) - 0.05x.$$

b. $P(x) = 5000x - (25,000 + 2,500x + 5x^{2/3} + 2x^2) - 0.05x$
$= 2,499.95x - 25,000 - 5x^{2/3} - 2x^2.$

Solutions to Chapter 1 – Review Exercises

1. a. $(-6, 3]$ b. $(-\infty, 4)$ c. $[2, \infty)$
2. $2x - 6 \leq 4$, $2x \leq 10$, $x \leq 5$.
3. $3 - x \geq 4 + x$, $-2x \geq 1$, $x \leq -\frac{1}{2}$.
4. $6x - 6 < x + 3$, $6x < x + 9$, $5x < 9$, $x < \frac{9}{5}$.
5. $2 - x \leq 4 - 4x$, $-x \leq 2 - 4x$, $3x \leq 2$, $x \leq \frac{2}{3}$.
6. $f(x) = (x - 2)^3$ 7.

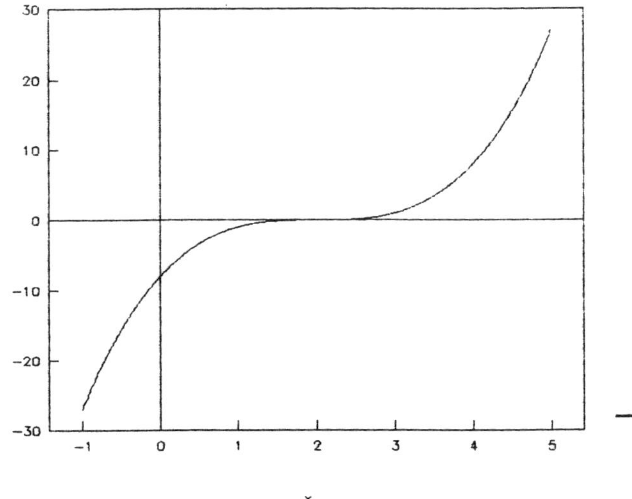

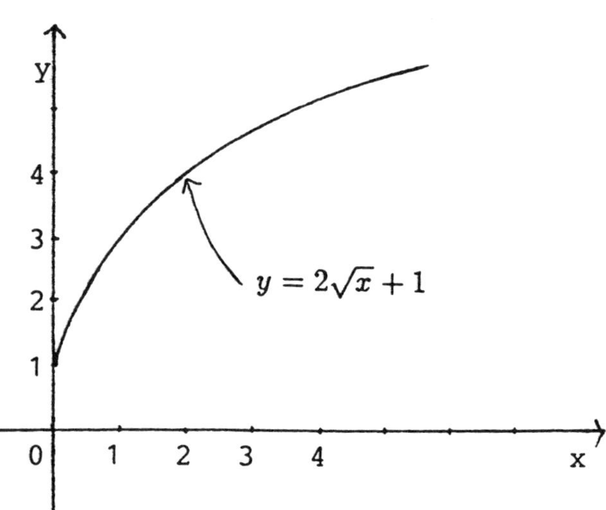

8.

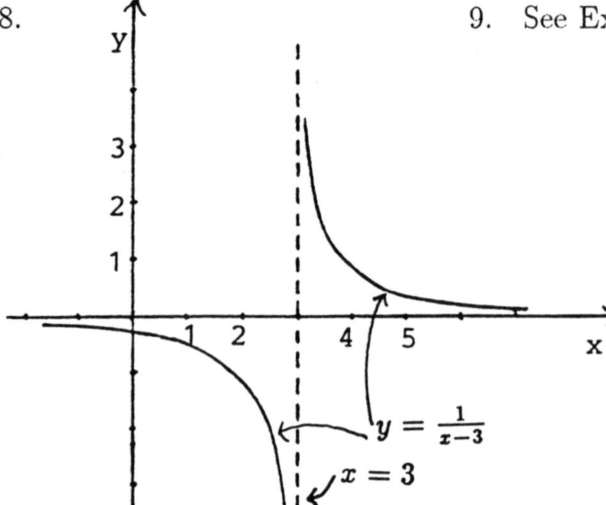

9. See Ex. 36 in Sec. 1.5.

10. Slope $= \frac{1-2}{4-(-3)} = \frac{-1}{7}$.
11. $x - 6y = 24$, $y = \frac{1}{6}x - 4$. The slope m_1 of this line is $\frac{1}{6}$.
 The required slope is $m_1 = \frac{1}{6}$.
12. From problem 11, we know that the slope m_1 of the line $x - 6y = 24$ is $\frac{1}{6}$. Let m_2 = slope of the required line. Since these two lines are perpendicular, $m_1 \cdot m_2 = -1 \Rightarrow \frac{1}{6} \cdot m_2 = -1 \Rightarrow m_2 = -6$. Thus an equation for the required line is: $y - (-3) = -6 \cdot (x - 2)$, $y = -6x + 9$.
13. $y = 5$.
14. $x = -6$.
15. The equation of the line is $x = 7$.
 Since $(b, 4)$ lies on the line, b must be 7.
16. The slope of the line $= \frac{4-(-6)}{-3-2} = -2$.
 The line is: $y - (-6) = -2(x - 2)$, $2x + y + 2 = 0$.
17. $-2y = 6 - x$ $y - \text{int} = -3$
 $y = -3 + \frac{x}{2}$
 $m = \frac{1}{2}$

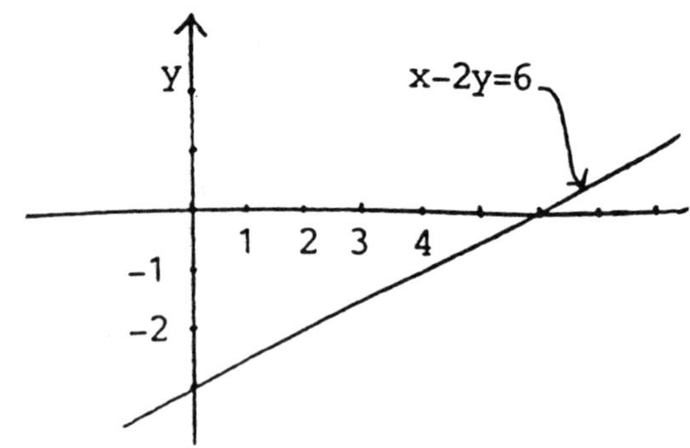

18. $-y = -4 - 3x \quad y-$ int $= 4.$
$y = 4 + 3x$
$m = 3$

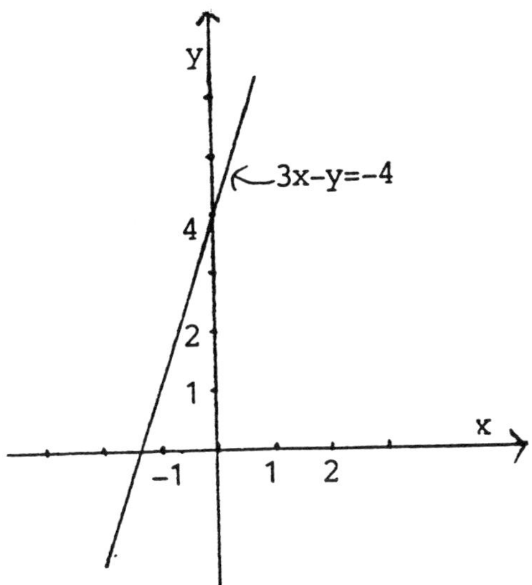

19. $2y = -2x + 8$
$y = -x + 4$
$m = -1, \quad b = 4.$

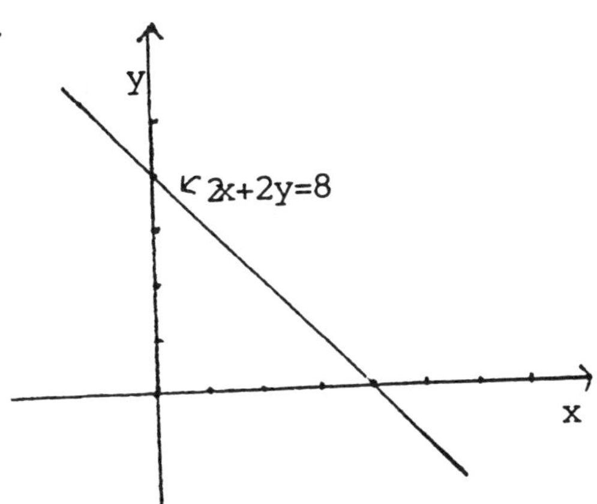

20. $3 - y = 4x + 6 \implies y = -4x - 3$
Slope $m = -4, \quad y$-intercept $b = -3.$

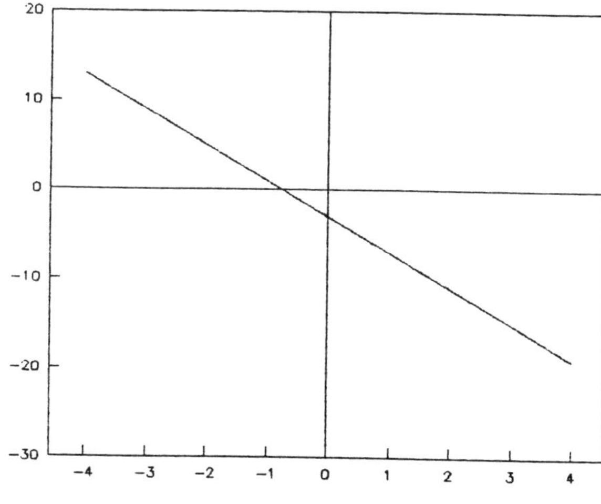

21. Domain is the set of all x except 0.
22. Domain is the set of all x such that $x^2 - 4 \neq 0$, so the domain is the set of all x except 2 and -2.
23. $f(x) = \sqrt{8 - x^3}$
    $8 - x^3 = 0$ for $x = 2$.
    $8 - x^3 > 0$ for $x < 2$.
    Domain is the set of all x such that $x \leq 2$.
24. Domain is the set of all x such that $(x^2 - 2x - 15) \neq 0$. Since $(x^2 - 2x - 15) = (x - 5)(x + 3)$, the domain is the set of all x except 5 and -3.
25. a. $f(6) = \frac{\sqrt{6-2}}{6+3} = \frac{2}{9}$.
    b. $f(11) = \frac{\sqrt{11-2}}{11+3} = \frac{3}{14}$.
26. a. $f(4) = 4^{3/2} + 2(4) = 8 + 8 = 16$
    b. $f(9) = 9^{3/2} + 2(9) = 27 + 18 = 45$.
27. $16^{3/4} = (16^{1/4})^3 = 2^3 = 8$.
28. $27^{-4/3} = \left(27^{1/3}\right)^{-4} = (3)^{-4} = \frac{1}{81}$
29. $\frac{(x^2y^4)^2}{xy^2} = \frac{x^4y^8}{xy^2} = x^3y^6$.
30. $\frac{3\sqrt{x}\,y}{x^2\sqrt{y}} = \frac{3x^{1/2}y}{x^2y^{1/2}} = 3y^{1/2}x^{-3/2}$
31.

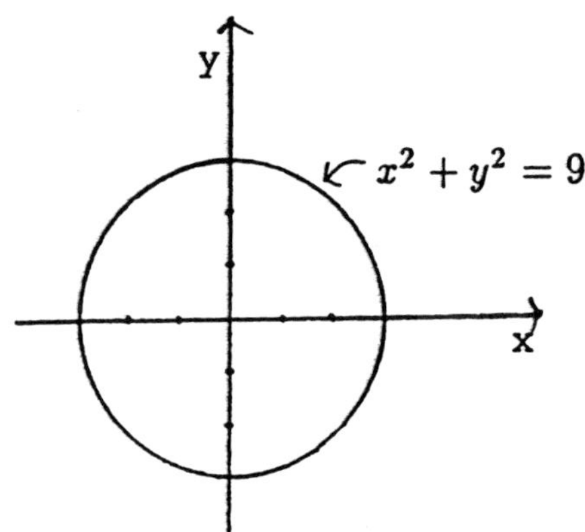

32.

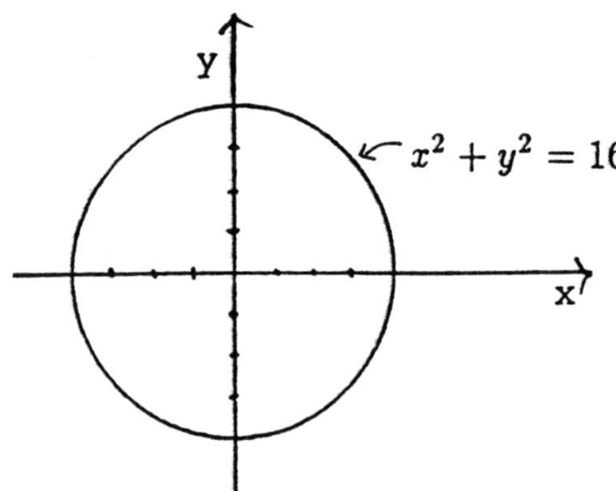

33. $x^2 + 2x + y^2 = 7 + 2y \implies x^2 + 2x + y^2 - 2y = 7$
implies $(x^2 + 2x + 1) + (y^2 - 2y + 1) = 7+1+1 \implies (x+1)^2 + (y-1)^2 = 3^2$.
The center is $(-1, 1)$, and the radius is 3.

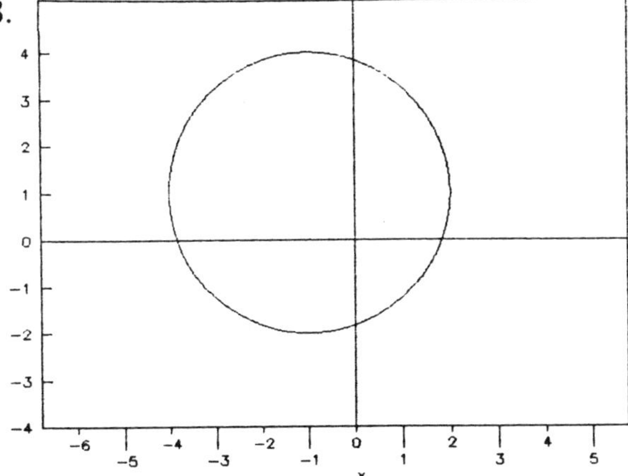

34. $x^2 + y^2 + 6y = 13$, $x^2 + (y^2 + 6y + 3^2) - 9 = 13$, $x^2 + (y+3)^2 = 22$. The center is $(0, -3)$, and the radius is $\sqrt{22} = 4.7$.

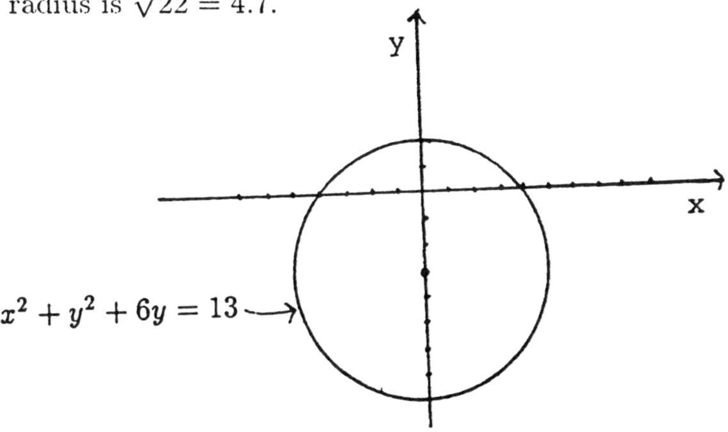

35. $x^2 + 4x + y^2 + 2y = 14$, $(x^2 + 4x + 2^2) - 4 + (y^2 + 2y + 1^2) - 1 = 14$,
$(x+2)^2 + (y+1)^2 = 19$. The center is $(-2, -1)$, and the radius is $\sqrt{19} = 4.36$.

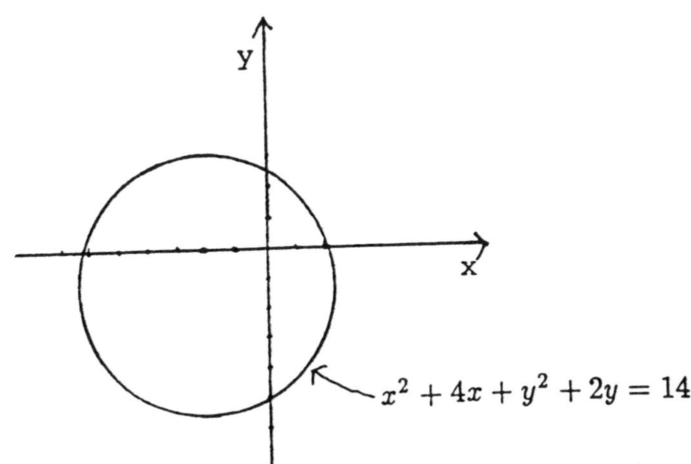

36. $x^2-2x+y^2-9y=22$, $(x^2-2x+(-1)^2)-1+\left(y^2-9y+\left(-\frac{9}{2}\right)^2\right)-20.25=22$, $(x-1)^2+(y-4.5)^2=43.25$. The center is $(1, 4.5)$, and the radius is $\sqrt{43.25}=6.58$.

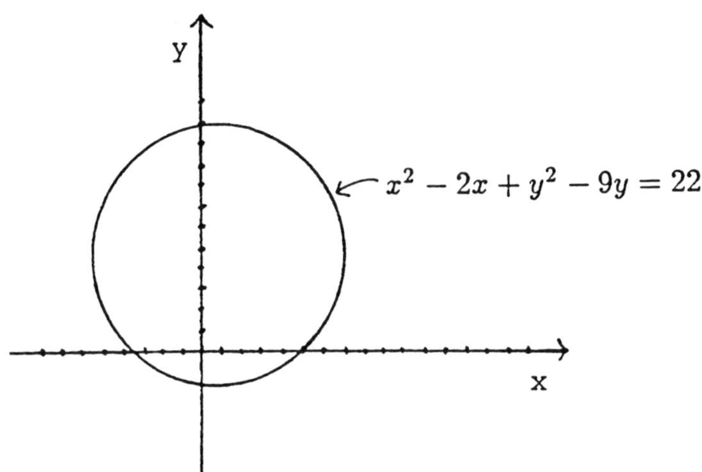

37. $f(x)=x^2-3x+2=(x-2)(x-1)$
$f(x)=0$ for $x=2$, $x=1$.
38. $f(x)=x^2-8x+16=(x-4)^2$.
$f(x)=0$ for $x=4$.
39. $f(x)=0$ for $x-3=0 \Rightarrow x=3$.
40. $f(x)=\sqrt{x}-x^{3/2}=x^{1/2}\cdot(1-x)$
$f(x)=0$ for $x=0$, $x=1$.
41. $f(x)=\frac{x^2+9}{x-2}$ $f(x)$ has no zeros.
42. $f(x)=6-2x-x^2=0$ for $x=\frac{-(-2)\pm\sqrt{(-2)^2-4(-1)(6)}}{2(-1)}=\frac{2\pm\sqrt{28}}{-2}=\frac{-2\pm2\sqrt{7}}{2}=-1\pm\sqrt{7}$.
43. $f(x)=g(x)$, $x-6=2x+2$, $x=-8$.
44. $f(x)=g(x)$, $\sqrt{x+2}=x$, $(x+2)^{1/2}=x$, $x+2=x^2$, $x^2-x-2=0$, $(x-2)(x+1)=0 \Rightarrow x=2$, $x=-1$ (extraneous)
45. $f(x)=g(x)$, $2-x^2=4x^2-3$, $5x^2=5$, $x^2=1 \Rightarrow x=1$, $x=-1$.
46. $f(x)=g(x) \implies \frac{1}{x+2}=x+2 \implies 1=(x+2)^2$
$\implies x+2=\pm1 \implies x=-1$ or $x=-3$.
47. $h_1(x)=(x^3+2)\cdot(x+4)=x^4+4x^3+2x+8$.
48. $h_2(x)=\frac{x^3+2}{x+4}$.
49. $h_1(0)=(0^3+2)\cdot(0+4)=8$.
50. $h_2(1)=\frac{1^3+2}{1+4}=\frac{3}{5}$.
51. $(g\circ f)(x)=g(f(x))=g(x^2-7)=\sqrt{(x^2-7)+3}=\sqrt{(x^2-4)}$.
52. $(g\circ u)(x)=g(u(x))=g(4x)=\sqrt{4x+3}$.
53. $(u\circ f)(x)=u(f(x))=u(x^2-7)=4(x^2-7)=4x^2-28$.

54. $(u \circ f \circ g)(x) = u(f(g(x))) = u\left(f\left(\sqrt{x+3}\right)\right) = u\left(\left(\sqrt{x+3}\right)^2 - 7\right)$
$= u(x-4) = 4(x-4).$

55. $f(x) = |x+3| = \begin{cases} -x-3, & x \le -3 \\ x+3, & x \ge -3 \end{cases}$

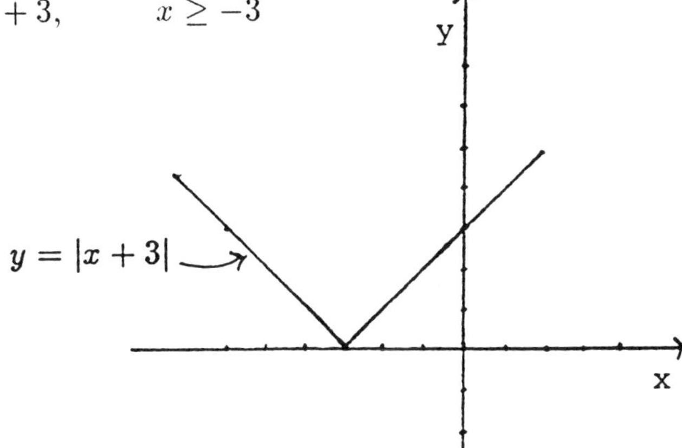

56. $f(x) = |x^2 - 2x - 3| = \begin{cases} x^2 - 2x - 3 & \text{for } x \le -1 \text{ or } x \ge 3 \\ -x^2 + 2x + 3 & \text{for } -1 \le x \le 3. \end{cases}$

x	$f(x)$
−3	12
−2	5
−1	0
0	3
1	4
2	3
3	0
4	5
5	12

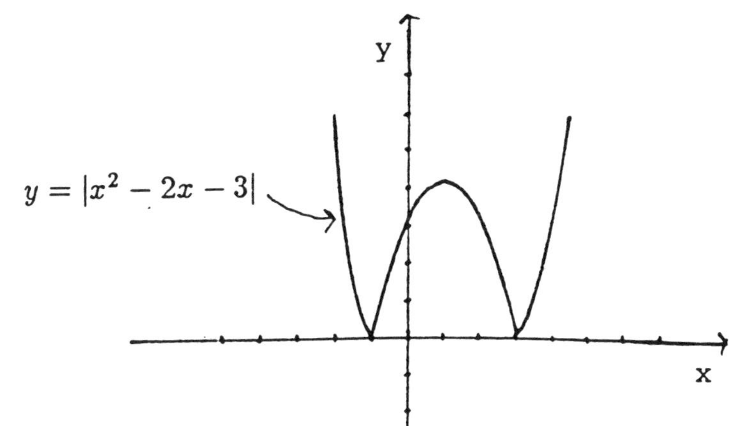

57. $f(x) = \begin{cases} x - 4, & x < 2 \\ -\frac{1}{2}x^2, & x \geq 2 \end{cases}$

x	$f(x)$
-1	-5
0	-4
1	-3
2	-2
37	$-\frac{9}{2}$
4	-8

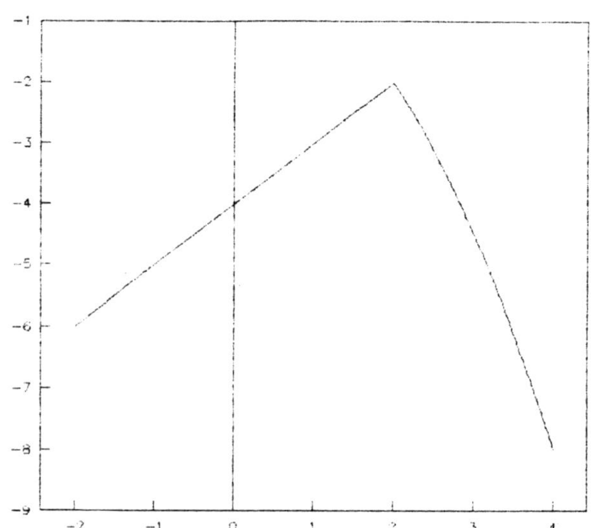

58. $f(x) = \begin{cases} \sqrt{2 - x}, & x \leq 1 \\ 2x^2 - 1, & x > 1 \end{cases}$

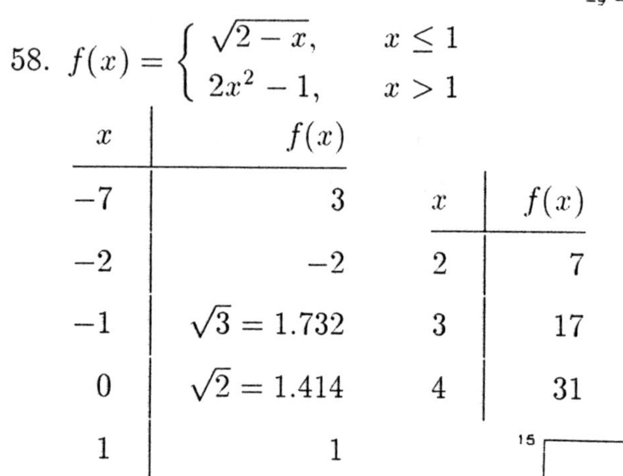

x	$f(x)$
-7	3
-2	-2
-1	$\sqrt{3} = 1.732$
0	$\sqrt{2} = 1.414$
1	1

x	$f(x)$
2	7
3	17
4	31

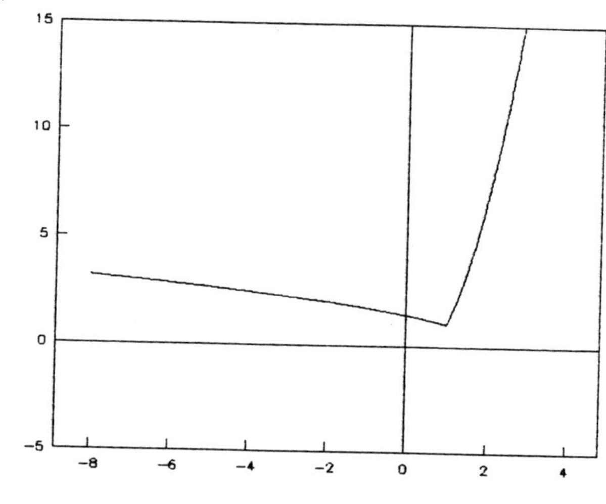

59. $D(p) = S(p)$. $2000 - 6p = 4p \Rightarrow 10p = 2000, \quad p = 200$.
The equilibrium price p_0 is \$200.

60. $T(x) = 10\left(1 + \frac{4}{\sqrt{x+2}}\right)$.

x	$T(x)$
1	33.1
2	30
4	26.33
6	24.14
8	22.65
14	20

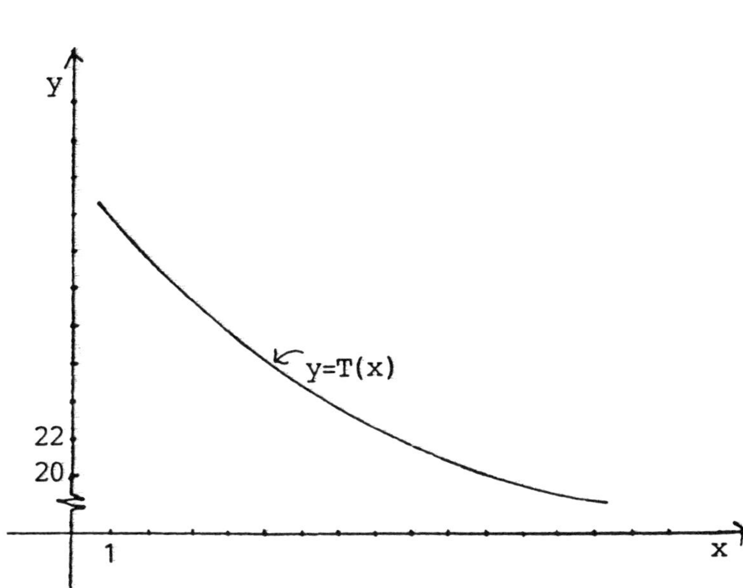

61. Figures b and c are graphs of functions.
62. $f(x) = 3(x-2)^2 + 4$. The lowest point is $(2, 4)$.
63. False.
64. $y - (-3) = -4(x-2)$, $y = -4x + 5$.
65. Let $x =$ the number of people served.
Let $C(x) =$ the corresponding total cost of the meal.
For $x \leq 50$, $C(x) = 20x$.
For $x > 50$, $C(x) = 20 \cdot 50 + 15(x - 50) = 1000 + 15x - 750 = 250 + 15x$.
66. Let x be the number of years after 1989, and let P be the corresponding cotton production in the United States in milions of bales.
$P = mx + b$. $\quad m = \frac{18-12.2}{2} = 2.9$.
$b = 12.2$
$P = 2.9x + 12.2$.
When $x = 6$, $\quad P = 2.9(6) + 12.2 = 29.6$ million bales.
67. Let x be the number of years after 1989, and let P_2 be the corresponding cotton production in India in millions of bales.
$P_2 = m_2 x + b_2$.
$m_2 = \frac{9.5-10.7}{2} = -.6$.
$b_2 = 10.7$. $\quad P_2 = -.6x + 10.7$
When $x = 7$, $\quad P_2 = -.6(7) + 10.7 = 6.5$ million bales.

68. $(c \circ n)(t) = c(n(t)) = \sqrt{1 + 0.2n(t)} = \sqrt{1 + 0.2(10,000 + 50t^{2/3})}$
$= \sqrt{1 + 2000 + 10t^{2/3}} = \sqrt{2001 + 10t^{2/3}}$.

69. Let x by the number of years after 1940, and let T be the average daily temperature in degrees Fahrenheit in this certain city.
$T = mx + b$.
$m = \frac{59.2-58}{40} = .03$.
$b = 58$. $\quad T = .03x + 58$.
When $x = 60$, $\quad t = .03(60) + 58 = 59.8°\text{F}$.

70. Cab fare is $2 + 1.20x$ where x = number of miles.
$10 = 2 + 1.2x \quad 8 = 1.20x \quad 6.7 = x$.

71. $D = (1 - r)T$.
a. Since $r = 0.3$ and $T = 30,000$, $D = (1 - 0.3) \cdot 30,000 = \$21,000$.
b. Since $D = 16,000$ and $T = 20,000$, $16,000 = (1 - r) \cdot 20,000$.
$0.8 = (1 - r)$. $\therefore r = 0.2 = 20\%$.
c. Since $D = 30,000$ and $r = 0.4$, $30,000 = (1 - 0.4) \cdot T$.
$30,000 = 0.6T$. $\therefore T = \$50,000$.

72. $V(n) = V_s + (V_0 - V_s)\left(1 - \frac{n}{N}\right)$
$= V_s + (V_0 - V_s) - (V_0 - V_s)\frac{n}{N}$
$= V_0 - \frac{V_0 - V_s}{N} \cdot n = m \cdot n + b$.
$\therefore m = -\frac{V_0 - V_s}{N}$, $b = V_0$.

73. $V(n) = -\frac{V_0 - V_s}{N} \cdot n + V_0$.
Since $V_0 = 30,000$, $V_s = 5,000$, and $N = 5$,

$$V(n) = -\frac{25,000}{5} \cdot n + 30,000 = -5,000 \cdot n + 30,000$$

For $n = 3$, $V(3) = -5,000 \cdot 3 + 30,000 = 15,000$.

74. $D(p) = \frac{400}{p^2+2p+5}$.
$D = 20$, $20 = \frac{400}{p^2+2p+5}$.
$p^2 + 2p + 5 = 20 \Rightarrow p^2 + 2p - 15 = 0$
$(p + 5)(p - 3) = 0$. $\therefore p = 3$ dollars.

75. $Y = (400 - 2x) \cdot x$.

76. We know that $R(0) = 0$ and $R(6) = 19$. Thus the linear function is
$R(w) = \frac{19-0}{6-0}w = \frac{19}{6}w$. $\quad \therefore R(15) = \frac{19}{6} \cdot 15 = \frac{95}{2}$.

77. $C(x) = C_0 + 16x + .02x^2$ and $R(x) = 24x$.
From $C(x) = R(x)$, we get $C_0 + 16x + 0.02x^2 = 24x$; $0.02x^2 - 8x + C_0 = 0$.
Since $C(x) = R(x)$ at $x = 100$, $\; 0.02 \cdot 100^2 - 8 \cdot 100 + C_0 = 0 \;\; \therefore C_0 = \$600$.

78. $C = (100 + 0.2 \cdot (800 - 500))x = (100 + 0.2 \cdot 300)x = 160x$.

79. $V(n) = (100,000)\left(1 - \frac{n}{20}\right)$.
(a) $V(5) = (100,000)\left(1 - \frac{5}{20}\right) = (100,000)\left(\frac{3}{4}\right) = 75,000$.
(b) $V(n) = (100,000)\left(1 - \frac{n}{30}\right)$.
(c) $V(5) = (100,000)\left(1 - \frac{5}{30}\right) = (100,000)\left(\frac{5}{6}\right) = 83,333.33$.

80. From $S(p_0) = D(p_0)$, we have
$2p_0 - 8 = \frac{154}{P_0}$.
$2p_0^2 - 8p_0 - 154 = 0$
$p_0^2 - 4p_0 - 77 = 0$
$(p_0 - 11)(p_0 + 7) = 0 \therefore p_0 = 11, -7$.
Since p_0 can't be negative,$p_0 = 11$.

81. Production level of sugar follows a linear function of times in U.S.A., Eastern Europe, South Africa and Australia.

82. Let t be the number of years after 1970, let P_1 be the corresponding domestic automobile production level for Japan in millions of automobiles, and let P_2 be the corresponding domestic automobile production level for the United States in millions of automobiles.
   a. $P_1 = m_1 t + b_1$.
   $m_1 = \frac{13-4.5}{20} = .425$.
   $b_1 = 4.5$.
   $P_1 = .425t + 4.5$.
   b. $P_2 = m_2 t + b_2$.
   $m_2 = \frac{8.3-7.3}{20} = .05$.
   $b_2 = 7.3$.
   $P_2 = .05t + 7.3$.
   c. $P_1 = P_2 \implies .425t + 4.5 = .05t + 7.3 \implies .375t = 2.8 \implies t \approx 7.5$.
   d. When $t = 30$, $P_1 = .425(30) + 4.5 = 17.25$ million automotiles and $P_2 = .05(30) + 7.3 = 8.8$ million automobiles.

83. Let t be the number of years after 1982, let C_1 be the avaerage manufacturing costs for small cars in Japan, and let C_2 be the average manufacturing costs for small cars in the United States.
   a. $C_1 = m_1 t + b_1$.
   $m_1 = \frac{7200-4200}{10} = 300$.
   $b_1 = 4200$.
   $C_1 = 300t + 4200$.
   $C_2 = m_2 t + b_2$.
   $m_2 = \frac{7400-6400}{10} = 100$.
   $b_2 = 6400$.
   $C_2 = 100t + 6400$.
   b. The linear model for Japan has a larger slope, so the average manufacturing costs for small cars are increasing more rapidly in Japan.
   c. $C_1 = C_2 \implies 300t + 4200 = 100t + 6400 \implies 200t = 2200 \implies t = 11$. Thus 11 years after 1982, that is in 1993, the average manufacturing costs for small cars will be the same in Japan and the United States.
   d. When $t = 15$, $C_1 = 300(15) + 4200 = \$8700$, and $C_2 = 100(15) + 6400 = \$7,900$.

CHAPTER 2

Solutions to Exercise Set 2.1

1. The tangent line appears to go through points $(1, 8)$ and $(0, 10)$, so the slope $m_{\tan}$ is $\frac{10-8}{0-1} = -2$.
2. The tangent line appears to go through points $(1, 1)$ and $(0, -2)$, so the slope $m_{\tan}$ is $\frac{-2-1}{0-1} = 3$.
3. The tangent line appears to go through points $(4, 2)$ and $(0, 1)$, so the slope $m_{\tan}$ is $\frac{1-2}{0-4} = \frac{1}{4}$.
4. The tangent line appears to go through points $(1, -1)$ and $(0, 0)$, so the slope $m_{\tan}$ is $\frac{0-(-1)}{0-1} = -1$.
5. The tangent line appears to go through points $\left(1, \frac{1}{2}\right)$ and $(0, 1)$, so the slope $m_{\tan}$ is $\frac{1-\frac{1}{2}}{0-1} = -\frac{1}{2}$.
6. The tangent line appears to go through points $(0, -3)$ and $(2, 1)$, so the slope $m_{\tan}$ is $\frac{1-(-3)}{2-0} = 2$.
7. $m = \lim_{h\to 0} \frac{f(1+h)-f(1)}{h} = \lim_{h\to 0} \frac{(1+h)^2-1^2}{h} = \lim_{h\to 0} \frac{1+2h+h^2-1}{h}$
$= \lim_{h\to 0} \frac{2h+h^2}{h} = \lim_{h\to 0} \frac{h(2+h)}{h} = \lim_{h\to 0}(2+h) = 2.$
8. $m = \lim_{h\to 0} \frac{f(2+h)-f(2)}{h} = \lim_{h\to 0} \frac{3(2+h)^2-1-(3\cdot 2^2-1)}{h} = \lim_{h\to 0} \frac{3(4+4h+h^2)-1-11}{h}$
$= \lim_{h\to 0} \frac{12+12h+3h^2-12}{h} = \lim_{h\to 0} \frac{3h(4+h)}{h} = \lim_{h\to 0} 3(4+h) = 12.$
9. $m = \lim_{h\to 0} \frac{f(1+h)-f(1)}{h} = \lim_{h\to 0} \frac{2-(1+h)^2-(2-1^2)}{h} = \lim_{h\to 0} \frac{2-1-2h-h^2-1}{h}$
$= \lim_{h\to 0} \frac{-h(2+h)}{h} = \lim_{h\to 0} -(2+h) = -2.$

10. $m = \lim_{h\to 0} \frac{f(-1+h)-f(-1)}{h} = \lim_{h\to 0} \frac{(-1+h)^2+(-1+h)-((-1)^2-1)}{h}$

$= \lim_{h\to 0} \frac{1-2h+h^2-1+h-0}{h} = \lim_{h\to 0} \frac{h(-2+h+1)}{h} = \lim_{h\to 0}(-1+h) = -1.$

11. $m = \lim_{h\to 0} \frac{f(1+h)-f(1)}{h} = \lim_{h\to 0} \frac{(1+h)^3-1^3}{h} = \lim_{h\to 0} \frac{1+3h+3h^2+h^3-1^3}{h}$

$= \lim_{h\to 0} \frac{h(3+3h+h^2)}{h} = \lim_{h\to 0}(3+3h+h^2) = 3.$

12. $m = \lim_{h\to 0} \frac{f(2+h)-f(2)}{h} = \lim_{h\to 0} \frac{(2+h)^2+(2+h)-2-(2^2+2-2)}{h} = \lim_{h\to 0} \frac{4+4h+h^2+2+h-6}{h}$

$= \lim_{h\to 0} \frac{h(5+h)}{h} = \lim_{h\to 0}(5+h) = 5.$

13. $m = \lim_{h\to 0} \frac{f(-1+h)-f(-1)}{h} = \lim_{h\to 0} \frac{(-1+h)^3+(-1+h)-(-2)}{h} = \lim_{h\to 0} \frac{-1+3h-3h^2+h^3-1+h+2}{h}$

$= \lim_{h\to 0} \frac{4h-3h^2+h^3}{h} = \lim_{h\to 0} \frac{h(4-3h+h^2)}{h} = \lim_{h\to 0}(4-3h+h^2) = 4.$

14. $m = \lim_{h\to 0} \frac{f(2+h)-f(2)}{h} = \lim_{h\to 0} \frac{(2+h)^4-2^4}{h}$

$= \lim_{h\to 0} \frac{16+32h+24h^2+8h^3+h^4-16}{h} = \lim_{h\to 0} \frac{h(32+24h+8h^2+h^3)}{h}$

$= \lim_{h\to 0}(32+24h+8h^2+h^3) = 32.$

15. $m = \lim_{h\to 0} \frac{f(1+h)-f(1)}{h} = \lim_{h\to 0} \frac{((1+h)^4+(1+h))-(1^4+1)}{h}$

$= \lim_{h\to 0} \frac{(1+4h+6h^2+4h^3+h^4+1+h)-(1+1)}{h}$

$= \lim_{h\to 0} \frac{h(5+6h+4h^2+h^3)}{h} = \lim_{h\to 0}(5+6h+4h^2+h^3) = 5.$

16. $m = \lim_{h\to 0} \frac{f(1+h)-f(1)}{h} = \lim_{h\to 0} \frac{\frac{1}{1+h}-\frac{1}{1}}{h} = \lim_{h\to 0} \frac{1}{h}\left\{\frac{1}{1+h}-1\right\}$

$= \lim_{h\to 0} \frac{1}{h}\left\{\frac{1}{1+h}-\frac{1+h}{1+h}\right\} = \lim_{h\to 0} \frac{1}{h}\left\{\frac{1-1-h}{1+h}\right\} = \lim_{h\to 0} \frac{1}{h}\left(\frac{-h}{1+h}\right)$

$= \lim_{h\to 0} \frac{h}{h}\left(\frac{-1}{1+h}\right) = -1.$

17. $m = \lim_{h\to 0} \frac{f(-2+h)-f(-2)}{h} = \lim_{h\to 0} \frac{\frac{1}{(-2+h)+3}-\frac{1}{(-2+3)}}{h} = \lim_{h\to 0} \frac{\frac{1}{1+h}-1}{h}$

$= \lim_{h\to 0} \frac{1}{h}\left\{\frac{1}{1+h}-1\right\} = \lim_{h\to 0} \frac{1}{h}\left\{\frac{1}{1+h}-\frac{1+h}{1+h}\right\} = \lim_{h\to 0} \frac{1}{h}\left\{\frac{1-1-h}{1+h}\right\}$

$= \lim_{h\to 0} \frac{1}{h}\left(\frac{-h}{1+h}\right) = \lim_{h\to 0} \frac{h}{h}\left(\frac{-1}{1+h}\right) = -1.$

18. $m = \lim_{h\to 0} \frac{f(2+h)-f(2)}{h} = \lim_{h\to 0} \frac{\frac{2+h}{1+(2+h)}-\frac{2}{1+2}}{h} = \lim_{h\to 0} \frac{\frac{2+h}{3+h}-\frac{2}{3}}{h}$

$= \lim_{h\to 0} \frac{1}{h}\left\{\frac{2+h}{3+h}-\frac{2+\frac{2}{3}h}{3+h}\right\} = \lim_{h\to 0} \frac{1}{h}\left\{\frac{h-\frac{2}{3}h}{3+h}\right\} = \lim_{h\to 0} \frac{h}{h}\left\{\frac{1-\frac{2}{3}}{3+h}\right\}$

$= \lim_{h\to 0} \frac{\frac{1}{3}}{3+h} = \frac{1}{9}.$

19. $m = \lim_{h\to 0} \frac{f(3+h)-f(3)}{h} = \lim_{h\to 0} \frac{\sqrt{(3+h)+1}-\sqrt{3+1}}{h}$
$= \lim_{h\to 0} \frac{\sqrt{4+h}-2}{h} = \lim_{h\to 0} \frac{\sqrt{4+h}-2}{h} \cdot \frac{\sqrt{4+h}+2}{\sqrt{4+h}+2}$
$= \lim_{h\to 0} \frac{\left(\sqrt{4+h}\right)^2-2^2}{h\left(\sqrt{4+h}+2\right)} = \lim_{h\to 0} \frac{(4+h)-4}{h\left(\sqrt{4+h}+2\right)} = \lim_{h\to 0} \frac{1}{\sqrt{4+h}+2} = \frac{1}{4}.$

20. $m = \lim_{h\to 0} \frac{f(4+h)-f(4)}{h} = \lim_{h\to 0} \frac{\left(2-\sqrt{4+h}\right)-\left(2-\sqrt{4}\right)}{h}$
$= \lim_{h\to 0} \frac{2-\sqrt{4+h}}{h} = \lim_{h\to 0} \frac{2-\sqrt{4+h}}{h} \cdot \frac{2+\sqrt{4+h}}{2+\sqrt{4+h}}$
$= \lim_{h\to 0} \frac{(2)^2-\left(\sqrt{4+h}\right)^2}{h\left(2+\sqrt{4+h}\right)} = \lim_{h\to 0} \frac{4-(4+h)}{h\left(2+\sqrt{4+h}\right)} = \lim_{h\to 0} \frac{-1}{2+\sqrt{4+h}} = -\frac{1}{4}.$

21. $m = \lim_{h\to 0} \frac{f(4+h)-f(4)}{h} = \lim_{h\to 0} \frac{(4+h)^{3/2}-4^{3/2}}{h}$
$= \lim_{h\to 0} \frac{\sqrt{(4+h)^3}-8}{h} = \lim_{h\to 0} \frac{\sqrt{(4+h)^3}-8}{h} \cdot \frac{\sqrt{(4+h)^3}+8}{\sqrt{(4+h)^3}+8}$
$= \lim_{h\to 0} \frac{(4+h)^3-64}{h\left(\sqrt{(4+h)^3}+8\right)} = \lim_{h\to 0} \frac{(64+48h+12h^2+h^3)-64}{h\left(\sqrt{(4+h)^3}+8\right)}$
$= \lim_{h\to 0} \frac{h(48+12h+h^2)}{h\left(\sqrt{(4+h)^3}+8\right)} = \lim_{h\to 0} \frac{48+12h+h^2}{\sqrt{(4+h)^3}+8} = \frac{48}{16} = 3.$

22. $m = \lim_{h\to 0} \frac{f(1+h)-f(1)}{h} = \lim_{h\to 0} \frac{4(1+h)^2-3-(4\cdot 1^2-3)}{h} = \lim_{h\to 0} \frac{4(1+2h+h^2)-3-1}{h}$
$= \lim_{h\to 0} \frac{4+8h+4h^2-4}{h} = \lim_{h\to 0} \frac{h(8+4h)}{h} = \lim_{h\to 0}(8+4h) = 8.$

An equation for the line is $y-1=8(x-1)$, $y-1=8x-8$, $y=8x-7$.

23. $m = \lim_{h\to 0} \frac{f(-2+h)-f(-2)}{h} = \lim_{h\to 0} \frac{9-(-2+h)^2-\left(9-(-2)^2\right)}{h} = \lim_{h\to 0} \frac{9-4+4h-h^2-5}{h}$
$= \lim_{h\to 0} \frac{h(4-h)}{h} = \lim_{h\to 0}(4-h) = 4.$

An equation for the line is $y-5=4(x+2)$, $y-5=4x+8$, $y=4x+13$.

24. $m = \lim_{h\to 0} \frac{f(1+h)-f(1)}{h} = \lim_{h\to 0} \frac{\frac{2}{3+(1+h)}-\frac{2}{3+1}}{h} = \lim_{h\to 0} \frac{\frac{2}{4+h}-\frac{2}{4}}{h}$
$= \lim_{h\to 0} \frac{1}{h}\left\{\frac{2}{4+h} - \frac{2+\frac{1}{2}h}{4+h}\right\} = \lim_{h\to 0} \frac{1}{h}\left\{\frac{2-2-\frac{1}{2}h}{4+h}\right\} = \lim_{h\to 0} \frac{h}{h}\left(\frac{-\frac{1}{2}}{4+h}\right)$
$= \lim_{h\to 0} \left(\frac{-\frac{1}{2}}{4+h}\right) = -\frac{1}{8}.$

An equation for the tangent line is $y-\frac{1}{2} = -\frac{1}{8}(x-1)$, $y-\frac{1}{2} = -\frac{1}{8}x+\frac{1}{8}$, $y = -\frac{1}{8}x+\frac{5}{8}$.

25. slope $= \lim_{h\to 0} \frac{f(x+h)-f(x)}{x} = \lim_{h\to 0} \frac{(m(x+h)+b)-(mx+b)}{h} = \lim_{h\to 0} \frac{mh}{h} = m.$

26. $m = \lim_{h\to 0} \frac{S(4+h)-S(4)}{h} = \lim_{h\to 0} \frac{\frac{1}{4}(4+h)^2+4-\left(\frac{1}{4}(4)^2+4\right)}{h} = \lim_{h\to 0} \frac{\frac{1}{4}(16+8h+h^2)+4-8}{h}$
$= \lim_{h\to 0} \frac{4+2h+\frac{1}{4}h^2-4}{h} = \lim_{h\to 0} \frac{h(2+\frac{1}{4}h)}{h} = \lim_{h\to 0}\left(2+\frac{1}{4}h\right) = 2.$

27. $m = \lim_{h\to 0} \frac{D(6+h)-D(6)}{h} = \lim_{h\to 0} \frac{\frac{10}{(6+h)+4} - \frac{10}{6+4}}{h} = \lim_{h\to 0} \frac{\frac{10}{10+h} - \frac{10}{10}}{h}$
$= \lim_{h\to 0} \frac{1}{h}\left(\frac{10}{10+h} - \frac{10+h}{10+h}\right) = \lim_{h\to 0} \frac{1}{h}\left(\frac{10-10-h}{10+h}\right) = \lim_{h\to 0} \frac{h}{h}\left(\frac{-1}{10+h}\right) = -\frac{1}{10}.$

28. a. $m = \lim_{h\to 0} \frac{f(-1+h)-f(-1)}{h} = \lim_{h\to 0} \frac{4-(-1+h)^2-(4-(-1)^2)}{h} = \lim_{h\to 0} \frac{4-(1-2h+h^2)-3}{h}$
$= \lim_{h\to 0} \frac{4-1+2h-h^2-3}{h} = \lim_{h\to 0} \frac{h(2-h)}{h} = 2.$

b. $m = \lim_{h\to 0} \frac{f(2+h)-f(2)}{h} = \lim_{h\to 0} \frac{2(2+h)+1-(2(2)+1)}{h} = \lim_{h\to 0} \frac{4+2h+1-4-1}{h}$
$= \lim_{h\to 0} \frac{h}{h} \cdot 2 = 2.$

29. $f(x) = -x + 3$ for $x < 3$, $f(x) = x - 3$ for $x \geq 3$.

a. for $a < 3$, the slope is -1.

b. for $a > 3$, the slope is 1.

30. a. $S(p_0) = D(p_0)$, $\frac{1}{4}p_0^2 = 12 - 2p_0$, $p_0^2 + 8p_0 - 48 = 0$,

$(p_0 + 12) \cdot (p_0 - 4) = 0.$

The equilibrium price $p_0 = 4$.

b. $m = \lim_{h\to 0} \frac{S(4+h)-S(4)}{h} = \lim_{h\to 0} \frac{\frac{1}{4}(4+h)^2 - \frac{1}{4}(4)^2}{h} = \lim_{h\to 0} \frac{\frac{1}{4}(16+8h+h^2)-4}{h}$
$= \lim_{h\to 0} \frac{4+2h+\frac{1}{4}h^2-4}{h} = \lim_{h\to 0} \frac{h\left(2+\frac{1}{4}h\right)}{h} = 2.$

c. $m = \lim_{h\to 0} \frac{D(4+h)-D(4)}{h} = \lim_{h\to 0} \frac{12-2(4+h)-(12-2(4))}{h} = \lim_{h\to 0} \frac{12-8-2h-12+8}{h}$
$= \lim_{h\to 0} \frac{h}{h}(-2) = -2.$

31. $m = \lim_{h\to 0} \frac{f(1+h)-f(1)}{h} = \lim_{h\to 0} \frac{a+b(1+h)+c(1+h)^2-(a+b(1)+c(1)^2)}{h}$
$= \lim_{h\to 0} \frac{a+b+bh+c+2ch+ch^2-a-b-c}{h} = \lim_{h\to 0} \frac{h(b+2c+ch)}{h}$
$= \lim_{h\to 0}(b + 2c + ch) = b + 2c.$

32. a.

t	$N(t)$
0	0
10	200
20	300
25	312.5
30	300
40	200
50	0

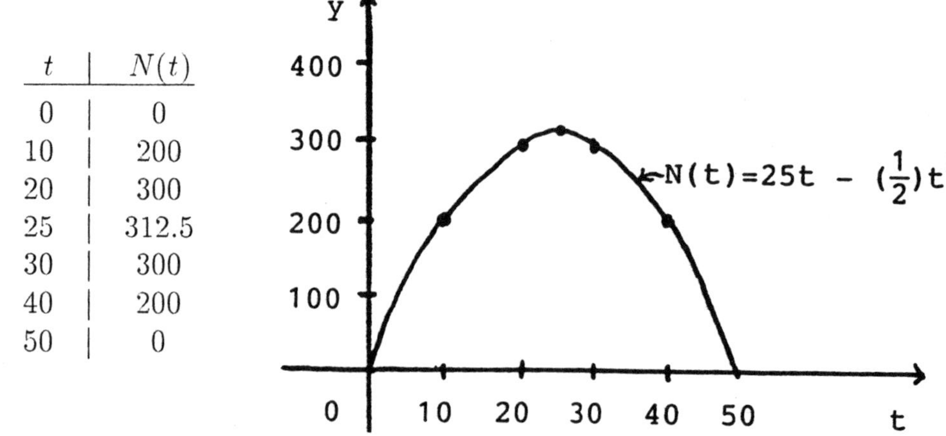

b. $m = \lim\limits_{h\to 0} \frac{N(1+h)-N(1)}{h} = \lim\limits_{h\to 0} \frac{25(1+h)-\left(\frac{1}{2}\right)(1+h)^2-\left(25(1)-\left(\frac{1}{2}\right)(1)^2\right)}{h}$

$= \lim\limits_{h\to 0} \frac{25+25h-\frac{1}{2}-h-\frac{1}{2}h^2-\frac{49}{2}}{h} = \lim\limits_{h\to 0} \frac{h\cdot\left(25-1-\frac{1}{2}h\right)}{h} = \lim\limits_{h\to 0} \left(24-\frac{1}{2}h\right) = 24.$

c. $m = \lim\limits_{h\to 0} \frac{N(a+h)-N(a)}{h} = \lim\limits_{h\to 0} \frac{25(a+h)-\frac{1}{2}(a+h)^2-\left(25(a)-\frac{1}{2}(a)^2\right)}{h}$

$= \lim\limits_{h\to 0} \frac{25a+25h-\frac{1}{2}a^2-ah-\frac{1}{2}h^2-25a+\frac{1}{2}a^2}{h} = \lim\limits_{h\to 0} \frac{h\left(25-a-\frac{1}{2}h\right)}{h} = 25-a.$

d. The tangent line is horizontal when $25-t=0$, $t=25$.

33. a. The tangent line to the curve labeled Type I is almost horizontal for quite a while and then becomes almost vertical. The slope of the tangent line to the curve labeled Type III is negative and rather large in absolute value for quite some time, and then this slope decreases in absolute value and approaches zero as t becomes large.

b. The anwers in part a. distinguish one curve from the other in the sense that in Type I a large percentage of the population lives for quite a few years whereas in Type III a large number of the population dies rather soon.

c. As the time variable t increases, the slope of the tangent line to the curve labeled Type I decreases in absolute value and eventually approaches zero, and the cloper of the tangent line to the curve labeled Type II is constant.

34. a. The slope of the tangent line is given by

$$m = \lim_{h \to 0} \frac{C(4+h)-C(4)}{h} = \lim_{h \to 0} \frac{\left(2{,}000+20(4+h)+2(4+h)^2\right)-2{,}112}{h}$$
$$= \lim_{h \to 0} \frac{2{,}000+80+20h+32+16h+2h^2-2{,}112}{h}$$
$$= \lim_{h \to 0} \frac{h(36+2h)}{h} = \lim_{h \to 0}(36+2h) = 36.$$

b. This tangent line has the equation

$$y - 2{,}112 = 36(x-4) \implies y = 36x + 1{,}968.$$

35. The cost function is given by $C(x) = 2{,}000 + 20x + 2x^2$.

a. The average cost function is given by

$$c(x) = \frac{C(x)}{x} = \frac{2{,}000}{x} + 20 + 2x.$$

b. The slope of the tangent line to the graph of $y = c(x)$ at the point where $x = 10$ is

$$m = \lim_{h \to 0} \frac{c(10+h)-c(10)}{h}$$
$$= \lim_{h \to 0} \frac{\frac{2{,}000}{10+h}+20+2(10+h)-240}{h}$$
$$= \lim_{h \to 0} \frac{2{,}000-220(10+h)+2(10+h)^2}{h(10+h)}$$
$$= \lim_{h \to 0} \frac{2{,}000-2200-220h+200+40h+2h^2}{h(10+h)}$$
$$= \lim_{h \to 0} \frac{h(-180+2h)}{h(10+h)} = \lim_{h \to 0} \frac{-180+2h}{10+h} = \frac{-180}{10} = -18.$$

Because this slope is negative, the average cost per racket is decreasing at this level of production.

c. The slope here is given by

$$\lim_{h \to 0} \frac{c(100+h)-c(100)}{h}$$
$$= \lim_{h \to 0} \frac{\frac{2{,}000}{100+h}+20+2(100+h)-240}{h}$$
$$= \lim_{h \to 0} \frac{2{,}000-220(100+h)+2(100+h)^2}{h(100+h)}$$
$$= \lim_{h \to 0} \frac{2{,}000-22{,}000-220h+20{,}000+400h+2h^2}{h(100+h)}$$
$$= \lim_{h \to 0} \frac{h(180+2h)}{h(100+h)} = \lim_{h \to 0} \frac{180+2h}{(100+h)} = 1.8.$$

Because this slope is positive, the average cost per racket is increasing at this level of production.

d. The slope of the tangent line to the graph of $y = C(x)$ at the point $(x_0, C(x_0))$ is given by

$$\lim_{h\to 0} \frac{C(x_0+h)-C(x_0)}{h}$$
$$= \lim_{h\to 0} \frac{(2{,}000+20(x_0+h)+2(x_0+h)^2)-(2{,}000+20x_0+2x_0^2)}{h}$$
$$= \lim_{h\to 0} \frac{20h+2(x_0^2+2hx_0+h^2)-2x_0^2}{h}$$
$$= \lim_{h\to 0} \frac{h(20+4x_0+2h)}{h} = \lim_{h\to 0}(20+4x_0+2h)$$
$$= 20 + 4x_0.$$

The number x_0 for which this slope is zero is given by $20 + 4x_0 = 0$ $\Longrightarrow x_0 = -5$.

Solutions to Exercise Set 2.2

1. $\lim_{x\to 3}(2x-6) = 2\cdot 3 - 6 = 0.$
2. $\lim_{x\to 3} \frac{2x-1}{x+5} = \frac{6-1}{3+5} = \frac{5}{8}.$
3. $\lim_{x\to 1}(x^2-4x) = 1 - 4 = -3.$
4. $\lim_{x\to 2} \frac{x^2-4}{x-2} = \lim_{x\to 2} \frac{(x+2)(x-2)}{(x-2)} = \lim_{x\to 2}(x+2) = 4.$
5. $\lim_{x\to -3} \frac{x^2-9}{x+3} = \lim_{x\to -3} \frac{(x-3)(x+3)}{(x+3)} = \lim_{x\to -3}(x-3) = -6.$
6. $\lim_{x\to 0} \frac{x^3-x}{x} = \lim_{x\to 0} \frac{x(x^2-1)}{x} = \lim_{x\to 0}(x^2-1) = -1.$
7. $\lim_{x\to 4} \frac{x-4}{x^2-x-12} = \lim_{x\to 4} \frac{x-4}{(x-4)(x+3)} = \lim_{x\to 4} \frac{1}{x+3} = \frac{1}{7}.$
8. $\lim_{x\to 6} \frac{36-x^2}{x-6} = \lim_{x\to 6} \frac{(6+x)(6-x)}{x-6} = \lim_{x\to 6}(-(6+x)) = -12.$
9. $\lim_{x\to 1} \frac{x^3-1}{x-1} = \lim_{x\to 1} \frac{(x-1)(x^2+x+1)}{x-1} = \lim_{x\to 1}(x^2+x+1) = 3.$
10. $\lim_{x\to 0} |x| = 0.$
11. $\lim_{h\to 0} \frac{(4+h)^2-16}{h} = \lim_{h\to 0} \frac{16+8h+h^2-16}{h} = \lim_{h\to 0}(8+h) = 8.$
12. $\lim_{x\to -1} \frac{x^2-2x-3}{x+1} = \lim_{x\to -1} \frac{(x-3)(x+1)}{(x+1)} = \lim_{x\to -1}(x-3) = -4.$

13. $\lim_{x\to 4} \frac{x-4}{\sqrt{x}-2} = \lim_{x\to 4} \frac{(\sqrt{x}+2)(\sqrt{x}-2)}{\sqrt{x}-2} = \lim_{x\to 4}(\sqrt{x}+2) = 4.$

14. $\lim_{x\to -3} \frac{x^2-6x-27}{x^2+3x} = \lim_{x\to -3} \frac{(x-9)(x+3)}{x(x+3)} = \lim_{x\to -3} \frac{x-9}{x} = \frac{-12}{-3} = 4.$

15. $\lim_{x\to -3} \frac{x^2-4x-21}{x+3} = \lim_{x\to -3} \frac{(x-7)(x+3)}{x+3} = \lim_{x\to -3}(x-7) = -10.$

16. $\lim_{x\to 4} \frac{|x-4|+2x}{x+3} = \frac{8}{7}.$

17. $\lim_{x\to 1} \frac{1-x^4}{x^2-1} = \lim_{x\to 1} \frac{(1-x^2)(1+x^2)}{(x^2-1)} = \lim_{x\to 1}\left(-(1+x^2)\right) = -2.$

18. $\lim_{x\to -1} \frac{x^3+2x^2-5x-6}{x+1} = \lim_{x\to -1} \frac{(x+1)(x^2+x-6)}{x+1} = \lim_{x\to -1}(x^2+x-6)$

$= (1-1-6) = -6.$

19. $\lim_{x\to 0} \frac{\sqrt{x+1}-1}{x} = \lim_{x\to 0} \frac{\sqrt{x+1}-1}{x} \cdot \frac{\sqrt{x+1}+1}{\sqrt{x+1}+1}$

$= \lim_{x\to 0} \frac{(x+1)-1}{x(\sqrt{x+1}+1)} = \lim_{x\to 0} \frac{1}{\sqrt{x+1}+1} = \frac{1}{2}.$

20. $\lim_{x\to 2} \frac{\sqrt{x+2}-2}{x-2} = \lim_{x\to 2} \frac{\sqrt{x+2}-2}{x-2} \cdot \frac{\sqrt{x+2}+2}{\sqrt{x+2}+2}$

$= \lim_{x\to 2} \frac{(x+2)-4}{(x-2)\cdot(\sqrt{x+2}+2)} = \lim_{x\to 2} \frac{1}{\sqrt{x+2}+2} = \frac{1}{4}.$

21. $\lim_{x\to 1} \frac{1-\sqrt{x}}{x+1} = 0.$

22. $\lim_{x\to 4} \frac{4-x}{\sqrt{x}-2} = \lim_{x\to 4} \frac{4-x}{\sqrt{x}-2} \cdot \frac{\sqrt{x}+2}{\sqrt{x}+2} = \lim_{x\to 4} \frac{(4-x)\cdot(\sqrt{x}+2)}{x-4}$

$= -\lim_{x\to 4}(\sqrt{x}+2) = -4.$

23. $\lim_{x\to 3} \frac{\sqrt{x+1}-\sqrt{x}}{1-x} = \frac{\sqrt{3+1}-\sqrt{3}}{1-3} = \frac{2-\sqrt{3}}{-2} = -1 + \frac{\sqrt{3}}{2}.$

24. $\lim_{x\to 4} \frac{\sqrt{x}-2}{x-4} = \lim_{x\to 4} \frac{\sqrt{x}-2}{(\sqrt{x}+2)(\sqrt{x}-2)} = \lim_{x\to 4} \frac{1}{(\sqrt{x}+2)} = \frac{1}{4}.$

25. $\lim_{x\to 1} \frac{1-\sqrt{x}}{x^{3/2}-\sqrt{x}} = \lim_{x\to 1} \frac{1-\sqrt{x}}{\sqrt{x}(x-1)} = \lim_{x\to 1} \frac{1-\sqrt{x}}{\sqrt{x}(\sqrt{x}+1)\cdot(\sqrt{x}-1)}$

$= \lim_{x\to 1} \frac{-1}{\sqrt{x}(\sqrt{x}+1)} = -\frac{1}{2}.$

26. $\lim_{x\to 1} \frac{\sqrt{x}-x}{x^2-x} = \lim_{x\to 1} \frac{\sqrt{x}\cdot(1-\sqrt{x})}{x\cdot(x-1)} = \lim_{x\to 1} \frac{1-\sqrt{x}}{\sqrt{x}(\sqrt{x}-1)\cdot(\sqrt{x}+1)}$

$= \lim_{x\to 1} \frac{-1}{\sqrt{x}(\sqrt{x}+1)} = -\frac{1}{2}.$

Solutions to Exercise Set 2.3

1. a. $f(1)$ doesn't exist.

   b. $\lim_{x\to 1^-} f(x) = 2.$

   c. $\lim_{x\to 1} f(x) = 2.$

   d. $\lim_{x\to 0} f(x) = 0.$

2. a. $f(1) = 0.$

   b. $\lim_{x\to 1^-} f(x) = -1.$

   c. $\lim_{x\to 1+} f(x) = 0.$

   d. $\lim_{x\to 1} f(x)$ doesn't exist.

3. a. $\lim_{x\to 0^-} f(x) = 2.$

   b. $\lim_{x\to 0+} f(x) = +\infty.$

   c. $\lim_{x\to -2} f(x) = 0.$

4. a. $\lim_{x\to -2+} f(x) = 0.$

   b. $\lim_{x\to 2^-} f(x) = 2.$

   c. $\lim_{x\to 2+} f(x) = -1.$

   d. $f(2) = 2.$

5. $\lim_{x\to 0^-} \frac{|x|}{x} = \lim_{x\to 0^-} \frac{-x}{x} = \lim_{x\to 0^-} (-1) = -1.$

6. $\lim_{x\to 0+} \frac{|x|}{x} = \lim_{x\to 0+} \frac{x}{x} = \lim_{x\to 0+} (1) = 1.$

7. $\lim_{x\to 0+} \sqrt{\frac{x}{x+2}} = \sqrt{\frac{0}{0+2}} = 0.$

8. $\lim_{x\to 2^-} \frac{|1-x^2|}{1+x^2} = \lim_{x\to 2^-} \frac{(x^2-1)}{1+x^2} = \frac{2^2-1}{1+2^2} = \frac{3}{5}.$

9. $\lim_{x\to 1^-} \frac{|x-2|}{x+2} = \lim_{x\to 1^-} \frac{2-x}{x+2} = \frac{2-1}{1+2} = \frac{1}{3}.$

10. $\lim_{x\to 6+} \frac{\sqrt{x^2-5x-6}}{x-3} = \lim_{x\to 6+} \frac{\sqrt{(x-6)(x+1)}}{x-3} = \frac{0}{3} = 0.$

11. $f(x) = \frac{1}{1-x^2}$ is discontinuous at $x = 1$ and $x = -1$ because $f(1)$ and $f(-1)$ don't exist.

12. $f(x) = \frac{x+2}{x^2-x-2} = \frac{x+2}{(x-2)(x+1)}$ is discontinuous at $x = 2$ and $x = -1$ because $f(2)$ and $f(-1)$ don't exist.

13. $f(x) = \frac{x^2-x-6}{x+3}$ is discontinuous at $x = -3$ because $f(-3)$ doesn't exist.

14. $f(x) = \frac{x^3+4x^2-7x-10}{x-2}$ is discontinuous at $x = 2$ because $f(2)$ doesn't exist.

15. $f(x) = \frac{x+3}{x^3+3x^2-x-3} = \frac{(x+3)}{(x-1)(x+3)(x+1)}$ is discontinuous at $x = 1, -3$, and -1 since $f(1)$, $f(-1)$ and $f(-3)$ don't exist.

16. $f(x) = \frac{|x+2|}{x+2}$ is discontinuous at $x = -2$ since $f(-2)$ doesn't exist.

17. $f(x) = \begin{cases} 1-x, & x \leq 2 \\ x-1, & x > 2 \end{cases}$ is discontinuous at $x = 2$ since $\lim_{x \to 2^+} f(x) = 1$ and $\lim_{x \to 2^-} f(x) = -1$, that is, $\lim_{x \to 2} f(x)$ doesn't exist.

18. $f(x) = \begin{cases} x^2 - 4, & x < 0 \\ 9 - \sqrt{x+16}, & x \geq 0 \end{cases}$

$\lim_{x \to 0^-} f(x) = \lim_{x \to 0^-} (x^2 - 4) = 0^2 - 4 = -4.$

$\lim_{x \to 0^+} f(x) = \lim_{x \to 0^+} \left(9 - \sqrt{x+16}\right) = 9 - 4 = 5.$

Since $\lim_{x \to 0^-} f(x) \neq \lim_{x \to 0^+} f(x)$, the function f is discontinuous at $x = 0$.

The function f is continuous for all other x.

19. $f(x) = \begin{cases} x^2 - 2x - 15 & , \quad x < -3 \\ 9 - x^2 & , \quad -3 < x < 5 \\ x^2 - 2x - 15 & , \quad x > 5 \end{cases}$

The function f is continuous everywhere except for $x = -3$ and $x = 5$, because $f(-3)$ and $f(5)$ are not defined.

20. $f(x) = \begin{cases} \dfrac{\sqrt{x^2+1}}{x+2} & , \quad x < 0 \\ \dfrac{1+x}{2-3x} & , \quad x \geq 0 \end{cases}$

$\lim_{x \to 0^-} f(x) = \lim_{x \to 0^-} \frac{\sqrt{x^2+1}}{x+2} = \frac{\sqrt{0^2+1}}{0+2} = \frac{1}{2}.$

$\lim_{x\to0+} f(x) = \lim_{x\to0+} \frac{1+x}{2-3x} = \frac{1+0}{2-3(0)} = \frac{1}{2}$.

Because $\lim_{x\to0^-} f(x) = \lim_{x\to0+} f(x) = f(0)$, the function f is continuous at $x = 0$.

The function f is also continuous for all other x.

21. From $\lim_{x\to2+} f(x) = \lim_{x\to2^-} f(x)$ we have $8 = 2^k$.

    Therefore $k = 3$.

22. From $\lim_{x\to1^-} f(x) = \lim_{x\to1+} f(x)$ we have $k = \frac{1}{\sqrt{2k}}$. Then $k^2 = \frac{1}{2k} \Rightarrow 2k^3 = 1 \Rightarrow$ $k^3 = \frac{1}{2} \Rightarrow k = \sqrt[3]{\frac{1}{2}} = \frac{1}{\sqrt[3]{2}}$.

23. From $\lim_{x\to2+} f(x) = \lim_{x\to2^-} f(x)$ we have $2k+5 = (2-k)(2+k)$. Then $2k+5 =$ $4-k^2 \Rightarrow k^2+2k+1 = 0 \Rightarrow k = -1$.

24. $S(p)$ is discontinuous at $p = 20$ since $\lim_{p\to20^-} S(p) = 10 \neq \lim_{p\to20+} S(p) = \frac{10}{3}$.

25. $p(q)$ is discontinuous at $q = 50$ and 100 since $\lim_{q\to50} p(q)$ and $\lim_{q\to100} p(q)$ don't exist.

26. $R(q) = \begin{cases} 40q & 0 \le q < 50 \\ 35q & 50 \le q < 100 \\ 30q & 100 \le q \end{cases}$.

    The answer is 'Yes' since $\lim_{q\to50} R(q)$ doesn't exist.

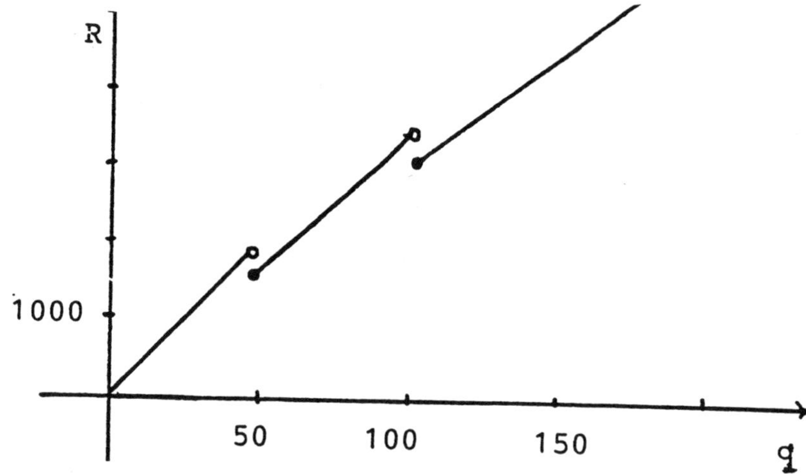

27. Let x be income in dollars. Then the tax function

$$T(x) \text{ is } \begin{cases} 2097 + 0.24(x - 15{,}000), & 15{,}000 \le x < 18{,}200 \\ A + 0.28\cdot(x - 18{,}200), & 18{,}200 \le x \le 23{,}500. \end{cases}$$

For $T(x)$ to be continuous at $x = 18,200$, it is required that $\lim_{x \to 18,200} T(x)$ exists and equals $T(18,200)$.

$2097 + 0.24(18,200 - 15,000) = A$. This implies that $A = \$2865$.

28. S is discontinuous at $x = 5$ and 10 since $\lim_{x \to 5} S(x)$ and $\lim_{x \to 10} S(x)$ don't exist.

29. From $\lim_{x \to 3^-} f(x) = \lim_{x \to 3^+} f(x)$, $9a + 2 = 20$. Hence $a = 2$.

30. Let $P(x)$ be the postage required to mail a letter weighing x ounces where $0 < x < 11$.

a.
$$P(x) = \begin{cases} 29 \text{ cents} & \text{for } 0 < x < 1 \\ 52 \text{ cents} & \text{for } 1 \le x < 2 \\ 75 \text{ cents} & \text{for } 2 \le x < 3 \\ 98 \text{ cents} & \text{for } 3 \le x < 4 \\ \$1.21 & \text{for } 4 \le x < 5 \\ \$1.44 & \text{for } 5 \le x < 6 \\ & \vdots \\ \$2.59 & \text{for } 10 \le x < 11 \end{cases}$$

The domain of this function is the set of all real numbers x such that $0 < x < 11$.

b. The numbers x at which the limit of this function fails to exist are $x = 1, 2, 3, 4, 5, 6, 7, 8, 9, 10$.

c. $\lim_{x \to 7^-} P(x) = \lim_{x \to 7^-} 1.67 = 1.67$.

d. $\lim_{x \to 7^+} P(x) = \lim_{x \to 7^+} 1.90 = 1.90$.

31. Let $P(x)$ be the price in dollars for a movie ticket for a person x years old.

$$P(x) = \begin{cases} 2.50 & \text{for } 0 < x < 12 \\ 4.50 & \text{for } x \ge 12 \end{cases}$$

$\lim_{x \to 12^-} P(x) = 2.50 < P(12)$.

Solutions to Exercise Set 2.4

1. $f'(x) = \lim_{h\to 0} \frac{f(x+h)-f(x)}{h}$
$= \lim_{h\to 0} \frac{[2(x+h)+5]-[2x+5]}{h} = \lim_{h\to 0} \frac{2x+2h+5-2x-5}{h} = 2.$

2. $f'(x) = \lim_{h\to 0} \frac{f(x+h)-f(x)}{h} = \lim_{h\to 0} \frac{[3-(x+h)]-[3-x]}{h} = \lim_{h\to 0} \frac{-h}{h} = -1.$

3. $f'(x) = \lim_{h\to 0} \frac{f(x+h)-f(x)}{h} = \lim_{h\to 0} \frac{\left[(x+h)^2+4\right]-[x^2+4]}{h}$
$= \lim_{h\to 0} \frac{x^2+2hx+h^2+4-x^2-4}{h} = \lim_{h\to 0} \frac{h(2x+h)}{h} = 2x.$

4. $f'(x) = \lim_{h\to 0} \frac{f(x+h)-f(x)}{h} = \lim_{h\to 0} \frac{\left((x+h)^3+7\right)-(x^3+7)}{h}$
$= \lim_{h\to 0} \frac{(x^3+3x^2h+3xh^2+h^3+7)-(x^3+7)}{h}$
$= \lim_{h\to 0} \frac{h(3x^2+3xh+h^2)}{h} = \lim_{h\to 0}(3x^2+3xh+h^2) = 3x^2.$

5. $f'(x) = \lim_{h\to 0} \frac{f(x+h)-f(x)}{h} = \lim_{h\to 0} \frac{(x+h)^4-x^4}{h} = \lim_{h\to 0} \frac{x^4+4hx^3+6h^2x^2+4h^3x+h^4-x^4}{h}$
$= \lim_{h\to 0} \frac{h(4x^3+6hx^2+4h^2x+h^3)}{h} = 4x^3.$

6. $f'(x) = \lim_{h\to 0} \frac{\sqrt{(x+h)+2}-\sqrt{x+2}}{h}$
$= \lim_{h\to 0} \frac{\sqrt{(x+h)+2}-\sqrt{x+2}}{h} \cdot \frac{\sqrt{(x+h)+2}+\sqrt{x+2}}{\sqrt{(x+h)+2}+\sqrt{x+2}}$
$= \lim_{h\to 0} \frac{((x+h)+2)-(x+2)}{h\left(\sqrt{(x+h)+2}+\sqrt{x+2}\right)} = \lim_{h\to 0} \frac{1}{\sqrt{(x+h)+2}+\sqrt{x+2}} = \frac{1}{2\sqrt{x+2}}.$

7. $f'(x) = \lim_{h\to 0} \frac{\frac{2}{(x+h)+3}-\frac{2}{x+3}}{h}$
$= \lim_{h\to 0} \frac{\frac{2}{(x+h)+3}-\frac{2}{x+3}}{h} \cdot \frac{((x+h)+3)(x+3)}{((x+h)+3)(x+3)}$
$= \lim_{h\to 0} \frac{2(x+3)-2((x+h)+3)}{h((x+h)+3)(x+3)} = \lim_{h\to 0} \frac{-2}{((x+h)+3)(x+3)} = \frac{-2}{(x+3)^2}.$

8. $f'(x) = \lim_{h\to 0} \frac{\frac{1}{\sqrt{(x+h)+1}}-\frac{1}{\sqrt{x+1}}}{h}$
$= \lim_{h\to 0} \frac{\frac{1}{\sqrt{(x+h)+1}}-\frac{1}{\sqrt{x+1}}}{h} \cdot \frac{\sqrt{(x+h)+1}\cdot\sqrt{x+1}}{\sqrt{(x+h)+1}\cdot\sqrt{x+1}}$
$= \lim_{h\to 0} \frac{\sqrt{x+1}-\sqrt{x+h+1}}{h\left(\sqrt{x+h+1}\cdot\sqrt{x+1}\right)} = \lim_{h\to 0} \frac{\sqrt{x+1}-\sqrt{x+h+1}}{h\left(\sqrt{x+h+1}\cdot\sqrt{x+1}\right)} \cdot \frac{\sqrt{x+1}+\sqrt{x+h+1}}{\sqrt{x+1}+\sqrt{x+h+1}}$
$= \lim_{h\to 0} \frac{(x+1)-(x+h+1)}{h\left(\sqrt{x+h+1}\cdot\sqrt{x+1}\right)\left(\sqrt{x+1}+\sqrt{x+h+1}\right)}$
$= \lim_{h\to 0} \frac{-1}{\left(\sqrt{x+h+1}\cdot\sqrt{x+1}\right)\left(\sqrt{x+1}+\sqrt{x+h+1}\right)} = \frac{-1}{(x+1)\cdot\left(2\sqrt{x+1}\right)} = -\frac{1}{2}(x+1)^{-3/2}.$

9. $f'(x) = 3\cdot 2x+4 = 6x+4.$

10. $\frac{dy}{dx} = 4 \cdot 3x^2 - 1 = 12x^2 - 1.$

11. $\frac{dy}{dx} = 3 \cdot (4x^3) + 0 - 4 = 12x^3 - 4.$

12. $f'(x) = -3 + 4 \cdot 2x + 5x^4 = 5x^4 + 8x - 3.$

13. $f'(x) = 3x^2 - 5x^4.$

14. $f'(x) = 2x - 2x^{-3} - 7x^{-2}.$

15. $f'(x) = 3 \cdot 4x^3 - 6 = 12x^3 - 6.$

16. $\frac{dy}{dx} = 4 \cdot \left((-2)x^{-3}\right) - 6 \cdot (3x^2) + 4\left((-1)x^{-2}\right)$

$= -8x^{-3} - 18x^2 - 4x^{-2}.$

17. $\frac{dy}{dx} = 2 \cdot 5x^4 - 3 \cdot (-2)x^{-3} = 10x^4 + 6x^{-3}.$

18. $f'(x) = \frac{1}{2}x^{\frac{1}{2}-1} = \frac{1}{2}x^{-\frac{1}{2}} = \frac{1}{2\sqrt{x}}.$

19. $f'(x) = 2 \cdot \left(\frac{2}{3}\right) x^{\frac{2}{3}-1} + 3 \cdot \left(\frac{5}{3}\right) x^{\frac{5}{3}-1} = \frac{4}{3}x^{-\frac{1}{3}} + 5x^{\frac{2}{3}}.$

20. $y = 3\sqrt{x} - 5x^{-3/2} = 3x^{1/2} - 5x^{-3/2}.$

$\frac{dy}{dx} = 3\left(\frac{1}{2}x^{-1/2}\right) - 5\left(\left(-\frac{3}{2}\right)x^{-5/2}\right) = \frac{3}{2}x^{-1/2} + \frac{15}{2}x^{-5/2}.$

21. $\frac{dy}{dx} = 9 \cdot (-2)x^{-3} + 3\left(\frac{1}{2}\right)x^{-1/2} = -18x^{-3} + \frac{3}{2}x^{-1/2}.$

22. $f'(x) = -4(-2)x^{-3} + 10 \cdot \left(\frac{2}{3}\right)x^{-1/3} = 8x^{-3} + \frac{20}{3}x^{-1/3}.$

23. $y = 5x^9 - 10\sqrt{x} + 3x^{11/2} = 5x^9 - 10x^{1/2} + 3x^{11/2}$. $\frac{dy}{dx} = 5 \cdot 9x^8 - 10 \cdot \frac{1}{2}x^{-1/2} +$ $3 \cdot \frac{11}{2}x^{9/2} = 45x^8 - 5x^{-1/2} + \frac{33}{2}x^{9/2}.$

24. $f(x) = (x-3)^2 = x^2 - 6x + 9$. $f'(x) = 2x - 6.$

25. $f'(x) = 4 \cdot \left(-\frac{2}{3}\right)x^{-\frac{2}{3}-1} - 7\left(-\frac{1}{4}\right)x^{-\frac{1}{4}-1} = -\frac{8}{3}x^{-\frac{5}{3}} + \frac{7}{4}x^{-\frac{5}{4}}.$

26. $f'(x) = 3\left(\frac{1}{2}\right)x^{-\frac{1}{2}} - 5\left(-\frac{4}{3}\right)x^{-\frac{4}{3}-1} = \frac{3}{2}x^{-\frac{1}{2}} + \frac{20}{3}x^{-\frac{7}{3}}.$

27. $f'(x) = 12x^2 - 4x$. $f'(2) = 12(2)^2 - 4(2) = 48 - 8 = 40.$

28. $f'(x) = 3x^2 - \frac{1}{2}x^{-1/2}$. $f'(4) = 3(4)^2 - \frac{1}{2}(4)^{-1/2} = 48 - \frac{1}{4} = \frac{191}{4}.$

29. $f'(x) = 12x^3 - 12x + 4$. $f'(-3) = 12(-3)^3 - 12(-3) + 4 = -284.$

30. $f'(x) = 12x^2 - 12x^3$. $f'(-1) = 12(-1)^2 - 12(-1)^3 = 24.$

31. $f'(x) = -3x^{-1/2} + \frac{10}{3}x^{-1/3}$. $f'(8) = -3(8)^{-1/2} + \frac{10}{3}(8)^{-1/3}$

$= -\frac{3}{2\sqrt{2}} + \frac{10}{3} \cdot \frac{1}{2} = -\frac{3}{2\sqrt{2}} + \frac{5}{3}$.

32. $f'(x) = 12x^2 - 10x^{-3}$. $f'(-2) = 12(-2)^2 - 10(-2)^{-3} = 48 + \frac{10}{8} = \frac{197}{4}$.

33. $f'(x) = 8x - 2$. At $(1, 3)$, $f'(1) = 8(1) - 2 = 6$.

The line tangent to the graph of $f(x)$ at the point $(1, 3)$ is $y - 3 = f'(1)(x - 1)$, $y - 3 = 6(x - 1)$, $6x - y - 3 = 0$.

34. $\frac{dy}{dx} = \frac{1}{2}x^{-1/2}$. At $(4, 2)$, $\frac{dy}{dx} = \frac{1}{2}(4)^{-1/2} = \frac{1}{4}$.

The line tangent to the graph of $y = \sqrt{x}$ at $(4, 2)$ is $y - 2 = \frac{1}{4}(x - 4)$, $x - 4y + 4 = 0$.

35. $f'(x) = \frac{1}{2}x^{-1/2} + \frac{1}{2}x^{-3/2}$. The slope of the graph of $f(x)$ at $\left(4, \frac{3}{2}\right)$ is

$f'(4) = \frac{1}{2}(4)^{-1/2} + \frac{1}{2}(4)^{-3/2} = \frac{1}{4} + \frac{1}{16} = \frac{5}{16}$.

36. $\frac{dy}{dx} = 12x^3 - 12x^2$. The slope at $(1, -1) = 12(1)^3 - 12(1)^2 = 0$.

37. $f'(x) = 2x + a$.

Since the slope of the tangent to the graph of $f(x)$ at $(x, f(x))$ equals $f'(x)$, then $m = 9 = f'(2) = 2(2) + a = 4 + a$, $a = 5$.

38. When $x = -2$, the corresponding y-coordinate $= \frac{b}{(-2)^2} = \frac{b}{4}$.

Since the point $\left(-2, \frac{b}{4}\right)$ is on the tangent line $4y - bx - 21 = 0$, we have $4\left(\frac{b}{4}\right) - b(-2) - 21 = 0$, $3b = 21$, $b = 7$.

39. $f(x) = \sqrt{x + 5}$.

$$f'(x) = \lim_{h \to 0} \frac{\sqrt{(x+h)+5} - \sqrt{x+5}}{h} = \lim_{h \to 0} \frac{\sqrt{(x+h)+5} - \sqrt{x+5}}{h} \cdot \frac{\sqrt{(x+h)+5} + \sqrt{x+5}}{\sqrt{(x+h)+5} + \sqrt{x+5}}$$

$$= \lim_{h \to 0} \frac{(x+h+5) - (x+5)}{h\left(\sqrt{x+h+5} + \sqrt{x+5}\right)} = \lim_{h \to 0} \frac{1}{\sqrt{x+h+5} + \sqrt{x+5}} = \frac{1}{2\sqrt{x+5}}.$$

$f'(a) = \frac{1}{4} \implies \frac{1}{2\sqrt{a+5}} = \frac{1}{4} \implies 2\sqrt{a+5} = 4 \implies \sqrt{a+5} = 2$

$\implies a + 5 = 4 \implies a = -1$.

40. $f(x) = A + x - x^2$ and $g(x) = 10 - 3x$.

a. $f'(x) = 1 - 2x$ and $g'(x) = -3$.

$f'(x) = g'(x) \implies 1 - 2x = -3 \implies x = 2$.

b. $f(2) = g(2) \implies A + 2 - 2^2 = 10 - 3(2) \implies A - 2 = 4 \implies A = 6$.

41. The graph of $y = P(x) = 32x - 2x^2 - 120$ is a parabola opening downward.

a. $32x - 2x^2 - 120 = -2(x^2 - 16x + 60) = -2(x - 10)(x - 6) = 0$ for $x = 6$, $x = 10$.

Therefore $P(x) > 0$ for $6 < x < 10$.

b. The vertex of this parabola is the highest point of this parabola. The x-coordinate of the vertex is given by $x = -\frac{32}{2(-2)} = 8$. Therefore the profit $P(x)$ increases as the production level x increases for $0 < x < 8$.

c. The largest possible profit $P(x)$ occurs when the level of production $x = 8$, and this largest possible profit is $P(8) = 32(8) - 2(8^2) - 120 = 8$.

d. Note that $P'(x) = 32 - 4x = 4(8 - x)$.

$P'(x) = 0$ for $x = 8$.

For $0 < x < 8$, $P'(x) > 0$, so $P(x)$ increases as x increases.

Then for $x > 8$, $P'(x) < 0$, so $P(x)$ decreases as x increases.

42. $P(x) = a\sqrt{x} = ax^{1/2}$.

a. $P'(x) = \frac{1}{2}ax^{-1/2}$.

b. $P'(4) = \frac{1}{2}a(4^{-1/2}) = \frac{1}{4}a$.

c. $y - 2a = \frac{1}{4}a(x - 4) \implies y - 2a = \frac{1}{4}ax - a \implies y = \frac{1}{4}ax + a$.

d. The rate at which biomass $P(x)$ increases with increasing precipitation x per unit time when $x = 4$ is $\frac{1}{4}a$ units of biomass per unit of precipitation x.

43. $f(t) = t^{-3/2}$.

a. $f'(t) = -\frac{3}{2}t^{-5/2}$.

b. $f'(2) = -\frac{3}{2} \cdot 2^{-5/2} = -3 \cdot 2^{-7/2} = -\frac{3}{8\sqrt{2}} = -\frac{3\sqrt{2}}{16}$.

c. $y - 2^{-3/2} = -\frac{3\sqrt{2}}{16}(t-2) \implies y = -\frac{3\sqrt{2}}{16}t + \frac{5\sqrt{2}}{8}$.

44. By the definition of the derivative,

$$r'(x) = (cf(x))' = \lim_{h\to 0} \frac{r(x+h) - r(x)}{h} = \lim_{h\to 0} \frac{cf(x+h) - cf(x)}{h}$$
$$= \lim_{h\to 0} c \cdot \frac{f(x+h) - f(x)}{h} = c \cdot \lim_{h\to 0} \frac{f(x+h) - f(x)}{h} = cf'(x).$$

Therefore the function $r(x) = cf(x)$ is differentiable, and $r'(x) = cf'(x)$.

45. a. By the definition of the derivative,

$$f'(x) = \lim_{h\to 0} \frac{f(x+h) - f(x)}{h} = \lim_{h\to 0} \frac{(x+h)^n - x^n}{h}.$$

b. $(x+h)^n = x^n + nx^{n-1}h + \frac{n(n-1)}{2} \cdot x^{n-2}h^2$

$+\frac{n(n-1)(n-2)}{6}x^{n-3}h^3 + \frac{n(n-1)(n-2)(n-3)}{24}x^{n-4}h^4$

$+ \cdots + nxh^{n-1} + h^n$.

c. $f'(x) = \lim_{h\to 0} \frac{f(x+h)-f(x)}{h} = \lim_{h\to 0} \Big\{ nx^{n-1} + h\big(\frac{n(n-1)}{2}x^{n-2}$

$+\frac{n(n-1)(n-2)}{6}x^{n-3}h + \frac{n(n-1)(n-2)(n-3)}{24}x^{n-4}h^2$

$+ \cdots + nxh^{n-3} + h^{n-2}\big)\Big\} = nx^{n-1}$.

Solutions to Exercise Set 2.5

1. $s(t) = 7 + 3t^2, \quad t_1 = 0, \quad t_2 = 3$

Average velocity $\bar{v} = \frac{s(t_2)-s(t_1)}{t_2-t_1} = \frac{s(3)-s(0)}{3-0} = \frac{34-7}{3} = 9$.

2. $s(t) = t^2 - 2t + 5, \quad t_1 = 1, \quad t_2 = 3$

Average velocity $\bar{v} = \frac{s(3)-s(1)}{3-1} = \frac{8-4}{2} = 2$.

3. $s(t) = \frac{10t}{1+t^2}, \quad t_0 = 0, \quad t_2 = 3$

Average velocity $\bar{v} = \frac{s(3)-s(0)}{3-0} = \frac{3-0}{3} = 1.$

4. $s(t) = 50 - \frac{3\sqrt{t}}{t+5}, \quad t_1 = 1, \quad t_2 = 9$

Average velocity $\bar{v} = \frac{s(9)-s(1)}{9-1} = \frac{49.3571-49.5}{8} = -.0179$

5. $s(t) = 40 + 64t - 3t^2, \quad t_1 = 1, \quad t_2 = 2$

Average velocity $\bar{v} = \frac{s(2)-s(1)}{2-1} = \frac{156-101}{1} = 55.$

6. $s(t) = 64 + 24t - 8t^2, \quad t_1 = 1, \quad t_2 = 3$

Average velocity $\bar{v} = \frac{s(3)-s(1)}{3-1} = \frac{64-80}{2} = -8.$

7. The velocity function $v(t) = s'(t) = 2t + 2.$

8. The velocity function $v(t) = s'(t) = 0 - 3t^2 = -3t^2.$

9. $v(t) = s'(t) = 3t^2 - 10t + 3.$

10. $v(t) = s'(t) = 8t^3 - 16t.$

11. $v(t) = s'(t) = 3 \cdot \frac{1}{2}t^{-1/2} + 5 \cdot \frac{3}{2}t^{1/2} = \frac{3}{2}t^{-1/2} + \frac{15}{2}t^{1/2}.$

12. $v(t) = s'(t) = \frac{2}{3}t^{-1/3} + 6 \cdot \frac{5}{2}t^{3/2} = \frac{2}{3}t^{-1/3} + 15t^{3/2}.$

13. $v(t) = s'(t) = 2 \cdot \frac{1}{2}t^{-1/2} + 5 \cdot \left(-\frac{2}{3}\right) t^{-5/3} = t^{-1/2} - \frac{10}{3}t^{-5/3}.$

14. $s(t) = t(t-1)(t+2) = t(t^2 + t - 2) = t^3 + t^2 - 2t.$

$v(t) = s'(t) = 3t^2 + 2t - 2.$

15. $s(t) = \frac{(t+6)(t-2)}{\sqrt{t}} = \frac{t^2+4t-12}{t^{1/2}} = \frac{t^2}{t^{1/2}} + 4\frac{t}{t^{1/2}} - \frac{12}{t^{1/2}}$

$= t^{3/2} + 4t^{1/2} - 12t^{-1/2}.$

$v(t) = s'(t) = \frac{3}{2}t^{1/2} + 4 \cdot \frac{1}{2}t^{-1/2} - 12\left(-\frac{1}{2}\right) t^{-3/2} = \frac{3}{2}t^{1/2} + 2t^{-1/2} + 6t^{-3/2}.$

16. $s(t) = \frac{t^2+9t+20}{t+4} = \frac{(t+4)(t+5)}{t+4} = t + 5.$

$v(t) = s'(t) = 1.$

17. $s(t) = (t^2+2)(t+1) = t^3 + t^2 + 2t + 2.$

$v(t) = s'(t) = 3t^2 + 2t + 2.$

$v(3) = s'(3) = 3 \cdot 3^2 + 2 \cdot 3 + 2 = 35.$

18. $s(t) = 6 + 5t - t^2$.

$v(t) = s'(t) = 5 - 2t$.

$v(t) = 0 \implies t = 2.5$.

$s(2.5) = 6 + 5(2.5) - (2.5)^2 = 12.25$.

19. $s(t) = 6t - t^2 = t(6 - t)$.

$v(t) = s'(t) = 6 - 2t$.

a. The initial velocity $v(0) = 6 - 2(0) = 6$.

b. The particle changes direction when $v(t) = 0$.

$v(t) = 0 \implies t = 3$.

c. The particle crosses the origin the second time when $t = 6$.

$v(6) = 6 - 2(6) = -6$.

20. The height of the rocket in feet after t seconds is $s(t) = 32t - 16t^2$.

a. The velocity at time t is given by

$v(t) = s'(t) = 32 - 32t$ feet per second.

b. The rocket reaches its maximum height when $v(t) = 0 \implies t = 1$ second.

c. The maximum height is $s(1) = 16$ feet.

d. Then the rocket strikes the ground, $s(t) = 0 \implies t(32 - 16t) = 0 \implies$

$t = 2$ seconds.

$v(2) = 32 - 32(2) = -32$ feet per second.

When the rocket strikes the ground, it is moving downward with a speed of 32 feet per second.

21. Let $y = s(t)$ denote the height of the water balloon t seconds after Carol dropped it. Let s_0 denote the initial height of the water balloon, and let v_0 denote the initial velocity of the water baloon. We have the formula

$$s(t) = -16t^2 + v_0 t + s_0.$$

Since Carol dropped the water baloon, it follows that $v_0 = 0$. Therefore

$$s(t) = -16t^2 + s_0.$$

a. When the water balloon strikes the ground, $s(t) = 0 \implies 16t^2 = s_0$ $\implies t^2 = \frac{s_0}{16} \implies t = \frac{\sqrt{s_0}}{4}$. Since the water balloon strikes the ground with a velocity of -96 ft/sec, it follows that

$$-96 = -32 \cdot \frac{\sqrt{s_0}}{4} \implies \sqrt{s_0} = 12 \implies s_0 = 144.$$

Therefore Carol dropped the water balloon from a height of 144 feet.

b. The number of seconds the water balloon falls is $\frac{\sqrt{144}}{4} = 3$.

22. The height of the flare in feet above the ground after t seconds is given by the formula

$$s(t) = 100 + 256t - 16t^2.$$

a. The flare reaches its maximum height when $v(t) = 0$.

$$v(t) = s'(t) = 256 - 32t$$

$$v(t) = 0 \implies t = 8.$$

The flare reaches its maximum height after 8 seconds.

b. This maximum height is

$$s(8) = 100 + 256(8) - 16(8^2) = 1124 \text{ feet}.$$

23. The position of the particle along the line after t seconds is given by the formula

$$s(t) = 6 + 5t - t^2.$$

The velocity of the particle after t seconds is given by the formula

$$v(t) = s'(t) = 5 - 2t.$$

Note that $v(t) = 0$ when $t = 2.5$.

When $0 < t < 2.5$, $v(t) > 0 \implies s(t)$ increases as t increases.

When $2.5 < t < 6$, $v(t) < 0 \implies s(t)$ decreases as t increases.

Note that $s(0) = 6$ and $s(6) = 0$.

Therefore the maximum distance of the particle from the position $s = 0$ in the time interval $[0, 6]$ is $s(2.5) = 12.25$.

Solutions to Exercise Set 2.6

1. a. $R'(x) = 10$.

   b. $C'(x) = 6$.

   c. $P(x) = 10x - 50 - 6x = 4x - 50$. $P'(x) = 4$.

2. a. $R'(x) = 100$.

   b. $C'(x) = 40 + 2x$.

   c. $P(x) = 100x - 400 - 40x - x^2 = -x^2 + 60x - 400$. $P'(x) = -2x + 60$.

3. a. $R'(x) = 100 - 4x$.

   b. $C'(x) = 20 + \frac{3}{2}x^{1/2}$.

   c. $P(x) = (100x - 2x^2) - (20x + x^{3/2} + 400) = -2x^2 - x^{3/2} + 80x - 400$.

   $P'(x) = -4x - \frac{3}{2}x^{1/2} + 80$.

4. a. $R'(x) = 10x^{-1/2}$.

   b. $C'(x) = 10 + \frac{1}{2}x^{-1/2}$.

   c. $P(x) = 20\sqrt{x} - 10x - \sqrt{x} - 30 = 19\sqrt{x} - 10x - 30$.

   $P'(x) = \frac{19}{2}x^{-1/2} - 10$.

5. a. $R'(x) = 40 + 25x^{-1/2}$.

b. $C'(x) = 20 + \frac{2}{3}x^{-1/3}$.

c. $P(x) = 40x + 50\sqrt{x} - 150 - 20x - x^{2/3} = 20x + 50\sqrt{x} - x^{2/3} - 150$.

$P'(x) = 20 + 25x^{-1/2} - \frac{2}{3}x^{-1/3}$.

6. a. $R(x) = 40(20 - x) = 800 - 40x$.

$R'(x) = -40$.

b. $C'(x) = 20 + 3 \cdot \frac{4}{3}x^{1/3} = 20 + 4x^{1/3}$.

c. $P(x) = (800 - 40x) - (250 + 20x + 3x^{4/3}) = -3x^{4/3} - 60x + 550$.

$P'(x) = -4x^{1/3} - 60$.

7. a. $R'(x) = 400 - \frac{20}{3}x^{-1/3}$.

b. $C'(x) = -5000x^{-2}$.

c. $P(x) = 400x - 10x^{2/3} - 5000x^{-1} - 40$.

$P'(x) = 400 - \frac{20}{3}x^{-1/3} + 5000x^{-2}$.

8. a. $R'(x) = (100x - 4000 - 30x^{-1})' = 100 + 30x^{-2}$.

b. $C'(x) = 2x^{-1/3} + 20x^{-3}$.

c. $P(x) = 100x - 4000 - 30x^{-1} - 200 - 3x^{2/3} + 10x^{-2}$

$= 100x - 30x^{-1} - 3x^{2/3} + 10x^{-2} - 4200$.

$P'(x) = 100 + 30x^{-2} - 2x^{-1/3} - 20x^{-3}$.

9. a. $P'(x) = 160 - 2x$. $P'(40) = 160 - 80 = 80$.

b. $P'(x) = 160 - 2x > 0 \implies -2x > -160, \quad x < 80$.

10. $MR(x) = 200 - 8x$. $MC(x) = 40$.

$MR(x) = MC(x) \implies 200 - 8x = 40, -8x = -160, \quad x = 20$.

11. a. $R(x) = x(600 - 3x) = 600x - 3x^2$.

b. $P(x) = 600x - 3x^2 - 4000 - 150x - 0.5x^2 = -3.5x^2 + 450x - 4000$.

c. $MC(x) = 150 + x$, $MR(x) = 600 - 6x$.

d. $MC(x) = MR(x) \implies 150 + x = 600 - 6x, \quad 7x = 450,\ x \approx 64.$

12. a. $R(x) = x(500 - x) = 500x - x^2.$

b. $P(x) = 500x - x^2 - 150 - 4x - x^2 = -2x^2 + 496x - 150.$

c. $MR(x) = 500 - 2x,\ MC(x) = 4 + 2x.$

d. $MC(x) = MR(x) \implies 4 + 2x = 500 - 2x, \quad 4x = 496, \quad x = 124.$

e. $MP(x) = -4x + 496.$

$MP(125) = -4(125) + 496 = -4 < 0.$

The level of production should be decreased to $x = 124$ because a negative derivative means profit decreases as x increases.

13. a. $N(3) - N(2) = \{20(3) - (3)^2\} - \{20(2) - (2)^2\} = 15.$

b. $N'(t) = 20 - 2t. \quad N'(2) = 20 - 4 = 16.$

14. a. The rate of basic skills per hour $= \frac{5}{2}t^{-1/2} - \frac{1}{4}.$

b. Yes, this rate can become negative when $\frac{5}{2}t^{-1/2} - \frac{1}{4} < 0, \quad \frac{5}{2t^{1/2}} < \frac{1}{4},$

$\frac{5}{2} \cdot 4 < t^{1/2}, \quad t > 100.$

15. a. $P'(t) = 1000t^{-1/2}.$

b. No, the population of the city in 1980 does not affect its growth rate.

16. $C'(x) = V'(x)$ which does not depend on F.

17. Let $R(x)$ = weekly revenue function.

$R(x) = x \cdot p(x) = x \cdot (a - bx) = ax - bx^2.$

$R'(x) = a - 2bx$ is a linear function with slope $m = -2b$.

18. a. $MU(x) = U'(x) = 6 - 2x.$

b. $MU(x) = 0 \implies 6 - 2x = 0,\ x = 3.$

c. Since $MU(x)$ is negative for $x > 3$, $U(x)$ decreases as x increases for $x > 3$. Thus if the person has already eaten 4 ice cream cones that day, her satisfaction will go down by eating another one.

19. $y(0) = A(0)^2 + B(0) + C = 40, \quad C = 40.$

$y(5) = A(5)^2 + B(5) + 40 = 315, \quad 25A + 5B = 275, \quad 5A + B = 55.$

$v(5) = 2A(5) + B = 105, 10A + B = 105, B = 105 - 10A.$

$5A + (105 - 10A) = 55 \implies A = 10.$

$B = 105 - 10(10) = 5.$

20. a. $P(x) = 30x - 400 - 10x - 20\sqrt{x} = 20x - 20\sqrt{x} - 400.$

$P(16) = 20(16) - 20 \cdot \sqrt{16} - 400 = -160.$

The manufacturer does not earn a profit at $x_0 = 16$.

b. $P(100) = 20(100) - 20\sqrt{100} - 400 = 1400.$

The manufacturer earns a profit at $x_0 = 100$.

c. $C(x) = R(x), \quad 400 + 10x + 20\sqrt{x} = 30x, \quad 20\sqrt{x} = 20x - 400,$

$\sqrt{x} = x - 20, \quad x = x^2 - 40x + 400, \quad x^2 - 41x + 400 = 0,$

$(x - 25)(x - 16) = 0$ implies $x_0 = 25$, $x = 16$ extraneous.

d. $P(x) = 20x - 20\sqrt{x} - 400.$

$P(16) = 20(16) - 20\sqrt{16} - 400 = -160.$

$P(25) = 20(25) - 20\sqrt{25} - 400 = 0.$

$P(49) = 20(49) - 20\sqrt{49} - 400 = 440.$

$P(100) = 20(100) - 20\sqrt{100} - 400 = 1400.$

21. a. $R(x) = x \cdot p(x) = x(1600 - 5x) = -5x^2 + 1600x.$

b. $P(x) = R(x) - C(x) = (-5x^2 + 1600x) - (25x^2 + 400x + 10,000)$

$= -30x^2 + 1200x - 10,000.$

c. $MP(x) = P'(x) = -60x + 1200.$

d. $MP(x) = 0 \implies x = 20.$

e. $MP(x) = MR(x) - MC(x).$

$MP(20) = 0 \implies MR(20) = MC(20).$

f. For $0 < x < 20$, $MP(x) > 0 \implies P(x)$ increases as x increases.

For $20 < x < 320$, $MP(x) < 0 \implies P(x)$ decreases as x increases.

Therefore the maximum possible profit occurs when $x = 20$.

22. The total cost of producing x bats per week is given by

$$C(x) = A + Bx + Dx^2.$$

a. The fixed cost of production is $C(0) = A$.

b. $MC(x) = B + 2Dx$, which does not depend on A.

c. The demand function does not depend on A, so the revenue function does not depend on A. Therefore, in view of part b, the marginal profit function does not depend on A.

23. a. $R(x) = x \cdot p = x(10,000 - 5x) = -5x^2 + 10,000x$.

b. $MR(x) = R'(x) = -10x + 10,000$.

c. $MR(500) = 5,000$.

$MR(1,000) = 0$.

$MR(1,500) = -5,000$.

d. For $0 < x < 1,000$, $MR(x) > 0 \implies R(x)$ increases as x increases.

For $1,000 < x < 2,000$, $MR(x) < 0 \implies R(x)$ decreases as x increases.

Therefore the revenues are the largest when the production level x is 1,000.

24. $P(x) = R(x) - C(x) \implies MP(x) = MR(x) - MC(x)$.

Let x_0 be a production level such that $MP(x_0) < 0$. It is possible for $MR(x_0)$ to be positive by having $MC(x_0)$ large enough so that $MC(x_0)$ is larger than $MR(x_0)$.

25. Let x_0 be a production level such that $MP(x_0) > 0$. Suppose it would be true that $MR(x_0) < 0$. Then it would follow that $MC(x_0) < 0$, which is not likely.

26. a. $B(x) = \sqrt{x+9} - 3$

$$MB(x) = B'(x) = \lim_{h\to 0} \frac{B(x+h)-B(x)}{h} = \lim_{h\to 0} \frac{\left(\sqrt{x+h+9}-3\right)-\left(\sqrt{x+9}-3\right)}{h}$$

$$= \lim_{h\to 0} \frac{\sqrt{x+h+9}-\sqrt{x+9}}{h} \cdot \frac{\sqrt{x+h+9}+\sqrt{x+9}}{\sqrt{x+h+9}+\sqrt{x+9}}$$

$$= \lim_{h\to 0} \frac{(x+h+9)-(x+9)}{h\left(\sqrt{x+h+9}+\sqrt{x+9}\right)} = \lim_{h\to 0} \frac{1}{\sqrt{x+h+9}+\sqrt{x+9}} = \frac{1}{2\sqrt{x+9}}.$$

b.

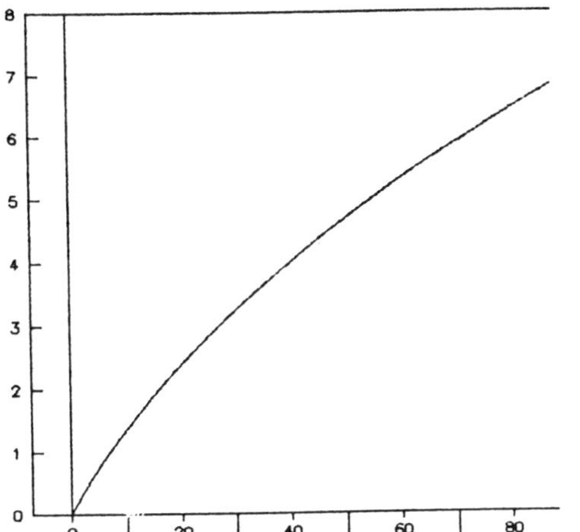

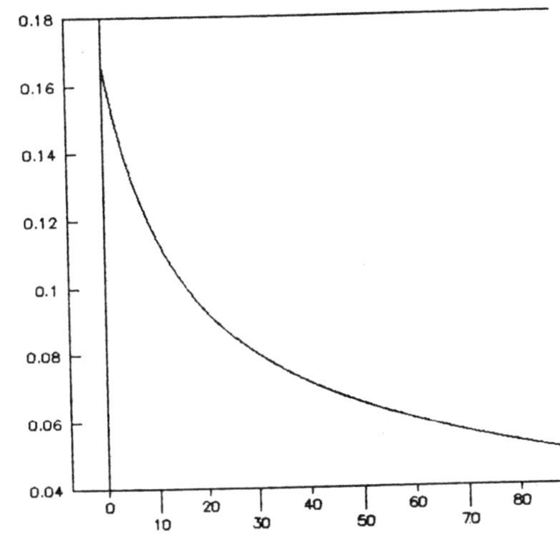

c. The marginal utility $MB(x) = \frac{1}{2\sqrt{x+9}}$ is positive for all $x > 0$. d. It is probably not true that the plant biomass $B(x)$ constantly increases as the amount x of precipitation increases because after the x gets beyond a certain amount, the plant biomass $B(x)$ is not going to keep increasing.

Solutions to Exercise Set 2.7

1. $f'(x) = 1(x+1) + (x-1)1 = x + 1 + x - 1 = 2x$.
2. $f'(x) = 2x(2-x) + (x^2-1)(-1) = 4x - 2x^2 - x^2 + 1 = -3x^2 + 4x + 1$.
3. $f'(x) = (6x-8)(x^2+2) + (3x^2-8x)(2x) = 12x^3 - 24x^2 + 12x - 16$.

4. $f(x) = (x^2 - 6x + 10)(7 - x)$.

$f'(x) = (2x - 6)(7 - x) + (x^2 - 6x + 10)(-1) = (-2x^2 + 20x - 42) + (-x^2 + 6x - 10)$

$= (-3x^2 + 26x - 52)$.

5. $f(x) = (x^3 - x)(x^3 - x)$.

$f'(x) = (3x^2 - 1)(x^3 - x) + (x^3 - x)(3x^2 - 1) = 3x^5 - 3x^3 - x^3 + x + 3x^5 - x^3 - 3x^3 + x$

$= 6x^5 - 8x^3 + 2x$.

6. $f(x) = (x^2 - 3x^{-2})(x^2 - 3x^{-2})$.

$f'(x) = (2x + 6x^{-3})(x^2 - 3x^{-2}) + (x^2 - 3x^{-2})(2x + 6x^{-3})$

$= 2x^3 - 6x^{-1} + 6x^{-1} - 18x^{-5} + 2x^3 + 6x^{-1} - 6x^{-1} - 18x^{-5} = 4x^3 - 36x^{-5}$.

7. $f(x) = \sqrt{x}\left(x^3 + x^{-2/3}\right)$.

$f'(x) = \frac{1}{2}x^{-1/2}\left(x^3 + x^{-2/3}\right) + x^{1/2}\left(3x^2 - \frac{2}{3}x^{-5/3}\right)$

$= \frac{1}{2}x^{5/2} + \frac{1}{2}x^{-7/6} + 3x^{5/2} - \frac{2}{3}x^{-7/6}$

$= \frac{7}{2}x^{5/2} - \frac{1}{6}x^{-7/6}$.

8. $f'(x) = (2x - 1)(x^{2/3} + 2x^{1/3}) + (x^2 - x)\left(\frac{2}{3}x^{-1/3} + \frac{2}{3}x^{-2/3}\right)$

$= 2x^{5/3} + 4x^{4/3} - x^{2/3} - 2x^{1/3} + \frac{2}{3}x^{5/3} + \frac{2}{3}x^{4/3} - \frac{2}{3}x^{2/3} - \frac{2}{3}x^{1/3}$

$= \frac{8}{3}x^{5/3} + \frac{14}{3}x^{4/3} - \frac{5}{3}x^{2/3} - \frac{8}{3}x^{1/3}$.

9. $f'(x) = (-6x^{-3} - x^{-2})(x - 4) + (3x^{-2} + x^{-1})(1)$

$= -6x^{-2} + 24x^{-3} - x^{-1} + 4x^{-2} + 3x^{-2} + x^{-1} = 24x^{-3} + x^{-2}$.

10. $f(x) = \left(x^{4/3} - x^{-2}\right)^2$.

$f'(x) = \left(\frac{4}{3}x^{1/3} + 2x^{-3}\right)\left(x^{4/3} - x^{-2}\right) + \left(x^{4/3} - x^{-2}\right)\left(\frac{4}{3}x^{1/3} + 2x^{-3}\right)$

$= 2\left(x^{4/3} - x^{-2}\right)\left(\frac{4}{3}x^{1/3} + 2x^{-3}\right)$

$= 2\left(\frac{4}{3}x^{5/3} - \frac{4}{3}x^{-5/3} + 2x^{-5/3} - 2x^{-5}\right)$

$= 4\left(\frac{2}{3}x^{5/3} + \frac{1}{3}x^{-5/3} - x^{-5}\right)$

$= \frac{4}{3}\left(2x^{5/3} + x^{-5/3} - 3x^{-5}\right)$.

11. $f'(x) = \frac{(x-2)(1)-(x+2)(1)}{(x-2)^2} = \frac{x-2-x-2}{(x-2)^2} = -\frac{4}{(x-2)^2}$.

12. $f'(x) = \frac{(3-x)(2x)-(x^2-6)(-1)}{(3-x)^2} = \frac{6x-2x^2+x^2-6}{(3-x)^2} = \frac{-x^2+6x-6}{(3-x)^2}.$

13. $g(x) = (8x+2)(x+1) \implies g'(x) = 8(x+1)+(8x+2)(1) = 8x+8+8x+2 = 16x+10.$

$$f'(x) = \frac{(x-3)(16x+10)-(8x+2)(x+1)(1)}{(x-3)^2}$$

$$= \frac{16x^2+10x-48x-30-8x^2-8x-2x-2}{(x-3)^2} = \frac{8x^2-48x-32}{(x-3)^2}.$$

14. $f(x) = \frac{x^4+4}{1-x^3}.$

$$f'(x) = \frac{(1-x^3)(4x^3)-(x^4+4)(-3x^2)}{(1-x^3)^2}$$

$$= \frac{(4x^3-4x^6)-(-3x^6-12x^2)}{(1-x^3)^2} = \frac{4x^3-x^6+12x^2}{(1-x^3)^2}.$$

15. $f'(x) = \frac{(x+2)(2x)-(x^2-4)(1)}{(x+2)^2} = \frac{2x^2+4x-x^2+4}{(x+2)^2}$

$$= \frac{x^2+4x+4}{x^2+4x+4} = 1.$$

16. $g(x) = (x^2+7)(x+2) \implies g'(x) = 2x(x+2)+(x^2+7)(1) = 2x^2+4x+x^2+7$

$$= 3x^2+4x+7.$$

$$f'(x) = \frac{(x^2-3x)(3x^2+4x+7)-(x^2+7)(x+2)(2x-3)}{(x^2-3x)^2} = \frac{x^4-6x^3-13x^2+28x+42}{(x^2-3x)^2}.$$

17. $h(x) = (1+x)(1+x) \implies h'(x) = 1(1+x)+(1+x)(1) = 1+x+1+x = 2+2x.$

$$f'(x) = \frac{(1+x)^2(-1)-(1-x)(2+2x)}{\{(1+x)^2\}^2} = \frac{-1-2x-x^2-2-2x+2x+2x^2}{\{(1+x)^2\}^2}$$

$$= \frac{x^2-2x-3}{\{(1+x)^2\}^2} = \frac{x-3}{(1+x)^3}.$$

18. $f(x) = \frac{\sqrt{x}+1}{1+x^2}.$

$$f'(x) = \frac{(1+x^2)\left(\frac{1}{2}x^{-1/2}\right)-\left(x^{1/2}+1\right)(2x)}{(1+x^2)^2}$$

$$= \frac{\left(\frac{1}{2}x^{-1/2}+\frac{1}{2}x^{3/2}\right)-\left(2x^{3/2}+2x\right)}{(1+x^2)^2}$$

$$= \frac{\frac{1}{2}x^{-1/2}-\frac{3}{2}x^{3/2}-2x}{(1+x^2)^2}.$$

19. $f'(x) = \frac{(x^{1/2}+1)\left(2x-\frac{2}{3}x^{-1/3}\right)-(x^2-x^{2/3})\left(\frac{1}{2}x^{-1/2}\right)}{\left(\sqrt{x}+1\right)^2}$

$$= \frac{2x^{3/2}-\frac{2}{3}x^{1/6}+2x-\frac{2}{3}x^{-1/3}-\frac{1}{2}x^{3/2}+\frac{1}{2}x^{1/6}}{\left(\sqrt{x}+1\right)^2}$$

$$= \frac{\frac{3}{2}x^{3/2}+2x-\frac{1}{6}x^{1/6}-\frac{2}{3}x^{-1/3}}{\left(\sqrt{x}+1\right)^2}.$$

20. $f'(x) = \frac{(x^2-x^{1/2}+1)(1+2x)-(x-4+x^2)\left(2x-\frac{1}{2}x^{-1/2}\right)}{\left(x^2-\sqrt{x}+1\right)^2}$

$= \frac{x^2-x^{1/2}+1+2x^3-2x^{3/2}+2x-2x^2+8x-2x^3+\frac{1}{2}x^{1/2}-2x^{-1/2}+\frac{1}{2}x^{3/2}}{\left(x^2-\sqrt{x}+1\right)^2}$

$= \frac{-x^2-\frac{3}{2}x^{3/2}+10x-\frac{1}{2}x^{1/2}-2x^{-1/2}+1}{\left(x^2-\sqrt{x}+1\right)^2}.$

21. $f(x) = x^{1/3}(x^2+2)$.

$f'(x) = \frac{1}{3}x^{-2/3}(x^2+2) + x^{1/3}(2x)$

$f'(x) = \left(\frac{1}{3}x^{4/3} + \frac{2}{3}x^{-2/3}\right) + (2x^{4/3})$

$= \frac{7}{3}x^{4/3} + \frac{2}{3}x^{-2/3}.$

22. $f(x) = \frac{x^2}{1+x^3}.$

$f'(x) = \frac{(1+x^3)(2x)-(x^2)(3x^2)}{(1+x^3)^2}$

$= \frac{(2x+2x^4)-(3x^4)}{(1+x^3)^2} = \frac{2x-x^4}{(1+x^3)^2}.$

23. $f(x) = g(x)h(x)$ where $g(x) = \frac{1}{x+1}$ and $h(x) = \frac{x-3}{x}$.

$g'(x) = \frac{(x+1)(0)-1(1)}{(x+1)^2} = \frac{-1}{(x+1)^2}.$

$h'(x) = \frac{x\cdot 1-(x-3)\cdot 1}{x^2} = \frac{3}{x^2}.$

$f'(x) = g'(x)\cdot h(x) + g(x)\cdot h'(x)$

$= \frac{-1}{(x+1)^2}\cdot\frac{(x-3)}{x} + \frac{1}{x+1}\cdot\frac{3}{x^2}$

$= \frac{-(x-3)\cdot x}{x^2(x+1)^2} + \frac{3(x+1)}{x^2(x+1)^2} = \frac{-x^2+6x+3}{x^2(x+1)^2}.$

24. $f(x) = \frac{x^{1/2}-x^{7/2}}{1+x}.$

$f'(x) = \frac{(1+x)\left(\frac{1}{2}x^{-1/2}-\frac{7}{2}x^{5/2}\right)-(x^{1/2}-x^{7/2})\cdot 1}{(1+x)^2}$

$= \frac{\frac{1}{2}x^{-1/2}+\frac{1}{2}x^{1/2}-\frac{7}{2}x^{5/2}-\frac{7}{2}x^{7/2}-x^{1/2}+x^{7/2}}{(1+x)^2}$

$= \frac{\frac{1}{2}x^{-1/2}-\frac{1}{2}x^{1/2}-\frac{7}{2}x^{5/2}-\frac{5}{2}x^{7/2}}{(1+x)^2}$

$= \frac{\frac{1}{2}x^{-1/2}(1-x-7x^3-5x^4)}{(1+x)^2}$

$= \frac{1-x-7x^3-5x^4}{2\sqrt{x}(1+x)^2}.$

25. $f(x) = g(x)h(x)$ where $g(x) = ax+b$ and $h(x) = cx^2+d$.

$g'(x) = a.$

$h'(x) = 2cx.$

$$f'(x) = a(cx^2 + d) + (ax + b)(2cx)$$
$$= acx^2 + ad + 2acx^2 + 2bcx$$
$$= 3acx^2 + 2bcx + ad.$$

26. $f'(x) = \frac{x \cdot C'(x) - C(x) \cdot 1}{x^2} = \frac{x \cdot C'(x) - C(x)}{x^2}.$

27. $f'(x) = \frac{(2x-4)3 - 3x(2)}{(2x-4)^2} = \frac{-12}{(2x-4)^2}.$

$f'(x) = -3 \implies \frac{-12}{(2x-4)^2} = -3, \quad -12 = -3(2x-4)^2,$

$4x^2 - 16x + 16 = 4 \implies 4x^2 - 16x + 12 = 0 \implies 4(x^2 - 4x + 3) = 0 \implies$

$4(x-3)(x-1) = 0 \implies x = 3$ or $x = 1$. The points are $\left(3, \frac{9}{2}\right)$ and $\left(1, -\frac{3}{2}\right)$.

28. $\frac{dy}{dx} = \frac{(ax+2)(0) - 1 \cdot a}{(ax+2)^2} = \frac{-a}{(ax+2)^2}.$

When $x = 0$, $y = \frac{1}{2}$, and $\frac{dy}{dx} = \frac{-a}{4}$.

The tangent line to the graph at $\left(0, \frac{1}{2}\right)$ has the equation

$y - \frac{1}{2} = \frac{-a}{4}(x - 0)$, $4y - 2 = -ax$, $4y + ax - 2 = 0$.

Thus $a = 3$.

29. $\frac{d}{dx}(fgh) = \frac{d}{dx}(f(gh)) = f'(gh) + f(gh)' = f'(gh) + f(g'h + gh')$

$$= (f')gh + f(g')h + fg(h').$$

30. $f'(x) = 1(x+1)(x+2) + x \cdot 1(x+2) + x(x+1)1 = 3x^2 + 6x + 2.$.

31. $f'(x) = 1(1-x)(x^3 - x) + (x-3)(-1)(x^3 - x) + (x-3)(1-x)(3x^2 - 1).$

$$= -5x^4 + 16x^3 - 6x^2 - 8x + 3.$$

32. $f'(x) = \frac{1}{2}x^{-1/2}(1+x)(1-x) + x^{1/2}(1)(1-x) + x^{1/2}(1+x)(-1).$

33. $f'(x) = (1-2x)(x^3 - x^{-2})(x^{-1} - x^{-3}) + (x - x^2)(3x^2 + 2x^{-3})(x^{-1} - x^{-3})$

$$+ (x - x^2)(x^3 - x^{-2})(-x^{-2} + 3x^{-4}).$$

34. $C(x) = 500 + (1 + x^{1/2})(20x + x^2)$.

a. $MC(x) = C'(x) = \frac{1}{2}x^{-1/2}(20x + x^2) + (1 + x^{1/2})(20 + 2x)$

$= 10x^{1/2} + \frac{1}{2}x^{3/2} + 20 + 20x^{1/2} + 2x + 2x^{3/2}$

$= 20 + 30x^{1/2} + 2x + \frac{5}{2}x^{3/2}$.

b. $MC(16) = 20 + 30(16)^{1/2} + 2(16) + \frac{5}{2}(16)^{3/2}$

$= 20 + 30(4) + 2(16) + \frac{5}{2}(64)$

$= 20 + 120 + 32 + 160 = 332$.

35. $R(x) = \frac{10000x^{3/2}}{2+x^{1/2}}$

a. $MR(x) = R'(x)$

$= \left[(2 + x^{1/2})(15000x^{1/2}) - (10000x^{3/2})(\frac{1}{2}x^{-1/2})\right] / (2 + x^{1/2})^2$

$= \frac{30000x^{1/2}+15000x-5000x}{(2+x^{1/2})^2} = \frac{30000x^{1/2}+10000x}{(2+x^{1/2})^2}$.

b. $MR(9) = \frac{30000(9)^{1/2}+10000(9)}{(2+9^{1/2})^2} = \frac{90000+90000}{25} = \frac{180000}{25} = 7200$.

36. $f(x) = \frac{x}{1+x^2}$.

$f'(x) = \frac{(1+x^2)\cdot 1-x(2x)}{(1+x^2)^2} = \frac{1-x^2}{(1+x^2)^2}$.

$f'(2) = \frac{1-2^2}{(1+2^2)^2} = \frac{-3}{25}$.

An equation for the tangent line is $y - \frac{2}{5} = -\frac{3}{25}(x - 2)$,

$25y - 10 = -3x + 6$, $25y = -3x + 16$, $3x + 25y - 16 = 0$.

37. $C(x) = 400 + 20x + x^2$.

a. $MC(x) = C'(x) = 20 + 2x$.

b. $c(x) = \frac{C(x)}{x} = \frac{400+20x+x^2}{x} = 400x^{-1} + 20 + x$.

c. $c'(x) = -400x^{-2} + 1$.

38. $c(x) = \frac{C(x)}{x} \implies c'(x) = \frac{xC'(x)-C(x)\cdot 1}{x^2} = \frac{x\cdot MC(x)-C(x)}{x^2}$.

39. $W(t) = \frac{50t^2}{10+t^2}$

$W'(t) = \frac{(10+t^2)100t-(50t^2)(2t)}{(10+t^2)^2} = \frac{1000t+100t^3-100t^3}{(10+t^2)^2}$

$= \frac{1000t}{(10+t^2)^2}$.

40. $s(t) = \frac{t^2-6t+4}{t+3}$.

a. $v(t) = s'(t) = \frac{(t+3)(2t-6)-(t^2-6t+4)(1)}{(t+3)^2}$

$= \frac{2t^2-18-t^2+6t-4}{(t+3)^2}$

$= \frac{t^2+6t-22}{(t+3)^2}$.

b. $v(3) = \frac{3^2+6(3)-22}{(3+3)^2} = \frac{9+18-22}{36} = \frac{5}{36}$.

41. $U(x) = \frac{x}{x^2+9}$.

a. $MU(x) = U'(x) = \frac{(x^2+9)(1)-x(2x)}{(x^2+9)^2} = \frac{9-x^2}{(x^2+9)^2}$.

b. $MU(x) = 0 \implies 9 - x^2 = 0 \implies x = 3$.

42. a. $R(x) = xp(x) = \frac{8,000x}{10+x}$.

b. $MR(x) = R'(x) = \frac{(10+x)(8,000)-(8,000x)(1)}{(10+x)^2} = \frac{80,000}{(10+x)^2}$.

c. $MR(10) = \frac{80,000}{(10+10)^2} = 200$.

$MR(90) = \frac{80,000}{(10+90)^2} = 8$.

d. $MR(x) > 0$ for all $x > 0$.

43. $P(t) = 10 + \frac{100t}{t^2+9}, \qquad t > 0$.

a. $P'(t) = 0 + \frac{(t^2+9)(100)-(100t)(2t)}{(t^2+9)^2}$

$= \frac{900-100t^2}{(t^2+9)^2}$.

b. $P'(0) = \frac{900}{(9)^2} = \frac{100}{9} > 0$.

Therefore the population increases initially.

c. $P'(t) = 0$ when $t = 3$.

For $t > 3$, $P'(t) < 0$, so the population is declining in size.

d. As t becomes very large, $\frac{100t}{t^2+9}$ becomes close to zero,

so $P(t)$ becomes close to 10. Therefore the eventual size of the population is 10.

Solutions to Exercise Set 2.8

1. $f'(x) = 3(x+4)^2 \cdot \frac{d}{dx}(x+4) = 3(x+4)^2 \cdot 1 = 3(x+4)^2$.

2. $f'(x) = 4(3x-2)^3 \cdot \frac{d}{dx}(3x-2) = 4(3x-2)^3 \cdot 3 = 12(3x-2)^3$.

3. $f(x) = (3x^2-2x)^5$.

   $f'(x) = 5(3x^2-2x)^4 \cdot \frac{d}{dx}(3x^2-2x) = 5(3x^2-2x)^4(6x-2) = 10(3x^2-2x)^4(3x-1)$.

4. $f'(x) = 9(x^2+8x+8)^8 \cdot \frac{d}{dx}(x^2+8x+8) = 9(x^2+8x+8)^8 \cdot (2x+8)$.

5. $f(x) = \sqrt{x^4-2x^2} = (x^4-2x^2)^{1/2}$.

   $f'(x) = \frac{1}{2}(x^4-2x^2)^{-1/2} \cdot \frac{d}{dx}(x^4-2x^2) = \frac{1}{2}(x^4-2x^2)^{-1/2} \cdot (4x^3-4x)$

   $= 2x(x^4-2x^2)^{-1/2}(x^2-1)$.

6. $f(x) = x^3(x^2-7)^4$.

   $f'(x) = (3x^2)(x^2-7)^4 + (x^3)\left(4(x^2-7)^3(2x)\right)$

   $= x^2(x^2-7)^3\left(3(x^2-7)+8x^2\right)$

   $= x^2(x^2-7)^3(11x^2-21)$.

7. $f(x) = \frac{x}{(x^2-9)^3}$.

   $f'(x) = \frac{(x^2-9)^3(1)-(x)\left(3(x^2-9)^2(2x)\right)}{(x^2-9)^6}$

   $= \frac{(x^2-9)^2\left((x^2-9)-6x^2\right)}{(x^2-9)^6} = \frac{-5x^2-9}{(x^2-9)^4}$.

8. $f'(x) = \frac{(x^2-6x+3)^2 \cdot \frac{d}{dx}(x+3)-(x+3)\cdot\frac{d}{dx}(x^2-6x+3)^2}{(x^2-6x+3)^4}$

   $= \frac{(x^2-6x+3)^2-(x+3)\cdot 2(x^2-6x+3)\cdot\frac{d}{dx}(x^2-6x+3)}{(x^2-6x+3)^4}$

   $= \frac{(x^2-6x+3)-2(x+3)(2x-6)}{(x^2-6x+3)^3}$.

9. $f'(x) = 4\left(3\sqrt{x}-2\right)^3 \cdot \frac{d}{dx}\left(3\sqrt{x}-2\right) = 4\left(3\sqrt{x}-2\right)^3 \cdot 3 \cdot \left(\frac{1}{2}x^{-1/2}\right)$

   $= 6x^{-1/2}\left(3\sqrt{x}-2\right)^3$.

10. $f'(x) = -3(x^6-x^2+2)^{-4} \cdot \frac{d}{dx}(x^6-x^2+2) = -3(x^6-x^2+2)^{-4} \cdot (6x^5-2x)$.

11. $f'(x) = 4(x^{2/3}-x^{1/4})^3 \cdot \frac{d}{dx}(x^{2/3}-x^{1/4}) = 4(x^{2/3}-x^{1/4})^3 \cdot \left(\frac{2}{3}x^{-1/3}-\frac{1}{4}x^{-3/4}\right)$.

12. $f'(x) = \frac{1}{2}(x^2-6x+6)^{-1/2} \cdot \frac{d}{dx}(x^2-6x+6) = \frac{1}{2}(x^2-6x+6)^{-1/2} \cdot (2x-6)$.

13. $f'(x) = 4\left(\frac{x-3}{x+3}\right)^3 \cdot \frac{d}{dx}\left(\frac{x-3}{x+3}\right)$
$= 4\left(\frac{x-3}{x+3}\right)^3 \cdot \frac{(x+3)\frac{d}{dx}(x-3)-(x-3)\frac{d}{dx}(x+3)}{(x+3)^2}$
$= 4\frac{(x-3)^3}{(x+3)^3} \cdot \frac{(x+3)\cdot 1-(x-3)\cdot 1}{(x+3)^2} = 4\frac{(x-3)^3}{(x+3)^3} \cdot \frac{6}{(x+3)^2} = \frac{24(x-3)^3}{(x+3)^5}.$

14. $f'(x) = 6\left(\frac{1+\sqrt{x}}{1-\sqrt{x}}\right)^5 \cdot \frac{d}{dx}\left(\frac{1+\sqrt{x}}{1-\sqrt{x}}\right)$
$= 6\frac{(1+\sqrt{x})^5}{(1-\sqrt{x})^5} \cdot \frac{(1-\sqrt{x})\frac{d}{dx}(1+\sqrt{x})-(1+\sqrt{x})\frac{d}{dx}(1-\sqrt{x})}{(1-\sqrt{x})^2}$
$= 6\frac{(1+\sqrt{x})^5}{(1-\sqrt{x})^5} \cdot \frac{(1-\sqrt{x})\cdot\left(\frac{1}{2\sqrt{x}}\right)-(1+\sqrt{x})\left(-\frac{1}{2\sqrt{x}}\right)}{(1-\sqrt{x})^2}$
$= 6\frac{(1+\sqrt{x})^5}{(1-\sqrt{x})^5} \cdot \frac{\frac{1}{2\sqrt{x}}+\frac{1}{2\sqrt{x}}}{(1-\sqrt{x})^2} = 6\frac{(1+\sqrt{x})^5}{(1-\sqrt{x})^7\cdot\sqrt{x}}.$

15. $f'(x) = 3\left(\frac{x+2}{\sqrt[3]{x}}\right)^2 \cdot \frac{d}{dx}\left(\frac{x+2}{\sqrt[3]{x}}\right) = 3\left(\frac{x+2}{\sqrt[3]{x}}\right)^2 \cdot \frac{\sqrt[3]{x}\cdot 1-(x+2)\cdot\frac{1}{3}x^{-2/3}}{(\sqrt[3]{x})^2}$
$= 3\frac{(x+2)^2}{(\sqrt[3]{x})^2} \cdot \frac{1}{3} \cdot \frac{3\sqrt[3]{x}-(x+2)x^{-2/3}}{(\sqrt[3]{x})^2} = \frac{(x+2)^2\left(3\sqrt[3]{x}-(x+2)x^{-2/3}\right)}{x\sqrt[3]{x}}$
$= 2(x+2)^2(x^{-1}-x^{-2}).$

16. $f(x) = \sqrt{x^{-2/3}-2x^{1/3}}.$

$f'(x) = \frac{1}{2}\left(x^{-2/3}-2x^{1/3}\right)^{-1/2}\left(-\frac{2}{3}x^{-5/3}-\frac{2}{3}x^{-2/3}\right)$

$= -\frac{1}{3}\left(x^{-2/3}-2x^{1/3}\right)^{-1/2}\left(x^{-5/3}+x^{-2/3}\right).$

17. $f'(x) = -3(4x^{-1/4}+6)^{-4} \cdot \frac{d}{dx}(4x^{-1/4}+6)$

$= -3\left(4x^{-1/4}+6\right)^{-4}\left(4\cdot\left(-\frac{1}{4}\right)x^{-5/4}\right)$

$= 3x^{-5/4}\left(4x^{-1/4}+6\right)^{-4}.$

18. $f'(x) = \frac{1}{4}(3-2x-x^4)^{-3/4} \cdot \frac{d}{dx}(3-2x-x^4) = \frac{1}{4}(3-2x-x^4)^{-3/4}(-2-4x^3)$

$= -\frac{1}{2}(3-2x-x^4)^{-3/4}(1+2x^3).$

19. $f'(x) = \frac{(1-x)\frac{d}{dx}\left[(x^3+1)^3+1\right]-\left[(x^3+1)^3+1\right]\frac{d}{dx}(1-x)}{(1-x)^2}$
$= \frac{(1-x)\left[3(x^3+1)^2(3x^2)\right]-\left[(x^3+1)^3+1\right](-1)}{(1-x)^2}$
$= \frac{9x^2(1-x)(x^3+1)^2+(x^3+1)^3+1}{(1-x)^2}.$

20. $f'(x) = \frac{\left[3+(x^2+1)^3\right]\cdot 1-(x+2)\left(3(x^2+1)^2(2x)\right)}{[3+(x^2+1)^3]^2} = \frac{(x^2+1)^3-6x(x+2)(x^2+1)^2+3}{[3+(x^2+1)^3]^2}.$

21. $f(x) = \frac{1}{1+(x^2+2)^2} = \left[1 + (x^2+2)^2\right]^{-1}$.

$f'(x) = (-1) \cdot \left[1 + (x^2+2)^2\right]^{-2} \cdot (2(x^2+2)(2x))$

$= \frac{-4x^3-8x}{[1+(x^2+2)^2]^2}$.

22. $f'(x) = 6\left(\frac{1+x}{1-x}\right)^5 \cdot \frac{d}{dx}\left(\frac{1+x}{1-x}\right) = 6\left(\frac{1+x}{1-x}\right)^5 \cdot \frac{(1-x)\cdot 1-(1+x)(-1)}{(1-x)^2} = \frac{12(1+x)^5}{(1-x)^7}$.

23. $f'(x) = 3\left(\frac{x^{2/3}}{1+\sqrt{x}}\right)^2 \cdot \frac{d}{dx}\left(\frac{x^{2/3}}{1+\sqrt{x}}\right) = 3\left(\frac{x^{2/3}}{1+\sqrt{x}}\right)^2 \cdot \frac{\left(1+\sqrt{x}\right)\frac{2}{3}x^{-1/3}-x^{2/3}\cdot\frac{1}{2}x^{-1/2}}{\left(1+\sqrt{x}\right)^2}$

$= 3\frac{x^{4/3}}{\left(1+\sqrt{x}\right)^2} \cdot \frac{\frac{2}{3}x^{-1/3}\left(1+\sqrt{x}\right)-\frac{1}{2}x^{1/6}}{\left(1+\sqrt{x}\right)^2} = \frac{3x^{4/3}\left[\frac{2}{3}x^{-1/3}\left(1+\sqrt{x}\right)-\frac{1}{2}x^{1/6}\right]}{\left(1+\sqrt{x}\right)^4}$

$= \frac{4x+x^{3/2}}{2\left(1+\sqrt{x}\right)^4}$

24. $f(x) = \frac{x}{\left(1+\sqrt{x}\right)^4}$.

$f'(x) = \frac{\left(1+\sqrt{x}\right)^4(1)-(x)\left(4\left(1+\sqrt{x}\right)^3\left(\frac{1}{2}x^{-1/2}\right)\right)}{\left(1+\sqrt{x}\right)^8}$

$= \frac{\left(1+\sqrt{x}\right)^3\left(\left(1+\sqrt{x}\right)-2\sqrt{x}\right)}{\left(1+\sqrt{x}\right)^8}$

$= \frac{1-\sqrt{x}}{\left(1+\sqrt{x}\right)5}$.

25. $f'(x) = 3\left(\sqrt{x}-\frac{1}{\sqrt{x}}\right)^2 \cdot \frac{d}{dx}\left(\sqrt{x}-\frac{1}{\sqrt{x}}\right) = 3\left(\sqrt{x}-\frac{1}{\sqrt{x}}\right)^2 \cdot \frac{d}{dx}\left(x^{1/2}-x^{-1/2}\right)$

$= 3\left(\sqrt{x}-\frac{1}{\sqrt{x}}\right)^2\left(\frac{1}{2}x^{-1/2}+\frac{1}{2}x^{-3/2}\right) = \frac{3}{2}\left(\sqrt{x}-\frac{1}{\sqrt{x}}\right)^2\left(\frac{1}{\sqrt{x}}+\frac{1}{x\sqrt{x}}\right)$

$= \frac{3}{2}\left[x^{1/2}-x^{-1/2}-x^{-3/2}+x^{-5/2}\right]$

26. $f'(x) = \frac{(x+5)^3\frac{1}{2}(x^2+1)^{-1/2}\cdot 2x-\sqrt{x^2+1}\cdot 3(x+5)^2}{(x+5)^6} = \frac{x(x+5)(x^2+1)^{-1/2}-3\sqrt{x^2+1}}{(x+5)^4}$.

27. $f'(x) = \frac{\sqrt{x^3+6}\cdot 5(x-4)^4-(x-4)^5\cdot\frac{d}{dx}\sqrt{x^3+6}}{\left(\sqrt{x^3+6}\right)^2}$

$= \frac{\sqrt{x^3+6}\cdot 5(x-4)^4-(x-4)^5\cdot\frac{1}{2}\cdot\frac{1}{\sqrt{x^3+6}}\cdot(3x^2)}{\left(\sqrt{x^3+6}\right)^2} = \frac{5(x^3+6)(x-4)^4-\frac{3}{2}x^2(x-4)^5}{\left(\sqrt{x^3+6}\right)^3}$.

28. $f'(x) = \left(\frac{d}{dx}\left(x^{1/4}-x^{3/4}\right)^3\right) \cdot \left(x^{-2/3}+x^{-4/3}\right)^5 + \left(x^{1/4}-x^{3/4}\right)^3$

$\cdot\frac{d}{dx}\left(x^{-2/3}+x^{-4/3}\right)^5$

$= 3\left(x^{1/4}-x^{3/4}\right)^2 \cdot \left(\frac{1}{4}x^{-3/4}-\frac{3}{4}x^{-1/4}\right) \cdot \left(x^{-2/3}+x^{-4/3}\right)^5$

$+ \left(x^{1/4}-x^{3/4}\right)^3 \cdot 5 \cdot \left(x^{-2/3}+x^{-4/3}\right)^4\left(-\frac{2}{3}x^{-5/3}-\frac{4}{3}x^{-7/3}\right)$

$= \frac{3}{4}\left(x^{1/4}-x^{3/4}\right)^2\left(x^{-3/4}-3x^{-1/4}\right)\left(x^{-2/3}+x^{-4/3}\right)^5$

$-\frac{10}{3}\left(x^{1/4}-x^{3/4}\right)^3\left(x^{-2/3}+x^{-4/3}\right)^4\left(x^{-5/3}+2x^{-7/3}\right)$.

29. $f'(x) = \left(\frac{d}{dx}\left(x^{1/3}-4\right)^3\right)\cdot\left(x-\sqrt{x}\right)^{-2/3} + \left(x^{1/3}-4\right)^3\cdot\frac{d}{dx}\left(x-\sqrt{x}\right)^{-2/3}$

$= 3\left(x^{1/3}-4\right)^2\left(\frac{1}{3}x^{-2/3}\right)\left(x-\sqrt{x}\right)^{-2/3} + \left(x^{1/3}-4\right)^3$

$\cdot\left(-\frac{2}{3}\right)\left(x-\sqrt{x}\right)^{-5/3}\left(1-\frac{1}{2}x^{-1/2}\right)$

$= x^{-2/3}\left(x^{1/3}-4\right)^2\left(x-\sqrt{x}\right)^{-2/3} - \frac{1}{3}\left(x^{1/3}-4\right)^3\left(x-\sqrt{x}\right)^{-5/3}\left(2-x^{-1/2}\right).$

30. $f(x) = \frac{\sqrt{1-x^2}}{1+\sqrt{x}}.$

$f'(x) = \frac{(1+\sqrt{x})\cdot\left(\frac{1}{2}(1-x^2)^{-1/2}(-2x)\right)-\sqrt{1-x^2}\left(\frac{1}{2}x^{-1/2}\right)}{\left(1+\sqrt{x}\right)^2}$

$= \frac{(1-x^2)^{-1/2}\left(-(1+\sqrt{x})(x)-\frac{1}{2}(1-x^2)(x^{-1/2})\right)}{\left(1+\sqrt{x}\right)^2}$

$= \frac{-x-x^{3/2}-\frac{1}{2}x^{-1/2}+\frac{1}{2}x^{3/2}}{(1-x^2)^{1/2}\left(1+\sqrt{x}\right)^2}$

$= \frac{-\frac{1}{2}x^{3/2}-x-\frac{1}{2}x^{-1/2}}{(1-x^2)^{1/2}\left(1+\sqrt{x}\right)^2}.$

31. $f'(x) = 3(1-x^2)^2(-2x) = -6x(1-x^2)^2.$

At $P = (1,0)$, $f'(1) = -6(1)(1-1^2)^2 = 0.$

The tangent line to the graph of $y = f(x)$ at P is $y - 0 = f'(1)(x-1)$, that is,

$y = 0.$

32. $f'(x) = 2\left(\frac{x}{x+1}\right)\cdot\frac{(x+1)-x}{(x+1)^2} = \frac{2x}{(x+1)^3}.$

At $P = (0,0)$, $f'(0) = \frac{2\cdot 0}{(0+1)^3} = 0.$

The line tangent to the graph of $y = f(x)$ at P is $y - 0 = f'(0)(x-0)$, that is,

$y = 0.$

33. $f(x) = x\sqrt{1+x^3}$; $P = (2,6)$

$f'(x) = 1\sqrt{1+x^3} + x\left(\frac{1}{2}(1+x^3)^{-1/2}(3x^2)\right).$

$f'(2) = 1\sqrt{1+(2)^3} + 2\left(\frac{1}{2}(1+2^3)^{-1/2}\left(3(2)^2\right)\right)$

$= 3 + \frac{1}{3}\cdot 12$

$= 7.$

The tangent line has the equation

$$y - 6 = 7(x-2) \implies y = 7x - 8.$$

34. $f'(x) = \frac{1}{2}\left(\frac{1-x}{1+x}\right)^{-1/2} \cdot \frac{-(1+x)-(1-x)}{(1+x)^2} = -\frac{(1-x)^{-1/2}}{(1+x)^{3/2}}$.

At $P = (0,1)$, $f'(0) = \frac{-(1-0)^{-1/2}}{(1+0)^{3/2}} = -1$.

The tangent line to the graph of $y = f(x)$ at the point P is $y - 1 = f'(0)(x - 0)$,

$y - 1 = (-1)x$, that is, $x + y - 1 = 0$.

35. $MC(x) = C'(x) = \frac{1}{2}(40 + 16x^2)^{-1/2} \cdot 16(2x) = 16x(40 + 16x^2)^{-1/2}$.

36. a. The marginal revenue $MR(x) = R'(x) = 20 \cdot \frac{1}{2}(100x - x^2)^{-1/2} \cdot (100 - 2x)$

$= \frac{10(100-2x)}{\sqrt{100x-x^2}}$.

b. Solve the inequality $MR(x) \geq 0$, that is, $\frac{10(100-2x)}{\sqrt{100x-x^2}} \geq 0$ for x.

$100 - 2x \geq 0, \quad x \leq 50$.

Thus $MR(x) \geq 0$ for x such that $0 < x \leq 50$.

37. a. The marginal weekly profit

$MP(x) = P'(x) = \frac{1}{3}(x^3 + 10x + 125)^{-2/3}(3x^2 + 10)$.

b. Use the fact that if $P'(x) > 0$ (or $P'(x) < 0$) for all x, $P(x)$ is always increasing (or decreasing) with x.

From part (a), $P'(x) = \frac{3x^2+10}{3\left(\sqrt[3]{(x^3+10x+125)}\right)^2} > 0$ for all $x \geq 0$.

Therefore the profit, $P(x)$, is always increasing with increasing sales x.

38. a. The velocity $v(t) = s'(t) = 3(t^3 - 9t + 2)^2(3t^2 - 9)$.

b. Let $v(t) < 0$. $3(t^3 - 9t + 2)^2(3t^2 - 9) < 0$.

Since $3(t^3 - 9t + 2)^2$ is always non-negative, it is equivalent to solve

$3t^2 - 9 < 0$ for $t \geq 0$. $\quad 3t^2 < 9$, $t^2 < 3$, $-\sqrt{3} < t < \sqrt{3}$.

But only $t \geq 0$ is considered, so for the values of

$0 \leq t < \sqrt{3}$, $v(t)$ is negative.

39. a. $N'(t) = \frac{d}{dt}\left(20 - 30(9 + t^2)^{-1/2}\right) = -30 \cdot \left(-\frac{1}{2}\right)(9 + t^2)^{-3/2} \cdot (2t)$

$N'(t) = 30t(9 + t^2)^{-3/2} = \frac{30t}{\left(\sqrt{9+t^2}\right)^3}$.

b. Since $N'(t) = \frac{30t}{\left(\sqrt{9+t^2}\right)^3} > 0$ for $t > 0$, the number of items $N(t)$ a cashier can ring up per minute is always increasing with t.

Thus he or she will never stop improving.

40. a. The equilibrium price is the price p when demand $D(p)$ equals supply $S(p)$.

$$D(2) = \left(\frac{20}{2+2}\right)^{2/3} = 5^{2/3} \qquad \text{and}$$
$$S(2) = (2^2 + 1)^{2/3} = 5^{2/3}.$$

$D(2) = S(2)$, so $p = 2$ is the equilibrium price.

b. $D'(p) = \frac{2}{3}\left(\frac{20}{2+p}\right)^{-1/3} \cdot \left(\frac{-20}{(2+p)^2}\right) = -\frac{2\cdot 20^{2/3}}{3(2+p)^{5/3}}$.

$D'(2) = \frac{-2\cdot 20^{2/3}}{3(2+2)^{5/3}} = -\frac{2\cdot 5^{2/3}}{3\cdot 4} = -\frac{5^{2/3}}{6} = -0.487.$

c. $S'(p) = \frac{2}{3}(p^2+1)^{-1/3}\cdot(2p) = \frac{4}{3}p(p^2+1)^{-1/3}$.

$S'(2) = \frac{4}{3}2(2^2+1)^{-1/3} = \frac{8}{3\cdot 5^{1/3}} = 1.56.$

41. The average monthly cost for the manufacturer $= \frac{C(x)}{x} = \frac{500+\sqrt{40+16x^2}}{x}$.

The marginal average monthly cost for the manufacturer $= \left(\frac{C(x)}{x}\right)'$

$= \frac{\frac{1}{2}(40+16x^2)^{-1/2}(16\cdot 2x)\cdot x - \left(500+\sqrt{40+16x^2}\right)\cdot 1}{x^2} = \frac{16}{\sqrt{40+16x^2}} - \frac{500+\sqrt{40+16x^2}}{x^2}$.

$= \frac{-\left(500\sqrt{40+16x^2}+40\right)}{x^2\sqrt{40+16x^2}}$

42. $r(c) = 10 + .2c + .01c^2$.

$f = (r^2+5)^{1/3} = \left((10+.2c+.01c^2)^2+5\right)^{1/3}$.

$\frac{df}{dc} = \frac{1}{3}\left((10+.2c+.01c^2)^2+5\right)^{-2/3}\cdot\left(2(10+.2c+.01c^2)(.2+.02c)\right)$.

Solutions to the Exercise Set 2.9

1. $\lim\limits_{x\to\infty} \frac{3x^2+2}{10x^2-3x} = \lim\limits_{x\to\infty} \frac{3+\frac{2}{x^2}}{10-\frac{3}{x}} = \frac{3+0}{10-0} = \frac{3}{10}.$

2. $\lim\limits_{x\to\infty} \frac{2x^4-6x}{7-3x^4} = \lim\limits_{x\to\infty} \frac{2x^4-6x}{7-3x^4} \cdot \frac{\frac{1}{x^4}}{\frac{1}{x^4}}$
$= \lim\limits_{x\to\infty} \frac{\frac{2x^4}{x^4}-\frac{6x}{x^4}}{\frac{7}{x^4}-\frac{3x^4}{x^4}} = \lim\limits_{x\to\infty} \frac{2-\frac{6}{x^3}}{\frac{7}{x^4}-3} = \frac{2-0}{0-3} = -\frac{2}{3}.$

3. $\lim\limits_{x\to\infty} \frac{x(4-x^3)}{3x^4+2x^2} = \lim\limits_{x\to\infty} \frac{4x-x^4}{3x^4+2x^2} = \lim\limits_{x\to\infty} \frac{\frac{4}{x^3}-1}{3+\frac{2}{x^2}} = -\frac{1}{3}.$

4. $\lim\limits_{x\to\infty} \frac{(x-3)(x+4)}{2x^2+2} = \lim\limits_{x\to\infty} \frac{x^2+x-12}{2x^2+2} = \lim\limits_{x\to\infty} \frac{1+\frac{1}{x}-\frac{12}{x^2}}{2+\frac{2}{x^2}} = \frac{1}{2}.$

5. $\lim\limits_{x\to-\infty} \frac{2x^2+7}{3-4x^2} = \lim\limits_{x\to-\infty} \frac{2x^2+7}{3-4x^2} \cdot \frac{\frac{1}{x^2}}{\frac{1}{x^2}}$
$= \lim\limits_{x\to-\infty} \frac{2+\frac{7}{x^2}}{\frac{3}{x^2}-4} = \frac{2+0}{0-4} = -\frac{1}{2}.$

6. $\lim\limits_{x\to-\infty} \frac{2x+6}{x^2+1} = \lim\limits_{x\to-\infty} \frac{\frac{2}{x}+\frac{6}{x^2}}{1+\frac{1}{x^2}} = \frac{0+0}{1+0} = \frac{0}{1} = 0.$

7. $\lim\limits_{x\to\infty} \frac{3x^2+7x}{1-x^4} = \lim\limits_{x\to\infty} \frac{\frac{3}{x^2}+\frac{7}{x^3}}{\frac{1}{x^4}-1} = \frac{0}{-1} = 0.$

8. $\lim\limits_{x\to\infty} \frac{x^4-4x^2}{x^3+7x^2} = \lim\limits_{x\to\infty} \frac{x^4-4x^2}{x^3+7x^2} \cdot \frac{\frac{1}{x^3}}{\frac{1}{x^3}}$
$= \lim\limits_{x\to\infty} \frac{\frac{x^4}{x^3}-\frac{4x^2}{x^3}}{\frac{x^3}{x^3}+\frac{7x^2}{x^3}} = \lim\limits_{x\to\infty} \frac{x-\frac{4}{x}}{1+\frac{7}{x}} = \infty$
because $\left(x-\frac{4}{x}\right) \to \infty$ and $\left(1+\frac{7}{x}\right) \to 1$ as $x \to \infty$.

9. $\lim\limits_{x\to+\infty} \frac{\sqrt{x-1}}{x^2} = \lim\limits_{x\to\infty} \sqrt{\frac{x-1}{x^4}} = \lim\limits_{x\to\infty} \sqrt{\frac{1}{x^3}-\frac{1}{x^4}} = 0.$

10. $\lim\limits_{x\to\infty} \frac{x^{2/3}+x^{4/3}}{x^2} = \lim\limits_{x\to\infty} \left(\frac{1}{x^{4/3}}+\frac{1}{x^{2/3}}\right) = 0.$

11. $\lim\limits_{x\to\infty} \frac{x+7-x^3}{10x^2+18} = \lim\limits_{x\to\infty} \frac{\frac{1}{x^2}+\frac{7}{x^3}-1}{\frac{10}{x}+\frac{18}{x^3}} = -\infty.$

12. $\lim\limits_{x\to\infty} \frac{3x^{2/3}-x^{5/2}}{6+x^{3/2}} = \lim\limits_{x\to\infty} \frac{\left(3\cdot\frac{1}{x^{11/6}}-1\right)}{\left(\frac{6}{x^{5/2}}+\frac{1}{x}\right)} = -\infty.$

13. $C(x) = 3{,}000 + 24x.$

    a. The average cost per item $c(x) = \frac{C(x)}{x} = \frac{3{,}000}{x} + 24.$

    b. $\lim\limits_{x\to\infty} c(x) = 0 + 24 = 24.$

14. $C(x) = 500 + 30x + 16\sqrt{x}.$

    a. $c(x) = \frac{C(x)}{x} = \frac{500}{x} + 30 + \frac{16}{\sqrt{x}}.$

b. $\lim_{x\to\infty} c(x) = 0 + 30 + 0 = 30.$

15. $P(x) = \frac{1{,}000+30x+\sqrt{x}}{10+2x}.$

$$\lim_{x\to\infty} P(x) = \lim_{x\to\infty} \frac{1{,}000+30x+\sqrt{x}}{10+2x} \cdot \frac{\frac{1}{x}}{\frac{1}{x}} = \lim_{x\to\infty} \frac{\frac{1{,}000}{x}+30+\frac{1}{\sqrt{x}}}{\frac{10}{x}+2} = 15.$$

We assume from this positive limit that the company is profitable at high production levels.

16. $B(x) = \frac{40x}{\sqrt{1+x^2}} + 20.$

a.
$$B'(x) = \frac{(1+x^2)^{1/2}\cdot(40) - 40x\cdot\left(\frac{1}{2}(1+x^2)^{-1/2}\cdot 2x\right)}{1+x^2} = 40\cdot\frac{(1+x^2)^{1/2} - x^2(1+x^2)^{-1/2}}{1+x^2}\cdot\frac{(1+x^2)^{1/2}}{(1+x^2)^{1/2}} = 40\cdot\frac{(1+x^2)-x^2}{(1+x^2)^{3/2}} = \frac{40}{(1+x^2)^{3/2}}.$$

b. $\lim_{x\to\infty} B'(x) = 0.$

Therefore in the long-term the rate of change of overall plant size $B(x)$ with respect to the precipitation level x approaches zero.

17. a. $\lim_{t\to\infty} MP(t) = \lim_{t\to\infty} P'(t) = D.$

b. It is not necessarily true that the population function $P(t)$ will level off in the long-run. For example, suppose $P(t) = \sqrt{250{,}000t^2 + 5{,}000t + 10{,}000} + 110{,}000$. Then

$$P'(t) = \frac{1}{2}\cdot\left(250{,}000t^2 + 5{,}000t + 10{,}000\right)^{-1/2}\cdot(500{,}000t + 5{,}000) = \frac{250{,}000t + 2{,}500}{\sqrt{250{,}000t^2 + 5{,}000t + 10{,}000}}.$$

Therefore,

$$\lim_{t\to\infty} P'(t) = \lim_{t\to\infty} \frac{250{,}000 + \frac{2{,}500}{t}}{\sqrt{250{,}000 + \frac{5{,}000}{t} + \frac{10{,}000}{t^2}}} = \frac{250{,}000}{500} = 500.$$

However, $\lim_{t\to\infty} P(t) = \infty.$

c. If $D > 0$, then in the long-run the population size will be increasing as time t increases at the rate approximately equal to D. If $D < 0$, then in the long-run the population size $P(t)$ will be decreasing as time t increases at the rate approximately equal to D.

18. Let $C(t)$ denote the cost of getting rid of the toxic waste t years after it is created. The ecologist's statement probably means that $\lim_{t\to\infty} C(t)$ is much larger than $C(0)$.

Solutions to the Review Exercises - Chapter 2

1. $\lim_{x\to 2}(3x-1) = 3\cdot 2 - 1 = 5.$
2. $\lim_{x\to 2}\frac{2x-1}{x+6} = \frac{2\cdot 2-1}{2+6} = \frac{3}{8}.$
3. $\lim_{x\to 3}(x^4-4) = 3^4 - 4 = 77.$
4. $\lim_{x\to 5}\frac{x^2-25}{x-5} = \lim_{x\to 5}\frac{(x-5)(x+5)}{x-5} = \lim_{x\to 5}(x+5) = 10.$
5. $\lim_{x\to 3}\frac{x^2-9}{x-3} = \lim_{x\to 3}\frac{(x-3)(x+3)}{x-3} = \lim_{x\to 3}(x+3) = 6.$
6. $\lim_{x\to 3}\frac{x-3}{x^2-9} = \lim_{x\to 3}\frac{x-3}{(x-3)(x+3)} = \lim_{x\to 3}\frac{1}{x+3} = \frac{1}{3+3} = \frac{1}{6}.$
7. $\lim_{x\to -2}\frac{x^2+x-2}{x+2} = \lim_{x\to -2}\frac{(x+2)(x-1)}{x+2} = \lim_{x\to -2}(x-1) = -2-1 = -3.$
8. $\lim_{x\to 1}\frac{x^2+6x-7}{x-1} = \lim_{x\to 1}\frac{(x-1)(x+7)}{x-1} = \lim_{x\to 1}(x+7) = 1+7 = 8.$
9. $\lim_{x\to 1}\frac{x^2+2x-3}{x^2+x-2} = \lim_{x\to 1}\frac{(x-1)(x+3)}{(x-1)(x+2)} = \lim_{x\to 1}\frac{x+3}{x+2} = \frac{1+3}{1+2} = \frac{4}{3}.$
10. $\lim_{x\to -2}\frac{x^2+4x+4}{x^2+x-2} = \lim_{x\to -2}\frac{(x+2)(x+2)}{(x+2)(x-1)} = \lim_{x\to -2}\frac{(x+2)}{(x-1)} = \frac{0}{-3} = 0.$
11. $\lim_{t\to -1}\sqrt{\frac{1-t^2}{1-t}} = \sqrt{\frac{1-(-1)^2}{1-(-1)}} = \sqrt{\frac{0}{2}} = 0.$
12. $\lim_{x\to 8}\sqrt{\frac{x-7}{x+2}} = \sqrt{\frac{8-7}{8+2}} = \sqrt{\frac{1}{10}} = \frac{1}{\sqrt{10}}.$
13. $\lim_{x\to 0}\frac{(2+x)^2-4}{x} = \lim_{x\to 0}\frac{4+4x+x^2-4}{x} = \lim_{x\to 0}\frac{x(4+x)}{x} = \lim_{x\to 0}(4+x) = 4+0 = 4.$
14. $\lim_{x\to 0}\frac{3x+5x^2}{x} = \lim_{x\to 0}\frac{x(3+5x)}{x} = \lim_{x\to 0}(3+5x) = 3+5\cdot 0 = 3.$

15. The denominator $x-2$ has only the zero $x=2$. Therefore $f(x)$ is discontinuous only at $x=2$.

16. $f(x)=\frac{x^2-7}{x^2-3x-4}=\frac{x^2-7}{(x-4)(x+1)}$

$f(x)$ is discontinuous at $x=4$ and $x=-1$ because $f(4)$ and $f(-1)$ are not defined since the denominator (x^2-3x-4) is zero for $x=4$ and $x=-1$.

17. The denominator of $f(x)$ is $x^2-x-6=(x-3)(x+2)$; it has zeros $x=-2$ and $x=3$. Therefore $f(x)$ is discontinuous only at $x=-2$ and $x=3$.

18. $f(x)=\frac{x}{|x|}=\begin{cases}-1 & x<0\\ 1 & x>0\end{cases}$.

Since $\lim\limits_{x\to 0^-} f(x)=-1$ and $\lim\limits_{x\to 0+} f(x)=1$, $\lim\limits_{x\to 0} f(x)$ does not exist.

Therefore $f(x)$ is discontinuous only at $x=0$.

19. $f(x)=\begin{cases}2-x^2 & , \quad x\le 2\\ -x & , \quad x>2\end{cases}$

$\lim\limits_{x\to 2^-} f(x)=\lim\limits_{x\to 2^-}(2-x^2)=2-2^2=-2.$

$\lim\limits_{x\to 2+} f(x)=\lim\limits_{x\to 2+}(-x)=-2.$

$f(2)=2-2^2=-2.$

Because $\lim\limits_{x\to 2^-} f(x)=\lim\limits_{x\to 2+} f(x)=f(2)$, the function $f(x)$ is continuous at $x=2$. Therefore $f(x)$ is continuous everywhere.

20. For all $x\ge 0$ and $x\ne 4$, $f(x)$ is continuous.

For $x=4$, $\lim\limits_{x\to 4^-} f(x)=\lim\limits_{x\to 4^-}\sqrt{x}=\sqrt{4}=2$ and $\lim\limits_{x\to 4+} f(x)=\lim\limits_{x\to 4+}(x-1)=3\ne\lim\limits_{x\to 4^-} f(x)$, so $f(x)$ is discontinuous at $x=4$.

21. $f'(x)=2x-1.$

22. $f'(x)=-3x^2.$

23. $f(x)=\sqrt{x}+4.$

$f'(x)=\frac{1}{2}x^{-1/2}=\frac{1}{2\sqrt{x}}.$

24. $f'(x)=\frac{1}{2}x^{-1/2}.$

25. $f'(x) = \left(\frac{1}{x^2}\right)' = (x^{-2})' = -2x^{-3} = -\frac{2}{x^3}.$

26. $f'(x) = (9x^{-3})' = 9(-3)x^{-4} = -\frac{27}{x^4}.$

27. $f'(x) = \left((x-2)^{-1}\right)' = (-1)(x-2)^{-2}\frac{d}{dx}(x-2) = -(x-2)^{-2}\cdot 1 = -(x-2)^{-2}.$

28. $f(x) = x^{1/3} - x.$

$f'(x) = \frac{1}{3}x^{-2/3} - 1.$

29. $f'(x) = m.$

30. $f'(x) = \frac{1}{2}(1-x)^{-1/2}(-1) = -\frac{1}{2}(1-x)^{-1/2}.$

31. $f(x) = \frac{1}{3x-7}$

$f'(x) = \frac{(3x-7)(0)-(1)(3)}{(3x-7)^2} = \frac{-3}{(3x-7)^2}.$

32. $f(x) = \frac{1}{1+\sqrt{x}}$

$f'(x) = \frac{(1+\sqrt{x})(0)-(1)(\frac{1}{2}x^{-1/2})}{(1+\sqrt{x})^2} = \frac{-1}{2\sqrt{x}(1+\sqrt{x})^2}.$

33. $f'(x) = \frac{(x+3)\cdot 1-(x-1)\cdot 1}{(x+3)^2} = \frac{4}{(x+3)^2}.$

34. $f(x) = \frac{\sqrt{x}}{1+x}$

$f'(x) = \frac{(1+x)(\frac{1}{2}x^{-1/2})-(\sqrt{x})(1)}{(1+x)^2} = \frac{\frac{1}{2}x^{-1/2}-\frac{1}{2}x^{1/2}}{(1+x)^2}$

$= \frac{1}{2}\cdot\frac{x^{-1/2}-x^{1/2}}{(1+x)^2}.$

35. $f(x) = (x^2+6)^{-1}.$

$f'(x) = (-1)(x^2+6)^{-2}\cdot\frac{d}{dx}(x^2+6) = -(x^2+6)^{-2}\cdot(2x) = -\frac{2x}{(x^2+6)^2}.$

36. $f'(x) = \left((x+3)^{-3}\right)' = -3(x+3)^{-4}.$

37. $f(x) = (x^2+4x+4)^2.$

$f'(x) = 2(x^2+4x+4)\cdot\frac{d}{dx}(x^2+4x+4) = 2(x^2+4x+4)\cdot(2x+4).$

38. $f'(x) = \frac{1}{2}(9+x^2)^{-1/2}\cdot(2x) = \frac{x}{\sqrt{9+x^2}}.$

39. $f(x) = (x^2+3)^{-2/3}$

$f'(x) = -\frac{2}{3}(x^2+3)^{-5/3}\cdot\frac{d}{dx}(x^2+3) = -\frac{2}{3}\left(x^2+3\right)^{-5/3}\cdot(2x).$

40. $f(x) = x^{-4} + (x-6)^{-3/4}$

$f'(x) = -4x^{-5} - \frac{3}{4}(x-6)^{-7/4}.$

41. $f'(x) = \left((1+x^2)^{-1/2}\right)' = -\frac{1}{2}(1+x^2)^{-3/2} \cdot \frac{d}{dx}(1+x^2) = -\frac{1}{2}(1+x^2)^{-3/2} \cdot (2x) = -\frac{x}{(1+x^2)^{3/2}}$.

42. $f(x) = \frac{(x+1)^2}{x^3-3}$.

$$f'(x) = \frac{(x^3-3)(2(x+1)) - (x+1)^2 \cdot (3x^2)}{(x^3-3)^2}$$
$$= \frac{(x+1)\left(2(x^3-3) - (x+1) \cdot (3x^2)\right)}{(x^3-3)^2} = \frac{(x+1)(-x^3-3x^2-6)}{(x^3-3)^2}.$$

43. $f(x) = x^4\sqrt{1+x^2}$.

$$f'(x) = (4x^3)(1+x^2)^{1/2} + (x^4)\left(\tfrac{1}{2}(1+x^2)^{-1/2} \cdot 2x\right)$$
$$= x^3(1+x^2)^{-1/2}\left(4(1+x^2) + x^2\right)$$
$$= x^3(1+x^2)^{-1/2}(4+5x^2).$$

44. $f'(x) = 3(x-2)^2(x^2+9)^4 + (x-2)^3 \cdot 4(x^2+9)^3 \cdot (2x)$

$$= 3(x-2)^2(x^2+9)^4 + 8x(x-2)^3(x^2+9)^3.$$

45. $f(x) = (1-x^2)^3(6+2x+5x^3)^{-3}$.

$$f'(x) = \left(3(1-x^2)^2(-2x)\right) \cdot (6+2x+5x^3)^{-3} + (1-x^2)^3\left(-3(6+2x+5x^3)^{-4}(2+15x^2)\right)$$

$$= -6x(1-x^2)^2(6+2x+5x^3)^{-3} - 3(1-x^2)^3(6+2x+5x^3)^{-4}(2+15x^2).$$

46. $f'(x) = 2x\sqrt{3-x} + (x^2-9)\frac{1}{2}(3-x)^{-1/2}(-1) = 2x\sqrt{3-x} - \frac{1}{2}(x^2-9)(3-x)^{-1/2}$.

47. $D(p) = \frac{200}{40+p^2}$.

$$D'(p) = \frac{(40+p^2)(0) - (200)(2p)}{(40+p^2)^2} = \frac{-400p}{(40+p^2)^2}.$$

The slope of the demand curve at the point where $p = 10$ is

$$D'(10) = \frac{-400(10)}{(40+10^2)^2} = \frac{-4000}{19600} - \frac{-10}{49}.$$

48. $f'(x) = \frac{2}{3}(9-x^2)^{-1/3} \cdot (-2x) = -\frac{4}{3}x(9-x^2)^{-1/3}$.

The slope of the tangent line to the graph of $f(x)$ at $(1,4) = f'(1)$

$= -\frac{4}{3} \cdot 1 \cdot (9-1^2)^{-1/3} = -\frac{4}{3} \cdot 2^{-1} = -\frac{2}{3}$.

49. The tangent line is $y - 4 = f'(1)(x-1)$, $y - 4 = -\frac{2}{3}(x-1)$

$2x + 3y - 14 = 0$.

50. $f'(x) = 2x.$

Consider a line tangent to the graph of $f(x)$ at the point $(a, f(a))$. Then the equation of such a line is $y - f(a) = f'(a)(x - a)$,

$y = f'(a)x - af'(a) + f(a)$. Since the line passes through $(0, -2)$,

$$-2 = f'(a)0 - af'(a) + f(a)$$

$$-2 = -a \cdot (2a) + (a^2 - 1)$$

$$a^2 = 1, \qquad a = \pm 1.$$

The points are $(1, 0)$ and $(-1, 0)$. The corresponding tangent lines therefore are

$$y - 0 = f'(1)(x - 1) \implies y = 2(x - 1), \quad y = 2x - 2$$

and $\quad y - 0 = f'(-1)\,(x - (-1)) \implies y = -2(x + 1), \quad y = -2x - 2.$

51. a. $S(p_0) = D(p_0) \implies \frac{12}{p_0 - 1} = p_0 - 5$

$12 = (p_0 - 5)(p_0 - 1), \qquad 12 = p_0^2 - 6p_0 + 5$

$p_0^2 - 6p_0 - 7 = 0, \qquad (p_0 - 7)(p_0 + 1) = 0, \qquad p_0 = 7, \qquad p_0 = -1.$

But $p_0 > 5$ so the equilibrium price $p_0 = 7$.

b. $D'(p) = \left(12(p - 1)^{-1}\right)' = 12(-1)(p - 1)^{-2} = -12(p - 1)^{-2}.$

The slope of $D(p)$ at $(7, D(7)) = D'(7) = -12(7 - 1)^{-2} = -\frac{12}{36} = -\frac{1}{3}.$

c. $S'(p) = 1$. The slope of $S(p)$ at $(7, S(7)) = S'(7) = 1.$

52.

$$y = T(x) = \begin{cases} 10 + .02x, & 0 \le x < 5{,}000 \\ A + .03(x - 5{,}000), & 5{,}000 \le x < 10{,}000 \\ B + .04(x - 10{,}000), & 10{,}000 \le x \end{cases}$$

$$\lim_{x \to 5000^-} T(x) = \lim_{x \to 5000^-} (10 + .02x) = 110.$$

$$\lim_{x \to 5000+} T(x) = \lim_{x \to 5000+} (A + .03(x - 5{,}000)) = A = T(5{,}000).$$

See 2nd Ed p. 80-81 for a better solution methodology

$\therefore A = 110.$

$$\lim_{x \to 10000^-} T(x) = \lim_{x \to 10000^-} (110 + .03(x - 5{,}000)) = 260$$
$$\lim_{x \to 10000^+} T(x) = \lim_{x \to 10000^+} (B + .04(x - 10{,}000)) = B = T(10{,}000)$$

$\therefore B = 260.$

53. For all $p \geq 50$ such that $p \neq 200$ and $p \neq 400$, $D(p)$ is continuous.

For $p = 200$: $\lim_{p \to 200^-} D(p) = \lim_{p \to 200^-} (2p - 5) = 2 \cdot 200 - 5 = 395.$
$\lim_{p \to 200^+} D(p) = \lim_{p \to 200^+} (p + 400) = 200 + 400 = 600 \neq \lim_{p \to 200^-} D(p).$
$D(p)$ is discontinuous at $p = 200$.

For $p = 400$: $\lim_{p \to 400^-} D(p) = \lim_{p \to 400^-} (p + 400) = 400 + 400 = 800.$
$\lim_{p \to 400^+} D(p) = \lim_{p \to 400^+} \left(\frac{1}{2}p + 800\right) = \frac{1}{2}(400) + 800 = 1000 \neq \lim_{p \to 400^-} D(p).$
$D(p)$ is discontinuous at $p = 400$.

54. a.

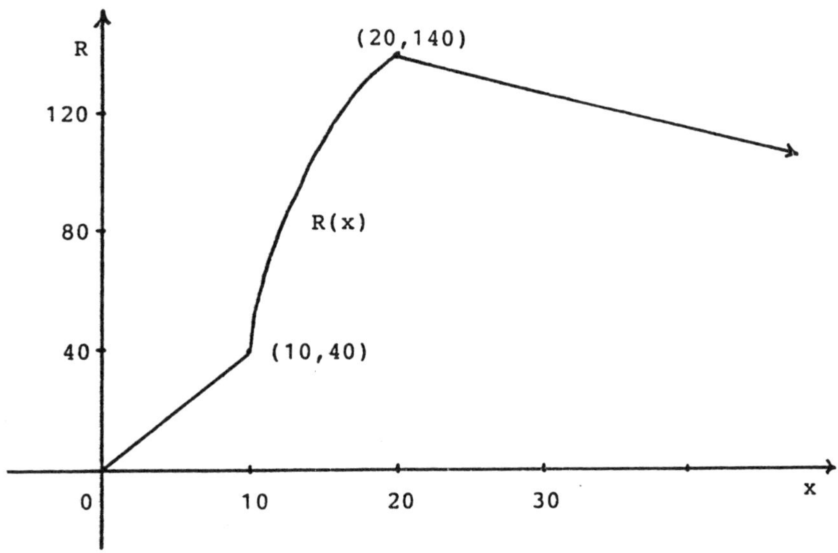

b. $R'(x) = \begin{cases} 4, & 0 < x < 10 \\ -2(x-20), & 10 < x < 20 \\ -1, & 20 < x. \end{cases}$

Only for x such that $10 < x < 20$ can $R'(x)$ equal 10. So let

$R'(x) = -2(x-20) = 10$, $x - 20 = -5$, $x = 15$. When $x = 15$, $R'(x) = 10$.

c. When $0 < x < 10$, $R'(x) = 4$. Also, $R'(18) = 4$.

55. a. $MC(x) = C'(x) = 30 + \frac{1}{2} \cdot 2x = 30 + x$.

b. $MC(x) = 30 + x = 60 \implies x = 30$.

56. a. $MR(x) = R'(x) = 450 - \frac{2x}{4} = 450 - \frac{1}{2}x$.

b. $MR(40) = 450 - \frac{1}{2}(40) = 430$.

c. $MR(x) = 450 - \frac{1}{2}x > 0$ for all values of $x < 900$.

d. $MR(x) = 450 - \frac{1}{2}x < 0$ for all values of $x > 900$.

57. a. $MC(x) = C'(x) = 40 + \frac{1}{3} \cdot 3x^2 = x^2 + 40$.

b. $MC(x) = x^2 + 40 = 76$, $x^2 = 36$, $x = \pm 6$.

But x must not be negative. So for $x = 6$, $MC(x) = 76$.

58. a. $MR(x) = R'(x) = 750 - \frac{2x}{6} - \frac{2 \cdot 3x^2}{3} = 750 - \frac{1}{3}x - 2x^2$.

b. $MR(10) = 750 - \frac{1}{3}(10) - 2(10)^2 = 546.67$.

59. a. $x = 500 - 2p$, $2p = 500 - x$, $p = 250 - \frac{1}{2}x$.

b. $R(x) = xp(x) = x\left(250 - \frac{1}{2}x\right) = 250x - \frac{1}{2}x^2$.

c. $MR(x) = \frac{dR}{dx} = 250 - x$.

d. $MR(x) = 0 \implies 250 - x = 0$, $x = 250$.

60. a. $C'(x) = MC(x) = 50$.

b. $P(x) = R(x) - C(x) = 250x - \frac{1}{2}x^2 - 40 - 50x = -\frac{1}{2}x^2 + 200x - 40$.

c. $MP(x) = -x + 200$.

d. $MP(x) = 0 \implies x = 200$.

61. a. $c(x) = \frac{C(x)}{x} = \frac{400+50x}{x} = \frac{400}{x} + 50.$

b. $c'(x) = -\frac{400}{x^2}.$

62. a. $MU(x) = U'(x) = 20x(x^2+9)^{-1/2}.$

b. $MU(4) = 20(4)(4^2+9)^{-1/2} = \frac{80}{5} = 16.$

63. $u(x) = \sqrt{x(x+24)} = \sqrt{x^2+24x}.$

$mu(x) = u'(x) = \frac{1}{2}(x^2+24x)^{-1/2} \cdot (2x+24) = \frac{x+12}{\sqrt{x^2+24x}}.$

64. a. $u'(1) = \frac{1+12}{\sqrt{1^2+24(1)}} = \frac{13}{5}$

b. $u'(3) = \frac{3+12}{\sqrt{3^2+24(3)}} = \frac{15}{9} = \frac{5}{3}.$

65. Because $u'(1) > u'(3)$, the employee earning 1,000 per month would appreciate more an increase of an equal amount in salary.

CHAPTER 3

Solutions to Exercise Set 3.1

1. $f'(x) = 3 > 0$ for all x, so f is increasing on $(-\infty, \infty)$.
2. $f'(x) = -1 < 0$ for all x, so f is decreasing on $(-\infty, \infty)$.
3. $f'(x) = -2x$, so $f'(x) = 0$ at $x = 0$.

   If $x < 0$, $f'(x) > 0$; and hence f is increasing on $(-\infty, 0)$.

   If $x > 0$, $f'(x) < 0$; and hence f is decreasing on $(0, \infty)$.
4. $f'(x) = 2x - 7$, so $f'(x) = 0$ at $x = \frac{7}{2}$.

Interval I	Test Number	Sign of $f'(x)$	Conclusion
$\left(-\infty, \frac{7}{2}\right)$	$x_1 = 1$	$f'(1) < 0$	decreasing
$\left(\frac{7}{2}, \infty\right)$	$x_2 = 5$	$f'(5) > 0$	increasing

5. $f'(x) = 1 - 3x^2$, so $f'(x) = 0$ at $x = \sqrt{\frac{1}{3}}$ and $-\sqrt{\frac{1}{3}}$.

Interval I	Test Number	Sign of $f'(x)$	Conclusion
$\left(-\infty, -\sqrt{\frac{1}{3}}\right)$	$x_1 = -2$	$f'(-2) < 0$	decreasing
$\left(-\sqrt{\frac{1}{3}}, \sqrt{\frac{1}{3}}\right)$	$x_2 = 0$	$f'(0) > 0$	increasing
$\left(\sqrt{\frac{1}{3}}, \infty\right)$	$x_3 = 2$	$f'(2) < 0$	decreasing

6. $f(x) = x^3 + 3x^2 - 24x + 10$.

   $f'(x) = 3x^2 + 6x - 24 = 3(x^2 + 2x - 8) = 3(x + 4)(x - 2)$.

   $f'(x) = 0$ for $x = -4$, $x = 2$.

Interval I	Test Number	Sign of $f'(x)$	Conclusion
$(-\infty, -4)$	$x_1 = -5$	$f'(-5) > 0$	increasing
$(-4, 2)$	$x_2 = -3$	$f'(-3) < 0$	decreasing
$(2, \infty)$	$x_3 = 3$	$f'(3) > 0$	increasing

7. $f'(x) = 3x^2 - 3 = 3(x+1)(x-1)$, so $f'(x) = 0$ at $x = 1$ and -1.

Interval I	Test Number	Sign of $f'(x)$	Conclusion
$(-\infty, -1)$	$x_1 = -3$	$f'(-3) > 0$	increasing
$(-1, 1)$	$x_2 = 0$	$f'(0) < 0$	decreasing
$(1, \infty)$	$x_3 = 2$	$f'(2) > 0$	increasing

8. $f'(x) = 6x^2 - 6x - 12 = 6(x^2 - x - 2) = 6(x-2)(x+1) = 0$ for $x = 2$ and -1.

Interval I	Test Number	Sign of $f'(x)$	Conclusion
$(-\infty, -1)$	$x_1 = -5$	$f'(-5) > 0$	increasing
$(-1, 2)$	$x_2 = 0$	$f'(0) < 0$	decreasing
$(2, \infty)$	$x_3 = 5$	$f'(5) > 0$	increasing

9. $f(x) = x^3 - 12x^2 - 48x + 12$.

$f'(x) = 3x^2 - 24x - 48 = 3(x^2 - 8x - 16)$.

$f'(x) = 0$ for $x = \frac{8 \pm \sqrt{8^2 - 4 \cdot 1 \cdot (-16)}}{2} = \frac{8 \pm \sqrt{128}}{2} = \frac{8 \pm 8\sqrt{2}}{2} = 4 \pm 4\sqrt{2}$.

Interval I	Test Number	Sign of $f'(x)$	Conclusion
$(-\infty, 4 - 4\sqrt{2})$	$x_1 = -2$	$f'(-2) > 0$	increasing
$(4 - 4\sqrt{2}, 4 + 4\sqrt{2})$	$x_2 = 0$	$f'(0) < 0$	decreasing
$(4 + 4\sqrt{2}, \infty)$	$x_3 = 10$	$f'(10) > 0$	increasing

10. $f'(x) = 3x^2 + 6x - 9 = 3(x^2 + 2x - 3) = 3(x+3)(x-1) = 0$ for $x = -3$ and 1.

Interval I	Test Number	Sign of $f'(x)$	Conclusion
$(-\infty, -3)$	$x_1 = -4$	$f'(-4) > 0$	increasing
$(-3, 1)$	$x_2 = 0$	$f'(0) < 0$	decreasing
$(1, \infty)$	$x_3 = 2$	$f'(2) > 0$	increasing

11. $f'(x) = 6x^2 - 6x - 36 = 6(x^2 - x - 6) = 6(x-3)(x+2) = 0$ for $x = 3$ and -2.

Interval I	Test Number	Sign of $f'(x)$	Conclusion
$(-\infty, -2)$	$x_1 = -3$	$f'(-3) > 0$	increasing
$(-2, 3)$	$x_2 = 0$	$f'(0) < 0$	decreasing
$(3, \infty)$	$x_3 = 4$	$f'(4) > 0$	increasing

12. $f(x) = x^4 - 18x^2 + 8$.

$f'(x) = 4x^3 - 36x = 4x(x^2 - 9)$.

$f'(x) = 0$ for $x = 0$, $x = -3$, $x = 3$.

Interval I	Test Number	Sign of $f'(x)$	Conclusion
$(-\infty, -3)$	$x_1 = -4$	$f'(-4) < 0$	decreasing
$(-3, 0)$	$x_2 = -1$	$f'(-1) > 0$	increasing
$(0, 3)$	$x_3 = 1$	$f'(1) < 0$	decreasing
$(3, \infty)$	$x_4 = 4$	$f'(4) > 0$	increasing

13. $f'(x) = 12x^3 + 12x^2 - 24x = 12x(x^2 + x - 2) = 12x(x+2)(x-1) = 0$ for $x = 0$, -2, and 1.

Interval I	Test Number	Sign of $f'(x)$	Conclusion
$(-\infty, -2)$	$x_1 = -3$	$f'(-3) < 0$	decreasing
$(-2, 0)$	$x_2 = -1$	$f'(-1) > 0$	increasing
$(0, 1)$	$x_3 = \frac{1}{2}$	$f'\left(\frac{1}{2}\right) < 0$	decreasing
$(1, \infty)$	$x_4 = 2$	$f'(2) > 0$	increasing

14. $f'(x) = -\frac{1}{(x-3)^2}$ fails to exist only at $x = 3$.

If $x < 3$, $f'(x) < 0$; so f is decreasing on $(-\infty, 3)$.

If $x > 3$, $f'(x) < 0$; so f is decreasing on $(3, \infty)$.

15. $f'(x) = \frac{4}{3}x^{1/3} = 0$ for $x = 0$.

If $x < 0$, $f'(x) < 0$; so f is decreasing on $(-\infty, 0)$.

If $x > 0$, $f'(x) > 0$; so f is increasing on $(0, \infty)$.

16. $f'(x) = \frac{x}{\sqrt{x^2+2}}$ exists for all x. On $(-\infty, 0)$, $f'(x) < 0$, so $f(x)$ is decreasing. On $(0, \infty)$, $f'(x) > 0$, so $f(x)$ is increasing.

17. $f(x) = |4 - x^2|$.

$$f(x) = \begin{cases} x^2 - 4, & -\infty < x \leq -2 \\ 4 - x^2, & -2 < x \leq 2 \\ x^2 - 4, & 2 < x < \infty. \end{cases}$$

If $x < -2$, $f'(x) = 2x < 0$, so f is decreasing.

If $-2 < x < 0$, $f'(x) = -2x > 0$, so f is increasing.

If $0 < x < 2$, $f'(x) = -2x < 0$, so f is decreasing.

If $x > 2$, $f'(x) = 2x > 0$, so f is increasing.

18. $f'(x) = \frac{2}{3}(x+3)^{-1/3} = \frac{2}{3\cdot(\sqrt[3]{x+3})}$ fails to exist only at $x = -3$.

If $x < -3$, $f'(x) < 0$; so f is decreasing on $(-\infty, -3)$.

If $x > -3$, $f'(x) > 0$; so f is increasing on $(-3, \infty)$.

19. $f'(x) = \frac{1}{3}(8-x^3)^{-2/3}\cdot(-3x^2) = \frac{-x^2}{(8-x^3)^{2/3}}$ fails to exist at $x = 2$, and $f'(x) = 0$ at $x = 0$.

Interval I	Test Number	Sign of $f'(x)$	Conclusion
$(-\infty, 0)$	$x_1 = -1$	$f'(-1) < 0$	decreasing
$(0, 2)$	$x_2 = 1$	$f'(1) < 0$	decreasing
$(2, \infty)$	$x_3 = 3$	$f'(3) < 0$	decreasing

20. $f(x) = |9 - x^2| = \begin{cases} x^2 - 9 & \text{for } x \le -3 \\ 9 - x^2 & \text{for } -3 \le x \le 3 \\ x^2 - 9 & \text{for } x \ge 3 \end{cases}.$

So $f'(x) = \begin{cases} 2x & \text{for } x < -3 \\ -2x & \text{for } -3 < x < 3 \\ 2x & \text{for } x > 3. \end{cases}$

$f'(x)$ doesn't exist at $x = 3$ and -3, and $f'(x) = 0$ at $x = 0$.

If $x < -3$, $f'(x) = 2x < 0$, so f is decreasing.

If $-3 < x < 0$, $f'(x) = -2x > 0$, so f is increasing.

If $0 < x < 3$, $f'(x) = -2x < 0$, so f is decreasing.

If $x > 3$, $f'(x) = 2x > 0$, so f is increasing.

21. $f'(x) = 6x^2 - 6ax$.

$f'(3) = 6\cdot 3^2 - 6\cdot a\cdot(3) = 0$, $54 - 18a = 0$.

$\therefore a = 3$.

22. $P(t) = 50(1 + t^2 - 12t + 36)^{-1}$.

$P'(t) = -50(t^2 - 12t + 37)^{-2}(2t - 12) = 0$ for $t = 6$.

If $t < 6$, $P'(t) > 0$, so P is increasing.

If $t > 6$, $P'(t) < 0$, so P is decreasing.

23. The numbers are x and $50 - x$. Let $f(x) = x(50 - x) = 50x - x^2$.

    Then $f'(x) = 50 - 2x = 0$ for $-2x = -50$, $x = 25$.

    If $x < 25$, $f'(x) > 0$; so $f(x)$ is increasing on $(-\infty, 25)$.

    If $x > 25$, $f'(x) < 0$; so $f(x)$ is decreasing on $(25, \infty)$.

24. a. $C'(x) = 5 + 0.04x > 0$ for all $x > 0$.

    So, $C(x)$ is increasing on $(0, \infty)$.

    b. $c(x) = \frac{C(x)}{x} = \frac{100+5x+0.02x^2}{x} = \frac{100}{x} + 5 + 0.02x$.

    $c'(x) = -\frac{100}{x^2} + 0.02 = 0$ for $x^2 = 5000 \implies x = 70.71$.

    $c'(x) > 0$ for $x > 70.71$. So $c(x)$ is increasing for $x > 70.71$.

25. $c(x) = \frac{C(x)}{x} = \frac{20+200x+0.01x^3}{x} = 0.01x^2 + 200 + \frac{20}{x}$.

    $c'(x) = 0.02x - \frac{20}{x^2} = \frac{0.02x^3-20}{x^2} = \frac{0.02(x^3-1000)}{x^2}$.

    $c'(10) = 0$.

    a. If $x > 10$, $c'(x) > 0$, so $c(x)$ is increasing.

    b. If $0 < x < 10$, $c'(x) < 0$, so $c(x)$ is decreasing.

26. a. $C(x) = 450{,}000 + 5x = 450{,}000 + 5(10{,}000 - 200p)$.

    Therefore $C(p) = 500{,}000 - 1000p$.

    b. $C'(p) = -1000 < 0$; $C(p)$ is decreasing.

    c. $R(p) = x \cdot p = (10{,}000 - 200p)p = -200p^2 + 10{,}000p$.

    d. $R'(p) = -400p + 10{,}000 = 0$ for $p = 25$.

    If $0 < p < 25$, $R'(p) > 0$, so $R(p)$ is increasing.

    If $p > 25$, $R'(p) < 0$, so $R(p)$ is decreasing.

    e. $P(p) = R(p) - C(p)$

    $= -200p^2 + 10{,}000p - (500{,}000 - 1000p)$

    $= -200p^2 + 11{,}000p - 500{,}000$.

    f. $P'(p) = -400p + 11{,}000$.

    $P'(p) = 0$ at $p = 27.5$.

    P is increasing on $(0, 27.5)$ since $P'(p) > 0$ on $(0, 27.5)$.

    g. P is decreasing on $(27.5, \infty)$.

27. $R(x) = 1500x - 60x^2 - x^3$.

$R'(x) = -3x^2 - 120x + 1500 = -3(x^2 + 40x - 500)$

$= -3(x + 50)(x - 10)$.

$R'(x) = 0$ at $x = 10$ and -50.

Since $x > 0$, $R'(x) > 0$ on $(0, 10)$ and $R'(x) < 0$ on $(10, \infty)$.

a. $(0, 10)$.

b. $MR(x) = R'(x) = -3(x^2 + 40x - 500)$.

$(MR)'(x) = -3(2x + 40)$. $\therefore MR'(x) = 0$ at $x = -20$.

MR is decreasing on $(-20, \infty)$. Since $x > 0$, MR is decreasing on $(0, \infty)$.

28. $U(x) = \sqrt{40 - 10x + x^2}$

a. $MU(x) = U'(x) = \frac{2x-10}{2\sqrt{40-10x+x^2}}$.

b. $U' > 0$ on $(5, \infty)$ and $U' < 0$ on $(0, 5)$, so U is increasing on $(5, \infty)$.

c. $(0, 5)$.

29. $x = 1000 - 2p \implies p = 500 - \frac{x}{2}$. $C(x) = 400 + 2x$.

a. $R(x) = xp = x\left(500 - \frac{x}{2}\right) = 500x - \frac{1}{2}x^2$.

b. $R'(x) = 500 - x \implies R'(x) = 0$ at $x = 500$.

$\therefore R$ is increasing on $(0, 500)$ since $R'(x) > 0$ on $(0, 500)$.

c. $P(x) = R(x) - C(x) = 500x - \frac{1}{2}x^2 - (400 + 2x)$

$= -\frac{1}{2}x^2 + 498x - 400$.

d. $P'(x) = -x + 498$, so $P'(x) = 0$ at $x = 498$.

$\therefore P$ is increasing on $(0, 498)$ since $P'(x) > 0$ there.

e. $(498, 1000)$.

30. a. $R(p) = (500 - 10(p - 30)) \cdot p = (500 - 10p + 300) \cdot p = 800p - 10p^2$.

b. $R'(p) = 800 - 20p$, so $R'(p) = 0$ at $p = 40$.

$\therefore R$ is increasing on $(0, 40)$ since $R'(p) > 0$ there.

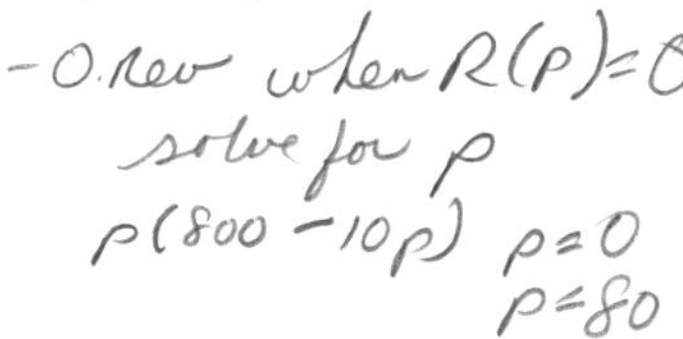

c. $(40, 80)$.

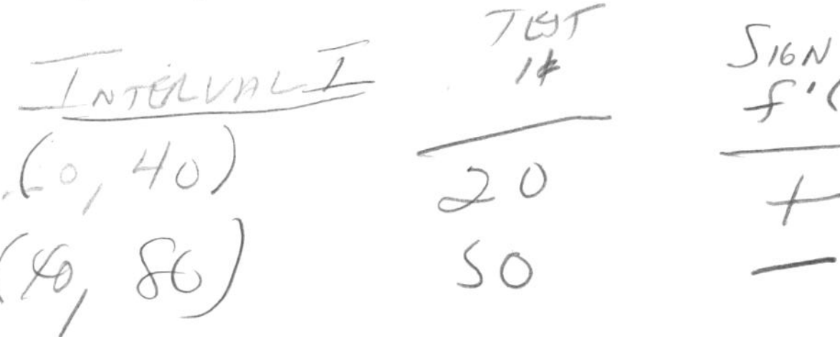

INTERVAL I
TEST #
SIGN OF f'(p)
(0, 40)
20
+
(40, 80)
50
—

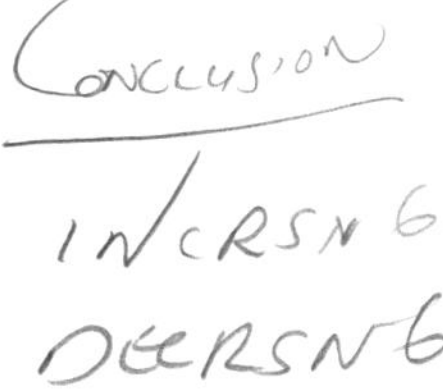

31. $P(t) = \frac{Kt}{100+t^2}, \quad t > 0.$

$P'(t) = \frac{(100+t^2)\cdot K-(Kt)\cdot(2t)}{(100+t^2)^2} = \frac{K(100-t^2)}{(100+t^2)^2}.$

$P'(t) = 0$ for $t = 10$.

If $0 < t < 10$, $P'(t) > 0$, so $P(t)$ is increasing.

If $10 < t < 20$, $P'(t) < 0$, so $P(t)$ is decreasing.

32. $N(t) = 20,000 + 20t^2 - 2t^3, t > 0.$

$N'(t) = 40t - 6t^2 = 2t(20 - 3t).$

$N'(t) = 0$ for $t = \frac{20}{3}$.

a. If $t < \frac{20}{3}$, $N'(t) > 0$, so $N(t)$ is increasing, and hence the epidemic is growing.

b. It $t > \frac{20}{3}$, $N'(t) < 0$, so $N(t)$ is decreasing, and hence the epidemic is declining.

33.

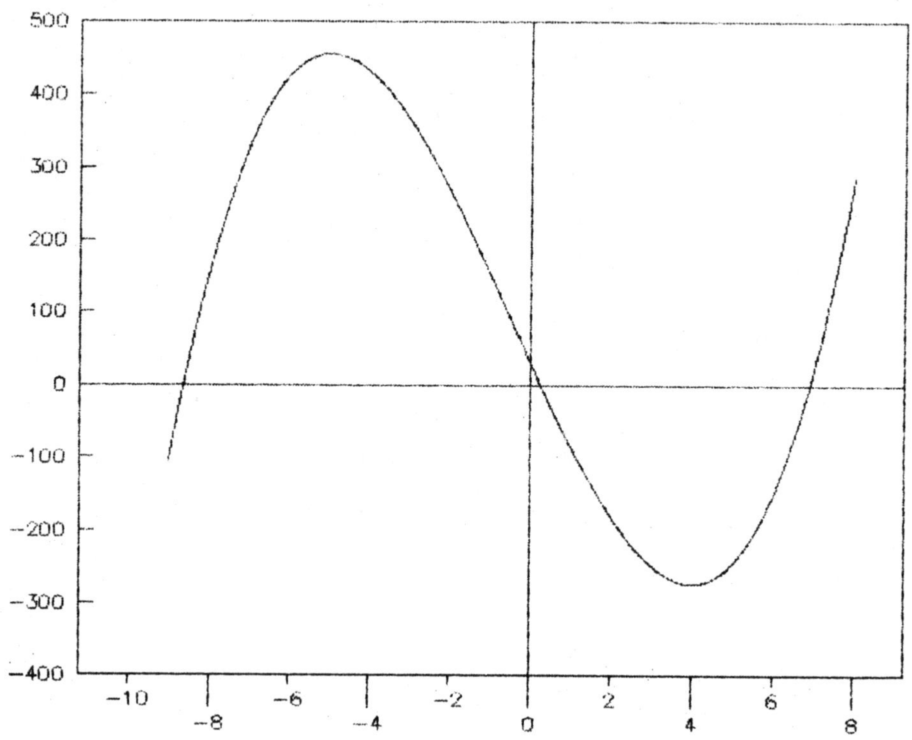

34. $C(x) = 500 + 20x + 5x^2$, $x \geq 0$.

$C'(x) = 20 + 10x > 0$ for all $x > 0$.

Therefore $C(x)$ is increasing for all $x > 0$.

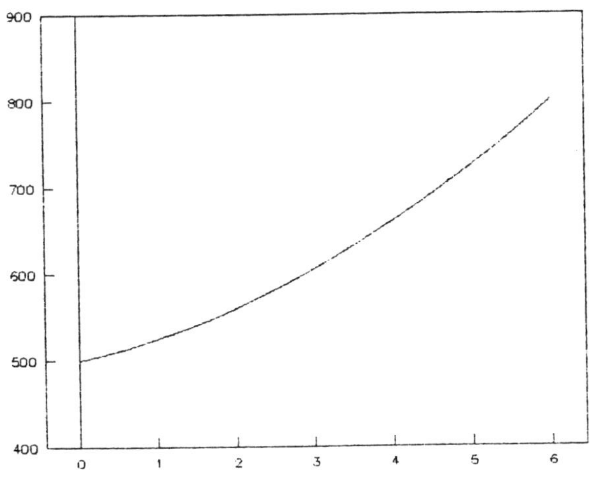

35. $c(x) = \frac{C(x)}{x} = \frac{500}{x} + 20 + 5x$.

$c'(x) = \frac{-500}{x^2} + 5$.

$c'(x) = 0$ for $x = 10$.

If $x < 10$, $c'(x) < 0$, so $c(x)$ is decreasing.

If $x > 10$, $c'(x) > 0$, so $c(x)$ is increasing.

From problem 34, we have that the total cost $C(x)$ keeps increasing as x increases.

36.

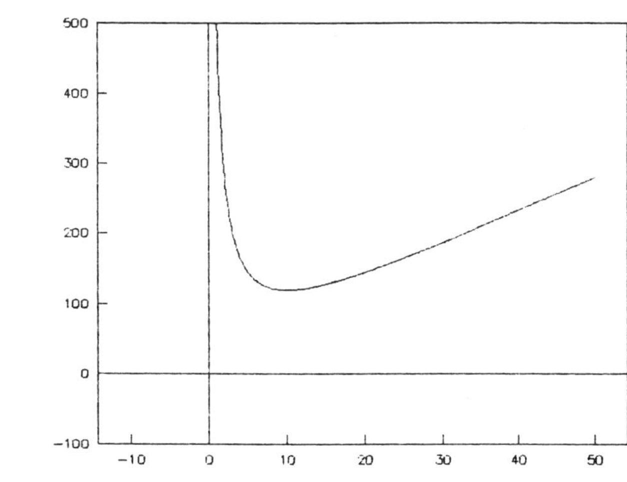

Solutions to Exercise Set 3.2

1. $fx) = x^2 - 2x$.
   $f'(x) = 2x - 2$.
   $f'(x) = 0$ for $x = 1$.
   For $x < 1$, $f'(x) < 0$. For $x > 1$, $f'(x) > 0$.
   $f(1)$ is a relative minimum.

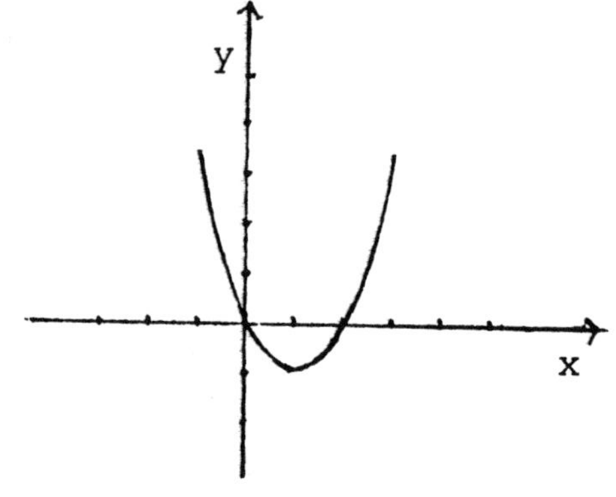

x	-1	0	1	2	3
$f(x)$	3	0	-1	0	3

2. $f(x) = x^2 + 3x + 2$.
   $f'(x) = 2x + 3$.
   $f'(x) = 0$ for $x = -\frac{3}{2}$.
   For $x < -\frac{3}{2}$, $f'(x) < 0$, so $f(x)$ is decreasing.
   For $x > -\frac{3}{2}$, $f'(x) > 0$, so $f(x)$ is increasing.
   $f\left(-\frac{3}{2}\right)$ is a relative minimum.

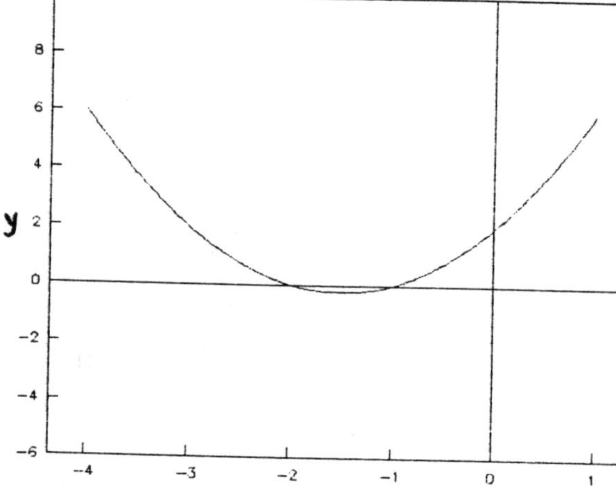

x	-4	-3	-2	$-\frac{3}{2}$	-1	0	1
$f(x)$	6	2	0	$-\frac{1}{4}$	0	2	6

3. $f(x) = 3x^2 - 6x + 3$.
   $f'(x) = 6x - 6$.
   $f'(x) = 0$ for $x = 1$.
   For $x < 1$, $f'(x) < 0$.
   For $x > 1$, $f'(x) > 0$.
   $f(1)$ is a relative minimum.

x	-1	0	1	2	3
$f(x)$	12	3	0	3	12

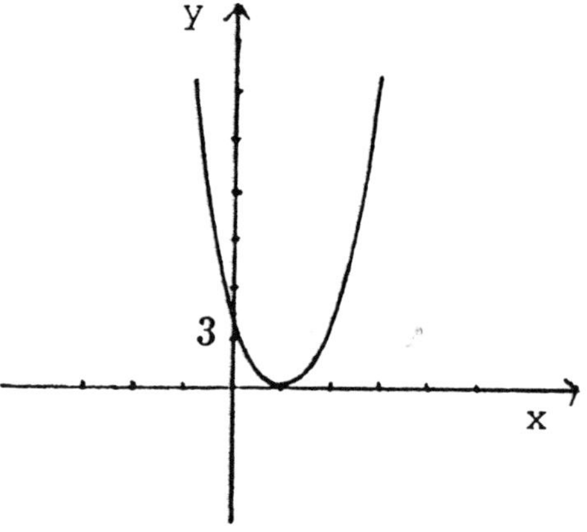

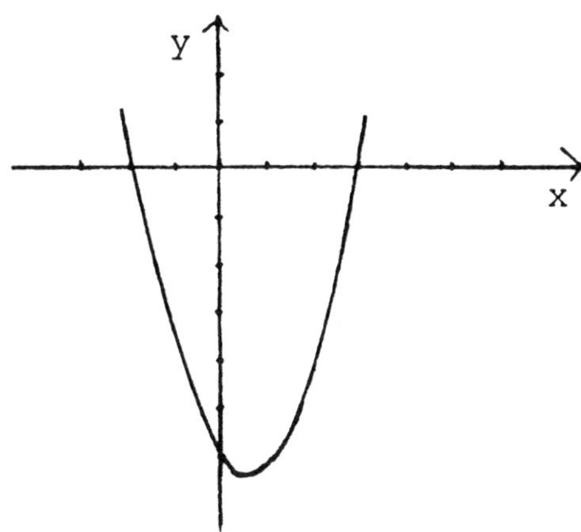

4. $f(x) = x^2 - x - 6.$

   $f'(x) = 2x - 1.$

   $f'(x) = 0$ for $x = \frac{1}{2}.$

   For $x < \frac{1}{2}$, $f'(x) < 0.$

   For $x > \frac{1}{2}$, $f'(x) > 0.$

   $f\left(\frac{1}{2}\right)$ is a relative minimum.

x	-1	0	$\frac{1}{2}$	1	2
$f(x)$	-4	-6	-6.25	-6	-4

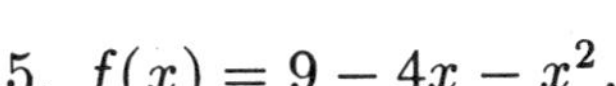

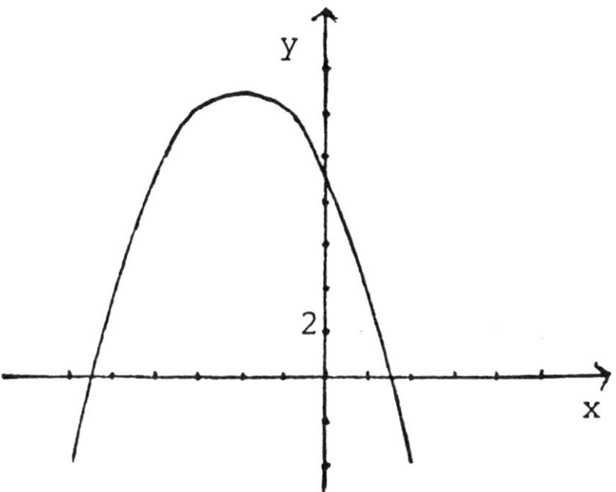

5. $f(x) = 9 - 4x - x^2.$

   $f'(x) = -4 - 2x.$

   $f'(x) = 0$ for $x = -2.$

   For $x < -2$, $f'(x) > 0.$

   For $x > -2$, $f'(x) < 0.$

   $f(-2)$ is a relative maximum.

x	-4	-3	-2	-1	0
$f(x)$	9	12	13	12	9

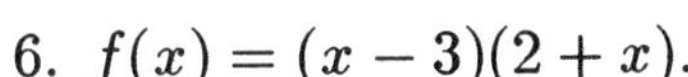

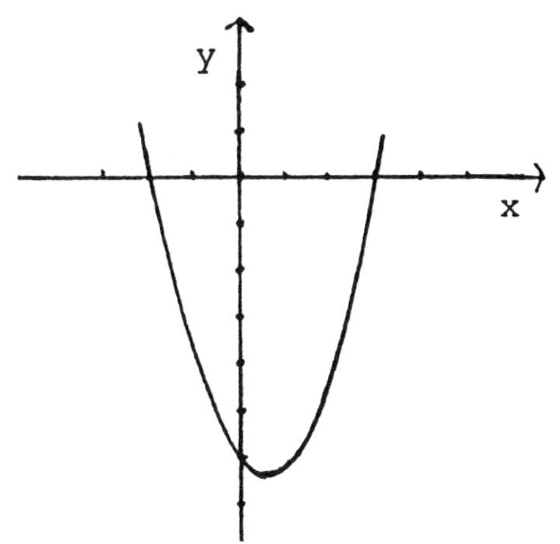

6. $f(x) = (x - 3)(2 + x).$

   $f'(x) = 1 \cdot (2 + x) + (x - 3) \cdot 1$

   $= 2 + x + x - 3 = 2x - 1.$

   $f'(x) = 0$ for $x = \frac{1}{2}.$

   For $x < \frac{1}{2}$, $f'(x) < 0.$

   For $x > \frac{1}{2}$, $f'(x) > 0.$

   $f\left(\frac{1}{2}\right)$ is a relative minimum.

x	-1	0	$\frac{1}{2}$	1	2
$f(x)$	-4	-6	-6.25	-6	-4

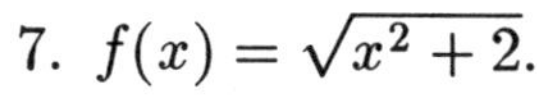

7. $f(x) = \sqrt{x^2 + 2}$.

$f'(x) = \frac{1}{2}(x^2 + 2)^{-1/2} \cdot (2x) = \frac{x}{\sqrt{x^2+2}}$.

$f'(x) = 0$ for $x = 0$.

For $x < 0$, $f'(x) < 0$, so $f(x)$ is decreasing.

For $x > 0$, $f'(x) > 0$, so $f(x)$ is increasing.

$f(0)$ is a relative minimum.

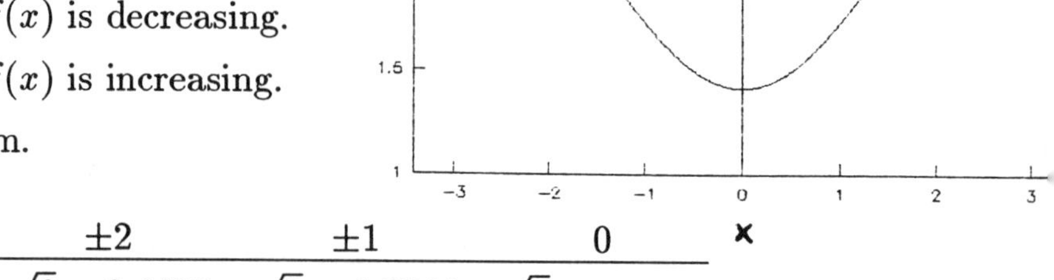

x	± 3	± 2	± 1	0
$f(x)$	$\sqrt{11} = 3.3166$	$\sqrt{6} = 2.4495$	$\sqrt{3} = 1.7321$	$\sqrt{2} = 1.4142$

8. $f(x) = x^3 - 3x^2 + 6$.

$f'(x) = 3x^2 - 6x$.

$f'(x) = 0$ for $x = 0$ and $x = 2$.

For $x < 0$, $f'(x) > 0$. For $0 < x < 2$, $f'(x) < 0$.

For $x > 2$, $f'(x) > 0$.

$f(0)$ is a relative maximum

and $f(2)$ is a relative minimum.

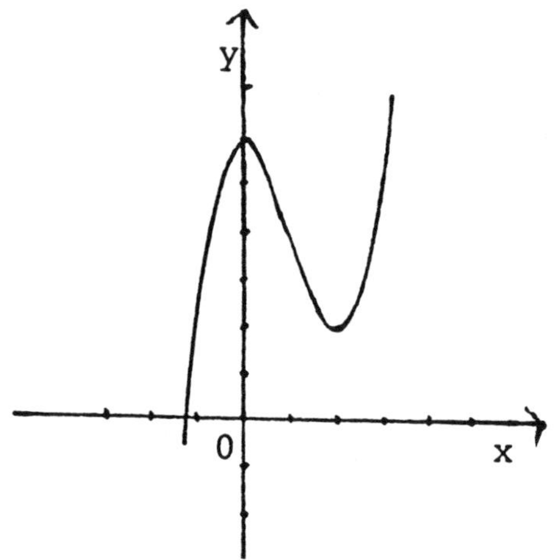

x	-1	0	1	2	3
$f(x)$	2	6	4	2	6

9. $f(x) = x^3 - 3x + 7$.

$f'(x) = 3x^2 - 3$.

$f'(x) = 0$ for $x = 1$ and $x = -1$

For $x < -1$, $f'(x) > 0$.

For $-1 < x < 1$, $f'(x) < 0$.

For $x > 1$, $f'(x) > 0$.

$f(-1)$ is a relative maximum

and $f(1)$ is a relative minimum.

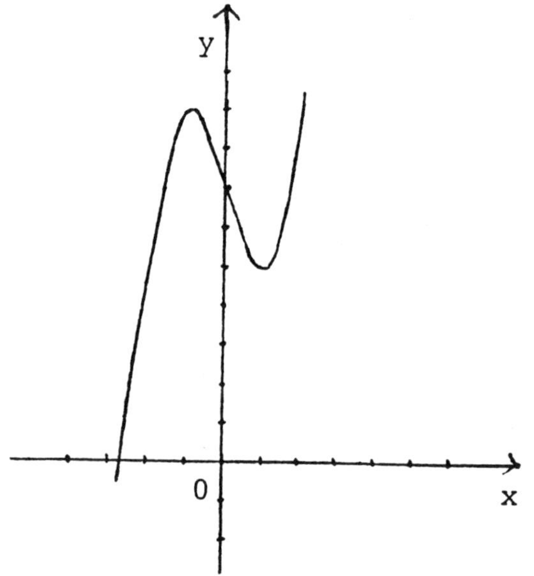

x	-2	-1	0	1	2
$f(x)$	5	9	7	5	9

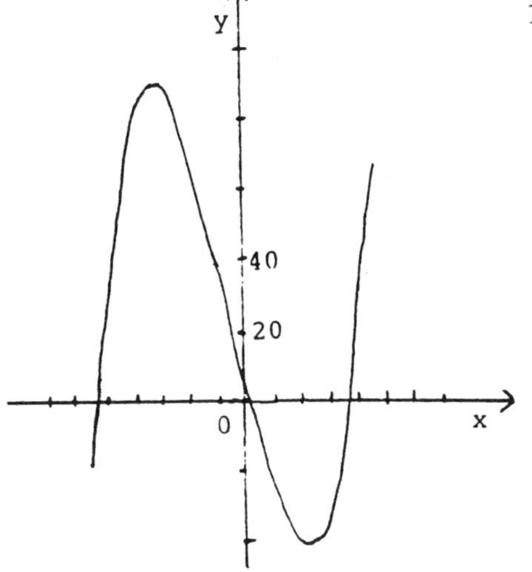

10. $f(x) = 2x^3 + 3x^2 - 36x + 4$.

$f'(x) = 6x^2 + 6x - 36 = 6(x+3)(x-2)$.

$f'(x) = 0$ for $x = -3$ and $x = 2$.

For $x < -3$, $f'(x) > 0$.

For $-3 < x < 2$, $f'(x) < 0$.

For $x > 2$, $f'(x) > 0$.

$f(-3)$ is a relative maximum and

$f(2)$ is a relative minimum.

x	-5	-4	-3	-2	-1	0	1	2	3	4
$f(x)$	68	9	85	72	41	4	-27	-40	-23	36

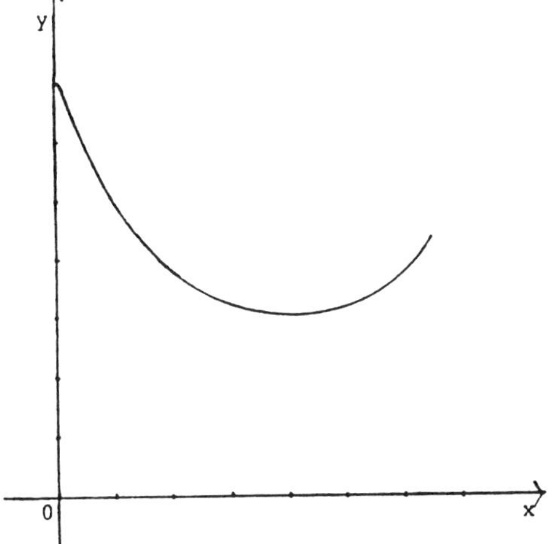

11. $f(x) = x^{3/2} - 3x + 7$.

$f'(x) = \frac{3}{2}x^{1/2} - 3$ for $x > 0$.

$f'(x) = 0$ for $x = 4$.

For $0 < x < 4$, $f'(x) < 0$.

For $x > 4$, $f'(x) > 0$.

$f(4)$ is a relative minimum.

x	0	1	2	3	4	5	6
$f(x)$	7	5	3.8	3.2	3	3.1	3.7

12. $f(x) = x^{4/3} - 4x - 3$.

$f'(x) = \frac{4}{3}x^{1/3} - 4$.

$f'(x) = 0$ for $x = 27$.

For $x < 27$, $f'(x) < 0$.

For $x > 27$, $f'(x) > 0$.

$f(27)$ is a relative minimum.

x	-8	0	8	27	64
$f(x)$	45	-3	-19	-30	-3

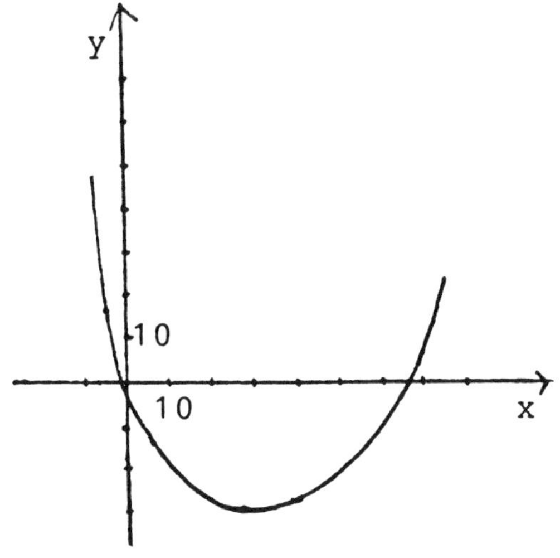

13. $f(x) = 4x^3 + 9x^2 - 12x + 7$.

$f'(x) = 12x^2 + 18x - 12 = 6(2x^2 + 3x - 2) = 6(2x - 1)(x + 2)$.

$f'(x) = 0$ for $x = -2$ and $x = \frac{1}{2}$.

For $x < -2$, $f'(x) > 0$.

For $-2 < x < \frac{1}{2}$, $f'(x) < 0$.

For $x > \frac{1}{2}$, $f'(x) > 0$.

$f(-2)$ is a relative maximum and

$f\left(\frac{1}{2}\right)$ is a relative minimum.

x	-3	-2	-1	0	$\frac{1}{2}$	1
$f(x)$	16	35	24	7	3.75	8

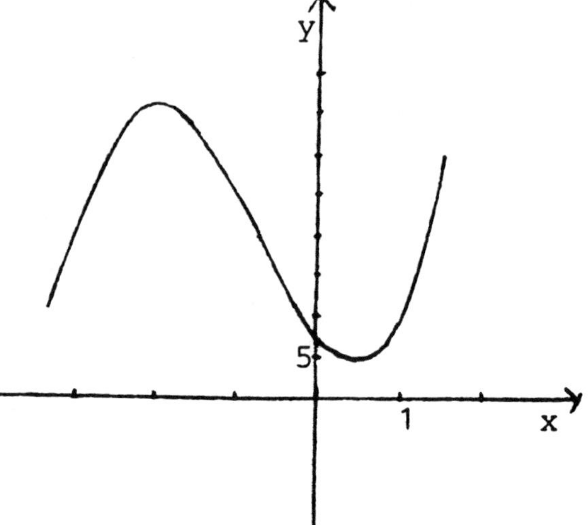

14. $f(x) = x^4 + 4x^3 - 8x^2 - 48x + 9$.

$f'(x) = 4x^3 + 12x^2 - 16x - 48 = 4(x^3 + 3x^2 - 4x - 12) = 4(x+3)(x+2)(x-2)$.

$f'(x) = 0$ for $x = -3$, $x = -2$, and $x = 2$.

For $x < -3$, $f'(x) < 0$.

For $-3 < x < -2$, $f'(x) > 0$.

For $-2 < x < 2$, $f'(x) < 0$.

For $x > 2$, $f'(x) > 0$.

$f(-3)$ and $f(2)$ are the relative minimums.

$f(-2)$ is a relative maximum.

x	-4	-3	-2	-1	0	1	2	3
$f(x)$	73	54	57	46	9	-42	-71	-18

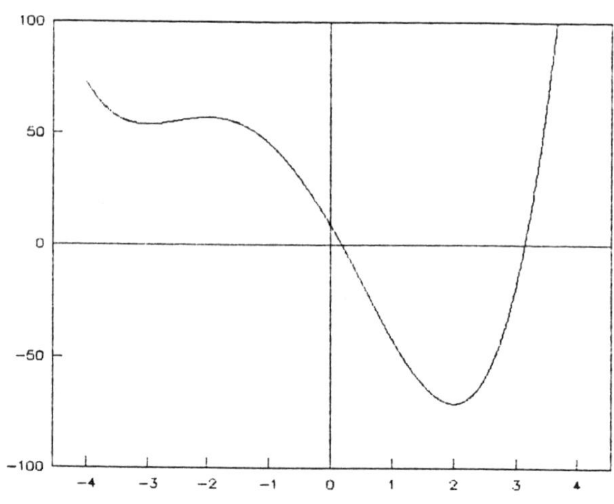

15. $f(x) = 3x^4 - 8x^3 - 18x^2 + 36$.

$f'(x) = 12x^3 - 24x^2 - 36x = 12x(x^2 - 2x - 3) = 12x\,((x-3)(x+1))$. $f'(x) = 0$ for all $x = -1$, $x = 0$, $x = 3$.

For $x < -1$, $f'(x) < 0$, so $f(x)$ is decreasing.

For $-1 < x < 0$, $f'(x) > 0$, so $f(x)$ is increasing.

For $0 < x < 3$, $f'(x) < 0$, so $f(x)$ is decreasing.

For $x > 3$, $f'(x) > 0$, so $f(x)$ is increasing.

$f(-1)$ is a relative minimum,

$f(0)$ is a relative maximum,

$f(3)$ is a relative minimum.

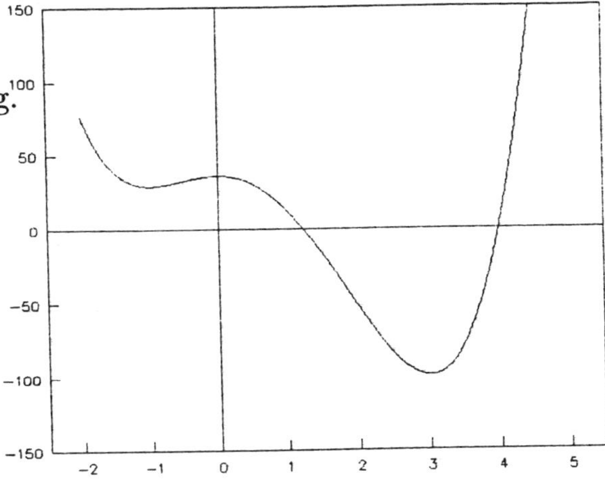

x	-2	-1	0	1	2	3	4
$f(x)$	76	29	36	13	-52	-99	4

16. $f(x) = \frac{x-1}{x+1}$.

$f'(x) = \frac{(x+1)\cdot 1-(x-1)\cdot 1}{(x+1)^2} = \frac{2}{(x+1)^2}$.

$f'(x)$ does not exist for $x = -1$.

$f'(x) > 0$ for $x \neq 1$, so $f(x)$ is increasing on $(-\infty, -1)$ and $(-1, \infty)$.

x	-4	-3	-2	$-\frac{3}{2}$	$-\frac{1}{2}$	$\frac{1}{2}$	1	2	3	4
$f(x)$	$\frac{5}{3}$	2	3	5	-3	$-\frac{1}{3}$	0	$\frac{1}{3}$	$\frac{1}{2}$	$\frac{3}{5}$

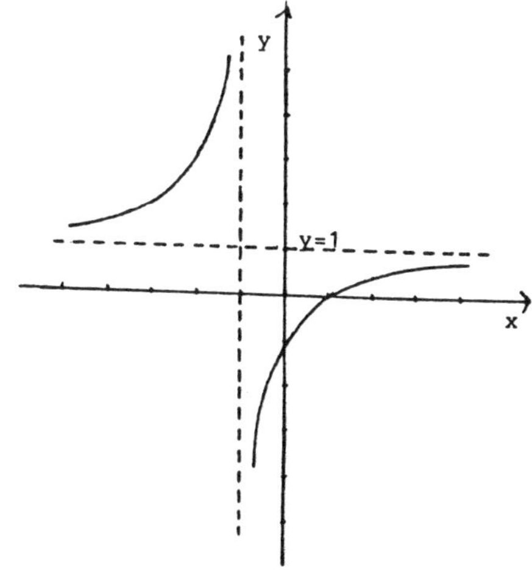

17. $f(x) = 9x - x^{-1}$.

$f'(x) = 9 + x^{-2}$.

$f'(x)$ does not exist for $x = 0$.

$f'(x) > 0$ for $x \neq 0$, so $f(x)$ is increasing on $(-\infty, 0)$ and $(0, \infty)$.

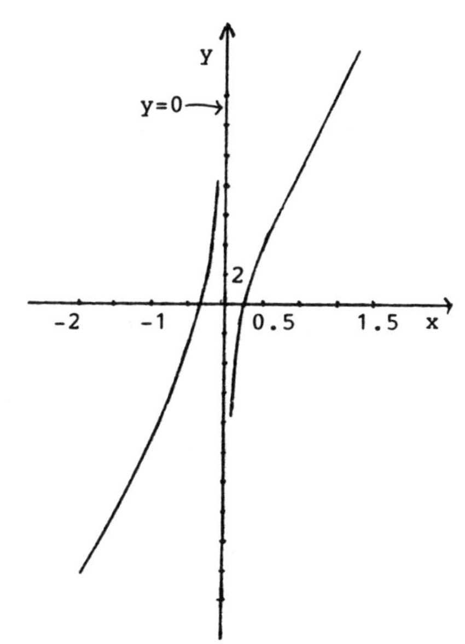

x	-2	-1	$-\frac{1}{2}$	$-\frac{1}{4}$	$-\frac{1}{8}$
$f(x)$	-17.5	-8	$-\frac{5}{2}$	$\frac{7}{4}$	$\frac{55}{8}$
x	$\frac{1}{8}$	$\frac{1}{4}$	$\frac{1}{2}$	1	2
$f(x)$	$-\frac{55}{8}$	$-\frac{7}{4}$	$\frac{5}{2}$	8	17.5

18. $f(x) = x^{1/3}(x-7)^2$.

$$\begin{aligned} f'(x) &= \tfrac{1}{3}x^{-2/3} \cdot (x-7)^2 + x^{1/3} \cdot 2(x-7) \\ &= \tfrac{1}{3}x^{-2/3} \cdot (x-7)\left((x-7) + 6x\right) \\ &= \tfrac{7}{3}x^{-2/3} \cdot (x-7) \cdot (x-1). \end{aligned}$$

$f'(x) = 0$ for $x = 1$, $x = 7$.

$f'(x)$ does not exist for $x = 0$.

For $x < 0$, $f'(x) > 0$.

For $0 < x < 1$, $f'(x) > 0$.

For $1 < x < 7$, $f'(x) < 0$.

For $x > 7$, $f'(x) > 0$.

$f(1)$ is a relative maximum.

$f(7)$ is a relative minimum.

x	-1	1	4	7	8	10
$f(x)$	-64	36	14.29	0	2	19.39

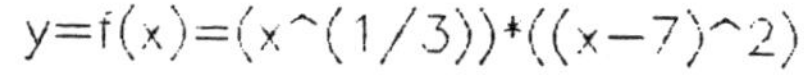

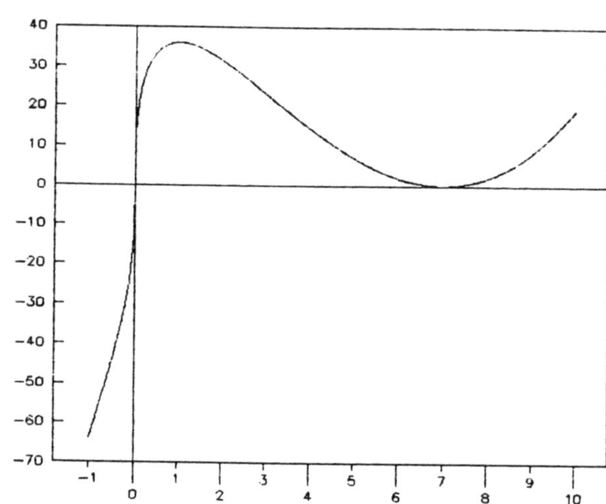

19. $f(x) = x^{5/3} - 5x^{2/3} + 3.$

$f'(x) = \frac{5}{3}x^{2/3} - \frac{10}{3}x^{-1/3} = \frac{5}{3}x^{-1/3}(x-2).$

$f'(x) = 0$ for $x = 2$.

$f'(x)$ does not exist for $x = 0$.

For $x < 0$, $f'(x) > 0$.

For $0 < x < 2$, $f'(x) < 0$.

For $x > 2$, $f'(x) > 0$.

$f(0)$ is a relative maximum.

$f(2)$ is a relative minimum.

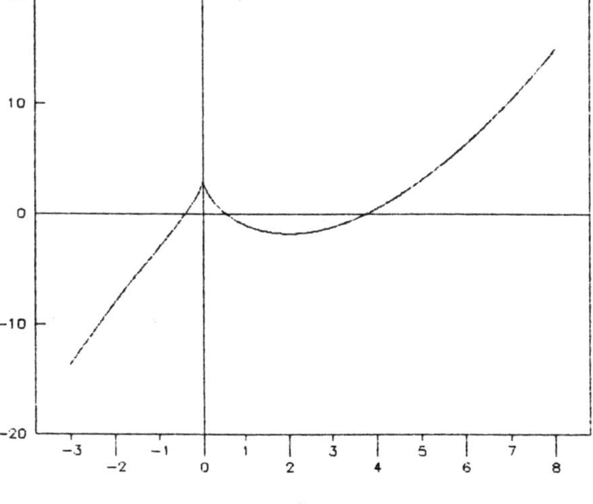

x	-3	-2	-1	0	
$f(x)$	-13.64	-8.11	-3	3	
x	1	2	4	6	8
$f(x)$	-1	-1.76	$.48$	6.3	15

20. $f(x) = \frac{x-1}{x^2+2}.$

$f'(x) = \frac{(x^2+2)\cdot 1 - (x-1)\cdot 2x}{(x^2+2)^2} = \frac{x^2+2-2x^2+2x}{(x^2+2)^2} = \frac{-x^2+2x+2}{(x^2+2)^2}.$

$f'(x) = 0$ for $(-x^2+2x+2) = 0 \implies x = \frac{-2\pm\sqrt{2^2-4\cdot(-1)\cdot 2}}{2\cdot(-1)} = \frac{2\pm\sqrt{12}}{2} = \frac{2\pm 2\sqrt{3}}{2}$

$= 1 \pm \sqrt{3}.$

For $x < 1-\sqrt{3}$, $f'(x) < 0$. For $1-\sqrt{3} < x < 1+\sqrt{3}$, $f'(x) > 0$.

For $x > 1+\sqrt{3}$, $f'(x) < 0$.

$f(1-\sqrt{3})$ is a relative minimum. $f(1+\sqrt{3})$ is a relative maximum.

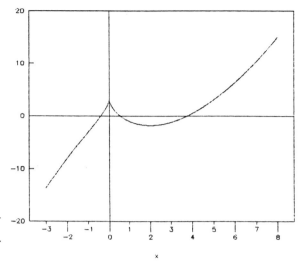

x	-10	-5	-1	$1-\sqrt{3} = -.73$	0
$f(x)$	$-.1$	$-.22$	$-.67$	$-.68$	$-.5$

x	1	2	$1+\sqrt{3} = 2.73$	3	5	10
$f(x)$	0	$.17$	$.183$	$.182$	$.15$	$.09$

21. $f(x) = \sqrt{x^2+4x+6}$.

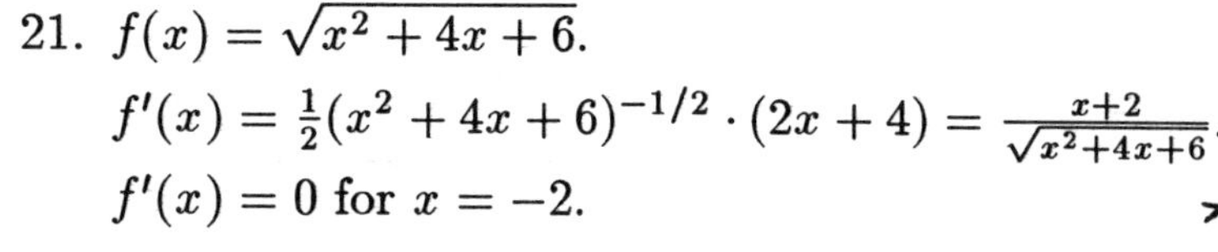

$f'(x) = \frac{1}{2}(x^2+4x+6)^{-1/2} \cdot (2x+4) = \frac{x+2}{\sqrt{x^2+4x+6}}$.

$f'(x) = 0$ for $x = -2$.

For $x < -2$, $f'(x) < 0$, so $f(x)$ is decreasing.

For $x > -2$, $f'(x) > 0$, so $f(x)$ is increasing.

$f(-2)$ is a relative minimum.

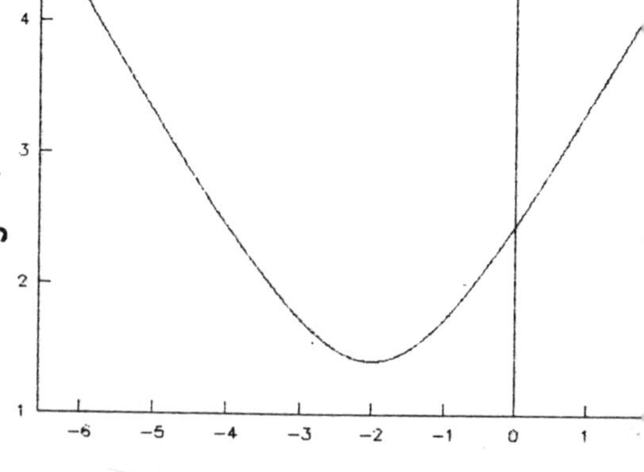

x	-4	-3	-2	-1	0
$f(x)$	$\sqrt{6} = 2.4495$	$\sqrt{3} = 1.7321$	$\sqrt{2} = 1.4142$	$\sqrt{3} = 1.7321$	$\sqrt{6} = 2.4495$

22. $f(x) = (x^2+2x+2)^{1/3}$.

$f'(x) = \frac{1}{3}(x^2+2x+2)^{-2/3} \cdot (2x+2)$.

$f'(x) = 0$ for $x = -1$.

For $x < -1$, $f'(x) < 0$, so $f(x)$ is decreasing.

For $x > -1$, $f'(x) > 0$, so $f(x)$ is increasing.

$f(-1)$ is a relative minimum.

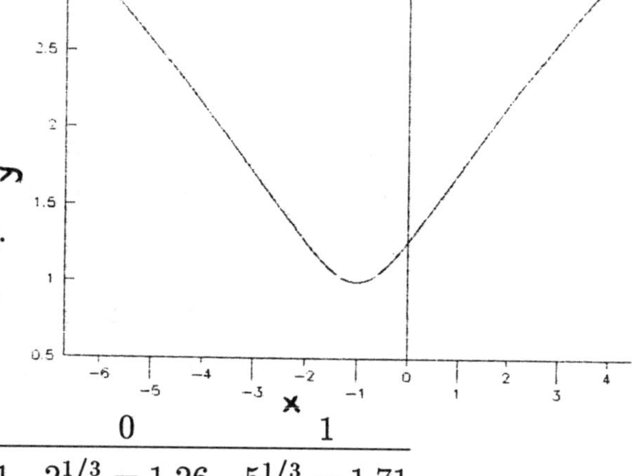

x	-3	-2	-1	0	1
$f(x)$	$5^{1/3} = 1.71$	$2^{1/3} = 1.26$	$1^{1/3} = 1$	$2^{1/3} = 1.26$	$5^{1/3} = 1.71$

23. $P'(x) = -96 + 36x - 3x^2 = -3\{(x-4)(x-8)\} = 0$ for $x = 4$ and $x = 8$. For $0 < x < 4$, $P'(x) < 0$; for $4 < x < 8$, $P'(x) > 0$; for $8 < x < \infty$, $P'(x) < 0$. Thus, $P(4)$ is a relative minimum and $P(8)$ is a relative maximum.

24. $P'(x) = -120 \cdot (3x-27)^{-1/3}$ for $x > 0$ and $x \neq 9$. For $0 < x < 9$, $P'(x) > 0$; for $x > 9$, $P'(x) < 0$. Thus $P(9)$ is a relative maximum.

25. $f'(x) = a$ for all x. If $a = 0$, $f(x) = b$ for all x.

If $a > 0$, $f'(x) > 0$ for all x, so $f(x)$ is always increasing.

If $a < 0$, $f'(x) < 0$ for all x, so $f(x)$ is always decreasing.

26. $f(x) = x^2 + bx + c$.

$f'(x) = 2x + b = 0$ for $x = -\frac{b}{2}$.

For $-\infty < x < -\frac{b}{2}$, $f'(x) < 0$; for $x > -\frac{b}{2}$, $f'(x) > 0$. Thus $f(x)$ has a relative minimum at $x = -\frac{b}{2}$ and has no relative maximum.

27. $f(x) = x^2 - ax + b$.

$f'(x) = 2x - a = 0$ for $x = 2$, so $a = 4$.

28. The total cost function $C(x) = 20 + 200x + .01x^3$.

The average cost function $c(x) = \frac{C(x)}{x} = \frac{20}{x} + 200 + .01x^2$.

a. $C'(x) = 200 + .03x^2 > 0$ for all $x > 0$.

Therefore $C(x)$ is increasing for all $x > 0$.

b. $c'(x) = -\frac{20}{x^2} + .02x$.

$c'(x) = 0 \implies \frac{20}{x^2} = .02x \implies .02x^3 = 20 \implies x^3 = 1000 \implies x = 10$.

For $0 < x < 10$, $c'(x) < 0$, so $c(x)$ is decreasing.

For $x > 10$, $c'(x) > 0$, so $c(x)$ is increasing.

Therefore $c(x)$ has a minimum value for $x = 10$.

29. Let x be one of these numbers. Then the other number is $50 - x$.

Let $f(x)$ be the product of these two numbers.

$f(x) = x \cdot (50 - x) = -x^2 + 50x$.

$f'(x) = -2x + 50$.

$f'(x) = 0$ for $x = 25$.

For $0 < x < 25$, $f'(x) > 0$, so $f(x)$ is increasing.

For $25 < x < 50$, $f'(x) < 0$, so $f(x)$ is decreasing.

Therefore the product is a maximum when the two numbers are both 25.

30. $f(x) = 2x^3 - ax^2 + 6$.

$f'(x) = 6x^2 - 2ax = 2x(3x - a)$.

$f'(x) = 0$ for $x = \frac{a}{3}$, $x = 0$.

a. Since $\frac{a}{3} = 3$, we have that $a = 9$. Therefore

$$f(x) = 2x^3 - 9x^2 + 6.$$

b. $f'(x) = 6x^2 - 18x = 6x(x - 3)$.

For $x < 0$, $f'(x) > 0$, so $f(x)$ is increasing.

For $0 < x < 3$, $f'(x) < 0$, so $f(x)$ is decreasing.

For $x > 3$, $f'(x) > 0$, so $f(x)$ is increasing.

Therefore the relative maximum value of $f(x)$ is $f(0) = 6$.

c. The relative minimum value of $f(x)$ is $f(3) = -21$.

31. $P(t) = \frac{Kt}{100+t^2}, \quad t > 0.$

$P'(t) = K \cdot \frac{(100+t^2)(1)-(t)(2t)}{(100+t^2)^2} = K \cdot \frac{100-t^2}{(100+t^2)^2}.$

$P'(t) = 0$ for $t = 10$.

For $0 < t < 10$, $P'(t) > 0$, so $P(t)$ is increasing.

For $10 < t < 20$, $P'(t) < 0$, so $P(t)$ is decreasing.

The population has a maximum value when $t = 10$.

$P(10) = .05K$.

32. $N(t) = 20,000 + 20t^2 - 2t^3, \quad t > 0.$

$N'(t) = 40t - 6t^2 = 2t(20 - 3t).$

$N'(t) = 0$ for $t = \frac{20}{3}$.

For $0 < t < \frac{20}{3}$, $\quad N'(t) > 0$, so $N(t)$ is increasing.

For $t > \frac{20}{3}$, $\quad N'(t) < 0$, so $N(t)$ is decreasing.

Therefore the maximum value of $N(t)$ occurs when $t = \frac{20}{3}$.

$N\left(\frac{20}{3}\right)$ is approximately 20,296.

Solutions to Exercise Set 3.3

1. $f(x) = x^2 + 6x + 9.$

   $f'(x) = 2x + 6. \qquad f''(x) = 2.$

2. $f(x) = 7x^2 - 5x + 14.$

   $f'(x) = 14x - 5. \qquad f''(x) = 14.$

3. $f(x) = x^3 - 4x^2 + 10x - 7.$

   $f'(x) = 3x^2 - 8x + 10. \qquad f''(x) = 6x - 8.$

4. $f'(x) = 1 - 3x^2$. $f''(x) = -6x$.

5. $f(x) = 4x^4 - 3x^3 + 9x^2 - x + 6.$

   $f'(x) = 16x^3 - 9x^2 + 18x - 1. \qquad f''(x) = 48x^2 - 18x + 18.$

6. $f'(x) = 4x^3 - 8x$. $f''(x) = 12x^2 - 8$.

7. $f(x) = 9x^3 - x^6.$

   $f'(x) = 27x^2 - 6x^5. \qquad f''(x) = 54x - 30x^4.$

8. $f'(x) = -2x^{-3} - 6x^2$. $f''(x) = 6x^{-4} - 12x$.

9. $f'(x) = 18x^5 - 24x^3$. $f''(x) = 90x^4 - 72x^2$.

10. $f'(x) = 4 + x^{-2} - 6x^{-3}$. $f''(x) = -2x^{-3} + 18x^{-4}$.

11. $f(x) = 3x^{5/3} - 2x^{-2/3} + x^{-2}$.

$f'(x) = 5x^{2/3} + \frac{4}{3}x^{-5/3} - 2x^{-3}$. $\qquad f''(x) = \frac{10}{3}x^{-1/3} - \frac{20}{9}x^{-8/3} + 6x^{-4}$.

12. $f(x) = 4(2x-3)^{5/3}$.

$f'(x) = 4 \cdot \frac{5}{3}(2x-3)^{2/3} \cdot (2) = \frac{40}{3}(2x-3)^{2/3}$.

$f''(x) = \frac{40}{3} \cdot \frac{2}{3}(2x-3)^{-1/3}(2) = \frac{160}{9}(2x-3)^{-1/3}$.

13. $f'(x) = \left((x-1)^{-1}\right)' = -(x-1)^{-2} \cdot 1 = -\frac{1}{(x-1)^2}$.

$f''(x) = \left(-(x-1)^{-2}\right)' = 2(x-1)^{-3} = \frac{2}{(x-1)^3}$.

14. $f'(x) = \frac{1\cdot(x+3) - x\cdot 1}{(x+3)^2} = \frac{3}{(x+3)^2}$.

$f''(x) = \left(3(x+3)^{-2}\right)' = -6(x+3)^{-3} = -\frac{6}{(x+3)^3}$.

15. $f'(x) = \frac{1}{2}(x+2)^{-1/2}$. $f''(x) = \frac{1}{2}\left(-\frac{1}{2}\right)(x+2)^{-3/2} = -\frac{1}{4}(x+2)^{-3/2}$.

16. $f'(x) = \frac{2}{3}(x-6)^{-1/3}$. $f''(x) = -\frac{2}{9}(x-6)^{-4/3}$.

17. $f(x) = 2(x^3 - 3x)^3$.

$f'(x) = 2 \cdot 3(x^3-3x)^2(3x^2-3) = 18(x^3-3x)^2(x^2-1)$.

$$\begin{aligned} f''(x) &= 18\left[2(x^3-3x)\cdot(3x^2-3)\cdot(x^2-1) + (x^3-3x)^2\cdot(2x)\right] \\ &= 36(x^3-3x)\left[3(x^2-1)^2 + (x^3-3x)\cdot x\right] \\ &= 36(x^3-3x)\cdot\left[3x^4 - 6x^2 + 3 + x^4 - 3x^2\right] \\ &= 36(x^3-3x)(4x^4 - 9x^2 + 3) \\ &= 36(4x^7 - 9x^5 + 3x^3 - 12x^5 + 27x^3 - 9x) \\ &= 36(4x^7 - 21x^5 + 30x^3 - 9x). \end{aligned}$$

18. $f(x) = \sqrt{x^2+4x+3}$.

$f'(x) = \frac{1}{2}\left(x^2+4x+3\right)^{-1/2}\cdot(2x+4) = \left(x^2+4x+3\right)^{-1/2}\cdot(x+2)$.

$$\begin{aligned} f''(x) &= -\tfrac{1}{2}\left(x^2+4x+3\right)^{-3/2}\cdot(2x+4)\cdot(x+2) + (x^2+4x+3)^{-1/2}(1) \\ &= \left(x^2+4x+3\right)^{-3/2}\left[-(x+2)^2 + (x^2+4x+3)\right] \\ &= \left(x^2+4x+3\right)^{-3/2}\left[-(x^2+4x+4) + (x^2+4x+3)\right] \\ &= \left(x^2+4x+3\right)^{-3/2}(-1). \end{aligned}$$

19. $f(x) = \frac{2x+1}{9-3x}$.

$f'(x) = \frac{(9-3x)(2)-(2x+1)(-3)}{(9-3x)^2} = \frac{21}{(9-3x)^2}$.

$f''(x) = 21(-2)(9-3x)^{-3}(-3) = \frac{126}{(9-3x)^3}$.

20. $f(x) = x\sqrt{4+x^3}$.

$f'(x) = (1)\cdot(4+x^3)^{1/2} + x\left(\frac{1}{2}\right)\left(4+x^3\right)^{-1/2}(3x^2)$
$= \left(4+x^3\right)^{-1/2}\left((4+x^3)+\frac{3}{2}x^3\right)$
$= \left(4+x^3\right)^{-1/2}\left(4+\frac{5}{2}x^3\right)$.

$f''(x) = \left[-\frac{1}{2}\left(4+x^3\right)^{-3/2}(3x^2)\right]\left(4+\frac{5}{2}x^3\right) + \left(4+x^3\right)^{-1/2}\left(\frac{15}{2}x^2\right)$
$= \left(4+x^3\right)^{-3/2}\left[\left(-\frac{3}{2}x^2\right)\left(4+\frac{5}{2}x^3\right) + \left(4+x^3\right)\left(\frac{15}{2}x^2\right)\right]$
$= \left(4+x^3\right)^{-3/2}\left(24x^2+\frac{15}{4}x^5\right)$.

21. $f'(x) = 1\cdot(x-1)^{2/3} + x\cdot\frac{2}{3}(x-1)^{-1/3}\cdot 1 = (x-1)^{2/3} + \frac{2}{3}x(x-1)^{-1/3}$.

$f''(x) = \frac{2}{3}(x-1)^{-1/3} + \frac{2}{3}\left[1\cdot(x-1)^{-1/3} + x\cdot\left(-\frac{1}{3}\right)(x-1)^{-4/3}\cdot 1\right]$
$= \frac{4}{3}(x-1)^{-1/3} - \frac{2}{9}x(x-1)^{-4/3}$.

22. $f'(x) = \frac{1\cdot(1+x^2)-x\cdot 2x}{(1+x^2)^2} = \frac{1-x^2}{(1+x^2)^2}$.

$f''(x) = \frac{-2x(1+x^2)^2-(1-x^2)\cdot 2(1+x^2)\cdot 2x}{(1+x^2)^4} = \frac{-2x(1+x^2)-4x(1-x^2)}{(1+x^2)^3}$.
$= \frac{-2x-2x^3-4x+4x^3}{(1+x^2)^3} = \frac{2x(x^2-3)}{(1+x^2)^3}$.

23. $f'(x) = 3(1-x^2)^2\cdot(-2x) = -6x(1-x^2)^2$.

$f''(x) = -6\left[1\cdot(1-x^2)^2 + x\cdot 2(1-x^2)(-2x)\right]$
$= -6(1-x^2)^2 + 24x^2(1-x^2)$.

24. $f'(x) = \frac{1}{2}(1+x^3)^{-1/2}\cdot(3x^2) = \frac{3}{2}x^2(1+x^3)^{-1/2}$.

$f''(x) = \frac{3}{2}\left[2x\cdot(1+x^3)^{-1/2} + x^2\cdot\left(-\frac{1}{2}\right)(1+x^3)^{-3/2}\cdot(3x^2)\right]$
$= 3x(1+x^3)^{-1/2} - \frac{9}{4}x^4(1+x^3)^{-3/2}$.

25. $f(x) = x^3 - 3x + 2$.

$f'(x) = 3x^2 - 3$.

$f''(x) = 6x$.

$f''(x) = 0$ for $x = 0$.

For $x < 0$, $f''(x) < 0$, so the graph is concave down.

For $x > 0$, $f''(x) > 0$, so the graph is concave up.

The point $(0, 2)$ is an inflection point.

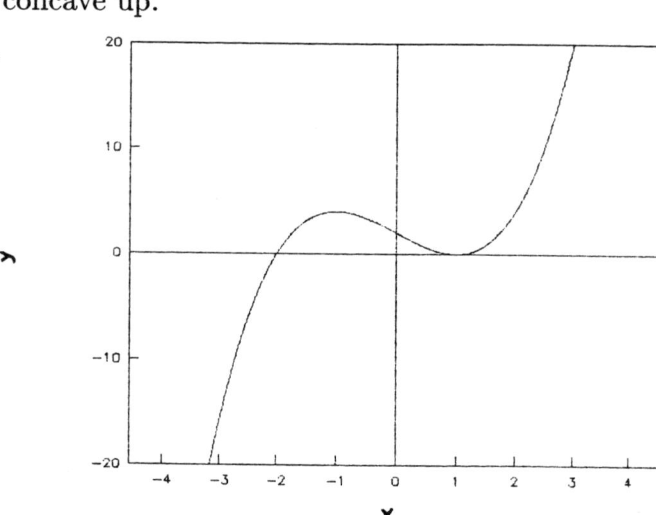

26. $f'(x) = -3x^2$. $f''(x) = -6x$. $f''(x) = 0$ for $x = 0$.

Intervals	Sign of $f''(x)$	Concavity of the graph of $f(x)$
$(-\infty, 0)$	+	concave up
$(0, \infty)$	−	concave down

The point $(0, 9)$ is an inflection point.

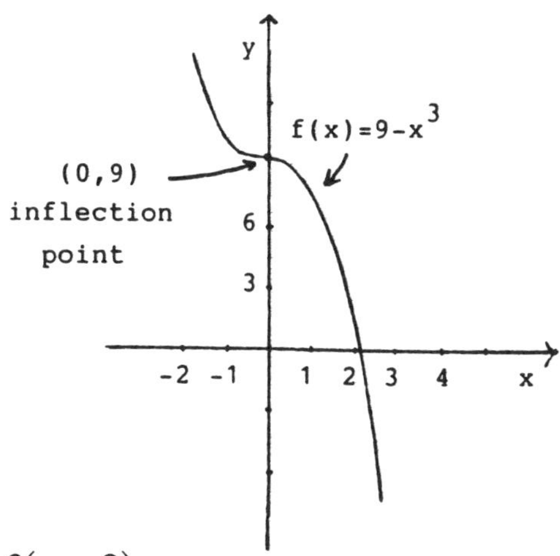

27. $f'(x) = 3x^2 - 18x + 12$. $f''(x) = 6x - 18 = 6(x - 3)$.
$f''(x) = 0$ for $x = 3$.

Interval	Test number t	$f''(t)$	Sign of $f''(x)$	Concavity of the graph of $f(x)$
$(-\infty, 3)$	0	−18	−	concave down
$(3, \infty)$	4	6	+	concave up

The point $(3, -24)$ is an inflection point.

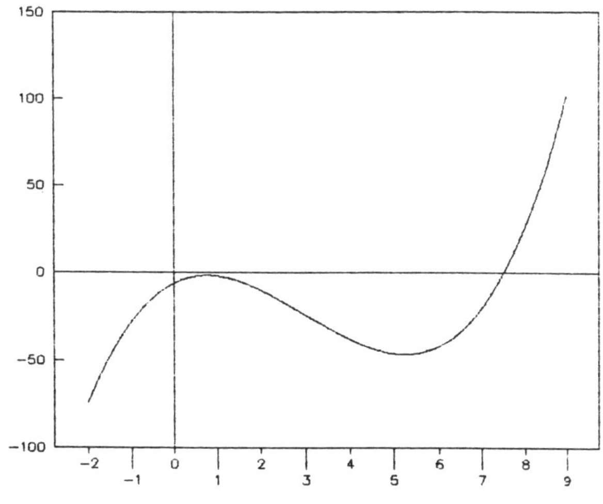

28. $f'(x) = 2x - 3x^2 + 3$. $f''(x) = 2 - 6x = 2(1 - 3x)$.
$f''(x) = 0$ for $x = \frac{1}{3}$.

Intervals	Sign of $f''(x)$	Concavity of the graph of $f(x)$
$(-\infty, \frac{1}{3})$	$+$	concave up
$(\frac{1}{3}, \infty)$	$-$	concave down

The point $\left(\frac{1}{3}, -\frac{133}{27}\right)$ is an inflection point.

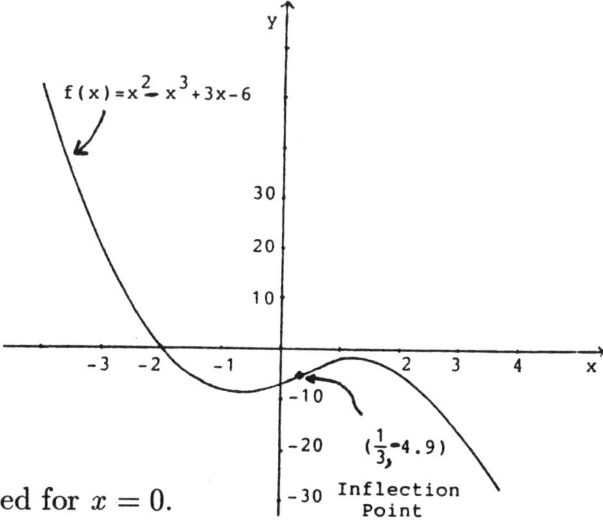

29. $f'(x) = -\frac{1}{x^2}$. $f''(x) = \frac{2}{x^3}$. $f''(x)$ is undefined for $x = 0$.

Intervals	Sign of $f''(x)$	Concavity of the graph of $f(x)$
$(-\infty, 0)$	$-$	concave down
$(0, \infty)$	$+$	concave up

The function $f(x)$ is not defined for $x = 0$. There is no inflection point.

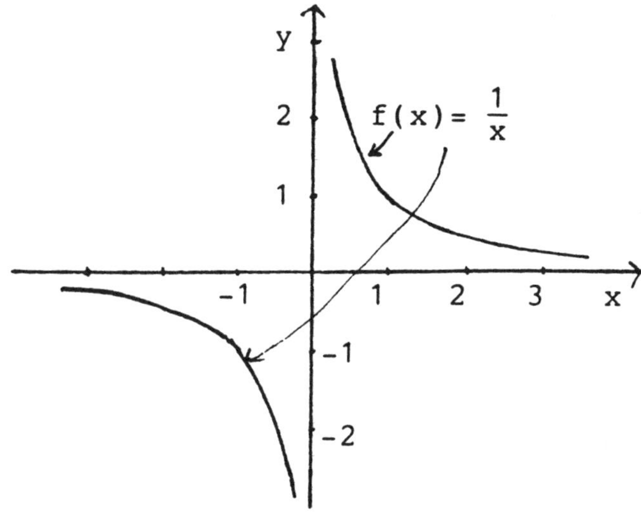

30. $f'(x) = \frac{-1}{(x-4)^2}$. $f''(x) = \frac{2}{(x-4)^3}$. $f''(x)$ is undefined for $x = 4$.
On $(-\infty, 4)$, $f''(x) < 0$, the graph of $f(x)$ is concave down.
On $(4, \infty)$, $f''(x) > 0$, the graph of $f(x)$ is concave up.
$f(x)$ is undefined for $x = 4$. Therefore, it has no inflection point.

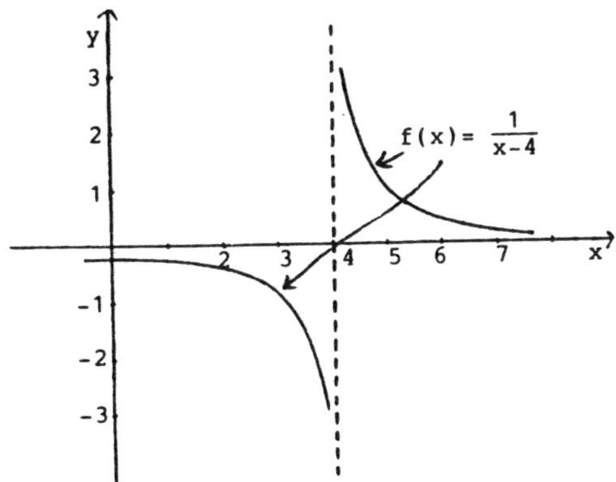

31. $f'(x) = \frac{(x+1)-x}{(x+1)^2} = \frac{1}{(x+1)^2}$. $f''(x) = -\frac{2}{(x+1)^3}$.
$f''(x)$ is undefined for $x = -1$.
On $(-\infty, -1)$, $f''(x) > 0$, the graph of $f(x)$ is concave up.
On $(-1, \infty)$, $f''(x) < 0$, the graph of $f(x)$ is concave down.
$f(x)$ is undefined for $x = -1$. Therefore, it has no inflection point.

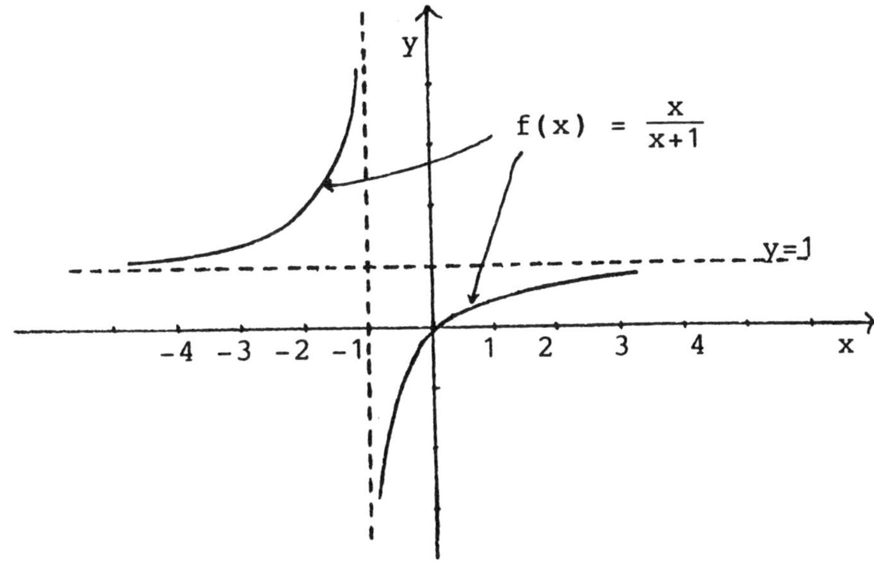

32. $f'(x) = \frac{1}{3}(x+2)^{-2/3}$. $f''(x) = -\frac{2}{9}(x+2)^{-5/3}$.
$f''(x)$ is undefined for $x = -2$. On $(-\infty, -2)$, $f''(x) > 0$, the graph of $f(x)$ is concave up.
On $(-2, \infty)$, $f''(x) < 0$, the graph of $f(x)$ is concave down.
The point $(-2, 0)$ is an inflection point.

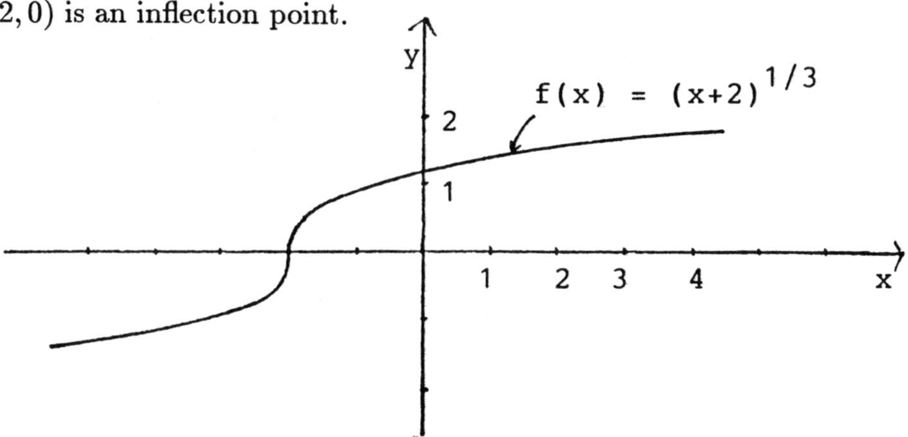

33. $f'(x) = 3(2x+1)^2 \cdot 2 = 6 \cdot (2x+1)^2$.
$f''(x) = 2 \cdot 6(2x+1) \cdot 2 = 48x + 24$.
$f''(x) = 0$ for $x = -\frac{1}{2}$.

Intervals	Sign of $f''(x)$	Concavity of the graph of $f(x)$
$\left(-\infty, -\frac{1}{2}\right)$	$-$	concave down
$\left(-\frac{1}{2}, \infty\right)$	$+$	concave up

The point $\left(-\frac{1}{2}, 0\right)$ is an inflection point.

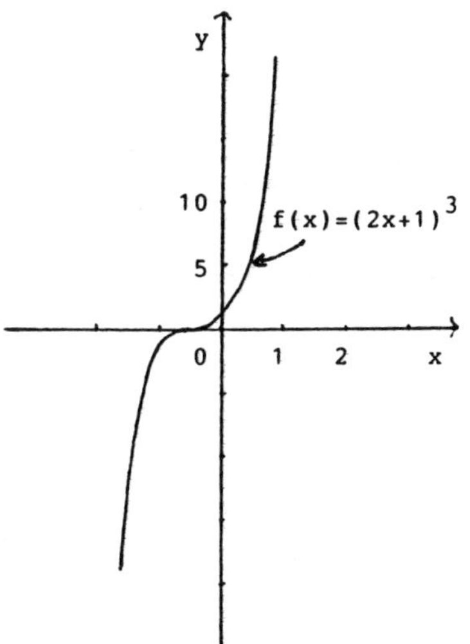

34. $f'(x) = \frac{1}{2}(x+4)^{-1/2}$. $f''(x) = -\frac{1}{4}\cdot(x+4)^{-3/2}$.
$f(x)$ is only defined for $x \geq -4$.
On $(-4,\infty)$, $f''(x) < 0$, the graph of $f(x)$ is concave down.

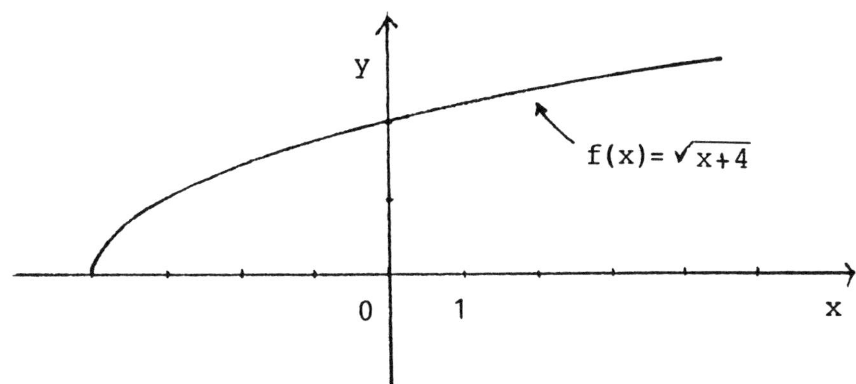

35. $f'(x) = 6x^2 - 6x + 18$. $f''(x) = 12x - 6$.
$f''(x) = 0$ for $x = \frac{1}{2}$.

Intervals	Sign of $f''(x)$	Concavity of the graph of $f(x)$
$(-\infty, \frac{1}{2})$	$-$	concave down
$(\frac{1}{2}, \infty)$	$+$	concave up

The point $(\frac{1}{2}, -\frac{7}{2})$ is an inflection point.

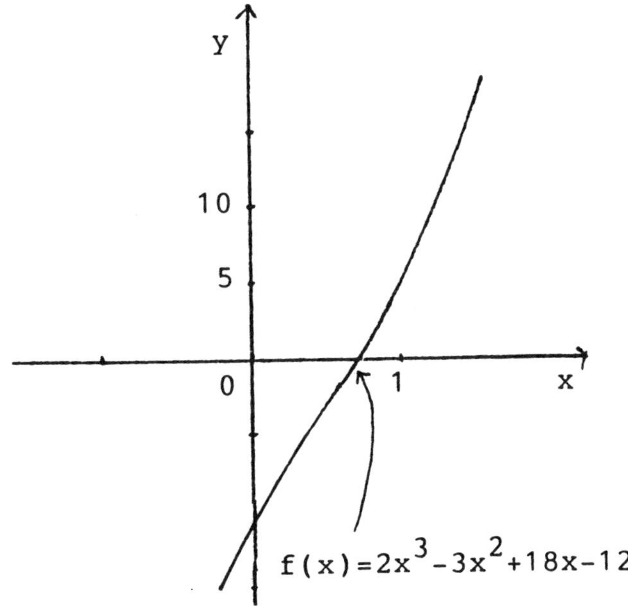

36. $f'(x) = \frac{1\cdot(x-4)-(x+4)\cdot 1}{(x-4)^2} = \frac{-8}{(x-4)^2}$. $f''(x) = \frac{16}{(x-4)^3}$.

$f''(x)$ is undefined for $x = 4$.

On $(-\infty, 4)$, $f''(x) < 0$, the graph of $f(x)$ is concave down.

On $(4, \infty)$, $f''(x) > 0$, the graph of $f(x)$ is concave up.

$f(x)$ is undefined for $x = 4$, therefore it has no inflection point.

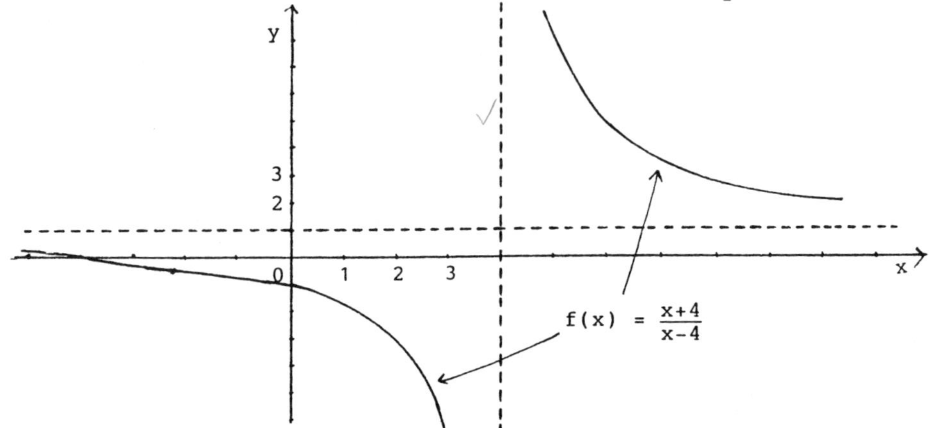

37. $f(x) = (x-2)^{1/3}$. $f'(x) = \frac{1}{3}(x-2)^{-2/3}$. $f''(x) = -\frac{2}{9}(x-2)^{-5/3}$.

$f''(x)$ is undefined for $x = 2$.

On $(-\infty, 2)$, $f''(x) > 0$, the graph of $f(x)$ is concave up.

On $(2, \infty)$, $f''(x) < 0$, the graph of $f(x)$ is concave down.

The point $(2, 0)$ is an inflection point.

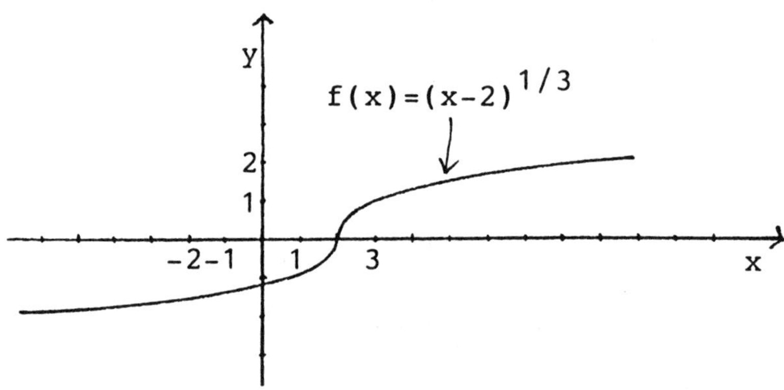

38. $f(x) = |4 - x^2| = \begin{cases} 4 - x^2, & -2 \le x \le 2, \\ x^2 - 4, & x < -2 \text{ or } x > 2. \end{cases}$

For $x \ne \pm 2$: $f'(x) = \begin{cases} -2x & \text{for } -2 < x < 2 \\ 2x & \text{for } x < -2 \text{ or } x > 2 \end{cases}$

$f''(x) = \begin{cases} -2 & \text{for } -2 < x < 2 \\ 2 & \text{for } x < -2 \text{ or } x > 2. \end{cases}$

On $(-2, 2)$, $f''(x) = -2 < 0$, the graph of $f(x)$ is concave down.

On $(-\infty, -2)$ and $(2, \infty)$, $f''(x) = 2 > 0$, the graph of $f(x)$ is concave up.

The inflection points are $(-2, 0)$ and $(2, 0)$.

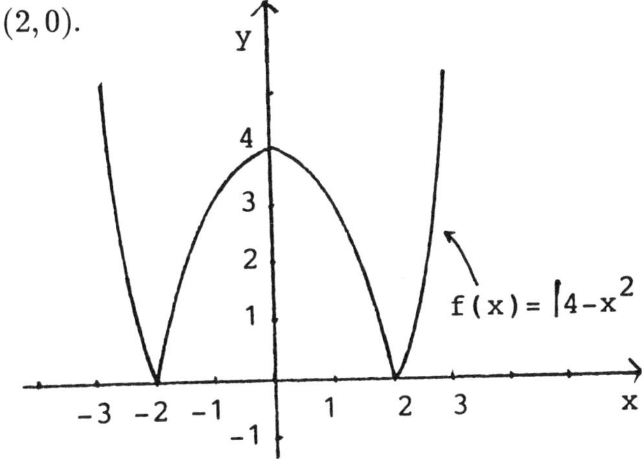

39. $f(x) = \frac{x}{1-x}$. $f'(x) = \frac{1\cdot(1-x) - x\cdot(-1)}{(1-x)^2} = \frac{1}{(1-x)^2}$. $f''(x) = \frac{2}{(1-x)^3}$.

$f''(x)$ is undefined for $x = 1$.

On $(-\infty, 1)$, $f''(x) > 0$, the graph of $f(x)$ is concave up.

On $(1, \infty)$, $f''(x) < 0$, the graph of $f(x)$ is concave down.

$f(x)$ is undefined for $x = 1$. Therefore, it has no inflection point.

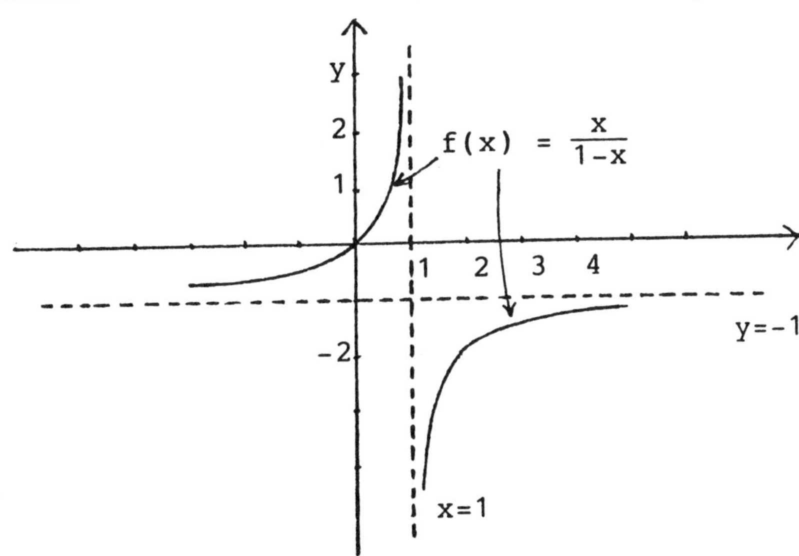

40. $f(x) = x^{5/3} - 5x^{2/3} + 3$. $f'(x) = \frac{5}{3}x^{2/3} - \frac{10}{3}x^{-1/3}$.
$f''(x) = \frac{10}{9}x^{-1/3} + \frac{10}{9}x^{-4/3} = \frac{10}{9}x^{-4/3}(x+1)$.
$f''(x) = 0$ for $x = -1$, and $f''(x)$ is undefined for $x = 0$.

Intervals	Sign of $f''(x)$	Concavity of the graph of $f(x)$
$(-\infty, -1)$	$-$	concave down
$(-1, 0)$	$+$	concave up
$(0, \infty)$	$+$	concave up

The inflection point is $(-1, -3)$.

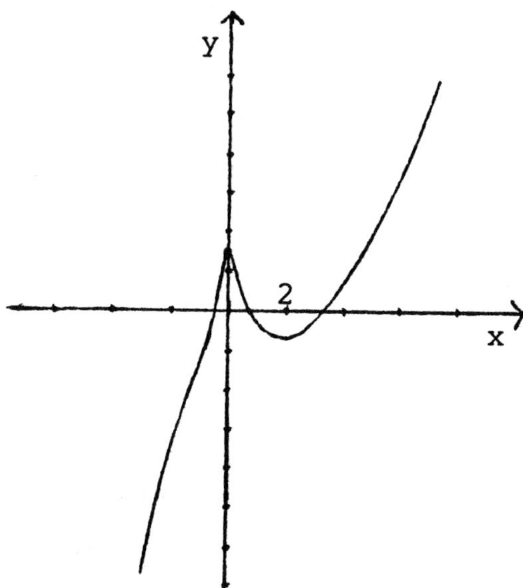

41. $f'(x) = 2x - \frac{2}{x^2}$. $f''(x) = 2 + \frac{4}{x^3}$.
$f'(1) = 2(1) - \frac{2}{1^2} = 0$ and $f''(1) = 2 + \frac{4}{1^3} = 6 > 0$.
So at $x = 1$, $f(x)$ has a relative minimum.

42. $f'(x) = 6x^2 - 6x$. $f''(x) = 12x - 6$. $f'(1) = 0$ and $f''(1) = 6 > 0$.
So $f(x)$ has a relative minimum.

43. $f'(x) = \frac{x-1}{x+1}$. $f''(x) = \frac{1\cdot(x+1)-(x-1)\cdot 1}{(x+1)^2} = \frac{2}{(x+1)^2}$.
$f'(1) = \frac{1-1}{1+1} = 0$ and $f''(1) = \frac{2}{(1+1)^2} = \frac{1}{2} > 0$.
So $f(x)$ has a relative minimum at $x = 1$.

44. $f'(x) = (x-1)(x+2)$ and $f''(x) = (x+2) + (x-1) = 2x + 1$.
$f'(-2) = 0$ and $f''(-2) = -3 < 0$.
So $f(x)$ has a relative maximum at $x = -2$.

45. $f'(x) = (x^3 - 4x)^{4/3}$. $f''(x) = \frac{4}{3}(x^3 - 4x)^{1/3}(3x^2 - 4)$.
$f'(2) = (2^3 - 4\cdot 2)^{4/3} = 0$ and $f''(2) = \frac{4}{3}(2^3 - 4\cdot 2)^{1/3}(3\cdot 2^2 - 4) = 0$.
The test failed.

46. a. The derivative of the inflation function is decreasing, and the second derivative of the inflation function is negative.

b. The derivative of the inflation function was constant last year, and the second derivative of the inflation function was zero last year.

c. The derivative of the radioactive isotope function is constant, and the second derivative of the radioactive isotope function is zero.

d. The derivative of the population function is positive and is increasing, and the second derivative of the population function is positive.

e. The derivative of the real estate price function is negative and is increasing in absolute value, and the second derivative of the real estate price function is negative.

f. The derivative of the home price function is positive and is decreasing, and the second derivative of the home price function is negative.

g. The derivative of her weight function is negative and is increasing in absolute value, and the second derivative of her weight function is negative.

h. The derivative of his weight function is positive and is decreasing and the second derivative is negative.

47. a. Since $R'(x)$ is decreasing for all $x > 0$, the graph of $y = R(x)$ is concave down on $(0, \infty)$.

b. Since $C'(x)$ is increasing for all $x > 0$, the graph of $y = C(x)$ is concave up on $(0, \infty)$.

48. a. On Interval #2, U is increasing, and the graph is concave down.

b. On Interval #1, because the graph is rising and is concave up.

c. On Interval #3, because the graph is falling and is concave down.

d. On Interval #4, because the graph is falling and is concave up.

49. a. For curve $U_1(x)$ marginal utility is increasing, because the graph of $U_1(x)$ is concave up.

b. The investor whose amount of annual return is $Y = U_1(x)$ knows how to make money work for him, because $U_1(x) > U_2(x)$ for $x > b$.

c. (i) At $x = a$, the investor whose amount of annual return is $Y = U_2(x)$ is more inclined to risk an additional dollar on his investments, because $U_2'(a) > U_1'(a)$.

At $x = b$, the investor whose amount of annual return is $Y = U_1(x)$ is more inclined to risk an additional dollar on his investments, because $U_1'(b) > U_2'(b)$.

50. a. $f'(x) = 3x^2$. $f''(x) = 6x$. Thus $f'(0) = 0$.

$f''(0) = 0$, but $f(0) = 0$ is neither a relative maximum nor a relative minimum because $f'(x) > 0$ for $x \neq 0$ so $f(x)$ is increasing for all x.

b. $f'(x) = 4x^3$. $f''(x) = 12x^2$. Thus $f'(0) = 0$. $f''(0) = 0$.

$f'(x) < 0$ for $x < 0$ and $f'(x) > 0$ for $x > 0$.

Therefore $f(0) = 0$ is a relative minimum.

c. $f'(x) = -4x^3$. $f''(x) = -12x^2$. $f'(0) = 0 = f''(0)$.

$f'(x) > 0$ for $x < 0$ and $f'(x) < 0$ for $x > 0$.

Thus $f(0) = 0$ is a relative maximum.

Solutions to Exercise Set 3.4

1. $f(x) = \frac{3}{x+5}$

$\lim_{x\to\infty} f(x) = 0$ and $\lim_{x\to-\infty} f(x) = 0$.

The line $y = 0$ is a horizontal asymptote.

2. $f(x) = \frac{1+x}{3-x}$.

$\lim_{x\to\pm\infty} f(x) = \lim_{x\to\pm\infty} \frac{1+x}{3-x} \cdot \frac{\frac{1}{x}}{\frac{1}{x}} = \lim_{x\to\pm\infty} \frac{\frac{1}{x}+1}{\frac{3}{x}-1} = \frac{0+1}{0-1} = -1.$

The line $y = -1$ is a horizontal asymptote.

3. $f(x) = \frac{3x^2}{1+x^2}$.

$\lim_{x\to\pm\infty} f(x) = \lim_{x\to\pm\infty} \frac{3x^2}{1+x^2} \cdot \frac{\frac{1}{x^2}}{\frac{1}{x^2}} = \lim_{x\to\pm\infty} \frac{3}{\frac{1}{x^2}+1} = \frac{3}{0+1} = 3.$

The line $y = 3$ is a horizontal asymptote.

4. $f(x) = \frac{\sqrt{x}}{1+x^2}$.

$\lim_{x\to\infty} f(x) = \lim_{x\to\infty} \frac{\sqrt{x}}{1+x^2} \cdot \frac{\frac{1}{x^2}}{\frac{1}{x^2}} = \lim_{x\to\infty} \frac{x^{-3/2}}{\frac{1}{x^2}+1} = \frac{0}{0+1} = 0.$

The line $y = 0$ is a horizontal asymptote.

5. $f(x) = \frac{x^3}{9-x^3}$.

$\lim_{x\to\pm\infty} f(x) = \lim_{x\to\pm\infty} \frac{x^3}{9-x^3} \cdot \frac{\frac{1}{x^3}}{\frac{1}{x^3}} = \lim_{x\to\pm\infty} \frac{1}{\frac{9}{x^3}-1} = \frac{1}{0-1} = -1.$

The line $y = -1$ is a horizontal asymptote.

6. $f(x) = \frac{4x^2}{1+3x+2x^2}$.

$\lim_{x\to\pm\infty} f(x) = \lim_{x\to\pm\infty} \frac{4x^2}{1+3x+2x^2} \cdot \frac{\frac{1}{x^2}}{\frac{1}{x^2}} = \lim_{x\to\pm\infty} \frac{4}{\frac{1}{x^2}+\frac{3}{x}+2} = \frac{4}{0+0+2} = 2.$

The line $y = 2$ is a horizontal asymptote.

7. $f(x) = \sqrt{\frac{1+7x^2}{x^2+3}}$.

$\lim_{x\to\pm\infty} f(x) = \lim_{x\to\pm\infty} \sqrt{\frac{1+7x^2}{x^2+3} \cdot \frac{1/x^2}{1/x^2}} = \lim_{x\to\pm\infty} \sqrt{\frac{(1/x^2)+7}{1+(3/x^2)}} = \sqrt{7}.$

The line $y = \sqrt{7}$ is a horizontal asymptote.

8. $f(x) = \frac{4x-\sqrt{x}}{x^{2/3}+x}$.

$\lim_{x\to\infty} f(x) = \lim_{x\to\infty} \frac{4x-\sqrt{x}}{x^{2/3}+x} \cdot \frac{\frac{1}{x}}{\frac{1}{x}} = \lim_{x\to\infty} \frac{4-\frac{1}{\sqrt{x}}}{\frac{1}{x^{1/3}}+1} = \frac{4-0}{0+1} = 4.$

The line $y = 4$ is a horizontal asymptote.

9. $f(x) = 6 - 4x^{-2/3}$.

$\lim_{x\to\pm\infty} f(x) = 6 - 0 = 6.$

The line $y = 6$ is a horizontal asymptote.

10. $f(x) = \frac{x^{2/3}-x^{-1/3}}{x^{1/4}+x^{1/2}} = \frac{x^{2/3}-x^{-1/3}}{x^{1/4}+x^{1/2}} \cdot \frac{\frac{1}{x^{2/3}}}{\frac{1}{x^{2/3}}} = \frac{1-\frac{1}{x}}{\frac{1}{x^{5/12}}+\frac{1}{x^{1/6}}}$.

As $x \to \infty$, $\left(1 - \frac{1}{x}\right) \to 0$, $\left(\frac{1}{x^{5/12}} + \frac{1}{x^{1/6}}\right) > 0$, and $\left(\frac{1}{x^{5/12}} + \frac{1}{x^{1/6}}\right) \to 0$.

Therefore $\lim_{x\to\infty} f(x) = \infty$.

There is no horizonal asymptote.

11. $x = 4$ is a vertical asymptote since $x - 4 = 0$ for $x = 4$.

12. $x = -2$ is a vertical asymptote since $x + 2 = 0$ for $x = -2$.

13. $x = 2$ and $x = -2$ are vertical asymptotes since $x^2 - 4 = 0$ for both $x = 2$ and $x = -2$.

14. $x = 1$ and $x = -1$ are vertical asymptotes since $x^2 - 1 = 0$ for both $x = 1$ and $x = -1$.

15. $x = -1$ and $x = -4$ are the vertical asymptotes since $x^2+5x+4 = (x+4)(x+1)$.

16. $x = 1$ and $x = 3$ are the vertical asymptotes since $x^2 - 4x + 3 = (x - 3)(x - 1)$.

17. $f(x) = \frac{x+3}{x^2+x-6} = \frac{(x+3)}{(x+3)(x-2)} = \frac{1}{(x-2)}$ for $x \neq -3$, $x \neq 2$. Thus $x = 2$ is a vertical asymptote.

18. $f(x) = \frac{x^2+x-6}{x^2+3x-10} = \frac{(x+3)(x-2)}{(x+5)(x-2)} = \frac{(x+3)}{(x+5)}$ for $x \neq 2$, $x \neq -5$. Thus $x = -5$ is a vertical asymptote.

19. $f(x) = \frac{x^2+5x+6}{x^2-5x+6} = \frac{(x+2)(x+3)}{(x-2)(x-3)}$. Thus $x = 2$ and $x = 3$ are vertical asymptotes.

20. $f(x) = \frac{x^3+2x^2-5x-6}{x^2-x-2} = \frac{(x+1)(x+3)(x-2)}{(x-2)(x+1)} = x+3$ for $x \neq 2$ and $x \neq -1$.

Thus there is no vertical asymptote.

21. $f(x) = 3 + x + \frac{\sqrt{x}}{1+\sqrt{x}}$.

There is no vertical asymptote and there is no horizontal asymptote.

Note that $1 = \frac{1+\sqrt{x}}{1+\sqrt{x}} = \frac{1}{1+\sqrt{x}} + \frac{\sqrt{x}}{1+\sqrt{x}} \implies \frac{\sqrt{x}}{1+\sqrt{x}} = 1 - \frac{1}{1+\sqrt{x}}$.

Therefore $f(x) = 4 + x - \frac{1}{1+\sqrt{x}}$. Observe that $\frac{1}{1+\sqrt{x}} \to 0$ as $x \to \infty$.

Thus the line $y = 4 + x$ is an oblique asymptote.

22. $f(x) = 6 - x + \frac{x+3}{1-x^2}$.

$\lim_{x\to\pm\infty} \frac{x+3}{1-x^2} = \lim_{x\to\pm\infty} \frac{x+3}{1-x^2} \cdot \frac{\frac{1}{x^2}}{\frac{1}{x^2}} = \lim_{x\to\pm\infty} \frac{\frac{1}{x}+\frac{3}{x^2}}{\frac{1}{x^2}-1} = \frac{0+0}{0-1} = 0.$

Thus $\lim_{x\to\infty} f(x) = -\infty$ and $\lim_{x\to-\infty} f(x) = +\infty$.

Therefore there is no horizontal asymptote.

The lines $x = 1$ and $x = -1$ are vertical asymptotes.

The line $y = 6 - x$ is an oblique asymptote.

23. $f(x) = \frac{2x^3+3x+5}{x^2+1}$.

For $x \neq 0$, $f(x) = \frac{2x^3+3x+5}{x^2+1} \cdot \frac{\frac{1}{x^2}}{\frac{1}{x^2}} = \frac{2x+\frac{3}{x}+\frac{5}{x^2}}{1+\frac{1}{x^2}}$.

$\lim_{x\to\infty} f(x) = \infty$ and $\lim_{x\to-\infty} f(x) = -\infty$.

There is no horizontal asmptote.

$$\begin{array}{r} 2x \\ x^2+1\,\overline{)\,2x^3+3x+5} \\ 2x^3+2x \\ \hline x+5 \end{array}$$

Thus
$$f(x) = 2x + \frac{x+5}{x^2+1}$$

$$\lim_{x\to\pm\infty} \frac{x+5}{x^2+1} = \lim_{x\to\pm\infty} \frac{x+5}{x^2+1} \cdot \frac{\frac{1}{x^2}}{\frac{1}{x^2}} = \lim_{x\to\pm\infty} \frac{\frac{1}{x}+\frac{5}{x^2}}{1+\frac{1}{x^2}}$$
$$= \frac{0+0}{1+0} = 0.$$

The line $y = 2x$ is an oblique asymptote. There are no vertical asymptotes.

24. $f(x) = \frac{x^2+x+3}{x+1}$.

Similarly as in problem 23, there is no horizontal asymptote.

The line $x = -1$ is a vertical asymptote.

$$\begin{array}{r} x \\ x+1 \overline{) x^2+x+3} \\ \underline{x^2+x} \\ 3 \end{array}$$

Thus

$$f(x) = x + \frac{3}{x+1}.$$
$$\lim_{x\to\pm\infty} \frac{3}{x+1} = 0.$$

The line $y = x$ is an oblique asymptote.

25. $f(x) = \frac{2x-2x^3+1}{1-x^2}$.

Because the numerator is a polynomial of higher degree than the polynomial in the denominator, there is no horizontal asymptote.

The lines $x = 1$ and $x = -1$ are vertical asymptotes.

$$\begin{array}{r} 2x \\ -x^2+1 \overline{) -2x^3+2x+1} \\ \underline{-2x^3+2x} \\ 1 \end{array}$$

Therefore

$$f(x) = 2x + \frac{1}{-x^2+1}.$$
$$\lim_{x\to\pm\infty} \frac{1}{-x^2+1} = 0.$$

The line $y = 2x$ is an oblique asymptote.

26. $f(x) = \frac{x^3+x^2-3x-3}{x+1}$.

There is no horizontal asymptote.

$$\begin{array}{r} x^2 - 3 \\ x+1 \overline{)x^3 + x^2 - 3x - 3} \\ x^3 + x^2 \qquad\qquad \\ \hline -3x - 3 \\ -3x - 3 \\ \hline 0 \end{array}$$

Therefore

$$f(x) = \frac{(x+1)(x^2-3)}{x+1} = x^2 - 3.$$

There is no vertical asymptote, and there is no oblique asymptote.

27. It is possible for a function to have both an oblique asymptote and a horizontal asymptote. For example, let

$$f(x) = \begin{cases} \frac{3x^2}{1+x^2} & , \quad x \le -1 \\ \frac{x^2+x+3}{x+1} & , \quad x > -1. \end{cases}$$

As shown in Problem 3, $\lim_{x \to -\infty} f(x) = 3$, so the line $y = 3$ is a horizontal asymptote. As shown in Problem 24, we have that

$$f(x) = x + \frac{3}{x+1} \qquad \text{for} \quad x > -1.$$

So the line $y = x$ is an oblique asymptote.

28. The function in Problem 25 is such that the lines $x = 1$ and $x = -1$ are vertical asymptotes, and the line $y = 2x$ is an oblique asymptote.

29. a. $C(x) = 80x + 2500$.

b. $c(x) = \frac{C(x)}{x} = 80 + \frac{2500}{x}$.

c. The line $x = 0$ is a vertical asymptote.

$\lim_{x \to \infty} c(x) = 80 \implies$ the line $y = 80$ is a horizontal asymptote.

30. $C(x) = 500 + 20\sqrt{x} + 40x$.

$c(x) = \frac{C(x)}{x} = \frac{500}{x} + \frac{20}{\sqrt{x}} + 40$.

The line $x = 0$ is a vertical asymptote.

$\lim_{x\to\infty} c(x) = 40 \implies$ the line $y = 40$ is a horizontal asymptote.

31. $C(x) = 2000 + 500x^{1/3} + Ax$.

$\lim_{x\to\infty} c(x) = 35 \implies \lim_{x\to\infty} \frac{2000+500x^{1/3}+Ax}{x} = 35$

$\implies \lim_{x\to\infty} \left(\frac{2000}{x} + \frac{500}{x^{2/3}} + A\right) = 35$.

$\therefore A = 35$.

32. $\lim_{x\to\infty} c(x) = \lim_{x\to\infty} \frac{22+10x+20x^2}{1+x^2} = \lim_{x\to\infty} \frac{\frac{22}{x^2}+\frac{10}{x}+20}{\frac{1}{x^2}+1} = 20$.

Thus $y = 20$ is a horizontal asymptote.

33. $\lim_{x\to\infty} d(x) = \lim_{x\to\infty} \frac{72+96x^{1/3}}{10x^{1/5}+6x^{1/3}} = \lim_{x\to\infty} \frac{\frac{72}{x^{1/3}}+96}{\frac{10}{x^{2/15}}+6} = 16$.

Thus $y = 16$ is a horizontal asymptote.

34. $S(p) = \frac{100}{200-p}$, $0 \le p < 200$.

$p = 200$ is a vertical asymptote since $200 - p = 0$ for $p = 200$.

The domain of S is $0 \le p < 200$.

$S'(p) = \frac{100}{(200-p)^2} > 0$ on $(0, 200)$, so S is increasing on $(0, 200)$.

$S''(p) = \frac{-100\cdot 2(200-p)(-1)}{(200-p)^4} = \frac{200}{(200-p)^3} > 0$ on $(0, 200)$, so S is concave up on $(0, 200)$.

$\lim_{p\to 200^-} S(p) = \lim_{p\to 200^-} \frac{100}{200-p} = \infty$.

p	0	50	100	150
$S(p)$	$\frac{1}{2}$	$\frac{2}{3}$	1	2

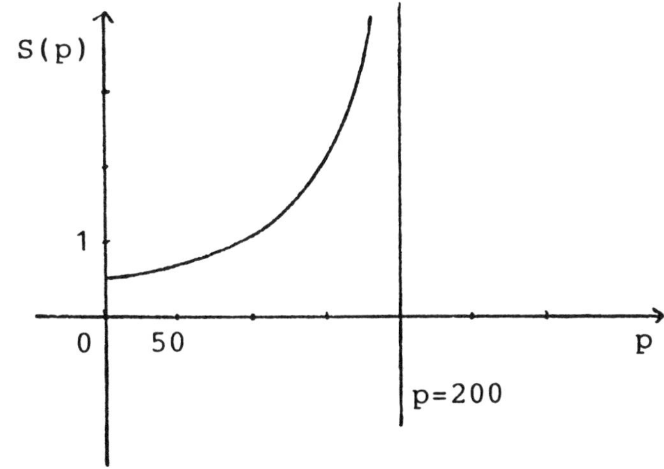

Solutions to Exercise Set 3.5

1. $f(x) = 9 - x^2$.
   $f(-x) = 9 - (-x)^2 = 9 - x^2 = f(x)$.
   Therefore f is an even function.
2. $f(x) = 3 + 5x^4$.
   $f(-x) = 3 + 5(-x)^4 = 3 + 5x^4 = f(x)$
   Therefore f is an even function.
3. $f(x) = x - x^5$.
   $f(-x) = (-x) - (-x)^5 = -x + x^5 = -f(x)$.
   Therefore f is an odd function.
4. $f(x) = 3 + x + 5x^3$.
   $f(-1) = -3$ and $f(1) = 9$, so $f(-1)$ is neither $f(1)$ nor $-f(1)$.
   Therefore f is neither even nor odd.
5. $f(x) = x(9 - x^2)$.
   $f(-x) = (-x) \cdot (9 - (-x)^2) = -x(9 - x^2) = -f(x)$.
   Therefore f is an odd function.
6. $f(x) = x^4(5 + x^2)$.
   $f(-x) = (-x)^4 \cdot (5 + (-x)^2) = x^4(5 + x^2) = f(x)$.
   Therefore f is an even function.
7. $f(x) = \frac{1+x^2}{9-x}$.
   $f(-1) = \frac{1+(-1)^2}{9-(-1)} = \frac{2}{10} = \frac{1}{5}$ and $f(1) = \frac{1+1^2}{9-1} = \frac{2}{8} = \frac{1}{4}$, so $f(-1)$ is neither $f(1)$ nor $-f(1)$.

   Therefore f is neither even nor odd.
8. $f(x) = \frac{x^2+2x+1}{\sqrt{3+x^2}}$.
   $f(-1) = \frac{(-1)^2+2(-1)+1}{\sqrt{3+(-1)^2}} = 0$ and $f(1) = \frac{1^2+2(1)+1}{\sqrt{3+1^2}} = \frac{4}{2} = 2$, so $f(-1)$ is neither $f(1)$ nor $-f(1)$.

   Therefore f is neither even nor odd.
9. $f(x) = \frac{1-x^{2/3}}{1+x^{4/3}}$.
   $f(-x) = \frac{1-(-x)^{2/3}}{1+(-x)^{4/3}} = \frac{1-x^{2/3}}{1+x^{4/3}} = f(x)$.
   Therefore f is an even function.
10. $f(x) = 3(x^3 + x^5)^{1/3}$.

$$f(-x) = 3\left((-x)^3 + (-x)^5\right)^{1/3} = 3\left(-(x^3 + x^5)\right)^{1/3}$$
$$= 3 \cdot (-1)(x^3 + x^5)^{1/3} = -f(x).$$

Therefore f is an odd function.

11. $f(x) = x^2 - 2x - 8$.

(1) The domain of f is all real numbers since $f(x)$ is a polynomial.

(2) f is neither even nor odd. (3) $f(x) = x^2 - 2x - 8 = (x-4)(x+2) = 0$ has two solutions $x = 4$ and $x = -2$.

(4,5) No vertical asymptote and no horizontal asymptote since $f(x)$ is a polynomial. (6) $f'(x) = 2x - 2 = 0$ for $x = 1$.

$f'(x) > 0$ for $x > 1$, so f is increasing on $(1, \infty)$.

$f'(x) < 0$ for $x < 1$, so f is decreasing on $(-\infty, 1)$.

Hence, $f(1) = -9$ is a relative minimum.

(7) $f''(x) = 2 > 0$ for all real numbers, so f is concave up on $(-\infty, \infty)$.

(8)

x	-2	-1	0	1	2	3	4
$y = f(x)$	0	-5	-8	-9	-8	-5	0

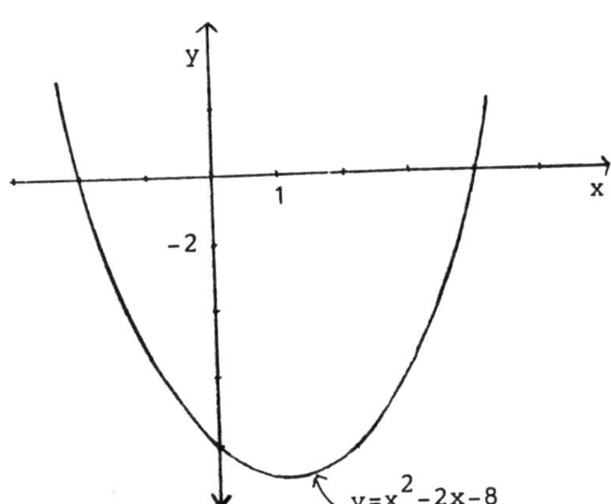

12. $f(x) = x^2 - 3x + 10$.

(1) The domain of f is all real numbers since $f(x)$ is a polynomial. (2) f is neither even nor odd. (3) The function f has no real zeros since the roots of the quadratic equation

$f(x) = 0$ are $x = \frac{3 \pm \sqrt{(-3)^2 - 4 \cdot 1 \cdot 10}}{2 \cdot 1} = \frac{3 \pm \sqrt{-31}}{2}$.

(4,5) Since $f(x)$ is a polynomial, there are no asymptotes.

(6) $f'(x) = 2x - 3$.

$f'(x) = 0$ for $x = \frac{3}{2}$.

For $x < \frac{3}{2}$, $f'(x) < 0$, so $f(x)$ is decreasing.

For $x > \frac{3}{2}$, $f'(x) > 0$, so $f(x)$ is increasing.

Hence $f\left(\frac{3}{2}\right) = 7.75$ is a relative minimum.

(7) $f''(x) = 2 > 0$ for all real numbers x,

so the graph of x is concave up on $(-\infty, \infty)$.

(8)

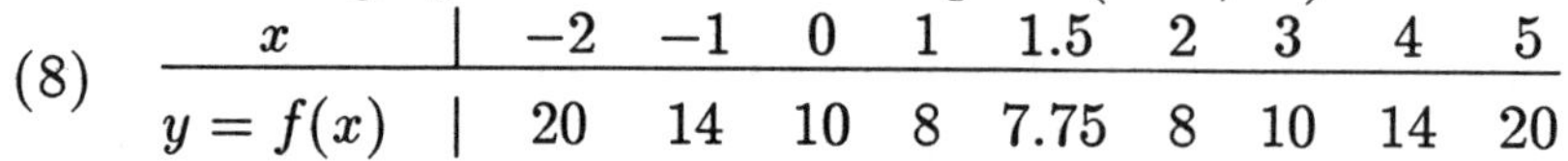

x	-2	-1	0	1	1.5	2	3	4	5
$y = f(x)$	20	14	10	8	7.75	8	10	14	20

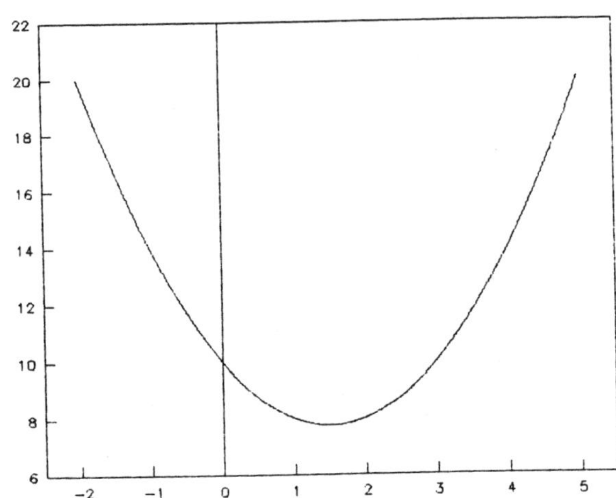

13. $f(x) = 2x^3 - 3x^2$.

(1) The domain of f is all real numbers since $f(x)$ is a polynomial.

(2) f is neither even nor odd.

(3) $f(x) = 2x^3 - 3x^2 = x^2(2x - 3) = 0$ for $x = 0$, $x = \frac{3}{2}$.

(4,5) Since $f(x)$ is a polynomial, there are no asymptotes.

(6) $f'(x) = 6x^2 - 6x = 6x(x - 1) = 0$ for $x = 0$ and $x = 1$.

Interval	Test number	$f'(x)$	Conclusion
$(-\infty, 0)$	$x = -1$	$f'(-1) > 0$	$f$increasing
$(0, 1)$	$x = \frac{1}{2}$	$f'\left(\frac{1}{2}\right) < 0$	f decreasing
$(1, \infty)$	$x = 2$	$f'(2) > 0$	f increasing

From these tests $f(0) = 0$ is a relative maximum for f

and $f(1) = -1$ is a relative minimum for f.

(7) $f''(x) = 12x - 6 = 0$ for $x = \frac{1}{2}$.

Since $f''(x) > 0$ for $x > \frac{1}{2}$, f is concave up on $\left(\frac{1}{2}, \infty\right)$.

Since $f''(x) < 0$ for $x < \frac{1}{2}$, f is concave down on $\left(-\infty, \frac{1}{2}\right)$.

Thus $\left(\frac{1}{2}, -\frac{1}{2}\right)$ is an inflection point for f.

(8)

x	-1	0	$\frac{1}{2}$	1	2
$y = f(x)$	-5	0	$-\frac{1}{2}$	-1	4

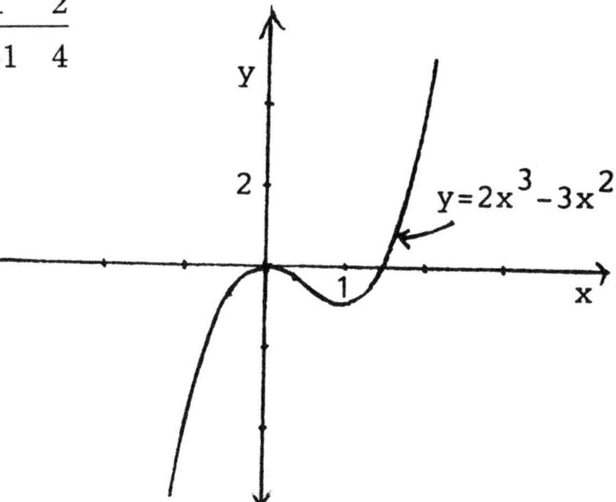

14. $f(x) = x^3 - 12x + 6$.

(1) The domain of f is all real numbers since $f(x)$ is a polynomial.

(2) f is neither even nor odd.

(3) It is difficult to find the zeros of $f(x)$, so we skip this step.

(4,5) Since $f(x)$ is a polynomial, there are no asymptotes.

(6) $f'(x) = 3x^2 - 12 = 3(x+2)(x-2) = 0$ for $x = 2$ and $x = -2$.
$f(x)$ is increasing on $(-\infty, -2)$ and $(2, \infty)$, $f(x)$ is decreasing on $(-2, 2)$.
$f(-2) = 22$ is a relative maximum.
$f(2) = -10$ is a relative minimum for f.

(7) $f''(x) = 6x$, so that $f(x)$ is concave up on $(0, \infty)$ and
$f(x)$ is concave down on $(-\infty, 0)$. Thus $(0, 6)$ is an inflection point.

(8)

x	-3	-2	0	2	3
$f(x)$	15	22	6	-10	-3

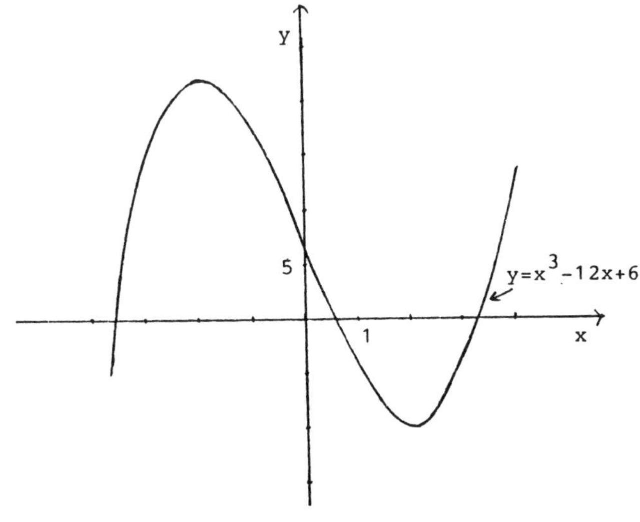

15. $f(x) = x^3 + x^2 - 8x + 8$.

(1) The domain of f is all real numbers since $f(x)$ is a polynomial.

(2) f is neither even nor odd.

(3) It is difficult to find the zeros of $f(x)$, so we skip this step. (4,5) Since $f(x)$ is a polynomial, there are no asymptotes. (6) $f'(x) = 3x^2 + 2x - 8 = (x+2)(3x-4) = 0$ for $x = \frac{4}{3}$ and $x = -2$.

Interval	Test number	$f'(x)$	Conclusion
$(-\infty, -2)$	$x = -3$	$f'(-3) > 0$	f increasing
$(-2, \frac{4}{3})$	$x = 0$	$f'(0) < 0$	f decreasing
$(\frac{4}{3}, \infty)$	$x = 3$	$f'(3) > 0$	f increasing

From these tests, $f(-2) = 20$ is a relative maximum for f and $f\left(\frac{4}{3}\right) = 1.48$ is a relative minimum for f.

(7) $f''(x) = 6x + 2 = 0$ for $x = -\frac{1}{3}$.
$f(x)$ is concave up on $\left(-\frac{1}{3}, \infty\right)$ since $f''(x) > 0$ on $\left(-\frac{1}{3}, \infty\right)$.
$f(x)$ is concave down on $\left(-\infty, -\frac{1}{3}\right)$ since $f''(x) < 0$ on $\left(-\infty, -\frac{1}{3}\right)$.
$\therefore \left(-\frac{1}{3}, 10.74\right)$ is an inflection point for f.

(8)

x	-3	-2	$-\frac{1}{3}$	0	$\frac{4}{3}$	2
$f(x)$	14	20	10.74	8	1.48	4

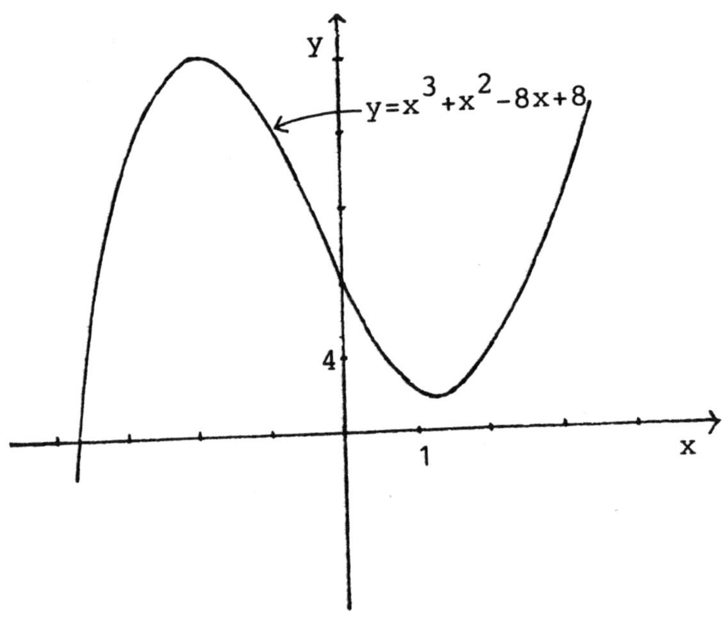

16. $f(x) = 9x - \frac{1}{x}$.

(1) The domain of f is all real numbers except $x = 0$.

(2) For $x \neq 0$, $f(-x) = 9(-x) - \frac{1}{-x} = -9x + \frac{1}{x} = -f(x)$,
so f is an odd function. (3) $f(x) = 9x - \frac{1}{x} = \frac{9x^2-1}{x} = \frac{(3x+1)(3x-1)}{x} = 0$ for $x = \pm\frac{1}{3}$.

(4) The line $x = 0$ is a vertical asymptote.

(5) There is no horizontal asymptote since $\lim_{x\to\infty} f(x) = \infty$ and $\lim_{x\to-\infty} f(x) = -\infty$.

(6) $f'(x) = 9 + \frac{1}{x^2} > 0$ for all real numbers except $x = 0$,
so that f is increasing on both $(-\infty, 0)$ and $(0, \infty)$.

(7) $f''(x) = -\frac{2}{x^3}$, from which $f''(x) > 0$ on $(-\infty, 0)$
and $f''(x) < 0$ on $(0, \infty)$. Thus f is concave up on $(-\infty, 0)$
and f is concave down on $(0, \infty)$. There is no inflection point.

(8)

x	-2	-1	$-\frac{1}{3}$	$+\frac{1}{3}$	$-\frac{1}{4}$	$\frac{1}{4}$	1	2
$f(x)$	-17.5	-8	0	0	$\frac{7}{4}$	$-\frac{7}{4}$	8	17.5

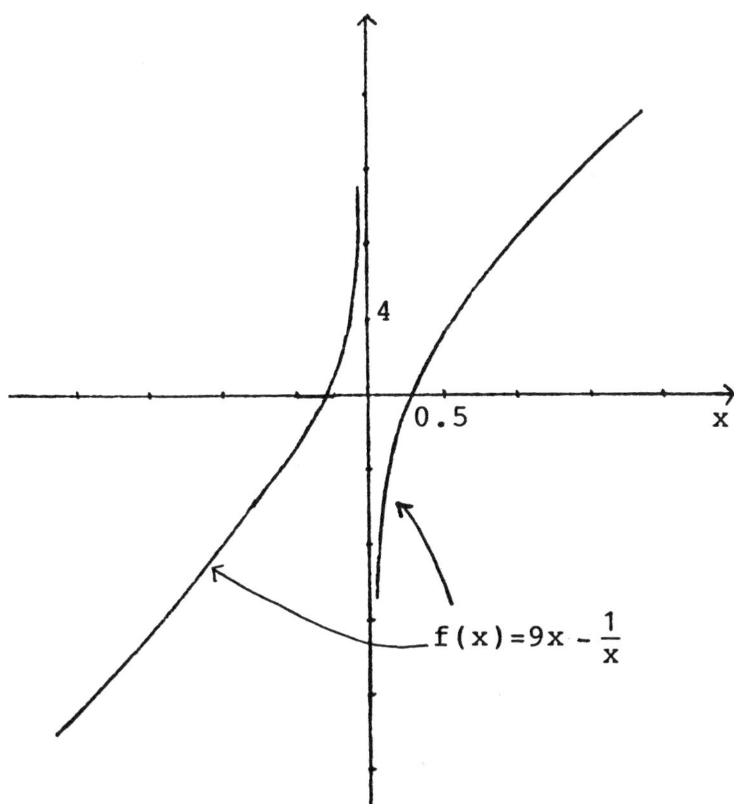

17. $y = f(x) = \frac{x+4}{x-4}$.

(1) The domain of f is all real numbers except $x = 4$ since $(x - 4) = 0$.

(2) f is neither even nor odd.

(3) $f(x) = 0$ for $x + 4 = 0 \implies x = -4$.

(4) $x = 4$ is a vertical asymptote.

(5) $y = 1$ is a horizontal asymptote since $\lim\limits_{x\to\pm\infty} \frac{x+4}{x-4} = 1$.

(6) $f'(x) = \frac{(x-4)-(x+4)}{(x-4)^2} = \frac{-8}{(x-4)^2} < 0$ for all real numbers except $x = 4$, so that f is decreasing on both $(-\infty, 4)$ and $(4, \infty)$.

(7) $f''(x) = \frac{8\cdot 2\cdot (x-4)}{(x-4)^4} = \frac{16}{(x-4)^3}$, from which $f''(x) < 0$ on $(-\infty, 4)$ and $f''(x) > 0$ on $(4, \infty)$. Hence $f(x)$ is concave up on $(4, \infty)$ and concave down on $(-\infty, 4)$. There is no inflection point.

(8)

x	-10	-4	0	1	3	3.5	4.5	5	10
$f(x)$	$\frac{3}{7}$	0	-1	$-\frac{5}{3}$	-7	-15	17	9	$\frac{7}{3}$

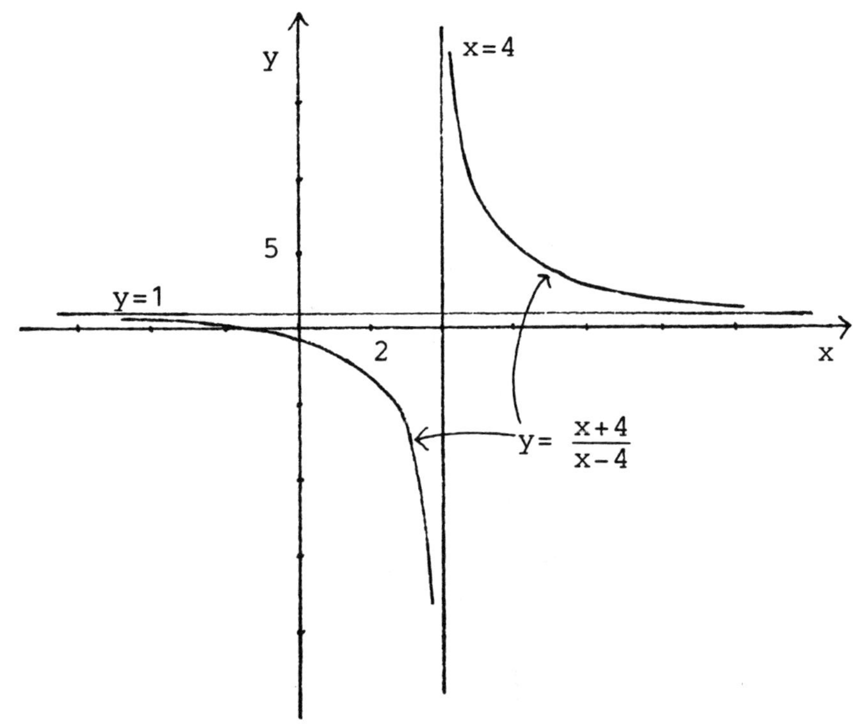

18. $f(x) = 4 - x^{2/3}$.

(1) The domain of f consists of all real numbers.

(2) Since $f(-x) = f(x)$ for all x, the function f is an even function.

(3) $f(x) = 0 \implies x^{2/3} = 4 \implies x = -8$ or $x = 8$.

(4,5) There are no asymptotes.

(6) $f'(x) = -\frac{2}{3}x^{-1/3}$. $f'(x)$ is undefined for $x = 0$.

For $x < 0$, $f'(x) > 0$, so $f(x)$ is increasing.

For $x > 0$, $f'(x) < 0$, so $f(x)$ is decreasing.

Hence $f(0) = 4$ is a relative maximum.

(7) $f''(x) = \frac{2}{9}x^{-4/3}$. $f''(x)$ is undefined for $x = 0$.

For $x < 0$, $f''(x) > 0$; so the graph of f is concave up on the interval $(-\infty, 0)$.

For $x > 0$, $f''(x) > 0$; so the graph of f is also concave up on the interval $(0, \infty)$.

There is no inflection points.

(8)

x	± 27	± 8	± 1	0
$y = f(x)$	-5	0	3	4

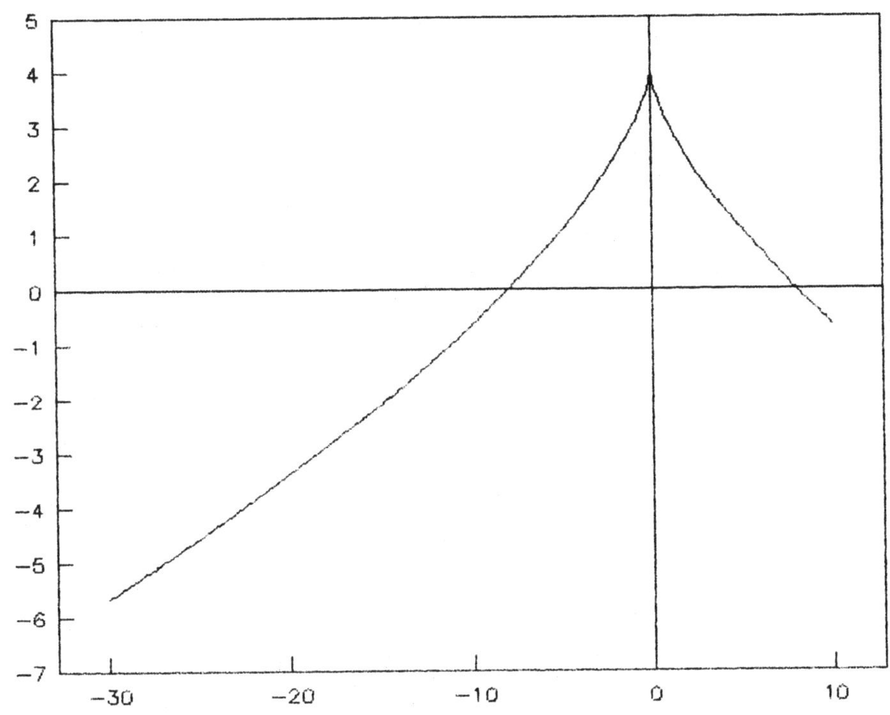

19. $f(x) = |4 - x^2|$.

This was discussed and drawn for Problem 38 in Exercise Set 3.3.

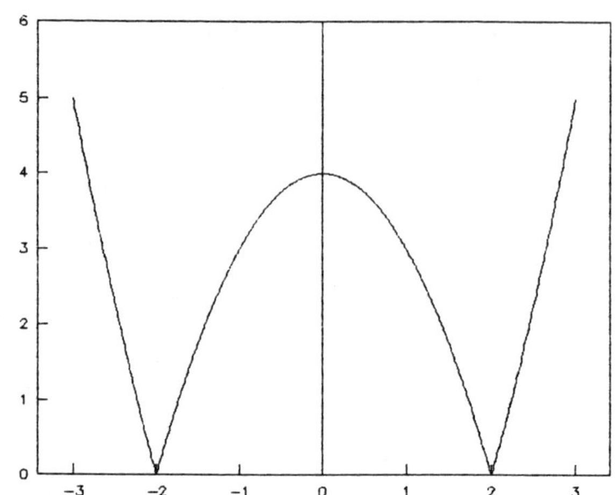

20. $f(x) = x^{5/3} - x^{2/3}$.

(1) The domain is all x.

(2) f is neither even nor odd. (3) $f(x) = x^{2/3}(x-1) = 0$ for $x = 0$ and $x = 1$.

(4,5) There is no vertical asymptote and no horizontal asymptote.

(6) $f'(x) = \frac{5}{3}x^{2/3} - \frac{2}{3}x^{-1/3} = \frac{1}{3}x^{-1/3}(5x-2)$.

$f'(x) = 0$ for $x = \frac{2}{5}$, and $f'(x)$ does not exist for $x = 0$.

Interval	Test number	$f'(x)$	Conclusion
$(-\infty, 0)$	-1	$f'(-1) > 0$	increasing
$\left(0, \frac{2}{5}\right)$	$\frac{1}{4}$	$f'\left(\frac{1}{4}\right) < 0$	decreasing
$\left(\frac{2}{5}, \infty\right)$	1	$f'(1) > 0$	increasing

$f(0)$ is a relative maximum and $f\left(\frac{2}{5}\right)$ is a relative minimum.

(7) $f''(x) = \frac{10}{9}x^{-1/3} + \frac{2}{9}x^{-4/3} = \frac{2}{9}x^{-4/3}(5x+1)$.

$f''(x) = 0$ for $x = -\frac{1}{5}$, and $f''(x)$ does not exist for $x = 0$.

Interval	Test number	$f''(x)$	Conclusion
$\left(-\infty, -\frac{1}{5}\right)$	$-\frac{1}{4}$	$f''\left(-\frac{1}{4}\right) < 0$	concave down
$\left(-\frac{1}{5}, 0\right)$	$-\frac{1}{10}$	$f''\left(-\frac{1}{10}\right) > 0$	concave up
$(0, \infty)$	1	$f''(1) > 0$	concave up

$\left(-\frac{1}{5}, -0.41\right)$ is the inflection point.

(8)

x	-2	-1	$-\frac{1}{5}$	0	$\frac{2}{5}$	1	2
$f(x)$	-4.76	-2	-0.41	0	-0.33	0	1.59

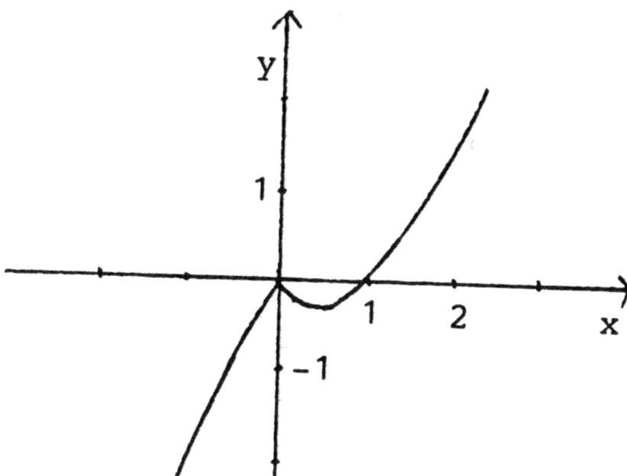

21. $y = \frac{1}{x(x-4)}$.

(1) The domain is all x except $x = 0$ and $x = 4$.

(2) The function f is neither even nor odd. The straight line $x = 2$ is an axis of symmetry for the graph of f.

(3) $y = \frac{1}{x(x-4)} \neq 0$ for all x in the domain of f.

(4) Vertical asymptotes are $x = 0$ and $x = 4$.

(5) Horizontal asymptote is $y = 0$ since $\lim_{x\to\pm\infty} \frac{1}{x(x-4)} = 0$.

(6) $y' = -1(x^2 - 4x)^{-2} \cdot (2x - 4) = (-2x + 4) \cdot (x^2 - 4x)^{-2}$.

$y' = 0$ for $x = 2$, and y' does not exist for $x = 0$ and $x = 4$.

Interval	Test Number	$f'(x)$	Conclusion
$(-\infty, 0)$	-1	$f'(-1) > 0$	increasing
$(0, 2)$	1	$f'(1) > 0$	increasing
$(2, 4)$	3	$f'(3) < 0$	decreasing
$(4, \infty)$	5	$f'(5) < 0$	decreasing

$f(2)$ is a relative maximum.

(7) $y'' = (-2)(x^2 - 4x)^{-2} + (-2x + 4)(-2)(x^2 - 4x)^{-3}(2x - 4)$

$= -2(x^2 - 4x)^{-3}\{(x^2 - 4x) + (-2x + 4)(2x - 4)\}$

$= -2(x^2 - 4x)^{-3}\{-3x^2 + 12x - 16\}$.

y'' has no zeros, and y'' does not exist for $x = 0$ and $x = 4$.

Interval	Test Number	$f''(x)$	Conclusion
$(-\infty, 0)$	-1	$f''(-1) > 0$	concave up
$(0, 4)$	1	$f''(1) < 0$	concave down
$(4, \infty)$	5	$f''(5) > 0$	concave up

There is no inflection point.

(8)

x	-5	-2	-1	1	2	3	5	7	10
y	$\frac{1}{45}$	$\frac{1}{12}$	$\frac{1}{5}$	$-\frac{1}{3}$	$-\frac{1}{4}$	$-\frac{1}{3}$	$\frac{1}{5}$	$\frac{1}{21}$	$\frac{1}{60}$

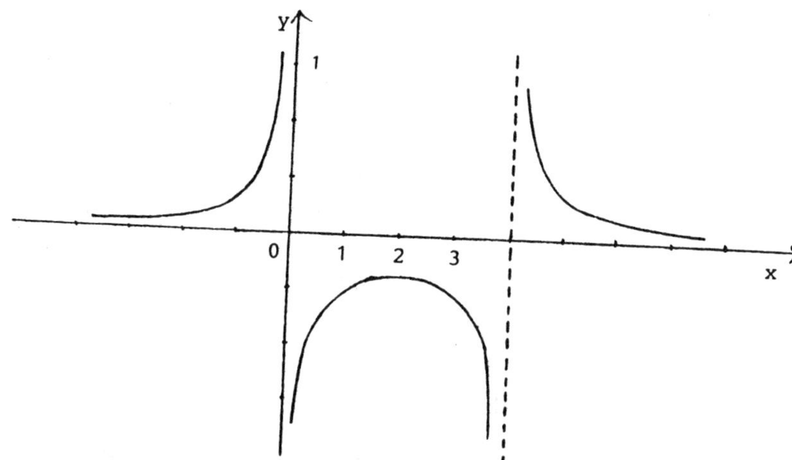

22. $y = x^2 - \frac{9}{x^2}$.

(1) The domain is all x except $x = 0$.

(2) f is an even function.

(3) $y = x^2 - \frac{9}{x^2} = 0$ for $x^4 = 9 \implies x = \pm\sqrt[4]{9} = \pm 1.732$.

(4) Vertical asymptote is $x = 0$.

(5) Since $\lim_{x\to\pm\infty} y = \infty$, there is no horizontal asymptote.

(6) $y' = 2x + \frac{18}{x^3}$.

y' does not exist for $x = 0$.

On $(-\infty, 0)$, $y' < 0$, so y is decreasing.

On $(0, \infty)$, $y' > 0$, so y is increasing.

(7) $y'' = 2 - \frac{54}{x^4}$.

$y'' = 0$ for $2x^4 = 54 \implies x^4 = 27 \implies x = \pm\sqrt[4]{27} = \pm 2.28$.

y'' does not exist for $x = 0$.

Interval	Test Number	$f''(x)$	Conclusion
$(-\infty, -2.28)$	-3	$f''(-3) > 0$	concave up
$(-2.28, 0)$	-1	$f''(-1) < 0$	concave down
$(0, 2.28)$	1	$f''(1) < 0$	concave down
$(2.28, \infty)$	3	$f''(3) > 0$	concave up

The inflection points are $(-2.28, 3.46)$ and $(2.28, 3.46)$.

(8)

x	± 3	± 2.28	± 2	± 1.732	± 1
y	8	3.46	$\frac{7}{4}$	0	-8

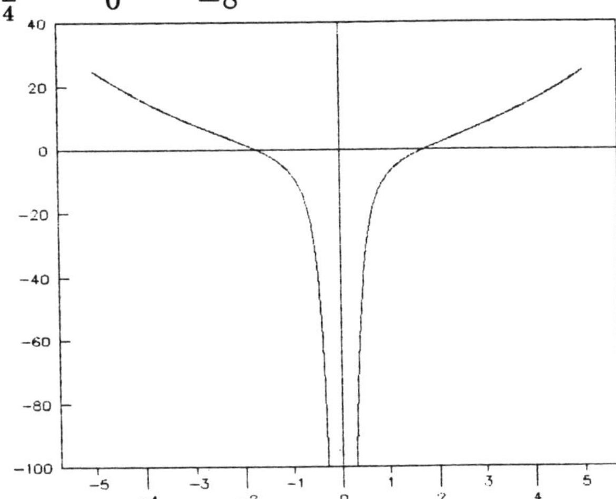

23. $f(x) = x(2x+1)^2$.

(1) The domain of f is all real numbers since $f(x)$ is a polynomial.

(2) f is neither even nor odd.

(3) $f(x) = 0$ for $x = 0$ and $x = -\frac{1}{2}$.

(4,5) There are no asymptotes.

(6) $f'(x) = (1)(2x+1)^2 + (x)(2(2x+1)(2)) = (2x+1)((2x+1)+(4x))$
$= (2x+1)(6x+1) = 12x^2 + 8x + 1.$

$f'(x) = 0$ for $x = -\frac{1}{6}$ and $x = -\frac{1}{2}$.

For $-\infty < x < -\frac{1}{2}$, $f'(x) > 0$, so $f(x)$ is increasing.

For $-\frac{1}{2} < x < -\frac{1}{6}$, $f'(x) < 0$, so $f(x)$ is decreasing.

For $-\frac{1}{6} < x < \infty$, $f'(x) > 0$, so $f(x)$ is increasing.

$f\left(-\frac{1}{2}\right) = 0$ is a relative maximum.

$f\left(-\frac{1}{6}\right) = -\frac{2}{27} = -.0741$ is a relative minimum.

(7) $f''(x) = 24x + 8$.

$f''(x) = 0$ for $x = -\frac{1}{3}$.

For $x < -\frac{1}{3}$, $f''(x) < 0$; the graph of f is concave down on the interval $\left(-\infty, -\frac{1}{3}\right)$.

For $x > -\frac{1}{3}$, $f''(x) > 0$; the graph of f is concave up on the interval $\left(-\frac{1}{3}, \infty\right)$.

The point $\left(-\frac{1}{3}, f\left(-\frac{1}{3}\right) = -\frac{1}{27}\right)$ is an inflection point.

(8)

x	-3	-2	-1	$-\frac{1}{2}$	$-\frac{1}{6}$	0	1	2
$y = f(x)$	-75	-18	-1	0	$-\frac{2}{27}$	0	9	50

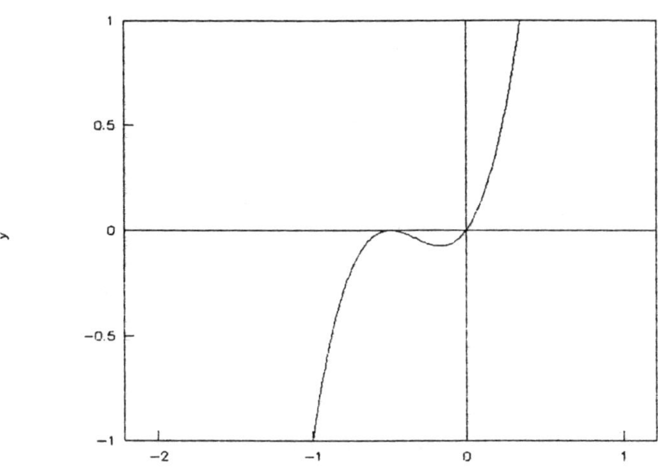

24. $f(x) = (x-3)^{2/3} + 1$.

(1) The domain is all x.

(2) f is neither even nor odd. The straight line $x = 3$ is an axis of symmetry for the graph of f. (3) $f(x) = (x-3)^{2/3} + 1 \geq 1$ for all x.

(4) There is no vertical asymptote.

(5) There is no horizonatal asymptote.

(6) $f'(x) = \frac{2}{3}(x-3)^{-1/3}$.

$f'(x)$ does not exist for $x = 3$.

For $x < 3$, $f'(x) < 0$, so $f(x)$ is decreasing.

For $x > 3$, $f'(x) > 0$, so $f(x)$ is increasing.

$f(3)$ is a relative minimum.

(7) $f''(x) = -\frac{2}{9}(x-3)^{-4/3}$.

$f''(x)$ does not exist for $x = 3$.

For $x < 3$, $f''(x) < 0$, so $f(x)$ is concave down.

For $x > 3$, $f''(x) < 0$, so $f(x)$ is concave down.

There is no inflection point.

(8)

x	0	1	2	3	4	5	6
y	3.08	2.59	2	1	2	2.59	3.08

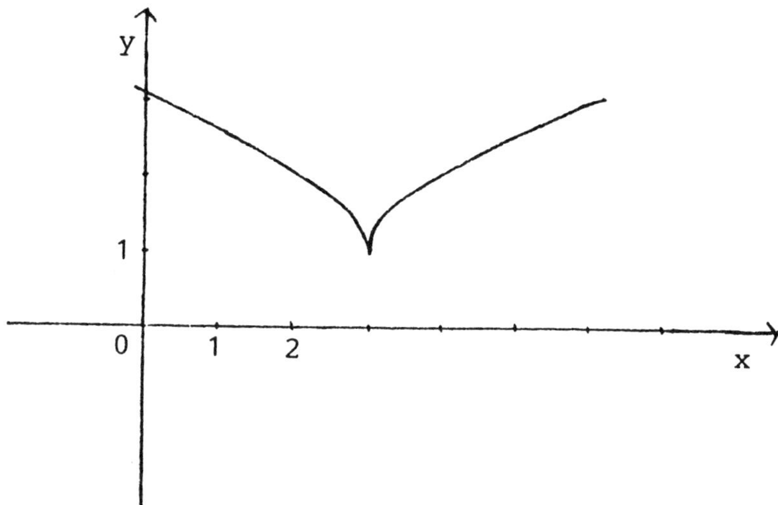

25. $y = \frac{1}{3}x^3 - x^2 - 3x + 4.$

(1) The domain is all x.

(2) f is neither even nor odd.

(3) It is difficult to determine for what x we have $y = 0$.

(4,5) There is no vertical asymptote and no horizontal asymptote.

(6) $y' = x^2 - 2x - 3 = (x - 3)(x + 1)$.

$y' = 0$ for $x = -1$ and $x = 3$.

Interval	Test Number	$f'(x)$	Conclusion
$(-\infty, -1)$	-2	$f'(-2) > 0$	increasing
$(-1, 3)$	0	$f'(0) < 0$	decreasing
$(3, \infty)$	4	$f'(4) > 0$	increasing

$f(-1)$ is a relative maximum and $f(3)$ is a relative minimum.

(7) $y'' = 2x - 2$. $y'' = 0$ for $x = 1$.

Interval	Test Number	$f''(x)$	Conclusion
$(-\infty, 1)$	0	$f''(0) < 0$	concave down
$(1, \infty)$	2	$f''(2) > 0$	concave up

$(1, \frac{1}{3})$ is the inflection point.

(8)

x	-3	-2	-1	0	1	2	3	4	5
y	-5	$\frac{10}{3}$	$\frac{17}{3}$	4	$\frac{1}{3}$	$-\frac{10}{3}$	-5	$-\frac{8}{3}$	$\frac{17}{3}$

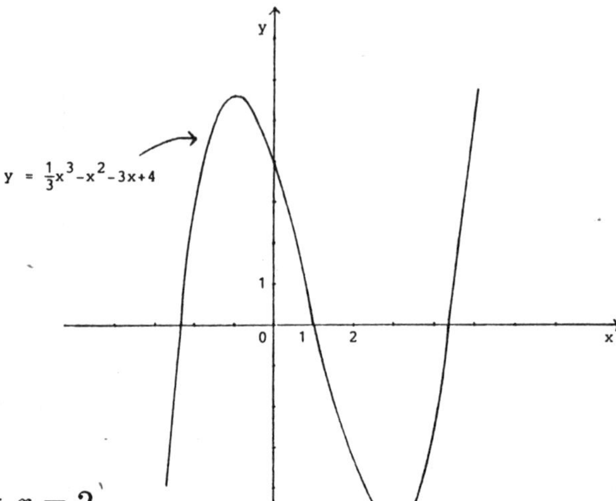

26. $f(x) = \frac{x^2-4x+5}{x-2}$.

(1) The domain is all x except $x = 2$.

(2) f is neither even nor odd.

(3) $f(x)$ has no zeros because $(x^2 - 4x + 5) > 0$ for all x.

(4) The vertical asymptote is $x = 2$.

(5) $\lim_{x\to+\infty} f(x) = +\infty$ and $\lim_{x\to-\infty} f(x) = -\infty$.
There is no horizontal asymptote.

$$\begin{array}{r} x-2 \\ x-2\overline{)x^2-4x+5} \\ x^2-2x \\ \hline -2x+5 \\ -2x+4 \\ \hline 1 \end{array}$$

Therefore

$$f(x) = x - 2 + \frac{1}{x-2}.$$

$$\lim_{x\to\pm\infty} \frac{1}{x-2} = 0.$$

The line $y = x - 2$ is an oblique asymptote.

(6) $f'(x) = \frac{(2x-4)(x-2)-(x^2-4x+5)(1)}{(x-2)^2} = \frac{(2x^2-8x+8)-x^2+4x-5}{(x-2)^2}$
$= \frac{x^2-4x+3}{(x-2)^2} = \frac{(x-3)(x-1)}{(x-2)^2}$.

$f'(x) = 0$ for $x = 3$ and $x = 1$.

$f'(x)$ does not exist for $x = 2$.

Interval	Test Number	$f'(x)$	Conclusion
$(-\infty, 1)$	0	$f'(0) > 0$	increasing
$(1, 2)$	$\frac{3}{2}$	$f'\left(\frac{3}{2}\right) < 0$	decreasing
$(2, 3)$	$\frac{5}{2}$	$f'\left(\frac{5}{2}\right) < 0$	decreasing
$(3, \infty)$	4	$f'(4) > 0$	increasing

$f(1)$ is a relative maximum, $f(3)$ is a relative minimum.

(7) $f''(x) = \frac{(x-2)^2(2x-4)-(x^2-4x+3)(2)(x-2)}{(x-2)^4}$
$= \frac{(x-2)(2x-4)-2(x^2-4x+3)}{(x-2)^3}$
$= \frac{2x^2-8x+8-2x^2+8x-6}{(x-2)^3} = \frac{2}{(x-2)^3}$.

$f''(x)$ does not exist for $x = 2$.

For $x < 2$, $f''(x) < 0$, so $f(x)$ is concave downward.

For $x > 2$, $f''(x) > 0$, so $f(x)$ is concave upward.

There is no inflection point.

(8)

x	-1	0	1	$\frac{3}{2}$	3	$\frac{5}{2}$	5	10
$f(x)$	$-\frac{10}{3}$	$-\frac{5}{2}$	-2	$-\frac{5}{2}$	2	$\frac{5}{2}$	$\frac{10}{3}$	$\frac{65}{8}$

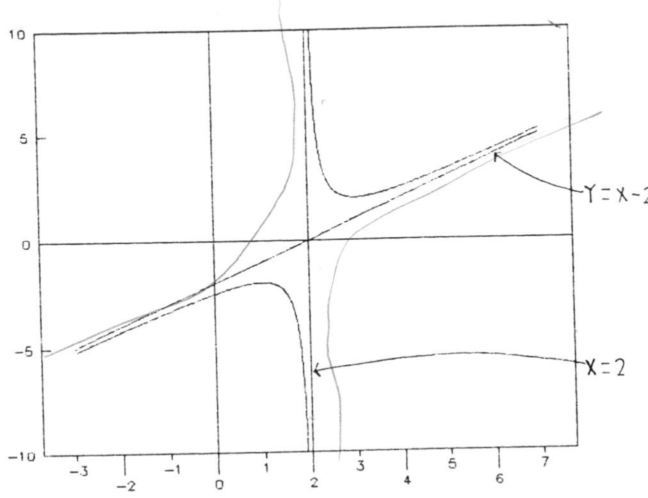

27. $y = f(x) = 4x^2(1 - x^2)$.

(1) The domain is all x.

(2) f is an even function. (3) $f(x) = 0$ for $x = 0$, $x = 1$, $x = -1$.

(4,5) There are no asymptotes.

(6) $f'(x) = 4x^2(-2x) + (1 - x^2)(8x) = -8x^3 + 8x - 8x^3$
$= -16x^3 + 8x = 8x(-2x^2 + 1)$.
$f'(x) = 0$ for $x = 0$, $x = \pm\sqrt{\frac{1}{2}}$.

Interval	Test Number	$f'(x)$	Conclusion
$\left(-\infty, -\sqrt{\frac{1}{2}}\right)$	-1	$f'(-1) > 0$	increasing
$\left(-\sqrt{\frac{1}{2}}, 0\right)$	$-\frac{1}{2}$	$f'\left(-\frac{1}{2}\right) < 0$	decreasing
$\left(0, \sqrt{\frac{1}{2}}\right)$	$\frac{1}{2}$	$f'\left(\frac{1}{2}\right) > 0$	increasing
$\left(\sqrt{\frac{1}{2}}, \infty\right)$	1	$f'(1) < 0$	decreasing

$f\left(-\sqrt{\frac{1}{2}}\right)$ and $f\left(\sqrt{\frac{1}{2}}\right)$ are relative maxima and $f(0)$ is a relative minimum.

(7) $f''(x) = -48x^2 + 8$. $f''(x) = 0$ for $x = \pm\sqrt{\frac{1}{6}}$.

Interval	Test Number	$f''(x)$	Conclusion
$\left(-\infty, -\sqrt{\frac{1}{6}}\right)$	$-\frac{1}{2}$	$f''\left(-\frac{1}{2}\right) < 0$	concave downward
$\left(-\sqrt{\frac{1}{6}}, \sqrt{\frac{1}{6}}\right)$	0	$f''(0) > 0$	concave upward
$\left(\sqrt{\frac{1}{6}}, \infty\right)$	$\frac{1}{2}$	$f''\left(\frac{1}{2}\right) < 0$	concave downward

$\left(-\sqrt{\frac{1}{6}}, \frac{5}{9}\right)$ and $\left(\sqrt{\frac{1}{6}}, \frac{5}{9}\right)$ are inflection points.

(8)

x	$\pm\frac{3}{2}$	± 1	$\pm\sqrt{\frac{1}{2}}$	0
$f(x)$	$-\frac{45}{4}$	0	1	0

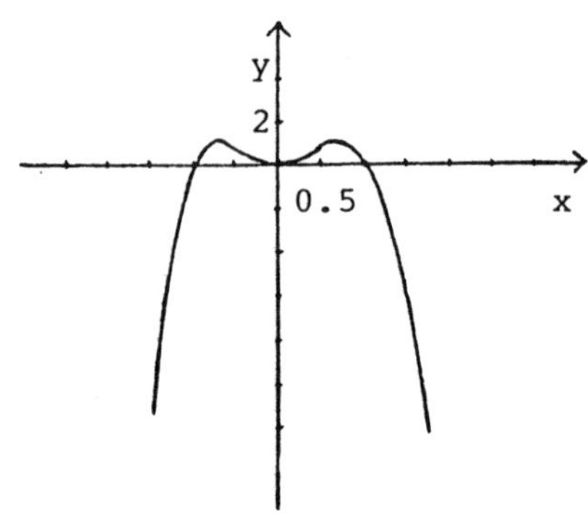

28. $f(x) = \frac{x^2}{9-x^2}$.

(1) The domain is all x except $x = \pm 3$.

(2) f is an even function.

(3) $f(x) = 0$ for $x = 0$.

(4) The vertical asymptotes are $x = 3$, $x = -3$.

(5) Since $\lim_{x \to \pm\infty} f(x) = -1$, the line $y = -1$ is a horizontal asymptote.

(6) $f'(x) = \frac{(9-x^2)\cdot 2x - x^2\cdot(-2x)}{(9-x^2)^2} = \frac{18x - 2x^3 + 2x^3}{(9-x^2)^2} = \frac{18x}{(9-x^2)^2}$.

$f'(x) = 0$ for $x = 0$.

$f'(x)$ does not exist for $x = \pm 3$.

Interval	Test Number	$f'(x)$	Conclusion
$(-\infty, -3)$	-4	$f'(-4) < 0$	decreasing
$(-3, 0)$	-2	$f'(-2) < 0$	decreasing
$(0, 3)$	2	$f'(2) > 0$	increasing
$(3, \infty)$	4	$f'(4) > 0$	increasing

$f(0)$ is a relative minimum.

(7) $f''(x) = \frac{(9-x^2)^2(18) - 18x(2)(9-x^2)(-2x)}{(9-x^2)^4}$

$= \frac{(9-x^2)(18) + 72x^2}{(9-x^2)^3} = \frac{54x^2 + 162}{(9-x^2)^3}$.

Interval	Test Number	$f''(x)$	Conclusion
$(-\infty, -3)$	-4	$f''(-4) < 0$	concave downward
$(-3, 3)$	0	$f''(0) > 0$	concave upward
$(3, \infty)$	4	$f''(4) < 0$	concave downward

There is no inflection point.

(8)

x	± 10	± 4	± 2	± 1	0
$f(x)$	-1.1	-2.29	$\frac{4}{5}$	$\frac{1}{8}$	0

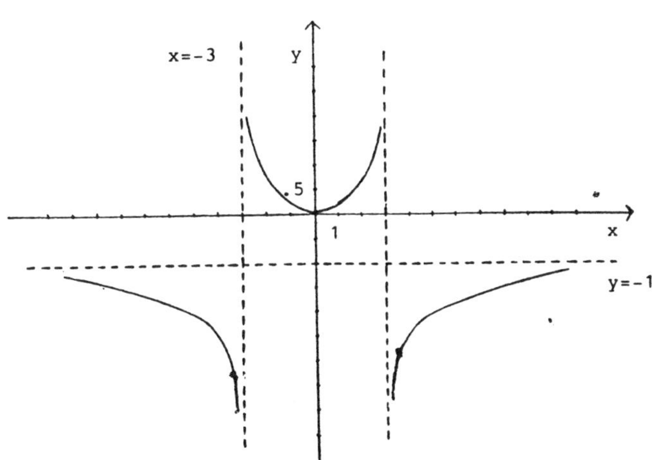

29. $f(x) = 16 - 20x^3 + 3x^5$.

(1) The domain is all x.

(2) This function f is neither even nor odd.

Note that

$$f(-x) - 16 = \left(16 - 20(-x)^3 + 3(-x)^5\right) - 16 = 20x^3 - 3x^5 = -f(x) + 16$$
$$= -\left[f(x) - 16\right].$$

Therefore the graph of f is symmetric with respect to the point (0,16).

(3) It is difficult to find the zeros of $f(x)$. (4,5) There are no asymptotes.

(6) $f'(x) = -60x^2 + 15x^4 = 15x^2(-4 + x^2)$.

$f'(x) = 0$ for $x = 0$, $x = 2$, $x = -2$.

Interval	Test Number	$f'(x)$	Conclusion
$(-\infty, -2)$	-3	$f'(-3) > 0$	increasing
$(-2, 0)$	-1	$f'(-1) < 0$	decreasing
$(0, 2)$	1	$f'(1) < 0$	decreasing
$(2, \infty)$	3	$f'(3) > 0$	increasing

$f(-2)$ is a relative maximum. $f(2)$ is a relative minimum.

(7) $f''(x) = -120x + 60x^3 = 60x(-2 + x^2)$.

$f''(x) = 0$ for $x = 0$, $x = \sqrt{2}$, $x = -\sqrt{2}$.

Interval	Test Number	$f''(x)$	Conclusion
$\left(-\infty, -\sqrt{2}\right)$	-2	$f''(-2) < 0$	concave downward
$\left(-\sqrt{2}, 0\right)$	-1	$f''(-1) > 0$	concave upward
$\left(0, \sqrt{2}\right)$	1	$f''(1) < 0$	concave downward
$\left(\sqrt{2}, \infty\right)$	2	$f''(2) > 0$	concave upward

The inflection points occur when $x = 0$, $x = \sqrt{2}$, $x = -\sqrt{2}$.

(8)

x	-3	-2	-1	0	1	2	3
$f(x)$	-173	80	33	16	-1	-48	205

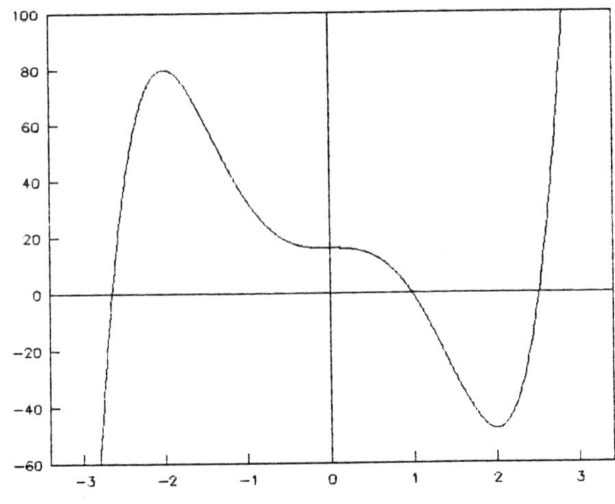

30. $f(x) = x^4 - 8x^2 + 10$.

(1) The domain is all x.

(2) This function f is even.

(3) $f(x) = 0 \implies x^2 = \frac{8 \pm \sqrt{(-8)^2 - (4)(1)(10)}}{(2)(1)} = \frac{8 \pm \sqrt{24}}{2} = \frac{8 \pm 2\sqrt{6}}{2} = 4 \pm \sqrt{6}$

$\implies x^2 = 6.4495$ or $x^2 = 1.5505 \implies x = 2.5396$,

$x = -2.5396$, $x = 1.2452$, $x = -1.2452$.

(4,5) There are no asymptotes.

(6) $f'(x) = 4x^3 - 16x = 4x(x^2 - 4)$.

$f'(x) = 0$ for $x = 0$, $x = 2$, $x = -2$.

Interval	Test Number	$f'(x)$	Conclusion
$(-\infty, -2)$	-3	$f'(-3) < 0$	decreasing
$(-2, 0)$	-1	$f'(-1) > 0$	increasing
$(0, 2)$	1	$f'(1) < 0$	decreasing
$(2, \infty)$	3	$f'(3) > 0$	increasing

$f(-2)$, $f(2)$ are relative minima. $f(0)$ is a relative maximum.

(7) $f''(x) = 12x^2 - 16 = 4(3x^2 - 4)$.

$f''(x) = 0$ for $x = \pm\sqrt{\frac{4}{3}}$.

Interval	Test Number	$f''(x)$	Conclusion
$\left(-\infty, -\sqrt{\frac{4}{3}}\right)$	-2	$f''(-2) > 0$	concave upward
$\left(-\sqrt{\frac{4}{3}}, \sqrt{\frac{4}{3}}\right)$	0	$f''(0) < 0$	concave downward
$\left(\sqrt{\frac{4}{3}}, \infty\right)$	2	$f''(2) > 0$	concave upward

The inflection points occur when $x = -\sqrt{\frac{4}{3}}$, $x = \sqrt{\frac{4}{3}}$.

(8)

x	± 3	± 2	± 1	0
$f(x)$	19	-6	3	10

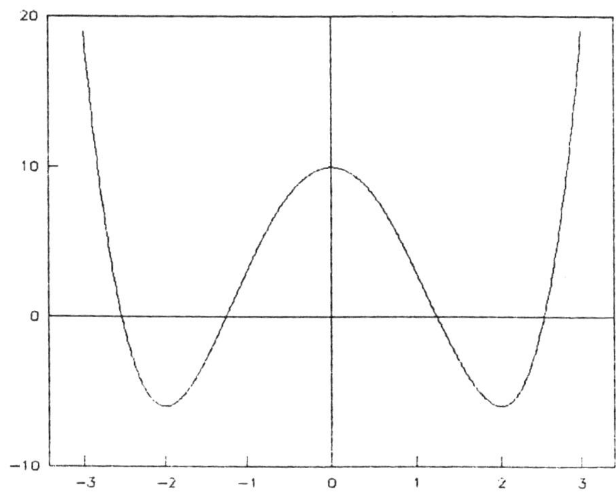

31. The function $f(x) = 9x - \frac{1}{x}$ in Problem 16 is an odd function, but the graph of f does not contain the origin.
32. The function $f(x) = x^2 - \frac{9}{x^2}$ in Problem 22 is an even function, but the graph of f does not contain the origin.
33. Let $g(x) = x^2 + 1$ for all real numbers x. Let $h(x) = \sqrt{g(x)}$ for all real numbers x. For every real number x, $h(-x) = \sqrt{g(-x)} = \sqrt{g(x)} = h(x)$. Therefore h is an even function.
34. Let $h(x) = x^2 + 1$ for all real numbers x. Let $g(x) = [h(x)]^3$.
For every real number x, $g(-x) = [h(-x)]^3 = [h(x)]^3 = g(x)$.
Therefore g is an even function and is not an odd function.
35. Let $g(x)$ and $h(x)$ be two odd functions defined, for simplicity, on the entire x-axis, and let $f(x) = g(x) \cdot h(x)$ for all real numbers x.
For any real number x, $f(-x) = g(-x) \cdot h(-x) = [-g(x)] \cdot [-h(x)] = g(x) \cdot h(x) = f(x)$. Therefore f is an even function.
36. Let $g(x)$ be an odd function defined on the entire x-axis, and let $h(x)$ be an even function also defined on the entire x-axis. Let $f(x) = g(x) \cdot h(x)$ for all real numbers x. For any real number x, $f(-x) = g(-x) \cdot h(-x) = [-g(x)] \cdot h(x) = -[g(x) \cdot h(x)] = -f(x)$. Therefore f is an odd function.
37. Let $f(x) = (x - a)^2 + b$. For any real number h, $f(a + h) = f(a - h)$, so the graph of f is symmetric with respect to the vertical line $x = a$.
38. $f(x) = x^2 - 4x - 2 = (x^2 - 4x + 4) - 6 = (x - 2)^2 - 6$. Therefore as shown in Problem 37 the graph of f is symmetric with respect to the vertical line $x = 2$.
39. $f(x) = x^5 + 2$. Let $g(x) = f(x) - 2 = x^5$. The graph of g is symmetric with respect to the origin, so the graph of f is symmetric with respect to the point (0,2).
40. The graph of a function $y = f(x)$ (not identically zero) cannot be symmetric with respect to the x-axis because for each x in the domain of the function there is a unique corresponding point on the graph of f.

Solutions to Exercise Set 3.6

1. $f(x) = 9 - x^2, \qquad x \text{ in } [-1,3].$
   $f'(x) = -2x.$
   $f'(x) = 0$ for $x = 0.$
   $f(-1) = 9 - (-1)^2 = 8, \quad f(3) = 9 - (3)^2 = 0, \quad f(0) = 9.$
   The maximum value is $f(0) = 9$, and the minimum value is $f(3) = 0$.
2. $f(x) = 9 - x^2 + 8x, \qquad x \text{ in } [0,5].$
   $f'(x) = -2x + 8.$
   $f'(x) = 0$ for $x = 4.$
   $f(0) = 9, \quad f(5) = 24, \quad f(4) = 25.$
   The maximum value is $f(4) = 25$, and the minimum value is $f(0) = 9$.
3. $f'(x) = 2x - 2$. $f'(x) = 0$ for $x = 1$.
   $f(-1) = (-1)^2 - 2(-1) + 3 = 6$, $f(1) = 1^2 - 2 \cdot 1 + 3 = 2$,
   and $f(3) = 9 - 6 + 3 = 6$.
   The maximum value of f is $f(-1) = f(3) = 6$ and the minimum value is $f(1) = 2$.
4. $f(x) = 7 - x^3 + 3x, \qquad x \text{ in } [-2,2].$
   $f'(x) = -3x^2 + 3 = 3(-x^2 + 1).$
   $f'(x) = 0$ for $x = 1, \quad x = -1.$
   $f(-2) = 9, \quad f(2) = 5, \quad f(-1) = 5, \quad f(1) = 9.$
   The maximum value is $f(-2) = f(1) = 9$, and the minimum value is $f(2) = f(-1) = 5$.
5. $f(x) = 3 + x + x^2 + 2x^3, \qquad x \text{ in } [-2,3].$
   $f'(x) = 1 + 2x + 6x^2.$
   $f'(x) > 0$ for all x.
   $f(-2) = -11, \qquad f(3) = 69.$
   The maximum value if $f(3) = 69$, and the minimum value is $f(-2) = -11$.
6. $f'(x) = 3 - 3x^2 = 3(1 - x^2)$. $f'(x) = 0$ for $x = \pm 1$.
   $f(-1) = -3 + 1 = -2$, $f(1) = 3 - 1 = 2$.
   The maximum value of f is $f(1) = 2$ and the minimum value is $f(-1) = -2$.

7. $f'(x) = 2x(x-1) + x^2 = x(3x-2)$. $f'(x) = 0$ for $x = 0$ and $x = \frac{2}{3}$.
$f(0) = 0$, $f\left(\frac{2}{3}\right) = \frac{4}{9}\left(-\frac{1}{3}\right) = -\frac{4}{27}$, and $f(3) = 9 \cdot 2 = 18$.
The maximum value of f is $f(3) = 18$ and the minimum value is $f\left(\frac{2}{3}\right) = -\frac{4}{27}$.

8. If $x > 2$, $f(x) = x - 2$, and $f'(x) = 1$.
If $x < 2$, $f(x) = 2 - x$, and $f'(x) = -1$.
For the case $x = 2$, $f'(x)$ does not exist. $x = 2$ is the only critical number.
$f(0) = 2$, $f(2) = 0$, and $f(5) = 3$.
The maximum value of f is $f(5) = 3$ and the minimum value is $f(2) = 0$.

9. $f'(x) = 3x^2 - 4x$. $f'(x) = 0$ for $x = 0$ and $x = \frac{4}{3}$.
$f(-1) = (-1)^3 - 2(-1)^2 = -3$, $f(0) = 0$, $f\left(\frac{4}{3}\right) = \left(\frac{4}{3}\right)^3 - 2\left(\frac{4}{3}\right)^2 = -\frac{32}{27}$, $f(2) = 0$.
The maximum value of f is $f(0) = f(2) = 0$ and the minimum value is $f(-1) = -3$.

10. $f(x) = 5x^3 - x^5, \quad x$ in $[-1, 1]$.
$f'(x) = 15x^2 - 5x^4 = 5x^2(3 - x^2)$.
$f'(x) = 0$ for $x = 0, \quad x = \sqrt{3}, \quad x = -\sqrt{3}$.
$f(-1) = -4, \quad f(1) = 4, \quad f(0) = 0$.
The maximum value is $f(1) = 4$, and the minimum value is $f(-1) = -4$.

11. $f'(x) = 1 - \frac{1}{x^2}$. $f'(x) = 0$ for $x = \pm 1$.
$f\left(\frac{1}{2}\right) = \frac{5}{2}$, $f(1) = 2$, and $f(2) = \frac{5}{2}$.
The maximum value of f is $f\left(\frac{1}{2}\right) = f(2) = \frac{5}{2}$ and the minimum value is $f(1) = 2$.

12. $f'(x) = 2(\sqrt{x} + x)\left(\frac{1}{2\sqrt{x}} + 1\right)$.
The only critical number for f is $x = 0$.
$f(0) = 0$, and $f(4) = 36$.
The maximum value of f is $f(4) = 36$ and the minimum value is $f(0) = 0$.

13. $f'(x) = \frac{2}{3x^{1/3}}$.
The only critical number for f is $x = 0$.
$f(0) = 2$. There is no maximum value for $f(x)$ since $f(x)$ is increasing.
The minimum value is $f(0) = 2$.

14. $f'(x) = \frac{-2}{3x^{1/3}}$.

The only critical number for f is $x = 0$.

$f(-1) = 2$, $f(0) = 3$, $f(8) = -1$.

The maximum value is $f(0) = 3$ and the minimum value is $f(8) = -1$.

15. $f'(x) = \frac{4-2x}{(x^2-4x)^2}$. $f'(x) = 0$ for $x = 2$.

$f(2) = -\frac{1}{4}$, $f(1) = -\frac{1}{3}$, and $f(3) = -\frac{1}{3}$.

The maximum value of f is $f(2) = -\frac{1}{4}$, and the minimum value is

$f(1) = f(3) = -\frac{1}{3}$.

16. $f'(x) = \frac{-2}{(x-1)^2}$.

There are no critical numbers for f.

$f(-3) = \frac{-2}{-4} = \frac{1}{2}$, $f(0) = -1$.

The maximum value of f is $f(-3) = \frac{1}{2}$ and the minimum value is $f(0) = -1$.

17. $f(x) = 2x^3 - 15x^2 + 24x + 10, \qquad x \text{ in } [0, 4]$.

$f'(x) = 6x^2 - 30x + 24 = 6(x^2 - 5x + 4) = 6(x-1)(x-4)$.

$f'(x) = 0$ for $x = 1, \quad x = 4$.

$f(0) = 10, \quad f(4) = -6, \quad f(1) = 21$.

The maximum value is $f(1) = 21$, and the minimum value is $f(4) = -6$.

18. $f'(x) = \frac{8}{3}x^{-2/3} - \frac{8}{3}x^{1/3} = \frac{8}{3}x^{-2/3}(1-x)$.

$f'(x) = 0$ for $x = 1$ and $f'(x)$ does not exist for $x = 0$.

$f(-1) = -10$, $f(0) = 0$, $f(1) = 6$, $f(8) = -16$.

The maximum value of f is $f(1) = 6$ and the minimum value is $f(8) = -16$.

19. $f'(x) = \left(\sqrt{x} - x^{3/2}\right)' = \frac{1}{2}x^{-1/2} - \frac{3}{2}x^{1/2} = \frac{1}{2}x^{-1/2}(1-3x)$.

$f'(x) = 0$ for $x = \frac{1}{3}$ and $f'(x)$ does not exist for $x = 0$.

$f(0) = 0$, $f\left(\frac{1}{3}\right) = \sqrt{\frac{1}{3}} \cdot \frac{2}{3} = \frac{2\sqrt{3}}{9}$, $f(4) = -6$.

The maximum value of f is $f\left(\frac{1}{3}\right) = \frac{2\sqrt{3}}{9}$ and the minimum value is $f(4) = -6$.

20. $f'(x) = 15x^4 - 15x^2 = 15x^2(x^2 - 1)$.

$f'(x) = 0$ for $x = 0$, $x = -1$, and $x = 1$.

$f(-2) = -56$, $f(-1) = 2$, $f(0) = 0$, $f(1) = -2$, $f(2) = 56$.

The maximum value of f is $f(2) = 56$

and the minimum value of f is $f(-2) = -56$.

21. $P(t) = 100{,}000 + 48t^{3/2} - 4t^2$, t in $[0, \infty)$.
 $P'(t) = 72t^{1/2} - 8t = 8t^{1/2}(9 - t^{1/2})$. $P'(t) = 0$ for $t = 0$, $t = 81$.
 On $(0, 81)$, $P'(t) > 0$, $P(t)$ is increasing.
 On $(81, \infty)$, $P'(t) < 0$, $P(t)$ is decreasing.
 Therefore the maximum population occurs when $t = 81$ which is the year 2066.
22. Let $R(x)$ = revenue from producing x pairs of shoes per week.
 Let $P(x)$ = profit from producing x pairs of shoes per week.
 Then $R(x) = 60x$ and $P(x) = R(x) - C(x) = -400 + 40x - x^2$.
 $P'(x) = 40 - 2x$. $P'(x) = 0$ for $x = 20$.
 For $0 < x < 20$, $P'(x) > 0$, $P(x)$ is increasing; and for $20 < x < \infty$, $P'(x) < 0$, $P(x)$ is decreasing. So the maximum profit occurs when $x = 20$.
23. $P'(x) = -96 + 36x - 3x^2 = -3(x-4)(x-8)$.
 $P'(x) = 0$ for $x = 4$, $x = 8$.
 $P(0) = 160$, $P(4) = 0$, $P(8) = 32$, $P(10) = 0$.
 The maximum value of P is $P(0) = 160$ which occurs at $x = 0$, and the minimum value is $P(4) = P(10) = 0$ which occurs at $x = 4$ and $x = 10$.
24. $P'(x) = -40(3x - 27)^{-1/3} \cdot 3 = -120(3x - 27)^{-1/3}$.
 $P'(x)$ does not exist for $x = 9$.
 $P(0) = -40$, $P(9) = 500$, $P(15) = 87.9$
 The maximum value of P is $P(9) = 500$ and the minimum value is $P(0) = -40$.
25. Let $R(p)$ = revenue from the sales per week at price p.
 $R(p) = xp = (400 - 4p)p = 400p - 4p^2$, p in $[0, 100]$.
 $R'(p) = 400 - 8p$.
 $R'(p) = 0$ for $p = 50$.
 $R(0) = 0$, $R(50) = 10{,}000$, $R(100) = 0$.
 The maximum revenue of 10,000 occurs when $p = 50$.
26. a. The peak demand of $E(t) = 4$ which occurs when $t = 13$ (1:00 p.m.).
 b. $R(t) = E(t) - S(t) = 3 - 2\left(\frac{t-13}{7}\right)^4 + \left(\frac{t-13}{7}\right)^2$.
 $R'(t) = -8\left(\frac{t-13}{7}\right)^3 + 2\left(\frac{t-13}{7}\right) = -2\left(\frac{t-13}{7}\right)\left[4\left(\frac{t-13}{7}\right)^2 - 1\right]$.
 $R'(t) = 0$ for $t = 13$, $\frac{t-13}{7} = \pm\frac{1}{2} \implies t = \frac{19}{2}$, $t = \frac{33}{2}$.
 $R(6) = 2$, $R\left(\frac{19}{2}\right) = \frac{25}{8}$, $R(13) = 3$, $R\left(\frac{33}{2}\right) = \frac{25}{8}$.

The resulting peak demand of $R(t) = \frac{25}{8}$ which occurs when $t = \frac{19}{2}$ (9:30 a.m.) and $t = \frac{33}{2}$ (4:30 p.m.).

c. The peak demand can be reduced by the amount (peak demand in part a.) - (peak demand in part b.) $= 4 - \frac{25}{8} = \frac{7}{8}$.

Solutions to Exercise Set 3.7

1. $p(t) = \sqrt{24 + 10t - t^2}$, t in $[0, 10]$.
   $p'(t) = \frac{1}{2}(24 + 10t - t^2)^{-1/2}(10 - 2t) = \frac{5-t}{(24+10t-t^2)^{1/2}}$.
   $p'(t) = 0$ for $t = 5$.
   $p(0) = \sqrt{24}$, $p(5) = 7$, $p(10) = \sqrt{24}$.
   Therefore this percentage will be the largest when $t = 5$ which is the year 1990.
2. Average cost $c(x) = \frac{C(x)}{x} = \frac{100}{x} + 5 + 0.02x$ for $0 \le x < \infty$.
   $c'(x) = -\frac{100}{x^2} + 0.02 = \frac{0.02x^2 - 100}{x^2}$.
   $c'(x) = 0$ for $x^2 = 5000$ or $x = 50\sqrt{2}$.
   For $0 < x < 50\sqrt{2}$, $c'(x) < 0$, $c(x)$ is decreasing.
   For $50\sqrt{2} < x < \infty$, $c'(x) > 0$, $c(x)$ is increasing.
   Therefore average cost per item is a minimum for $x = 50$ hairdryers.
3. $R'(t) = 0.08 - 0.04t = 0.04(2 - t)$.
   $R'(t) = 0$ for $t = 2$.
   For $0 < t < 2$, $R'(t) > 0$, $R(t)$ is increasing.
   For $2 < t < \infty$, $R'(t) < 0$, $R(t)$ is decreasing.
   Therefore efficiency is maximum when $t = 2$ hours.
4. Let two nonnegative numbers be x and y. Then $y = 10 - x$. The product $P = xy = x(10 - x) = 10x - x^2$.
   $P'(x) = 10 - 2x$. $P'(x) = 0$ for $x = 5 \implies y = 10 - 5 = 5$.
   Two numbers are 5 and 5.

5. Let two nonnegative numbers be x and y.

$x + y = 36$, so $y = 36 - x$.

$S(x) = x + (36 - x)^2 = x + 1296 - 72x + x^2 = x^2 - 71x + 1296.$

$S'(x) = 2x - 71 = 0$ for $x = 35.5$.

$S(0) = 1296$, $S(35.5) = 35.75$, $S(36) = 36$.

The maximum of $S(x)$ occurs when $x = 0$, so the numbers are 0 and 36.

6. $y + 2x = 20$, so $y = 20 - 2x$.

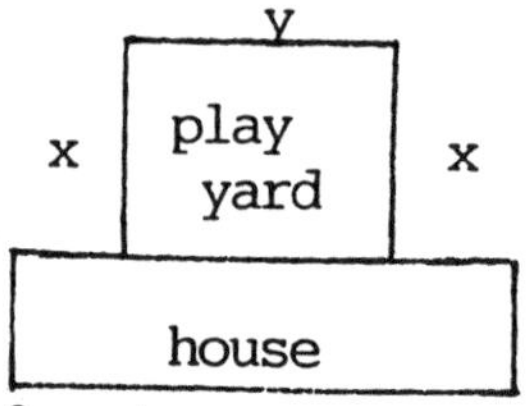

The area of the play yard $f(x) = x \cdot y = x(20 - 2x) = 20x - 2x^2$.

$f'(x) = 20 - 4x = 0$ for $x = 5$.

$f(0) = 0$, $f(5) = 50$, $f(10) = 0$.

The maximum area occurs when $x = 5$ meters in which case $y = 10$ meters.

7. Let N be the number of people in the population.

Let $I(t)$ be the number of people infected after time t.

Let $R(t)$ be the rate at which the disease is spreading after time t.

There is a positive constant k such that

$$\begin{aligned}
R(t) &= k \cdot I(t) \cdot \{N - I(t)\}. \\
R'(t) &= k \cdot \{I'(t) \cdot [N - I(t)] + I(t) \cdot [-I'(t)]\} \\
&= k\{I'(t)N - 2 \cdot I'(t)I(t)\} \\
&= k \cdot I'(t) \cdot \{N - 2 \cdot I(t)\}.
\end{aligned}$$

$$R'(t) = 0 \quad \text{for} \quad N - 2 \cdot I(t) = 0 \implies I(t) = \frac{N}{2}.$$

The disease is spreading most rapidly when $I(t) = \frac{N}{2}$, because

$R'(t) > 0$ when $I(t) < \frac{N}{2}$ and $R'(t) < 0$ when $I(t) > \frac{N}{2}$,

and because $I(t)$ increases as t increases.

8. $3\ell + 3\ell + 3w = 120$, $6\ell + 3w = 120$, $w = 40 - 2\ell$.

The area of the rectangular pen $A(\ell) = 3\ell \cdot w = 3\ell(40 - 2\ell) = 120\ell - 6\ell^2$.

$A'(\ell) = 120 - 12\ell = 0$ for $\ell = 10$.

For $0 < \ell < 10$, $A'(\ell) > 0$. For $10 < \ell < 20$, $A'(\ell) < 0$.

So the maximum area occurs when $\ell = 10$ meters, in which case $w = 20$ meters.

9. $f(x) = x^3 - 9x^2 + 7x - 6.$

   $f'(x) = 3x^2 - 18x + 7$

   $f''(x) = 6x - 18 = 0$ for $x = 3.$

   $f'(1) = -8,\ f'(3) = -20,\ f'(4) = -17.$

   The maximum slope if $f'(1) = -8$. The minimum slope is $f'(3) = -20$.

10. Average cost $c(x) = \frac{C(x)}{x} = \frac{16+4x+x^2}{x}.$

    $c'(x) = \frac{(4+2x)x-1(16+4x+x^2)}{x^2} = \frac{x^2-16}{x^2}.$

    $c'(x) = 0$ for $x = 4.$

    For $0 < x < 4$, $c'(x) < 0$, $c(x)$ is decreasing.

    For $4 < x < \infty$, $c'(x) > 0$, $c(x)$ is increasing.

    Therefore average cost per item is a minimum for $x = 4$.

11. $R(p) = p \cdot x = p(2000 - 100p) = 2000p - 100p^2.$

    $R'(p) = 2000 - 200p = 0$ for $p = 10.$

    $R(0) = 0,\ R(10) = 10,000,\ R(20) = 0.$

    The maximum revenue occurs for $p = 10$.

12. $C(p) = 500 + 10(2000 - 100p) = 500 + 20,000 - 1000p = 20,500 - 1000p.$

    $$P(p) = R(p) - C(p) = 2000p - 100p^2 - 20,500 + 1000p$$
    $$= -100p^2 + 3000p - 20,500.$$

    $P'(p) = -200p + 3000 = 0$ for $p = 15.$

    $P(0) = -20,500,\ P(15) = 2000,\ P(20) = -500.$

    The maximum profit occurs for $p = 15$.

13. Let p be the daily rental fee in dollars.

    Let x be the number of daily rentals of jackhammers per year.

    Let R be the revenue in dollars.

$$x = 500 - 10(p - 30) = 800 - 10p.$$

$$R = x \cdot p = (800 - 10p)p = 800p - 10p^2.$$

$$R' = 800 - 20p = 0 \quad \text{for} \quad p = 40.$$

$$R(30) = 15,000, \quad R(40) = 16,000, \quad R(80) = 0.$$

The maximum revenue is for $p = \$40$.

14. The length of the open box $= 20 - 2x$.

The width of the open box $= 20 - 2x$.

The height of the open box $= x$.

The volume $V(x) = (20 - 2x)(20 - 2x)(x) = x(400 - 40x - 40x + 4x^2)$

$= 400x - 40x^2 - 40x^2 + 4x^3 = 400x - 80x^2 + 4x^3$.

$V'(x) = 400 - 160x + 12x^2 = 4(100 - 40x + 3x^2) = 4(3x - 10)(x - 10)$.

$V'(x) = 0$ for $x = \frac{10}{3}$, $x = 10$.

$V(0) = 0$, $V\left(\frac{10}{3}\right) > 0$, $V(10) = 0$.

The maximum volume occurs at $x = \frac{10}{3}$ cm.

15. $\frac{6-x}{y} = \frac{6}{8} \implies 6y = 48 - 8x$, $y = 8 - \frac{4}{3}x$.

The area of the rectangle $A(x) = x \cdot y = x\left(8 - \frac{4}{3}x\right) = 8x - \frac{4}{3}x^2$.

$A'(x) = 8 - \frac{8}{3}x = 0$ for $x = 3$.

$A(0) = 0$, $A(3) > 0$, $A(6) = 0$.

The maximum area occurs when $x = 3$ cm and $y = 4$ cm.

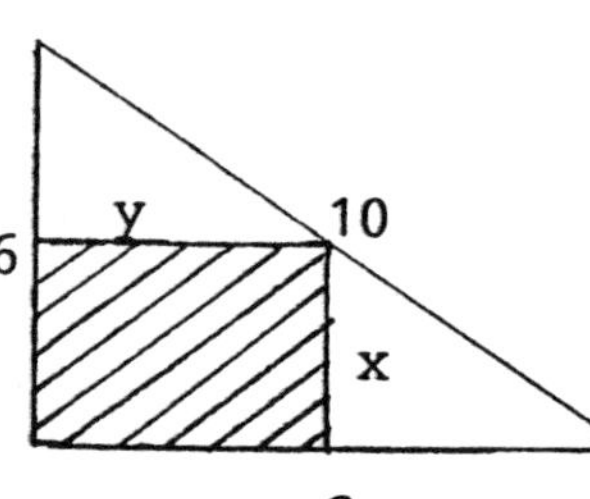

16. $w + 2h = 10$, $w = 10 - 2h$.

$2\ell + 2h = 16$, $\ell = 8 - h$.

Volume $V(h) = w \cdot \ell \cdot h$

$= (10 - 2h)(8 - h)h$

$= (80 - 10h - 16h + 2h^2)h$

$= 80h - 26h^2 + 2h^3$.

$V'(h) = 80 - 52h + 6h^2 = 2(3h - 20)(h - 2)$.

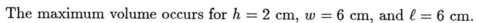

$V'(h) = 0$ for $h = 2$.

$V(0) = 0$, $V(2) > 0$, $V(5) = 0$.

The maximum volume occurs for $h = 2$ cm, $w = 6$ cm, and $\ell = 6$ cm.

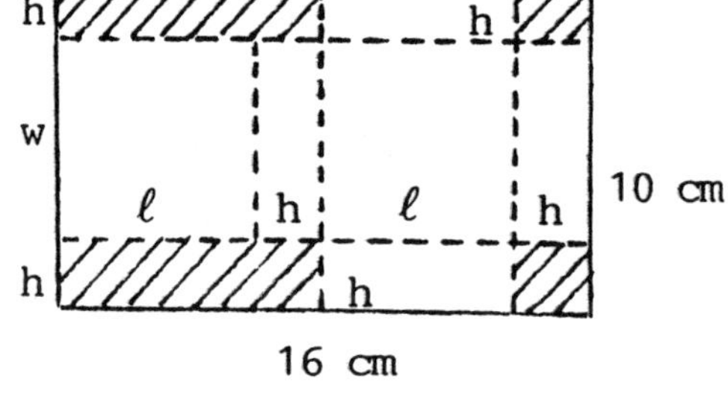

17. (Area of window) $= w \cdot h + \frac{1}{2}r^2\pi$.

$r = \frac{w}{2} \implies$ (Area of window) $= wh + \frac{1}{2}\pi\left(\frac{w}{2}\right)^2$

$= wh + \frac{1}{8}\pi w^2$.

$2h + w + \frac{1}{2}w\pi = 10 \implies 4h + (2 + \pi)w = 20$.

$h = \frac{20-(2+\pi)w}{4}$.

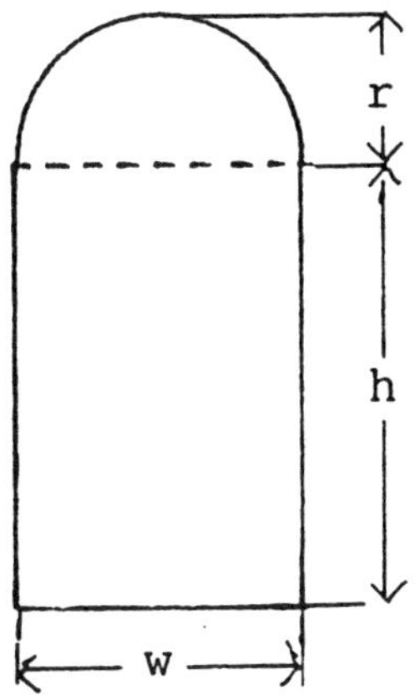

Therefore, (Area of window) $= w \cdot \left(\frac{20-(2+\pi)w}{4}\right) + \frac{1}{8}\pi w^2$

$= 5w - \frac{(2+\pi)}{4}w^2 + \frac{1}{8}\pi w^2$

$= w^2\left(\frac{\pi}{8} - \frac{1}{2} - \frac{\pi}{4}\right) + 5w$

$= w^2\left(-\frac{1}{2} - \frac{\pi}{8}\right) + 5w = f(w).$

$f'(w) = \left(-1 - \frac{\pi}{4}\right)w + 5 = 0$ for $w = \frac{5}{1+\frac{\pi}{4}} = \frac{20}{4+\pi}$.

$f'(w) > 0$ for $0 < w < \frac{20}{4+\pi}$, and $f'(w) < 0$ for $\frac{20}{4+\pi} < w < \frac{20}{2+\pi}$.

So maximum area occurs when $w = \frac{20}{(4+\pi)}$ and

$$h = \left(20 - (2+\pi)\cdot\frac{20}{(4+\pi)}\right)\cdot\frac{1}{4} = 5\left(1 - \frac{2+\pi}{4+\pi}\right) = \frac{10}{4+\pi} \text{ meters.}$$

18. $x =$ (daily room rate).

$y =$ (the number of rooms rented per day).

$R =$ (daily revenue).

$y = 200 - 4\cdot(x-40) = 200 - 4x + 160 = 360 - 4x.$

$R = x\cdot y = x\cdot(360-4x) = -4x^2 + 360x.$

$\frac{dR}{dx} = -8x + 360 = 0$ for $x = 45$.

$\frac{dR}{dx} > 0$ for $40 < x < 45$ and $\frac{dR}{dx} < 0$ for $45 < x < 90$.

So the maximum revenue occurs when daily room rate is \$45.

19. $f(t) = \frac{6t}{t^2+2t+1}$, $t > 0$.

$f'(t) = \frac{6(t^2+2t+1)-6t(2t+2)}{(t^2+2t+1)^2} = \frac{6(t+1)-12t}{(t+1)^3} = \frac{-6t+6}{(t+1)^3} = \frac{-6(t-1)}{(t+1)^3}$.

So $f'(t) = 0$ for $t = 1$.

Interval	Test Number	$f'(t)$	Conclusion
$(0,1)$	$\frac{1}{2}$	$f'\left(\frac{1}{2}\right) > 0$	increasing
$(1,\infty)$	2	$f'(2) < 0$	decreasing

Thus, $f(t)$ has a maximum when $t = 1$ minute.

20. $U(x) = \frac{6x-x^2}{10x^2-60x+100}$, $x > 0$.

$U'(x) = \frac{(6-2x)(10x^2-60x+100)-(6x-x^2)(20x-60)}{100(x^2-6x+10)^2}$

$= \frac{(6-2x)(x^2-6x+10)-(6x-x^2)(2x-6)}{10(x^2-6x+10)^2}$

$= \frac{(6-2x)(x^2-6x+10+6x-x^2)}{10(x^2-6x+10)^2}$

$= \frac{2(3-x)}{(x^2-6x+10)^2} = 0$ for $x = 3$.

$U'(x) > 0$ for $0 < x < 3$ and $U'(x) < 0$ for $3 < x$.

So $U(x)$ has a maximum for $x = 3$.

Hence the saturation quantity for $U(x)$ is 3.

21. $U(x) = 2 - (x-8)^{2/3}$, $x \geq 0$.

$U'(x) = -\frac{2}{3}(x-8)^{-1/3} = \frac{-2}{3\sqrt[3]{x-8}}$.

$U'(x) > 0$ for $0 \leq x < 8$, and $U'(x) < 0$ for $x > 8$.

Hence $U(x)$ has a maximum when $x = 8$.

22. Let D be the length of the route.

Number of hours truck travels is $\frac{D}{\nu}$.

The cost of making the trip is $(C(\nu) + A) \cdot \frac{D}{\nu} = f(\nu)$.

$f(\nu) = (k\nu^{3/2} + A) \cdot \frac{D}{\nu} = \left(k\nu^{1/2} + \frac{A}{\nu}\right) \cdot D$.

$f'(\nu) = \left(k \cdot \frac{1}{2\sqrt{\nu}} - \frac{A}{\nu^2}\right) \cdot D = \left(\frac{k\nu^{3/2}-2A}{2\nu^2}\right) \cdot D = 0$ for $\nu = \left(\frac{2A}{k}\right)^{2/3}$.

$f'(\nu) < 0$ for $\nu < \left(\frac{2A}{k}\right)^{2/3}$ and $f'(\nu) > 0$ for $\nu > \left(\frac{2A}{k}\right)^{2/3}$.

So $f(\nu)$ has a minimum at $\nu = \left(\frac{2A}{k}\right)^{2/3}$.

23.

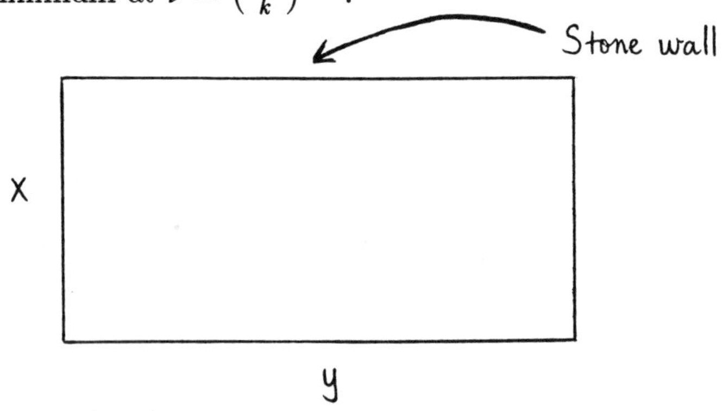

Stone wall

x

y

The cost $C = 50y + 2(20x) + 20y = 70y + 40x$.

$x \cdot y = 800 \implies y = \frac{800}{x}$.
$C = 70\left(\frac{800}{x}\right) + 40x = \frac{56000}{x} + 40x$.
$\frac{dC}{dx} = \frac{-56000}{x^2} + 40 \implies x^2 = 1400 \implies x = 10\sqrt{14}$.
For $0 < x < 10\sqrt{14}$, $\frac{dC}{dx} < 0$, so C decreases as x increases.
For $x > 10\sqrt{14}$, $\frac{dC}{dx} > 0$, so C increases as x increases.
Therefore the minimum cost C occurs when $x = 10\sqrt{14}$ and $y = \frac{800}{10\sqrt{14}} = \frac{800\sqrt{14}}{140} = \frac{40\sqrt{14}}{7}$.

24. Let p denote the price of a ticket, and let x denote the corresponding number of tickets sold.

$$x = mp + b.$$
$$m = -100.$$
$$4,000 = (-100)20 + b \implies b = 6,000.$$
$$x = -100p + 6,000.$$

The revenue $R = px = -100p^2 + 6000p$.
$\frac{dR}{dp} = -200p + 6000$.
$\frac{dR}{dp} = 0$ for $p = 30$.
For $p < 30$, $\frac{dR}{dp} > 0$, so R increases as p increases.
For $30 < p < 60$, $\frac{dR}{dp} < 0$, so R decreases as p increases.
Therefore the maximum revenue occurs when the price per ticket is \$30.

25. Let P denote the annual profit for the bank.
$P = .10(50,000r) - (100,000r)r = -100,000r^2 + 5,000r$.
$\frac{dP}{dr} = -200,000r + 5,000$.
$\frac{dP}{dr} = 0$ for $r = .025$.
For $0 < r < .025$, $\frac{dP}{dr} > 0$, so P increases as r increases.
For $r > .025$, $\frac{dP}{dr} < 0$, so P decreases as r increases.
Therefore the profit on P is maximized when the interest rate paid on savings deposits is 2.5% per year.

26. Again let P denote the annual profit for the bank.

$P = .10(100,000r) - (100,000r)r = -100,000r^2 + 10,000r.$

$\frac{dP}{dr} = -200,000r + 10,000.$

$\frac{dP}{dr} = 0$ for $r = .05.$

By proceeding as in Problem 25, we can show that the profit P is maximized when the interest rate paid on savings deposits is 5% per year.

27. The cost of enrolling n students in the summer program is

$$C(n) = 3n^2 + 2,000n + 7,500.$$

The average cost per child is

$$c(n) = \frac{C(n)}{n} = 3n + 2,000 + \frac{7,500}{n}.$$

$$c'(n) = 3 - \frac{7500}{n^2}.$$

$$c'(n) = 0 \text{ for } \frac{7500}{n^2} = 3 \implies n^2 = 2500 \implies n = 50.$$

For $0 < n < 50$, $c'(n) < 0$, so $c(n)$ decreases as n increases.

For $n > 50$, $c'(n) > 0$, so $c(n)$ increases as n increases.

Therefore the smallest average cost per child occurs when the number of children is 50.

28. The revenue $R = p \cdot n = 5,000n - 27n^2$.

The profit $P = R - C = (5,000n - 27n^2) - (3n^2 + 2,000n + 7,500)$

$= -30n^2 + 3,000n - 7,500.$

$\frac{dP}{dn} = -60n + 3,000.$

$\frac{dP}{dn} = 0$ for $n = 50.$

For $0 < n < 50$, $\frac{dP}{dn} > 0$, so P increases as n increases.

For $n > 50$, $\frac{dP}{dn} < 0$, so P decreases as n increases.

Therefore the maximum profit occurs when the number of children is 50.

29. The revenue R is given by

$$R = p \cdot x = 10x - .002x^2.$$

The profit P is given by

$$P = R - C = (10x - .002x^2) - (2,000 + 2x)$$
$$= -.002x^2 + 8x - 2,000.$$
$$\frac{dP}{dx} = -.004x + 8.$$
$$\frac{dP}{dx} = 0 \text{ for } x = 2,000.$$

For $0 < x < 2,000$, $\frac{dP}{dx} > 0$, so P increases as x increases.

For $2,000 < x < 5,000$, $\frac{dP}{dx} < 0$, so P decreases as x increases.

Therefore the maximum profit occurs when $2,000$ sets of cards are produced.

30. 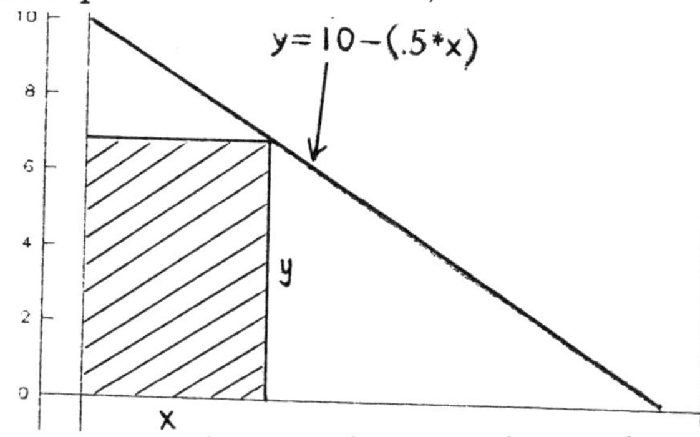

Suppose the rectangle is in the first quadrant as shown above.

The area of this rectangle is $A = x \cdot y = 10x - \frac{1}{2}x^2$.

$\frac{dA}{dx} = 10 - x$.

$\frac{dA}{dx} = 0$ for $x = 10$.

For $0 < x < 10$, $\frac{dA}{dx} > 0$, so A increases as x increases.

For $10 < x < 20$, $\frac{dA}{dx} < 0$, so A decreases as x increases.

Therefore the maximum area occurs when $x = 10$.

This maximum area is $10(10) - \frac{1}{2}(10)^2 = 50$.

31. Let x be the number of additional trees per acre.

Let y be the yield per tree $\Longrightarrow y = 300 - 5x$.

(Yield per acre) $= (x + 30)y = (x + 30)(300 - 5x) = f(x)$.

$f(x) = (30 + x)(300 - 5x) = 9000 + 150x - 5x^2$.

$f'(x) = 150 - 10x$. $f'(x) = 0$ for $x = 15$.

$f(0) = 9000$, $f(15) = 10,125$, $f(60) = 0$.

So, 15 additional trees per acre should be planted to maximize yield.

32. $d^2 + w^2 = 20^2$.

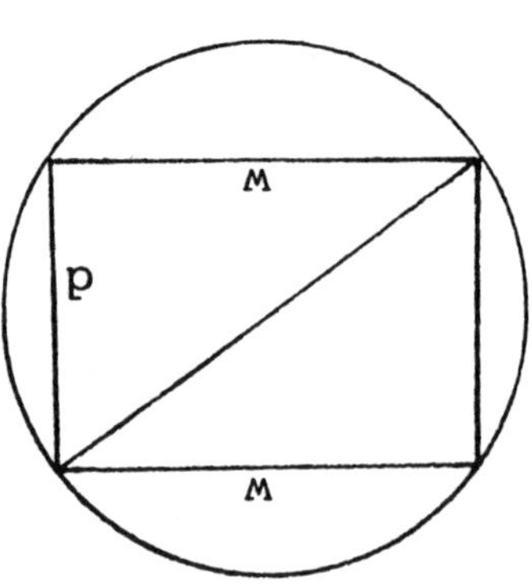

$d^2 = 400 - w^2$.

Strength $= S = k \cdot w \cdot d^2 = kw(400 - w^2)$

$= k(400w - w^3)$.

$\frac{dS}{dw} = k(400 - 3w^2) = 0$ for $w = \sqrt{\frac{400}{3}} = \frac{20}{\sqrt{3}}$.

For $0 < w < \frac{20}{\sqrt{3}}$, $\frac{dS}{dw} > 0$, S is increasing.

For $\frac{20}{\sqrt{3}} < w < 20$, $\frac{dS}{dw} < 0$, S is decreasing.

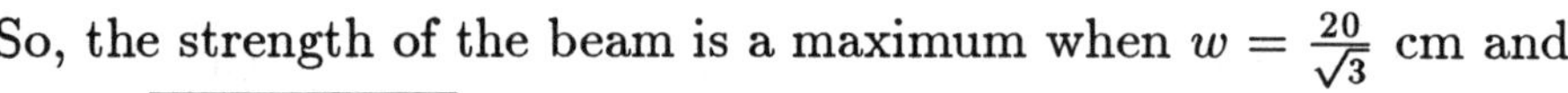

So, the strength of the beam is a maximum when $w = \frac{20}{\sqrt{3}}$ cm and

$d = \sqrt{400 - \left(\frac{20}{\sqrt{3}}\right)^2} = 20\sqrt{\frac{2}{3}}$ cm.

33. $v(\rho) = \frac{100}{1+\rho^2}$.

$\rho(t) = -10(t-3)(t-7)$, t in $[3,7]$.

$\widehat{v}(t) = v\left(\rho(t)\right)$.

$\widehat{v}'(t) = v'(\rho) \cdot \rho'(t) = \frac{-100(2\rho)}{(1+\rho^2)^2} \cdot [-10(t-3+t-7)]$

$= \frac{-20000(t-3)(t-7)(2t-10)}{(1+100(t-3)^2(t-7)^2)^2}$.

$\widehat{v}'(t) = 0$ for $t = 3$, $t = 7$, and $t = 5$.

$\widehat{v}(3) = v\left(\rho(3)\right) = 100$, $\widehat{v}(5) = \frac{100}{1601} = .0625$, $\widehat{v}(7) = 100$.

The maximum velocity is 100 km/hr when $t = 3$ p.m. and $t = 7$ p.m., and the minimum velocity is 0.0625 km/hr when $t = 5$ p.m.

34. $4\ell + 2r\pi = 50$.

$r = \frac{1}{\pi}(25 - 2\ell)$.

Combined area $A = \ell^2 + \pi r^2 = \ell^2 + \frac{1}{\pi}(25 - 2\ell)^2$

$= \ell^2 + \frac{1}{\pi}(625 - 100\ell + 4\ell^2)$

$= \left(1 + \frac{4}{\pi}\right)\ell^2 - \frac{100}{\pi}\ell + \frac{625}{\pi}$.

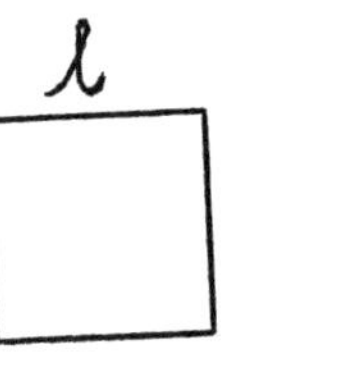

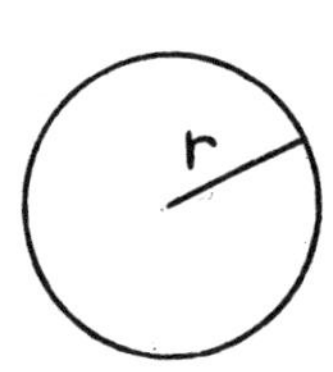

$\frac{dA}{d\ell} = \left(2 + \frac{8}{\pi}\right)\ell - \frac{100}{\pi} = 0$ for $\ell = \frac{50}{\pi+4}$.

$\frac{dA}{d\ell} < 0$ for $\ell < \frac{50}{\pi+4}$ and $\frac{dA}{dt} > 0$ for $\ell > \frac{50}{\pi+4}$.

So the maximum combined area will be obtained at either $\ell = 0$ or $\ell = \frac{25}{2}$. $A(0) = \frac{625}{\pi}$ and $A\left(\frac{25}{2}\right) = \frac{625}{4}$. So the maximum combined area is obtained when $\ell = 0$, i.e., only the circle is made.

35. Let $y =$ the number of days it takes to complete the job.

Let $x =$ the number of overtime hours worked during each of the first $(y-1)$ days.

Let $z =$ the number of regular hours worked during the yth day.

Let $w =$ the number of overtime hours worked during the yth day.

$$8(y-1) + x(y-1) + z + w = 200.$$

$$\text{Cost } C = 80(y-1) + 15x(y-1) + 10z + 15w + 200(y-1).$$

$$\begin{aligned} 8(y-1) + x(y-1) + z + w &= 200 \\ (8+x)(y-1) &= 200 - z - w \\ y - 1 &= \frac{200-z-w}{8+x} \end{aligned}$$

$$\begin{aligned} C &= 80\left(\frac{200-z-w}{8+x}\right) + 15x\left(\frac{200-z-w}{8+x}\right) + 10z + 15w + 200\left(\frac{200-z-w}{8+x}\right) \\ &= (280+15x)\left(\frac{200-z-w}{8+x}\right) + 10z + 15w. \end{aligned}$$

Think of z and w as being given and that C then depends only on x. Then

$$\begin{aligned} \frac{d}{dx}C &= (200-z-w)\left(\frac{15(8+x) - 1(280+15x)}{(8+x)^2}\right) \\ &= (200-z-w)\frac{120 + 15x - 280 - 15x}{(8+x)^2} \\ &= (200-z-w)\left(\frac{-160}{(8+x)^2}\right). \end{aligned}$$

Since $z \le 8$ and $w \le 4$, it follows that $\frac{d}{dx}C < 0$ for $0 \le x \le 4$. Therefore the minimum value of C occurs when $x = 4$.

$$y - 1 = \frac{50}{3} - \frac{z+w}{12}.$$

For $(y-1)$ to be an integer, we must have $z = 8$ and $w = 0$. Thus, the minimum cost is obtained by putting in 17 days, and having the worker put in 4 hours of overtime during each of the first 16 days and no overtime on the 17th day.

36. Time it takes the swimmer $= \frac{\sqrt{100^2+x^2}}{3} + \frac{300-x}{5} = f(x)$.

$f'(x) = \frac{1}{3} \cdot \frac{2x}{2\sqrt{10000+x^2}} - \frac{1}{5} = \frac{1}{15} \cdot \frac{5x-3\sqrt{10000+x^2}}{\sqrt{10000+x^2}}$.

$f'(x) = 0$ for $5x - 3\sqrt{10000+x^2} = 0 \implies 5x = 3\sqrt{10000+x^2} \implies 25x^2 = 9(10000+x^2) \implies 16x^2 = 90000 \implies x^2 = 5625 \implies x = 75$.

$f'(x) < 0$ on $0 < x < 75$ and $f'(x) > 0$ on $75 < x < 300$. So $f(x)$ is a minimum when $x = 75$.

Hence the swimmer can reach the distressed person most quickly by swimming to the point 75 m from the point on the shoreline closest to him.

37. The total cost of laying the cable as in the figure

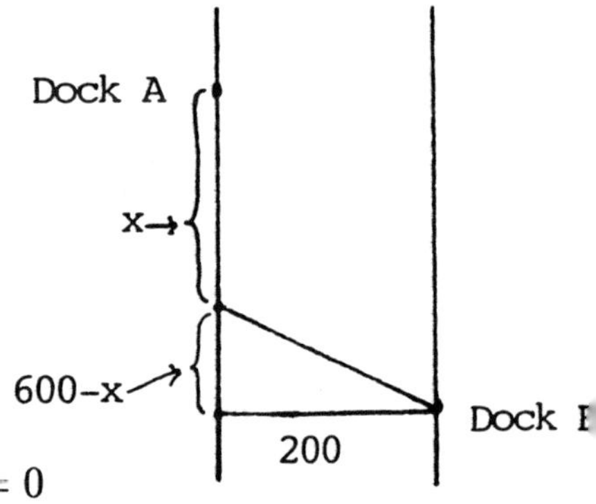

$$= 30x + 50 \cdot \sqrt{(600-x)^2 + 200^2}$$
$$= 30x + 50 \cdot \sqrt{x^2 - 1200x + 400,000} = f(x).$$

$f'(x) = 30 + 25 \cdot \frac{2x-1200}{\sqrt{x^2-1200x+400,000}}$

$= \frac{30\sqrt{x^2-1200x+400,000}+50x-30,000}{\sqrt{x^2-1200x+400,000}}$.

$f'(x) = 0$ for $30\sqrt{x^2 - 1200x + 400,000} + 50x - 30,000 = 0$

$\implies 3\sqrt{x^2 - 1200x + 400,000} = 3000 - 5x \implies 9(x^2 - 1200x + 400,000)$

$= 9,000,000 - 30,000x + 25x^2 \implies 9x^2 - 10800x + 3,600,000$

$= 9,000,000 - 30000x + 25x^2 \implies 16x^2 - 19,200x + 5,400,000 = 0$

$\implies x^2 - 1200x + 337,500 = 0$

$$\Longrightarrow x = \frac{1200 \pm \sqrt{(1200)^2 - 4(337{,}500)}}{2} = \frac{1200 \pm \sqrt{1{,}440{,}000 - 1{,}350{,}000}}{2}$$

$$= \frac{1200 \pm \sqrt{90{,}000}}{2} = \frac{1200 \pm 300}{2} = \frac{900}{2} \text{ or } \frac{1500}{2}$$

$$= 450 \text{ or } 750.$$

$f'(x) < 0$ for $0 < x < 450$ and $f'(x) > 0$ for $450 < x < 600$. Hence the total cost of laying the cable is minimized when 450 m of cable are laid on the land.

38. Let $f(x)$ be the survival rate for seedlings at x meters from the trunk of the tree. Then

$$f(x) = k \cdot d(x) \cdot p(x) = k \frac{0.1x}{1 + (0.2x)^2},$$

where k is a constant and $0 \leq x \leq 10$.

$$f'(x) = k \frac{0.1(1 + 0.04x^2) - 0.1x(0.08x)}{(1 + 0.04x^2)^2}$$

$$= k \frac{0.1 - 0.004x^2}{(1 + 0.04x^2)^2} = \frac{k}{1000} \cdot \frac{100 - 4x^2}{(1 + 0.04x^2)^2} = 0 \quad \text{for } x = 5.$$

$f'(x) > 0$ for $0 < x < 5$ and $f'(x) < 0$ for $5 < x < 10$.

Hence the survival rate is a maximum when the distance from the trunk of the tree is 5 m.

39. Here $f(x) = k \cdot d(x) \cdot p(x) = k \cdot \frac{1}{x} \cdot \frac{1}{(x-10)^2}$,

where k is a constant and $1 \leq x \leq 9$.

$f'(x) = -\frac{k(3x^2 - 40x + 100)}{x^2(x-10)^4} = -\frac{k(3x-10)(x-10)}{x^2(x-10)^4} = 0$ for $x = \frac{10}{3}$.

$f'(x) < 0$ for $0 < x < \frac{10}{3}$ and $f'(x) > 0$ for $\frac{10}{3} < x < 9$.

Hence the survival rate is a minimum when the distance from the parent tree is $\frac{10}{3}$ m.

40.

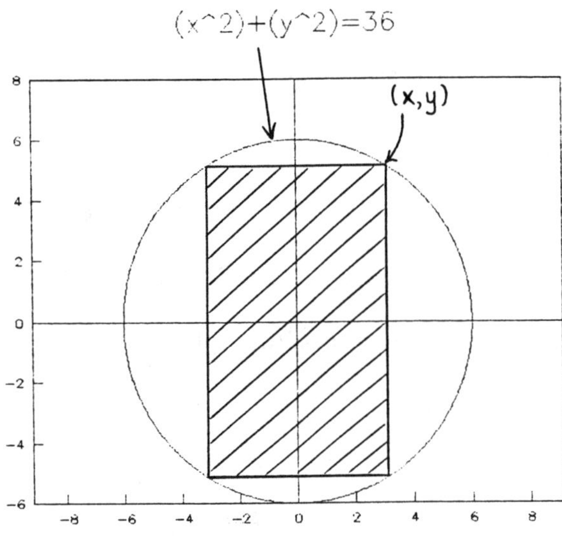

The area of the indicated rectangle is

$$A = (2x)(2y) = 4xy.$$

$$y = \sqrt{36 - x^2}.$$

$$A = 4x\sqrt{36 - x^2}.$$

$$\frac{dA}{dx} = (4)(36 - x^2)^{1/2} + (4x)\left(\frac{1}{2}(36 - x^2)^{-1/2}(-2x)\right)$$

$$= 4(36 - x^2)^{-1/2}\left((36 - x^2) - x^2\right) = 4(36 - x^2)^{-1/2}(36 - 2x^2)$$

$$= 8(36 - x^2)^{-1/2}(18 - x^2).$$

$$\frac{dA}{dx} = 0 \text{ for } x^2 = 18 \implies x = 3\sqrt{2}.$$

For $0 < x < 3\sqrt{2}$, $\frac{dA}{dx} > 0$, so A increases as x increases.

For $3\sqrt{2} < x < 6$, $\frac{dA}{dx} < 0$, so A decreases as x increases.

Therefore the maximum area A occurs when $x = 3\sqrt{2}$ and $y = \sqrt{36 - (3\sqrt{2})^2} = \sqrt{18} = 3\sqrt{2}$.

This maximum possible area is $A = 4\left(3\sqrt{2}\right)\left(3\sqrt{2}\right) = 72$.

Solutions to Exercise Set 3.8

1. $P(x) = \left(100x - \frac{1}{4}x^2\right) - (500 + 4x)$

   $= -\frac{1}{4}x^2 + 96x - 500,\ 0 \le x \le 200.$

   $P'(x) = -\frac{1}{2}x + 96 = 0$ for $x = 192$.

   Since $P'(x) > 0$ for $x < 192$ and $P'(x) < 0$ for $x > 192$, $P(x)$ has a maximum at $x = 192$.

2. $P(x) = 80x - \left(1000 + 30x + \frac{1}{2}x^2\right)$

   $= -\frac{1}{2}x^2 + 50x - 1000,\ 0 \le x \le 500.$

   $P'(x) = -x + 50 = 0$ for $x = 50$.

   Since $P'(x) > 0$ for $x < 50$ and $P'(x) < 0$ for $x > 50$, $P(x)$ has a maximum at $x = 50$.

3. $P(x) = 30x - (500 + 20x + x^2)$

   $= -x^2 + 10x - 500,\ 0 \le x \le 100.$

   $P'(x) = -2x + 10 = 0$ for $x = 5$.

   Since $P'(x) > 0$ for $x < 5$ and $P'(x) < 0$ for $x > 5$, $P(x)$ has a maximum at $x = 5$.

4. $P(x) = 100x - \frac{1}{10}x^2 - \left(200 + 50x + \frac{1}{10}x^2\right)$

   $= -\frac{1}{5}x^2 + 50x - 200,\ 0 \le x \le 40.$

   $P'(x) = -\frac{2}{5}x + 50 = 0$ for $x = 125$.

   Since $P'(x) > 0$ for $x < 125$, $P(x)$ has a maximum at endpoint $x = 40$.

5. Maximum Profit in 1: $P(192) = 8716$

   Maximum Profit in 2: $P(50) = 250$

   Maximum Profit in 3: $P(5) = -475$.

   Maximum Profit in 4: $P(40) = 1480$.

Companies in Exercises 1, 2, and 4 are profitable.

6. Let s be the speed in kilometers per hour.

   Let C be the cost in dollars per kilometer.

$$C = \frac{s}{100} + \frac{1}{s} \cdot 200$$
$$C' = \frac{1}{100} - \frac{200}{s^2} = 0 \quad \text{for } s = 141.42.$$

   For $0 < s < 141.42$, $C' < 0$, so C is decreasing; for $s > 141.42$, $C' > 0$, so C is increasing.

   Hence the cost per kilometer is minimized when the speed is 141.42 km/hour.

7. Let x be the number of times per year the store orders bulbs.

   Yearly cost $= 3000 \cdot \frac{1}{2x}(.48) + 20x + 150 = 720\frac{1}{x} + 20x + 150 = f(x)$.

   $f'(x) = -720\frac{1}{x^2} + 20$. $f'(x) = 0$ for $x = 6$.

   For $0 < x < 6$, $f'(x) < 0$; for $x > 6$, $f'(x) > 0$.

   The minimum value of $f(x)$ occurs for $x = 6$. Hence the store should order bulbs 6 times a year.

8. Let x be the number of passengers. Let y be the cost per person. Let $R(x)$ be the revenue.

   $y = 400 - 1.5(x - 100) = -1.5x + 550$.

   $R(x) = x \cdot y = -1.5x^2 + 550x$.

   $R'(x) = -3x + 550 = 0$ for $x = \frac{550}{3} = 183.333\ldots$.

   $R'(x) > 0$ for $x < \frac{550}{3}$ and $R'(x) < 0$ for $x > \frac{550}{3}$.

   The maximum value of $R(x)$ occurs when $x = \frac{550}{3} \approx 183$ people.

9. Let x be the number of times per year the merchant orders shirts.

   Yearly costs $= 2000 \cdot \frac{1}{2x} \cdot 3 + 50x + 400 = 3000 \cdot \frac{1}{x} + 50x + 400 = f(x)$.

   $f'(x) = -\frac{3000}{x^2} + 50 = 0$ for $x^2 = 60 \implies x = 7.746$.

For $x < 7.746$, $f'(x) < 0$; for $x > 7.746$, $f'(x) > 0$.

$f(x)$ is a minimum for $x = 7.746$.

$f(7) = 1178.57$ and $f(8) = 1175$, so $f(8) < f(7)$.

Thus the merchant should order shirts 8 times a year.

10. As in Example 2, let x be the number of times per year the merchant orders brooms. Here yearly cost $= 1200 \cdot \frac{1}{2x} \cdot 2 + 30x + 144$

$$= \frac{1200}{x} + 30x + 144 = f(x).$$

$f'(x) = -\frac{1200}{x^2} + 30$.

$f'(x) = 0$ for $x^2 = 40 \implies x = 6.325$.

For $x < 6.325$, $f'(x) < 0$; for $x > 6.325$, $f'(x) > 0$.

$f(6) = 524$ and $f(7) = 525.43$, so $f(6) < f(7)$. Now the merchant should order brooms 6 times a year.

11. The profit $P(x) = R(x) - C(x) = (200x - 4x^2) - (900 + 40x) = -4x^2 + 160x - 900$.

$P'(x) = -8x + 160 = 0$ for $x = 20$.

$P(0) = -900$, $P(20) = 700$, $P(50) = -2900$.

The maximum profit occurs when $x = 20$.

12. Let $x =$ the price of a book. Let $y =$ the number of copies sold.

$R(x) =$ revenue. $P(x) =$ profit.

$y = 20,000 + 2,000(16 - x) = 20,000 + 32,000 - 2,000x = 52,000 - 2,000x$.

$R(x) = x \cdot y = 52,000x - 2,000x^2$.

$$\begin{aligned} P(x) &= R(x) - C(x) = (52,000x - 2000x^2) - 4y \\ &= 52,000x - 2,000x^2 - 4(52,000 - 2,000x) \\ &= 52,000x - 2,000x^2 - 208,000 + 8,000x \\ &= -2,000x^2 + 60,000x - 208,000. \end{aligned}$$

a. $R'(x) = 52,000 - 4,000x$

$R'(x) = 0$ for $x = 13$.

$R(0) = 0,\ R(13) = 338,000,\ R(26) = 0$.

The maximum revenue occurs when $x = \$13$.

b. Since $C = 4y$, the marginal cost per book is $\frac{dC}{dy} = \$4$.

c. $P'(x) = -4,000x + 60,000$.

$P'(x) = 0$ for $x = 15$.

$P'(x) > 0$ for $0 < x < 15$, $P'(x) < 0$ for $15 < x < 26$.

The maximum profit occurs when the price is \$15 per book.

13. a. $R(x) = x \cdot p = x(600 - 3x) = -3x^2 + 600x$.

b. $P(x) = R(x) - C(x)$

$$= (-3x^2 + 600x) - (400 + 150x + 0.5x^2)$$

$$= -3.5x^2 + 450x - 400.$$

c. $C'(x) = 150 + x$. $R'(x) = -6x + 600$.

d. $P'(x) = -7x + 450$.

$P'(x) = 0$ for $x = \frac{450}{7} = 64.29$.

For $0 < x < \frac{450}{7}$, $P'(x) > 0$; for $\frac{450}{7} < x < 100$, $P'(x) < 0$.

$P(64) = 14,064$, $P(65) = 14,062.50$.

Hence, the maximum profit occurs when the weekly production level is 64 dishwashers.

e. $C'(64) = 214$. $R'(64) = 216$.

14. $S(p) = \frac{100}{200-p}$.

a. Marginal supply $= S'(p) = \frac{(200-p)\cdot 0 - 100(-1)}{(200-p)^2} = \frac{100}{(200-p)^2}$.

$0.25 = \frac{100}{(200-p)^2} \implies (200-p)^2 = 400 \implies 200 - p = 20 \implies p = 180$.

b. S is increasing because $S'(p) > 0$ for $0 < p < 180$.

c. It would be unrealistic to let p have the range $0 \le p < 200$ because the supply of uranium could increase beyond bound.

15. $c(x) = \frac{C(x)}{x}$. $c'(x) = \frac{x \cdot C'(x) - C(x)}{x^2}$.
When the average cost is a minimum, $[x \cdot C'(x) - C(x)] = 0 \implies C'(x) = \frac{C(x)}{x}$, so marginal cost = average cost.

16. $P(x) = R(x) - C(x) - tx$, x = output. When the profit is a maximum, $P'(x) = R'(x) - C'(x) - t = 0$, so $R'(x) = C'(x) + t$.

17. $P(x) = p \cdot x - C(x) - t \cdot x$ where x = number of items
$C(x)$ = total cost of producing these x items
$P(x)$ = profit from sales of these items
t = per item tax imposed on the manufacturer
Suppose C is differentiable and the range of C' contains the open interval $(0, p)$. Suppose $C'(x)$ is an increasing function of x. The output level which maximizes profit is given by $P'(x) = p - C'(x) - t = 0 \implies C'(x) = p - t$. Let t_1 and t_2 be per item taxes such that $t_2 > t_1$, and let x_1 and x_2 be the output levels for tax t_1 and tax t_2, respectively, which maximize profit. Then

$$C'(x_1) = p - t_1$$
$$C'(x_2) = p - t_2.$$

Thus, $C'(x_2) - C'(x_1) = t_1 - t_2 < 0 \implies C'(x_2) < C'(x_1) \implies x_2 < x_1$.

18. $P(x) = x(600 - 3x) - (400 + 150x + 0.5x^2) - 30x$
$= 600x - 3x^2 - 400 - 150x - 0.5x^2 - 30x$
$= -3.5x^2 + 420x - 400.$

a. $P'(x) = -7x + 420$.
$P'(x) = 0$ for $x = 60$.
Reasoning as before, we find that the profit is maximized when the output is 60 dishwashers per week.

b. The resulting price per dishwasher is

$$p = 600 - 3(60) = 600 - 180 = \$420.$$

In problem 13, the price per dishwasher for maximizing profit was $600 - 3(64) = 600 - 192 = 408$. The resulting change in price per dishwasher is $420 - 408 = \$12$.

c. The amount of tax absorbed by the manufacturer is $30 - 12 = \$18$.

19. Now, $P(x) = -3.5x^2 + 485x - 400$.

$$P'(x) = -7x + 485 = 0 \text{ for } x = \frac{485}{7} = 69.3.$$

$P(69) = 16,401.5, \quad P(70) = 16,400.$

Therefore the manufacturer should manufacture 69 dishwashers per week.

20. $E(p) = \frac{-pQ'(p)}{Q(p)} = \frac{-p\cdot\left(-\frac{3}{2}\right)}{-\frac{3}{2}p+10}.$

$E(3) = \frac{-3\cdot\left(-\frac{3}{2}\right)}{-\frac{3}{2}(3)+10} = \frac{\frac{9}{2}}{\frac{11}{2}} = \frac{9}{11}.$

21. $E(p) = \frac{-pQ'(p)}{Q(p)} = \frac{-p\cdot\left(-\frac{5}{4}\right)}{-\frac{5}{4}p+40}.$

$E(20) = \frac{-20\cdot\left(-\frac{5}{4}\right)}{-\frac{5}{4}\cdot(20)+40} = \frac{25}{15} = \frac{5}{3}.$

22. $E(p) = \frac{-pQ'(p)}{Q(p)} = \frac{-p(-2p)}{169-p^2} = \frac{2p^2}{169-p^2}.$

$E(6) = \frac{2(6)^2}{169-(6)^2} = \frac{72}{133}.$

23. $E(p) = \frac{-pQ'(p)}{Q(p)} = \frac{-p\cdot\left(\frac{1}{2}\right)(1000-4p)^{-1/2}(-4)}{(1000-4p)^{1/2}} = \frac{2p}{(1000-4p)}.$

$E(40) = \frac{2(40)}{1000-4\cdot 40} = \frac{80}{840} = \frac{2}{21}.$

24. $E(p) = \frac{-pQ'(p)}{Q(p)} = -p\left(\frac{-2}{p^3}\Big/\frac{1}{p^2}\right) = 2.$

25. $E(p) = \frac{-p\left(-\frac{1}{2}p^{-3/2}\right)}{p^{-1/2}} = \frac{1}{2}.$

26. $E(p) = \frac{-p(-80)}{(1+p)^3}\Big/\frac{40}{(1+p)^2} = \frac{2p}{1+p}. \quad E(5) = \frac{5}{3}.$

27. $E(p) = \frac{-p(-20)}{(p+2)^3}\Big/\frac{10}{(p+2)^2} = \frac{2p}{p+2}. \quad E(2) = 1.$

28. $E(p) = \frac{-p\left(-\frac{1}{2}p^{-3/2}\right)}{p^{-1/2}} = \frac{1}{2}.$

The curve is inelastic for all p.

29. $E(p) = -p\left(\frac{-2}{p^3}\Big/\frac{1}{p^2}\right) = 2.$

The curve is elastic for all $p > 0$.

30. $E(p) = \frac{-p\left(-\frac{1}{2}\right)}{500-\frac{p}{2}} = \frac{p}{1000-p}.$

$E(p) < 1 \implies \frac{p}{1000-p} < 1 \implies p < 1000 - p \implies 2p < 1000 \implies p < 500.$

The curve for is inelastic $p < 500$ and is elastic for $500 < p < 1000$.

31. $E(p) = \frac{-p\left((1+p)\cdot\frac{1}{2\sqrt{p}} - \sqrt{p}\right)}{(1+p)^2} \Big/ \frac{\sqrt{p}}{1+p}$

$= \frac{-p\left(\frac{1+p}{2\sqrt{p}} - \sqrt{p}\right)}{(1+p)^2} \cdot \frac{1+p}{\sqrt{p}}$

$= \frac{-\sqrt{p}\cdot\left(\frac{1+p}{2\sqrt{p}} - \sqrt{p}\right)}{(1+p)} = \frac{p - \frac{1+p}{2}}{1+p}$

$= \frac{2p-(1+p)}{2(1+p)} = \frac{p-1}{2(p+1)}$.

Then $E(p) < 1 \implies p - 1 < 2(p+1) \implies p > -3$. So the curve is inelastic for all $p > 1$.

32. a. $E(p) = \frac{-p(6-2p)}{40+6p-p^2} = \frac{-6p+2p^2}{40+6p-p^2}$.

b. $E(4) = \frac{8}{48} = 0.167$.

33. $Q = 200 - p \implies p = 200 - Q$.

$R(Q) = pQ = 200Q - Q^2$.

$P(Q) = 200Q - Q^2 - (2Q + 100) = -Q^2 + 198Q - 100$.

$P'(Q) = -2Q + 198 = 0$ for $Q = 99$.

a. The profit is maximized when $Q = 99$.

The resulting price is $p = 200 - 99 = 101$.

b. $E(p) = \frac{-p\cdot Q'(p)}{Q(p)} = \frac{-p(-1)}{200-p}$. $E(101) = \frac{101}{99}$. The demand is elastic.

34. a. $x = Q(p) = 45 - .2p^2$.

b. $E(p) = \frac{-pQ'(p)}{Q(p)} = \frac{-p(-.4p)}{45-.2p^2} = \frac{.4p^2}{45-.2p^2}$.

$E(4) = \frac{.4(4)^2}{45-.2(4)^2} = \frac{6.4}{41.8} = .1531 < 1$.

Therefore the demand is inelastic when $p = 4$.

c. $R(p) = p \cdot (x) = 45p - .2p^3$.

$R(10) = 45(10) - .2(10)^3 = 250$ and $R(11) = 45(11) - .2(11)^3 = 228.8$.

35. a. $x = Q(p) = \sqrt{1000 - 10p}$.

b. $E(p) = \frac{-pQ'(p)}{Q(p)} = \frac{-p\left(\frac{1}{2}\right)(1000-10p)^{-1/2}\cdot(-10)}{\sqrt{1000-10p}} = \frac{5p}{1000-10p}$.

$E(50) = \frac{5(50)}{1000-10(50)} = \frac{250}{500} = \frac{1}{2} < 1$.

Therefore the demand is inelastic when $p = 50$.

c. $E(70) = \frac{5(70)}{1000-10(70)} = \frac{350}{300} = \frac{7}{6} > 1$.

Therefore the demand is elastic when $p = 70$.

d. $E(p) = 1 \implies 5p = 1000 - 10p \implies 15p = 1000 \implies p = 66.67$.

36. a. $p=\sqrt{8,000-4x^2} \implies p^2=8,000-4x^2 \implies 4x^2=8,000-p^2 \implies x^2=2,000-\frac{1}{4}p^2 \implies x=\sqrt{2,000-\frac{1}{4}p^2}=Q(p)$.

b. $E(p)=\frac{-pQ'(p)}{Q(p)}=\frac{-p\cdot\left(\frac{1}{2}\right)\left(2,000-\frac{1}{4}p^2\right)^{-1/2}\left(-\frac{1}{2}p\right)}{\left(2,000-\frac{1}{4}p^2\right)^{1/2}}=\frac{\frac{1}{4}p^2}{2,000-\frac{1}{4}p^2}$.

$E(40)=\frac{\frac{1}{4}(40)^2}{2,000-\frac{1}{4}(40)^2}=\frac{400}{1,600}=\frac{1}{4}<1$.

Therefore the demand is inelastic when $p=40$.

c. $E(p)=1 \implies \frac{1}{4}p^2=2,000-\frac{1}{4}p^2 \implies \frac{1}{2}p^2=2,000 \implies p^2=4,000 \implies p=63.25$.

Solutions to Exercise Set 3.9

1. $x^2+y^2=9$. $\quad 2x+2y\frac{dy}{dx}=0, \quad 2y\frac{dy}{dx}=-2x,$
$\frac{dy}{dx}=\frac{-2x}{2y}=\frac{-x}{y}$.

2. $x^2-y^2=16$. $\quad 2x-2y\frac{dy}{dx}=0, \quad -2y\frac{dy}{dx}=-2x,$
$\frac{dy}{dx}=\frac{-2x}{-2y}=\frac{x}{y}$.

3. $xy^2=6$. $\quad x\cdot 2y\frac{dy}{dx}+y^2\cdot 1=0, \quad 2xy\frac{dy}{dx}=-y^2,$
$\frac{dy}{dx}=\frac{-y^2}{2xy}=\frac{-y}{2x}$.

4. $4x^2+2y^2=4$. $\quad 8x+4y\cdot\frac{dy}{dx}=0, \quad 4y\cdot\frac{dy}{dx}=-8x,$
$\frac{dy}{dx}=\frac{-8x}{4y}=\frac{-2x}{y}$.

5. $x^2+2xy+y^2=8$. $\quad 2x+2x\cdot\frac{dy}{dx}+y\cdot 2+2y\frac{dy}{dx}=0,$
$2x\frac{dy}{dx}+2y\frac{dy}{dx}=-2x-2y, \quad (2x+2y)\frac{dy}{dx}=-(2x+2y),$
$\frac{dy}{dx}=\frac{-(2x+2y)}{(2x+2y)}=-1$.

6. $x^2+3xy+6y+y^2=10$.
$2x+3y+3x\frac{dy}{dx}+6\frac{dy}{dx}+2y\frac{dy}{dx}=0$.
$(3x+6+2y)\frac{dy}{dx}=-2x-3y$.
$\frac{dy}{dx}=\frac{-2x-3y}{3x+6+2y}$.

7. $x^3+2xy^2+x^2y+2y^3=5$.
$3x^2+2y^2+2x\left(2y\frac{dy}{dx}\right)+2xy+x^2\frac{dy}{dx}+6y^2\frac{dy}{dx}=0$.
$(4xy+x^2+6y^2)\frac{dy}{dx}=-3x^2-2y^2-2xy$.
$\frac{dy}{dx}=\frac{-3x^2-2y^2-2xy}{4xy+x^2+6y^2}$.

8. $3x^2 - xy + 5xy^3 = 5.$
$6x - y - x\frac{dy}{dx} + 5y^3 + 5x\left(3y^2\frac{dy}{dx}\right) = 0.$
$(-x + 15xy^2)\frac{dy}{dx} = -6x + y - 5y^3.$
$\frac{dy}{dx} = \frac{-6x+y-5y^3}{-x+15xy^2}.$

9. $x^3 + 2x^2y + 5xy^2 - y^2 = 10.$
$3x^2 + 4xy + 2x^2\frac{dy}{dx} + 5y^2 + 5x\left(2y\frac{dy}{dx}\right) - 2y\frac{dy}{dx} = 0.$
$(2x^2 + 10xy - 2y)\frac{dy}{dx} = -3x^2 - 4xy - 5y^2.$
$\frac{dy}{dx} = \frac{-3x^2-4xy-5y^2}{2x^2+10xy-2y}.$

10. $xy + 3\sqrt{x+y} - y^2 = 5.$
$y + x\frac{dy}{dx} + 3 \cdot \frac{1}{2}(x+y)^{-1/2}\left(1 + \frac{dy}{dx}\right) - 2y\frac{dy}{dx} = 0.$
$\left(x + \frac{3}{2}(x+y)^{-1/2} - 2y\right)\frac{dy}{dx} = -y - \frac{3}{2}(x+y)^{-1/2}.$
$\frac{dy}{dx} = \frac{-y-\frac{3}{2}(x+y)^{-1/2}}{x+\frac{3}{2}(x+y)^{-1/2}-2y}.$

11. $xy - y^4 + \sqrt{xy} = 10.$
$y + x\frac{dy}{dx} - 4y^3\frac{dy}{dx} + \frac{1}{2}(xy)^{-1/2} \cdot \left(y + x\frac{dy}{dx}\right) = 0.$
$\left(x - 4y^3 + \frac{1}{2}(xy)^{-1/2} \cdot (x)\right)\frac{dy}{dx} = -y - \frac{1}{2}(xy)^{-1/2} \cdot (y).$
$\frac{dy}{dx} = \frac{-y-\frac{1}{2}(xy)^{-1/2}(y)}{x-4y^3+\frac{1}{2}(xy)^{-1/2}(x)}.$

12. $\sqrt{x+y} = x - y.$
$\frac{1}{2}(x+y)^{-1/2} \cdot \left(1 + \frac{dy}{dx}\right) = 1 - \frac{dy}{dx}.$
$\left(\frac{1}{2}(x+y)^{-1/2} + 1\right)\frac{dy}{dx} = 1 - \frac{1}{2}(x+y)^{-1/2}.$
$\frac{dy}{dx} = \frac{1-\frac{1}{2}(x+y)^{-1/2}}{\frac{1}{2}(x+y)^{-1/2}+1}.$

13. $\sqrt{x} - xy + x^{2/3}y^{4/3} = 30.$
$\frac{1}{2}x^{-1/2} - y - x\frac{dy}{dx} + \frac{2}{3}x^{-1/3}y^{4/3} + x^{2/3} \cdot \frac{4}{3}y^{1/3}\frac{dy}{dx} = 0.$
$\left(-x + \frac{4}{3}x^{2/3}y^{1/3}\right)\frac{dy}{dx} = -\frac{1}{2}x^{-1/2} + y - \frac{2}{3}x^{-1/3}y^{4/3}.$
$\frac{dy}{dx} = \frac{-\frac{1}{2}x^{-1/2}+y-\frac{2}{3}x^{-1/3}y^{4/3}}{-x+\frac{4}{3}x^{2/3}y^{1/3}}.$

14. $\frac{x+y}{x-y} = 3.$
$x + y = 3(x - y).$
$2x - 4y = 0$
$y = \frac{1}{2}x.$
$\frac{dy}{dx} = \frac{1}{2}.$

15. $xy = 9.$ $\quad x\frac{dy}{dx} + y = 0,$ $\quad \frac{dy}{dx} = -\frac{y}{x}.$
At $(3,3),$ $\frac{dy}{dx} = \frac{-3}{3} = -1.$

16. $x^2 + y^2 = 4.$ $\quad 2x + 2y\frac{dy}{dx} = 0,$ $\quad 2y\frac{dy}{dx} = -2x,$ $\frac{dy}{dx} = \frac{-2x}{2y} = \frac{-x}{y}.$
At $(\sqrt{2}, \sqrt{2}),$ $\frac{dy}{dx} = \frac{-\sqrt{2}}{\sqrt{2}} = -1.$

17. $x^3 + y^3 = 16.$ $\quad 3x^2 + 3y^2\frac{dy}{dx} = 0,$ $\quad \frac{dy}{dx} = \frac{-3x^2}{3y^2} = \frac{-x^2}{y^2}.$
At $(2,2),$ $\frac{dy}{dx} = \frac{-4}{4} = -1.$

18. $x^2 \cdot y^2 = 16.$ $\quad x^2 \cdot 2y\frac{dy}{dx} + y^2 \cdot 2x = 0,$
$2x^2y\frac{dy}{dx} = -2xy^2,$ $\quad \frac{dy}{dx} = \frac{-2xy^2}{2x^2y} = \frac{-y}{x}.$
At (-1,4), $\frac{dy}{dx} = \frac{-4}{-1} = 4.$

19. $\frac{x+y}{x-y} = 4.$ $\quad \frac{(x-y)\left(1+\frac{dy}{dx}\right)-(x+y)\left(1-\frac{dy}{dx}\right)}{(x-y)^2} = 0,$
$2x\frac{dy}{dx} - 2y = 0,$ $\quad \frac{dy}{dx} = \frac{2y}{2x} = \frac{y}{x}.$
At (5,3), $\frac{dy}{dx} = \frac{3}{5}.$

20. $(y-x)^2 = x.$ $\quad 2(y-x)\left(\frac{dy}{dx} - 1\right) = 1,$ $\quad (2y-2x)\left(\frac{dy}{dx} - 1\right) = 1,$
$2y\frac{dy}{dx} - 2y - 2x\frac{dy}{dx} + 2x = 1,$ $\quad 2y\frac{dy}{dx} - 2x\frac{dy}{dx} = 1 - 2x + 2y,$
$\frac{dy}{dx}(2y - 2x) = 1 - 2x + 2y,$ $\quad \frac{dy}{dx} = \frac{1-2x+2y}{2y-2x}.$
At (9,12), $\frac{dy}{dx} = \frac{1-2(9)+2(12)}{2(12)-2(9)} = \frac{1-18+24}{24-18} = \frac{7}{6}.$

21. $\sqrt{x} + \sqrt{y} = 4.$ $\quad \frac{1}{2}x^{-1/2} + \frac{1}{2}y^{-1/2}\frac{dy}{dx} = 0,$ $\quad \frac{1}{2y^{1/2}} \cdot \frac{dy}{dx} = \frac{-1}{2x^{1/2}},$
$\frac{dy}{dx} = \frac{-2y^{1/2}}{2x^{1/2}} = \frac{-y^{1/2}}{x^{1/2}}.$
At (4,4), $\frac{dy}{dx} = \frac{-\sqrt{4}}{\sqrt{4}} = -1.$
Hence, -1 is the desired slope. The equation for the line tangent to the graph is $(y-4) = -1(x-4)$ where $y = -x + 8.$

22. $x^2 + y^2 = 8.$ $\quad 2x + 2y\frac{dy}{dx} = 0,$ $\quad 2y\frac{dy}{dx} = -2x,$ $\quad \frac{dy}{dx} = \frac{-x}{y}.$
At (-2,2), $\frac{dy}{dx} = \frac{-(-2)}{2} = 1.$
Hence, 1 is the desired slope. The equation for the line tangent to the graph is $(y-2) = 1(x+2)$ where $y = x + 4.$

23. $y^3 - x^2 = 7.$ $\quad 3y^2\frac{dy}{dx} - 2x = 0,$ $\quad \frac{dy}{dx} = \frac{2x}{3y^2}.$
At (1,2), $\frac{dy}{dx} = \frac{2(1)}{3(2^2)} = \frac{1}{6}.$ Hence, $\frac{1}{6}$ is the desired slope.

24. $\frac{x^2}{16} + \frac{y^2}{9} = 1. \qquad \frac{2x}{16} + \frac{2y}{9} \cdot \frac{dy}{dx} = 0,$

$\frac{2y}{9} \cdot \frac{dy}{dx} = -\frac{x}{8}, \qquad \frac{dy}{dx} = -\frac{9x}{16y}.$

At $\left(2, \frac{3\sqrt{3}}{2}\right)$, $\frac{dy}{dx} = \frac{-9(2)}{16\left(\frac{3\sqrt{3}}{2}\right)} = \frac{-18}{24\sqrt{3}} = -\frac{3}{4\sqrt{3}} = -\frac{\sqrt{3}}{4}.$

Hence, $-\frac{\sqrt{3}}{4}$ is the desired slope.

25. $A = \pi r^2. \qquad \frac{dA}{dt} = \pi \cdot \frac{dr^2}{dt} = \pi \cdot 2r\frac{dr}{dt} = 2\pi r\frac{dr}{dt}.$

When $r = 10$ and $\frac{dr}{dt} = 2$,

$\frac{dA}{dt} = 2\pi \cdot 10 \cdot 2 = 40\pi \cdot m^2/\text{sec}.$

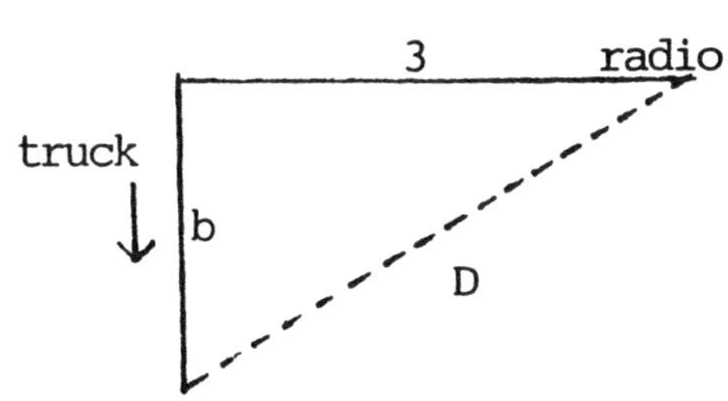

26. $D = \sqrt{9 + b^2}, \qquad \frac{db}{dt} = 80$

$\frac{dD}{dt} = \frac{1}{2}(9 + b^2)^{-1/2}(2b)\frac{db}{dt}.$

When $D = 5$, $9 + b^2 = 5^2 \implies b^2 = 16 \implies b = 4.$

So $\frac{dD}{dt} = \frac{1}{2}(9 + 4^2)^{-1/2}(2 \cdot 4) \cdot 80 = 64$ km/hr.

27. $N(t) = I(t) + S(t), \quad \frac{dI}{dt} = 24$, and $\frac{dS}{dt} = -20.$

$\frac{dN}{dt} = \frac{dI}{dt} + \frac{dS}{dt} = 24 + (-20) = 4$ persons/day.

28. Let $x =$ the number of years from now

$y =$ the average yield per tree in pounds of apples

$z =$ the price per pound of apples in cents

$R =$ the revenue in cents.

Then $y = 200 + 20x, \qquad z = 60 + 10x,$

$$\begin{aligned} R &= 100 \cdot y \cdot z = 100(200 + 20x)(60 + 10x) \\ &= 100(12,000 + 1200x + 2000x + 200x^2) \\ &= 20,000(x^2 + 16x + 60). \end{aligned}$$

$\frac{dR}{dx} = 20,000(2x + 16).$

When $x = 1$, $\frac{dR}{dx} = 20,000(2 \cdot 1 + 16) = 360,000$ cents/yr.

29. $2x\frac{dx}{dt} + 5\frac{dx}{dt}p + 5x\frac{dp}{dt} + 2p\frac{dp}{dt} = 0.$

$2(20)\frac{dx}{dt} + 5(10)\frac{dx}{dt} + 5(20)(-1) + 2(10)(-1) = 0.$

$90\frac{dx}{dt} - 120 = 0 \implies \frac{dx}{dt} = \frac{4}{3}.$

30. $2x\frac{dx}{dt} + 4\frac{dx}{dt}p + 4x\frac{dp}{dt} + 4\frac{dp}{dt} = 0.$

$2(40)\frac{dx}{dt} + 4(100)\frac{dx}{dt} + 4(40)(-.5) + 4(-.5) = 0.$

$480\frac{dx}{dt} - 82 = 0.$

$\frac{dx}{dt} = \frac{82}{480} = \frac{41}{240}.$

31. $5\frac{dx}{dt} + 5\frac{dx}{dt}p + 5x\frac{dp}{dt} + 2p\frac{dp}{dt} = 0.$

$5\frac{dx}{dt} + 5(40)\frac{dx}{dt} + 5(20)(.20) + 2(40)(.20) = 0.$

$205\frac{dx}{dt} + 36 = 0.$

$\frac{dx}{dt} = \frac{-36}{205}.$

32. Let the demand function be denoted by $x = Q(p)$.

The elasticity of demand at price p is given by

$$E(p) = \frac{-pQ'(p)}{Q(p)} = -\frac{p\frac{dx}{dp}}{x}.$$

$$2x\frac{dx}{dp} + 5\frac{dx}{dp}p + 5x + 2p = 0.$$

$$2(20)\frac{dx}{dp} + 5(10)\frac{dx}{dp} + 5(20) + 2(10) = 0.$$

$$90\frac{dx}{dp} + 120 = 0.$$

$$\frac{dx}{dp} = -\frac{4}{3}.$$

$$E(10) = \frac{-10\left(-\frac{4}{3}\right)}{20} = \frac{2}{3}.$$

33. $2x\frac{dx}{dp} + 4\frac{dx}{dp}p + 4x + 4 = 0.$

$2(40)\frac{dx}{dp} + 4(100)\frac{dx}{dp} + 4(40) + 4 = 0.$

$480\frac{dx}{dp} + 164 = 0.$

$\frac{dx}{dp} = -\frac{164}{480} = -\frac{41}{120}.$

$E(p) = -\frac{p\frac{dx}{dp}}{x}.$

$E(100) = -\frac{100\left(-\frac{41}{120}\right)}{40} = \frac{4100}{4800} = \frac{41}{48}.$

Solutions to Exercise Set 3.10

1. $f(9.2) \approx f(8) + f'(8)(9.2 - 8).$

$f(8) = 3(8^2) + 7 = 199.$

$f'(x) = 6x \implies f'(8) = 48.$

$f(9.2) \approx 199 + 48(1.2) = 256.6.$

2. $f(2.7) \approx f(3) + f'(3)(2.7 - 3)$.
   $f(3) = 3^3 + 5 = 32$.
   $f'(x) = 3x^2 \implies f'(3) = 27$.
   $f(2.7) \approx 32 + 27(-.3) = 23.9$.
3. $f(-5.4) \approx f(-6) + f'(-6)\,(-5.4 - (-6))$.
   $f(-6) = 2(-6)^2 + 3(-6) = 54$.
   $f'(x) = 4x + 3 \implies f'(-6) = 4(-6) + 3 = -21$.
   $f(-5.4) \approx 54 + (-21)(.6) = 41.4$.
4. $f(6.2) \approx f(6) + f'(6)(6.2 - 6)$.
   $f(6) = \sqrt{6+3} = 3$.
   $f'(x) = \frac{1}{2}(x+3)^{-1/2} \implies f'(6) = \frac{1}{6}$.
   $f(6.2) \approx 3 + \frac{1}{6}(.2) = \frac{91}{30}$.
5. $f(2.2) \approx f(2) + f'(2)(2.2 - 2)$.
   $f(2) = \sqrt{\frac{2+2}{2-1}} = 2$.
   $f'(x) = \frac{1}{2}\left(\frac{x+2}{x-1}\right)^{-1/2} \cdot \frac{(x-1)(1)-(x+2)(1)}{(x-1)^2} = \frac{1}{2}\left(\frac{x+2}{x-1}\right)^{-1/2} \cdot \frac{-3}{(x-1)^2}$.
   $f'(2) = \frac{1}{2}\left(\frac{2+2}{2-1}\right)^{-1/2} \frac{-3}{(2-1)^2} = \frac{1}{2} \cdot \frac{1}{2} \cdot (-3) = -\frac{3}{4}$.
   $f(2.2) \approx 2 + \left(-\frac{3}{4}\right)(.2) = 1.85$.
6. $f(6) \approx f(5) + f'(5)(6 - 5)$.
   $f(5) = \sqrt{5^2 + 11} = 6$.
   $f'(x) = \frac{1}{2}\left(x^2 + 11\right)^{-1/2} \cdot (2x) = \left(x^2 + 11\right)^{-1/2}(x)$.
   $f'(5) = (5^2 + 11)^{-1/2}(5) = \frac{5}{6}$.
   $f(6) \approx 6 + \frac{5}{6}(1) = \frac{41}{6}$.
7. $f(120) \approx f(125) + f'(125)(120 - 125)$.
   $f(125) = 5 + 125^{2/3} = 5 + 25 = 30$.
   $f'(x) = \frac{2}{3}x^{-1/3} \implies f'(125) = \frac{2}{3}(125)^{-1/3} = \frac{2}{3}\left(\frac{1}{5}\right) = \frac{2}{15}$.
   $f(120) \approx 30 + \frac{2}{15}(-5) = \frac{88}{3}$.
8. $f(25) \approx f(27) + f'(27)(25 - 27)$.
   $f(27) = 27^{1/3} + 3(27) = 3 + 81 = 84$.
   $f'(x) = \frac{1}{3}x^{-2/3} + 3 \implies f'(27) = \frac{1}{3}(27)^{-2/3} + 3 \implies \frac{1}{27} + 3 = \frac{82}{27}$.
   $f(25) \approx 84 + \frac{82}{27}(-2) = \frac{2104}{27}$.

9. Let $f(x) = \sqrt{x}$.
   $f'(x) = \frac{1}{2\sqrt{x}}$.
   Take $\sqrt{38.6} \approx f(36) + f'(36)(38.6 - 36) = 6 + \frac{1}{12}(2.6) = 6.2167$.
10. Let $f(x) = \sqrt[3]{x}$.
    $f'(x) = \frac{1}{3}x^{-2/3}$.
    Take $\sqrt[3]{124.3} \approx f(125) + f'(125)(124.3 - 125) = 5 + \frac{1}{75}(-.7) = 4.9907$.
11. Let $f(x) = \frac{1}{\sqrt{x}}$.
    $f'(x) = -\frac{1}{2}x^{-3/2}$.
    Take $\frac{1}{\sqrt{17.2}} \approx f(16) + f'(16)(17.2 - 16) = \frac{1}{4} - \frac{1}{128}(1.2) = .2406$.
12. Let $f(x) = x^{3/5}$.
    $f'(x) = \frac{3}{5}x^{-2/5}$.
    Take $(31)^{3/5} \approx f(32) + f'(32)(31 - 32) = 8 + \frac{3}{20}(-1) = 7.85$.
13. Let $f(x) = x^{3/4}$.
    $f'(x) = \frac{3}{4}x^{-1/4}$.
    Take $(83.2)^{3/4} \approx f(81) + f'(81)(83.2 - 81) = 27 + \frac{1}{4}(2.2) = 27.55$.
14. Let $f(x) = \sqrt{x}$.
    $f'(x) = \frac{1}{2\sqrt{x}}$.
    Take $\sqrt{123.5} \approx f(121) + f'(121)(123.5 - 121) = 11 + \frac{1}{22}(2.5) = 11.1136$.
15. $R(p) = 800p - 3p^2$.
    a. $R(100) = 800(100) - 3(100)^2 = 50,000$.
    b. $R(105) - R(100) \approx R'(100)(105 - 100)$.
       $R'(p) = 800 - 6p \implies R'(100) = 800 - 6(100) = 200$.
       $R(105) - R(100) \approx 200(5) = 1,000$.
    c. $R(120) - R(100) \approx R'(100)(120 - 100) = 200(20) = 4,000$.
16. Let $f(x) = x^3$.

    Our estimate of the error in the calculation of the volume of the cube is

$$f'(10) \cdot \left(\frac{1}{4}\right) = (3 \cdot 10^2)\left(\frac{1}{4}\right) = 75.$$

    Our estimate of the percent error in this calculation is

$$\frac{75}{f(10)} \cdot 100 = \frac{75}{1000} \cdot 100 = 7.5\% .$$

17. $R(p) = x \cdot p = 1000p - 2p^{7/4}$.

The percent change is approximately

$$\frac{R'(81) \cdot [85 - 81]}{R(81)} \cdot 100.$$

$R(81) = 1000(81) - 2(81)^{7/4} = 81,000 - 2(2187) = 76,626.$

$R'(p) = 1000 - \frac{7}{2}p^{3/4}.$

$R'(81) = 1000 - \frac{7}{2}(81)^{3/4} = 905.5.$

Therefore the percent change is approximately

$$\frac{905.5(4)}{76,626} \cdot 1000 = 4.73\% \ .$$

18. $T(n) = \frac{1}{4}\sqrt{n^2 + 4n + 4}$.

a. $T(6) = \frac{1}{4}\sqrt{6^2 + 4(6) + 4} = 2.$

b. $T(8) \approx T(6) + T'(6) \cdot (8 - 6).$

$T'(n) = \frac{1}{4} \cdot \frac{1}{2}(n^2 + 4n + 4)^{-1/2} \cdot (2n + 4) = \frac{1}{4}(n^2 + 4n + 4)^{-1/2}(n + 2).$

$T'(6) = \frac{1}{4} \cdot \frac{1}{8} \cdot 8 = \frac{1}{4}.$

$T(8) \approx 2 + \frac{1}{4} \cdot 2 = 2.5.$

19. $f(K) = 500\sqrt{K}$.

$f'(K) = 500 \cdot \frac{1}{2\sqrt{K}} = \frac{250}{\sqrt{K}}.$

$f(640,000) = 400,000.$

$f'(640,000) = .3125.$

The approximate percentage increases in the daily output is

$$\frac{f'(640,000) \cdot (700,000 - 640,000)}{f(640,000)} \cdot 100$$
$$= \frac{.3125(60,000)}{400,000} \cdot 100 = 4.6875\% \ .$$

20. $C(x) = 4,000 + 40x + 3x^2$.

$C'(x) = 40 + 6x.$

$C(100) = 38,000.$

$C'(100) = 640$.

The approximate percentage increase in the tota cost is

$$\frac{C'(100)\cdot(110-100)}{C(100)}\cdot 100 = \frac{640(10)}{38,000}\cdot 100 = 16.8421\% \ .$$

21. $x = 2,000 - 40p \implies 40p = 2,000 - x \implies p = 50 - \frac{1}{40}x = 50 - .025x$.
   a. $R(x) = x \cdot p = 50x - .025x^2$.
   b. $MR(x) = R'(x) = 50 - .05x$.
      $MR(100) = 50 - .05(100) = 45$.
   c. $R(80) \approx R(100) + R'(100)\cdot(80-100) = 4,750 + 45(-20) = 3,850$.
   d. $R(80) = 50 \cdot 80 - .025(80)^2 = 3,840$.
   e. The approximation to $R(80)$ obtained in part c. comes from the linear approximation

   $$L(x) = R(100) + R'(100)\cdot(x-100)$$

   to $R(x)$ for x near 100.
22. The linear approximation to $C(x_0+1)$ based at production level x_0 is obtained by using the linear approximation

   $$L(x) = C(x_0) + C'(x_0)\cdot(x - x_0)$$

   to values of $C(x)$ for x near x_0.
23. The volume V of a sphere of radius r is given by the formula

   $$V = f(r) = \frac{4}{3}\pi r^3.$$

   When the radius of the sphere is increased by 5% from a given value to r_0, the corresponding percentage increase in the volume is given by

   $$\frac{f'(r_0)\cdot[1.05r_0 - r_0]}{f(r_0)}\cdot 100 = \frac{4\pi r_0^2\cdot(.05r_0)}{\frac{4}{3}\pi r_0^3}\cdot 100 = 15\% \ .$$

24. The drag D as a function of the speed v is given by the formula

$$D = cv^2 = f(v)$$

where c is a constant. When the speed is increased by 50% from a given value to v_0, the corresponding percentage increase in the drag is given by

$$\frac{f'(v_0) \cdot (1.5v_0 - v_0)}{f(v_0)} \cdot 100 = \frac{2cv_0 \cdot (.5v_0)}{cv_0^2} \cdot 100 = 100\% .$$

25. a. $V(r) = 3 \cdot \pi r^2$ liters/min.
    b. $V(4) = 3\pi(4)^2 = 48\pi$ liters/min.
    c. $V(4.5) - V(4) \approx V'(4) \cdot (4.5 - 4) = 6\pi(4) \cdot (.5) = 12\pi$ liters/min.

26. Let V denote the volume of the roller bearing, let ℓ denote the length, and let r denote the radius of the base. The volume is given by the formula $V = \pi r^2 \ell$. Since the length is given as 5 cm and is assumed to be precise, the volume is given by $V = 5\pi r^2 = f(r)$. Our estimate of the deviation from the intended volume is

$$f'(2) \cdot [2.05 - 2] = 20\pi \cdot (.05) = \pi.$$

27. The area A of the plot is given by the formula

$$A = s^2 = f(s).$$

Our approximation to the maximum absolute variation in the area of the plot is given by

$$f'(200) \cdot (2) = 800 \text{ ft}^2.$$

28. $f(9) \approx f(10) + f'(10) \cdot [9 - 10] = 6 - 3 \cdot (-1) = 9.$

Solutions to Exercise Set 3.11

1. The Mean Value Theorem does not hold for this function, because $f(x)$ is not differentiable at the point (0,0).

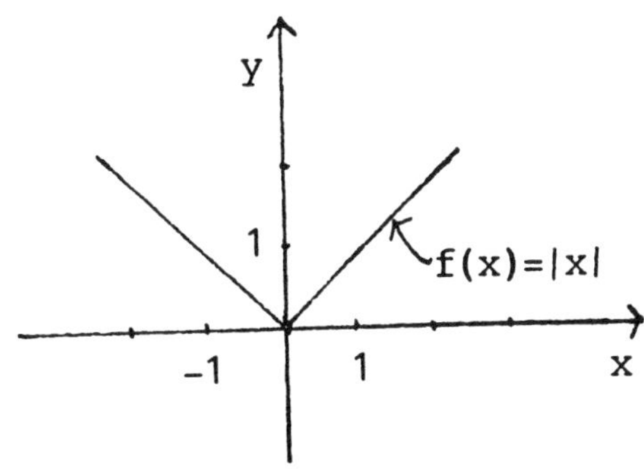

2. $f(x) = 2x^2 - 7, \qquad x \text{ in } [-1, 5].$
   $f'(x) = 4x.$
   $f'(c) = \frac{f(5)-f(-1)}{5-(-1)} \implies 4c = \frac{43-(-5)}{6} \implies 4c = 8 \implies c = 2.$
3. $f'(x) = -2x.$
   $f'(c) = -2c = \frac{f(2)-f(0)}{2-0} = \frac{4-2^2-(4-0^2)}{2} = -2.$
   $c = 1.$
4. $f'(x) = 2x + 2.$
   $f'(c) = 2c + 2 = \frac{f(4)-f(0)}{4-0} = \frac{(4^2+2\cdot4)-(0^2+2\cdot0)}{4} = 6 \implies c = 2.$
5. $f(x) = \sqrt{x+5}, \qquad x \text{ in } [-5, 4].$
   $f'(x) = \frac{1}{2\sqrt{x+5}}.$
   $f'(c) = \frac{f(4)-f(-5)}{4-(-5)} \implies \frac{1}{2\sqrt{c+5}} = \frac{3-0}{9} \implies 2\sqrt{c+5} = 3 \implies \sqrt{c+5} = \frac{3}{2}$
   $\implies c + 5 = \frac{9}{4} \implies c = -\frac{11}{4}.$
6. $f'(x) = -\frac{1}{x^2}.$
   $f'(c) = -\frac{1}{c^2} = \frac{f(3)-f(1)}{3-1} = \frac{\frac{1}{3}-1}{2} = -\frac{1}{3} \implies c^2 = 3,\ c = \sqrt{3}.$
7. $f'(x) = 2x + 2.$
   $f'(c) = 2c + 2 = \frac{f(0)-f(-3)}{-3-0} = \frac{-3-(9-6-3)}{-3} = 1 \implies 2c + 2 = 1,\ c = -\frac{1}{2}.$
8. $f'(x) = 3x^2 - 2.$
   $f'(c) = 3c^2 - 2 = \frac{f(2)-f(0)}{2-0} = \frac{(8-4+4)-4}{2} = 2 \implies 3c^2 = 4;\ c = \frac{2}{\sqrt{3}}.$
9. By the Mean Value Theorem there is at least one time t_0 during the 2 hours that the speed $v(t_0)$ equals the average speed during the 2 hours.

   $$v(t_0) = \frac{120}{2} = 60 > 55 \text{ mph.}$$

   Therefore the driver violates the speed limit at least once.
10. a. No. The revenue function R might not be differentiable. Even if R is differentiable, the Mean Value Theorem only says that $R'(x) = 0$ for at least one x in (100,300).

    b. Yes. By the Mean Value Theorem for at least one sale level c in (100,300)

    $$R'(c) = \frac{R(300) - R(100)}{300 - 100} = 0.$$

11. Let x_1, x_2 be in (a,b) such that $x_1 < x_2$.

Applying the Mean Value Theorem on the closed interval $[x_1, x_2]$ we have for at least one c in (x_1, x_2) that

$$f'(c) = \frac{f(x_2) - f(x_1)}{x_2 - x_1}.$$

Since $f'(x) < 0$ for all x in (a,b), $f'(c) < 0$.

So $\frac{f(x_2)-f(x_1)}{x_2-x_1} < 0 \implies f(x_2) - f(x_1) < 0$ since $x_2 - x_1 > 0$

$\implies f(x_2) < f(x_1)$. So, $f(x)$ is decreasing on (a,b).

12. a. Define a function $d(x) = f(x) - \left[f(a) + \frac{f(b)-f(a)}{b-a}(x-a)\right]$.

b. The first part $f(x)$ of $d(x)$ is differentiable on (a,b).
The part in the brackets is a linear function, therefore, it is differentiable. Thus $d(x)$ is differentiable on (a,b).

c. $d(a) = f(a) - \left[f(a) + \frac{f(b)-f(a)}{b-a}(a-a)\right] = 0$
$d(b) = f(b) - \left[f(a) + \frac{f(b)-f(a)}{b-a}(b-a)\right]$
$= f(b) - [f(a) + f(b) - f(a)] = 0.$

d. Suppose $d(x)$ is different from 0 for at least one x in $[a,b]$. By Theorem 5 in Section 3.5, $d(x)$ has both a maximum and a minimum value on $[a,b]$. Since $d(a) = d(b) = 0$, this maximum and minimum cannot both be obtained at the end points of the interval.

e. From part d. the maximum or minimum value of $d(x)$ must occur at a c in (a,b). Therefore $d'(c) = 0$.

f. $d'(x) = f'(x) - \frac{f(b)-f(a)}{b-a}$, and $d'(c) = 0$.
$d'(c) = f'(c) - \frac{f(b)-f(a)}{b-a} = 0.$
So, $f'(c) = \frac{f(b)-f(a)}{b-a}$.

13. Suppose f is an even function on the interval $[-a,a]$ where $a > 0$. Suppose f is differentiable on the open interval $(-a,a)$, and suppose that the derivative function f' is continuous at $x = 0$.

If h is any positive real number such that $h < a$, then there is at least one real number c satisfying $-h < c < h$ such that

$$f'(c) = \frac{f(h) - f(-h)}{h - (-h)} = 0.$$

Therefore there are real numbers x arbitrarily close to 0 such that $f'(x) = 0$. In view of the supposed continuity of $f'(x)$ at $x = 0$, then $f'(0) = 0$. Therefore $f'(0) = \frac{f(a)-f(-a)}{a-(-a)}$.

14. Let $f(x) = \frac{1}{x^2}$ for all $x \neq 0$. Then, f is an even function, but $f'(0)$ does not exist.

15. Let x_1 and x_2 be any two real numbers in the interval (a, b) such that $x_1 < x_2$. Then there is a real number c satisfying $x_1 < c < x_2$ such that

$$f'(c) = \frac{f(x_2) - f(x_1)}{x_2 - x_1}.$$

Since $f'(c) = 0$, it follows that $[f(x_2) - f(x_1)] = 0 \implies f(x_2) = f(x_1)$. Therefore f is a constant function on (a, b).

Solutions to Review Exercises - Chapter 3

1. $f'(x) = 4 - 2x = 0$ for $x = 2$.

   For $x < 2$, $f'(x) > 0$, so $f(x)$ is increasing. For $x > 2$, $f'(x) < 0$, so $f(x)$ is decreasing. $f(2)$ is a relative maximum.

2. $f(x) = x^2 - 6x - 16$.

   $f'(x) = 2x - 6$.

   $f'(x) = 0$ for $x = 3$.

   For $x < 3$, $f'(x) < 0$, so $f(x)$ is decreasing.

   For $x > 0$, $f'(x) > 0$, so $f(x)$ is increasing.

   $f(3) = -25$ is a relative minimum.

3. $f'(x) = 2x - 2 = 0$ for $x = 1$.

   For $x < 1$, $f'(x) < 0$, so $f(x)$ is decreasing. For $x > 1$, $f'(x) > 0$, so $f(x)$ is increasing. $f(1)$ is a relative minimum.

4. $f'(x) = \frac{-1(x)-1(1-x)}{x^2} = -\frac{1}{x^2}$.

   $f'(x) < 0$ for all $x \neq 0$, so $f'(x)$ is decreasing on $(-\infty, 0)$ and on $(0, \infty)$.

5. $f'(x) = \frac{1(x+3)-1(x-3)}{(x+3)^2} = \frac{6}{(x+3)^2}$.

   $f'(x) > 0$ for all $x \neq -3$, so $f'(x)$ is increasing on $(-\infty, -3)$ and on $(-3, \infty)$.

6. $f(x) = \frac{x}{x^2+1}$.

$f'(x) = \frac{(x^2+1)\cdot 1 - x(2x)}{(x^2+1)^2} = \frac{1-x^2}{(x^2+1)^2}$.

$f'(x) = 0$ for $x = 1$, $x = -1$.

For $-\infty < x < -1$, $f'(x) < 0$, so $f(x)$ is decreasing.

For $-1 < x < 1$, $f'(x) > 0$, so $f(x)$ is increasing.

For $1 < x < \infty$, $f'(x) < 0$, so $f(x)$ is decreasing.

$f(-1) = -\frac{1}{2}$ is a relative minimum, and $f(1) = \frac{1}{2}$ is a relative maximum.

7. $f'(x) = 1 \cdot (16 - x^2)^{1/2} + x \cdot \left\{\frac{1}{2}(16 - x^2)^{-1/2} \cdot (-2x)\right\}$

$= (16 - x^2)^{-1/2} \cdot \{(16 - x^2) - x^2\}$

$= (16 - x^2)^{-1/2} \cdot (16 - 2x^2)$.

$f'(x) = 0$ for $x = \pm\sqrt{8}$.

$f'(x)$ does not exist for $x = \pm 4$.

Interval	Test Number	Sign of $f'(x)$	Conclusion
$(-4, -\sqrt{8})$	-3	$f'(-3) < 0$	decreasing
$(-\sqrt{8}, \sqrt{8})$	0	$f'(0) > 0$	increasing
$(\sqrt{8}, 4)$	3	$f'(3) < 0$	decreasing

$f(-\sqrt{8})$ is a relative minimum. $f(\sqrt{8})$ is a relative maximum.

8. $f'(x) = 4x^3 - 4x = 4x(x^2 - 1) = 0$ for $x = 0$ and ± 1.

Interval	Test Number	Sign of $f'(x)$	Conclusion
$(-\infty, -1)$	-2	$f'(-2) < 0$	decreasing
$(-1, 0)$	$-\frac{1}{2}$	$f'\left(-\frac{1}{2}\right) > 0$	increasing
$(0, 1)$	$\frac{1}{2}$	$f'\left(\frac{1}{2}\right) < 0$	decreasing
$(1, \infty)$	2	$f'(2) > 0$	increasing

$f(-1)$ and $f(1)$ are relative minima. $f(0)$ is a relative maximum.

9. $f'(x) = 3x^2 - 6x = 3x(x - 2) = 0$ for $x = 0$ and $x = 2$.

Interval	Test Number	Sign of $f'(x)$	Conclusion
$(-\infty, 0)$	-1	$f'(-1) > 0$	increasing
$(0, 2)$	1	$f'(1) < 0$	decreasing
$(2, \infty)$	3	$f'(3) > 0$	increasing

$f(0)$ is a relative maximum. $f(2)$ is a relative minimum.

10. $f'(x) = 2x - 2x^{-2} = 2x^{-2}(x^3 - 1) = 0$ for $x = 1$. $f'(x)$ does not exist for $x = 0$.

Interval	Test Number	Sign of $f'(x)$	Conclusion
$(-\infty, 0)$	-1	$f'(-1) < 0$	decreasing
$(0, 1)$	$\frac{1}{2}$	$f'\left(\frac{1}{2}\right) < 0$	decreasing
$(1, \infty)$	2	$f'(2) > 0$	increasing

$f(1)$ is a relative minimum.

11. $f(x) = 4 - 2x^{2/3}$.

$f'(x) = -2 \cdot \frac{2}{3}x^{-1/3} = -\frac{4}{3}x^{-1/3}$.

$f'(x)$ is undefined for $x = 0$.

For $x < 0$, $f'(x) > 0$, so $f(x)$ is increasing.

For $x > 0$, $f'(x) < 0$, so $f(x)$ is decreasing.

$f(0) = 4$ is a relative maximum.

12. $f(x) = |x^2 - 4x - 5| = |(x - 5)(x + 1)|$.

$$f(x) = \begin{cases} x^2 - 4x - 5, & x < -1 \\ -x^2 + 4x + 5, & -1 < x < 5 \\ x^2 - 4x - 5, & x > 5. \end{cases}$$

For $x < -1$, $f'(x) = 2x - 4 < 0$, so $f(x)$ is decreasing.

For $-1 < x < 5$, $f'(x) = -2x + 4$. $f'(x) = 0$ for $x = 2$.

For $-1 < x < 2$, $f'(x) > 0$, so $f(x)$ is increasing.

For $2 < x < 5$, $f'(x) < 0$, so $f(x)$ is decreasing.

For $x > 5$, $f'(x) = 2x - 4 > 0$, so $f(x)$ is increasing.

$f(-1) = 0$ is a relative minimum, $f(2) = 9$ is a realtive maximum, and $f(5) = 0$ is a relative minimum.

13. $f'(x) = -2x(x^2 - 4)^{-2} = 0$ for $x = 0$.

$f(-1) = -\frac{1}{3}$, $f(1) = -\frac{1}{3}$, $f(0) = -\frac{1}{4}$.

The maximum is $f(0) = -\frac{1}{4}$. The minimum is $f(-1) = f(1) = -\frac{1}{3}$.

14. $f'(x) = 3x^2 - 6x = 3x(x - 2) = 0$ for $x = 0$ and $x = 2$.

$f(-1) = -3$, $f(1) = -1$, $f(0) = 1$.

The maximum is $f(0) = 1$. The minimum is $f(-1) = -3$.

15. $f(x) = x\sqrt{1-x^2}$, x in $[-1, 1]$.
$$f'(x) = 1 \cdot (1-x^2)^{1/2} + x\left(\tfrac{1}{2}(1-x^2)^{-1/2}(-2x)\right)$$
$$= (1-x^2)^{-1/2}\left((1-x^2) - x^2\right) = (1-x^2)^{-1/2} \cdot (1-2x^2).$$
$f'(x) = 0$ for $1 - 2x^2 = 0 \implies x^2 = \frac{1}{2} \implies x = \frac{-1}{\sqrt{2}}$ or $x = \frac{1}{\sqrt{2}}$.
$f'(-1) = 0$, $f(1) = 0$, $f\left(-\frac{1}{\sqrt{2}}\right) = -\frac{1}{2}$, $f\left(\frac{1}{\sqrt{2}}\right) = \frac{1}{2}$.
The maximum value is $f\left(\frac{1}{\sqrt{2}}\right) = \frac{1}{2}$, and the minimum value is $f\left(-\frac{1}{\sqrt{2}}\right) = -\frac{1}{2}$.

16. $f'(x) = 1 + \frac{2}{3}x^{-1/3}$.
$f'(x) = 0$ for $\frac{2}{3}x^{-1/3} = -1 \implies x^{-1/3} = -\frac{3}{2} \implies x^{1/3} = -\frac{2}{3} \implies x = \left(-\frac{2}{3}\right)^3 = -\frac{8}{27}$. $f'(x)$ does not exist for $x = 0$.
$f(-1) = 0$, $f(1) = 2$, $f\left(-\frac{8}{27}\right) = \frac{4}{27}$, $f(0) = 0$.
The maximum is $f(1) = 2$. The minimum is $f(-1) = f(0) = 0$.

17. $f'(x) = 1 - \left\{\frac{1}{2}(1-x^2)^{-1/2} \cdot (-2x)\right\} = 1 + x \cdot (1-x^2)^{-1/2}$.
$f'(x) = 0$ for $\frac{x}{\sqrt{1-x^2}} = -1 \implies \frac{x^2}{1-x^2} = 1 \implies x^2 = 1 - x^2 \implies x^2 = \frac{1}{2} \implies x = \pm\sqrt{\frac{1}{2}}$.
$f(-1) = -1$, $f(1) = 1$, $f\left(-\sqrt{\frac{1}{2}}\right) = -\sqrt{2}$, $f\left(\sqrt{\frac{1}{2}}\right) = 0$.
The maximum is $f(1) = 1$. The minimum is $f\left(-\sqrt{\frac{1}{2}}\right) = -\sqrt{2}$.

18. $f'(x) = \frac{1}{3}x^{-2/3} - 1$.
$f'(x) = 0$ for $\frac{1}{3} = x^{2/3} \implies x = \pm\left(\frac{1}{3}\right)^{3/2}$. $f'(x)$ does not exist for $x = 0$.
$f(-1) = 0$, $f(1) = 0$, $f\left(\left(\frac{1}{3}\right)^{3/2}\right) = \frac{2}{3^{3/2}}$, $f\left(-\left(\frac{1}{3}\right)^{3/2}\right) = -\frac{2}{3^{3/2}}$, $f(0) = 0$.
The maximum is $f\left(\left(\frac{1}{3}\right)^{3/2}\right) = \frac{2}{3^{3/2}}$. The minimum is $f\left(-\left(\frac{1}{3}\right)^{3/2}\right) = -\frac{2}{3^{3/2}}$.

19. $f'(x) = 1 + 8x^{-3}$.
$f'(x) = 0$ for $x^{-3} = -\frac{1}{8} \implies x = -2$.
$f(-3) = -\frac{31}{9}$, $f(-1) = -5$, $f(-2) = -3$.
The maximum is $f(-2) = -3$. The minimum is $f(-1) = -5$.

20. $f'(x) = 1 \cdot (1+x^2)^{-1} + x \cdot \{(-1)(1+x^2)^{-2} \cdot (2x)\}$
$$= (1+x^2)^{-2}\{(1+x^2) - 2x^2\} = (1+x^2)^{-2}(1-x^2).$$
$f'(x) = 0$ for $x = \pm 1$.
$f(-2) = -\frac{2}{5}$, $f(2) = \frac{2}{5}$, $f(-1) = -\frac{1}{2}$, $f(1) = \frac{1}{2}$.
The maximum is $f(1) = \frac{1}{2}$. The minimum is $f(-1) = -\frac{1}{2}$.

21. $f'(x) = 4x^3 - 4x = 4x(x^2 - 1) = 0$ for $x = 0$ and $x = \pm 1$.
$f(-2) = 8$, $f(2) = 8$, $f(0) = 0$, $f(-1) = -1$, $f(1) = -1$.
The maximum is $f(-2) = f(2) = 8$. The minimum is $f(-1) = f(1) = -1$.

22. $f'(x) = \frac{(1)(x^2+3)-(x-1)(2x)}{(x^2+3)^2} = \frac{x^2+3-2x^2+2x}{(x^2+3)^2} = \frac{-x^2+2x+3}{(x^2+3)^2}$
$= \frac{(-x+3)(x+1)}{(x^2+3)^2}$.
$f'(x) = 0$ for $x = 3$ and $x = -1$.
$f(-2) = -\frac{3}{7}$, $f(4) = \frac{3}{19}$, $f(3) = \frac{2}{12} = \frac{1}{6}$, $f(-1) = -\frac{2}{4} = -\frac{1}{2}$.
The maximum is $f(3) = \frac{1}{6}$. The minimum is $f(-1) = -\frac{1}{2}$.

23. Vertical asymptote: $x = 7$.
$\lim_{x \to \pm\infty} f(x) = 0 \implies$ horizontal asymptote is $y = 0$.

24. Vertical asymptote: $x = -2$.
$\lim_{x \to \pm\infty} f(x) = \lim_{x \to \pm\infty} \frac{1-\frac{2}{x}}{1+\frac{2}{x}} = \frac{1-0}{1+0} = 1 \implies$ horizontal asymptote is $y = 1$.

25. $y = \frac{x^2}{4-x^2}$.
$\lim_{x \to \pm\infty} \frac{x^2}{4-x^2} = \lim_{x \to \pm\infty} \frac{x^2}{4-x^2} \cdot \frac{\frac{1}{x^2}}{\frac{1}{x^2}} = \lim_{x \to \pm\infty} \frac{1}{\frac{4}{x^2}-1} = \frac{1}{0-1} = -1.$
Therefore the line $y = -1$ is a horizontal asymptote.
The lines $x = 2$ and $x = -2$ are vertical asymptotes.

26. $f(x) = \frac{x^2+5}{3-x^2}$.
$\lim_{x \to \pm\infty} f(x) = -1 \implies$ the line $y = -1$ is a horizontal asymptote.
The lines $x = \sqrt{3}$ and $x = -\sqrt{3}$ are vertical asymptotes.

27. Vertical asymptotes: $x = \pm 3$.
$\lim_{x \to \pm\infty} f(x) = \lim_{x \to \pm\infty} \frac{1+\frac{9}{x^2}}{1-\frac{9}{x^2}} = \frac{1+0}{1-0} = 1 \implies$ horizontal asymptote is $y = 1$.

28. $y = \frac{(x-1)}{(x-1)(x+2)} = \frac{1}{(x+2)}$. The only vertical asymptote is $x = -2$.
$\lim_{x \to \pm\infty} f(x) = \lim_{x \to \pm\infty} \frac{\frac{1}{x}-\frac{1}{x^2}}{1+\frac{1}{x}-\frac{2}{x^2}} = \frac{0-0}{1+0-0} = 0 \implies$ horizontal asymptote is $y = 0$.

29. For $x < 0$, $f(x) = \frac{x}{-x} = -1$; for $x > 0$, $f(x) = \frac{x}{x} = 1$. So there is no vertical asymptote.
Horizontal asymptotes: $y = -1$ and $y = 1$.

30. $f(x) = \frac{9+2x^4}{x^2+3x+2} = \frac{9+2x^4}{(x+1)(x+2)}$.

For $x \neq 0$, $f(x) = \frac{\frac{9}{x^2}+2x^2}{1+\frac{3}{x}+\frac{2}{x^2}} \to \infty$ as $x \to \pm\infty$.

Therefore there is no horizontal asymptote.

The lines $x = -1$ and $x = -2$ are vertical asymptotes.

31. Vertical asymptote: $x = 0$.

$\lim_{x\to\pm\infty} f(x) = 4 \implies$ horizontal asymptote is $y = 4$.

32. There is no vertical asymptote because $(x^2+6)^2 > 0$ for all x.

$\lim_{x\to\pm\infty} f(x) = \lim_{x\to\pm\infty} \frac{x^3}{x^4+12x^2+36} = \lim_{x\to\pm\infty} \frac{\frac{1}{x}}{1+\frac{12}{x^2}+\frac{36}{x^4}} = \frac{0}{1+0+0} = 0 \implies$ horizontal asymptote is $y = 0$.

33. $\lim_{x\to+\infty} f(x) = \lim_{x\to+\infty} \frac{2+\frac{6}{x}}{\frac{9}{x}-1} = \frac{2+0}{0-1} = -2$.

34. $\lim_{x\to\infty} \frac{x^2+3x+5}{3x^2+6} = \lim_{x\to\infty} \frac{x^2+3x+5}{3x^2+6} \cdot \frac{\frac{1}{x^2}}{\frac{1}{x^2}} = \lim_{x\to\infty} \frac{1+\frac{3}{x}+\frac{5}{x^2}}{3+\frac{6}{x^2}}$

$= \frac{1+0+0}{3+0} = \frac{1}{3}$.

35. $\lim_{x\to\infty} \frac{7}{3x-x^{-3/2}} = 0$ because $(3x) \to \infty$ and $x^{-3/2} \to 0$ as $x \to \infty$.

36. $\lim_{x\to+\infty} f(x) = \lim_{x\to+\infty} \frac{7}{3-\frac{1}{x^3}} = \frac{7}{3}$.

37. $\lim_{x\to 4^-} \frac{x^2+5\sqrt{x}}{x^2-16} = -\infty$ because as $x \to 4^-$, $(x^2+5\sqrt{x}) \to 26 > 0$ and $(x^2-16) \to 0^-$.

38. $\lim_{x\to 3^+} \left|\frac{x^2-4x-3}{x-3}\right| = \lim_{x\to 3^+} \frac{-x^2+4x+3}{x-3} = +\infty$ because as $x \to 3^+$, $(-x^2+4x+3) \to 6 > 0$ and $(x-3) \to 0^+$.

39. As $x \to 2^-$, (x^2-4) approaches 0 through negative values. Hence,

$\lim_{x\to 2^-} \frac{6}{x^2-4} = -\infty$.

40. $\lim_{x\to\infty} \frac{(x-6)(x+3)}{2x^3+8} = \lim_{x\to\infty} \frac{x^2-3x-18}{2x^3+8} = \lim_{x\to\infty} \frac{x^2-3x-18}{2x^3+8} \cdot \frac{\frac{1}{x^3}}{\frac{1}{x^3}}$

$= \lim_{x\to\infty} \frac{\frac{1}{x}-\frac{3}{x^2}-\frac{18}{x^3}}{2+\frac{8}{x^3}} = \frac{0-0-0}{2+0} = \frac{0}{2} = 0$.

41. $\lim_{x\to 1^-} \frac{5x^2}{1-x} = +\infty$ because as $x \to 1^-$, $5x^2 \to 5 > 0$ and $(1-x) \to 0^+$.

42. $\lim_{x\to+\infty} \frac{1-x^{-2}}{2+x^{-2}} = \lim_{x\to+\infty} \frac{1-\frac{1}{x^2}}{2+\frac{1}{x^2}} = \frac{1-0}{2+0} = \frac{1}{2}$.

43. $y = 4x - x^2$.

(1) The domain is all x.

(2) $y = 0$ for $x = 0$ and $x = 4$.

(3),(4) There are no asymptotes.

(5) $y' = 4 - 2x = 0$ for $x = 2$.

Interval	Test Number	Sign of $f'(x)$	Conclusion
$(-\infty, 2)$	-1	$f'(-1) > 0$	increasing
$(2, \infty)$	3	$f'(3) < 0$	decreasing

$f(2)$ is the relative maximum.

(6) $f''(x) = -2$

$f''(x) < 0$ for all x, so the curve is concave down on $(-\infty, \infty)$.

(7)

x	-1	0	1	2	3	4
y	-5	0	3	4	3	0

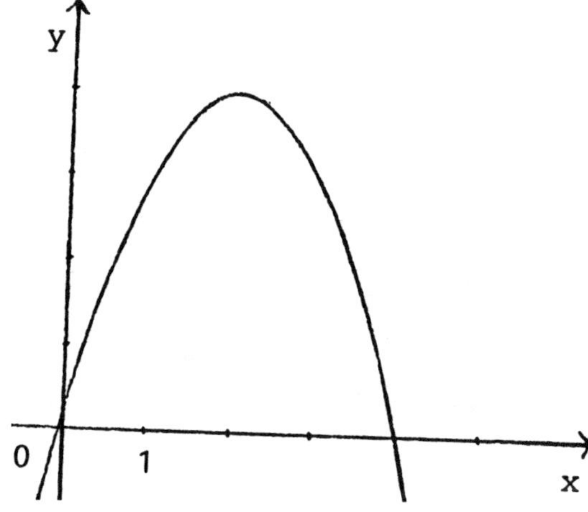

44. $y = x(x-1)(x+3) = x^3 + 2x^2 - 3x$.

(1) The domain is all x.

(2) $y = 0$ for $x = 0$, $x = 1$, and $x = -3$.

(3),(4) There are no asymptotes.

(5) $y' = 3x^2 + 4x - 3$.

$y' = 0$ for $x = \frac{-4 \pm \sqrt{4^2 - 4(3)(-3)}}{2 \cdot (3)} = \frac{-4 \pm 2\sqrt{13}}{6} = \frac{-2 \pm \sqrt{13}}{3}$

$= 0.535$ and -1.869.

Interval	Test Number	Sign of $f'(x)$	Conclusion
$(-\infty, -1.869)$	-2	$f'(-2) > 0$	increasing
$(-1.869, 0.535)$	0	$f'(0) < 0$	decreasing
$(0.535, \infty)$	2	$f'(2) > 0$	increasing

$f(-1.869)$ is the relative maximum. $f(0.535)$ is the relative minimum.

(6) $y'' = 6x + 4 = 0$ for $x = -\frac{2}{3}$.

Interval	Test Number	$f''(x)$	Conclusion
$(-\infty, -\frac{2}{3})$	-1	$f''(-1) < 0$	concave down
$(-\frac{2}{3}, \infty)$	0	$f''(0) > 0$	concave up

The inflection point occurs when $x = -\frac{2}{3}$.

(7)

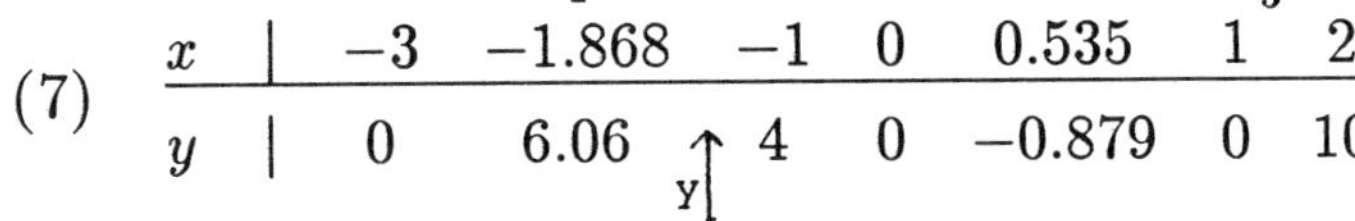

x	-3	-1.868	-1	0	0.535	1	2
y	0	6.06	4	0	-0.879	0	10

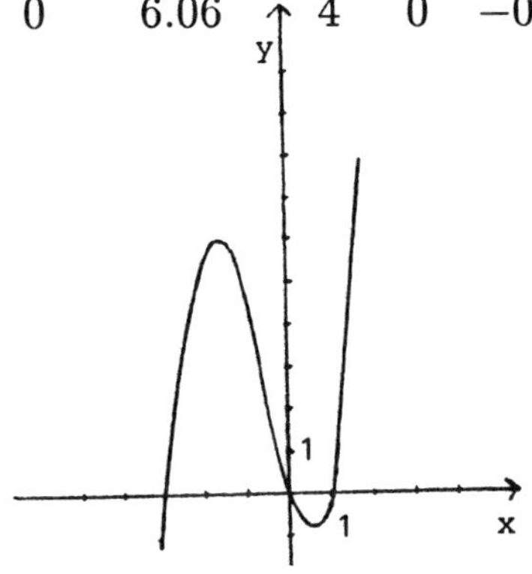

45. $f(x) = x^2 - 2x - 3 = (x + 1)(x - 3)$.

(1) The domain is all x.

(2) $f(x) = 0$ for $x = -1$, $x = 3$.

(3),(4) There are no asymptotes.

(5) $f'(x) = 2x - 2 = 0$ for $x = 1$.

For $x < 1$, $f'(x) < 0$, $f(x)$ is decreasing.

For $x > 1$, $f'(x) > 0$, $f(x)$ is increasing.

$f(1) = -4$ is the relative minimum.

(6) $f''(x) = 2 > 0$ for all x. So the graph is concave down on $(-\infty, \infty)$.

(7)

x	-2	-1	0	1	2	3
$f(x)$	5	0	-3	-4	-3	0

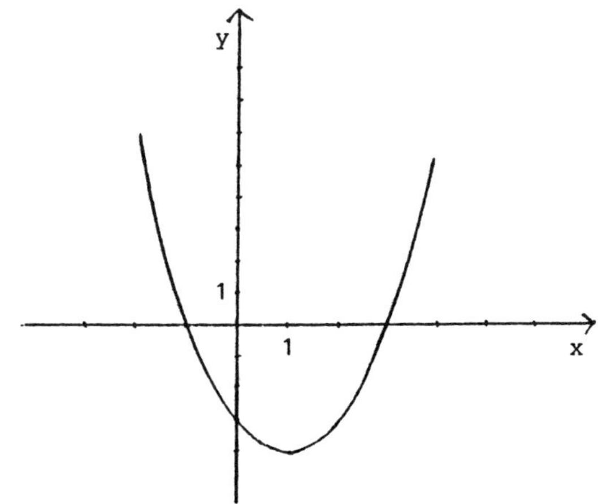

46. $f(x) = \frac{1-x}{x}$.

(1) The domain is all $x \neq 0$.

(2) $f(x) = 0$ for $x = 1$.

(3) The vertical asymptote is $x = 0$.

(4) $\lim_{x\to\pm\infty} \frac{1-x}{x} = -1$. So the horizontal asymptote is $y = -1$.

(5) $f'(x) = -\frac{1}{x^2} < 0$ for all $x \neq 0$.

So $f(x)$ is decreasing on $(-\infty, 0)$ and $(0, \infty)$, and there is no relative extremum.

(6) $f''(x) = \frac{2}{x^3}$.

For $x < 0$, $f''(x) < 0$, the graph of $f(x)$ is concave down.

For $x > 0$, $f''(x) > 0$, the graph of $f(x)$ is concave up.

There is no inflection point.

(7)

x	-2	-1	$-\frac{1}{2}$	$\frac{1}{2}$	1	2
$f(x)$	$-\frac{3}{2}$	-2	-3	1	0	$-\frac{1}{2}$

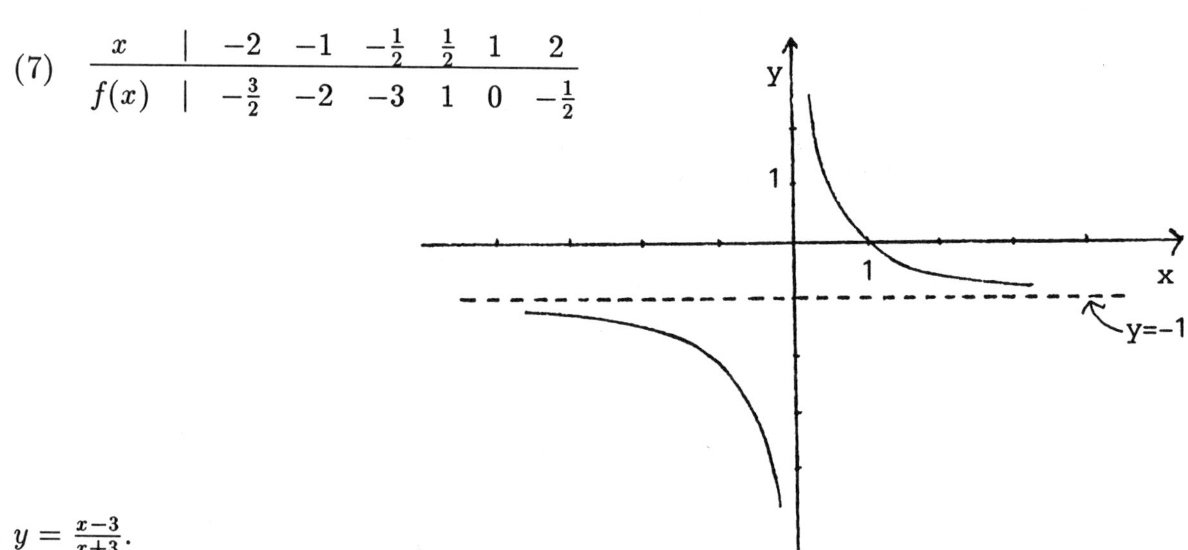

47. $y = \frac{x-3}{x+3}$.

(1) The domain is all $x \neq -3$.

(2) $y = 0$ for $x = 3$.

(3) The vertical asymptote is $x = -3$.

(4) $\lim_{x\to\pm\infty} \frac{x-3}{x+3} = 1$. So the horizontal asymptote is $y = 1$.

(5) $y' = \frac{6}{(x+3)^2} > 0$ for all $x \neq -3$.

So the graph of y is increasing on $(-\infty, -3)$ and $(-3, \infty)$, and there is no relative extremum.

(6) $y'' = \frac{-12}{(x+3)^3}$.

For $x < -3$, $y'' > 0$, the graph is concave up.

For $x > -3$, $y'' < 0$, the graph is concave down.

There is no inflection point.

(7)

x	-5	-4	-3.5	-2.5	-2	-1	0	1	3
y	4	7	13	-11	-5	-2	-1	$-\frac{1}{2}$	0

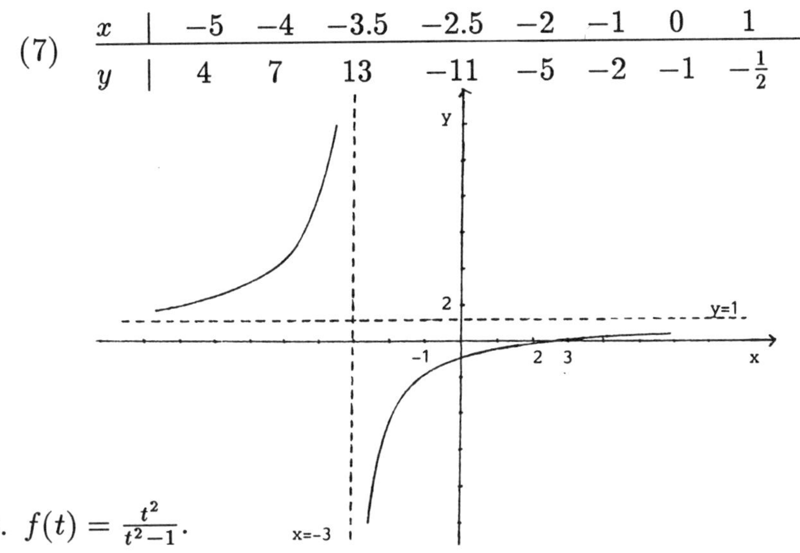

48. $f(t) = \frac{t^2}{t^2-1}$.

(1) The domain is all t except ± 1.

(2) $f(t) = 0$ for $t = 0$.

(3) The vertical asymptotes are $t = -1$ and $t = 1$.

(4) $\lim\limits_{t\to\pm\infty} \frac{t^2}{t^2-1} = 1$, so the horizontal asymptote is $y = 1$.

(5) $f'(t) = \left(1 + \frac{1}{t^2-1}\right)' = -\frac{2t}{(t^2-1)^2}$.

$f'(t) = 0$ for $t = 0$.

$(t^2-1)^2 > 0$ for all $t \neq \pm 1$.

For $t < 0$, $t \neq -1$, $f'(t) > 0$. So $f(t)$ is increasing on $(-\infty, -1)$ and $(-1, 0)$.

For $t > 0$, $t \neq 1$, $f'(t) < 0$. So $f(t)$ is decreasing on $(0, 1)$ and $(1, \infty)$.

$f(0) = 0$ is the relative maximum.

(6) $f''(t) = \frac{-2(t^2-1)^2 + 2t\cdot 2(t^2-1)2t}{(t^2-1)^4} = \frac{2(3t^2+1)}{(t^2-1)^3}$.

$f''(t)$ does not exist for $t = \pm 1$. $3t^2 + 1 > 0$ for all t.

Interval	Sign of $f''(x)$	Conclusion
$(-\infty, -1)$	$+$	concave up
$(-1, 1)$	$-$	concave down
$(1, \infty)$	$+$	concave up

$f(t)$ has no inflection points.

(7)

t	-4	-3	-2
$f(t)$	1.07	1.125	1.33

t	-1.5	-0.5	0
$f(t)$	1.8	-0.33	0

t	0.5	1.5	2
$f(t)$	-0.33	1.8	1.33

t	3	4
$f(t)$	1.125	1.07

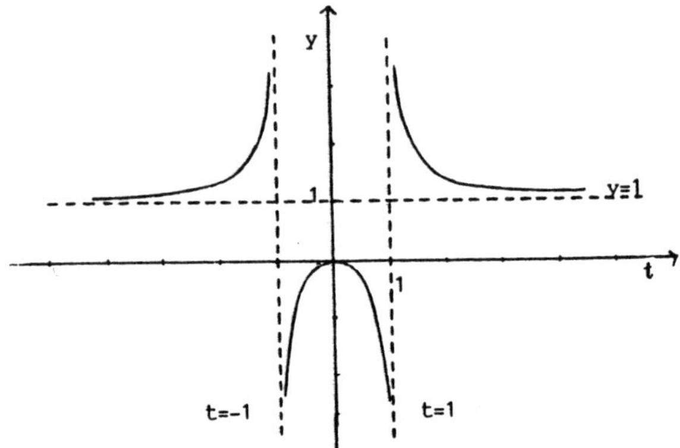

49. $f(x) = x\sqrt{16-x^2}$.

(1) The domain is all x such that $-4 \le x \le 4$.

(2) $f(x) = 0$ for $x = 0$ and $x = \pm 4$.

(3),(4) There are no asymptotes.

(5) $f'(x) = 1 \cdot (16-x^2)^{1/2} + x\left[\frac{1}{2}(16-x^2)^{-1/2} \cdot (-2x)\right]$
$= (16-x^2)^{-1/2} \cdot (16-2x^2)$
$f'(x) = 0$ for $x = \pm 2\sqrt{2}$. $f'(x)$ does not exist for $x = \pm 4$.

Interval	Sign of $f'(x)$	Conclusion
$(-4, -2\sqrt{2})$	$-$	decreasing
$(-2\sqrt{2}, 2\sqrt{2})$	$+$	increasing
$(2\sqrt{2}, 4)$	$-$	decreasing

$f\left(-2\sqrt{2}\right) = -8$ is the relative minimum.
$f\left(2\sqrt{2}\right) = 8$ is the relative maximum.

(6) $f''(x) = -\frac{1}{2}(16-x^2)^{-3/2}(-2x)(16-2x^2) + (16-x^2)^{-1/2}(-4x)$

$= x(16-x^2)^{-3/2}\left[(16-2x^2) - 4(16-x^2)\right]$

$= x(16-x^2)^{-3/2}(-48+2x^2).$

$f''(x) = 0$ for $x = 0$. $f''(x)$ does not exist for $x = \pm 4$.

Interval	Sign of $f''(x)$	Conclusion
$(-4, 0)$	$+$	concave up
$(0, 4)$	$-$	concave down

The inflection point occurs when $x = 0$.

(7)

x	-4	-3	$-2\sqrt{2}$
$f(x)$	0	-7.94	-8

x	-2	-1	0
$f(x)$	-6.93	-3.87	0

x	1	2	$-2\sqrt{2}$
$f(x)$	$+3.87$	$+6.93$	$+8$

x	3	4
$f(x)$	$+7.94$	0

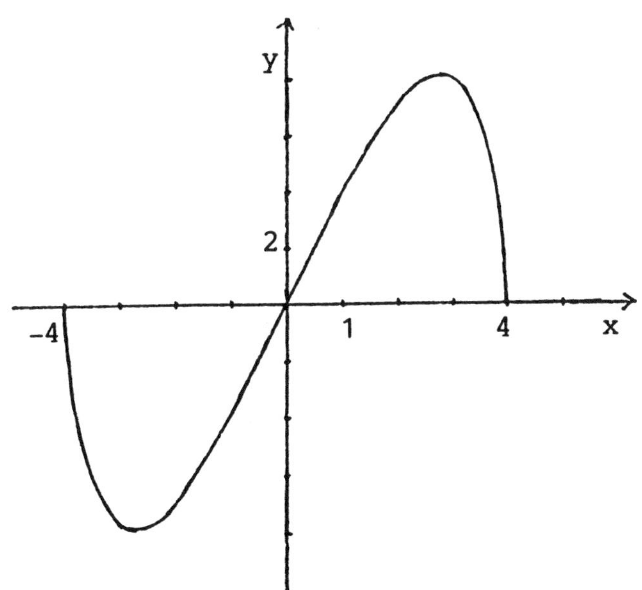

50. $y = \frac{\sqrt{x}}{1+\sqrt{x}}$.

(1) The domain is all $x \geq 0$.

(2) $y = 0$ for $x = 0$.

(3) There is no vertical asymptote.

(4) $\lim_{x\to+\infty} \frac{\sqrt{x}}{1+\sqrt{x}} = 1$. So the horizontal asymptote is $y = 1$.

(5) $y' = \left(1 - \frac{1}{1+\sqrt{x}}\right)' = \frac{1}{\left(1+\sqrt{x}\right)^2} \cdot \frac{1}{2\sqrt{x}} = \frac{1}{2\sqrt{x}\left(1+\sqrt{x}\right)^2}$.
$y' > 0$ for all $x > 0$, and y' does not exist for $x = 0$.
So y is increasing on $[0, \infty)$.

(6) $y'' = \frac{1}{2} \cdot \frac{-\left[\frac{1}{2}x^{-1/2}\left(1+\sqrt{x}\right)^2+\sqrt{x}\cdot 2\left(1+\sqrt{x}\right)\cdot\frac{1}{2}x^{-1/2}\right]}{x\left(1+\sqrt{x}\right)^4}$
$= -\frac{x^{-1/2}\left(1+\sqrt{x}+2\sqrt{x}\right)}{4x\left(1+\sqrt{x}\right)^3} = -\frac{3\sqrt{x}+1}{4\left[\sqrt{x}\left(1+\sqrt{x}\right)\right]^3}$.
$y'' < 0$ for all $x > 0$.
The graph of $y = \frac{\sqrt{x}}{1+\sqrt{x}}$ is concave down on $[0, \infty)$.

(7)

x	0	$\frac{1}{2}$	1	2
y	0	0.41	0.5	0.59

x	3	4
y	0.63	0.67

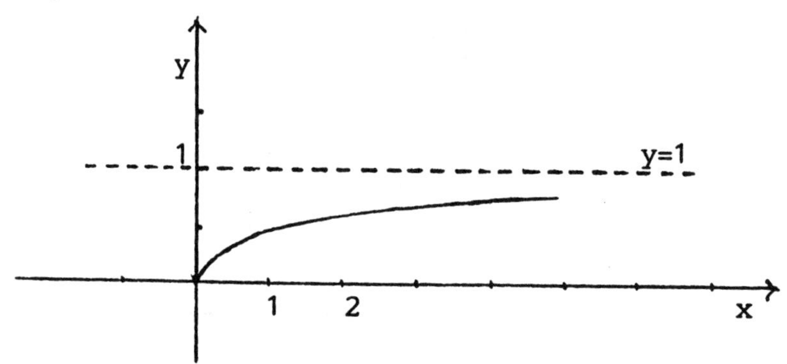

51. $y = \frac{x^2}{x^2+9}$.

(1) The domain is all x.

(2) $y = 0$ for $x = 0$.

(3),(4) There is no vertical asymptote.
Since $\lim_{x\to\pm\infty} \frac{x^2}{x^2+9} = 1$, the horizontal asymptote is $y = 1$.

(5) $y' = \frac{18x}{(x^2+9)^2} = 0$ for $x = 0$.

For $x < 0$, $y' < 0$, the graph is decreasing.

For $x > 0$, $y' > 0$, the graph is increasing.

$y = 0$ is the relative minimum when $x = 0$.

(6) $y'' = 18 \cdot \frac{1\cdot(x^2+9)^2 - x\cdot 2(x^2+9)\cdot 2x}{(x^2+9)^4} = \frac{18\cdot(x^2+9-4x^2)}{(x^2+9)^3} = \frac{54(3-x^2)}{(x^2+9)^3}$.

$y'' = 0$ for $x = \pm\sqrt{3}$.

Interval	Sign of $f''(x)$	Conclusion
$(-\infty, -\sqrt{3})$	$-$	concave down
$(-\sqrt{3}, \sqrt{3})$	$+$	concave up
$(\sqrt{3}, \infty)$	$-$	concave down

$(-\sqrt{3}, 0.25)$ and $(\sqrt{3}, 0.25)$ are the inflection points.

(7)

x	-3	-2	-1	0
y	0.5	0.31	0.1	0

x	1	2	3
y	0.1	0.31	0.5

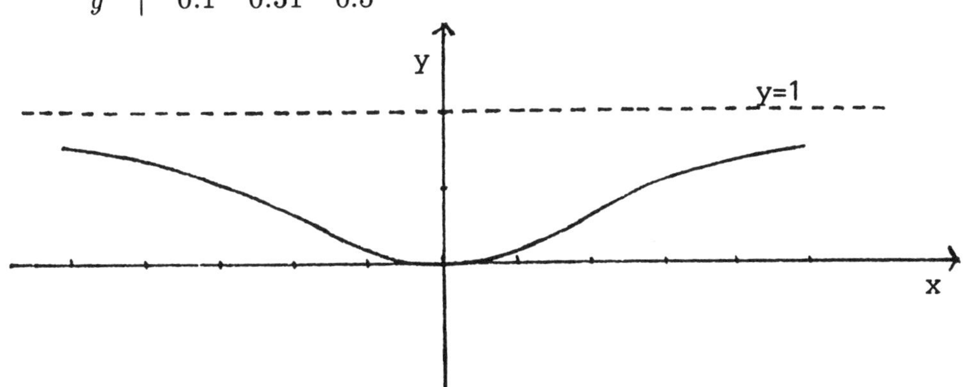

52. $f(x) = x^2 + \frac{2}{x} = \frac{x^3+2}{x}$.

(1) The domain is all $x \neq 0$.

(2) $f(x) = 0$ for $x = -\sqrt[3]{2}$.

(3) The vertical asymptote is $x = 0$.

(4) $\lim_{x\to\pm\infty} \left(x^2 + \frac{2}{x}\right) = \infty$. There is no horizontal asymptote.

(5) $f'(x) = 2x - \frac{2}{x^2} = \frac{2(x^3-1)}{x^2}$.

$f'(x) = 0$ for $x = 1$.

For $x < 1$, $x \neq 0$, $f'(x) < 0$; $f(x)$ is decreasing on $(-\infty, 0)$ and $(0, 1)$.

For $x > 1$, $f'(x) > 0$; $f(x)$ is increasing on $(1, \infty)$.

$f(1) = 3$ is the relative minimum.

(6) $f''(x) = 2 + \frac{4}{x^3} = \frac{2(x^3+2)}{x^3} = 0$ for $x = -\sqrt[3]{2}$. $f''(x)$ does not exist for $x = 0$.

Interval	Sign of $f''(x)$	Conclusion
$(-\infty, -\sqrt[3]{2})$	$+$	concave up
$(-\sqrt[3]{2}, 0)$	$-$	concave down
$(0, \infty)$	$+$	concave up

$(-\sqrt[3]{2}, 0)$ is the inflection point.

(7)

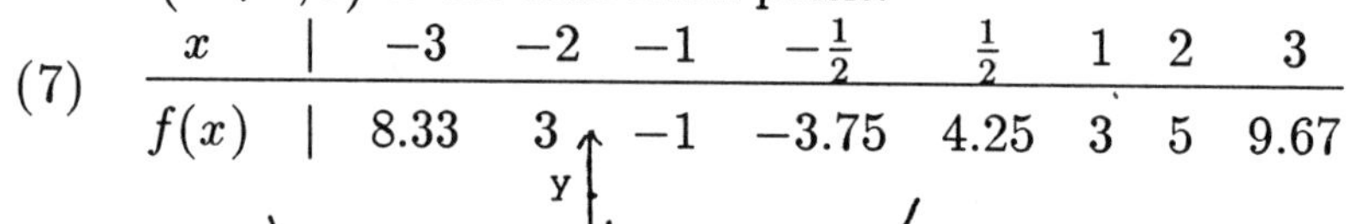

x	-3	-2	-1	$-\frac{1}{2}$	$\frac{1}{2}$	1	2	3
$f(x)$	8.33	3	-1	-3.75	4.25	3	5	9.67

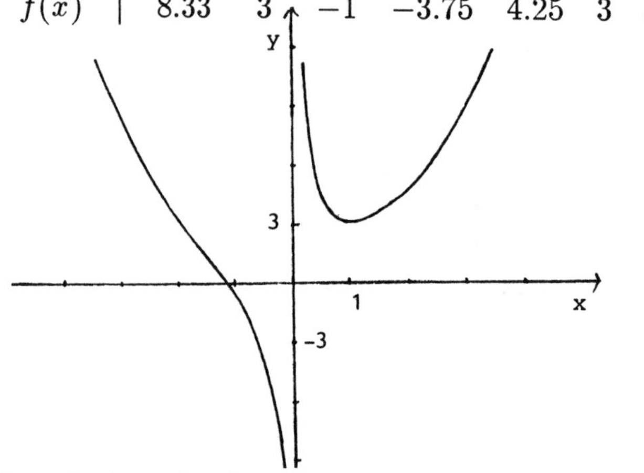

53. Let $\ell =$ the length of a side of the square.

    $A =$ the area of the square.

    Then $A = \ell^2$. $\frac{dA}{dt} = 2\ell \cdot \frac{d\ell}{dt}$.

    When $\ell = 10$ and $\frac{d\ell}{dt} = 2$, $\frac{dA}{dt} = 2 \cdot 10 \cdot 2 = 40\text{cm}^2/\text{sec}$.

54. Let $x =$ the number of trees per acre

    $y =$ the average yield per tree

    $z =$ the total crop.

    Then $y = 300 - 10(x - 25) = 550 - 10x$.

    $z = x \cdot y = x \cdot (550 - 10x) = 550x - 10x^2 = f(x)$.

    $\frac{dz}{dx} = 550 - 20x = 0$ for $x = 27.5$.

    For $0 < x < 27.5$, $\frac{dz}{dx} > 0$.

    For $27.5 < x < 55$, $\frac{dz}{dx} < 0$.

So $f(x)$ is maximum when $x = 27.5$.

$f(27) = 7560, \qquad f(28) = 7560.$

Thus either 27 or 28 trees per acre gives the largest crop.

55. Let $r =$ the radius of the disc of oil,

$\ell =$ the thickness of the layer of oil.

Then $V = \pi r^2 \ell. \qquad \ell = \frac{V}{\pi r^2}, \qquad \frac{d\ell}{dt} = -\frac{2V}{\pi r^3} \cdot \frac{dr}{dt}.$

When $t = 9$, $r(9) = 6a$ and $\frac{dr}{dt} = \frac{a}{\sqrt{9}} = \frac{a}{3}$,

$\frac{d\ell}{dt} = -\frac{2V}{\pi(6a)^3} \cdot \frac{a}{3} = -\frac{V}{324\pi a^2}$ m/sec.

56. a. The total weekly revenue $R(x) = x \cdot p(x) = x \cdot \left(90 - \frac{x}{500}\right) = 90x - \frac{x^2}{500}$.

b. $R'(x) = 90 - \frac{x}{250}$.

c. $R'(x) = 0 \implies \frac{x}{250} = 90 \implies x = 22500$ lb.

When $0 < x < 22500$, $R'(x) > 0$

When $22500 < x < 50 \times 2,000$, $R'(x) < 0$.

So the total weekly revenue is maximum for sales level $x = 22500$ lb. or 11.25 tons.

57. The weekly revenue $R(x) = x \cdot p(x) = x\left(5 + \frac{200}{x}\right) = 5x + 200$.

The weekly profit $P(x) = R(x) - C(x) = (5x + 200) - (5000 + 2x) = 3x - 4800$.

Since $P(x)$ is increasing with x for $0 \leq x \leq 1000$, the sales level $x = 1000$ maximizes profit.

58. Let $D =$ the number of miles you drive.

Then the total cost $f(v) = \frac{D}{v}C(v) = \frac{D}{v}(10 + 0.004v^2) = \frac{10D}{v} + 0.004Dv$.

$f'(v) = -\frac{10D}{v^2} + 0.004D = \frac{(0.004v^2 - 10)D}{v^2}$.

$f'(v) = 0$ for $0.004v^2 = 10 \implies v = 50$ m.p.h.

For $0 \leq v \leq 50$, $f'(v) < 0$. For $v > 50$, $f'(v) > 0$.

So the most economical speed is $v = 50$ m.p.h.

59. Since the time the car and the truck travel is the same, then

$\frac{20-b}{80} = \frac{a}{60} \implies a = \frac{3(20-b)}{4}$.

$\ell = \sqrt{a^2 + b^2}$.

$\frac{d\ell}{dt} = \frac{1}{2}(a^2 + b^2)^{-1/2} \cdot \left(2a\frac{da}{dt} + 2b\frac{db}{dt}\right)$

$= (a^2 + b^2)^{-1/2}(a \cdot 60 - b \cdot 80)$.

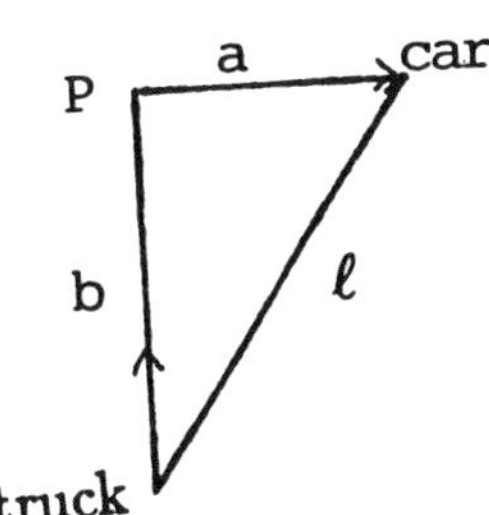

When $b=0$, $a=\frac{3\cdot 20}{4}=15$, $\frac{d\ell}{dt}=(15^2+0)^{-1/2}(15\cdot 60-0)=60$ km/hr.

60. $\frac{b}{a+b}=\frac{1.6}{8}$.

$8b=1.6(a+b)$.

$b=\frac{1}{4}a$.

$\frac{db}{dt}=\frac{1}{4}\cdot\frac{da}{dt}=\frac{1}{4}(1.2)=0.3$ m/sec.

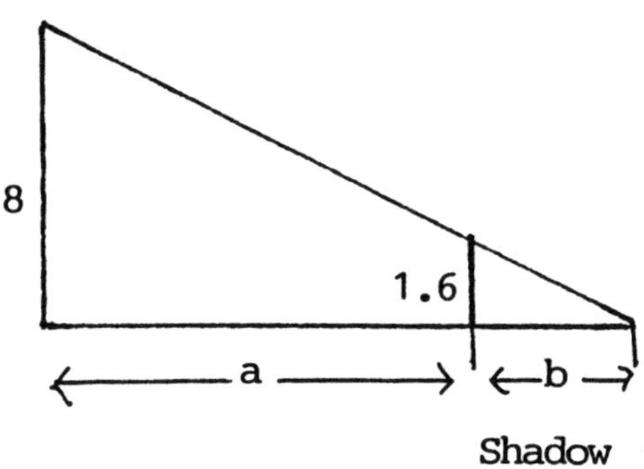

61. Let $x=$ the number of times per year the store orders dishwashers

$C(x)=$ the yearly costs.

Then $C(x)=150\cdot\frac{1}{2x}\cdot 15+25x+2\cdot 150=\frac{1125}{x}+25x+300$.

$C'(x)=-\frac{1125}{x^2}+25=\frac{25(x^2-45)}{x^2}=0$ for $x=\sqrt{45}=6.7$.

$C(6)=637.5$, $C(7)=635.7$.

So the store should order 7 times/yr to minimize cost.

62. The revenue $R(x)=xp(x)=x(800-2x)=800x-2x^2$.

The profits $P(x)=R(x)-C(x)=(800x-2x^2)-(100+6x+x^2)$

$=-3x^2+794x-100$.

$P'(x)=-6x+794=0$ for $x=132.3$.

For $0<x<132.3$, $P'(x)>0$. For $132.3<x<200$, $P'(x)<0$.

$P(132)=52436$, $P(133)=52435$.

So the maximum profits occur when output level $x=132$ items.

63. $P(x)=R(x)-C(x)-30x=-3x^2+794x-100-30x$

$=-3x^2+764x-100$.

$P'(x)=-6x+764=0$ for $x=127.3$.

For $0<x<127.3$, $P'(x)>0$. For $127.3<x<200$, $P'(x)<0$.

$P(127)=48541$, $P(128)=48540$.

So the optimal level would be changed to $x=127$ items/week.

64. Let $x =$ the subscription rate

$y =$ the number of subscriptions

$R =$ the revenue.

Then $y = 2000 - 50(x - 40) = 4000 - 50x$.

$R = x \cdot y = x(4000 - 50x) = 4000x - 50x^2$.

$R' = 4000 - 100x = 0$ for $x = 40$.

For $0 < x < 40$, $R' > 0$. For $40 < x < 80$, $R' < 0$.

So the maximum revenue occurs when the subscription rate $x = \$40$.

65. The revenue $R(x) = x \cdot p = x \cdot \sqrt{100 - x^2}$.

$R'(x) = \sqrt{100 - x^2} \cdot 1 + x \cdot \frac{1}{2}(100 - x^2)^{-1/2}(-2x) = \frac{(100-x^2)-x^2}{\sqrt{100-x^2}} = \frac{100-2x^2}{\sqrt{100-x^2}}$.

$R'(x) = 0$ for $x = \sqrt{50} \approx 7.07$, $p = \sqrt{100 - \left(\sqrt{50}\right)^2} = \sqrt{50} \approx 7.07$.

For $0 < x < 7.07$, $R'(x) > 0$; and for $7.07 < x < 10$, $R'(x) < 0$.

So the maximum revenue occurs when the price $p = 7.07$.

66. $(x + 2)^2 + (y - 3)^2 = 4$.

$2(x + 2) + 2(y - 3)\frac{dy}{dx} = 0$, $\quad \frac{dy}{dx} = -\frac{x+2}{y-3}$.

At $\left(-1, 3 + \sqrt{3}\right)$, $\frac{dy}{dx} = -\frac{-1+2}{3+\sqrt{3}-3} = -1/\sqrt{3}$.

The tangent line is $y - \left(3 + \sqrt{3}\right) = -\frac{1}{\sqrt{3}}(x - 1)$.

67. a. $v = s'(t) = 96 - 32t$.

b. $s'(t) = 0$ for $t = \frac{96}{32} = 3$ sec. $s(3) = 144$.

For $0 < t < 3$, $s'(t) > 0$; and for $3 < t < 6$, $s'(t) < 0$.

So the maximum height $s = 144$ ft. occurs when $t = 3$ sec after the ball is thrown.

c. $v' = -32 < 0$ for all $t \geq 0$.

So v is decreasing on (0,6) and has a maximum when $t = 0$ at height $s = 0$.

d. $s(t) = 0$ for $t = 0$ and $t = 6$.

So when $t = 6$ sec., $v = 96 - 32 \cdot 6 = -96$ ft/sec.

So the speed is 96 ft/sec. when it strikes the ground.

68. $P(x) = 40x - 300 - 0.5x^2$.

a. $P'(x) = 40 - x = 0$ for $x = 40$.

In $(0, 40)$, $P'(x) > 0$, so $P(x)$ is increasing.

b. In $(0, 40)$, $P'(x) > 0$; and in $(40, \infty)$, $P'(x) < 0$.

So the profit is maximum when $x = 40$.

69. a. $c(x) = \frac{C(x)}{x} = \frac{30x+200+0.5x^2}{x} = 30 + \frac{200}{x} + 0.5x$.

b. $c'(x) = -\frac{200}{x^2} + 0.5 = \frac{0.5x^2-200}{x^2} = 0$ for $x = \sqrt{\frac{200}{0.5}} = 20$.

For $0 < x < 20$, $c'(x) < 0$; and for $x > 20$, $c'(x) > 0$.

So average cost will be a minimum for $x = 20$ tires/day.

70. $C(x) = 8x + \frac{3200}{x}$.

$C'(x) = 8 - \frac{3200}{x^2} = \frac{8(x^2-400)}{x^2} = 0$ for $x = 20$.

For $0 < x < 20$, $C'(x) < 0$; and for $x > 20$, $C'(x) > 0$.

So the cost is minimum for lot size $x = 20$ bicycles.

71. $x^2y + xy^2 = -4$. $\quad 2xy + x^2\frac{dy}{dx} + y^2 + x \cdot 2y\frac{dy}{dx} = 0$.

$(x^2 + 2xy)\frac{dy}{dx} = -y^2 - 2xy$. $\quad \frac{dy}{dx} = \frac{-y(y+2x)}{x(x+2y)}$.

72. $x^3y - xy^3 = 10$. $\quad 3x^2y + x^3\frac{dy}{dx} - y^3 - x \cdot 3y^2 \cdot \frac{dy}{dx} = 0$.

$(x^3 - 3xy^2)\frac{dy}{dx} = y^3 - 3x^2y$. $\quad \frac{dy}{dx} = \frac{y(y^2-3x^2)}{x(x^2-3y^2)}$.

73. $\sqrt[3]{xy} - y = -6$. $\quad \frac{1}{3}(xy)^{-2/3}\left(y + x\frac{dy}{dx}\right) - \frac{dy}{dx} = 0$.

$\left(\frac{1}{3}(xy)^{-2/3}x - 1\right)\frac{dy}{dx} = -\frac{1}{3}(xy)^{-2/3}y$.

$\frac{dy}{dx} = \frac{-\frac{1}{3}(xy)^{-2/3}y}{\frac{1}{3}(xy)^{-2/3}x-1} = \frac{-y}{x-3(xy)^{2/3}}$.

At $(3, 9)$, $\frac{dy}{dx} = \frac{-9}{3-3(3\cdot 9)^{2/3}} = \frac{3}{8}$.

74. $\sqrt{x} - \sqrt{y} = 1$. $\quad \frac{1}{2\sqrt{x}} - \frac{1}{2\sqrt{y}}\frac{dy}{dx} = 0$.

$\frac{dy}{dx} = \sqrt{\frac{y}{x}}$.

At $(25, 16)$, $\frac{dy}{dx} = \sqrt{\frac{16}{25}} = \frac{4}{5}$.

75. $u(x) = \sqrt{x(x+24)} = \sqrt{x^2 + 24x}$.

$u'(x) = \frac{1}{2}(x^2 + 24x)^{-1/2} \cdot (2x + 24) = \frac{x+12}{\sqrt{x^2+24x}}$.

a. $u(x)$ is increasing for all $x > 0$.

b. There are no intervals on which $u(x)$ is decreasing.

c. $u(x)$ does not have any relative extrema.

d. $u''(x) = \frac{(x^2+24x)^{1/2}\cdot 1-(x+12)\cdot\frac{1}{2}(x^2+24x)^{-1/2}\cdot(2x+24)}{x^2+24x}$

$= \frac{(x^2+24x)^{-1/2}\cdot\left[(x^2+24x)-(x+12)^2\right]}{x^2+24x}$

$= \frac{\left[(x^2+24x)-(x^2+24x+144)\right]}{(x^2+24x)^{3/2}}$

$= \frac{-144}{(x^2+24x)^{3/2}}.$

$u''(x) < 0$ for all $x > 0$, so the graph of u is concave down on the interval $(0, \infty)$.

e. The employee satisfaction continually increases as the salary increases, but the rate of increase is continuously decreasing.

76. $D(p) = \frac{500}{10+\sqrt{p(p+24)}} = \frac{500}{10+\sqrt{p^2+24p}}.$

a. $D'(p) = \frac{0-500\left[\frac{1}{2}(p^2+24p)^{-1/2}(2p+24)\right]}{\left(10+\sqrt{p^2+24p}\right)^2}$

$= \frac{-500(p^2+24p)^{-1/2}(p+12)}{\left(10+\sqrt{p^2+24p}\right)^2}.$

b. Demand decreases as price increases.

c. $D''(p) = \frac{-500}{\left(10+\sqrt{p^2+24p}\right)^4}\cdot\left\{\left(10+\sqrt{p^2+24p}\right)^2\left[\left(-\frac{1}{2}\right)(p^2+24p)^{-3/2}\cdot\right.\right.$

$(2p+24)(p+12)+(p^2+24p)^{-1/2}\cdot 1\Big]-(p^2+24p)^{-1/2}(p+12)\cdot$

$\left.2\left(10+\sqrt{p^2+24p}\right)\cdot\frac{1}{2}\cdot(p^2+24p)^{-1/2}(2p+24)\right\}$

$= \frac{-500}{\left(10+\sqrt{p^2+23p}\right)^4}\left\{\left(10+\sqrt{p^2+24p}\right)^2\cdot\left[-144(p^2+24p)^{-3/2}\right]\right.$

$\left.-2(p^2+24p)^{-1}\cdot(p+12)^2\cdot\left(10+\sqrt{p^2+24p}\right)\right\}.$

$D''(p) > 0$ for all $p > 0$, so the graph of D is concave up for all $p > 0$.

d. $MD(1) = D'(1) = \frac{-500(1+24)^{-1/2}(1+12)}{\left(10+\sqrt{1+24}\right)^2}$

$$= \frac{-500\cdot\left(\frac{1}{5}\right)(13)}{15^2} = \frac{-1300}{225} = -5.78.$$

$$MD(3) = D'(3) = \frac{-500(3^2+24\cdot 3)^{-1/2}\cdot(3+12)}{\left(10+\sqrt{3^2+24\cdot 3}\right)^2}$$

$$= \frac{(-500)\cdot\left(\frac{1}{9}\right)\cdot(15)}{19^2} = -2.31.$$

$MD(3) > MD(1)$. Note that $D''(p) > 0$ for all $p > 0$, so the marginal demand increases as the price p increases.

77. $D(p) = \frac{10\sqrt{p}}{16+p}, \qquad p > 0.$

$$D'(p) = \frac{(16+p)\cdot 10\cdot\frac{1}{2}\cdot p^{-1/2} - 10\sqrt{p}\cdot 1}{(16+p)^2}$$

$$= \frac{5(16+p)p^{-1/2} - 10p^{1/2}}{(16+p)^2}$$

$$= \frac{5p^{-1/2}[(16+p)-2p]}{(16+p)^2} = \frac{5p^{-1/2}[16-p]}{(16+p)^2}.$$

$D'(p) = 0$ for $p = 16$.

For $0 < p < 16$, $D'(p) > 0$, so the demand $D(p)$ increases as the price p increases.

For $p > 16$, $D'(p) < 0$, so the demand $D(p)$ decreases as the price p increases.

The demand $D(p)$ in units of thousands of ounces of cologne per month is a maximum when the price p is \$16.

CHAPTER 4

Solutions to Exercise Set 4.1

1. a. $9^{3/2} = 27.$ b. $16^{1/4} = 2.$

   c. $49^{3/2} = 7^3 = 343.$ (d) $4^{-3/2} = 2^{-3} = \frac{1}{8}.$

2. a. $36^{-3/2} = 6^{-3} = \frac{1}{216}.$ b. $81^{-3/4} = 3^{-3} = \frac{1}{27}.$

   c. $8^{5/3} = 2^5 = 32.$ (d) $125^{-2/3} = 5^{-2} = \frac{1}{25}.$

3. a. $\left(\frac{1}{4}\right)^{3/2} = \frac{1}{2^3} = \frac{1}{8}.$ b. $\left(\frac{27}{8}\right)^{2/3} = \frac{27^{2/3}}{8^{2/3}} = \frac{3^2}{2^2} = \frac{9}{4}.$

   c. $\left(\frac{1}{16}\right)^{-3/4} = 16^{3/4} = 2^3 = 8.$ d. $\left(\frac{81}{36}\right)^{-3/2} = \frac{36^{3/2}}{81^{3/2}} = \frac{6^3}{9^3} = \frac{8}{27}.$

4. $f(x) = 2^x.$

x	-2	-1	0	1	2
$f(x)$	$\frac{1}{4}$	$\frac{1}{2}$	1	2	4

5. $f(x) = 3^{-x}.$

x	-1	$-\frac{1}{2}$	0	$\frac{1}{2}$	1
$f(x)$	3	1.73	1	0.58	$\frac{1}{3}$

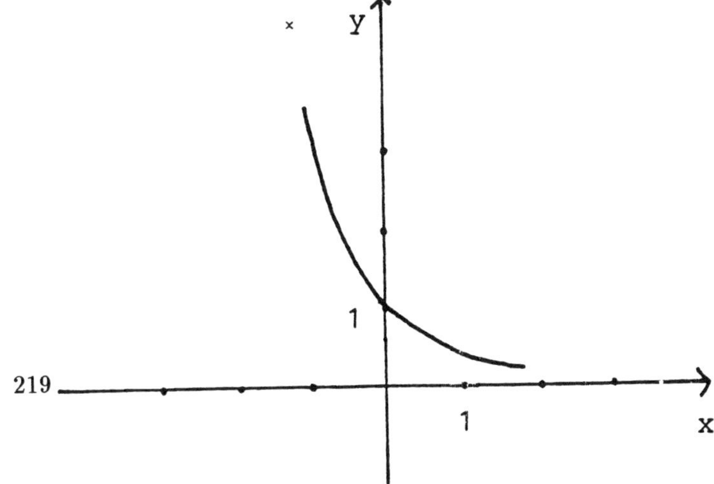

6. $f(x) = \left(\frac{1}{2}\right)^x$.

x	-1	$-\frac{1}{2}$	0	$\frac{1}{2}$	1
$f(x)$	2	1.41	1	0.71	0.5

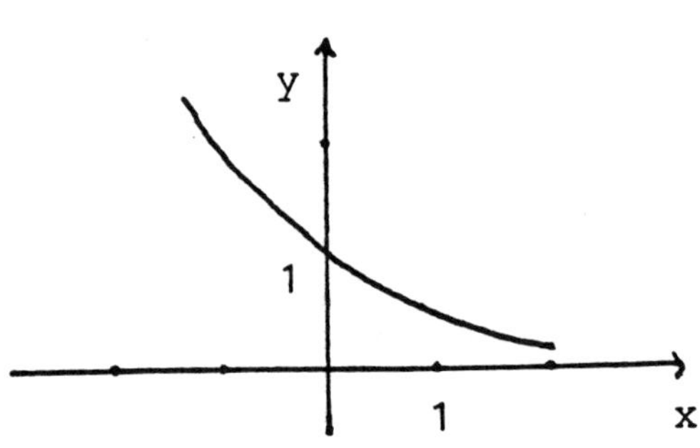

7. $f(x) = 10^{-x}$.

x	-1	0	1	2
$f(x)$	10	1	$\frac{1}{10}$	$\frac{1}{100}$

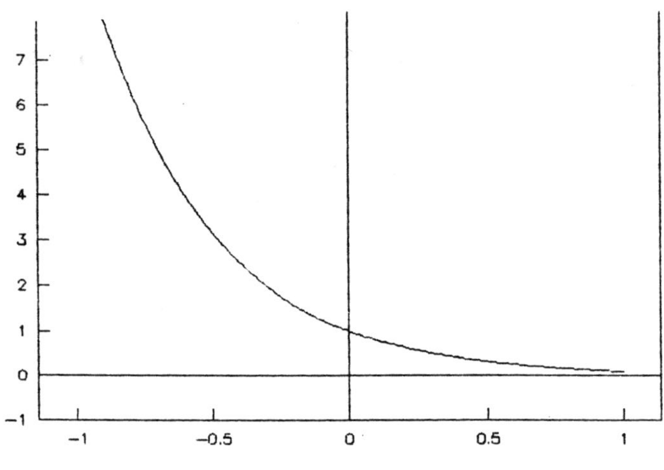

8. $f(x) = 2^{2x}$.

x	-2	-1	0	1	2
$f(x)$	$\frac{1}{16}$	$\frac{1}{4}$	1	4	16

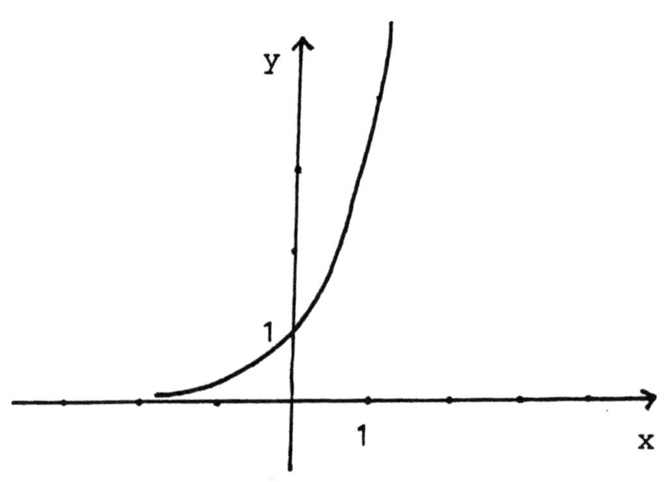

9. $f(x) = e^{-x}$.

x	-1	$-\frac{1}{2}$	0	$\frac{1}{2}$	1
$f(x)$	2.72	1.65	1	0.61	0.37

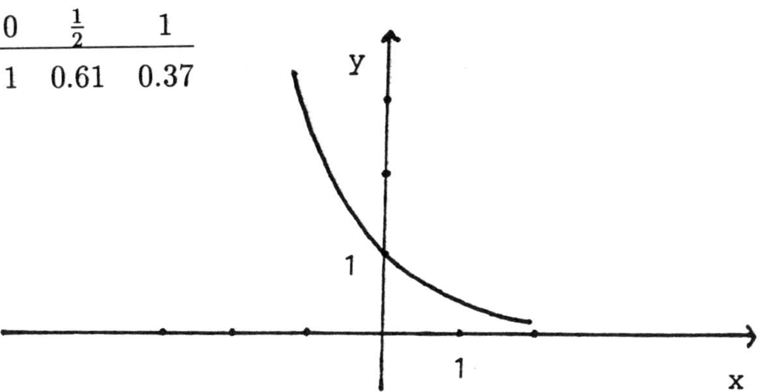

10. $f(x) = -2e^{x}$.

x	-1	$-\frac{1}{2}$	0	$\frac{1}{2}$	1
$f(x)$	-0.74	-1.21	-2	-3.30	-5.44

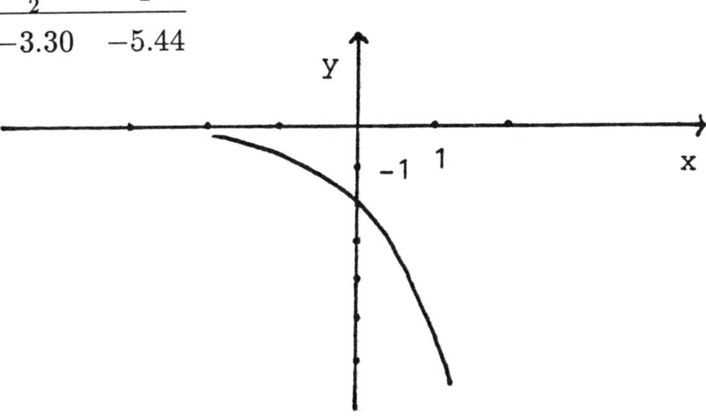

11. $f(x) = -3e^{-x}$.

x	-1	$-\frac{1}{2}$	0	$\frac{1}{2}$	1	2
$f(x)$	-8.15	-4.95	-3	-1.82	-1.10	-0.41

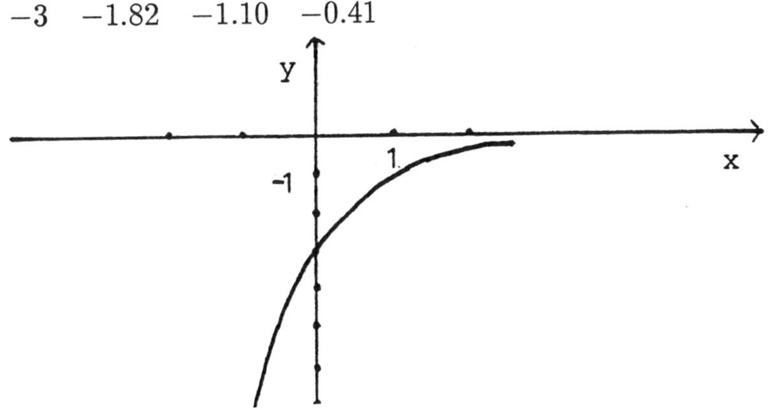

12. $2^3 \cdot 4^{-2} = 2^3 \cdot 2^{-4} = 2^{3-4} = \frac{1}{2}$.

13. $\frac{27^{4/3}\cdot 4^3}{9^{3/2}\cdot 32} = \frac{3^4\cdot(2^2)^3}{3^3\cdot 2^5} = \frac{3^4\cdot 2^6}{3^3\cdot 2^5} = 3\cdot 2 = 6.$

14. $\frac{3^{-6}\cdot 3^5}{3^2} = \frac{3^{-1}}{3^2} = \frac{1}{27}.$

15. $\frac{[2^3\cdot 3^{-2}]^2}{\frac{1}{3}(8)} = \frac{2^6\cdot 3^{-4}}{3^{-1}2^3} = \frac{2^{6-3}}{3^{-1+4}} = \frac{2^3}{3^3} = \frac{8}{27}.$

16. $e^2e^{-2} = e^{2-2} = e^0 = 1.$

17. $e^3\cdot e^x = e^{3+x}.$

18. $\frac{e^3e^{-2x}}{e^{-x}} = e^{3-2x-(-x)} = e^{3-x}.$

19. $\sqrt{16e^{4x}} = (16)^{1/2}(e^{4x})^{1/2} = 4\cdot e^{2x}.$

20. $(27e^{-6}e^{3x})^{1/3} = 3e^{-2}e^x = 3e^{x-2}.$

21. $\frac{(4e^{3x})^{2/3}}{\sqrt{8e^4}} = \frac{2^{4/3}e^{2x}}{2^{3/2}e^2} = 2^{\frac{4}{3}-\frac{3}{2}}\cdot e^{2x-2} = 2^{-1/6}e^{2x-2} = \frac{e^{2x-2}}{\sqrt[6]{2}}.$

22. We find the value of his investment after 3 years under the conditions given here.

a. annual compounding $P(3) = (1+0.10)^3\cdot 1000$
$= 1.331\cdot 1000 = \$1,331.$

b. quarterly compounding $P(3) = \left(1+\frac{0.10}{4}\right)^{3\cdot 4}\cdot 1000$
$= 1.344889\cdot 1000 = \$1,344.89.$

c. monthly compounding $P(3) = \left(1+\frac{0.10}{12}\right)^{3\cdot 12}\cdot 1000$
$= 1.34818\cdot 1000 = \$1,348.18.$

d. continuous compounding $P(3) = e^{3(0.10)}\cdot 1000$
$= 1.34986\cdot 1000 = \$1,349.86.$

23. a. $P(1) = \left(1+\frac{0.08}{2}\right)^{1\cdot 2}\cdot 500 = 1.0816\cdot 500 = \$540.80.$

b. $P(2) = \left(1+\frac{0.08}{2}\right)^{2\cdot 2}\cdot 500 = 1.16986\cdot 500 = \$584.93.$

c. $P(1.5) = \left(1+\frac{0.08}{2}\right)^{1.5\cdot 2}\cdot 500 = 1.124864\cdot 500 = \$562.43.$

24. a. $P(1) = e^{1\cdot 0.08}\cdot 500 = 1.08329\cdot 500 = \$541.64.$

b. $P(2) = e^{2\cdot 0.08}\cdot 500 = 1.17351\cdot 500 = \$586.76.$

c. $P(1.5) = e^{1.5\cdot 0.08}\cdot 500 = 1.12749\cdot 500 = \$563.75.$

25. $P_0 = \frac{P(t)}{e^{rt}} = \frac{1000}{e^{0.10\cdot 4}} = \frac{1000}{1.492} = \$670.32.$

26. $P_0 = \frac{20,000}{e^{(.08)10}} = \$8,986.58.$

27. $P_0 = \frac{P(t)}{e^{rt}} = \frac{10,000}{e^{0.08 \cdot 5}} = \frac{10,000}{1.492} = \$6703.20.$

28. $P_0 = \frac{P(t)}{e^{rt}} = \frac{5,000}{e^{0.06 \cdot 5}} = \frac{5,000}{1.34986} = \$3,704.09.$

29. $P(7) = e^{0.07 \cdot 7} \cdot 10,000 = 1.6323 \cdot 10,000 = 16,323.16.$

30. Consider first continuous compounding of interest at the nominal rate $r = 10\%$.

Then

$$P(7) = e^{0.10 \cdot 7} \cdot P_0 = 2.014 \cdot P_0,$$

so the money has doubled in 7 years.

Consider next continuous compounding of interest at the nominal rate $r = 7\%$.

Then

$$P(10) = e^{0.07 \cdot 10} \cdot P_0 = 2.014 \cdot P_0,$$

so the money doubled in 10 years.

31. $P_0 = \frac{P(t)}{e^{rt}} = \frac{10,000}{e^{0.05 \cdot 6}} = \frac{10,000}{1.34986} = \$7408.18.$

32. a. $P_0 = \frac{P(t)}{e^{rt}} = \frac{12 \cdot 20}{e^{0.04 \cdot 6}} = \frac{240}{1.27125} = \$188.79.$

b. $P_0 = \frac{P(t)}{e^{rt}} = \frac{12 \cdot 20}{e^{0.07 \cdot 6}} = \frac{240}{1.52196} = \$157.69.$

33. $e^{rt} P_0 = (1 + 0.08)^t P_0$. For $t = 1$, $e^r = 1.08$.

If $r = 0.05$, $e^{0.05} \approx 1.05$; if $r = 0.06$, $e^{0.06} \approx 1.06$;

if $r = 0.07$, $e^{0.07} \approx 1.07$; if $r = 0.08$, $e^{0.08} \approx 1.083$;

if $r = 0.073$, $e^{0.073} \approx 1.076$; if $r = 0.077$, $e^{0.077} \approx 1.08$.

So $r \approx 0.077$.

34. $P_0 = \frac{P(t)}{e^{rt}} = \frac{1000}{e^{(.10)5}} = \$606.53.$

35. a. $(1 + r) = \left(1 + \frac{0.10}{2}\right)^2.$

$1 + r = 1.1025, \quad r = 0.1025$ or 10.25%.

b. $(1 + r) = e^{\cdot 1}.$

$1 + r = 1.1052, \quad r = 0.1052$ or 10.52%.

Solutions to Exercise Set 4.2

1. a. $\log_{10} 100 = 2$ since $10^2 = 100$.

   b. $\log_{10} 10 = 1$ since $10^1 = 10$.

   c. $\log_2 16 = 4$ since $2^4 = 16$.

   d. $\log_4 64 = 3$ since $4^3 = 64$.

   e. $\log_3 81 = 4$ since $3^4 = 81$.

   f. $\ln e^2 = 2$ since $e^2 = e^2$.

2. a. $\log_8 2 = \frac{1}{3}$ since $8^{1/3} = 2$.

   b. $\log_2 \sqrt{2} = \frac{1}{2}$ since $2^{1/2} = \sqrt{2}$.

   c. $\log_9 \left(\frac{1}{3}\right) = -\frac{1}{2}$ since $9^{-1/2} = \frac{1}{3}$.

   d. $\log_{10} \left(\frac{1}{100}\right) = -2$ since $10^{-2} = \frac{1}{100}$.

   e. $\log_{10}(0.001) = -3$ since $10^{-3} = 0.001$.

   f. $\log_4 \left(\frac{1}{8}\right) = -\frac{3}{2}$ since $4^{-3/2} = \frac{1}{8}$.

3. a. $\ln e = 1$. b. $\ln 1 = 0$.

   c. $\ln(2.2) = 0.78846$. d. $\ln(0.3) = -1.20397$.

   e. $\ln(e^3) = 3$. f. $\ln\left(\frac{1}{e^2}\right) = -2$.

4. a. Let $y = \log_a x$.

   By the definition of the logarithm, we have $x = a^y$.

   b. Taking natural logs of both sides of the equation $x = a^y$ and using the property (L3) of logarithms, we obtain

   $\ln x = \ln(a^y) = y \ln a$.

   c. From part (b) $y = \frac{\ln x}{\ln a}$.

   d. Use part a. and c.; we have $\log_a x = \frac{\ln x}{\ln a}$.

   Therefore, $\ln x = (\ln a) \log_a x$.

5. Set $a = 10$ in the formulas in Ex. 4; then

   $\log_{10} x = \frac{\ln x}{\ln 10}$ and $\ln x = (\ln 10) \log_{10} x$.

6. a. $\log_{10} 5 = \frac{\ln 5}{\ln 10} = \frac{1.60944}{2.30259} = 0.69897.$

   b. $\log_3 8 = \frac{\ln 8}{\ln 3} = \frac{2.07944}{1.09861} = 1.89279.$

   c. $\log_2 12 = \frac{\ln 12}{\ln 2} = \frac{2.48490}{0.69315} = 3.58496.$

   d. $\log_8 4 = \frac{\ln 4}{\ln 8} = \frac{1.38629}{2.07944} = 0.66667.$

   e. $\log_3 \frac{1}{2} = \frac{\ln \frac{1}{2}}{\ln 3} = \frac{-0.69315}{1.09861} = -0.63093.$

   f. $\log_4 6 = \frac{\ln 6}{\ln 4} = \frac{1.79176}{1.38629} = 1.29248.$

7. $e^x = 1. \qquad x = \ln 1 = 0.$

8. $e^x = 3. \qquad x = \ln 3 = 1.09861.$

9. $e^{x^2-3} = 5 \implies x^2 - 3 = \ln 5 \implies x^2 = \ln 5 + 3 \implies x^2 = 4.6094 \implies x = \pm 2.147.$

10. $\ln x^2 = 2. \qquad x^2 = e^2, \qquad x = -e$ and $x = e.$

11. $e^{x+\ln x} = 2x \implies x + \ln x = \ln 2x \implies x + \ln x = \ln 2 + \ln x \implies x = \ln 2.$

12. $2^{x-2} = 8. \qquad 2^{x-2} = 2^3, \qquad x - 2 = 3, \qquad x = 5.$

13. $\ln x^2 = 8. \qquad x^2 = e^8, \qquad x = \pm(e^8)^{1/2} = \pm e^4.$

14. $e^{x^2-2x-3} = 1. \qquad x^2 - 2x - 3 = \ln 1 = 0.$

    $(x+1)(x-3) = 0 \implies x = -1, x = 3.$

15. $e^{\ln(2x+3)} = 7.$

    By identity (4), $e^{\ln(2x+3)} = 2x + 3.$

    Then $2x + 3 = 7$, $x = \frac{7-3}{2} = 2.$

16. $e^{2x} + e^x - 2 = 0. \qquad (e^x + 2)(e^x - 1) = 0. \qquad e^x = -2$ or $e^x = 1.$

    But e^x cannot be negative. So $x = \ln 1 = 0.$

17. $\log_{10} P = \frac{-A}{t+C} + B.$ Since $\log_{10} P = \frac{\ln P}{\ln 10}$,

    then $\frac{\ln P}{\ln 10} = \frac{-A}{t+C} + B. \qquad \ln P = (\ln 10)\left(\frac{-A}{t+C} + B\right).$

18. $\log_{10} P = B - \frac{A}{t+C}. \qquad P = 10^{\left(B - \frac{A}{t+C}\right)}.$

19. $p = 100e^{-x}.$

    $e^{-x} = \frac{p}{100}, \qquad e^x = \frac{100}{p}, \qquad x = \ln \frac{100}{p} = \ln 100 - \ln p.$

20. $P_T = e^{rT} P_0$.

$e^{rT} = \frac{P_T}{P_0}, \quad rT = \ln\left(\frac{P_T}{P_0}\right), \quad T = \frac{1}{r}\ln\left(\frac{P_T}{P_0}\right).$

21. $T = \frac{1}{0.1}\ln\left(\frac{1500}{1000}\right) = 10(0.40547) \approx 4.05$ yrs.

22. $T = \frac{1}{0.05}\ln\left(\frac{1500}{1000}\right) = 20(0.40547) \approx 8.11$ yrs.

23. $T = \frac{1}{0.1}\ln\left(\frac{2P_0}{P_0}\right) = 10\ln 2 \approx 6.93$ yrs.

24. $p(x) = 100e^{-0.02x}$. When $x = 100$, $p(100) = 100e^{-0.02(100)} = 100e^{-2} = \$13.53$.

25. $p(x) = 100e^{-0.02x}$. When $p = 50$, $\quad 50 = 100e^{-.02x}$,

$e^{-0.02x} = \frac{1}{2}, \quad e^{0.02x} = 2, \quad 0.02x = \ln 2, \quad x = \frac{\ln 2}{0.02} = 34.7 \approx 35$ items.

26. $f(t) = 10 + 3t + 6\ln t$.

   a. $f(1) = 10 + 3(1) + 6\ln 1 = 10 + 3 + 6(0) = 13$.

   b. $f(2) = 10 + 3(2) + 6\ln 2 = 16 + 6(0.69315) = 20.15888$.

   c. $f(e) = 10 + 3(e) + 6\ln e = 10 + 3e + 6(1) = 16 + 3e$.

27. $f(t) = a + b\ln t$.

Since $f(1) = 15, \quad 15 = a + b\ln 1, \quad 15 = a + b\cdot 0, \quad a = 15$.

Since $f(e) = 20, \quad 20 = a + b\ln e, \quad 20 = 15 + b(1), \quad b = 5$.

28. $f(t) = a + b(1 - e^{2t})$.

Since $f(0) = 10, \quad 10 = a + b(1 - e^{2(0)}), \quad 10 = a + b(1 - 1), \quad a = 10$.

Since $f(\ln 2) = 4, \quad 4 = a + b(1 - e^{2\ln 2}), \quad 4 = 10 + b(1 - e^{\ln 4}), \quad 4 = 10 + b(1 - 4)$,

$b = 2$.

29. Let $P(t)$ denote the value of the store after t years.

Let P_0 denote the original value of the store.

Let r denote the annual return.

$P(t) = e^{rt} P_0$.

$P_0 = \$1$ million.

$P(5) = \$2$ million $\implies 2 = e^{r(5)}1 \implies e^{5r} = 2 \implies 5r = \ln 2 \implies r = \frac{\ln 2}{5} =$

.1386 or 13.86%.

30. $2 = e^{.05t} \implies .05t = \ln 2 \implies t = \frac{\ln 2}{.05} = 13.86$ years.

31. $P(5) = 100,000e^{.15(5)} = \$211,700$.

32. $P_0 = 250,000e^{-.12(4)} = \$154,695.85$.

33. Let $P(t)$ denote the value of the gift after t years.

Let P_0 denote the initial value of the gift.

Let r denote the nominal interest rate.

$P(t) = e^{rt}P_0$.

$P(6) = 2P_0 \implies 2P_0 = e^{r(6)}P_0 \implies e^{6r} = 2 \implies 6r = \ln 2 \implies r = \frac{\ln 2}{6} =$ $.1155$ or 11.55%.

34. $N(t) = Ce^{kt}$, where t denotes the number of years after 1980, and where $N(t)$ is the population of the region in millions of people.

$N(0) = 10 \implies C = 10$.

a. $N(10) = 16 \implies 16 = 10e^{k(10)} \implies e^{10k} = 1.6 \implies 10k = \ln 1.6 \implies$ $k = \frac{\ln 1.6}{10} \implies k = .047$.

b. $N(20) = 10e^{.047(20)} = 25.6$ million.

35. $P(t) = 200,000e^{rt}$.

$P(4) = 140,000 \implies 140,000 = 200,000e^{r(4)} \implies e^{4r} = .7 \implies 4r =$ $\ln .7 \implies r = \frac{\ln .7}{4} = -.089$ or -8.9%.

Therefore the house depreciated at the rate of 8.9% per year.

36. True. Because the time $T = \frac{1}{r}\ln\left(\frac{2P_0}{P_0}\right) = \frac{1}{r}\ln 2$.

37. a. Let $u = \log_b x$ and $v = \log_b y$.

Then $x = b^u$ and $y = b^v$ by the definition of a logarithm.

b. Use the fact that $xy = b^u \cdot b^v = b^{u+v}$.

Then $\log_b(xy) = \log_b(b^u \cdot b^v) = \log_b(b^{u+v}) = u + v$.

c. From (a), $u = \log_b x$ and $v = \log_b y$.

Then by (b), $\log_b(xy) = \log_b x + \log_b y$.

38. a. Let $u = \log_b x$ and $v = \log_b y$.

Then $x = b^u$ and $y = b^v$ by the definition of a logarithm.

b. Use the fact that $x/y = b^u/b^v = b^{u-v}$.

Then $\log_b\left(\frac{x}{y}\right) = \log_b\left(\frac{b^u}{b^v}\right) = \log_b b^{u-v} = u - v$.

c. Since $u = \log_b x$ and $v = \log_b y$,

then by (b), $\log_b\left(\frac{x}{y}\right) = \log_b x - \log_b y$.

39. a. Let $u = \log_b x$. Then $x = b^u$ by the definition of a logarithm.

b. Then $x^r = (b^u)^r = b^{ru}$. $\log_b x^r = \log_b b^{ru} = ru$.

c. Since $u = \log_b x$, then by b. $\log_b x^r = r\log_b x$.

Solutions to Exercise Set 4.3

1. $y' = \frac{1}{2x} \cdot 2 = \frac{1}{x}$.
2. $f'(x) = \frac{1}{ax} \cdot a = \frac{1}{x}$.
3. $f(x) = 10\ln(x^2+5)$.

   $f'(x) = 10 \cdot \frac{1}{x^2+5} \cdot \frac{d}{dx}(x^2+5) = 10 \cdot \frac{1}{x^2+5} \cdot 2x = \frac{20x}{x^2+5}$.
4. $f'(x) = 4 \cdot \frac{1}{9-x^2} \cdot (-2x) = \frac{-8x}{9-x^2}$.
5. $y' = x \cdot \frac{1}{x} + \ln x$. $y' = 1 + \ln x$.
6. $y = \ln\sqrt{x+6}$.

   $y = \frac{1}{2}\ln(x+6)$.

   $\frac{dy}{dx} = \frac{1}{2} \cdot \frac{1}{x+6} \cdot \frac{d}{dx}(x+6) = \frac{1}{2} \cdot \frac{1}{x+6}$.
7. $f'(x) = \frac{1}{\sqrt{x^3-x}} \cdot \frac{1}{2}(x^3-x)^{-1/2}(3x^2-1) = \frac{3x^2-1}{2(x^3-x)}$.
8. $f'(x) = \frac{1}{(x^2-2x)^3} \cdot 3(x^2-2x)^2(2x-2)$.

   $f'(x) = \frac{3(2x-2)}{x^2-2x}$.
9. $f'(x) = 3\left[\ln(x^2-2x)\right]^2 \cdot \frac{(2x-2)}{x^2-2x}$.

   $f'(x) = \frac{3\ln^2(x^2-2x)(2x-2)}{x^2-2x}$.
10. $f'(x) = \frac{1}{\sqrt{1+\sqrt{x}}} \cdot \frac{1}{2}\left(1+\sqrt{x}\right)^{-1/2}\left(\frac{1}{2}x^{-1/2}\right)$.

    $f'(x) = \frac{1}{4\left(1+\sqrt{x}\right)\left(\sqrt{x}\right)}$. $f'(x) = \frac{1}{4\left(\sqrt{x}+x\right)}$.
11. $y' = \frac{1}{\ln t} \cdot \frac{1}{t} = \frac{1}{t\ln t}$.

12. $y = x\sqrt{1+\ln x}$.

$\frac{dy}{dx} = (1)\cdot\sqrt{1+\ln x} + (x)\cdot\left(\frac{1}{2}(1+\ln x)^{-1/2}\frac{1}{x}\right) = \sqrt{1+\ln x} + \frac{1}{2}(1+\ln x)^{-1/2}$.

13. $f(x) = \frac{x^3}{1+\ln x}$.

$f'(x) = \frac{(1+\ln x)\cdot(3x^2) - x^3\cdot\left(\frac{1}{x}\right)}{(1+\ln x)^2} = \frac{x^2(3(1+\ln x)-1)}{(1+\ln x)^2} = \frac{x^2(2+3\ln x)}{(1+\ln x)^2}$.

14. $f'(x) = x^2\cdot 2\ln x\cdot\frac{1}{x} + \ln^2 x\cdot 2x$

$= 2x\ln x + 2x\ln^2 x = 2x\ln x(1+\ln x)$.

15. $y' = 4\left(3\ln\sqrt{x}\right)^3\cdot\left(3\cdot\frac{1}{\sqrt{x}}\cdot\frac{1}{2}x^{-1/2}\right)$. $\quad y' = \frac{162\ln^3\sqrt{x}}{x}$.

16. $f'(x) = \frac{\ln(c+dx)\cdot\frac{1}{a+bx}\cdot b - \ln(a+bx)\cdot\frac{1}{c+dx}\cdot d}{\ln^2(c+dx)}$.

$f'(x) = \frac{b\ln(c+dx)}{(a+bx)(\ln^2(c+dx))} - \frac{d\ln(a+bx)}{(c+dx)(\ln^2(c+dx))}$.

17. $f'(x) = x^2\cdot\frac{1}{3x-6}\cdot 3 + \ln(3x-6)\cdot 2x = \frac{3x^2}{3x-6} + 2x\ln(3x-6)$.

18. $y = \ln\left(\frac{x^2+3}{x+5}\right) = \ln(x^2+3) - \ln(x+5)$.

$\frac{dy}{dx} = \frac{1}{x^2+3}\cdot(2x) - \frac{1}{x+5}$.

19. $f'(x) = \ln x\frac{1}{\sqrt{x}}\cdot\frac{1}{2}x^{-1/2} + \ln\sqrt{x}\cdot\frac{1}{x}$

$= \frac{\ln x}{2x} + \frac{\ln\sqrt{x}}{x} = \frac{\ln x + 2\ln\sqrt{x}}{2x} = \frac{\ln x+\ln x}{2x} = \frac{\ln x}{x}$.

20. $f'(x) = \frac{(1-\ln x)\left(\frac{1}{x}\right) - (1+\ln x)\left(-\frac{1}{x}\right)}{(1-\ln x)^2}$

$= \frac{\frac{1}{x}(1-\ln x+1+\ln x)}{(1-\ln x)^2} = \frac{2}{x(1-\ln x)^2}$.

21. $f(x) = \frac{x+\ln^2(3x)}{\sqrt{1+\ln x}}$.

$f'(x) = \frac{(1+\ln x)^{1/2}\cdot\left(1+2\ln(3x)\cdot\frac{1}{x}\right) - \left[x^2+\ln^2(3x)\right]\cdot\left[\frac{1}{2}(1+\ln x)^{-1/2}\cdot\frac{1}{x}\right]}{(1+\ln x)}$.

22. $f(x) = (x^2 - x\ln x)\sqrt{x+\ln x}$.

$f'(x) = \left(2x - 1\cdot\ln x - x\cdot\frac{1}{x}\right)(x+\ln x)^{1/2} + (x^2 - x\ln x)\left(\frac{1}{2}(x+\ln x)^{-1/2}\left(1+\frac{1}{x}\right)\right)$

$= (2x - \ln x - 1)(x+\ln x)^{1/2} + (x^2 - x\ln x)\left(\frac{1}{2}(x+\ln x)^{-1/2}\left(1+\frac{1}{x}\right)\right)$.

23. $f(x) = \ln(x^3+3)^{4/3}$. $\quad f(x) = \frac{4}{3}\ln(x^3+3)$.

$f'(x) = \frac{4}{3}\cdot\frac{1}{x^3+3}\cdot 3x^2 = \frac{4x^2}{x^3+3}$.

24. $f(x) = \ln\sqrt{x} + \ln(x^2+3)$.

$f'(x) = \frac{1}{\sqrt{x}}\cdot\frac{1}{2}x^{-1/2} + \frac{1}{x^2+3}\cdot 2x = \frac{1}{2x} + \frac{2x}{x^2+3}$.

25. $f(x) = \ln(x-6)^{2/3} - \ln\sqrt{1+x}$.

$f'(x) = \frac{1}{(x-6)^{2/3}}\cdot\frac{2}{3}(x-6)^{-1/3} - \frac{1}{\sqrt{1+x}}\cdot\frac{1}{2}(1+x)^{-1/2} = \frac{2}{3(x-6)} - \frac{1}{2(1+x)}$.

26. $f(x) = \ln\sqrt{x}(x^2+2) - \ln\sqrt[3]{x}(1+x^2)$.

$f(x) = \ln\sqrt{x} + \ln(x^2+2) - \ln\sqrt[3]{x} - \ln(1+x^2)$.

$f'(x) = \frac{1}{\sqrt{x}} \cdot \frac{1}{2}(x)^{-1/2} + \frac{1}{x^2+2} \cdot 2x - \frac{1}{\sqrt[3]{x}} \cdot \frac{1}{3}(x)^{-2/3} - \frac{1}{1+x^2} \cdot 2x$.

$f'(x) = \frac{1}{2x} + \frac{2x}{x^2+2} - \frac{1}{3x} - \frac{2x}{1+x^2}$.

27. $\ln y = \ln(x+3)^x$. $\quad \ln y = x\ln(x+3)$.

$\frac{1}{y} \cdot \frac{dy}{dx} = x \cdot \frac{1}{x+3} + \ln(x+3)$. $\quad \frac{1}{y} \cdot \frac{dy}{dx} = \frac{x}{x+3} + \ln(x+3)$.

$\frac{dy}{dx} = (x+3)^x \left[\frac{x}{x+3} + \ln(x+3)\right]$.

28. $\ln y = \sqrt{x+2}\ln x$.

$\frac{1}{y} \cdot \frac{dy}{dx} = \sqrt{x+2} \cdot \frac{1}{x} + \ln x \cdot \frac{1}{2}(x+2)^{-1/2}$.

$\frac{1}{y} \cdot \frac{dy}{dx} = \frac{\sqrt{x+2}}{x} + \frac{\ln x}{2\sqrt{x+2}}$.

$\frac{dy}{dx} = x^{\sqrt{x+2}}\left(\frac{\sqrt{x+2}}{x} + \frac{\ln x}{2(\sqrt{x+2})}\right)$.

29. $\ln y = \sqrt{x+1}\ln x$.

$\frac{1}{y} \cdot \frac{dy}{dx} = \sqrt{x+1} \cdot \frac{1}{x} + \ln x \cdot \frac{1}{2}(x+1)^{-1/2}$.

$\frac{dy}{dx} = x^{\sqrt{x+1}}\left(\frac{\sqrt{x+1}}{x} + \frac{\ln x}{2\sqrt{x+1}}\right)$.

30. $\ln y = \sqrt{x+1}\ln(x^2+3)$.

$\frac{1}{y} \cdot \frac{dy}{dx} = \sqrt{x+1} \cdot \frac{1}{x^2+3} \cdot 2x + \ln(x^2+3) \cdot \frac{1}{2}(x+1)^{-1/2}$.

$\frac{1}{y} \cdot \frac{dy}{dx} = \frac{2x\sqrt{x+1}}{x^2+3} + \frac{\ln(x^2+3)}{2\sqrt{x+1}}$.

$\frac{dy}{dx} = (x^2+3)^{\sqrt{x+1}}\left[\frac{2x\sqrt{x+1}}{x^2+3} + \frac{\ln(x^2+3)}{2\sqrt{x+1}}\right]$.

31. $y = \frac{x(x+1)(x+2)}{(x+3)(x+4)}$.

$\ln y = \ln x(x+1)(x+2) - \ln(x+3)(x+4)$.

$\ln y = \ln x + \ln(x+1) + \ln(x+2) - \ln(x+3) - \ln(x+4)$.

$\frac{1}{y} \cdot \frac{dy}{dx} = \frac{1}{x} + \frac{1}{x+1} + \frac{1}{x+2} - \frac{1}{x+3} - \frac{1}{x+4}$.

$\frac{dy}{dx} = \frac{x(x+1)(x+2)}{(x+3)(x+4)}\left[\frac{1}{x} + \frac{1}{x+1} + \frac{1}{x+2} - \frac{1}{x+3} - \frac{1}{x+4}\right]$.

32. $y = \frac{\sqrt{1+x^2}}{x(1-x^5)}$.

$\ln y = \ln\sqrt{1+x^2} - \ln x(1-x^5)$.

$\ln y = \ln\sqrt{1+x^2} - \ln x - \ln(1-x^5)$.

$\frac{1}{y} \cdot \frac{dy}{dx} = \frac{1}{\sqrt{1+x^2}} \cdot \frac{1}{2}(1+x^2)^{-1/2}(2x) - \frac{1}{x} - \frac{1}{1-x^5} \cdot (-5x^4)$.

$\frac{1}{y}\cdot\frac{dy}{dx}=\frac{2x}{2(1+x^2)}-\frac{1}{x}+\frac{5x^4}{1-x^5}$.

$\frac{dy}{dx}=\frac{\sqrt{1+x^2}}{x(1-x^5)}\left(\frac{x}{(1+x^2)}-\frac{1}{x}+\frac{5x^4}{1-x^5}\right)$.

33. $y=\frac{(x^2+3)(1-x)^{2/3}}{x\sqrt{1+x}}$.

$\ln y=\ln(x^2+3)+\frac{2}{3}\ln(1-x)-\ln x-\frac{1}{2}\ln(1+x)$.

$\frac{1}{y}\frac{dy}{dx}=\frac{2x}{x^2+3}+\frac{2}{3}\cdot\left(\frac{1}{1-x}\cdot(-1)\right)-\frac{1}{x}-\frac{1}{2}\cdot\frac{1}{1+x}$.

$\frac{dy}{dx}=\frac{(x^2+3)(1-x)^{2/3}}{x\sqrt{1+x}}\cdot\left(\frac{2x}{x^2+3}-\frac{2}{3}\cdot\frac{1}{1-x}-\frac{1}{x}-\frac{1}{2}\cdot\frac{1}{1+x}\right)$.

34. $\ln y=\ln x(x^3+6)^4-\ln\sqrt{1+x^2}$.

$\ln y=\ln x+4\ln(x^3+6)-\frac{1}{2}\ln(1+x^2)$.

$\frac{1}{y}\cdot\frac{dy}{dx}=\frac{1}{x}+4\cdot\frac{1}{x^3+6}\cdot 3x^2-\frac{1}{2}\cdot\frac{1}{1+x^2}\cdot 2x$.

$\frac{dy}{dx}=\frac{x(x^3+6)^4}{\sqrt{1+x^2}}\left[\frac{1}{x}+\frac{12x^2}{x^3+6}-\frac{x}{1+x^2}\right]$.

35. $\ln x+\ln y=x+y$. $\frac{1}{x}+\frac{1}{y}\cdot\frac{dy}{dx}=1+\frac{dy}{dx}$.

$\left(\frac{1}{y}-1\right)\frac{dy}{dx}=1-\frac{1}{x}$. $\frac{dy}{dx}=\frac{1-\frac{1}{x}}{\frac{1}{y}-1}$. $\frac{dy}{dx}=\frac{y(x-1)}{x(1-y)}$.

36. $x\ln y+y^2\ln x=1$.

$1\cdot\ln y+x\cdot\frac{1}{y}\frac{dy}{dx}+2y\frac{dy}{dx}\ln x+y^2\cdot\frac{1}{x}=0$.

$\left(\frac{x}{y}+2y\ln x\right)\frac{dy}{dx}=-\ln y-\frac{y^2}{x}$.

$\frac{dy}{dx}=\frac{-\ln y-\frac{y^2}{x}}{\frac{x}{y}+2y\ln x}=\frac{y(-x\ln y-y^2)}{x(x+2y^2\ln x)}$.

37. $\frac{dy}{dx}=x\cdot 2\ln x\cdot\frac{1}{x}+(\ln x)^2+\frac{\ln x-x\cdot\frac{1}{x}}{(\ln x)^2}$.

$\frac{dy}{dx}=2\ln x+\ln^2 x+\frac{\ln x-1}{(\ln x)^2}$. At $x=e$, $\frac{dy}{dx}=2+1=3$.

The equation is $(y-2e)=3(x-e)$. $y=3x-e$.

38. $y=x\ln x$.

$y'=1\cdot\ln x+x\cdot\frac{1}{x}=(\ln x)+1=0$ for $\ln x=-1$, $x=e^{-1}$.

For $0<x<e^{-1}$, $y'<0$; and for $x>e^{-1}$, $y'>0$.

So $y=e^{-1}\ln e^{-1}=-\frac{1}{e}$ is the relative minimum when $x=e^{-1}$.

39. $y=x-\ln x$. $y'=1-\frac{1}{x}=0$ for $x=1$.

For $0<x<1$, $y'<0$; and for $x>1$, $y'>0$.

So $y=1-\ln 1=1$ is the relative minimum when $x=1$.

40. $y = x^2 \ln x$. $\quad y' = 2x \ln x + x^2 \frac{1}{x} = x(2\ln x + 1)$.

$y' = 0$ for $\ln x = -\frac{1}{2} \implies x = e^{-1/2}$.

For $0 < x < e^{-1/2}$, $y' < 0$; and for $x > e^{-1/2}$, $y' > 0$.

So $y = (e^{-1/2})^2 \ln e^{-1/2} = e^{-1}\left(-\frac{1}{2}\right) = -\frac{1}{2e}$ is the relative minimum when $x = e^{-1/2}$.

41. $y = \ln(x^2 - x)$. $\quad y' = \frac{1}{x^2-x} \cdot (2x - 1)$.

y' does not exist for $x = 0$ and $x = 1$.

Since the domain of y is $x < 0$ and $x > 1$, y has no relative extrema.

42. $y = \ln x^3$.

x	$\frac{1}{3}$	$\frac{2}{3}$	1	$\frac{3}{2}$	2
y	-3.30	-1.22	0	1.22	2.08

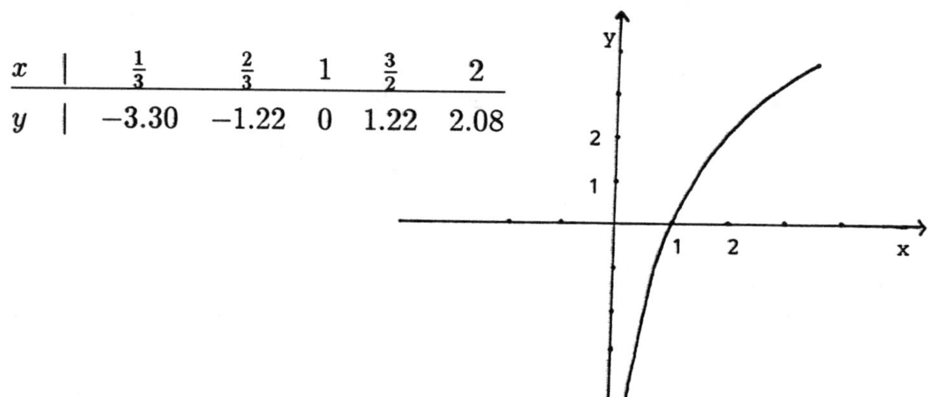

43. $y = \ln(4 - x)$.

x	-1	0	1	2	3	$3\frac{1}{3}$	$3\frac{2}{3}$
y	1.61	1.39	1.10	0.69	0	-0.41	-1.10

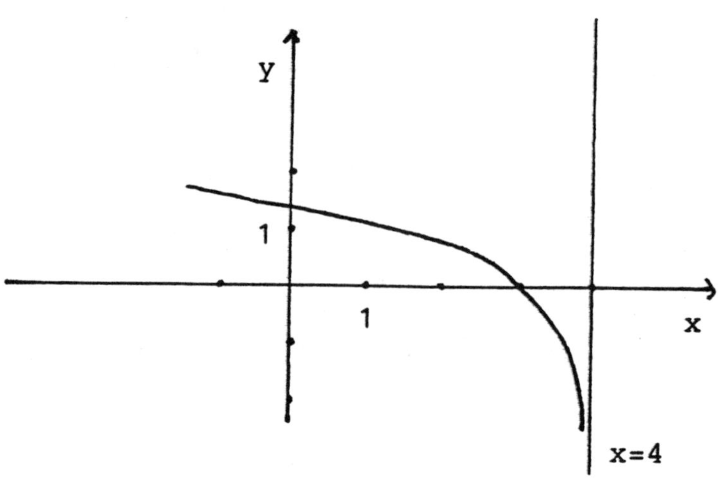

44. $C(x) = 2,000 + 200x + x^2$.

a. $MC(x) = C'(x) = 200 + 2x$. $\qquad MC(20) = 200 + 2(20) = 240$.

b. The company's rate of growth at production level $x = 20$ is

$$\frac{C'(20)}{C(20)} = \frac{240}{6,400} = .0375.$$

45. a. $C(x) = 20x + 300$.

$C'(x) = 20$.

$\frac{C'(20)}{C(20)} = \frac{20}{700} = .0286$ or 2.86%.

b. $R(x) = 60x - x^2$.

$R'(x) = 60 - 2x$.

$\frac{R'(20)}{R(20)} = \frac{20}{800} = .025$ or 2.5%.

c. $\frac{P'(20)}{P(20)} = \frac{R'(20)-C'(20)}{R(20)-C(20)} = \frac{20-20}{800-700} = 0$.

46. $C(x) = 200 + 10x$ and $R(x) = 50x$.

$P(x) = R(x) - C(x) = 50x - (200 + 10x) = 40x - 200$.

a. The break-even point x_0 is such that $R(x_0) = C(x_0) \implies 50x_0 = 200 + 10x_0 \implies x_0 = 5$.

b. The growth in profits is given by

$$G(x) = \frac{P'(x)}{P(x)} = \frac{40}{40x - 200}.$$

$$G'(x) = \frac{(0)(40x - 200) - 40(40)}{(40x - 200)^2} = \frac{-1600}{(40x - 200)^2}.$$

$G'(x)$ is undefined for $x = 5$.

For $0 < x < 5$, $G'(x) < 0$, so the rate of growth in profits is decreasing.

For $x > 5$, $G'(x) < 0$, so the rate of growth in profits is decreasing.

c. For $x > x_0 = 5$, $P'(x) = 40 > 0$, so the profits are increasing.

47. $P(t) = 750,000 + 30,000t + 10,000e^{-t}$.

$P'(t) = 30,000 + 10,000e^{-t}(-1)$.

a. $\frac{P'(2)}{P(2)} \cdot 100 = \frac{28646.64717}{811353.3528} \cdot 100 = 3.53\%$.

b. $\frac{P'(5)}{P(5)} \cdot 100 = \frac{29932.62053}{900067.3795} \cdot 100 = 3.33\%$.

48. $V(t) = 5000e^{\frac{1}{2}\sqrt{t}}$

$V'(t) = 5000e^{\frac{1}{2}\sqrt{t}} \cdot \frac{1}{2} \cdot \frac{1}{2\sqrt{t}}$.

a. $\frac{V'(5)}{V(5)} = \frac{1709.940516}{15294.17292} = .1118$ (relative rate of growth)

Relative percent rate of growth $= \frac{V'(5)}{V(5)} \cdot 100 = 11.18\%$.

b. $\frac{V'(10)}{V(10)} = \frac{1921.27655}{24302.43965} = .079$ (relative rate of growth)

Relative percent rate of growth $= \frac{V'(10)}{V(10)} \cdot 100 = 7.9\%$.

49. $V(t) = 10,000te^{-0.5t}$.

$V'(t) = 10,000\left((1)(e^{-0.5t} + t\left(e^{-0.5t}(-0.5)\right)\right)$

$= 10,000e^{-0.5t}(1 - 0.5t)$.

Relative rate of growth $= \frac{V'(4)}{V(4)} = \frac{-1353.352832}{5413.41133} = -.25$

The relative rate of decline is .25.

The relative percent rate of decline is 25% .

Solutions to Exercise Set 4.4

1. $f'(x) = 6e^{6x}$.
2. $f'(x) = -e^{9-x}$.
3. $f'(x) = 2x \cdot e^{x^2-4}$.
4. $f'(x) = 3x^{-1/2} \cdot e^{2\sqrt{x}}$.
5. $f(x) = \frac{e^x}{x^2+2}$.

   $f'(x) = \frac{(x^2+2)(e^x)-(e^x)(2x)}{(x^2+2)^2} = \frac{e^x(x^2-2x+2)}{(x^2+2)^2}$.
6. $f(x) = x^2e^{-x+5}$.

   $f'(x) = (2x)(e^{-x+5}) + (x^2)\left(e^{-x+5} \cdot (-1)\right) = xe^{-x+5}(2-x)$.

7. $f(x) = \sqrt{x + e^{-x}}$.

$f'(x) = \frac{1}{2}(x + e^{-x})^{-1/2}(1 - e^{-x})$.

8. $f(x) = \frac{e^x+1}{e^x-1}$.

$f'(x) = \frac{(e^x-1)(e^x)-(e^x+1)(e^x)}{(e^x-1)^2} = \frac{-2e^x}{(e^x-1)^2}$.

9. $f'(x) = e^{x^2-\sqrt{x}} \cdot (2x - \frac{1}{2}x^{-1/2})$.

10. $f'(x) = 1 \cdot (e^{-2\ln x}) + x \cdot (e^{-2\ln x}) \cdot (-2) \cdot \left(\frac{1}{x}\right) = e^{-2\ln x}(1-2) = -e^{-2\ln x}$.

11. $f(x) = \ln \frac{e^x+1}{x^3+1} = \ln(e^x + 1) - \ln(x^3 + 1)$.

$f'(x) = \frac{1}{e^x+1} \cdot e^x - \frac{1}{x^3+1} \cdot 3x^2$.

12. $y' = \frac{e^x(e^{-x}+1)-(e^x-1)(e^{-x})(-1)}{(e^{-x}+1)^2} = \frac{1+e^x+1-e^{-x}}{(e^{-x}+1)^2} = \frac{2+e^x-e^{-x}}{(e^{-x}+1)^2}$.

13. $y' = \frac{1}{2}(e^x - e^{-x})$.

14. $y' = (2x+1)(e^{x^2+3}) + (x^2 + x - 1)(e^{x^2+3})(2x)$

$= 2x \cdot e^{x^2+3} \cdot \left\{\left(1 + \frac{1}{2x}\right) + (x^2 + x - 1)\right\} = 2x \cdot e^{x^2+3}\left(x^2 + x + \frac{1}{2x}\right)$.

15. $f'(x) = 4(x - e^{-2x})^3\left(1 - e^{-2x}(-2)\right) = 4(x - e^{-2x})^3(1 + 2e^{-2x})$.

16. $y = e^{\sqrt{x}+\ln x} = e^{\sqrt{x}} \cdot e^{\ln x} = e^{\sqrt{x}} \cdot x$.

$\frac{dy}{dx} = \left(e^{\sqrt{x}} \cdot \frac{1}{2\sqrt{x}}\right)(x) + \left(e^{\sqrt{x}}\right)(1) = e^{\sqrt{x}}\left(\frac{1}{2}\sqrt{x} + 1\right)$.

17. $f'(x) = 1 \cdot (e^{x^{-2}}) + (x)(e^{x^{-2}})(-2x^{-3}) = e^{x^{-2}}(1 - 2x^{-2})$.

18. $y' = \frac{1}{(x+e^{x^2})} \cdot (1 + e^{x^2} \cdot 2x) = \frac{1+2x\cdot e^{x^2}}{(x+e^{x^2})}$.

19. $f(x) = \frac{xe^{-x}}{1+x^2}$.

$f'(x) = \frac{(1+x^2)\left((1)(e^{-x})+(x)\left(e^{-x}(-1)\right)\right)-(xe^{-x})(2x)}{(1+x^2)^2}$

$= \frac{e^{-x}(1+x^2-x-x^3-2x^2)}{(1+x^2)^2} = \frac{e^{-x}(1-x-x^2-x^3)}{(1+x^2)^2}$.

20. $y = \ln^2 x e^{\sqrt{1+x^2}}$.

$\frac{dy}{dx} = \left(2\ln x \cdot \frac{1}{x}\right)\left(e^{\sqrt{1+x^2}}\right) + (\ln^2 x)\left(e^{\sqrt{1+x^2}} \cdot \frac{1}{2}(1 + x^2)^{-1/2} \cdot (2x)\right)$

$= e^{\sqrt{1+x^2}} \cdot \ln x\left(\frac{2}{x} + \ln x \cdot (1 + x^2)^{-1/2} \cdot x\right)$.

21. $e^{xy}\left(x \cdot \frac{dy}{dx} + y \cdot 1\right) = 1 \implies e^{xy}x\frac{dy}{dx} = 1 - e^{xy}y$.

$\implies \frac{dy}{dx} = \frac{1-e^{xy}y}{e^{xy}x}$.

22. $e^{x-y} = y^2 e^{x^2}$.

$e^{x-y}\left(1 - \frac{dy}{dx}\right) = \left(2y\frac{dy}{dx}\right)\left(e^{x^2}\right) + (y^2)\left(e^{x^2} \cdot 2x\right)$.

$\frac{dy}{dx}\left(-e^{x-y} - 2y \cdot e^{x^2}\right) = 2xy^2e^{x^2} - e^{x-y}$.

$\frac{dy}{dx} = \frac{2xy^2e^{x^2} - e^{x-y}}{\left(-e^{x-y} - 2y \cdot e^{x^2}\right)}$.

23. $f'(x) = 1 \cdot (2^x) + x \cdot 2^x \cdot 1 \cdot \ln 2 = 2^x + x \cdot 2^x \cdot \ln 2$.

24. $f'(x) = 7^{x^3-6x} \cdot (3x^2 - 6) \cdot \ln 7 = (3x^2 - 6)7^{x^3-6x} \cdot \ln 7$.

25. $y' = \frac{1}{2}(1 + 4^x)^{-1/2}(4^x)(1)\ln 4 = \frac{4^x \ln 4}{2\sqrt{1+4^x}}$.

26. $f(x) = \log_{10}(x^2 - 10^x)$.

$f'(x) = \frac{1}{\ln 10} \cdot \frac{1}{x^2 - 10^x} \cdot (2x - \ln 10 \cdot 10^x)$.

27. $f'(x) = 2(\log_2 x^{1/2})\left\{\frac{1}{x^{1/2} \cdot \ln 2}\left(\frac{1}{2}x^{-1/2}\right)\right\}$

$= 2\left(\log_2 x^{1/2}\right)\frac{\frac{1}{2} \cdot x^{-1/2}}{x^{1/2} \cdot \ln 2} = \frac{\log_2 x^{1/2}}{x \ln 2}$.

28. $y' = \frac{1}{(x^2-4)^3 \cdot \ln 8} \cdot 3(x^2 - 4)^2 \cdot (2x) = \frac{6x(x^2-4)^2}{(x^2-4)^3 \ln 8} = \frac{6x}{(x^2-4)\ln 8}$.

29. $f(x) = (3 - x^2)e^x$.

$f'(x) = (3 - x^2)e^x + (-2x)e^x = e^x(3 - x^2 - 2x) = e^x(-x + 1)(x + 3)$.

$f'(x) = 0$ for $x = 1$ and $x = -3$.

Interval I	Test Number	$f'(x)$	Conclusion
$(-\infty, -3)$	-4	$f'(-4) < 0$	decreasing
$(-3, 1)$	-2	$f'(-2) > 0$	increasing
$(1, \infty)$	2	$f'(2) < 0$	decreasing

$\lim_{x \to -\infty} f(x) = 0$ and $f(1) = 2e > 0$, so the maximum is $f(1) = 2e$.

30. $y' = (2x)(e^{1-x^2}) + (x^2)(e^{1-x^2})(-2x) = 2xe^{1-x^2}(1 - x^2)$.

$y' = 0$ for $x = 0$, $x = 1$, and $x = -1$.

Interval I	Test Number	$f'(x)$	Conclusion
$(-\infty, -1)$	-2	$f'(-2) > 0$	increasing
$(-1, 0)$	$-\frac{1}{2}$	$f'\left(-\frac{1}{2}\right) < 0$	decreasing
$(0, 1)$	$\frac{1}{2}$	$f'\left(\frac{1}{2}\right) > 0$	increasing
$(1, \infty)$	2	$f'(2) < 0$	decreasing

$f(-1)$ and $f(1)$ are relative maxima. $f(0)$ is the relative minimum.

31. $y' = (1)(e^{1-x^3}) + (x)(e^{1-x^3})(-3x^2) = e^{1-x^3}(1 - 3x^3)$.

$y' = 0$ for $x = \sqrt[3]{\frac{1}{3}}$.

For $x < \sqrt[3]{\frac{1}{3}}$, $y' > 0$; for $x > \sqrt[3]{\frac{1}{3}}$, $y' < 0$.

So y has a relative maximum at $x = \sqrt[3]{\frac{1}{3}}$.

32. $y' = 1 \cdot (e^x) + x \cdot (e^x) = e^x(1 + x)$.

$y' = 0$ for $x = -1$. The point P is $(-1, -e^{-1})$.

33. $p(x) = 100e^{-.5x}$.

$R(x) = x \cdot p(x) = 100xe^{-.5x}$.

$R'(x) = (100)(e^{-.5x}) + (100x)\left(e^{-.5x}(-.5)\right) = 100e^{-.5x}(1 - .5x)$.

$R'(x) = 0$ for $(1 - .5x) = 0 \implies x = 2$.

For $0 < x < 2$, $R'(x) > 0$, so $R(x)$ increases as x increases.

For $x > 2$, $R'(x) < 0$, so $R(x)$ decreases as x increases.

Therefore the maximum revenue occurs when $x = 2$.

The corresponding price is $100e^{-.5(2)} = 36.79$.

34. $P'(t) = 100,000 \cdot \left\{5(e^{-0.05t}) + (1 + 5t)(e^{-0.05t})(-0.05)\right\}$

$= 100,000\{4.95e^{-0.05t} - 0.25te^{-0.05t}\} = 100,000e^{-0.05t}(4.95 - 0.25t)$.

The city's population will be a maximum when $P'(t) = 0$.

$P'(t) = 0 \implies 4.95 - 0.25t = 0 \implies t = 19.8$ years.

35. Revenue $R(x) = 10000x$. Cost $C(x) = 500 + 40x + e^{0.5x}$.

Profit $P(x) = 10000x - 500 - 40x - e^{0.5x} = 9960x - 500 - e^{0.5x}$.

$P'(x) = 9960 - e^{0.5x}(0.5) = 9960 - 0.5e^{0.5x}$.

The weekly production level x that maximizes profits occurs when

$P'(x) = 0 \implies 0.5e^{0.5x} = 9960 \implies \ln e^{0.5x} = \ln 19920$

$\implies 0.5x = \ln 19920 \implies x = 19.8$.

36. $V'(t) = 5000\left(2^{\sqrt{t}}\right)\left(\frac{1}{2}t^{-1/2}\right) \cdot \ln 2 = 2500 \ln 2 \left(2^{\sqrt{t}}\right)(t^{-1/2})$.

$V'(4) = 2500 \ln 2 \left(2^{\sqrt{4}}\right)\left(\frac{1}{\sqrt{4}}\right) = 2500 \ln 2(2^2)\left(\frac{1}{2}\right)$

$= 2500 \ln 2(2) = 5000 \ln 2 = 3465.74$.

37. Rate of growth $= \frac{V'(4)}{V(4)} = \frac{2500\ln 2\left(2^{\sqrt{4}}\right)\left(\frac{1}{\sqrt{4}}\right)}{5000\left(2^{\sqrt{4}}\right)} = \frac{\ln 2\left(\frac{1}{2}\right)}{2} = \frac{\ln 2}{4}$.

38. The present value of the revenue is

$$R(t) = e^{-0.125t} \cdot 100(1.5)^{\sqrt{t}}$$
$$= 100e^{-0.125t}\left(e^{\sqrt{t}\ln 1.5}\right) = 100e^{-0.125t+\sqrt{t}\ln 1.5}.$$
$$R'(t) = 100e^{-0.125t+\sqrt{t}\ln 1.5}\left(-0.125 + \tfrac{1}{2}t^{-1/2}\ln 1.5\right).$$
$R'(t) = 0$ for $-0.125 + \frac{1}{2}t^{-1/2}\ln 1.5 = 0 \implies \frac{1}{2}t^{-1/2}\ln 1.5 = 0.125 \implies$ $t^{-1/2} = \frac{0.125}{\frac{1}{2}\ln 1.5} \implies t^{1/2} = \frac{\frac{1}{2}\ln 1.5}{0.125} \implies t = \frac{\frac{1}{4}\ln^2 1.5}{(0.125)^2} \implies t = 2.63$ years.

39. The present value of the revenue is

$$R(t) = e^{-0.1t} \cdot 10,000(1.2)^{\sqrt{t}} = 10,000e^{-0.1t}(1.2)^{\sqrt{t}}.$$
$$R'(t) = 10,000\{e^{-0.1t}\cdot(-0.1)(1.2)^{\sqrt{t}} + e^{-0.1t}\cdot(1.2)^{\sqrt{t}}\cdot\left(\tfrac{1}{2}t^{-1/2}\right)\cdot\ln 1.2\}$$
$$= 10,000\left[e^{-0.1t}(1.2)^{\sqrt{t}}\{-0.1 + \tfrac{1}{2}t^{-1/2}\ln 1.2\}\right].$$
$R'(t) = 0$ for $\frac{1}{2}t^{-1/2}\ln 1.2 = 0.1 \implies t^{-1/2} = \frac{0.1}{\frac{1}{2}\ln 1.2} \implies t^{1/2} = \frac{\frac{1}{2}\ln 1.2}{0.1} \implies$ $t = \frac{\frac{1}{4}\ln^2 1.2}{(0.1)^2} = 0.83$ year.

40. Equilibrium: $S(x) = D(x) \implies 10e^{0.5x} = 20e^{-0.5x+5} \implies$

$e^{0.5x} = 2e^{-0.5x}\cdot e^5 \implies e^{0.5x}\cdot e^{0.5x} = 2e^5 \implies e^x = 2e^5 \implies x = \ln(2e^5)$.

41. a. $f = \lim_{t\to\infty} y(t) = \frac{a}{1+0} + \frac{f-a}{1+0}$, since b_1 and b_2 are positive.

b. If $b_1 = b_2$ and $c_1 = c_2 = c$, then $y(t) = \frac{f}{1+e^{-b_1(t-c)}}$.

$$y'(t) = \frac{-f\cdot e^{-b_1(t-c)}\cdot(-b_1)}{\left(1+e^{-b_1(t-c)}\right)^2} = \frac{b_1\cdot f\cdot e^{-b_1(t-c)}}{\left(1+e^{-b_1(t-c)}\right)^2}.$$
$$y''(t) = b_1 f\left\{\left(1+e^{-b_1(t-c)}\right)^2\cdot e^{-b_1(t-c)}\cdot(-b_1)\right.$$
$$\left. -e^{-b_1(t-c)}\cdot 2\left(1+e^{-b_1(t-c)}\right)\cdot e^{-b_1(t-c)}\cdot(-b_1)\right\}$$
$$\div\left(1+e^{-b_1(t-c)}\right)^4$$
$$= \frac{b_1 f\{\left(1+e^{-b_1(t-c)}\right)\cdot e^{-b_1(t-c)}\cdot(-b_1)+2b_1e^{-2b_1(t-c)}\}}{\left(1+e^{-b_1(t-c)}\right)^3}$$
$$= \frac{b_1^2 f\{-e^{-b_1(t-c)}-e^{-2b_1(t-c)}+2e^{-2b_1(t-c)}\}}{\left(1+e^{-b_1(t-c)}\right)^3}.$$
$$= \frac{b_1^2 f\cdot e^{-2b_1(t-c)}\{1-e^{b_1(t-c)}\}}{\left(1+e^{-b_1(t-c)}\right)^3}.$$

$y''(t) = 0$ for $\{1 - e^{b_1(t-c)}\} = 0 \implies e^{b_1(t-c)} = 1 \implies t = c$.

For $t < c$, $y''(t) > 0$; for $t > c$, $y''(t) < 0$.

So the maximum value of $y'(t)$ occurs when $t = c$.

42. $L(t) = a + b(1 - e^{-ct})$.

   a. $v(t) = L'(t) = bce^{-ct}$.

   b. No, because the derivative is never zero.

   c. $a(t) = L''(t) = -bc^2e^{-ct}$.

   d. If $b > 0$, then $-bc^2e^{-ct} < 0$ for all t.

43. $f(x) = \frac{1}{\sigma\sqrt{2\pi}}e^{-x^2/2\sigma^2}$.

   $f'(x) = \frac{1}{\sigma\sqrt{2\pi}}e^{-x^2/2\sigma^2}\left(\frac{-2x}{2\sigma^2}\right) = \frac{-1}{\sigma^3\sqrt{2\pi}}xe^{-x^2/2\sigma^2}$.

   $f'(x) = 0$ for $x = 0$.

   For $x < 0$, $f'(x) > 0$, so $f(x)$ is increasing.

   For $x > 0$, $f'(x) < 0$, so $f(x)$ is decreasing.

   So $f(0)$ is a relative maximum.

   $$f''(x) = -\frac{1}{\sigma^3\sqrt{2\pi}}\left\{1e^{-x^2/2\sigma^2} + x\left(e^{-x^2/2\sigma^2}\right)\left(-\frac{1}{2\sigma^2}(2x)\right)\right\}$$
   $$= -\frac{1}{\sigma^3\sqrt{2\pi}}\left\{e^{-x^2/2\sigma^2}\left(1 - \frac{x^2}{\sigma^2}\right)\right\}.$$

   $f''(x) = 0$ for $x = \pm\sigma$.

Interval I	Test Number	$f''(x)$	Conclusion
$(-\infty, -\sigma)$	-2σ	$f''(-2\sigma) > 0$	concave up
$(-\sigma, \sigma)$	0	$f''(0) < 0$	concave down
(σ, ∞)	2σ	$f''(2\sigma) > 0$	concave up

   The inflection points occur at $x = \pm\sigma$.

Solutions to Exercise Set 4.5

1. $N(t) = Ce^{4t}$.
2. $N(t) = Ce^{-3t}$.
3. $f(x) = Ce^{-2x}$.
4. $\frac{dy}{dt} = \pi y$. $\quad y = y(0)e^{\pi t}$.
5. $y = y(0)e^{-3t}$.

6. $f(t) = Ce^{6t}$.

7. $N(t) = N(0)e^{4t} \implies N(t) = 4e^{4t}$.

8. $y = -4e^{-2x}$.

9. $\frac{dN}{dt} + 10N = 0, \qquad N(0) = 6 \implies N(t) = 6e^{-10t}$.

10. $\frac{dy}{dt} - 5y = 0, \qquad y(0) = 4$.

    $y(t) = 4e^{5t}$.

11. A = amount of deposit today.

    $2500 = Ae^{0.1(8)}. \qquad A = 2500e^{-0.8}. \qquad A = \$1123.32.$

12. Let r = interest rate.

    $e^r = 1.08. \qquad r = \ln 1.08 = .07696.$

    The rate is about 7.7%.

13. $e^r = 1 + i. \qquad i = e^r - 1.$

14. True.

15. Let $P(t)$ denote the value of the deposit after t years.

    $P(t) = P_0 e^{.05t}$.

    $P(t) = 2P_0 \implies 2P_0 = P_0 e^{.05t} \implies e^{.05t} = 2 \implies .05t = \ln 2$

    $\implies t = \frac{\ln 2}{.05} = 13.86$ years.

16. False. Refer to Example 2, Section 4.5.

17. a. Let $Y(t)$ be the amount present after t days.

    $Y(t) = Y(0)e^{Kt}$.

    $Y(20) = \frac{1}{2}Y(0) \implies \frac{1}{2} = e^{20K} \implies 20K = \ln\frac{1}{2} = -\ln 2$

    $\implies K = \frac{-\ln 2}{20}$.

    $Y(10) = 50 \implies 50 = Y(0) \cdot e^{(-\ln 2/20)\cdot 10}$

    $\implies Y(0) = 50e^{\ln 2/2} = 70.71$ mg.

    b. $Y(30) = 70.71e^{(-\ln 2/20)\cdot 30} = 25$ mg.

18. When $Y(t) = 0.1Y(0)$, $0.1 = e^{(-\ln 2/20)t} \implies (-\ln 2/20)t = \ln 0.10 = -\ln 10 \implies t = 20\ln 10/\ln 2$.

19. Let $Y(t)$ be the number of bacteria in the colony after t hours.

$Y(0) = 50 \implies Y(t) = 50e^{Kt}$.

$Y(12) = 400 \implies 400 = 50e^{K(12)} \implies e^{12K} = 8 \implies 12K = \ln 8 \implies K = \ln 8/12$.

a. When $Y(t) = 100$, $100 = 50e^{(\ln 8/12)t}$

$\implies e^{(\ln 8/12)t} = 2 \implies (\ln 8/12)t = \ln 2$

$\implies t = 12\ln 2/\ln 2^3 = 12\ln 2/(3\ln 2) = 4$ hrs.

b. $Y(16) = 50e^{(\ln 8/12)\cdot 16} = 50e^{\ln 8(4/3)}$

$= 50e^{\ln(8)^{4/3}} = 50e^{\ln 16} = 50(16) = 800$.

20. $P(t) = e^{rt}P_0$. $\quad P'(t) = P_0e^{rt}\cdot r$.

The relative rate of growth $= \frac{P'(t)}{P(t)} = \frac{rP_0e^{rt}}{P_0e^{rt}} = r$.

21. Let $N(t) =$ the number of bacteria after t hours.

Then $N(t) = N(0)e^{Kt}$ and $N(0) = 100,000$.

Since $N(2) = 150,000$, $\quad 150,000 = 100,000e^{K\cdot 2}$, $\quad 1.5 = e^{2K}$, $\quad 2K = \ln 1.5$

$\implies K = \frac{1}{2}\ln 1.5$.

Then $N(t) = 100,000e^{\left(\frac{1}{2}\ln 1.5\right)t}$.

So $N(5) = 100,000e^{\left(\frac{1}{2}\ln 1.5\right)(5)} = 100,000e^{\ln(1.5)^{5/2}}$

$= 100,000(1.5)^{5/2} \approx 275,568$.

22. Let $R(t) =$ the amount of radium after t years.

Let $t_0 =$ the number of years it takes 10 grams to decay to 6 grams.

$R(0) = 10$, $\quad R(1600) = 5$. Then $R(t) = R(0)e^{kt} = 10e^{kt}$.

$5 = 10e^{1600k}$, $\quad k = \frac{1}{1600}\ln 0.5$. $\quad R(t) = 10e^{\left(\frac{1}{1600}\ln 0.5\right)t}$.

Since $R(t_0) = 6$, $\quad 10e^{\left(\frac{1}{1600}\ln 0.5\right)t_0} = 6$, $\quad e^{\ln(0.5)^{t_0/1600}} = \frac{3}{5}$,

$\left(\frac{1}{2}\right)^{t_0/1600} = \frac{3}{5}$, $\quad \frac{t_0}{1600} = \log_{1/2}\left(\frac{3}{5}\right) = \frac{\ln\left(\frac{3}{5}\right)}{\ln\left(\frac{1}{2}\right)} \approx 0.737$. $\quad t_0 \approx 1179$ years.

23. $P(t) = 1000e^{rt}$. $\quad P(3) = 1000 + 400 = 1400$.

Then $P(3) = 1400 = 1000e^{r3}$, $\quad 1.4 = e^{3r}$, $\quad r = \frac{1}{3}\ln 1.4 \approx 0.112 = 11.2\%$.

24. $3P = Pe^{.08t} \implies e^{.08t} = 3 \implies .08t = \ln 3 \implies t = \frac{\ln 3}{.08} = 13.73$ years.

25. $A'(t) = -0.10A(t), \qquad A(0) = 10$ mg.

$A(t) = A(0)e^{-0.1t} = 10e^{-0.1t}$. Then $A(6) = 10e^{-0.1(6)} \approx 5.5$ mg.

26. The amount of a radioactive isotope $A(t) = Ce^{Kt}$. Let t_0 be the half-life.

$A(2) = 0.8C. \quad 0.8C = Ce^{2K}, \qquad K = \frac{1}{2}\ln 0.8.$

Then $A(t) = Ce^{\left(\frac{1}{2}\ln 0.8\right)t}$.

$A(t_0) = \frac{1}{2}C, \quad \frac{1}{2}C = Ce^{\left(\frac{1}{2}\ln 0.8\right)t_0}, \quad \frac{1}{2} = e^{\ln(0.8)^{t_0/2}},$

$\frac{1}{2} = (0.8)^{t_0/2}, \quad t_0 = 2\log_{0.8}\left(\frac{1}{2}\right) = 2 \cdot \frac{\ln\frac{1}{2}}{\ln 0.8} \approx 6.2$ years.

27. The sales level t years after year one is $N(t) = 10,000e^{Kt}$.

$N(1) = 150,000. \quad 150,000 = 10,000e^{K\cdot 1}, \quad K = \ln 15.$

$N(t) = 10,000e^{(\ln 15)t} = 10,000 \cdot 15^t.$

$N(2) = 10,000 \cdot 15^2 = \$2,250,000.$

28. a. $\frac{dN}{dt} = bN - dN = (b-d)N.$

Then $N = N_0 e^{(b-d)t}$.

b. $b = 20/100 = 0.2, \qquad d = 10/100 = 0.1, \qquad N_0 = 200.$

Then after 20 years, $N = 200e^{(0.2-0.1)20} = 200e^2 \approx 1478.$

29. The amount of the substance $A(t) = Ce^{Kt}$.

Then $A(12) = 0.7C = Ce^{K(12)}. \quad 0.7 = e^{12K}, \quad K = \frac{1}{12}\ln 0.7.$

Thus $A(t) = Ce^{\left(\frac{1}{12}\ln 0.7\right)t}$.

Let t_0 be the half-life of the substance.

Then $A(t_0) = \frac{1}{2}C = Ce^{\left(\frac{1}{12}\ln 0.7\right)t_0}, \quad \frac{1}{2} = (0.7)^{t_0/12},$

$t_0 = 12\log_{0.7}\left(\frac{1}{2}\right) = 12 \cdot \frac{\ln\frac{1}{2}}{\ln 0.7} \approx 23.3$ hours.

30. Let $A(t)$ denote the amount of the foreign substance in the body after t hours.

There is a negative constant k such that

$$\frac{dA}{dt} = kA.$$

31. From Example 7 the amount of ^{14}C remaining t years after the fossil's death is $A(t) = A_0 e^{\left(\frac{-\ln 2}{5760} t\right)}$. Let t_0 be the fossil's age.

Then $A(t_0) = 0.8A_0 = A_0 e^{\left(\frac{-\ln 2}{5760} t_0\right)}$.

$\frac{-\ln 2}{5760} t_0 = \ln 0.8, \quad t_0 = -5760 \frac{\ln 0.8}{\ln 2} \approx 1854.31$ years.

32. The amount remaining after t days is $N(t) = 50e^{kt}$.

Then $N(8) = 25 = 50e^{k \cdot 8}$.

$k = \frac{1}{8} \ln \frac{1}{2} = -\frac{1}{8} \ln 2. \quad N(t) = 50e^{\left(-\frac{1}{8} \ln 2\right)t}$.

Thus $N(21) = 50e^{\left(-\frac{1}{8} \ln 2\right) \cdot 21} = 50 \cdot 2^{-21/8} \approx 8.11$ micrograms.

33. As in Exercise 32, let $N(t)$ be the number of micrograms of iodine in the bloodstream after t days. As shown in our discussion of Exercise 32,

$$N(t) = 50e^{\left(-\frac{1}{8} \ln 2\right)t}.$$

$$N(t) = 20 \implies 20 = 50e^{\left(-\frac{1}{8} \ln 2\right)t} \implies e^{\left(-\frac{1}{8} \ln 2\right)t} = .4 \implies$$

$$\left(-\frac{1}{8} \ln 2\right) t = \ln .4 \implies t = \frac{\ln .4}{-\frac{1}{8} \ln 2} = 10.58 \text{ days.}$$

34. The amount of the chemical remaining undissolved after t minutes is $N(t) = 20e^{kt}$.

Then $N(5) = 10 = 20e^{k \cdot 5}, \qquad k = \frac{1}{5} \ln \frac{1}{2}$.

$N(t) = 20e^{\left(\frac{1}{5} \ln \frac{1}{2}\right)t}$.

$N(t_0) = 0.1(20) = 20e^{\left(\frac{1}{5} \ln \frac{1}{2}\right)t_0}, \qquad \left(\frac{1}{5} \ln \frac{1}{2}\right) t_0 = \ln 0.1$.

$t_0 = \frac{5 \ln 0.1}{\ln \frac{1}{2}} \approx 16.6$ mins.

Solutions to Exercise Set 4.6

1. Let $P(t)$ denote the number of bats present after t years.

The logistic model is defined by the equation

$$P(t) = \frac{M}{1 + Ce^{-kt}}.$$

We are given that $M = 500$ bats.

$$P(0) = 10 \implies 10 = \frac{500}{1+C} \implies 10(1+C) = 500 \implies C = 49.$$

$$P\left(\frac{1}{2}\right) = 40 \implies 40 = \frac{500}{1+49e^{-k(.5)}} \implies 40\left(1+49e^{-.5k}\right) = 500 \implies$$

$$1960e^{-.5k} = 460 \implies e^{-.5k} = \frac{460}{1960} = .234693877 \implies$$

$$-.5k = \ln .234693877 \implies k = \frac{\ln .234693877}{-.5} \implies$$

$$k = 2.8989.$$

Therefore

$$P(t) = \frac{500}{1+49e^{-2.8989t}}.$$

a. $P(1) = \frac{500}{1+49e^{-2.8989(1)}} = 135.17$ or 135.

b. $P(3) = \frac{500}{1+49e^{-2.8989(3)}} = 495.94$ or 496.

2. Let $P(t)$ denote the number of fruit flies present after t days. Again

$$P(t) = \frac{M}{1+Ce^{-kt}}.$$

We are given that $M = 1000$ fruit flies.

$$P(0) = 50 \implies 50 = \frac{1000}{1+C} \implies 50(1+C) = 1000 \implies C = 19.$$

$$P(2) = 120 \implies 120 = \frac{1000}{1+19e^{-k(2)}} \implies 120(1+19e^{-2k}) = 1000 \implies$$

$$2280e^{-2k} = 880 \implies e^{-2k} = \frac{880}{2280} = .385964912 \implies$$

$$-2k = \ln .385964912 \implies k = \frac{\ln .385964912}{-2} \implies$$

$$k = .476.$$

Therefore

$$P(t) = \frac{1000}{1+19e^{-.476t}}.$$

$$P(4) = \frac{1000}{1+19e^{-.476(4)}} = 261.$$

3. Let $P(t)$ denote the number of bats persent after t years. Assuming the Law of Natural Growth, we have that

$$P(t) = P_0 e^{kt}.$$

$$P(0) = 10 \implies P_0 = 10.$$

$$P(.5) = 40 \implies 40 = 10e^{k(.5)} \implies e^{.5k} = 4 \implies .5k = \ln 4 \implies$$

$$k = \frac{\ln 4}{.5} = 2.7726.$$

a. $P(1) = 10e^{2.7726(1)} = 160.$

b. $P(3) = 10e^{2.7726(3)} = 40,961.$

4. Let $P(t)$ denote the number of fruit flies present after t days. Assuming the Law of Natural Growth, we have that

$$P(t) = P_0 e^{kt}.$$

$$P(0) = 50 \implies P_0 = 50.$$

$$P(2) = 120 \implies 120 = 50e^{k(2)} \implies e^{2k} = \frac{120}{50} = 2.4 \implies$$

$$2k = \ln 2.4 \implies k = \frac{\ln 2.4}{2} = .4377.$$

Therefore

$$P(t) = 50e^{.4377t}$$

$$P(4) = 50e^{.4377(4)} = 288.$$

5. $f(t) = \frac{100}{1+5e^{-2t}}.$

$$f'(t) = \frac{(1+5e^{-2t})\cdot 0 - 100\cdot\left(5e^{-2t}(-2)\right)}{(1+5e^{-2t})^2} = 1000\frac{e^{-2t}}{(1+5e^{-2t})^2}.$$

$$f''(t) = 1000\cdot\frac{(1+5e^{-2t})^2\cdot\left(e^{-2t}(-2)\right) - e^{-2t}\cdot\left(2(1+5e^{-2t})\left(5e^{-2t}(-2)\right)\right)}{(1+5e^{-2t})^4}$$

$$= 1000\cdot\frac{(1+5e^{-2t})(-2e^{-2t}) + 20e^{-4t}}{(1+5e^{-2t})^3}$$

$$= 1000\cdot\frac{2e^{-2t}(-1+5e^{-2t})}{(1+5e^{-2t})^3}.$$

$f''(t) = 0$ for $(-1 + 5e^{-2t}) = 0 \implies e^{-2t} = \frac{1}{5} \implies -2t = \ln\frac{1}{5}$

$\implies -2t = -\ln 5 \implies t = \frac{\ln 5}{2} \approx .8047.$

For $0 < t < \frac{\ln 5}{2}$, $f''(t) > 0$, so the graph is concave up.

For $t > \frac{\ln 5}{2}$, $f''(t) < 0$, so the graph is concave down.

$f\left(\frac{\ln 5}{2}\right) = \frac{100}{1+5e^{(-2)(\ln 5/2)}} = \frac{100}{1+5e^{-\ln 5}} = \frac{100}{1+5\cdot\frac{1}{5}} = 50.$

The inflection point is $\left(\frac{\ln 5}{2}, 50\right)$.

6.

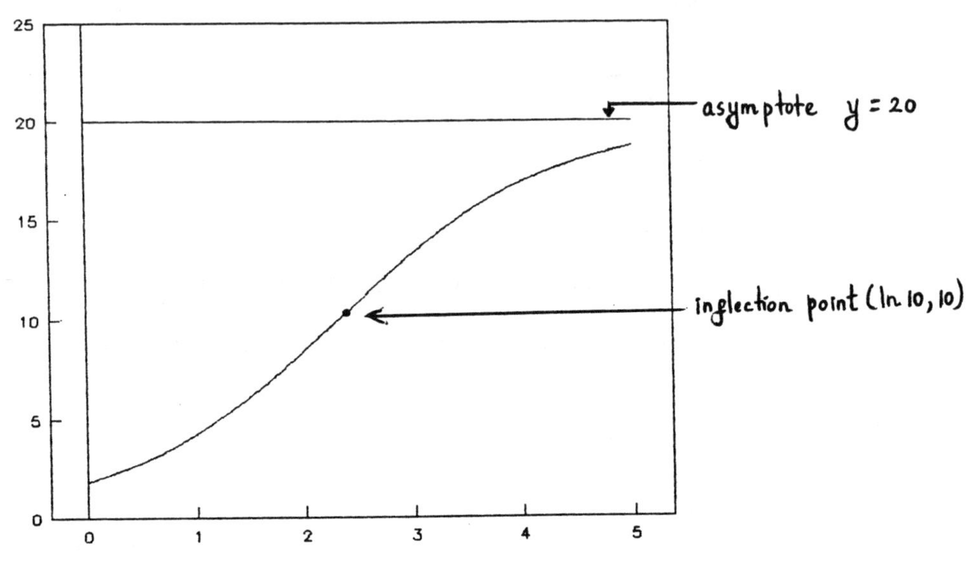

7. $f(x) = \frac{1}{\sigma\sqrt{2\pi}} e^{-\frac{1}{2}\left(\frac{x-\mu}{\sigma}\right)^2}$

$f(\mu - x) = \frac{1}{\sigma\sqrt{2\pi}} e^{-\frac{1}{2}\left(\frac{-x}{\sigma}\right)^2} = \frac{1}{\sigma\sqrt{2\pi}} e^{-\frac{1}{2\sigma^2}x^2}.$

$f(\mu + x) = \frac{1}{\sigma\sqrt{2\pi}} e^{-\frac{1}{2}\left(\frac{x}{\sigma}\right)^2} = \frac{1}{\sigma\sqrt{2\pi}} e^{-\frac{1}{2\sigma^2}x^2}.$

$f(\mu - x) = f(\mu + x)$.

8. (Percent of the scores below 116) = (percent of the total area lying between $x = 100$ and $x = 116$) + (percent of the total area lying to the left of $x = 100$) = (34 + 50)% = 84%.

9. (Percent of the scores above 116) = 100 - (percent of the total area lying to the oeft of 116) = 100 - 84 = 16%.

10. (Percent of the scores above 132) = 50 - (percent of the total area lying between $x = 100$ and $x = 132$) = 50 - 47.5 = 2.5%.

11. (Percent of the scores below 132) = 100 - (percent of the scores above 132)= 100 - 2.5 = 97.5%.

12. (Percent of the scores between 84 and 132) = (percent of the scores between 84 and 100) + percent of the scores between 100 and 132) = 34 + 47.5 = 81.5%.

13. (Percent of the scores between 68 and 100) = $\frac{1}{2}\cdot$ (percent of the total area between 68 and 132) = $\frac{1}{2} \cdot 95 = 45.5\%$.

14. (Percent of the scores between 68 and 132) = (percent of the total area between 68 and 132) = 95%.

15. (Percentage falling above $x = 0$) = (percent of the total area above $x = 0$) = 50%.

16. (Percentage falling above $x = -1$) = (percent of the total area between $x = -1$ and $x = 0$) + (percent of the total area above $x = 0$) = 34 + 50 = 84%.

17. (Percentage falling between $x = -1$ and $x = 1$) = (percent of the total area between 4x=-1andx=1) = 68%.

18. (Percentage falling between $x = 0$ and $x = 2$) = (percent of the total area between $x = 0$ and $x = 2$) = $\frac{1}{2}\cdot$ (percent of the total area between $x = -2$ and $x = 2$) = $\frac{1}{2} \cdot 95 = 47.5\%$.

19. (Percentage falling below $x = -2$) = 50 - (percent of the total area between $x = -2$ and $x = 0$) = 50 - 47.5 = 2.5%.

20. (Percentage falling between $x = -1$ and $x = 2$) = (percent of the total area between $x = -1$ and $x = 0$) + (percent of the total area between $x = 0$ and $x = 2$) = 34 + 47.5 = 81.5%.

21.

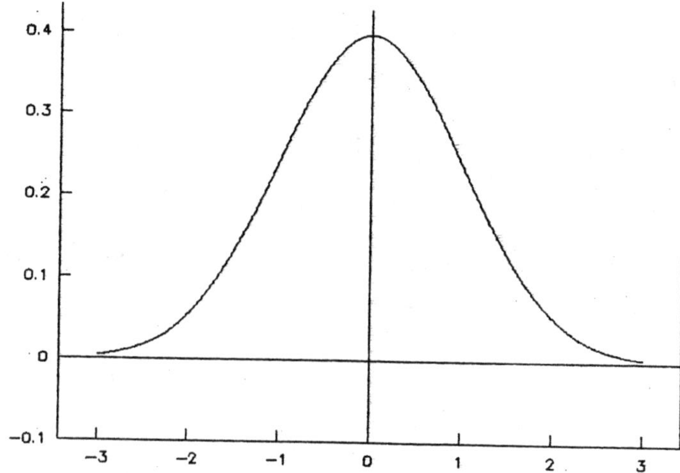

22. $g(x) = \frac{1}{\sqrt{2\pi}}e^{-x^2/2}$.

$g'(x) = \frac{1}{\sqrt{2\pi}}e^{-x^2/2} \cdot ((-2x)/2) = \frac{1}{\sqrt{2\pi}}e^{-x^2/2} \cdot (-x)$.

$$g''(x) = \frac{1}{\sqrt{2\pi}}\left((e^{-x^2/2}(-x)\right) \cdot (-x) + \left(e^{-x^2/2} \cdot (-1)\right)$$

$$= \frac{1}{\sqrt{2\pi}}e^{-x^2/2} \cdot (x^2 - 1).$$

$g''(x) = 0$ for $x = -1$ and $x = 1$.

For $x < -1$, $g''(x0 > 0$, so the graph of g is concave up.

For $-1 < x, 1$, $g''(x) < 0$, so the graph of g is concave down.

For $x > 1$, $g''(x) > 0$, so the graph of g is concave up.

Therefore the inflection points occur at $x = -1$ and $x = 1$.

Thus the inflection points are $\left(-1, \frac{1}{\sqrt{2\pi}}e^{-1/2}\right)$ and $\left(1, \frac{1}{\sqrt{2\pi}}e^{-1/2}\right)$.

23. Suppose a particle is launched vertically upward from a height of 125 feet above ground elvel with an initial velocity of 144 feet/second. Let $s(t)$ denote the height in feet of the particle above ground level after t seconds. Then the position function for the particle is given by

$$s(t) = -16t^2 + 144t + 125.$$

The velocity function for the particle is given by

$$v(t) = -32t + 144.$$

This velocity function for predicting the velocity of the particle t seconds after it is launched provides an examples of a mathematical model from physics.

24. The example of a mathematical model from physics described in Problem 23 provides an example of a mathematical model where we know that the prediction is very precise.

25. Suppose Linda tosses a pair of fair dice 25 times. Let n denote the number of the toss, and let $f(n)$ be the sum of the dots on the top faces of the two dice. This provides an example of a mathematical model discussed in probability. However, this example is not entirely precise in its predictions because Linda is probably not able to determine the outcome of a toss of the dice.

26. Let $f(x) = \frac{1}{\sigma\sqrt{2\pi}} e^{-\frac{1}{2}\left(\frac{x-\mu}{\sigma}\right)^2}$.

$$f'(x) = \frac{1}{\sigma\sqrt{2\pi}} \cdot \left(e^{-\frac{1}{2}\left(\frac{x-\mu}{\sigma}\right)^2} \cdot \left(-\frac{1}{2} \cdot 2 \cdot \frac{x-\mu}{\sigma} \cdot \frac{1}{\sigma}\right)\right)$$
$$= -\frac{1}{\sigma^3\sqrt{2\pi}} \cdot e^{-\frac{1}{2}\left(\frac{x-\mu}{\sigma}\right)^2} \cdot (x-\mu).$$
$$f''(x) = -\frac{1}{\sigma^3\sqrt{2\pi}} \cdot \left(\left(e^{-\frac{1}{2}\left(\frac{x-\mu}{\sigma}\right)^2} \cdot \left(-\frac{x-\mu}{\sigma^2}\right)\right) \cdot (x-\mu) + \left(e^{-\frac{1}{2}\left(\frac{x-\mu}{\sigma}\right)^2}\right) \cdot (1)\right)$$
$$= -\frac{1}{\sigma^3\sqrt{2\pi}} e^{-\frac{1}{2}\left(\frac{x-\mu}{\sigma}\right)^2} \cdot \left(-\left(\frac{x-\mu}{\sigma}\right)^2 + 1\right).$$
$$f''(x) = 0 \implies -\left(\frac{x-\mu}{\sigma}\right)^2 + 1 = 0 \implies (x-\mu)^2 = \sigma^2 \implies x - \mu = \pm\sigma$$
$$\implies x = \mu \pm \sigma.$$

$x < \mu - \sigma \implies f''(x) > 0 \implies$ the graph of f is concave up.

$\mu - \sigma < x < \mu + \sigma \implies f''(x) < 0 \implies$ the graph of f is concave down.

$x > \mu + \sigma \implies f''(x) > 0 \implies$ the graph of f is concave up.

Therefore the inflection points occur at $x = \mu \pm \sigma$.

Solutions to Review Exercises - Chapter 4

1. a. $\frac{1}{343}$.

   b. $8^{-4/3} = (8^{1/3})^{-4} = 2^{-4} = \frac{1}{2^4} = \frac{1}{16}$.

   c. $\frac{4}{9}$.

   d. $\frac{27}{8}$.

2. a. e^4.

   b. $\frac{(2^{-3} \cdot 3^{-2})^3}{\frac{1}{3}(16)^{3/4}} = \frac{2^{-9} \cdot 3^{-6}}{\frac{1}{3} \cdot 8} = \frac{3}{2^9 \cdot 3^6 \cdot 2^3} = \frac{1}{2^{12} \cdot 3^5} = \frac{1}{4096 \cdot 243}$
   $= \frac{1}{995328}$.

   c. $\frac{1}{3e \cdot (8e^{3x})^{2/3}} = \frac{1}{3e \cdot (2e^x)^2} = \frac{1}{3e \cdot 4e^{2x}} = \frac{1}{12e^{2x+1}}$.

   d. $(2e^{-1}2^{2x})^3 = 8e^{-3}2^{6x}$.

3. $P(3) = \left(1 + \frac{0.06}{2}\right)^6 \cdot 400$.

   $P(3) = (1.03)^6 \cdot 400 = \$477.62$.

4. $P(3) = e^{(0.06)(3)} \cdot 400 = 400 \cdot e^{(0.18)} = \$478.89$.

5. $P(t) = P_0 e^{0.1t}$. $P(5) = P_0 e^{0.1 \cdot 5}$. $10,000 = P_0 e^{0.5}$.

   $P_0 = 10,000e^{-0.5} \approx \$6,065.31$.

6. a. $10^y = \frac{1}{100}$. $y = -2$.

   b. $\log_3 243 = 5$ because $3^5 = 243$.

   c. $2^y = \frac{1}{4}$. $y = -2$.

   d. $e^y = e^{-3}$. $y = -3$.

7. a. $\ln e^{3x} = \ln 1$, $3x \ln e = 0$, $3x = 0$, $x = 0$.

   b. $\ln e^{x^2+x-2} = \ln 1$, $(x^2 + x - 2) \ln e = 0$,

   $x^2 + x - 2 = 0$, $(x+2)(x-1) = 0$; $x = -2, 1$.

   c. $e^{3x} = e^{2x+5} \implies 3x = 2x + 5 \implies x = 5$.

   d. $(e^x)^2 - e^x - 2 = 0$, $(e^x - 2)(e^x + 1) = 0$,

   $e^x = 2$ or $e^x = -1$.

   But e^x cannot be negative. So $e^x = 2$, $x = \ln 2$.

8. $\ln p = \ln(50e^{-2x})$, $\quad \ln p = \ln 50 + \ln e^{-2x}$,

$\ln p = \ln 50 - 2x$, $\quad x = -\frac{1}{2}\ln\frac{p}{50}$.

9. $R = 50e^{-2x}\cdot x$. $\quad R' = 50e^{-2x} + x(50e^{-2x})(-2)$,

$R' = 50e^{-2x}(1-2x)$, $\quad x = \frac{1}{2}$ when $R' = 0$.

$p = 50e^{-2(\frac{1}{2})} = 50e^{-1} = \frac{50}{e} \approx \$18.39$.

10. $x = -\frac{1}{2}\ln\frac{2}{50} \approx 1.6$ items.

11. $\frac{dy}{dx} = \frac{1}{6x}\cdot 6 = \frac{1}{x}$.

12. $f(x) = xe^{-3x}$.

$f'(x) = (1)(e^{-3x}) + (x)\left(e^{-3x}(-3)\right) = e^{-3x}(1-3x)$.

13. $\frac{dy}{dx} = -2xe^{6-x^2}$.

14. $f(x) = 4\ln(3x^2+2)$.

$f'(x) = 4\cdot\frac{1}{3x^2+2}\cdot\frac{d}{dx}(3x^2+2) = 4\cdot\frac{1}{3x^2+2}\cdot 6x = \frac{24x}{3x^2+2}$.

15. $f'(t) = (3t^2-3)e^{t^3-3t+2}$.

16. $y' = (t)(-1)(e^{1-t}) + e^{1-t} = e^{1-t}(1-t)$.

17. $f'(x) = \frac{1}{\sqrt{x+1}}\cdot\frac{1}{2}(x+1)^{-1/2} = \frac{1}{2(x+1)}$.

18. $f(x) = \sqrt{1-\ln^2 x}$.

$f'(x) = \frac{1}{2}\left(1-\ln^2 x\right)^{-1/2}\cdot\frac{d}{dx}(1-\ln^2 x) = \frac{1}{2}(1-\ln^2 x)^{-1/2}\left(-2\ln x\cdot\frac{1}{x}\right)$

$= -\frac{\ln x}{x\cdot\sqrt{1-\ln^2 x}}$.

19. $y' = x^2\cdot\frac{1}{2}x^{-1/2}e^{\sqrt{x}} + e^{\sqrt{x}}\cdot 2x = \frac{x^2e^{\sqrt{x}}}{2\sqrt{x}} + e^{\sqrt{x}}\cdot 2x$

$= \frac{x^2e^{\sqrt{x}}}{2\sqrt{x}} + 2xe^{\sqrt{x}} = \left(2x + \frac{1}{2}x^{3/2}\right)e^{\sqrt{x}}$.

20. $f'(x) = \frac{2e^{2x}}{e^{2x}-2}$.

21. $y = \ln\frac{x}{x+2} = \ln x - \ln(x+2)$.

$\frac{dy}{dx} = \frac{1}{x} - \frac{1}{x+2} = \frac{(x+2)-x}{x(x+2)} = \frac{2}{x(x+2)}$.

22. $y' = e^{\sqrt{x}}\cdot\frac{1}{x} + \ln x\cdot e^{\sqrt{x}}\cdot\frac{1}{2}x^{-1/2}$

$= \frac{e^{\sqrt{x}}}{x} + \frac{e^{\sqrt{x}}\ln x}{2\sqrt{x}} = \frac{(2+\sqrt{x}\ln x)e^{\sqrt{x}}}{2x}$.

23. $f'(x) = \frac{1}{\ln\sqrt{x}}\cdot\frac{1}{\sqrt{x}}\cdot\frac{1}{2}x^{-1/2} = \frac{1}{2x\ln\sqrt{x}}$.

24. $y' = x \cdot \frac{1}{\sqrt{x}-e^{-x}} \cdot \left(\frac{1}{2}x^{-1/2} + e^{-x}\right) + \ln\left(\sqrt{x} - e^{-x}\right)$
$= \frac{xe^{-x}+x\cdot\frac{1}{2}x^{-1/2}}{\sqrt{x}-e^{-x}} + \ln\left(\sqrt{x} - e^{-x}\right)$
$= \frac{xe^{-x}+\frac{1}{2}\sqrt{x}}{\sqrt{x}-e^{-x}} + \ln\left(\sqrt{x} - e^{-x}\right).$

25. $y' = x \cdot e^{x-\sqrt{x}}\left(1 - \frac{1}{2}x^{-1/2}\right) + e^{x-\sqrt{x}} = xe^{x-\sqrt{x}}\left(1 - \frac{1}{2\sqrt{x}}\right) + e^{x-\sqrt{x}}$
$= xe^{x-\sqrt{x}} - \frac{xe^{x-\sqrt{x}}}{2\sqrt{x}} + e^{x-\sqrt{x}} = \left(x - \frac{1}{2}\sqrt{x} + 1\right)e^{x-\sqrt{x}}.$

26. $y' = e^{t+\ln t} \cdot \left(1 + \frac{1}{t}\right) = \left(1 + \frac{1}{t}\right)e^{t+\ln t}.$

27. $y = \ln^2(1 - e^{-2x}).$
$\frac{dy}{dx} = 2\ln(1 - e^{-2x}) \cdot \frac{1}{1-e^{-2x}} \cdot (2e^{-2x}) = \frac{4\ln(1-e^{-2x})\cdot e^{-2x}}{1-e^{-2x}}.$

28. $y' = \frac{1}{2}\left(e^x - e^{-x}\right)^{-1/2}\left(e^x + e^{-x}\right).$

29. $f'(t) = e^{\sqrt{t}-\ln t} \cdot \left(\frac{1}{2}t^{-1/2} - \frac{1}{t}\right).$

30. $f'(x) = x^2 \cdot e^{1-\sqrt{x}}\left(-\frac{1}{2}x^{-1/2}\right) + e^{1-\sqrt{x}} \cdot 2x$
$= \frac{-x^2e^{1-\sqrt{x}}}{2\sqrt{x}} + 2xe^{1-\sqrt{x}} = xe^{1-\sqrt{x}}\left(2 - \frac{1}{2}\sqrt{x}\right).$

31. $x\ln y + y\ln x = 4$
$x \cdot \frac{1}{y} \cdot \frac{dy}{dx} + \ln y \cdot 1 + y\frac{1}{x} + \ln x \cdot \frac{dy}{dx} = 0$
$\frac{x}{y} \cdot \frac{dy}{dx} + \ln x\frac{dy}{dx} = -\ln y - \frac{y}{x}$
$\left(\frac{x}{y} + \ln x\right)\frac{dy}{dx} = -\ln y - \frac{y}{x}$
$\frac{dy}{dx} = \frac{-\ln y - \frac{y}{x}}{\frac{x}{y}+\ln x} = \frac{-y^2-xy\ln y}{x^2+xy\ln x}.$

32. $x^2 + e^{xy} - y = 2$
$2x + \left(x \cdot \frac{dy}{dx} + y\right)e^{xy} - \frac{dy}{dx} = 0.$
$2x + xe^{xy}\frac{dy}{dx} + ye^{xy} - \frac{dy}{dx} = 0.$
$\frac{dy}{dx}\left(xe^{xy} - 1\right) = -2x - ye^{xy}$
$\frac{dy}{dx} = \frac{-2x-ye^{xy}}{xe^{xy}-1} = \frac{2x+ye^{xy}}{1-xe^{xy}}.$

33. $e^{xy} = 3xy.$
$e^{xy}\left(1 \cdot y + x \cdot \frac{dy}{dx}\right) = 3 \cdot 1 \cdot y + 3x \cdot \frac{dy}{dx}.$
$\frac{dy}{dx}\left(xe^{xy} - 3x\right) = 3y - ye^{xy}.$
$\frac{dy}{dx} = \frac{3y-ye^{xy}}{xe^{xy}-3x} = \frac{y(3-e^{xy})}{x(e^{xy}-3)} = -\frac{y}{x}.$

34. $\ln xy = x - y$. $\frac{1}{xy}\left(x\frac{dy}{dx} + y\right) = 1 - \frac{dy}{dx}$.

$\frac{x}{xy} \cdot \frac{dy}{dx} + \frac{y}{xy} = 1 - \frac{dy}{dx}$, $\frac{dy}{dx}\left(\frac{1}{y} + 1\right) = 1 - \frac{1}{x}$,

$\frac{dy}{dx} = \frac{1-\frac{1}{x}}{\frac{1}{y}+1} = \frac{xy-y}{xy+x}$.

35. $y = x^2 \ln x$, $x > 0$.

$y' = 2x \ln x + x^2 \cdot \frac{1}{x} = x(2\ln x + 1)$.

$y' = 0$ for $2\ln x + 1 = 0$, $x = e^{-1/2}$.

For $0 < x < e^{-1/2}$, $y' < 0$; and for $x > e^{-1/2}$, $y' > 0$.

So $y = (e^{-1/2})^2 \ln e^{-1/2} = -\frac{1}{2e}$ is the relative minimum for $x = e^{-1/2}$.

36. $f(x) = e^{x^2/3}$ in $\left[-1, \sqrt{8}\right]$.

$f'(x) = \frac{2}{3}xe^{x^2/3} = 0$ for $x = 0$.

$f(-1) = e^{(-1)^2/3} = e^{1/3}$, $f(0) = 1$, $f\left(\sqrt{8}\right) = e^{\left(\sqrt{8}\right)^2/3} = e^{8/3}$.

So the minimum of $f(x)$ is $f(0) = 1$ and the maximum is $f\left(\sqrt{8}\right) = e^{8/3}$.

37. $y = \ln(1 + x^2)$. $y' = \frac{2x}{1+x^2}$. $y'' = \frac{2(1+x^2)-2x\cdot 2x}{(1+x^2)^2} = \frac{2-2x^2}{(1+x^2)^2}$.

$y'' = 0$ for $x = \pm 1$.

On $(-1, 1)$, $y'' > 0$. On $(-\infty, -1)$ and $(1, \infty)$, $y'' < 0$.

So y is concave up on $(-1, 1)$.

38. $y = x^2 - \ln x^2$, $x \neq 0$.

$y' = 2x - \frac{2x}{x^2} = \frac{2(x^2-1)}{x} = 0$ for $x = \pm 1$.

y' does not exist for $x = 0$.

On $(-\infty, -1)$, $y' < 0$. On $(-1, 0)$, $y' > 0$.

On $(0, 1)$, $y' < 0$. On $(1, \infty)$, $y' > 0$.

So $f(-1) = f(1) = 1$ are the relative minima, and there is no relative maximum.

39. $2^x = e^{rx}$, $\ln 2^x = \ln e^{rx}$, $x \ln 2 = rx$, $r = \ln 2$.

40. $y = xe^{2x}$. $y' = 1 \cdot e^{2x} + xe^{2x} \cdot 2 = e^{2x}(1 + 2x)$.

At $(\ln 2, 4\ln 2)$, $y' = e^{2\ln 2}(1 + 2\ln 2) = 4(1 + \ln 4)$.

So the tangent line at $(\ln 2, 4\ln 2)$ is

$y - 4\ln 2 = 4(1 + \ln 4)(x - \ln 2)$

$y - 4\ln 2 = 4(1+\ln 4)x - 4\ln 2 - 4\ln 4 \cdot \ln 2$

$y = 4(1+\ln 4)x - 8(\ln 2)^2.$

41. The volume of the water inside the jar is

$V(t) = 6^2\pi \cdot H(t).$

$V'(t) = 36\pi \cdot H'(t).$

Now $V'(t) = -\ln(t^2)$.

So $H'(t) = \frac{-1}{36\pi}V'(t) = \frac{-1}{36\pi}\ln(t^2)$.

42. $\frac{dy}{dx} = 2y. \qquad y = Ce^{2x}.$

43. $\frac{dy}{dx} + y = 0. \qquad \frac{dy}{dx} = -y. \qquad y = Ce^{-x}.$

44. $2y' + 4y = 0, \qquad y(0) = \pi. \qquad y' = -2y.$

$y = y(0)e^{-2x}. \qquad y = \pi e^{-2x}.$

45. The population t years after 1970 is $N(t) = 203e^{kt}$ million.

$N(10) = 227 = 203e^{k\cdot 10}. \qquad k = \frac{1}{10}\ln\frac{227}{203}.$

So $N(t) = 203e^{\left(\frac{1}{10}\ln\frac{227}{203}\right)t}$

Then in 1990, $N(20) = 203e^{\left(\frac{1}{10}\ln\frac{227}{203}\right)(20)} = 203\left(\frac{227}{203}\right)^2 \approx 254$ million.

46. $N(t) =$ the number of kilograms of raw sugar after t hours.

$N(0) = 100. \qquad N(t) = 100e^{kt}.$

$N(6) = 75 \implies 75 = 100e^{k6} \implies e^{6k} = 0.75 \implies 6k = \ln 0.75 \implies$

$k = \frac{1}{6}\ln(0.75)$. $N(t) = 100e^{\left(\frac{1}{6}\ln 0.75\right)t}$.

(a) When $N(t) = 50$, $50 = 100e^{\left(\frac{1}{6}\ln 0.75\right)t} \implies e^{\left(\frac{1}{6}\ln 0.75\right)t} = 0.5$

$\implies \left(\frac{1}{6}\ln 0.75\right)t = \ln 0.5 \implies t = 6\frac{\ln 0.5}{\ln 0.75} \implies t = 14.46$ hours.

(b) $N(t) = 10$, $10 = 100e^{\left(\frac{1}{6}\ln 0.75\right)t} \implies e^{\left(\frac{1}{6}\ln 0.75\right)t} = .1$

$\implies \left(\frac{1}{6}\ln 0.75\right)t = \ln 0.1 \implies t = 6\frac{\ln 0.1}{\ln 0.75} \implies t = 48$ hours.

47. $N(t) =$ the number of fruit flies after t days.

$N(0) = 100. \qquad N(t) = 100e^{kt}.$

$N(10) = 500 \implies 500 = 100e^{k\cdot 10} \implies e^{10k} = 5 \implies 10k = \ln 5$

$\implies k = \frac{1}{10}\ln 5.$

$N(t) = 100e^{\left(\frac{1}{10}\ln 5\right)t}. \qquad N(4) = 100e^{\left(\frac{1}{10}\ln 5\right)4} = 190.36$ flies.

48. $N(t) =$ the number of grams of a radioactive substance after t hours.

$N(0) = 100. \qquad N(t) = 100e^{kt}.$

$N(6) = 40 \implies 40 = 100e^{k6} \implies e^{k6} = 0.4 \implies 6k = \ln 0.4 \implies$

$k = \frac{1}{6}\ln(0.4)$. $N(t) = 100e^{\left(\frac{1}{6}\ln 0.4\right)t}$.

$N(t) = 50 \implies 50 = 100e^{\left(\frac{1}{6}\ln 0.4\right)t} \implies e^{\left(\frac{1}{6}\ln 0.4\right)t} = 0.5 \implies$

$\left(\frac{1}{6}\ln 0.4\right)t = \ln 0.5 \implies t = 6\frac{\ln 0.5}{\ln 0.4} \implies t = 4.54$ hours.

49. $f(x) = C \cdot e^{kx}. \qquad f'(x) = k \cdot C \cdot e^{kx}.$

$f'(x+a) = k \cdot C \cdot e^{k(x+a)} = k \cdot f(x+a).$

50. We assume here that k is a positive constant. Then for all $x > 0$, $f(x) = \ln(kx) = \ln k + \ln x$. Therefore $f'(x) = \frac{1}{x} = g'(x)$ for all $x > 0$. Hence the functions f and g have the same derivative.

51. Let $P(t)$ denote the value of the bond after t years. Let P_0 denote the present value of the bond.

$$P(t) = P_0 e^{.05t} \implies P_0 = e^{-.05t} \cdot P(t).$$

$$P(4) = 2000 \implies P_0 = e^{-.05(4)} \cdot 2000 = 1,637.46 \text{ dollars.}$$

52. The value of the account after 1 year is $De^{.05}$.

The value of the account after 2 years is $(De^{.05} + D)e^{.05} = D(e^{.10} + e^{.05})$.

The value of the account after 3 years is $\left(D(e^{.10} + e^{.05}) + D\right)e^{.05}$

$= D(e^{.15} + e^{.10} + e^{.05})$.

a. Continuing in this manner, we find that the value of the account after 5 years is

$$T = D(e^{.25} + e^{.20} + e^{.15} + e^{.10} + e^{.05}).$$

b. $T = 20,000 = D(e^{.25} + e^{.20} + e^{.15} + e^{.10} + e^{.05})$

$\implies 5.823704432D = 20,000 \implies D = \$3,434.24.$

c. The interest rate r is changed to 8%. Then

$$T = D(e^{.40} + e^{.32} + e^{.24} + e^{.16} + e^{.08}).$$

$$20,000 = D(6.396999551) \implies D = \$3,126.47.$$

CHAPTER 5

Solutions to Exercise Set 5.1

1. $\int 5dx = 5x + C.$
2. $\int 3x^2 dx = x^3 + C.$
3. $\int 2x^3 dx + \int 5x dx = \frac{1}{2}x^4 + \frac{5}{2}x^2 + C.$
4. $\int 4dx - \int 4x^3 dx = 4x - x^4 + C.$
5. $\int x^3 dx - \int 6x^2 dx + \int 2x dx - \int 1dx = \frac{1}{4}x^4 - 2x^3 + x^2 - x + C.$
6. $\int 9dx - \int x^3 dx + \int 5x^4 dx = 9x - \frac{1}{4}x^4 + x^5 + C.$
7. $\int x^{2/3} dx - \int 3x^{1/2} dx = \frac{3}{5}x^{5/3} - 2x^{3/2} + C.$
8. $\int e^{3x} dx = \frac{1}{3}e^{3x} + C.$
9. $\int \frac{4}{x} dx = 4\ln x + C, \qquad x > 0.$
10. $\int \frac{1}{x+2} dx = \ln(x+2) + C_1$ for $x > -2$.
    $= \ln(-(x+2)) + C_2$ for $x < -2$.
11. $\int e^{5x} dx = \frac{1}{5}e^{5x} + C.$
12. $\int 3e^{-x/2} dx = 3 \cdot \frac{1}{-\frac{1}{2}} e^{-x/2} + C = -6e^{-x/2} + C.$
13. $\int x^{1/3} dx - \int 2x^{-2/3} dx + \int x^{-5/3} dx = \frac{3}{4}x^{4/3} - 6x^{1/3} - \frac{3}{2}x^{-2/3} + C.$
14. $\int \left(\sqrt{x} - 4x^{2/3} + 5\right) dx = \frac{1}{\frac{3}{2}}x^{3/2} - 4 \cdot \frac{1}{\frac{5}{3}}x^{5/3} + 5x + C.$
    $= \frac{2}{3}x^{3/2} - 4 \cdot \frac{3}{5}x^{5/3} + 5x + C = \frac{2}{3}x^{3/2} - \frac{12}{5}x^{5/3} + 5x + C.$
15. $\int 4e^{-x} dx - \int \frac{1}{x} dx + \int \sqrt{x} dx = -4e^{-x} - \ln x + \frac{2}{3}x^{3/2} + C, \quad x > 0.$
16. $\int (x^2 + x - 2) dx = \frac{1}{3}x^3 + \frac{1}{2}x^2 - 2x + C.$
17. $\int 4x^{-6} dx - \int 6x^{-4} dx = -\frac{4}{5}x^{-5} + 2x^{-3} + C.$
18. $\int \sqrt{x} dx - \int x^{1/3} dx = \frac{2}{3}x^{3/2} - \frac{3}{4}x^{4/3} + C.$
19. $\int (x - 2x^{5/6} + x^{2/3}) dx = \frac{1}{2}x^2 - \frac{12}{11}x^{11/6} + \frac{3}{5}x^{5/3} + C.$

20. $\int \frac{(x-1)(x-2)}{(x-1)}dx = \int (x-2)dx = \frac{1}{2}x^2 - 2x + C.$

21. $\int (\sqrt{x} - e^{-x})\,dx = \frac{1}{\frac{3}{2}}x^{3/2} - \frac{1}{-1}e^{-x} + C = \frac{2}{3}x^{3/2} + e^{-x} + C.$

22. $\int x^{3/2}dx - \int 4x^{1/2}dx = \frac{2}{5}x^{5/2} - \frac{8}{3}x^{3/2} + C.$

23. $\int (6x + 3x^2 + 5)dx = 6 \cdot \left(\frac{1}{2}x^2\right) + 3 \cdot \left(\frac{1}{3}x^3\right) + 5x + C$

$$= 3x^2 + x^3 + 5x + C.$$

$f(x) = 3x^2 + x^3 + 5x$ and $g(x) = 3x^2 + x^3 + 5x + 1$ are functions such that $f'(x) = g'(x) = 6x + 3x^2 + 5$ for all x.

$[f(x) - g(x)] = -1$ for all x.

24. $f(x) = x^2 + Ax + 6$ and $g(x) = Bx^2 + 10x - 9.$

$f'(x) = 2x + A$ and $g'(x) = 2Bx + 10.$

$f'(x) = g'(x)$ for all $x \implies 2 = 2B$ and $A = 10.$

a. $A = 10.$

b. $2B = 2 \implies B = 1.$

25. $f(x) = Ae^{2x} + \sqrt{x} + 5$ and $g(x) = 6e^{2x} + 3Bx^{1/2} + C.$

$f'(x) = 2Ae^{2x} + \frac{1}{2\sqrt{x}}$ and $g'(x) = 12e^{2x} + \frac{3B}{2\sqrt{x}}.$

$f'(x) = g'(x)$ for all $x > 0 \implies 2A = 12$ and $1 = 3B.$

a. $2A = 12 \implies A = 6.$

b. $3B = 1 \implies B = \frac{1}{3}.$

c. C is any constant.

26. $MR(x) = R'(x) = A$ where A is a constant.

$R(x) = Ax + B$ where B is a constant. If $A > 0$, then the revenue $R(x)$ is increasing for all $x > 0$.

If $A < 0$, then the revenue $R(x)$ is decreasing for all $x > 0$.

27. $P'(t) = 4 + 2t$ for $t > 0.$

a. $P(t) = 4t + t^2 + C$ where C is the population of the deer at time $t = 0.$

b. The population is increasing because $P'(t) = 4 + 2t > 0$ for all $t > 0.$

28. $P'(t) = 4\sqrt{t} - 6$ for $t > 0$.

a. $P(t) = 4 \cdot \frac{1}{\frac{3}{2}} t^{3/2} - 6t + C = \frac{8}{3} t^{3/2} - 6t + C$ where C is the pheasant population when $t = 0$.

b. $P'(t) = 0 \implies 4\sqrt{t} = 6 \implies 16t = 36 \implies t = \frac{36}{16} = \frac{9}{4}$.

For $0 < t < \frac{9}{4}$, $P'(t) < 0$, so the pheasant population is decreasing.

For $t > \frac{9}{4}$, $P'(t) > 0$, so the pheasant population is increasing.

c. The population size is decreasing at the rate of about 6 pheasants per unit of time.

Solutions to Exercise Set 5.2

1. ii 2. i 3. iv 4. iii

5. vi 6. v

7. $F(x) = \int 4dx = 4x + C$.

$F(0) = 3 = 4(0) + C \implies C = 3$.

$F(x) = 4x + 3$.

8. $F(x) = \int (x - 2)dx = \frac{1}{2}x^2 - 2x + C$.

$F(0) = 6 = \frac{1}{2}(0)^2 - 2(0) + C \implies C = 6$.

$F(x) = \frac{1}{2}x^2 - 2x + 6$.

9. $F(x) = \int (x^2 - 2x)dx = \frac{1}{3}x^3 - x^2 + C$.

$F(0) = -3 = \frac{1}{3}(0)^3 - (0)^2 + C \implies C = -3$.

$F(x) = \frac{1}{3}x^3 - x^2 - 3$.

10. $F(x) = \int \sqrt{x}dx = \frac{2}{3}x^{3/2} + C$.

$F(4) = 7 = \frac{2}{3}(4)^{3/2} + C \implies 7 = \frac{16}{3} + C \implies C = \frac{5}{3}$.

$F(x) = \frac{2}{3}x^{3/2} + \frac{5}{3}$.

11. $F(x) = \int \frac{1}{x} dx = \ln x + C.$

$F(e) = 5 = \ln e + C \implies C = 4.$

$F(x) = \ln x + 4.$

12. $F(x) = \int (4e^x + 5) dx = 4e^x + 5x + C.$

$F(1) = 4e + 8 = 4e^1 + 5(1) + C \implies C = 3.$

$F(x) = 4e^x + 5x + 3.$

13. $F(x) = \int (x^{2/3} - x^{-1/3}) dx = \frac{3}{5}x^{5/3} - \frac{3}{2}x^{2/3} + C.$

$F(0) = 5 = \frac{3}{5} \cdot (0)^{5/3} - \frac{3}{2} \cdot (0)^{2/3} + C \implies C = 5.$

$F(x) = \frac{3}{5}x^{5/3} - \frac{3}{2}x^{2/3} + 5.$

14. $F(x) = \int (x^2 - 4 + 3x^{-7/2}) dx = \frac{1}{3}x^3 - 4x - \frac{6}{5}x^{-5/2} + C.$

$F(0) = -4 = \frac{1}{3}(0)^3 - 4(0) - \frac{6}{5}(0)^{-5/2} + C \implies C = -4.$

$F(x) = \frac{1}{3}x^3 - 4x - \frac{6}{5}x^{-5/2} - 4.$

15. $F(x) = \int (5e^{2x} + 4) dx = \frac{5}{2}e^{2x} + 4x + C.$

$F(0) = 10 = \frac{5}{2}e^{2(0)} + 4(0) + C \implies C = \frac{15}{2}.$

$F(x) = \frac{5}{2}e^{2x} + 4x + \frac{15}{2}.$

16. $F(x) = \int \left(\sqrt{x} - \frac{1}{x}\right) dx = \frac{2}{3}x^{3/2} - \ln x + C.$

$F(1) = 3 = \frac{2}{3}(1)^{3/2} - \ln 1 + C \implies C = \frac{7}{3}.$

$F(x) = \frac{2}{3}x^{3/2} - \ln x + \frac{7}{3}.$

17. $F(x) = \int \left(\frac{3}{x+5} + 2e^{-3x}\right) dx = 3\ln(x+5) - \frac{2}{3}e^{-3x} + C.$

$F(0) = 0 \implies 0 = 3\ln(0+5) - \frac{2}{3}e^{-3(0)} + C = 3\ln 5 - \frac{2}{3} + C$

$\implies C = \frac{2}{3} - 3\ln 5.$

$F(x) = 3\ln(x+5) - \frac{2}{3}e^{-3x} + \frac{2}{3} - 3\ln 5.$

18. $F'(x) = \frac{x-1}{x+1} = \frac{(x+1)-2}{x+1} = \frac{x+1}{x+1} - \frac{2}{x+1} = 1 - \frac{2}{x+1}.$

$F(x) = x - 2\ln(x+1) + C.$

$F(0) = 1 \implies 1 = 0 - 2\ln(0+1) + C \implies C = 1.$

$F(x) = x - 2\ln(x+1) + 1.$

19. $C(x) = \int MC(x)dx = \int 250dx = 250x + K.$

$C(0) = 700 = 250(0) + K \implies K = 700.$

$C(x) = 250x + 700$ dollars.

20. (a) We are given that $C(20) = 7000 = 250(20) + K \implies K = \$2000$

$=$ weekly fixed costs.

(b) $C(x) = 250x + 2000.$

21. $C(x) = \int MC(x)dx = \int \left(400 + \frac{1}{4}x\right) dx = 400x + \frac{1}{8}x^2 + K.$

(a) The manufacturer's cost for 10 dishwashers $= C(10) = 5200$

$= 400(10) + \frac{1}{8}(10)^2 + K \implies K = \$1187.50$, fixed monthly cost.

(b) $C(x) = 400x + \frac{1}{8}x^2 + 1187.5.$

22. $MC(x) = 50 + 2x \implies C(x) = 50x + x^2 + K.$

$c(x) = \frac{C(x)}{x} = 50 + x + \frac{K}{x} = \frac{1000}{x} + A + x.$

a. $A = 50.$

b. $K = 1000 \implies C(x) = 50x + x^2 + 1000.$

23. $MC(x) = 100 + 4x \implies C(x) = 100x + 2x^2 + K.$

$C(100) = 35,000 \implies 35,000 = 100(100) + 2(100)^2 + K \implies K = 5000.$

Therefore $C(x) = 100x + 2x^2 + 5000.$

24. $MR(x) = 400 - 2x \implies R(x) = 400x - x^2 + K.$

$R(0) = 0 \implies K = 0.$

Therefore $R(x) = 400x - x^2.$

25. $R(x) = \int MR(x)dx = \int \left(250 - \frac{1}{2}x\right) dx = 250x - \frac{1}{4}x^2 + K.$

$R(0) = 0 = 250(0) - \frac{1}{4}(0)^2 + K \implies K = 0.$

The total revenue function $R(x) = 250x - \frac{1}{4}x^2.$

26. $U(x) = \int MU(x)dx = \int \sqrt{x}dx = \frac{2}{3}x^{3/2} + K.$

$U(0) = 0 = \frac{2}{3}(0)^{3/2} + K \implies K = 0.$

The utility function $U(x) = \frac{2}{3}x^{3/2}.$

27. $P'(t) = 4 + 2t, \quad t > 0.$

$P(t) = 4t + t^2 + K.$

$P(0) = 12 \implies K = 12.$

Therefore $P(t) = 4t + t^2 + 12.$

28. $P'(t) = 4t - 6, \quad t > 0.$

$P(t) = 2t^2 - 6t + K.$

$P(0) = 40 \implies K = 40.$

Therefore $P(t) = 2t^2 - 6t + 40.$

29. I matches up with c.

II matches up with a.

III matches up with b.

30. a. $C(x) = \int MC(x)dx = \int \left(40 + \frac{1}{2}x\right) dx = 40x + \frac{1}{4}x^2 + K.$

$C(0) = 800 = 40(0) + \frac{1}{4}(0)^2 + K \implies K = 800.$

The manufacturer's monthly cost $C(x) = 40x + \frac{1}{4}x^2 + 800.$

b. $R(x) = \int MR(x)dx = \int 60dx = 60x + K.$

$R(0) = 0 = 60(0) + K \implies K = 0.$

The manufacturer's monthly revenue $R(x) = 60x.$

c. The manufacturer's monthly profit $P(x) = R(x) - C(x)$

$\implies P(x) = 60x - 40x - \frac{1}{4}x^2 - 800 \implies P(x) = -\frac{1}{4}x^2 + 20x - 800.$

d. $MP(x) = P'(x) = MR(x) - MC(x) \implies P'(x) = 60 - 40 - \frac{1}{2}x$

$\implies P'(x) = 20 - \frac{1}{2}x.$

31. $s(t) = \int v(t)dt = \int (3t^2 + 6t + 2)dt = t^3 + 3t^2 + 2t + C.$

$s(0) = 4 = (0)^3 + 3(0)^2 + 2(0) + C \implies C = 4.$

$s(t) = t^3 + 3t^2 + 2t + 4.$

(a) Its location after t seconds $s(t) = t^3 + 3t^2 + 2t + 4.$

(b) Its location after 4 seconds $s(4) = (4)^3 + 3(4)^2 + 2(4) + 4 \implies s(4) = 124.$

32. $s(t) = \int v(t)dt = \int(-32t)dt = -\frac{32}{2}t^2 + C.$

(a) $s(4) = -\frac{32}{2}(4)^2 + C = -256 \text{ ft} + C \implies$ it has fallen 256 feet.

(b) $s(8) = -\frac{32}{2}(8)^2 + C = -1024 \text{ ft} + C \implies$ it has fallen 1024 feet.

(c) $s(t) = -\frac{32}{2}t^2 + C \implies$ it has fallen $16t^2$ feet.

33. $N(t) = \int \left(20 + 24\sqrt{t}\right) dt = 20t + 16t^{3/2} + C.$

$N(0) = 2 = 20(0) + 16(0)^{3/2} + C \implies C = 2.$

(a) $N(t) = 20t + 16t^{3/2} + 2.$

(b) $N(16) = 20(16) + (16)(16)^{3/2} + 2 \implies N(16) = 1346.$

Solutions to Exercise Set 5.3

1. $\int x(3 + x^2)^3 dx.$

   Let $u = 3 + x^2$. $\quad du = 2x\,dx \implies x\,dx = \frac{1}{2}du.$

   $\int x(3 + x^2)^3 dx = \int u^3 \left(\frac{1}{2}du\right) = \frac{1}{2} \cdot \frac{1}{4}u^4 + C = \frac{1}{8}(3 + x^2)^4 + C.$

2. $\int x\sqrt{x^2 + 1}\, dx$

   Let $u(x) = x^2 + 1$. $\quad du = 2x\, dx.$

   $\int x\sqrt{x^2 + 1}dx = \frac{1}{2}\int \sqrt{u}\, du = \frac{1}{2} \cdot \frac{2}{3}u^{3/2} + C = \frac{1}{3}(x^2 + 1)^{3/2} + C.$

3. $\int 5x\sqrt{9 + x^2}\, dx.$

   Let $u = 9 + x^2$. $\quad du = 2x\,dx \implies x\,dx = \frac{1}{2}du.$

   $\int 5x\sqrt{9 + x^2}\, dx = \int 5u^{1/2}\left(\frac{1}{2}du\right) = \frac{5}{2} \cdot \frac{2}{3}u^{3/2} + C = \frac{5}{3}\left(9 + x^2\right)^{3/2} + C.$

4. $\int x^2(4 + x^3)^{-1/2}\, dx.$

   Let $u(x) = 4 + x^3$. $\quad du = 3x^2\, dx.$

   $\int x^2(4 + x^3)^{-1/2}\, dx = \frac{1}{3}\int u^{-1/2}\, du = \frac{1}{3}2u^{1/2} + C = \frac{2}{3}(4 + x^3)^{1/2} + C.$

5. $\int xe^{x^2}\, dx.$

   Let $u(x) = x^2$. $\quad du = 2x\, dx, \quad x\, dx = \frac{1}{2}du.$

   $\int xe^{x^2}\, dx = \frac{1}{2}\int e^u\, du = \frac{1}{2}e^u + C = \frac{1}{2}e^{x^2} + C.$

6. $\int \frac{x^2\,dx}{\sqrt{1+x^3}}$.

Let $u(x) = 1 + x^3$. $du = 3x^2\,dx$.

$\int \frac{x^2}{\sqrt{1+x^3}}\,dx = \frac{1}{3}\int \frac{1}{\sqrt{u}}\,du = \frac{1}{3}2u^{1/2} + C = \frac{2}{3}(1+x^3)^{1/2} + C.$

7. $\int \frac{\left(\sqrt{x}+5\right)^6}{\sqrt{x}}\,dx$.

Let $u = \sqrt{x} + 5$. $du = \frac{1}{2\sqrt{x}}\,dx \implies \frac{1}{\sqrt{x}}\,dx = 2\,du$.

$\int \frac{\left(\sqrt{x}+5\right)^6}{\sqrt{x}}\,dx = \int u^6(2\,du) = 2 \cdot \frac{1}{7}u^7 + C = \frac{2}{7}\left(\sqrt{x}+5\right)^7 + C.$

8. $\int \frac{x-1}{x^2-2x}\,dx$.

Let $u(x) = x^2 - 2x$. $du = (2x-2)dx$.

$\int \frac{x-1}{x^2-2x}dx = \int \frac{1}{u}\cdot\frac{1}{2}\,du = \frac{1}{2}\ln|u| + C = \frac{1}{2}\ln|x^2-2x| + C.$

9. $\int \frac{x}{1+3x^2}\,dx$.

Let $u = 1 + 3x^2$. $du = 6x\,dx \implies x\,dx = \frac{1}{6}\,du$.

$\int \frac{x}{1+3x^2}\,dx = \int \frac{1}{u}\left(\frac{1}{6}\,du\right) = \frac{1}{6}\ln|u| + C = \frac{1}{6}\ln(1+3x^2) + C.$

10. $\int \frac{\ln^3 x}{x}\,dx$.

Let $u = \ln x$. $du = \frac{1}{x}\,dx$.

$\int \frac{\ln^3 x}{x}\,dx = \int u^3\,du = \frac{1}{4}u^4 + C = \frac{1}{4}\ln^4 x + C.$

11. $\int e^{2x}(1+e^{2x})^3\,dx$.

Let $u(x) = 1 + e^{2x}$. $du = 2e^{2x}\,dx$.

$\int e^{2x}(1+e^{2x})^3\,dx = \int u^3 \cdot \frac{1}{2}\,du = \frac{1}{2}\cdot\frac{1}{4}u^4 + C = \frac{1}{8}(1+e^{2x})^4 + C.$

12. $\int \frac{e^x}{1+e^x}\,dx$.

Let $u(x) = 1 + e^x$. $du = e^x\,dx$.

$\int \frac{e^x}{1+e^x}\,dx = \int \frac{1}{u}\,du = \ln u + C = \ln(1+e^x) + C.$

13. $\int \frac{e^x}{\sqrt{e^x+1}}\,dx$.

Let $u(x) = e^x + 1$. $du = e^x\,dx$.

$\int \frac{e^x}{\sqrt{e^x+1}}\,dx = \int \frac{1}{\sqrt{u}}\,du = 2u^{1/2} + C = 2\sqrt{e^x+1} + C.$

14. $\int \frac{1}{x\ln x}\,dx$

Let $u(x) = \ln x$. $du = \frac{1}{x}\,dx$.

$\int \frac{1}{x\ln x}\,dx = \int \frac{1}{u}\,du = \ln|u| + C = \ln|\ln x| + C.$

15. $\int \frac{2x+3}{(x^2+3x+6)^3}\, dx$

Let $u(x) = x^2 + 3x + 6$. $du = (2x+3)dx$.

$\int \frac{2x+3}{(x^2+3x+6)^3}\, dx = \int \frac{1}{u^3}\, du = -\frac{1}{2}u^{-2} + C = -\frac{1}{2(x^2+3x+6)^2} + C.$

16. $\int \frac{3x+3}{(x^2+2x-3)^3}\, dx$

Let $u(x) = x^2 + 2x - 3$. $du = (2x+2)dx$.

$\int \frac{3x+3}{(x^2+2x-3)^3}\, dx = \int \frac{3}{u^3} \cdot \frac{1}{2}\, du = \frac{3}{2}\left(-\frac{1}{2}\right) u^{-2} + C = -\frac{3}{4(x^2+2x-3)^2} + C.$

17. $\int \left(1 - \frac{1}{x}\right)^3 \left(\frac{1}{x^2}\right) dx$

Let $u(x) = 1 - \frac{1}{x}$. $du = \frac{1}{x^2}\, dx$.

$\int \left(1 - \frac{1}{x}\right)^3 \left(\frac{1}{x^2}\right) dx = \int u^3\, du = \frac{1}{4}u^4 + C = \frac{1}{4}\left(1 - \frac{1}{x}\right)^4 + C.$

18. $\int \frac{x^2+2x+3x^{3/2}}{\sqrt{x}}\, dx = \int \left(\frac{x^2}{\sqrt{x}} + \frac{2x}{\sqrt{x}} + \frac{3x^{3/2}}{\sqrt{x}}\right) dx = \int (x^{3/2} + 2x^{1/2} + 3x)dx$

$= \frac{2}{5}x^{5/2} + 2 \cdot \frac{2}{3}x^{3/2} + 3 \cdot \frac{1}{2}x^2 + C = \frac{2}{5}x^{5/2} + \frac{4}{3}x^{3/2} + \frac{3}{2}x^2 + C.$

19. $\int \frac{x^3}{\sqrt{5+x^4}}\, dx$

Let $u(x) = 5 + x^4$. $du = 4x^3\, dx$.

$\int \frac{x^3}{\sqrt{5+x^4}}\, dx = \int \frac{1}{\sqrt{u}} \cdot \frac{1}{4}\, du = \frac{1}{4} \cdot 2u^{1/2} + C = \frac{1}{2}(5 + x^4)^{1/2} + C.$

20. $\int \frac{x}{x^2+5}\, dx$

Let $u(x) = x^2 + 5$. $du = 2x\, dx$.

$\int \frac{x}{x^2+5}\, dx = \int \frac{1}{u} \cdot \frac{1}{2}\, du = \frac{1}{2}\ln u + C = \frac{1}{2}\ln(x^2+5) + C.$

21. $\int \frac{4\ln x^2}{x}\, dx = \int \frac{8\ln x}{x}\, dx.$

Let $u = \ln x$. $du = \frac{1}{x}\, dx$.

$\int \frac{4\ln x^2}{x}\, dx = 8 \cdot \int u\, du = 8 \cdot \frac{1}{2}u^2 + C = 4(\ln x)^2 + C.$

22. $\int \frac{1}{\sqrt{x}(1-\sqrt{x})}\, dx$

Let $u(x) = 1 - \sqrt{x}$. $du = -\frac{1}{2\sqrt{x}}\, dx$.

$\int \frac{1}{\sqrt{x}(1-\sqrt{x})}\, dx = \int \frac{1}{u}(-2)du = -2\ln|u| + C = -2\ln|1 - \sqrt{x}| + C.$

23. $\int \frac{(x^{2/3}-5)^{2/3}}{\sqrt[3]{x}}\,dx$

Let $u(x) = x^{2/3} - 5$. $\quad du = \frac{2}{3}x^{-1/3}\,dx = \frac{2}{3} \cdot \frac{1}{\sqrt[3]{x}}\,dx$.

$\int \frac{(x^{2/3}-5)^{2/3}}{\sqrt[3]{x}}\,dx = \int u^{2/3} \cdot \frac{3}{2}\,du = \frac{3}{2} \cdot \frac{3}{5}u^{5/3} + C = \frac{9}{10}(x^{2/3} - 5)^{5/3} + C.$

24. $\int \frac{e^{-3x}-x^2}{x^3+e^{-3x}}\,dx$.

Let $u = x^3 + e^{-3x}$. $\quad du = (3x^2 - 3e^{-3x})dx = -3(e^{-3x} - x^2)dx$

$\implies (e^{-3x} - x^2)dx = -\frac{1}{3}\,du.$

$\int \frac{e^{-3x}-x^2}{x^3+e^{-3x}}\,dx = \int \frac{1}{u}\left(-\frac{1}{3}\,du\right) = -\frac{1}{3}\ln|u| + C = -\frac{1}{3}\ln|x^3 + e^{-3x}| + C.$

25. $\int \frac{(1+e^{\sqrt{x}})}{\sqrt{x}}\,e^{\sqrt{x}}\,dx$

Let $u(x) = 1 + e^{\sqrt{x}}$. $\quad du = \frac{1}{2} \cdot \frac{e^{\sqrt{x}}}{\sqrt{x}}\,dx$.

$\int \frac{(1+e^{\sqrt{x}})e^{\sqrt{x}}}{\sqrt{x}}\,dx = \int u \cdot 2\,du = u^2 + C = \left(1 + e^{\sqrt{x}}\right)^2 + C.$

26. $\int (x^{5/3} - 2x)^{3/2}(5x^{2/3} - 6)dx$

Let $u(x) = x^{5/3} - 2x$. $\quad du = \left(\frac{5}{3}x^{2/3} - 2\right)dx = \frac{1}{3}(5x^{2/3} - 6)dx.$

$\int (x^{5/3} - 2x)^{3/2}(5x^{2/3} - 6)dx = 3\int u^{3/2}du = 3 \cdot \frac{2}{5}u^{5/2} + C.$

$\qquad = \frac{6}{5}(x^{5/3} - 2x)^{5/2} + C.$

27. $MC(x) = 40 + \frac{20x}{1+x^2}$.

(a) $\lim_{x\to\infty} MC(x) = \lim_{x\to\infty}\left(40 + \frac{20x}{1+x^2}\right) = 40 + \lim_{x\to\infty} \frac{\frac{20x}{x^2}}{\frac{1}{x^2} + \frac{x^2}{x^2}}$

$\qquad = 40 + \lim_{x\to\infty} \frac{\frac{20}{x}}{\frac{1}{x^2}+1} = 40 + 0 = 40.$

(b) $C(x) = \int C'(x)dx = \int \left(40 + \frac{20x}{1+x^2}\right)dx = \int 40dx + \int \frac{20x}{1+x^2}dx$

$\qquad = 40x + \int \frac{20x}{1+x^2}dx.$

Let $u(x) = 1 + x^2$. $\quad du = 2xdx.$

$\int \frac{20x}{1+x^2}dx = \int \frac{10}{u}du = 10\ln u + C = 10\ln(1 + x^2) + C.$

Thus $C(x) = 40x + 10\ln(1 + x^2) + C.$

Since $C(0) = 500 = 40 \cdot 0 + 10\ln(1 + 0^2) + C, \quad 500 = C.$

Thus, $C(x) = 40x + 10\ln(1 + x^2) + 500.$

28. (a) $R(x) = 100x.$

(b) $R(0) = 100 \cdot 0 = 0.$

29. $MR(x) = R'(x) = \frac{10{,}000x}{100+0.2x^2}$.

(a) $\lim_{x\to\infty} MR(x) = \lim_{x\to\infty} \frac{10{,}000x}{100+0.2x^2} = \lim_{x\to\infty} \frac{\frac{10{,}000x}{x^2}}{\frac{100}{x^2}+\frac{0.2x^2}{x^2}} = \lim_{x\to\infty} \frac{\frac{10{,}000}{x}}{\frac{100}{x^2}+0.2} = 0.$

(b) $R(x) = \int MR(x)dx = \int \frac{10{,}000x}{100+0.2x^2}dx.$

Let $u(x) = 100 + 0.2x^2$. $du = 0.4x dx$.

$R(x) = \int \frac{10{,}000x}{100+0.2x^2}dx = \int \frac{10{,}000}{u} \cdot \frac{1}{0.4}du = 25{,}000 \ln u + C.$

$R(x) = 25{,}000 \ln(100 + 0.2x^2) + C$

Since $R(0) = 0$, $R(0) = 0 = 25{,}000 \ln(100 + 0.2 \cdot 0^2) + C.$

$0 = 25{,}000 \ln 100 + C.$ $C = -25{,}000 \ln 100 \approx -115{,}129.25.$

Thus $R(x) = 25{,}000 \ln(100 + 0.2x^2) - 115{,}129.25.$

30. $MU(x) = U'(x) = 4x(4 + x^2)^{-2/3}.$

$U(x) = \int MU(x)dx = \int 4x(4 + x^2)^{-2/3}dx.$

Let $v(x) = 4 + x^2$. $dv = 2xdx$.

$U(x) = \int 2v^{-2/3}dv = 2 \cdot 3 \cdot v^{1/3} + C = 6(4 + x^2)^{1/3} + C.$

Since $U(0) = 0$, $0 = 6(4+0)^{1/3} + C$, $C = -6\sqrt[3]{4}.$

Thus $U(x) = 6\sqrt[3]{4+x^2} - 6\sqrt[3]{4} = 6\left(\sqrt[3]{4+x^2} - \sqrt[3]{4}\right).$

31. $P'(t) = \frac{100e^{20t}}{1+e^{20t}}$, $P(0) = 40{,}000.$

$P(t) = \int P'(t)dt = \int \frac{100e^{20t}}{1+e^{20t}}dt.$

Let $u(t) = 1 + e^{20t}$, $du = 20e^{20t}dt.$

$P(t) = \int \frac{5}{u}du = 5 \ln u + C = 5 \ln(1 + e^{20t}) + C.$

Since $P(0) = 40{,}000$, $40{,}000 = 5 \ln(1 + e^{20\cdot 0}) + C.$

$C = -5 \ln 2 + 40{,}000.$

Thus $P(t) = 5 \ln(1 + e^{20t}) - 5 \ln 2 + 40{,}000$

$P(t) = 5 \ln \frac{1}{2}(1 + e^{20t}) + 40{,}000.$

32. $S'(t) = 5t/(t^2 + 1) + .10te^{-t^2}.$

a. $S'(0) = 0.$

b. Let $u_1 = t^2 + 1$. $du_1 = 2t\,dt \implies t\,dt = \frac{1}{2}\,du_1.$

$\int \frac{5t}{t^2+1}\,dt = \int 5 \cdot \frac{1}{u_1} \cdot \left(\frac{1}{2}du_1\right) = \frac{5}{2} \ln u_1 + C_1 = \frac{5}{2} \ln(t^2 + 1) + C_1.$

Let $u_2 = -t^2$. $du_2 = -2t\,dt \implies t\,dt = -\frac{1}{2}\,du_2$.

$\int .10te^{-t^2}\,dt = \int .10e^{u_2}\left(-\frac{1}{2}\,du_2\right) = -.10 \cdot \frac{1}{2}e^{u_2} + C_2 = -.05e^{-t^2} + C_2$.

Therefore

$$S(t) = \frac{5}{2}\ln(t^2+1) - .05e^{-t^2} + C.$$

$$S(0) = 10 \implies 10 = \frac{5}{2}\ln(0^2+1) - .05e^{-0^2} + C \implies C = 10.05.$$

Hence

$$S(t) = \frac{5}{2}\ln(t^2+1) - .05e^{-t^2} + 10.05.$$

33. $P'(t) = \sqrt{t}e^{t^{3/2}}$.

Let $u = t^{3/2}$. $du = \frac{3}{2}t^{1/2}\,dt \implies t^{1/2}\,dt = \frac{2}{3}du$.

$\int \sqrt{t}e^{t^{3/2}}dt = \int e^u\left(\frac{2}{3}\,du\right) = \frac{2}{3}e^u + C = \frac{2}{3}e^{t^{3/2}} + C$.

$P(0) = 20 \implies 20 = \frac{2}{3}e^{0^{3/2}} + C \implies C = \frac{58}{3}$.

Therefore $$P(t) = \frac{2}{3}e^{t^{3/2}} + \frac{58}{3}.$$

34. $D'(t) = \frac{4}{t} + te^{-t^2}$.

$D(t) = 4\ln t - \frac{1}{2}e^{-t^2} + C$.

$D(10) = 30 \implies 30 = 4\ln 10 - \frac{1}{2}e^{-10^2} + C \implies C = 30 - 4\ln 10 + \frac{1}{2}e^{-10^2}$

$\implies C = 20.79$.

Therefore $\quad D(t) = 4\ln t - \frac{1}{2}e^{-t^2} + 20.79.$

Solutions to Exercise Set 5.4

1. $f(x) = 2x + 5$.

$\Delta x = \frac{4-0}{4} = 1$, and use the left endpoint of each subinterval as t_j. Then

$$\text{Approx. Sum} = [f(0) + f(1) + f(2) + f(3)]\,\Delta x = [5 + 7 + 9 + 11] \cdot 1 = 32.$$

2. $f(x) = x^2 + 3$.

$\Delta x = \frac{2-0}{4} = \frac{1}{2}, \qquad t_1 = 0, \quad t_2 = \frac{1}{2}, \quad t_3 = 1, \quad t_4 = \frac{3}{2}.$

$$\begin{aligned}\text{Approx. Sum} &= \left[f(0) + f\left(\frac{1}{2}\right) + f(1) + f\left(\frac{3}{2}\right)\right]\Delta x \\ &= \left[3 + \left(\left(\frac{1}{2}\right)^2 + 3\right) + (1^2 + 3) + \left(\left(\frac{3}{2}\right)^2 + 3\right)\right]\frac{1}{2} \\ &= \left(12 + \frac{1}{4} + 1 + \frac{9}{4}\right)\frac{1}{2} = \frac{31}{4}.\end{aligned}$$

3. $f(x) = 4 - x^2$.

$\Delta x = \frac{2-(-2)}{8} = \frac{1}{2}, \qquad t_j =$ left endpoint of each interval.

$$\begin{aligned}\text{Approx. Sum} &= \left[f(-2) + f\left(-\frac{3}{2}\right) + f(-1) + f\left(-\frac{1}{2}\right) + f(0)\right. \\ &\quad \left. + f\left(\frac{1}{2}\right) + f(1) + f\left(\frac{3}{2}\right)\right]\Delta x \\ &= \left[(4-(-2)^2) + \left(4 - \left(-\frac{3}{2}\right)^2\right) + (4-(-1)^2) + \left(4 - \left(-\frac{1}{2}\right)^2\right)\right. \\ &\quad \left. + (4-0^2) + \left(4 - \left(\frac{1}{2}\right)^2\right) + (4-1^2) + \left(4 - \left(\frac{3}{2}\right)^2\right)\right] \cdot \frac{1}{2} \\ &= \left(0 + \frac{7}{4} + 3 + \frac{15}{4} + 4 + \frac{15}{4} + 3 + \frac{7}{4}\right) \cdot \frac{1}{2} = \frac{21}{2}.\end{aligned}$$

4. $f(x) = \frac{1}{x+2}$.

$\Delta x = \frac{3-0}{6} = \frac{1}{2}$, $t_j =$ left endpoint of each interval.

$$\begin{aligned}\text{Approx. Sum} &= \left[f(0) + f\left(\frac{1}{2}\right) + f(1) + f\left(\frac{3}{2}\right) + f(2) + f\left(\frac{5}{2}\right)\right]\Delta x \\ &= \left[\frac{1}{2} + \frac{1}{\frac{1}{2}+2} + \frac{1}{1+2} + \frac{1}{\frac{3}{2}+2} + \frac{1}{2+2} + \frac{1}{\frac{5}{2}+2}\right]\frac{1}{2} \\ &= \left(\frac{1}{2} + \frac{2}{5} + \frac{1}{3} + \frac{2}{7} + \frac{1}{4} + \frac{2}{9}\right)\frac{1}{2} = \frac{2509}{2520}.\end{aligned}$$

5. $f(x) = \frac{1}{1+x^2}$.

$\Delta x = \frac{1-(-1)}{4} = \frac{1}{2}$, $\quad t_j =$ left endpoint of each interval.

$$\begin{aligned}\text{Approx. Sum} &= \left[f(-1) + f\left(-\frac{1}{2}\right) + f(0) + f\left(\frac{1}{2}\right)\right]\Delta x \\ &= \left[\frac{1}{1+(-1)^2} + \frac{1}{1+\left(-\frac{1}{2}\right)^2} + \frac{1}{1+0^2} + \frac{1}{1+\left(\frac{1}{2}\right)^2}\right]\frac{1}{2} \\ &= \left(\frac{1}{2} + \frac{4}{5} + 1 + \frac{4}{5}\right)\frac{1}{2} = \frac{31}{20}.\end{aligned}$$

6. $f(x) = 2x + 5$.

$\Delta x = \frac{4-0}{4} = 1$, $\quad t_j =$ right endpoint of each interval.

$$\begin{aligned}\text{Approx. Sum} &= [f(1) + f(2) + f(3) + f(4)]\,\Delta x \\ &= [7 + 9 + 11 + 13]\cdot 1 = 40.\end{aligned}$$

7. $f(x) = x^2 + 3$.

$\Delta x = \frac{2-0}{4} = \frac{1}{2}$, $\quad t_j =$ right endpoint of each interval.

$$\begin{aligned}\text{Approx. Sum} &= \left[f\left(\frac{1}{2}\right) + f(1) + f\left(\frac{3}{2}\right) + f(2)\right]\Delta x \\ &= \left[\left(\left(\frac{1}{2}\right)^2 + 3\right) + (1^2+3) + \left(\left(\frac{3}{2}\right)^2 + 3\right) + (2^2+3)\right]\frac{1}{2} \\ &= \left(12 + \frac{1}{4} + 1 + \frac{9}{4} + 4\right)\frac{1}{2} = \frac{39}{4}.\end{aligned}$$

8. $f(x) = 9 - x^2$.

$\Delta x = \frac{3-(-3)}{6} = 1$, $\quad t_j =$ right endpoint of each interval.

$$\begin{aligned}\text{Approx. Sum} &= [f(-2) + f(-1) + f(0) + f(1) + f(2) + f(3)]\,\Delta x \\ &= [(9-(-2)^2) + (9-(-1)^2) + (9-0^2) + (9-1^2) + (9-2^2) + (9-3^2)]\cdot 1 \\ &= 45 - 4 - 1 - 1 - 4 = 35.\end{aligned}$$

9. $f(x) = \frac{1}{4-x}$.
$\Delta x = \frac{2-0}{4} = \frac{1}{2}$, $\quad t_j$ = right endpoint of each interval.

$$\begin{aligned}\text{Approx. Sum} &= \left[f\left(\frac{1}{2}\right) + f(1) + f\left(\frac{3}{2}\right) + f(2)\right]\Delta x\\ &= \left[\frac{1}{4-\frac{1}{2}} + \frac{1}{4-1} + \frac{1}{4-\frac{3}{2}} + \frac{1}{4-2}\right]\frac{1}{2}\\ &= \left(\frac{2}{7} + \frac{1}{3} + \frac{2}{5} + \frac{1}{2}\right)\frac{1}{2} = \frac{319}{420}.\end{aligned}$$

10. $f(x) = \frac{1}{2+x^2}$.
$\Delta x = \frac{2-(-2)}{4} = 1$, $\quad t_j$ = right endpoint of each interval.

$$\begin{aligned}\text{Approx. Sum} &= [f(-1) + f(0) + f(1) + f(2)]\,\Delta x\\ &= \left(\frac{1}{2+(-1)^2} + \frac{1}{2+0^2} + \frac{1}{2+1^2} + \frac{1}{2+2^2}\right)\cdot 1\\ &= \frac{1}{3} + \frac{1}{2} + \frac{1}{3} + \frac{1}{6} = \frac{4}{3}.\end{aligned}$$

11. The approximating sums in Exercises 1 and 2 are lower sums.

12. The approximating sum in Exercise 4 is an upper sum.

13. None of the approximating sums in Exercises 6-10 are lower sums.

14. The approximating sums in Exercises 6, 7 and 9 are upper sums.

15. $\sum_{j=1}^{6}(3j+2) = (3(1)+2) + (3(2)+2) + (3(3)+2) + (3(4)+2) + (3(5)+2)$

$$+(3(6)+2) = 5 + 8 + 11 + 14 + 17 + 20 = 75.$$

16. $\sum_{j=1}^{5}(2j^2-5) = (2(1)^2-5) + (2(2)^2-5) + (2(3)^2-5) + (2(4)^2-5) + (2(5)^2-5)$

$$= -3 + 3 + 13 + 27 + 45 = 85.$$

17. $\sum_{j=1}^{5}(2j-5)^2 = (2(1)-5)^2+(2(2)-5)^2+(2(3)-5)^2+(2(4)-5)^2+(2(5)-5)^2$

$= (-3)^2+(-1)^2+(1)^2+(3)^2+(5)^2$

$= 9+1+1+9+25 = 45.$

18. $\sum_{j=3}^{6}(j^2-4j+1) = \left(3^2-4(3)+1\right)+\left(4^2-4(4)+1\right)+\left(5^2-4(5)+1\right)+\left(6^2-4(6)+1\right)$

$= -2+1+6+13 = 18.$

19. $\sum_{j=4}^{7}(j-5) = (4-5)+(5-5)+(6-5)+(7-5)$

$= -1+0+1+2 = 2.$

20. $\sum_{j=1}^{10} j^3 = (1)^3+(2)^3+(3)^3+(4)^3+(5)^3+(6)^3+(7)^3+(8)^3+(9)^3+(10)^3$

$= 1+8+27+64+125+216+343+512+729+1000$

$= 3025.$

21. $\sum_{j=4}^{10}(j^3-10) = \left((4)^3-10\right)+\left((5)^3-10\right)+\left((6)^3-10\right)+\left((7)^3-10\right)+\left((8)^3-10\right)$

$+\left((9)^3-10\right)+\left((10)^3-10\right)$

$= 54+115+206+333+502+719+990 = 2919.$

22. $\sum_{j=2}^{6}(j+1)(j-1) = (2+1)(2-1)+(3+1)(3-1)+(4+1)(4-1)$

$+(5+1)(5-1)+(6+1)(6-1)$

$= 3+8+15+24+35 = 85.$

23. $\int_0^2 x dx = \text{Area of } R = \frac{1}{2} \cdot 2 \cdot 2 = 2.$

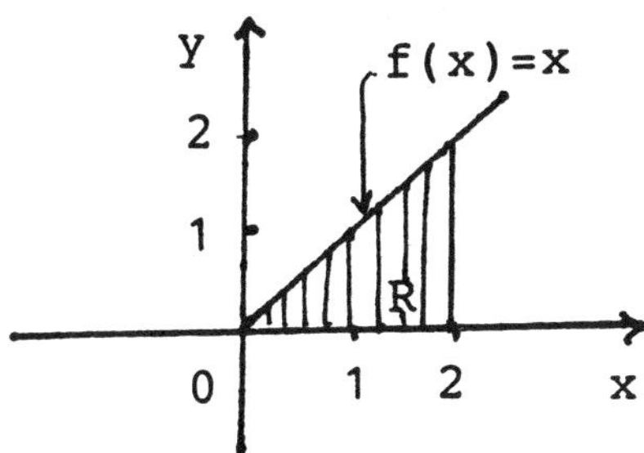

24. $\int_1^3 (x+4)dx = \frac{1}{2}(5+7) \cdot 2 = 12.$

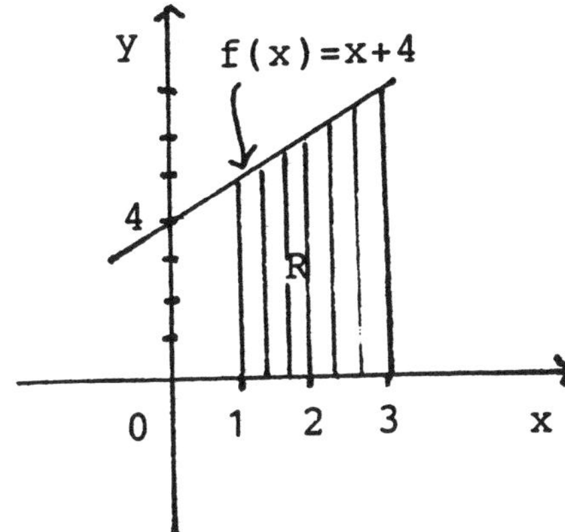

25. $\int_{-1}^3 4dx = 4 \cdot 4 = 16.$

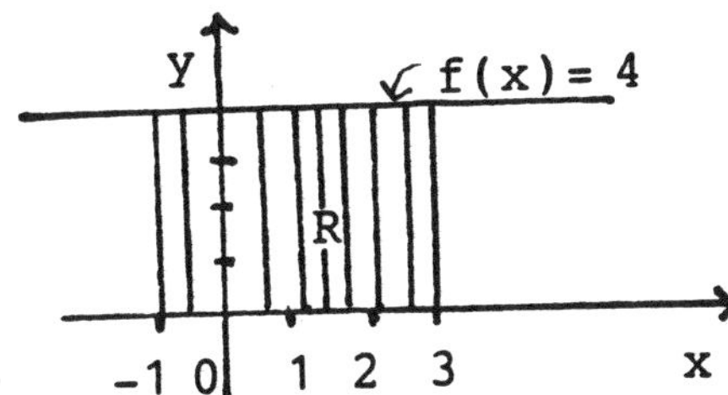

26. $\int_{-1}^{2} (x+1)dx = \frac{1}{2}3 \cdot 3 = \frac{9}{2}$

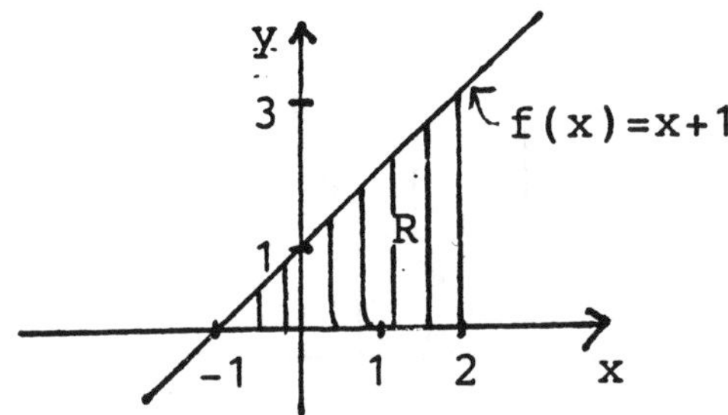

27. $\int_{0}^{9}(9-x)dx = \text{Area of } R = \frac{1}{2} \cdot 9 \cdot 9 = \frac{81}{2}.$

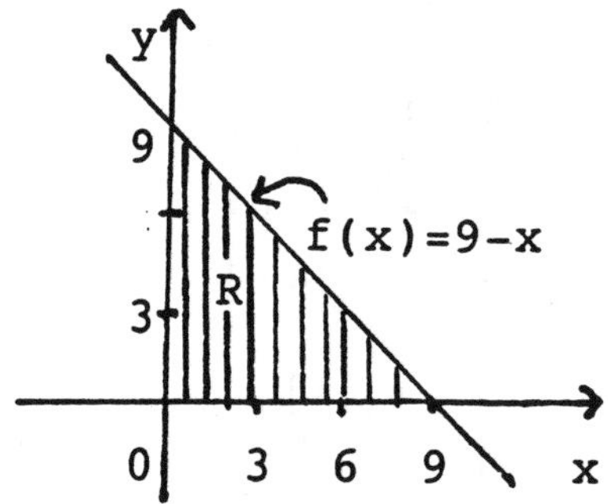

28. $\int_{3}^{6} 3dx = \text{Area of } R = 3 \cdot 3 = 9.$

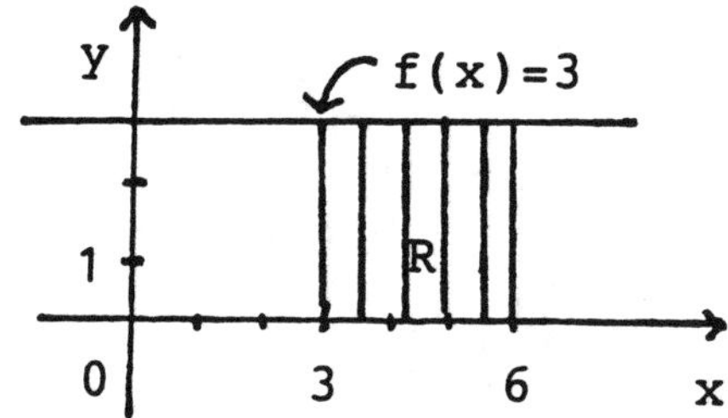

29. $\int_{-1}^{2} |x|dx = (\text{Area of } R_1) + (\text{Area of } R_2) = \frac{1}{2}1 \cdot 1 + \frac{1}{2}2 \cdot 2 = \frac{5}{2}$.

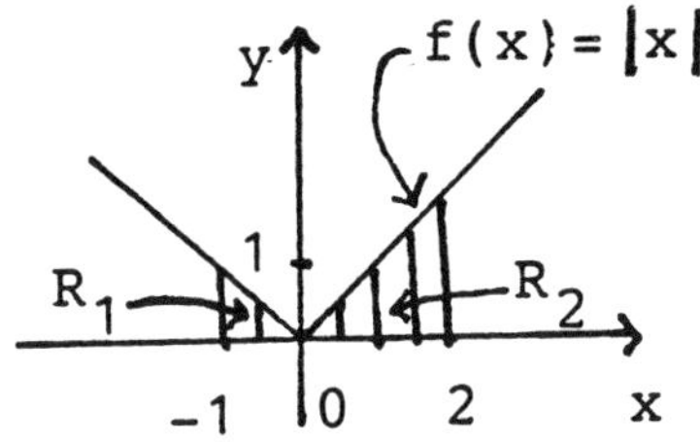

30. $\int_{2}^{4} (3x - 6)dx = \frac{1}{2} \cdot 2 \cdot 6 = 6$

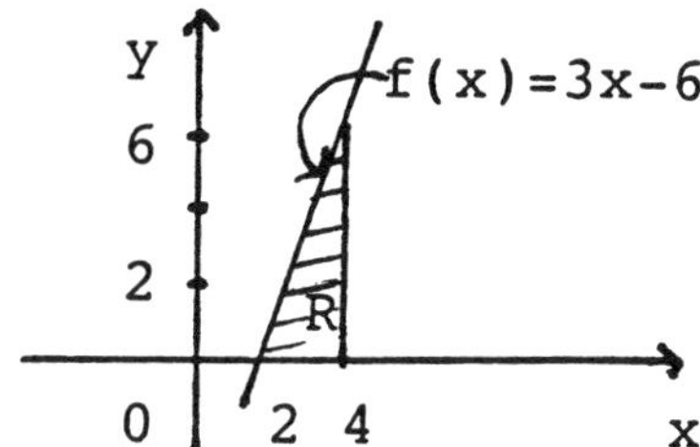

31. $\int_{0}^{4} |x - 2|dx = 2\left(\frac{1}{2} \cdot 2 \cdot 2\right) = 4$

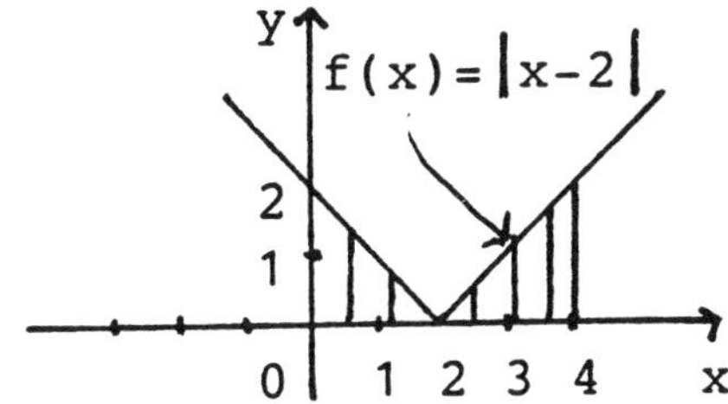

32. $\int_0^5 |2x-4|dx = \frac{1}{2}\cdot 2\cdot 4 + \frac{1}{2}\cdot 3\cdot 6 = 13.$

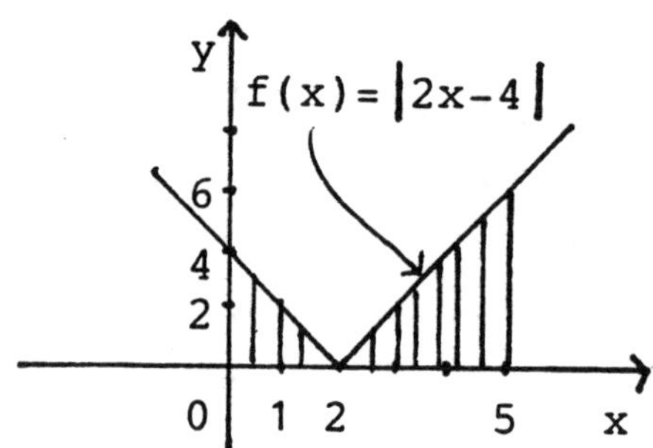

33. $\int_{-2}^{0} (-x)dx + \int_0^2 (x)dx.$

34. $\int_0^1 (-2x+2)dx + \int_1^6 (2x-2)dx.$

35. $\int_0^2 (-3x+6)dx + \int_2^4 (3x-6)dx.$

36. $\int_{-2}^{-1} (-x-1)dx + \int_{-1}^{2} (x+1)dx.$

37. $\int_{-2}^{-1} (-1+x^2)dx + \int_{-1}^{1} (1-x^2)dx + \int_1^2 (-1+x^2)dx.$

38. $\int_{-2}^{-1} (-x^3-1)dx + \int_{-1}^{3} (x^3+1)dx.$

39. $\int_0^2 \sqrt{x+2}dx \approx 3.441853$ using $n = 100$ subintervals. Note that the exact answer is

$$\frac{2}{3}(x+2)^{3/2}\Big|_0^2 = \frac{2}{3}\left(4^{3/2} - 2^{3/2}\right) = \frac{2}{3}\left(8 - 2\sqrt{2}\right) = 3.44771525.$$

40. For 100 subintervals, the appoximation is 12.62541.

For 1000 subintervals, the approximation is 12.50395.

The actual value is 12.49045772.

41. $\int_0^4 \ln(x+3)dx \approx 6.308563$ using $n = 100$ subintervals.
Using integration by parts (see Section 7.2), we find that the exact answer is

$$\int_0^4 \ln(x+3)dx = \{(x+3)\ln(x+3) - x\}\Big|_0^4 = 7\ln 7 - 3\ln 3 - 4$$
$$= 6.325534178.$$

42. $\int_0^2 \sqrt{4-x^2}dx \approx ?$

number of subintervals	approximation to integral
100	3.160416
1,000	3.143557
10,000	3.141794
100,000	3.141619

The exact answer is $\frac{1}{4}\cdot$ (the area of a circle with radius 2), which equals $\frac{1}{4}\cdot\pi\cdot 2^2 = \pi = 3.141592$.

43. $\int_0^4 \sqrt{5+x^2}dx \approx ?$

number of subintervals	approximation to integral
100	12.48087
1,000	12.52299
10,000	12.52723
100,000	12.52768

Using techniques available from Chapter 6 on trigonometric functions, we can show that

$$\int \sqrt{5+x^2}dx = \frac{1}{2}x\sqrt{5+x^2} + \frac{5}{2}\ln|\sqrt{5+x^2}+x| + C.$$

Hence the exact answer is

$$\frac{1}{2}\cdot 4\sqrt{21} + \frac{5}{2}\ln\left(\sqrt{21}+4\right) - \frac{5}{2}\ln\sqrt{5} = 2\sqrt{21} + \frac{5}{2}\ln\frac{\sqrt{21}+4}{\sqrt{5}}$$
$$= 12.52768916.$$

44. $\int_{-2}^{2} \sqrt{9-x^2}\,dx \approx$?

number of subintervals	approximation to integral
100	11.03945
1,000	11.03968
10,000	11.03970
100,000	11.04013
1,000,000	11.04305

Using techniques involving trigonometric functions, we can show that the exact answer is

$$\int_{-2}^{2} \sqrt{9-x^2}\,dx = 2 \cdot \int_{0}^{2} \sqrt{9-x^2}\,dx = \left(9 \arcsin \frac{x}{3} + x\sqrt{9-x^2}\right)\Big|_0^2$$
$$= 11.03968486.$$

Let $x = 3\sin\theta$.

$$\sqrt{9-x^2} = \sqrt{9-9\sin^2\theta} = 3 \cdot \sqrt{1-\sin^2\theta} = 3 \cdot \sqrt{\cos^2\theta} = 3\cos\theta.$$

$dx = 3\cos\theta d\theta$.

$$\begin{aligned}\int \sqrt{9-x^2}\,dx &= \int 3\cos\theta \cdot 3\cos\theta d\theta = 9 \cdot \int \frac{1+\cos 2\theta}{2} d\theta \\ &= \frac{9}{2}\left(\theta + \frac{\sin 2\theta}{2}\right) + C \\ &= \frac{9}{2}(\theta + \sin\theta\cos\theta) + C \\ &= \frac{9}{2}\left(\arcsin\frac{x}{3} + \frac{x}{3} \cdot \frac{\sqrt{9-x^2}}{3}\right) + C \\ &= \frac{9}{2}\arcsin\frac{x}{3} + \frac{1}{2}x\sqrt{9-x^2} + C\end{aligned}$$

$$\int_{-2}^{2} \sqrt{9-x^2}\,dx = \left(9 \arcsin \frac{x}{3} + x\sqrt{9-x^2}\right)\Big|_0^2$$
$$= 9\arcsin\frac{2}{3} + 2\sqrt{5} = 11.039685.$$

Solutions to Exercise Set 5.5

1. $\int_0^2 (x+3)dx = \left(\frac{x^2}{2}+3x\right)\Big]_0^2 = \left(\frac{4}{2}+6\right)-0 = 8.$

2. $\int_1^3 (4-2x)dx = \left(4x-x^2\right)\Big]_1^3 = (12-9)-(4-1) = 0.$

3. $\int_1^5 (x^2-6)dx = \left(\frac{x^3}{3}-6x\right)\Big]_1^5 = \left(\frac{125}{3}-30\right)-\left(\frac{1}{3}-6\right) = \frac{52}{3}.$

4. $\int_2^7 3dx = 3x]_2^7 = (21)-(6) = 15.$

5. $\int_{-3}^2 (9+x^2)dx = \left(9x+\frac{x^3}{3}\right)\Big]_{-3}^2 = \left(18+\frac{8}{3}\right)-(-27-9) = \frac{170}{3}.$

6. $\int_2^3 \left(6+\frac{1}{x^2}\right)dx = \left(6x-\frac{1}{x}\right)\Big]_2^3 = \left(18-\frac{1}{3}\right)-\left(12-\frac{1}{2}\right) = 6+\frac{1}{6} = \frac{37}{6}.$

7. $\int_0^1 e^{4x}dx = \frac{e^{4x}}{4}\Big]_0^1 = \frac{1}{4}(e^4-1).$

8. $\int_1^5 \frac{1}{x+3}dx = \ln(x+3)]_1^5 = \ln 8 - \ln 4 = \ln 2.$

9. $\int_1^4 \left(\sqrt{x}-\frac{3}{\sqrt{x}}\right)dx = \left(\frac{2}{3}x^{3/2}-6\sqrt{x}\right)\Big]_1^4 = \left(\frac{2}{3}\cdot 8-12\right)-\left(\frac{2}{3}-6\right) = -\frac{4}{3}.$

10. To find an antiderivative $\int xe^{x^2}dx$, we use the substitution $u = x^2$; $du = 2xdx$.

    Then $\int xe^{x^2}dx = \frac{1}{2}\int e^u du = \frac{1}{2}e^u + C = \frac{1}{2}e^{x^2}+C.$

    Therefore, by the Fundamental Theorem, one obtains

    $\int_0^2 xe^{x^2}dx = \frac{1}{2}e^{x^2}\Big]_0^2 = \frac{1}{2}(e^4-1).$

11. $\int_0^1 (x^4-6x^3+x)dx = \left(\frac{1}{5}x^5-\frac{3}{2}x^4+\frac{x^2}{2}\right)\Big]_0^1 = \left(\frac{1}{5}-1\right) = -\frac{4}{5}.$

12. $\int_0^2 \left(\frac{x-1}{x+1}\right)dx = \int_0^2 \left(1-\frac{2}{x+1}\right)dx = (x-2\ln(x+1))]_0^2$

    $= (2-2\ln 3)-0 = 2-2\ln 3.$

13. $\int_1^4 \left(\frac{x+3}{\sqrt{x}}\right)dx = \int_1^4 \left(\sqrt{x}+3\cdot x^{-1/2}\right)dx$

    $= \left(\frac{2}{3}x^{3/2}+6\cdot x^{1/2}\right)\Big]_1^4 = \left(\frac{16}{3}+12\right)-\left(\frac{2}{3}+6\right) = \frac{32}{3}.$

14. $\int_0^2 (\sqrt{x} - 2)(x+1)dx$

$= \int_0^2 \left(x^{3/2} - 2x + \sqrt{x} - 2\right) dx = \left(\frac{2}{5}x^{5/2} - x^2 + \frac{2}{3}x^{3/2} - 2x\right)\Big]_0^2$

$= \left(\frac{2}{5}2^{5/2} - 4 + \frac{2}{3}2^{3/2} - 4\right) = \left(\frac{8}{5}\sqrt{2} + \frac{4}{3}\sqrt{2} - 8\right) = \frac{44}{15}\sqrt{2} - 8.$

15. $\int_0^2 (x^3 - 6x^2 + 3x + 3)dx$

$= \left(\frac{x^4}{4} - 2x^3 + \frac{3}{2}x^2 + 3x\right)\Big]_0^2 = (4 - 16 + 6 + 6) = 0.$

16. $\int_{-1}^1 (4x^3 - 3x^4)dx = \left(x^4 - \frac{3}{5}x^5\right)\Big]_{-1}^1 = \left(1 - \frac{3}{5}\right) - \left(1 + \frac{3}{5}\right) = -\frac{6}{5}.$

17. $\int_1^2 \frac{x+6}{x^2+12x}dx = \int_1^2 \frac{1}{2} \cdot \frac{2x+12}{x^2+12x}dx = \frac{1}{2}\ln(x^2 + 12x)\Big]_1^2$

$= \left(\frac{1}{2}\ln 28 - \frac{1}{2}\ln 13\right) = \frac{1}{2}\ln\frac{28}{13}.$

18. $\int_0^1 xe^{3x^2}dx = \int_0^1 \frac{1}{6} \cdot 6x \cdot e^{3x^2}dx$

$= \frac{1}{6}e^{3x^2}\Big]_0^1 = \frac{1}{6}(e^3 - 1).$

19. $\int_2^4 \frac{x-1}{\sqrt{x}-1}dx = \int_2^4 \frac{(\sqrt{x}+1)(\sqrt{x}-1)}{(\sqrt{x}-1)}dx = \int_2^4 (\sqrt{x} + 1)\, dx$

$= \left(\frac{2}{3}x^{3/2} + x\right)\Big]_2^4 = \left(\frac{2}{3} \cdot 8 + 4\right) - \left(\frac{4}{3}\sqrt{2} + 2\right)$

$= \frac{22}{3} - \frac{4}{3}\sqrt{2} = \frac{2}{3}\left(11 - 2\sqrt{2}\right).$

20. $\int_0^3 x\left(\sqrt[3]{x} - 2\right) dx = \int_0^3 \left(x^{4/3} - 2x\right) dx = \left(\frac{3}{7}x^{7/3} - x^2\right)\Big]_0^3$

$= \left(\frac{3}{7}9 \cdot 3^{1/3} - 9\right) = \frac{27}{7}3^{1/3} - 9.$

21. $\int_0^1 \left(x^{3/5} - x^{5/3}\right) dx = \left(\frac{5}{8}x^{8/5} - \frac{3}{8}x^{8/3}\right)\Big]_0^1 = \frac{1}{4}.$

22. $\int_1^2 \frac{1-x}{x^3}dx = \int_1^2 \left(\frac{1}{x^3} - \frac{1}{x^2}\right) dx = \left(-\frac{1}{2x^2} + \frac{1}{x}\right)\Big]_1^2$

$= \left(-\frac{1}{8} + \frac{1}{2}\right) - \left(-\frac{1}{2} + 1\right) = -\frac{1}{8}.$

23. $\int_0^4 \frac{dx}{\sqrt{2x+1}} = \left(\sqrt{2x+1}\right)\Big]_0^4 = 3 - 1 = 2.$

24. $\int_1^4 \frac{e^{\sqrt{x}}}{\sqrt{x}} dx = \int_1^4 2 \cdot \frac{e^{\sqrt{x}}}{2\sqrt{x}} dx = 2 \cdot \left(e^{\sqrt{x}}\right)\Big]_1^4$

$= 2 \cdot (e^2 - e^1) = 2 \cdot (e^2 - e).$

25. Then $du = 2xdx$.

$\int \frac{x}{\sqrt{16+x^2}} dx = \int \frac{1}{\sqrt{u}} \cdot \frac{1}{2} du = u^{1/2} + C = \sqrt{16 + x^2} + C.$

Therefore, $\int_0^2 \frac{x}{\sqrt{16+x^2}} dx = \left(\sqrt{16 + x^2}\right)\Big|_0^2 = \sqrt{20} - 4.$

26. Let $u = x^3 + 3x + 7$. Then $du = (3x^2 + 3)dx \implies (x^2 + 1)dx = \frac{1}{3} du.$

$\int \frac{x^2+1}{x^3+3x+7} dx = \int \frac{1}{3u} du = \frac{1}{3} \ln u + C = \frac{1}{3} \ln(x^3 + 3x + 7) + C.$

$\int_0^1 \frac{x^2+1}{x^3+3x+7} dx = \frac{1}{3} \ln(x^3 + 3x + 7)\Big|_0^1 = \frac{1}{3}(\ln 11 - \ln 7) = \frac{1}{3} \ln \frac{11}{7}.$

27. Let $u = 9 - x^2$. Then $du = -2xdx \implies xdx = -\frac{1}{2} du.$

$\int x\sqrt{9 - x^2} dx = \int u^{1/2} \cdot \left(-\frac{1}{2}\right) du = -\frac{1}{2} \cdot \frac{2}{3} u^{3/2} + C$

$= -\frac{1}{3}(9 - x^2)^{3/2} + C.$

$\int_0^3 x\sqrt{9 - x^2} dx = \left(-\frac{1}{3}(9 - x^2)^{3/2}\right)\Big]_0^3 = -\frac{1}{3}(0 - 27) = 9.$

28. Let $u = 2x^2 - 1$. Then $du = 4xdx \implies xdx = \frac{1}{4} du.$

$\int \frac{x}{(2x^2-1)^3} dx = \int \frac{1}{u^3} \cdot \frac{1}{4} du = \frac{1}{4} \cdot \frac{u^{-2}}{-2} + C.$

$= -\frac{1}{8}(2x^2 - 1)^{-2} + C.$

$\int_1^2 \frac{x}{(2x^2-1)^3} dx = -\frac{1}{8}(2x^2 - 1)^{-2}\Big]_1^2 = -\frac{1}{8}(7^{-2} - 1)$

$= -\frac{1}{8}\left(\frac{-48}{49}\right) = \frac{6}{49}.$

29. Let $u = x^2 - 1$. Then $du = 2xdx \implies xdx = \frac{1}{2} du.$

$\int x(x^2 - 1)^{1/3} dx = \int u^{1/3} \cdot \frac{1}{2} du = \frac{1}{2} \cdot \frac{3}{4} \cdot u^{4/3} + C$

$= \frac{3}{8}(x^2 - 1)^{4/3} + C.$

$\int_1^2 x(x^2 - 1)^{1/3} dx = \frac{3}{8}(x^2 - 1)^{4/3}\Big]_1^2 = \frac{3}{8}(3^{4/3} - 0) = \frac{9}{8}\sqrt[3]{3}.$

30. Let $u = x - \sqrt{x}$. Then $du = \left(1 - \frac{1}{2\sqrt{x}}\right) dx.$

$\int \left(1 - \frac{1}{2\sqrt{x}}\right) e^{x-\sqrt{x}} dx = \int e^u du = e^u + C = e^{x-\sqrt{x}} + C.$

$\int_1^4 \left(1 - \frac{1}{2\sqrt{x}}\right) e^{x-\sqrt{x}} dx = e^{x-\sqrt{x}}\Big]_1^4 = e^2 - 1.$

31. $\int_0^8 \sqrt{x}\,dx = \frac{2}{3}x^{3/2}\Big]_0^8 = \frac{2}{3}\left(8^{3/2}\right) - \frac{2}{3}\left(0^{3/2}\right) = \frac{2}{3}\left(8^{3/2}\right) \approx 15.08.$

32. $\int_{-1}^{3} \frac{1}{x+2}\,dx = \ln(x+2)\Big]_{-1}^{3} = \ln 5 - \ln 1 = \ln 5 \approx 1.61.$

33. $\int_0^{\ln 5} e^{-x}\,dx = -e^{-x}\Big]_0^{\ln 5} = -e^{-\ln 5} + e^{-0} = -\frac{1}{5} + 1 = \frac{4}{5}.$

34. $\int_0^{\ln 2} \frac{e^x}{1+e^x}\,dx = \ln(1+e^x)\Big]_0^{\ln 2} = \ln(1+e^{\ln 2}) - \ln(1+e^0)$

$$= \ln(1+2) - \ln(1+1) = \ln(3) - \ln(2) = \ln\tfrac{3}{2} \approx 0.4055.$$

35. $\int_{-2}^{2}(-x^2+4)dx = \left(-\frac{1}{3}x^3 + 4x\right)\Big]_{-2}^{2} = \frac{16}{3} - \left(-\frac{16}{3}\right) = \frac{32}{3} \approx 10.67.$

36. $\int_0^{\ln 4}(-1+e^x)dx = (-x+e^x)\Big]_0^{\ln 4}$

$$= (-\ln 4 + e^{\ln 4}) - (-0 + e^0) = 3 - \ln 4 \approx 1.6137.$$

37. $\int_0^1(-x^3+x^2)dx = \left(-\frac{1}{4}x^4 + \frac{1}{3}x^3\right)\Big]_0^1 = \left(-\frac{1}{4} + \frac{1}{3}\right) - 0 = \frac{1}{12} \approx .0833.$

38. $\int_{-4}^{0} \frac{-x}{\sqrt{9+x^2}}\,dx.$

Let $u = 9 + x^2$. $du = 2x\,dx \implies x\,dx = \frac{1}{2}\,du.$

$\int \frac{-x}{\sqrt{9+x^2}}\,dx = \int u^{-1/2}\left(-\frac{1}{2}\,du\right) = -\frac{1}{2} \cdot \frac{u^{1/2}}{\frac{1}{2}} + C = -\sqrt{9+x^2} + C.$

$\int_{-4}^{0} \frac{-x}{\sqrt{9+x^2}}\,dx = -\sqrt{9+x^2}\Big]_{-4}^{0} = -3 - (-5) = 2.$

Solutions to Exercise Set 5.6

1. Since $f(x) > 0$ for all x in $[0,2]$,
   area $= \int_0^2(2x+5)dx = (x^2+5x)\Big]_0^2 = 14.$

2. Since $f(x) \geq 0$ for all x in $[-3,3]$, area $= \int_{-3}^{3}(9-x^2)dx = \left(9x - \frac{x^3}{3}\right)\Big]_{-3}^{3} =$ $(27-9) - (-27+9) = 18 - (-18) = 36.$

3. Since $f(x) > 0$ for all x in $[3, 5]$,

   area $= \int_3^5 \frac{1}{x-2} dx = \ln(x-2)]_3^5 = \ln 3 - \ln 1 = \ln 3.$

4. Clearly $e^{2x} > 0$ for all x.

   area $= \int_0^{\ln 2} e^{2x} dx = \frac{1}{2}e^{2x}\Big]_0^{\ln 2} = \frac{1}{2}(4-1) = \frac{3}{2}.$

5. Since $f(x) > 0$ on $[1, 2]$,

   area $= \int_1^2 \frac{x+1}{x^2+2x} dx = \frac{1}{2}\ln(x^2+2x)\Big]_1^2$

   $= \frac{1}{2}(\ln 8 - \ln 3) = \frac{1}{2}\ln\frac{8}{3}.$

6. Here $f(x) \geq 0$ on $[0, 1]$.

   area $= \int_0^1 xe^{1-x^2} dx = \left(-\frac{1}{2}e^{1-x^2}\right)\Big]_0^1$

   $= -\frac{1}{2}(1-e) = \frac{1}{2}(e-1).$

7. Since $f(x) > 0$ on $[-4, 4]$,

   area $= \int_{-4}^4 \sqrt{5+x}dx = \frac{2}{3}(5+x)^{3/2}\Big]_{-4}^4 = \frac{2}{3}(27-1) = \frac{52}{3}.$

8. Since $f(x) \geq 0$ on $[0, 4]$,

   area $= \int_0^4 \frac{x}{\sqrt{9+x^2}} dx = \sqrt{9+x^2}\Big]_0^4 = 5 - 3 = 2.$

9. $\int_0^{\ln 5} \frac{e^x}{1+e^x} x = \ln(1+e^x)]_0^{\ln 5} = \ln(1+e^{\ln 5}) - \ln(1+e^0)$

   $= \ln(1+5) - \ln(1+1) = \ln 6 - \ln 2 = \ln\frac{6}{2} = \ln 3 \approx 1.0986.$

10. $\int_1^4 \frac{1}{\sqrt{x}(1+\sqrt{x})} dx.$

    Let $u = (1+\sqrt{x})$. $du = \frac{1}{2\sqrt{x}} dx \implies \frac{1}{\sqrt{x}} dx = 2\, du.$

    $\int \frac{1}{\sqrt{x}(1+\sqrt{x})} dx = \int \frac{1}{u}(2\, du) = 2\ln u + C = 2\ln(1+\sqrt{x}) + C.$

    $\int_1^4 \frac{1}{\sqrt{x}(1+\sqrt{x})} dx = 2\ln(1+\sqrt{x})]_1^4 = 2\ln\left(1+\sqrt{4}\right) - 2\ln\left(1+\sqrt{1}\right)$

    $= 2\ln 3 - 2\ln 2 = 2\ln\frac{3}{2} \approx .8109.$

11. $\int_1^e \frac{\sqrt{1+\ln x}}{x}\,dx.$

Let $u = (1 + \ln x)$. $du = \frac{1}{x}\,dx$.

$\int \frac{\sqrt{1+\ln x}}{x}\,dx = \int u^{1/2}\,du = \frac{2}{3}u^{3/2} + C = \frac{2}{3}(1+\ln x)^{3/2} + C.$

$\int_1^e \frac{\sqrt{1+\ln x}}{x}\,dx = \frac{2}{3}(1+\ln x)^{3/2}\Big]_1^e = \frac{2}{3}(1+\ln e)^{3/2} - \frac{2}{3}(1+\ln 1)^{3/2}$

$= \frac{2}{3}(1+1)^{3/2} - \frac{2}{3}(1+0)^{3/2} = \frac{2}{3}\left(2^{3/2} - 1\right) \approx 1.219.$

12. $\int_0^8 \left(3x^{2/3} + \sqrt[3]{x}\right)dx = \left(3\left(\frac{3}{5}\right)x^{5/3} + \frac{3}{4}x^{4/3}\right)\Big]_0^8 = \left(\frac{9}{5}x^{5/3} + \frac{3}{4}x^{4/3}\right)\Big]_0^8$

$= \left(\frac{9}{5}(8^{5/3}) + \frac{3}{4}(8^{4/3})\right) - 0 = \frac{9}{5}(32) + \frac{3}{4}(16)$

$= \frac{288}{5} + 12 = \frac{348}{5} = 69.6.$

13. Since $f(x) > g(x)$ on $[-2, 2]$,

area $= \int_{-2}^{2} (f(x) - g(x))\,dx = \int_{-2}^{2} (9 - x^2 + 2)dx$

$= \left(9x - \frac{1}{3}x^3 + 2x\right)\Big]_{-2}^{2} = \left(22 - \frac{8}{3}\right) - \left(-22 + \frac{8}{3}\right)$

$= 44 - \frac{16}{3} = \frac{116}{3}.$

14. Since $g(x) > f(x)$ on $[1, 4]$,

area $= \int_1^4 (g(x) - f(x))\,dx = \int_1^4 (\sqrt{x} + x + 1)\,dx$

$= \left(\frac{2}{3}x^{3/2} + \frac{1}{2}x^2 + x\right)\Big]_1^4 = \left(\frac{2}{3}\cdot 8 + 8 + 4\right) - \left(\frac{2}{3} + \frac{3}{2}\right) = \frac{91}{6}.$

15. Since $f(x) \geq g(x)$ on $[0, 2]$,

area $= \int_0^2 (f(x) - g(x))\,dx = \int_0^2 (x + 1 + 2x - 1)dx$

$= \int_0^2 3x\,dx = \frac{3}{2}x^2\Big]_0^2 = 6.$

16. Since $f(x) > g(x)$ on $[-1, 1]$,

area $= \int_{-1}^{1} (f(x) - g(x))\,dx = \int_{-1}^{1} (2x + 3 - x^2 + 4)dx$

$= \int_{-1}^{1} (-x^2 + 2x + 7)dx = \left(-\frac{x^3}{3} + x^2 + 7x\right)\Big]_{-1}^{1}$

$= \left(-\frac{1}{3} + 8\right) - \left(\frac{1}{3} - 6\right) = \frac{40}{3}.$

17. Since $f(x) \geq g(x)$ on $[0,4]$,

$$\text{area} = \int_0^4 (f(x) - g(x))\,dx = \int_0^4 \left(\sqrt{x} + x^2\right) dx$$
$$= \left(\tfrac{2}{3}x^{3/2} + \tfrac{1}{3}x^3\right)\Big]_0^4 = \left(\tfrac{2}{3}\cdot 8 + \tfrac{1}{3}\cdot 64\right) = \tfrac{80}{3}.$$

18. Since $g(x) \geq f(x)$ on $[1,8]$,

$$\text{area} = \int_1^8 (g(x) - f(x))\,dx = \int_1^8 (x^{2/3} - x^{-2})dx$$
$$= \left(\tfrac{3}{5}x^{5/3} + x^{-1}\right)\Big]_1^8 = \left(\tfrac{3}{5}\cdot 32 + \tfrac{1}{8}\right) - \left(\tfrac{3}{5} + 1\right)$$
$$= \tfrac{88}{5} + \tfrac{1}{8} = \tfrac{709}{40}.$$

19. Since $f(x) \geq g(x)$ on $[0,3]$ and $g(x) \geq f(x)$ on $[-3,0]$,

$$\text{area} = \int_0^3 (f(x) - g(x))\,dx + \int_{-3}^0 (g(x) - f(x))\,dx$$
$$= \int_0^3 \left(x\sqrt{9-x^2} + x\right) dx + \int_{-3}^0 \left(-x - x\sqrt{9-x^2}\right) dx$$
$$= \left(\tfrac{x^2}{2} - \tfrac{1}{3}(9-x^2)^{3/2}\right)\Big]_0^3 + \left(-\tfrac{x^2}{2} + \tfrac{1}{3}(9-x^2)^{3/2}\right)\Big]_{-3}^0$$
$$= \left(\tfrac{9}{2} + \tfrac{1}{3}\cdot 27\right) + \left(\tfrac{1}{3}\cdot 27 + \tfrac{9}{2}\right) = (9+18) = 27.$$

20. $$f(x) = \begin{cases} 4 - x^2 & \text{on } [-2,2] \\ -4 + x^2 & \text{on } (-\infty,-2] \text{ and } [2,\infty). \end{cases}$$

Since $g(x) \geq f(x)$ on $[-3,3]$,

$$\text{area} = \int_{-3}^3 (g(x) - f(x))\,dx$$
$$= \int_{-3}^{-2} \left(5 - (-4+x^2)\right) dx + \int_{-2}^2 \left(5 - (4-x^2)\right) dx + \int_2^3 \left(5 - (-4+x^2)\right) dx$$
$$= \int_{-3}^{-2} (9-x^2)dx + \int_{-2}^2 (1+x^2)dx + \int_2^3 (9-x^2)dx$$
$$= \left(9x - \tfrac{1}{3}x^3\right)\Big]_{-3}^{-2} + \left(x + \tfrac{1}{3}x^3\right)\Big]_{-2}^2 + \left(9x - \tfrac{x^3}{3}\right)\Big]_2^3$$
$$= \left\{\left(-18 + \tfrac{8}{3}\right) - (-27+9)\right\} + \left\{\left(2 + \tfrac{8}{3}\right) - \left(-2 - \tfrac{8}{3}\right)\right\} + \left\{(27-9) - \left(18 - \tfrac{8}{3}\right)\right\}$$
$$= \tfrac{8}{3} + \tfrac{28}{3} + \tfrac{8}{3} = \tfrac{44}{3}.$$

21. The graphs intersect at two points. To find these points, we equate the two functions to obtain

$$4 - x^2 = x - 2 \implies x^2 + x - 6 = 0.$$

Thus $x^2 + x - 6 = 0$, so $x = -3$ or 2. Therefore, the points of intersection are $(-3, -5)$ and $(2, 0)$.

Since $4 - x^2 \geq x - 2$ on the interval $(-3, 2)$,

$$\begin{aligned} \text{area} &= \int_{-3}^{2} \left(4 - x^2 - (x-2)\right) dx \\ &= \int_{-3}^{2} (-x^2 - x + 6) dx = \left(-\tfrac{1}{3}x^3 - \tfrac{1}{2}x^2 + 6x\right)\Big]_{-3}^{2} \\ &= \left(-\tfrac{8}{3} - 2 + 12\right) - \left(9 - \tfrac{9}{2} - 18\right) \\ &= \left(-\tfrac{8}{3} + 10\right) - \left(-\tfrac{9}{2} - 9\right) = \tfrac{125}{6}. \end{aligned}$$

22. To find the intersection points of two graphs,

$$x^2 = x^3 \implies x^2(1-x) = 0 \implies x = 0 \text{ or } x = 1.$$

Thus the intersection points are (0,0) and (1,1).

Since $x^2 \geq x^3$ on [0,1],

$$\text{area} = \int_0^1 (x^2 - x^3) dx = \left(\tfrac{x^3}{3} - \tfrac{x^4}{4}\right)\Big]_0^1 = \left(\tfrac{1}{3} - \tfrac{1}{4}\right) = \tfrac{1}{12}.$$

23. First, let's find the intersection points.

$$x = x^3 \implies x(1 - x^2) = 0 \implies x(1+x)(1-x) = 0.$$

Thus the intersection points are (0,0), (1,1), and (-1,-1).

Since $x \geq x^3$ on [0,1] and $x^3 \geq x$ on [-1,0],

$$\begin{aligned} \text{area} &= \int_0^1 (x - x^3) dx + \int_{-1}^{0} (x^3 - x) dx \\ &= \left(\tfrac{x^2}{2} - \tfrac{x^4}{4}\right)\Big]_0^1 + \left(\tfrac{x^4}{4} - \tfrac{x^2}{2}\right)\Big]_{-1}^{0} \\ &= \left(\tfrac{1}{2} - \tfrac{1}{4}\right) + \left(0 - \left(\tfrac{1}{4} - \tfrac{1}{2}\right)\right) = \tfrac{1}{4} + \tfrac{1}{4} = \tfrac{1}{2}. \end{aligned}$$

24. Let's find the intersection points of the graphs.

$$9(9 - x^2) - x^2 + 9 = 0 \implies 90 - 10x^2 = 0$$

$$\implies 9 - x^2 = 0 \implies (3+x)(3-x) = 0 \implies x = \pm 3.$$

Since $(9-x^2) \geq \frac{1}{9}x^2 - 1$ on Let $u = 16 + x^2$. [-3,3],

$$\text{area} = \int_{-3}^{3} \left(9 - x^2 - \tfrac{1}{9}x^2 + 1\right) dx$$
$$= \int_{-3}^{3} \left(10 - \tfrac{10}{9}x^2\right) dx = \left(10x - \tfrac{10}{27}x^3\right)\Big]_{-3}^{3}$$
$$= (30-10) - (-30+10) = 20 - (-20) = 40.$$

25. To find the intersection points, we have

$$x^2 - 4 = 2 - x \implies x^2 + x - 6 = 0 \implies (x+3)(x-2) = 0$$
$$\implies x = -3 \text{ or } x = 2.$$

Since $2 - x \geq x^2 - 4$ on $(-3, 2)$,

$$\text{area} = \int_{-3}^{2} (2 - x - x^2 + 4) dx = \int_{-3}^{2} (6 - x - x^2) dx$$
$$= \left(6x - \tfrac{1}{2}x^2 - \tfrac{1}{3}x^3\right)\Big]_{-3}^{2} = \left(10 - \tfrac{8}{3}\right) - \left(-9 - \tfrac{9}{2}\right)$$
$$= 19 + \tfrac{11}{6} = \tfrac{125}{6}.$$

26. To find the intersection points, we have

$$\sqrt{x} = \sqrt[3]{x} \implies x^3 = x^2 \implies x = 0 \text{ or } 1.$$

Since $\sqrt[3]{x} \geq \sqrt{x}$ on [0,1],

$$\text{area} = \int_{0}^{1} \left(\sqrt[3]{x} - \sqrt{x}\right) dx = \int_{0}^{1} \left(x^{1/3} - x^{1/2}\right) dx$$
$$= \left(\tfrac{3}{4}x^{4/3} - \tfrac{2}{3}x^{3/2}\right)\Big]_{0}^{1} = \left(\tfrac{3}{4} - \tfrac{2}{3}\right) = \tfrac{1}{12}.$$

27. To find the intersection points, we have

$$x^{2/3} = x^2 \implies x^2 = x^6 \implies x^2(x^4 - 1) = 0$$
$$\implies x^2(x^2+1)(x+1)(x-1) = 0$$
$$\implies x = 0, -1, 1.$$

Since $x^{2/3} \geq x^2$ on both [-1,0] and [0,1],

$$\text{area} = \int_{-1}^{0} (x^{2/3} - x^2)dx + \int_{0}^{1}(x^{2/3} - x^2)dx$$
$$= \left(\tfrac{3}{5}x^{5/3} - \tfrac{x^3}{3}\right)\Big]_{-1}^{0} + \left(\tfrac{3}{5}x^{5/3} - \tfrac{x^3}{3}\right)\Big]_{0}^{1}$$
$$= \left\{0 - \left(-\tfrac{3}{5} + \tfrac{1}{3}\right)\right\} + \left\{\left(\tfrac{3}{5} - \tfrac{1}{3}\right) - 0\right\}$$
$$= \tfrac{3}{5} - \tfrac{1}{3} + \tfrac{3}{5} - \tfrac{1}{3} = \tfrac{6}{5} - \tfrac{2}{3} = \tfrac{8}{15}.$$

28. From the graphs we can see easily that there are two intersection points. Thus we have

$$x^{2/3} = 2 - x^2 \implies x^2 = (2 - x^2)^3$$
$$\implies x^2 = 8 - 3 \cdot 4 \cdot x^2 + 3 \cdot 2 \cdot x^4 - x^8$$
$$\implies x^8 - 6x^4 + 13x^2 - 8 = 0$$
$$\implies (x+1)(x-1)(x^6 + x^4 - 5x^2 + 8) = 0$$
$$\implies x = \pm 1$$

Since $(2 - x^2) \geq x^{2/3}$ on [-1,1],

$$\text{area} = \int_{-1}^{1} (2 - x^2 - x^{2/3})dx = \left(2x - \tfrac{1}{3}x^3 - \tfrac{3}{5}x^{5/3}\right)\Big]_{-1}^{1}$$
$$= \left(2 - \tfrac{1}{3} - \tfrac{3}{5}\right) - \left(-2 + \tfrac{1}{3} + \tfrac{3}{5}\right) = \left(4 - \tfrac{2}{3} - \tfrac{6}{5}\right)$$
$$= \left(4 - \tfrac{28}{15}\right) = \tfrac{32}{15}.$$

29. Note that $f(x) = x^3e^x \leq 0$ for $-2 \leq x \leq 0$, and $f(x) \geq 0$ for $0 \leq x \leq 2$.

Therefore the area described is given by

$$\int_{-2}^{0} [-f(x)]\,dx + \int_{0}^{2} f(x)\,dx = \int_{-2}^{0}(-x^3e^x)dx + \int_{0}^{2}(x^3e^x)dx.$$

30. $\sqrt{\frac{x}{2}} - 1 = 1 \implies \sqrt{\frac{x}{2}} = 2 \implies \frac{x}{2} = 4 \implies x = 8.$

For $3 \leq x < 8$, $\left(\sqrt{\frac{x}{2}} - 1\right) < 1 \implies \ln\left(\sqrt{\frac{x}{2}} - 1\right) < 0.$

For $8 < x \leq 16$, $\left(\sqrt{\frac{x}{2}} - 1\right) > 1 \implies \ln\left(\sqrt{\frac{x}{2}} - 1\right) > 0.$

Therefore the area described is given by

$$\int_{3}^{8} \left[-\ln\left(\sqrt{\frac{x}{2}} - 1\right)\right] dx + \int_{8}^{16} \left[\ln\left(\sqrt{\frac{x}{2}} - 1\right)\right] dx.$$

31. Note that $f(x) = e^{x^2} - e = 0$ for $e^{x^2} = e^1 \implies x^2 = 1 \implies x = \pm 1$.

We assume that the problem here is to find the area of the region bounded by the graph of $y = f(x)$ and the x-axis for $-1 \le x \le 1$.

Observe that $f(x) < 0$ for $-1 < x < 1$.

Therefore the desired area is given by

$$\int_{-1}^{1} [-f(x)]\, dx = \int_{-1}^{1} \left(-e^{x^2} + e\right) dx.$$

32. We assume that the problem here is to find the area of the region bounded by the graph of $y = f(x) = \sqrt{16 - x^4}$ and the x-axis for $-2 \le x \le 2$.

Therefore the desired area is given by

$$\int_{-2}^{2} f(x)\, dx = \int_{-2}^{2} \sqrt{16 - x^4}\, dx.$$

33.

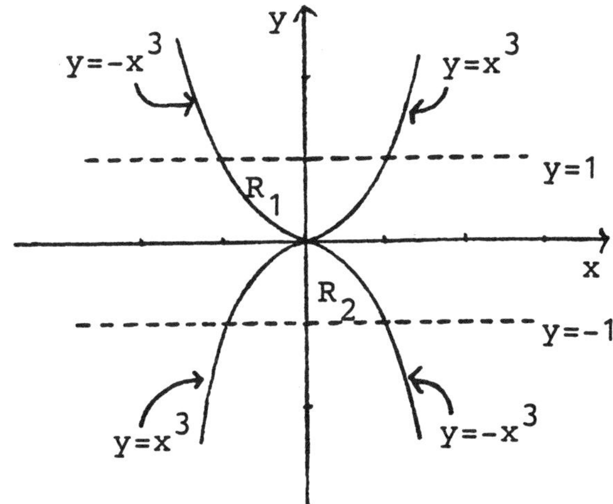

$$\begin{aligned}\text{Area} &= (\text{Area of } R_1) + (\text{Area of } R_2)\\ &= \int_{-1}^{0} \left(1 - (-x^3)\right) dx + \int_{0}^{1} (1 - x^3) dx + \int_{-1}^{0} \left(x^3 - (-1)\right) dx\\ &\quad + \int_{0}^{1} \left(-x^3 - (-1)\right) dx\\ &= \int_{-1}^{0} (1 + x^3) dx + \int_{0}^{1} (1 - x^3) dx + \int_{-1}^{0} (x^3 + 1) dx + \int_{0}^{1} (-x^3 + 1) dx\end{aligned}$$

$$= \left(x + \tfrac{x^4}{4}\right)\Big]_{-1}^{0} + \left(x - \tfrac{x^4}{4}\right)\Big]_{0}^{1} + \left(\tfrac{x^4}{4} + x\right)\Big]_{-1}^{0} + \left(-\tfrac{x^4}{4} + x\right)\Big]_{0}^{1}$$
$$= 0 - \left(-1 + \tfrac{1}{4}\right) + \left(1 - \tfrac{1}{4}\right) + 0 - \left(\tfrac{1}{4} - 1\right) + \left(-\tfrac{1}{4} + 1\right) - 0$$
$$= \tfrac{3}{4} + \tfrac{3}{4} + \tfrac{3}{4} + \tfrac{3}{4} = 3.$$

34.

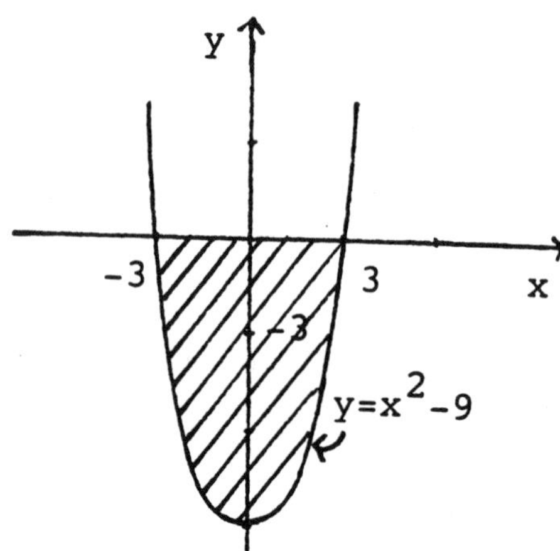

$$\text{Area} = \int_{-3}^{3} \left(0 - (x^2 - 9)\right) dx = \int_{-3}^{3} (-x^2 + 9) dx$$
$$= \left(-\tfrac{1}{3}x^3 + 9x\right)\Big]_{-3}^{3} = (-9 + 27) - (9 - 27) = 18 - (-18) = 36.$$

35.

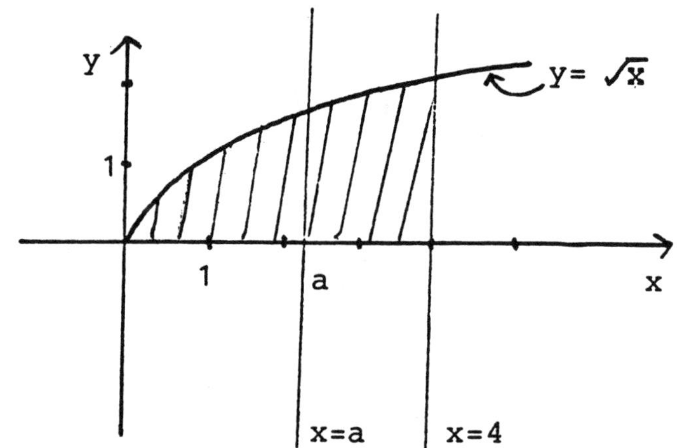

$$\int_0^a \sqrt{x}dx = \tfrac{1}{2}\int_0^4 \sqrt{x}dx \implies \tfrac{2}{3}x^{3/2}\Big]_0^a = \tfrac{1}{2}\left(\tfrac{2}{3}x^{3/2}\right)\Big]_0^4$$
$$\implies \tfrac{2}{3}a^{3/2} = \tfrac{1}{2}\left(\tfrac{2}{3}\cdot 8\right) \implies \tfrac{2}{3}a^{3/2} = \tfrac{8}{3} \implies a^{3/2} = 4 \implies a = 4^{2/3}.$$

Solutions to Exercise Set 5.7

1. $\int_0^{10}(20-0.5x)dx - 10\cdot D(10) = \left(20x - \frac{1}{4}x^2\right)\Big]_0^{10} - 10\cdot 15$
$= (200-25) - 150 = \$25.$

2. $\int_0^{10}(40-0.02x^2)dx - 10\cdot D(10)$
$= \left(40x - \frac{1}{150}x^3\right)\Big]_0^{10} - 10(40-2)$
$= \left(400 - \frac{100}{15}\right) - 380 = 20 - \frac{20}{3} = \frac{40}{3}$ dollars.

3. $\int_0^{5} 100(40-x^2)dx - 5\cdot D(5)$
$= 100\left(40x - \frac{1}{3}x^3\right)\Big]_0^{5} - 5\cdot(100(40-25))$
$= 100\left(200 - \frac{125}{3}\right) - 5\cdot(1500)$
$= \frac{47500}{3} - 7500 = \frac{25000}{3}$ dollars.

4. $\int_0^{15} \frac{500}{x+10}dx - 15\cdot D(15) = 500\cdot\ln(x+10)\Big]_0^{15} - 15\cdot\frac{500}{25}$
$= 500\cdot(\ln 25 - \ln 10) - 300 = 500\cdot\ln\frac{5}{2} - 300$ dollars.

5. $\int_0^{12} \frac{40}{\sqrt{2x+1}}dx - 12\cdot D(12) = 40\cdot\sqrt{2x+1}\Big]_0^{12} - 12\cdot\frac{40}{5}$
$= 40\cdot(5-1) - 96 = 160 - 96 = \$64.$

6. $16\cdot S(16) - \int_0^{16}(5+\sqrt{x})\,dx = 16\cdot(5+4) - \left(5x + \frac{2}{3}x^{3/2}\right)\Big]_0^{16}$
$= 144 - \left(5\cdot 16 + \frac{2}{3}\cdot 4^3\right) = 144 - \left(80 + \frac{128}{3}\right)$
$= 64 - \frac{128}{3} = \frac{64}{3}$ dollars.

7. $6\cdot S(6) - \int_0^{6}\frac{1}{100}x^2dx = 6\cdot\frac{36}{100} - \frac{1}{100}\left(\frac{x^3}{3}\right)\Big]_0^{6}$
$= \frac{216}{100} - \frac{1}{100}(72) = \frac{144}{100} = \$1.44.$

8. $4\cdot S(4) - \int_0^{4} x\sqrt{9+x^2}dx = 4\cdot(4\cdot 5) - \frac{1}{3}(9+x^2)^{3/2}\Big]_0^{4}$
$= 80 - \frac{1}{3}(5^3 - 3^3) = 80 - \frac{1}{3}(125-27)$
$= 80 - \frac{98}{3} = \frac{142}{3}$ dollars.

9. $20 \cdot S(20) - \int_0^{20} (e^{0.2x} - 1)dx$

$= 20 \cdot (e^4 - 1) - (5 \cdot e^{0.2x} - x)\Big]_0^{20}$

$= 20 \cdot (e^4 - 1) - \{5 \cdot e^4 - 20 - 5\}$

$= 20 \cdot e^4 - 20 + 25 - 5e^4 = 15 \cdot e^4 + 5$ dollars.

10. $10 \cdot S(10) - \int_0^{10} (x \cdot e^{0.01x^2})dx$

$= 10 \cdot (10 \cdot e) - (50 \cdot e^{0.01x^2})\Big]_0^{10}$

$= 100e - (50 \cdot e - 50) = 50e + 50$ dollars.

11. a. $\int_{400}^{500} MC(x)dx = \int_{400}^{500} (200 + .4x)dx.$

b. $C(500) - C(400) = \left(200x + .2x^2\right)\Big]_{400}^{500} = \left(200(500) + .2(500)^2\right)$

$- \left(200(400) + .2(400)^2\right) = 150,000 - 112,000 = 38,000.$

12. a. $\int_{20}^{40} \left(60 + \frac{40}{x+10}\right) dx.$

b. The value of the integral in part a. is

$$(60x + 40\ln(x+10))]_{20}^{40} = (60(40) + 40\ln(40+10)) - (60(20) + 40\ln(20+10))$$

$$= 1200 + 40\ln\frac{50}{30} = 1220.43.$$

13. $C(40) = C(30) + \int_{30}^{40} (200 + 0.4x)dx$

$= 7000 + (200x + 0.2x^2)\Big]_{30}^{40}$

$= 7000 + (8000 + 320) - (6000 + 180)$

$= 7000 + 2000 + 140 = \$9140.$

14. $C(20) = C(0) + \int_0^{20} \left(60 + \frac{40}{x+10}\right) dx$

$C(20) = 5000 + (60x + 40 \cdot \ln(x+10))]_0^{20}$

$C(20) = 5000 + (1200 + 40 \cdot \ln 30) - 40\ln 10$

$C(20) = 6200 + 40 \cdot \ln 3$ dollars.

15. $R(50) - R(30) = \int_{30}^{50} (400 - 0.5x)dx$
$= \left(400x - \frac{1}{4}x^2\right)\Big]_{30}^{50}$
$= \left(20000 - \frac{2500}{4}\right) - \left(12000 - \frac{900}{4}\right)$
$= 8000 - \frac{1600}{4} = \$7600.$

16. $\int_0^{30} \frac{100}{1+0.02x} dx = 100 \cdot \ln(1 + 0.02x) \cdot 50]_0^{30}$
$= 5000 \cdot (\ln 1.6 - 0) = 5000 \cdot \ln 1.6$ dollars.

17. The total nominal value of the revenue flow is

$$\int_0^3 A(t)dt = \int_0^3 (3000t + 2000)dt = \left(1500t^2 + 2000t\right)\Big]_0^3$$
$$= \left(1500(3)^2 + 2000(3)\right) - 0 = \$19,500.$$

18. The nominal value of the contract is

$$\int_0^8 A(t)dt = \int_0^8 \left(100\sqrt{t+1}\right) dt = 100 \cdot \frac{2}{3}(t+1)^{3/2}\Big]_0^8$$
$$= \frac{200}{3}(8+1)^{3/2} - \frac{200}{3}(0+1)^{3/2}$$
$$= \frac{200}{3}(27 - 1) = \$\frac{5200}{3}.$$

19. $\int_5^{10} 1000e^{-0.2t}dt = 1000 \cdot \left(-5 \cdot e^{-0.2t}\right)\Big]_5^{10}$

$= 1000 \cdot \left(-5 \cdot e^{-2} + 5e^{-1}\right) = 5000(e^{-1} - e^{-2})$ dollars.

20. $\int_0^5 A(t) \cdot e^{-rt}dt = \int_0^5 2000 \cdot e^{-0.08t}dt = 2000 \cdot \frac{e^{-0.08t}}{-0.08}\Big]_0^5$

$= 2000 \cdot \left(-\frac{25}{2}\right) \cdot \left(e^{-0.4} - 1\right)$

$= 25000 \cdot (1 - e^{-0.4})$ dollars.

21. $\int_0^{10} 5000 \cdot e^{-rt}dt = \int_0^{10} 5000 \cdot e^{-0.1t}dt$

$= 5000 \cdot (-10)(e^{-0.1t})\Big]_0^{10} = 50000(1 - e^{-1})$ dollars.

22. $\int_0^8 5000\sqrt{t+1}dt = 5000 \cdot \left(\frac{2}{3}(t+1)^{3/2}\right)\Big]_0^8$
$= 5000 \cdot \left(\frac{2}{3} \cdot 3^3 - \frac{2}{3}\right) = 5000\left(18 - \frac{2}{3}\right)$
$= 5000 \cdot \frac{52}{3} = \frac{260000}{3}$ dollars.

23. a. The present value of this payment stream is

$$\int_0^{20} 2,000e^{-.08t}\,dt = 2,000 \cdot \frac{1}{-.08}e^{-.08t}\Big]_0^{20}$$

$$= -25,000\left(e^{-.08(20)} - e^{-.08(0)}\right) = \$19,952.59.$$

b. We proceed as indicated in Problem 52 in the Review Exercises for Chapter 4. Let n be some given positive integer. Let $\Delta t = \frac{20}{n}$. Let t_0, t_1, t_2, t_3, ..., t_n be the real numbers such that

$$0 = t_0 < t_1 < t_2 < t_3 < \cdots < t_n = 20$$

and $(t_j - t_{j-1}) = \Delta t$ for $j = 1, 2, 3, \ldots, n$. Note that $t_j = j \cdot \Delta t$ for $j = 0, 1, 2, \ldots, n$. Suppose that at each instant of time t_0, t_1, t_2, t_3, ..., t_{n-1} Robert deposits $2,000 \cdot \Delta t$ dollars into his IRA account which is paying 8% annual interest compounded continuously. As in Problem 52 in the Review Exercises for Chapter 4, the value of Robert's IRA account after 20 years would be given by

$$T(n) = 2,000 \cdot \Delta t \cdot \left(e^{(n \cdot .08)\cdot\Delta t} + e^{((n-1).08)\cdot\Delta t} + \ldots\right.$$
$$\left. + e^{(3\cdot.08)\cdot\Delta t} + e^{(2\cdot.08)\cdot\Delta t} + e^{(1\cdot.08)\cdot\Delta t}\right)$$
$$= 2,000 \cdot \left(e^{.08\cdot(1\cdot\Delta t)} + e^{.08\cdot(2\cdot\Delta t)} + e^{.08\cdot(3\cdot\Delta t)}\right.$$
$$\left. + \cdots + e^{.08\cdot((n-1)\cdot\Delta t)} + e^{.08\cdot(n\cdot\Delta t)}\right) \cdot \Delta t$$
$$= 2,000 \cdot \left(e^{.08t_1} + e^{.08t_2} + e^{.08t_3} + \cdots + e^{.08t_{n-1}}\right.$$
$$\left. + e^{.08t_n}\right) \cdot \Delta t.$$

As n approaches ∞, this value $T(n)$ approaches the definite integral

$$\int_0^{20} 2,000e^{.08t}\,dt.$$

We take this definite integral to be the nominal value of Robert's IRA account after 20 years. This nominal value is

$$2,000 \cdot \frac{1}{.08} \cdot \left(e^{.08(20)} - e^{.08(0)}\right) = \$98,825.81.$$

24. a. The present value of this payment stream is

$$\int_0^{20} 2,000e^{-.10t}\,dt = 2,000 \cdot \frac{1}{-.10} e^{-.10t}\Big]_0^{20}$$

$$= -20,000\left(e^{-.10(20)} - e^{-.10(0)}\right) = \$17,293.29.$$

b. As in part b. of Problem 23, the nominal value of Robert's IRA account after 20 years is given by the definite integral

$$\int_0^{20} 2,000e^{.10t}\,dt = \frac{2,000}{.10}\left(e^{.10(20)} - e^{.10(0)}\right) = \$127,781.12.$$

25. Let R denote the payment rate.

$$1,000,000 = \int_0^{20} Re^{-.08t}\,dt = R \cdot \frac{1}{-.08} \cdot e^{-.08t}\Big]_0^{20}$$

$$= R \cdot \frac{1}{.08} \cdot (1 - e^{-1.6}) = 9.9762935R.$$

Therefore $R = \$100,237.63$.

26. $1,000,000 = R \cdot \frac{1}{.04} \cdot (1 - e^{-.08}) = 13.766776R \implies R = \$72,638.65.$

27. The present value of this payment stream is given by

$$\int_0^{60} 600e^{-.00833t}\,dt = -\frac{1}{.00833} \cdot 600 \cdot e^{-.00833t}\Big]_0^{60} = \$28,329.79.$$

28. The present value under this plan is

$$\int_0^{60} 400e^{-.00833t}\,dt + 4000e^{-.00833(60)}$$

$$= -\frac{1}{.00833}\cdot 400 \cdot e^{-.00833t}\Big]_0^{60} + 2426.12$$

$$= 18{,}886.53 + 2{,}426.12 = \$21{,}312.65.$$

29. Net present value for plan A is given by

$$PV_A = \int_0^5 (\$3 \text{ million})e^{(-.08t)}\,dt - \$10 \text{ million}$$

$$= \left\{3\left[\frac{-1}{.08}(e^{-.08t})\right]_0^5 - 10\right\} \text{ million dollars}$$

$$= \left\{3\left[12.5(1 - e^{-.4})\right] - 10\right\} \text{ million dollars}$$

$$= 2.362998274 \text{ million dollars}$$

$$= 2{,}362{,}998.27 \text{ dollars.}$$

Net present value for plan B is given by

$$PV_B = \int_0^5 (\$2 \text{ million})e^{-.08t}dt - \$6 \text{ million}$$

$$= \left\{2\left[\frac{-1}{.08}(e^{-.08t})\right]_0^5 - 6\right\} \text{ million dollars}$$

$$= \left\{2\left[12.5(1 - e^{-.4})\right] - 6\right\} \text{ million dollars}$$

$$= 2.241998849 \text{ million dollars}$$

$$= 2{,}241{,}998.85 \text{ dollars.}$$

Plan A would be chosen.

30. Net present value for Plan A is given by

$$PV_A = \int_0^{10} (\$3 \text{ million})e^{(-.10t)}\,dt - \$10 \text{ million}$$

$$= \left\{ 3\left[\frac{-1}{.10}\left(e^{-.10t}\right)\right]_0^{10} - 10 \right\} \text{ million dollars}$$

$$= \left\{3\left[10(1-e^{-1})\right] - 10\right\} \text{ million dollars}$$

$$= 8.963616765 \text{ million dollars}$$

$$= 8,963,616.77 \text{ dollars.}$$

Net present value for Plan B is given by

$$PV_B = \int_0^{10} (\$2 \text{ million})e^{(-.10t)}\,dt - \$6 \text{ million}$$

$$= \left\{ 2\left[\frac{-1}{.10}\left(e^{-.10t}\right)\right]_0^{10} - 6 \right\} \text{ million dollars}$$

$$= \left\{2\left[10(1-e^{-1})\right] - 6\right\} \text{ million dollars}$$

$$= 6.642411177 \text{ million dollars}$$

$$= 6,642,411.18 \text{ dollars.}$$

Plan A would be chosen.

31. The present value 30 years from now of the revenue stream which begins then is given by

$$PV_{30} = \int_0^{10} (30,000)e^{-.06t}\,dt = 30,000\left[\frac{-1}{.06}e^{-.06t}\right]_0^{10}$$

$$= 500,000(1 - e^{-.6}) = 225,594.18 \text{ dollars.}$$

The present value now of the amount PV_{30} is

$$PV_0 = 225,584.18e^{-.06(30)} = \$37,290.47.$$

Solutions to Exercise Set 5.8

1. $\int_1^4 \pi(2x+1)^2 dx$

$= \pi \int_1^4 (4x^2+4x+1)dx = \pi \left(\frac{4}{3}x^3 + 2x^2 + x\right)\Big]_1^4$

$= \pi \left(\frac{256}{3} + 32 + 4 - \frac{4}{3} - 3\right) = \pi \left(\frac{252}{3} + 33\right) = \pi \cdot \frac{351}{3} = 117\pi.$

2. $\int_1^5 \pi(4x-1)dx$

$= \pi(2x^2 - x)\Big]_1^5 = \pi(50-5-1) = 44\pi.$

3. $\int_0^3 \pi|x-1|^2 dx = \int_0^3 (x-1)^2 \cdot \pi\, dx$

$= \pi \int_0^3 (x^2-2x+1)dx = \pi \left(\frac{x^3}{3} - x^2 + x\right)\Big]_0^3$

$= \pi \cdot (9-9+3) = 3\pi.$

4. Volume $= \pi \int_0^2 [f(x)]^2\, dx = \pi \int_0^2 \left[\sqrt{4-x^2}\right]^2 dx = \pi \int_0^2 (4-x^2)dx$

$= \pi \left(4x - \frac{1}{3}x^3\right)\Big]_0^2 = \pi\left((4(2) - \frac{1}{3}(2)^3) - 0\right) = \frac{16}{3}\pi.$

5. Volume $= \pi \int_1^2 \left[\frac{\sqrt{x+1}}{x}\right]^2 dx = \pi \int_1^2 \frac{x+1}{x^2}\, dx = \pi \int_1^2 \left(\frac{1}{x} + \frac{1}{x^2}\right) dx$

$= \pi \left(\ln x - \frac{1}{x}\right)\Big]_1^2 = \pi\left((\ln 2 - \frac{1}{2}) - (\ln 1 - 1)\right) = \pi\left(\ln 2 + \frac{1}{2}\right).$

6. Let's find the intersections of the two graphs.

$x^2 = x^3 \implies x^2(1-x) = 0 \implies x = 0$ or $x = 1.$

$\int_0^1 \pi\left(f(x)^2 - g(x)^2\right) dx = \int_0^1 \pi(x^4 - x^6)dx$

$= \pi \cdot \left(\frac{1}{5}x^5 - \frac{1}{7}x^7\right)\Big]_0^1 = \pi\left(\frac{1}{5} - \frac{1}{7}\right) = \frac{2}{35}\pi.$

7. First, find the intersections of the two graphs.

$f(x) = g(x) \implies \frac{1}{4}x^3 = x \implies x^3 - 4x = 0 \implies x(x^2-4) = 0 \implies$

$x = 0$ or $x = 2$ or $x = -2.$

For $-2 < x < 0$, $[g(x)]^2 > [f(x)]^2$.

For $0 < x < 2$, $[g(x)]^2 > [f(x)]^2$.

Therefore the volume is

$$\pi \int_{-2}^{2} \left\{[g(x)]^2 - [f(x)]^2\right\} dx = \pi \int_{-2}^{2} \left(x^2 - \frac{1}{16}x^6\right) dx$$
$$= \pi \left(\frac{1}{3}x^3 - \frac{1}{112}x^7\right)\Big]_{-2}^{2}$$
$$= \pi \left\{\left(\frac{8}{3} - \frac{128}{112}\right) - \left(-\frac{8}{3} + \frac{128}{112}\right)\right\}$$
$$= 16\pi\left(\frac{1}{3} - \frac{1}{7}\right) = 16\pi \cdot \frac{4}{21} = \frac{64\pi}{21}.$$

8. Volume $= \pi \int_1^4 \left(x - \frac{1}{x^2}\right) dx = \pi\left(\frac{1}{2}x^2 + \frac{1}{x}\right)\Big]_1^4 = \pi\left(8 + \frac{1}{4} - \frac{1}{2} - 1\right) = \frac{27\pi}{4}.$

9. Volume $= \pi \int_0^4 (4-x)^2 dx = \frac{-\pi}{3}(4-x)^3\Big]_0^4 = \frac{64\pi}{3}.$

10. Volume $= \pi \int_0^4 x\,dx = \pi \cdot \frac{x^2}{2}\Big]_0^4 = 8\pi.$

11. Average value $= \dfrac{\int_1^5 f(x)dx}{5-1} = \dfrac{\int_1^5 \sqrt{2x-1}\,dx}{4} = \dfrac{\frac{1}{3}(2x-1)^{3/2}\big]_1^5}{4}$
$= \dfrac{\frac{1}{3}\cdot(27-1)}{4} = \dfrac{13}{6}.$

12. $\dfrac{\int_{-1}^3 (x^2-7)dx}{3-(-1)} = \dfrac{\left(\frac{1}{3}x^3-7x\right)\big]_{-1}^3}{4} = \dfrac{(9-21)-\left(-\frac{1}{3}+7\right)}{4} = -\dfrac{14}{3}.$

13. $\dfrac{\int_{-2}^2 (x+2)^{1/2}dx}{2-(-2)} = \dfrac{\frac{2}{3}(x+2)^{3/2}\big]_{-2}^2}{4} = \dfrac{\frac{16}{3}}{4} = \dfrac{4}{3}.$

14. $\dfrac{\int_0^1 \frac{1-x}{1+x}dx}{1-0} = \dfrac{\int_0^1\left(-1+\frac{2}{1+x}\right)dx}{1} = \dfrac{(-x+2\ln(1+x))]_0^1}{1} = \dfrac{(-1+2\ln 2)}{1} = -1+\ln 4.$

15. $\dfrac{\int_0^2 (x^2-x+1)dx}{2-0} = \dfrac{\left(\frac{1}{3}x^3-\frac{1}{2}x^2+x\right)\big]_0^2}{2} = \dfrac{\left(\frac{8}{3}-2+2\right)-0}{2} = \dfrac{4}{3}.$

16. The area of $R = \frac{1}{2}(4+10)\cdot 2 = 14.$

Average value $= \frac{14}{2} = 7.$

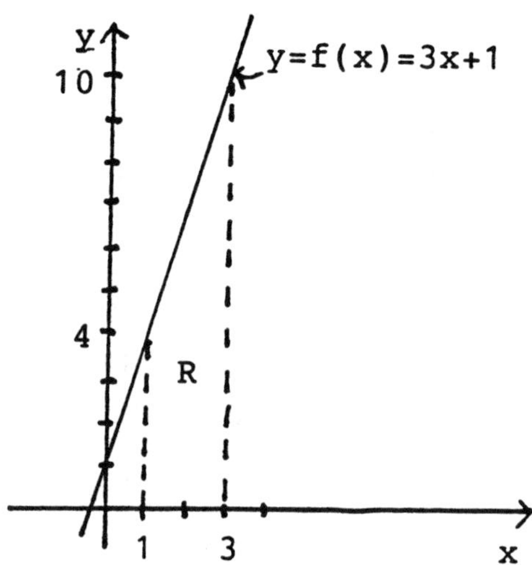

17. The area of $R = \frac{1}{2}\pi(3)^2 = \frac{9}{2}\pi$.
Average value $= \frac{\frac{9\pi}{2}}{6} = \frac{3\pi}{4}$.

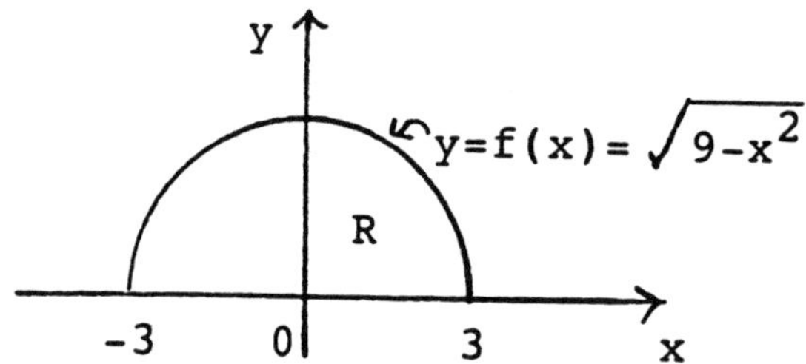

18. $\frac{\int_0^{20}(10+40x+0.3x^2)dx}{20-0} = \frac{\left(10x+20x^2+0.1x^3\right)\Big]_0^{20}}{20} = \frac{200+8000+800}{20} = 450.$

19. $\frac{\int_0^{10}(120x-4x^2)dx}{10-0} = \frac{\left(60x^2-\frac{4}{3}x^3\right)\Big]_0^{10}}{10} = \frac{6000-\frac{4000}{3}}{10} = \frac{1400}{3}.$

20. $\frac{\int_0^{4}\left(\frac{p}{1+p^2}\right)dp}{4-0} = \frac{\frac{1}{2}\ln\left(1+p^2\right)\Big]_0^{4}}{4} = \frac{\frac{1}{2}\ln 17-\frac{1}{2}\ln 1}{4} = \frac{1}{8}\ln 17.$

21. The average value of the marginal cost function $C'(x)$ on the interval $[0,x]$ is
$\frac{\int_0^{x}C'(u)du}{x} = \frac{C(x)-C(0)}{x} = \frac{C(x)}{x} = c(x)$ in the case $C(0)=0$.

22. $\frac{\int_6^{20}\left(4-2\left(\frac{t-13}{7}\right)^4\right)dt}{20-6}dt = \frac{\left(4t-\frac{2}{5}\left(\frac{t-13}{7}\right)^5\right)\Big]_6^{20}}{14} = \frac{\left(80-\frac{2}{5}\right)-\left(24+\frac{2}{5}\right)}{14} = \frac{138}{35}.$

23. Revolve R about the x-axis.

$$\text{Volume} = \pi \cdot \int_{-r}^{r} (f(x))^2\,dx = \pi \cdot \int_{-r}^{r} (r^2 - x^2)dx$$

$$= \pi \cdot \left(r^2 x - \tfrac{1}{3}x^3\right)\Big]_{-r}^{r}$$

$$= \pi\left[\left(r^3 - \tfrac{1}{3}r^3\right) - \left(-r^3 + \tfrac{1}{3}r^3\right)\right]$$

$$= \pi\left(\tfrac{4}{3}r^3\right).$$

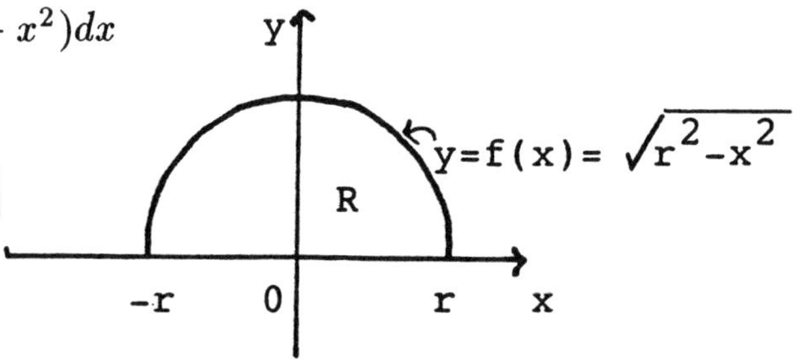

This is the formula for the volume of a sphere of radius r.

24. The average amount on deposit is given by

$$\frac{\int_0^{10} 1000e^{.08t}\,dt}{10} = \frac{1000 \cdot \frac{1}{.08}e^{.08t}\Big]_0^{10}}{10}$$

$$= 1,250 \cdot (e^{.8} - 1) = \$1,531.93.$$

25. Let $P(t)$ denote the population of the earth in billions of people t years from now. We have that

$$P(t) = P_0 e^{rt} \implies P(t) = 5e^{rt}.$$

$$P(50) = 10 \implies 10 = 5e^{r(50)} \implies e^{50r} = 2 \implies 50r = \ln 2$$

$$\implies r = \frac{\ln 2}{50}.$$

Therefore

$$P(t) = 5e^{\left(\frac{\ln 2}{50}\right)t}.$$

The average population of the earth during this 50-year period would be

$$\frac{\int_0^{50} P(t)dt}{50} = \frac{5 \cdot \frac{50}{\ln 2} \cdot e^{\left(\frac{\ln 2}{50}\right)t}\Big]_0^{50}}{50} = \frac{5}{\ln 2} \cdot \left(e^{\ln 2} - 1\right)$$

$$= \frac{5}{\ln 2} = 7.2135 \text{ billion.}$$

Solutions to the Review Exercises - Chapter 5

1. $\int (6x^2 - 2x + 1)dx = 2x^3 - x^2 + x + C.$
2. $\int (x^2 - 6x)^2 dx = \int (x^4 - 12x^3 + 36x^2)dx = \frac{1}{5}x^5 - 3x^4 + 12x^3 + C.$
3. $\int x\sqrt{3x^2+5}\,dx.$

   Let $u = 3x^2 + 5$. $du = 6x\,dx \implies x\,dx = \frac{1}{6}\,du.$

   $\int x\sqrt{3x^2+5}\,dx = \int u^{1/2}\left(\frac{1}{6}\,du\right) = \frac{1}{6}\cdot\frac{2}{3}u^{3/2} + C = \frac{1}{9}(3x^2+5)^{3/2} + C.$
4. $\int (3x^{1/2} + 3x^{-1/2})dx = 2x^{3/2} + 6x^{1/2} + C.$
5. $\int \left(t + \sqrt[3]{t}\right)^2 dt = \int (t^2 + 2t^{4/3} + t^{2/3})dt = \frac{1}{3}t^3 + \frac{6}{7}t^{7/3} + \frac{3}{5}t^{5/3} + C.$
6. $\int t\sqrt{9-t^2}\,dt = -\frac{1}{3}(9-t^2)^{3/2} + C.$
7. $\int \frac{x^3-7x^2+6}{x}dx = \int \left(x^2 - 7x + \frac{6}{x}\right)dx = \frac{1}{3}x^3 - \frac{7}{2}x^2 + 6\ln|x| + C.$
8. $\int \frac{x^3+x^2-x+2}{x+2}dx = \int (x^2 - x + 1)dx = \frac{1}{3}x^3 - \frac{1}{2}x^2 + x + C.$
9. $\int (2x-1)(2x+3)dx = \int (4x^2 + 4x - 3)dx = \frac{4}{3}x^3 + 2x^2 - 3x + C.$
10. $\int \frac{x}{4x^4+4x^2+1}dx = \int \frac{x}{(2x^2+1)^2}dx.$

    Let $u = 2x^2 + 1$. $du = 4x\ dx \implies x\ dx = \frac{1}{4}du.$

    So, $\int u^{-2}\frac{1}{4}du = \frac{1}{4}\frac{u^{-1}}{-1} + C = -\frac{1}{4}\left(\frac{1}{2x^2+1}\right) + C.$
11. Let $u = 1 - x^2$. $du = -2x\ dx \implies x\ dx = -\frac{1}{2}du.$

    $\int \frac{x}{1-x^2}dx = \int \frac{1}{u}\left(-\frac{1}{2}du\right) = -\frac{1}{2}\ln|u| + C = -\frac{1}{2}\ln|1-x^2| + C.$
12. $\int \frac{3x+3}{4x+2x^2}\,dx.$

    Let $u = 4x + 2x^2$. $du = (4+4x)dx \implies du = 4(x+1)dx \implies$

    $$(x+1)dx = \frac{1}{4}\,du \implies (3x+3)dx = \frac{3}{4}\,du.$$

    $\int \frac{3x+3}{4x+2x^2}\,dx = \int \frac{1}{u}\left(\frac{3}{4}\,du\right) = \frac{3}{4}\ln|u| + C = \frac{3}{4}\ln|4x+2x^2| + C.$
13. $\int \sqrt{e^x}dx = \int e^{x/2}dx = 2e^{x/2} + C.$
14. $\int (e^x+1)^3\,dx = \int \left((e^x)^3 + 3(e^x)^2\cdot(1) + 3(e^x)(1)^2 + (1)^3\right)dx$

    $= \int (e^{3x} + 3e^{2x} + 3e^x + 1)dx = \frac{1}{3}e^{3x} + \frac{3}{2}e^{2x} + 3e^x + x + C.$
15. Let $u = \ln x \implies du = \frac{1}{x}dx.$

    $\int \frac{1}{x\sqrt{\ln x}}dx = \int u^{-1/2}du = 2u^{1/2} + C = 2\sqrt{\ln x} + C.$

16. Let $u = e^x + e^{-x}$. $du = (e^x - e^{-x})dx$

$\int \frac{e^x - e^{-x}}{e^x + e^{-x}} dx = \int \frac{1}{u} du = \ln|u| + C = \ln(e^x + e^{-x}) + C.$

17. $\int \frac{x^3-1}{x+1} dx = \int \left(\frac{x^3+1}{x+1} - \frac{2}{x+1}\right) dx = \int \left(x^2 - x + 1 - \frac{2}{x+1}\right) dx$

$= \frac{1}{3}x^3 - \frac{1}{2}x^2 + x - 2\ln|x+1| + C.$

18. Let $u = \sqrt{x}$. $du = \frac{1}{2\sqrt{x}} dx \implies \frac{1}{\sqrt{x}} dx = 2du.$

$\int \frac{e^{\sqrt{x}}}{\sqrt{x}} dx = \int e^u 2du = 2e^u + C = 2e^{\sqrt{x}} + C.$

19. $\int e^{2x}(1 - e^{2x})^2\, dx.$

Let $u = 1 - e^{2x}$. $du = -2e^{2x}\, dx \implies e^{2x}\, dx = -\frac{1}{2}\, du.$

$\int e^{2x}(1 - e^{2x})^2\, dx = \int u^2 \left(-\frac{1}{2}\, du\right) = -\frac{1}{6}u^3 + C = -\frac{1}{6}(1 - e^{2x})^3 + C.$

20. Let $u = x^3 + x^2 - 7$. $du = (3x^2 + 2x)dx.$

$\int \frac{2x+3x^2}{x^3+x^2-7} dx = \int \frac{1}{u}\, du = \ln|u| + C = \ln|x^3 + x^2 - 7| + C.$

21. $v(t) = 2t - (t+1)^{-2}$. $v(t) = s'(t).$

(a) $s(t) = \int \left(2t - (t+1)^{-2}\right) dt = t^2 + \frac{1}{t+1} + C.$

$s(0) = 0 \implies 0 = 0^2 + \frac{1}{0+1} + C \implies C = -1.$

$s(t) = t^2 + \frac{1}{t+1} - 1.$

(b) $a(t) = v'(t) = 2 + \frac{2}{(t+1)^3}.$

22. $\lim_{t\to\infty} P(t) = \lim_{t\to\infty} \frac{200{,}000}{20+40e^{-0.2t}} = \frac{200{,}000}{20+40(0)} = 10{,}000.$

The horizontal asymptote is $y = 10{,}000$.

23. (a) $C(x) = \int MC(x)dx = \int(120 + 6x)dx = 120x + 3x^2 + K.$

$C(0) = 500 \implies K = 500.$

$C(x) = 120x + 3x^2 + 500.$

(b) $C(20) = 120(20) + 3(20)^2 + 500 = 4100.$

24. (a) $R(x) = \int MR(x)dx = \int 250dx = 250x + K.$

$R(0) = 0 \implies K = 0.$ $R(x) = 250x.$

(b) $R(20) = 250(20) = 5{,}000.$

25. $P(x) = R(x) - C(x) = 250x - (120x + 3x^2 + 500) = -3x^2 + 130x - 500.$

$P(20) = R(20) - C(20) = 5000 - 4100 = 900 > 0.$

The operation is profitable when $x = 20$.

26. $\frac{dP}{dt} = 8e^{0.5t}$.

$P = \frac{8}{0.5}e^{0.5t} + C = 16e^{0.5t} + C$.

$P(0) = 16 \implies 16 = 16e^0 + C \implies C = 0$.

Thus, $P = 16e^{0.5t}$.

27. $\frac{dV}{dt} = \frac{-80{,}000}{(t+1)^2}$. $\quad V = \frac{80{,}000}{(t+1)} + C$.

$V(0) = 100,000 \implies 100,000 = \frac{80{,}000}{0+1} + C \implies C = 20,000$

(a) $V(t) = \frac{80{,}000}{t+1} + 20,000$ dollars.

(b) $V(3) = \frac{80{,}000}{3+1} + 20,000 = 40,000$ dollars.

(c) $\lim_{t\to\infty} V(t) = 20,000$.

28. Let V = value of investment after t years.

$\frac{dV}{dt} = \frac{500e^{\sqrt{t}}}{\sqrt{t}}$. $\quad V = 1000e^{\sqrt{t}} + C$.

$V(0) = 1000 \implies 1000 = 1000e^0 + C \implies C = 0$.

$V = 1000e^{\sqrt{t}}$. $\quad V(4) = 1000e^2$ dollars.

29. (a) $C(x) = \int MC(x)dx = \int(70 + 2x)dx = 70x + x^2 + K$.

$C(10) = 1000 \implies 1000 = 70(10) + 10^2 + K \implies K = 200$.

$C(x) = 70x + x^2 + 200$ dollars.

(b) $C(0) = \$200$.

30. $C(x) = \int(40 + 2x)dx = 40x + x^2 + K$.

$R(x) = \int 120dx = 120x$.

$P(x) = R(x) - C(x) = -x^2 + 80x - K$.

$P(20) = 1050 \implies 1050 = -20^2+80(20)-K \implies -K = -150 \implies K = 150$.

Thus $C(0) = \$150$.

31. $\int_0^3(2x - 3)dx = (x^2 - 3x)\Big]_0^3 = \big(3^2 - 3(3)\big) - \big(0^2 - 3(0)\big) = 0$.

32. $\int_0^2 \frac{1}{x+5}\,dx = \ln(x+5)\Big]_0^2 = \ln(2+5) - \ln(0+5) = \ln 7 - \ln 5 = \ln\frac{7}{5}$.

33. $\int_0^1(x^2 - 6)^2\,dx = \int_0^1(x^4 - 12x^2 + 36)dx = \big(\frac{1}{5}x^5 - 4x^3 + 36x\big)\Big]_0^1$

$= \big(\frac{1}{5} - 4 + 36\big) = \frac{161}{5}$.

34. $\int_{-3}^{7} 6\,dx = 6x\big]_{-3}^{7} = 6(7) - 6(-3) = 60.$

35. $\int_{2}^{9} \sqrt{x+7}\,dx = \frac{2}{3}(x+7)^{3/2}\big]_{2}^{9} = \frac{2}{3}(9+7)^{3/2} - \frac{2}{3}(2+7)^{3/2} = \frac{2}{3}(64-27) = \frac{74}{3}.$

36. $\int_{0}^{1} 3e^{2x}dx = \frac{3}{2}e^{2x}\big]_{0}^{1} = \frac{3}{2}e^{2\cdot 1} - \frac{3}{2}e^{2\cdot 0} = \frac{3}{2}(e^2 - 1).$

37. $\int_{0}^{3} \frac{x}{x+3}dx = \int_{0}^{3}\left(1 - \frac{3}{x+3}\right)dx = (x - 3\ln(x+3))]_0^3$

$= [3 - 3\ln(3+3)] - [0 - 3\ln(0+3)] = 3 - 3(\ln 6 - \ln 3)$

$= 3 - 3\ln\frac{6}{3} = 3(1 - \ln 2).$

38. $\int_{-1}^{1} (x^2 - 3x + 9)dx = \left(\frac{1}{3}x^3 - \frac{3}{2}x^2 + 9x\right)\big]_{-1}^{1}$

$= \left(\frac{1}{3}\cdot 1^3 - \frac{3}{2}\cdot 1^2 + 9\cdot 1\right) - \left[\frac{1}{3}(-1)^3 - \frac{3}{2}(-1)^2 + 9(-1)\right]$

$= \frac{2}{3} + 18 = \frac{56}{3}.$

39. $\int_{3}^{4} \frac{x+2}{x-2}\,dx = \int_{3}^{4}\left(\frac{(x-2)+4}{x-2}\right)dx = \int_{3}^{4}\left(\frac{x-2}{x-2} + \frac{4}{x-2}\right)dx = \int_{3}^{4}\left(1 + \frac{4}{x-2}\right)dx$

$= (x + 4\ln(x-2))]_3^4 = (4 + 4\ln 2) - (3 + 4\ln 1) = 1 + 4\ln 2.$

40. $\int_{0}^{2} 3x^2e^{x^3}\,dx = e^{x^3}\Big]_{0}^{2} = e^{2^3} - e^{0^3} = e^8 - 1.$

41. $\int_{-1}^{3} (x-1)(x+5)dx = \int_{-1}^{3} (x^2 + 4x - 5)dx = \left(\frac{1}{3}x^3 + 2x^2 - 5x\right)\big]_{-1}^{3}$
$= \left(\frac{1}{3}\cdot 3^3 + 2\cdot 3^2 - 5\cdot 3\right) - \left(\frac{1}{3}\cdot(-1)^3 + 2\cdot(-1)^2 - 5(-1)\right)$
$= \frac{27}{3} + 3 + \frac{1}{3} - 7 = \frac{16}{3}.$

42. $\int_{4}^{9} (\sqrt{x} - 1)(\sqrt{x} + 1)\,dx = \int_{4}^{9}(x-1)dx = \left(\frac{1}{2}x^2 - x\right)\big]_{4}^{9}$
$= \left(\frac{1}{2}\cdot 9^2 - 9\right) - \left(\frac{1}{2}\cdot 4^2 - 4\right) = \frac{63}{2} - \frac{8}{2} = \frac{55}{2}.$

43. $\int_{1}^{3} |x-4|dx = \int_{1}^{3}(4-x)dx = \left(4x - \frac{1}{2}x^2\right)\big]_{1}^{3} = \left(4(3) - \frac{1}{2}(3)^2\right) - \left(4(1) - \frac{1}{2}(1)^2\right)$

$= \frac{15}{2} - \frac{7}{2} = 4.$

44. $\int_0^4 x\left(\sqrt{x}+x^{2/3}\right)dx = \int_0^4 \left(x^{3/2}+x^{5/3}\right)dx = \left(\frac{2}{5}x^{5/2}+\frac{3}{8}x^{8/3}\right)\Big]_0^4$
$= \left(\frac{2}{5}\cdot 4^{5/2}+\frac{3}{8}\cdot 4^{8/3}\right)-0 = \frac{64}{5}+\frac{3}{8}\cdot 4^{8/3}.$

45. $\int_{-8}^{-1}(x^{1/3}-x^{5/3})dx = \left(\frac{3}{4}x^{4/3}-\frac{3}{8}x^{8/3}\right)\Big]_{-8}^{-1}$
$= \left(\frac{3}{4}\cdot(-1)^{4/3}-\frac{3}{8}\cdot(-1)^{8/3}\right)-\left(\frac{3}{4}\cdot(-8)^{4/3}-\frac{3}{8}\cdot(-8)^{8/3}\right)$
$= \frac{3}{4}-\frac{3}{8}-\frac{3}{4}\cdot 16+\frac{3}{8}\cdot 2^8 = \frac{675}{8}.$

46. $\int_0^4 \frac{x}{\sqrt{x^2+9}}\,dx.$

Let $u = x^2+9$. $du = 2x\,dx \implies x\,dx = \frac{1}{2}\,du.$

$\int \frac{x}{\sqrt{x^2+9}}\,dx = \int u^{-1/2}\left(\frac{1}{2}\,du\right) = \frac{1}{2}\cdot\frac{u^{1/2}}{1/2}+C = \sqrt{x^2+9}+C.$

$\int_0^4 \frac{x}{\sqrt{x^2+9}}\,dx = \sqrt{x^2+9}\,\Big]_0^4 = 5-3 = 2.$

47. $\int_1^8 \frac{\sqrt[3]{x}}{5+x^{4/3}}dx.$

Let $u(x) = 5+x^{4/3}$. $du = \frac{4}{3}x^{1/3}dx.$ $\sqrt[3]{x}dx = \frac{3}{4}du.$

$\int \frac{\sqrt[3]{x}}{5+x^{4/3}}dx = \frac{3}{4}\int \frac{1}{u}du = \frac{3}{4}\ln u + C = \frac{3}{4}\ln(5+x^{4/3})+C.$

$\int_1^8 \frac{\sqrt[3]{x}}{5+x^{4/3}}dx = \frac{3}{4}\ln(5+x^{4/3})\Big]_1^8$
$= \frac{3}{4}\ln(5+8^{4/3})-\frac{3}{4}\ln(5+1^{4/3}) = \frac{3}{4}\ln(5+16)-\frac{3}{4}\ln 6$
$= \frac{3}{4}\ln\left(\frac{21}{6}\right) = \frac{3}{4}\ln\left(\frac{7}{2}\right).$

48. $\int_0^1 (\sqrt{x}+5)(x^{1/3}+x)dx = \int_0^1 (x^{5/6}+x^{3/2}+5x+5x^{1/3})dx$
$= \left(\frac{6}{11}x^{11/6}+\frac{2}{5}x^{5/2}+\frac{5}{2}x^2+\frac{15}{4}x^{4/3}\right)\Big]_0^1$
$= \left(\frac{6}{11}\cdot 1+\frac{2}{5}\cdot 1+\frac{5}{2}\cdot 1+\frac{15}{4}\cdot 1\right)-0 = \frac{1583}{220}.$

49. $\int_1^8 \frac{x^{2/3}+3x^{5/2}}{x}dx = \int_1^8 (x^{-1/3}+3x^{3/2})dx = \left(\frac{3}{2}x^{2/3}+\frac{6}{5}x^{5/2}\right)\Big]_1^8$
$= \left(\frac{3}{2}\cdot 8^{2/3}+\frac{6}{5}\cdot(8)^{5/2}\right)-\left(\frac{3}{2}\cdot 1+\frac{6}{5}\cdot 1\right) = \frac{9}{2}+\frac{6}{5}(8^{5/2}-1).$

50. $\int_0^{\ln 2} (e^x+e^{-x})^2dx = \int_0^{\ln 2} (e^{2x}+2+e^{-2x})dx = \left(\frac{1}{2}e^{2x}-\frac{1}{2}e^{-2x}+2x\right)\Big]_0^{\ln 2}$
$= \left(\frac{1}{2}e^{2\ln 2}-\frac{1}{2}e^{-2\ln 2}+2\ln 2\right)-\left(\frac{1}{2}e^{2\cdot 0}-\frac{1}{2}e^{-2\cdot 0}+2\cdot 0\right)$
$= \frac{1}{2}\cdot 4-\frac{1}{2}\cdot\frac{1}{4}+2\ln 2-0 = \frac{15}{8}+2\ln 2.$

51. $\int_{-4}^{4} \sqrt{16 - x^2}dx = \text{Area of } R$

$= \frac{1}{2}\pi \cdot 4^2 = 8\pi.$

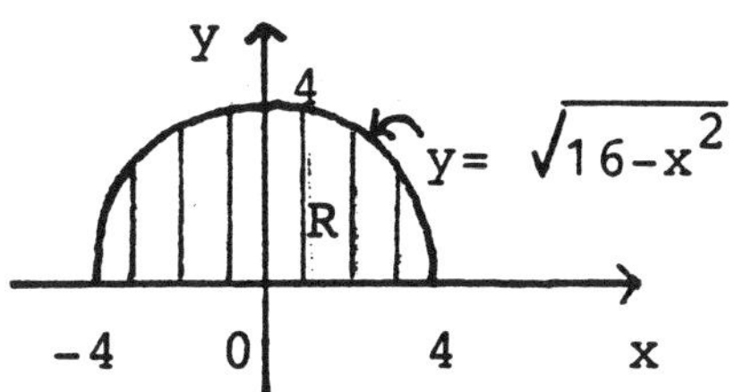

52. $\int_{-2}^{0} \sqrt{4 - x^2}dx = \text{Area of } R$

$= \frac{1}{4}\pi \cdot 2^2 = \pi.$

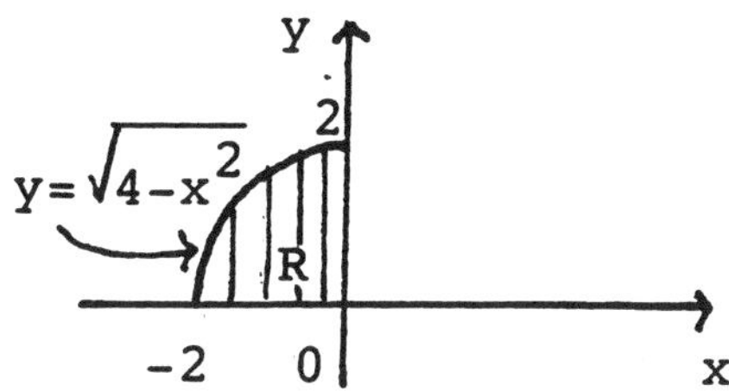

53. Area of $R = \int_0^9 (9 - \sqrt{x})\, dx = \left(9x - \frac{2}{3}x^{3/2}\right)\Big]_0^9 = (9{\cdot}9 - \frac{2}{3}{\cdot}9^{3/2}) - 0 = 81 - \frac{2}{3}{\cdot}27 =$ 63.

54. Area of $R = \int_0^5 (10 - 2x)dx = (10x - x^2)\big]_0^5 = (10 \cdot 5 - 5^2) - 0 = 25.$

55. From a sketch of the graphs it can be seen that there are two points of intersection. These points are (1,4) and (-1,4).

Area of $R = 2\int_0^1 \left[(3 + x^{2/3}) - (3x^2 + 1)\right] dx.$

Area of $R = 2\left(3x + \frac{3}{5}x^{5/3} - x^3 - x\right)\Big]_0^1$

$= 2\left(3 \cdot 1 + \frac{3}{5} \cdot 1 - 1 - 1\right) - 0$

$= 2\left(\frac{8}{5}\right) = \frac{16}{5}.$

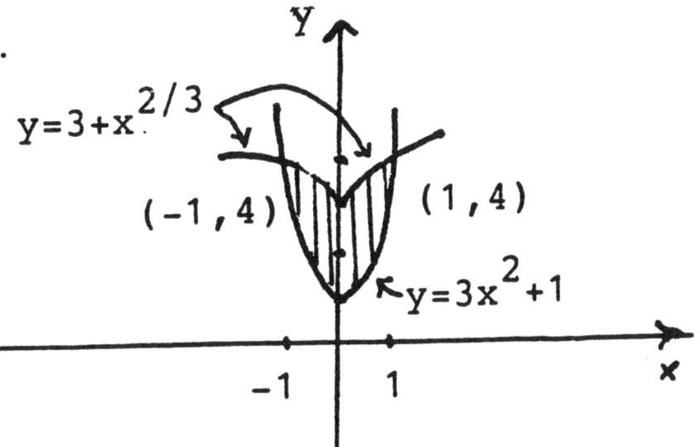

56. Area $R = \int_0^1 (e^x - e^{-x})dx = (e^x + e^{-x})]_0^1$
$= (e^1 + e^{-1}) - (e^0 + e^0) = e + \frac{1}{e} - 2.$

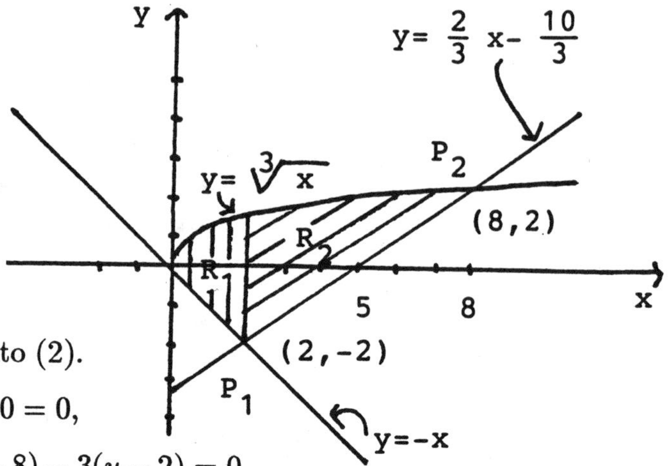

57. For intersection point P_1, solve
$\begin{cases} y = -x \\ y = \frac{2}{3}x - \frac{10}{3} \end{cases}.$
$x = 2, \quad y = -2, \quad P_1 = (2, -2).$

For intersection point P_2, solve
$\begin{cases} y = \sqrt[3]{x} & (1) \\ y = \frac{2}{3}x - \frac{10}{3} & (2) \end{cases}.$
From (1), $x = y^3$. Substitute into (2).

$y = \frac{2}{3}y^3 - \frac{10}{3}, \qquad 2y^3 - 3y - 10 = 0,$

$2y^3 - 3y - 16 + 6 = 0, \quad 2(y^3 - 8) - 3(y - 2) = 0,$

$2(y-2)(y^2 + 2y + 4) - 3(y - 2) = 0, \qquad (y - 2)(2y^2 + 4y + 5) = 0.$

$y = 2, \qquad x = 2^3 = 8.$ So $P_2 = (8, 2)$.

Area of R = (Area of R_1) + (Area of R_2)

$$= \int_0^2 [\sqrt[3]{x} - (-x)]\, dx + \int_2^8 [\sqrt[3]{x} - (\tfrac{2}{3}x - \tfrac{10}{3})]\, dx$$

$$= (\tfrac{3}{4}x^{4/3} + \tfrac{1}{2}x^2)]_0^2 + (\tfrac{3}{4}x^{4/3} - \tfrac{1}{3}x^2 + \tfrac{10}{3}x)]_2^8$$

$$= (\tfrac{3}{4} \cdot 2^{4/3} + \tfrac{1}{2} \cdot 2^2) - 0 + (\tfrac{3}{4} \cdot 8^{4/3} - \tfrac{1}{3} \cdot 8^2 + \tfrac{10}{3} \cdot 8) - (\tfrac{3}{4} \cdot 2^{4/3} - \tfrac{1}{3} \cdot 2^2 + \tfrac{10}{3} \cdot 2)$$

$$= 2 + 12 - \tfrac{64}{3} + \tfrac{80}{3} + \tfrac{4}{3} - \tfrac{20}{3} = 14.$$

58. $MC(x) = 120, \qquad C(40) = 6000.$

(a) $C(50) - C(40) = \int_{40}^{50} MC(x)dx.$

So $C(50) = C(40) + \int_{40}^{50} MC(x)dx = 6000 + \int_{40}^{50} 120dx.$

$C(50) = 6000 + 120x]_{40}^{50} = 6000 + (120 \cdot 50 - 120 \cdot 40) = \$7200.$

(b) $C(x) = \int MC(x)dx = \int 120dx = 120x + K.$

$C(40) = 6000 \implies 6000 = 120 \cdot 40 + K, \quad K = 1200.$

Thus $C(x) = 120x + 1200$ dollars.

59. $C(50) - C(40) = \int\limits_{40}^{50} MC(x)dx = \int\limits_{40}^{50} \left(110 + \frac{50}{x+10}\right) dx$

$$C(50) - C(40) = (110x + 50\ln(x+10))]_{40}^{50}$$
$$= 110 \cdot 50 + 50\ln(50+10) - 110 \cdot 40 - 50\ln(40+10)$$
$$= 1100 + 50\ln\tfrac{6}{5} = 50\left(22 + \ln\tfrac{6}{5}\right).$$

60. $R(49) - R(36) = \int\limits_{36}^{49} MR(x)dx = \int\limits_{36}^{49} (600 - 2\sqrt{x})\,dx$

$$= \left(600x - \tfrac{4}{3}x^{3/2}\right)\Big]_{36}^{49}$$
$$= \left(600 \cdot 49 - \tfrac{4}{3} \cdot 49^{3/2}\right) - \left(600 \cdot 36 - \tfrac{4}{3} \cdot 36^{3/2}\right)$$
$$= 7800 - \tfrac{4}{3}\left(7^3 - 6^3\right) = \tfrac{22{,}892}{3}$$

61. $D(x_0) = D(6) = 40(50 - 6^2) = 560.$

Consumers' surplus $= \int\limits_0^6 40(50 - x^2)dx - 6 \cdot (560)$

$$= 40\left(50x - \tfrac{1}{3}x^3\right)\Big]_0^6 - 3360$$
$$= 40\left(300 - \tfrac{1}{3} \cdot 216 - 0\right) - 3360 = 5760.$$

62. $D(x_0) = D(16) = \frac{40\sqrt{16}}{1+16^{3/2}} = \frac{40(4)}{1+64} = \frac{160}{65}.$

Consumers' surplus $= \int\limits_0^{16} \frac{40\sqrt{x}}{1+x^{3/2}}dx - 16 \cdot \frac{160}{65}.$

Let $u(x) = 1 + x^{3/2}.$ $\qquad du = \frac{3}{2}x^{1/2}dx.$

$\int \frac{40\sqrt{x}}{1+x^{3/2}}dx = 40 \cdot \frac{2}{3}\int \frac{1}{u}du = \frac{80}{3}\ln u + C = \frac{80}{3}\ln(1 + x^{3/2}) + C.$

$\int\limits_0^{16} \frac{40\sqrt{x}}{1+x^{3/2}}dx = \frac{80}{3}\ln(1 + x^{3/2})\Big]_0^{16}$

$$= \tfrac{80}{3}\left[\ln(1 + 16^{3/2}) - \ln(1+0)\right] = \tfrac{80}{3}\ln 65.$$

So consumers' surplus $= \frac{80}{3}\ln 65 - 16 \cdot \frac{160}{65} = 80\left(\frac{1}{3}\ln 65 - \frac{32}{65}\right).$

63. $S(x_0) = S(15) = \frac{1}{\sqrt{15+1}}e^{\sqrt{15+1}} = \frac{1}{4}e^4.$

Suppliers' surplus $= 15 \cdot \left(\frac{1}{4}e^4\right) - \int\limits_0^{15} \frac{e^{\sqrt{x+1}}}{\sqrt{x+1}}dx.$

Let $u(x) = \sqrt{x+1}.$ $\qquad du = \frac{1}{2\sqrt{x+1}}dx.$

$\int \frac{e^{\sqrt{x+1}}}{\sqrt{x+1}}dx = 2\int e^u du = 2e^u + C = 2e^{\sqrt{x+1}} + C.$

So the suppliers' surplus $= \frac{15}{4}e^4 - \left(2e^{\sqrt{x+1}}\right)\Big]_0^{15} = \frac{15}{4}e^4 - 2\left(e^{\sqrt{15+1}} - e^{\sqrt{0+1}}\right)$

$$= \tfrac{15}{4}e^4 - 2\left(e^4 - e\right) = \tfrac{7}{4}e^4 + 2e.$$

64. The total nominal value of revenue is

$$\int_0^4 A(t)dt = \int_0^4 \left(6\sqrt{t} + 3000\right) dt = \left(\tfrac{12}{3}t^{3/2} + 3000t\right)\Big]_0^4$$

$$= 4 \cdot 4^{3/2} + 3000(4) - 0 = 32 + 12000 = \$12,032.$$

65. The value of revenue $= \int_0^{10} A(t)dt = \int_0^{10} 100e^{0.2t}dt$

$$= 100 \cdot \tfrac{1}{0.2}e^{0.2t}\Big]_0^{10} = 500(e^2 - e^0) = 500(e^2 - 1) \text{ dollars.}$$

66. The present value of revenue $= \int_0^{10} A(t)e^{-rt}dt = \int_0^{10} 5000e^{-0.08t}dt$

$$= 5000\left(-\tfrac{1}{0.08}\right)e^{-0.08t}\Big]_0^{10} = -62500(e^{-0.8} - e^0)$$

$$= 62500(1 - e^{0.8}) \text{ dollars.}$$

67. The total value of the annuity $A = \int_0^5 10,000t\sqrt{1+t^2}dt.$

Let $u(t) = 1 + t^2$. $du = 2t\ dt$.

$$\int 10,000t\sqrt{1+t^2}dt = 5,000\int\sqrt{u}du = \tfrac{10,000}{3}u^{3/2} + C = \tfrac{10,000}{3}(1+t^2)^{3/2} + C.$$

$$A = \int_0^5 10,000t\sqrt{1+t^2}dt = \tfrac{10,000}{3}(1+t^2)^{3/2}\Big]_0^5 = 10,000(26^{3/2} - 1)/3 \text{ dollars.}$$

68. The points of intersection of the graphs of f and g are (0,0) and (1,1).

Volume of $R = \int_0^1 \left[\pi(x^{2/3})^2 - \pi \cdot x^2\right] dx$

$$= \pi\left(\tfrac{3}{7}x^{7/3} - \tfrac{1}{3}x^3\right)\Big]_0^1 = \pi\left(\tfrac{3}{7} - \tfrac{1}{3}\right) = \tfrac{2}{21}\pi.$$

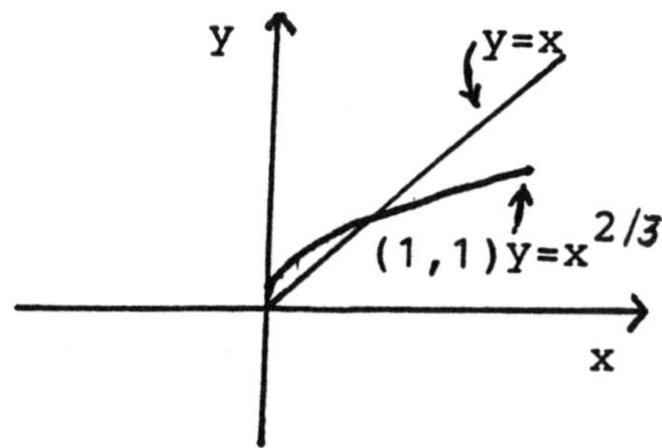

69. Average of $f = \frac{1}{3-0}\int_0^3 \frac{x}{x^2+7}dx = \frac{1}{3}\cdot\frac{1}{2}\cdot\ln(x^2+7)\Big]_0^3$

$$= \tfrac{1}{6}\left[\ln(3^2+7) - \ln(0+7)\right] = \tfrac{1}{6}(\ln 16 - \ln 7) = \tfrac{1}{6}\ln\left(\tfrac{16}{7}\right).$$

70. Average of $g = \frac{1}{4-1}\int_1^4 (1+\sqrt{x})(1-\sqrt{x})\,dx = \frac{1}{3}\int_1^4 (1-x)dx$

$= \frac{1}{3}\left(x - \frac{1}{2}x^2\right)\Big]_1^4 = \frac{1}{3}\left(4 - \frac{1}{2}\cdot 4^2 - 1 + \frac{1}{2}\right) = -\frac{3}{2}.$

71. $\ln(kx) = \ln x + C.$

$\ln x + \ln k = \ln x + C \implies C = \ln k.$

72. Suppose you manufacture x items per day. Suppose the cost per item of the material you use in the manufacture of these x items is given by the formula

$$\gamma = \frac{1}{12}x^2 - 10x + 325 = M(x)$$

where γ is in dollars. Suppose the labor costs per item you manufacture are \$50. Let your fixed daily costs be \$400. Then your total daily cost function is given by

$$C(x) = \frac{1}{12}x^3 - 10x^2 + 375x + 400.$$

Let I denote the open interval consisting of all real numbers x such that $40 < x < 50$. Throughout this interval I, the marginal cost $MC(x)$ increases as x increases, but the total daily cost function $C(x)$ decreases as x increases. Throughout the interval I, the cost per item $\gamma = M(x)$ of the material used in the manufacture of x items decreases as x increases, and γ is positive for all x in this interval.

73. Suppose you manufacture and sell x thousand items per day.

Suppose your total daily cost in thousands of dollars for manufacturing these x thousand items is given by

$$C(x) = -x^3 + 12x^2 + 20x + 15.$$

Suppose your revenue in thousands of dollars from the sale of these x thousand items is given by

$$R(x) = 60x.$$

Then your profit in thousands of dollars from the manufacture and sale of these x thousand items is given by

$$P(x) = R(x) - C(x) = x^3 - 12x^2 + 40x - 15.$$

Let I denote the open interval consisting of all real numbers x such that $4 < x < 5.633$. Throughout the interval I, the marginal profit $MP(x)$ increases as x increases, but the profit $P(x)$ decreases as x increases. Throughout the interval I, the total cost $C(x)$ is positive and increases as x increases, and the profit $P(x)$ is positive. The graphs of the cost function $C(x)$ and the profit function $P(x)$ are shown.

y=C(x)=(-(x^3))+(12*(x^2))+(20*x)+15

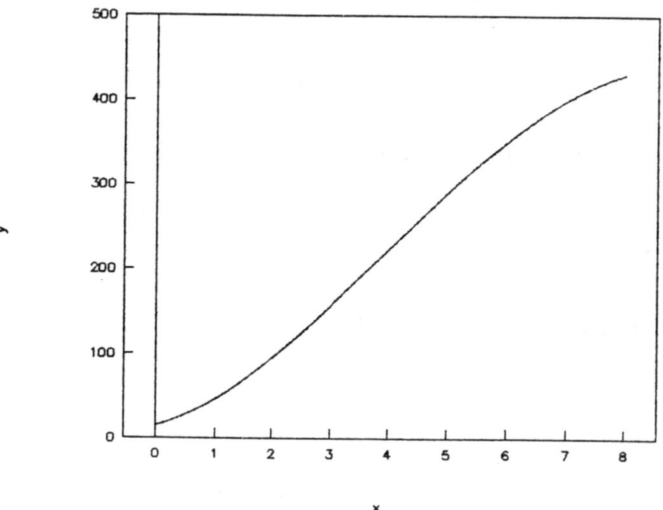

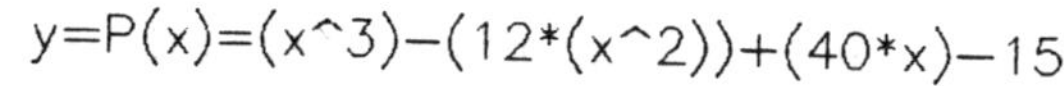

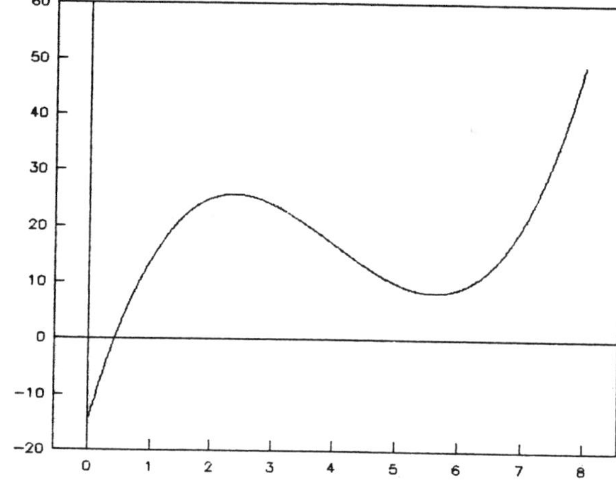

CHAPTER 6 (CHAPTER 7)

Solutions to Exercise Set 6.1 (7.1)

1.a. $\theta_r = \frac{\pi}{180} \cdot 90 = \frac{\pi}{2}$.

b. $\theta_r = \frac{\pi}{180} \cdot 45 = \frac{\pi}{4}$.

c. $\theta_r = \frac{\pi}{180} \cdot (-135) = -\frac{3}{4}\pi$.

d. $\theta_r = \frac{\pi}{180} \cdot 30 = \frac{\pi}{6}$.

e. $\theta_r = \frac{\pi}{180} \cdot 60 = \frac{\pi}{3}$.

f. $\theta_r = \frac{\pi}{180} \cdot (-150) = -\frac{5}{6}\pi$.

g. $\theta_r = \frac{\pi}{180} \cdot 180 = \pi$.

h. $\theta_r = \frac{\pi}{180} \cdot 210 = \frac{7}{6}\pi$.

2.a. $\theta_r = \frac{240}{180}\pi = \frac{4}{3}\pi$.

b. $\theta_r = \frac{225}{180}\pi = \frac{5}{4}\pi$.

c. $\theta_r = -\frac{270}{180}\pi = -\frac{3}{2}\pi$.

d. $\theta_r = \frac{300}{180}\pi = \frac{5}{3}\pi$.

e. $\theta_r = \frac{330}{180}\pi = \frac{11}{6}\pi$.

f. $\theta_r = -\frac{315}{180}\pi = -\frac{7}{4}\pi$.

g. $\theta_r = \frac{115}{180}\pi = \frac{23}{36}\pi$.

h. $\theta_r = \frac{235}{180}\pi = \frac{47}{36}\pi$.

3.a. $\theta_d = \frac{180}{\pi} \cdot \frac{\pi}{4} = 45°$.

b. $\theta_d = \frac{180}{\pi} \cdot \frac{3\pi}{2} = 270°$.

c. $\theta_d = \frac{180}{\pi} \cdot \left(-\frac{\pi}{12}\right) = -15°$.

d. $\theta_d = \frac{180}{\pi} \cdot \frac{7\pi}{6} = 210°$.

e. $\theta_d = \frac{180}{\pi} \cdot \frac{7\pi}{8} = \frac{315}{2}°$.

f. $\theta_d = \frac{180}{\pi} \cdot \left(-\frac{5\pi}{6}\right) = -150°$.

g. $\theta_d = \frac{180}{\pi} \cdot \frac{7\pi}{6} = 210°$.

h. $\theta_d = \frac{180}{\pi} \cdot \left(-\frac{3}{4}\pi\right) = -135°$.

4.a. Counterclockwise rotation through 720° corresponds to a radian measure $t = 720 \cdot \frac{\pi}{180} = 4\pi$.

b. $t = 480 \cdot \frac{\pi}{180} = \frac{8}{3}\pi$.

c. $t = 750 \cdot \frac{\pi}{180} = \frac{25}{6}\pi$.

d. $t = 1440 \cdot \frac{\pi}{180} = 8\pi$.

e. $t = 450 \cdot \frac{\pi}{180} = \frac{5}{2}\pi$.

f. $t = 390 \cdot \frac{\pi}{180} = \frac{13}{6}\pi$.

g. $t = 540 \cdot \frac{\pi}{180} = 3\pi$.

h. $t = 690 \cdot \frac{\pi}{180} = \frac{23}{6}\pi$.

5.a. Clockwise rotation through 45° corresponds to a radian measure $t = -45 \cdot \frac{\pi}{180} = -\frac{\pi}{4}$.

b. $t = -270 \cdot \frac{\pi}{180} = -\frac{3}{2}\pi$.

c. $t = -30 \cdot \frac{\pi}{180} = -\frac{\pi}{6}$.

d. $t = -150 \cdot \frac{\pi}{180} = -\frac{5}{6}\pi$.

e. $t = -390 \cdot \frac{\pi}{180} = -\frac{13}{6}\pi$.

f. $t = -135 \cdot \frac{\pi}{180} = -\frac{3}{4}\pi$.

g. $t = -330 \cdot \frac{\pi}{180} = -\frac{11}{6}\pi$.

h. $t = -540 \cdot \frac{\pi}{180} = -3\pi$.

6. $\left(3 + \frac{1}{3}\right) \cdot 2\pi = \frac{20}{3}\pi$.

7.a. $-6 \cdot \frac{\pi}{6} = -\pi$.

b. $-12 \cdot \frac{\pi}{6} = -2\pi$.

c. $-(12+3) \cdot \frac{\pi}{6} = -\frac{5}{2}\pi$.

d. $-(12+8) \cdot \frac{\pi}{6} = -\frac{10}{3}\pi$.

8. $2\pi + \frac{\pi}{2} = \frac{5}{2}\pi$.

9. $-\left(2 \cdot 2\pi + \frac{\pi}{2}\right) = -\frac{9}{2}\pi$.

10. $-\left(\pi + \frac{3}{4}\pi\right) = -\frac{7}{4}\pi$.

11. $(2\pi + \pi) = 3\pi$.

12. $-\left(2\pi + \pi + \frac{\pi}{4}\right) = -\frac{13}{4}\pi$.

13. $-\left(2\pi + \frac{\pi}{4}\right) = -\frac{9}{4}\pi$.

14.

$\left(-\frac{\sqrt{2}}{2}, \frac{\sqrt{2}}{2}\right)$

15.

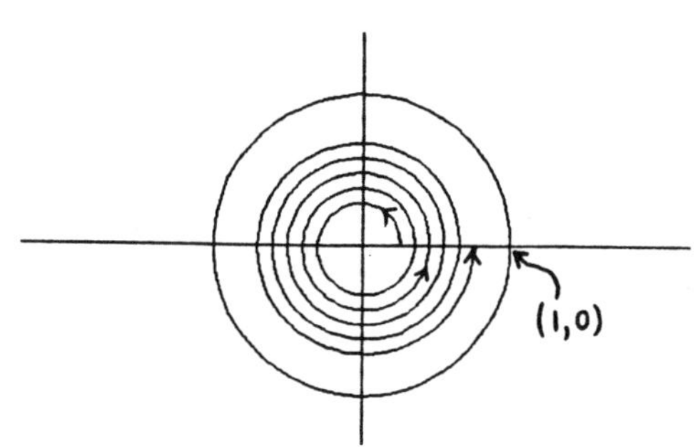

16.

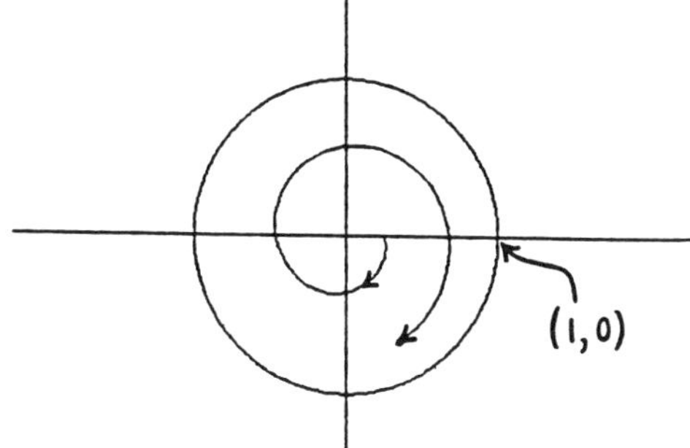
angle theta = (-(7/3)*pi)radians
(1,0)

17.

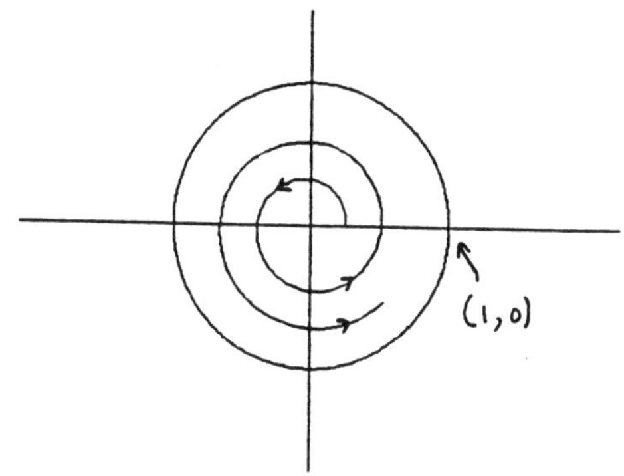
angle theta = ((15/4)*pi)radians
(1,0)

18.

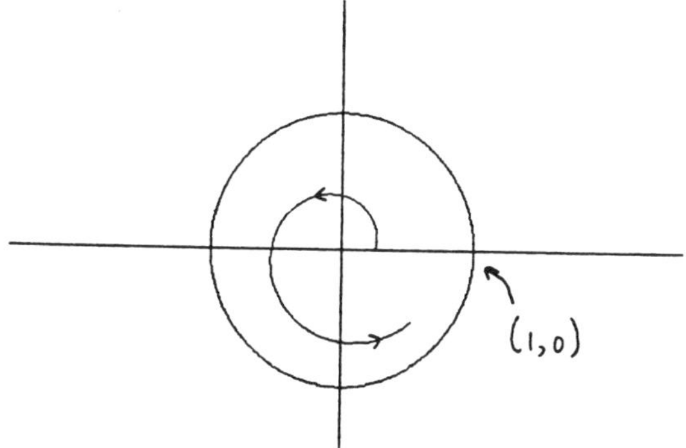
angle theta=((7/4)*pi)radians
(1,0)

19.

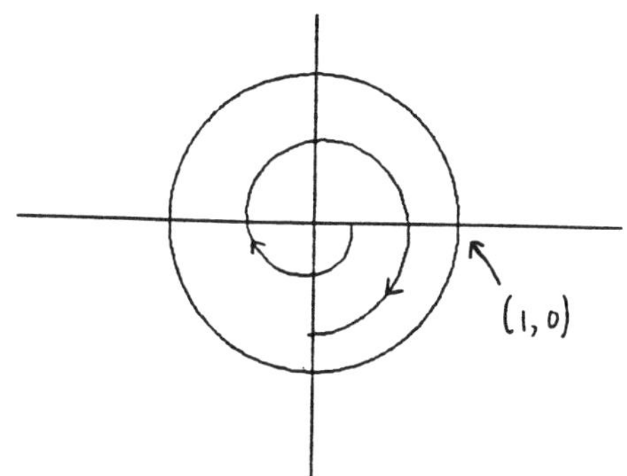
angle theta=((-(5/2))*pi)radians
(1,0)

20.

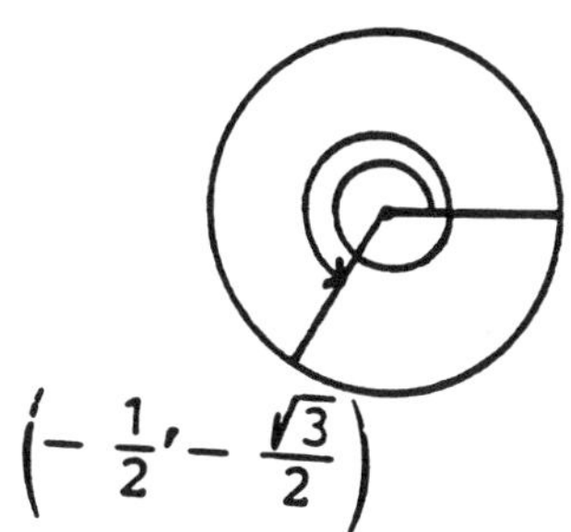
$\left(-\frac{1}{2}, -\frac{\sqrt{3}}{2}\right)$

21.

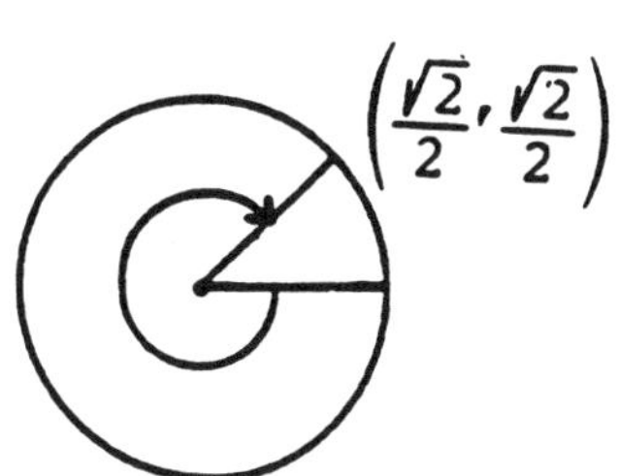
$\left(\frac{\sqrt{2}}{2}, \frac{\sqrt{2}}{2}\right)$

22.

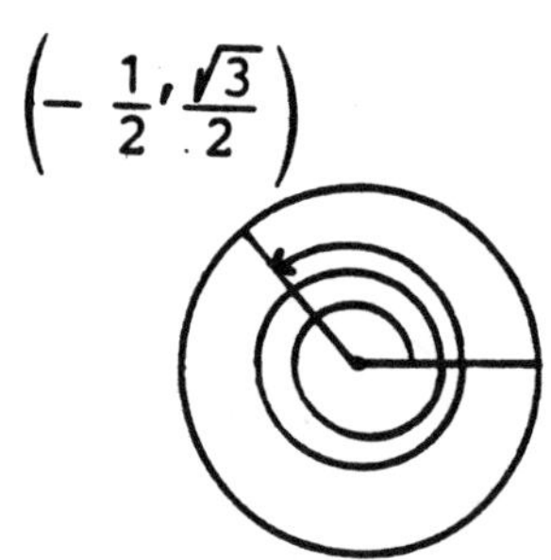

23.

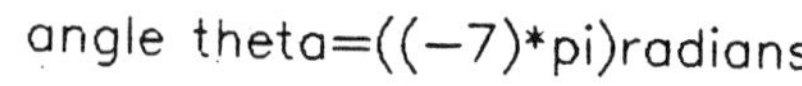

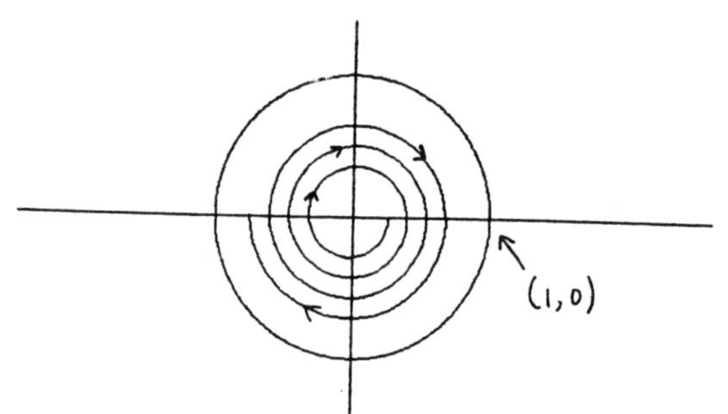

24.

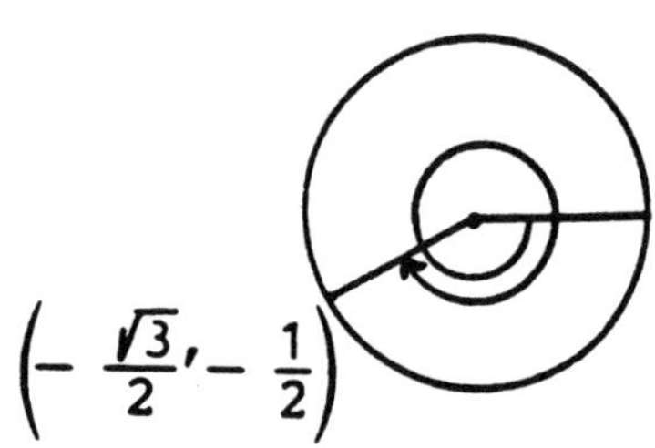

25. The radian measure of the angle is $(-3.25) \cdot 2\pi = -6.5\pi$.

26. The radian measure is $\frac{\pi}{180} \cdot 15 = \frac{\pi}{12}$.

27.a. The radian measure of -180 degrees is $-\pi$, corresponding to $\frac{1}{2}$ of a revolution in the clockwise direction.

b. The radian measure of -540 degrees is -3π, corresponding to $\frac{3}{2}$ revolutions in the clockwise direction.

c. The radian measure of 270 degrees in $\frac{3}{2}\pi$, corresponding to $\frac{3}{4}$ revolutions in the counterclockwise direction.

Solutions to Exercise Set 6.2 (7.2)

1.a. $\sin\theta = \frac{3}{5}$, b. $\cos\theta = \frac{4}{5}$. 2. a. $\sin\theta = \frac{4}{5}$ b. $\cos\theta = \frac{3}{5}$.

3. From the Pythagorean theorem, we have

$y^2 + 3^2 = \left(\sqrt{13}\right)^2 = 13 \implies y^2 = 4 \implies y = 2.$

$\therefore$ a. $\sin\theta = \frac{y}{\sqrt{13}} = \frac{2}{\sqrt{13}}$, b. $\cos\theta = \frac{3}{\sqrt{13}}$.

4. $h^2 = 2^2 + 3^2 = 13 \implies h = \sqrt{13}.$

a. $\sin\theta = \frac{3}{h} = \frac{3}{\sqrt{13}}$, b. $\cos\theta = \frac{2}{h} = \frac{2}{\sqrt{13}}$.

5. $x^2 + 7^2 = \left(\sqrt{65}\right)^2 = 65 \implies x^2 = 65 - 49 = 16 \implies x = 4.$

a. $\sin\theta = \frac{7}{\sqrt{65}}$, b. $\cos\theta = \frac{x}{\sqrt{65}} = \frac{4}{\sqrt{65}}$.

6. $h^2 = 4^2 + 7^2 = 16 + 49 = 65 \implies h = \sqrt{65}.$

a. $\sin\theta = \frac{4}{h} = \frac{4}{\sqrt{65}}$, b. $\cos\theta = \frac{7}{h} = \frac{7}{\sqrt{65}}$.

7. Since $t = \frac{3}{4}\pi$, $\cos t = -\frac{\sqrt{2}}{2}$ and $\sin t = \frac{\sqrt{2}}{2}$.

8. Since $t = \frac{7}{4}\pi$, $\cos t = \cos\left(2\pi - \frac{\pi}{4}\right) = \cos\left(-\frac{\pi}{4}\right) = \cos\frac{\pi}{4} = \frac{\sqrt{2}}{2}$

and $\sin t = \sin\left(2\pi - \frac{\pi}{4}\right) = \sin\left(-\frac{\pi}{4}\right) = -\sin\frac{\pi}{4} = -\frac{\sqrt{2}}{2}$.

9. Since $t = -\frac{5}{3}\pi$, $\cos t = \cos\left(\frac{\pi}{3} - 2\pi\right) = \cos\frac{\pi}{3} = \frac{1}{2}$

and $\sin t = \sin\left(\frac{\pi}{3} - 2\pi\right) = \sin\frac{\pi}{3} = \frac{\sqrt{3}}{2}$.

10. Since $t = 2\pi + \frac{\pi}{3} = \frac{7}{3}\pi$, $\cos t = \cos\left(2\pi + \frac{\pi}{3}\right) = \cos\frac{\pi}{3} = \frac{1}{2}$

and $\sin t = \sin\left(2\pi + \frac{\pi}{3}\right) = \sin\frac{\pi}{3} = \frac{\sqrt{3}}{2}$.

11. $\sin\left(-\frac{\pi}{4}\right) = -\sin\left(\frac{\pi}{4}\right) = -\frac{\sqrt{2}}{2}$.

12. $5\pi = (\pi + 4\pi) = (\theta + 2n\pi)$ where $\theta = \pi$, $n = 2$.

$cos(5\pi) = \cos(\pi) = -1.$

13. $\frac{9\pi}{2} = \left(\frac{\pi}{2} + 4\pi\right) = (\theta + 2n\pi)$ where $\theta = \frac{\pi}{2}$, $n = 2$.

$\cos\left(\frac{9\pi}{2}\right) = \cos\left(\frac{\pi}{2}\right) = 0.$

14. $\sin\left(\frac{11}{3}\pi\right) = \sin\left(2\pi + \frac{5}{3}\pi\right) = \sin\frac{5}{3}\pi = \sin\left(2\pi - \frac{\pi}{3}\right)$

$= \sin\left(-\frac{\pi}{3}\right) = -\sin\frac{\pi}{3} = -\frac{\sqrt{3}}{2}.$

15. $\sin\left(-\frac{5}{3}\pi\right) = \sin\left(-\frac{5}{3}\pi + 2\pi\right) = \sin\frac{\pi}{3} = \frac{\sqrt{3}}{2}$.

16. $-\frac{5\pi}{2} = \left(\frac{3\pi}{2} - 4\pi\right) = (\theta + 2n\pi)$ where $\theta = \frac{3\pi}{2}$, $n = -2$.

$\cos\left(-\frac{5\pi}{2}\right) = \cos\left(\frac{3\pi}{2}\right) = 0.$

17. $\cos\left(\frac{11}{4}\pi\right) = \cos\left(2\pi + \frac{3}{4}\pi\right) = \cos\frac{3}{4}\pi = -\frac{\sqrt{2}}{2}.$

18. $\sin\left(-\frac{9}{4}\pi\right) = \sin\left(-2\pi - \frac{1}{4}\pi\right) = \sin\left(-\frac{\pi}{4}\right) = -\sin\frac{\pi}{4} = -\frac{\sqrt{2}}{2}.$

19. $f(t)$ has amplitude 2 since $|f(t)| \leq 2$ for all t and $\left|f\left(\frac{\pi}{2}\right)\right| = \left|f\left(\frac{3\pi}{2}\right)\right| = 2$, and $f(t)$ has period 2π.

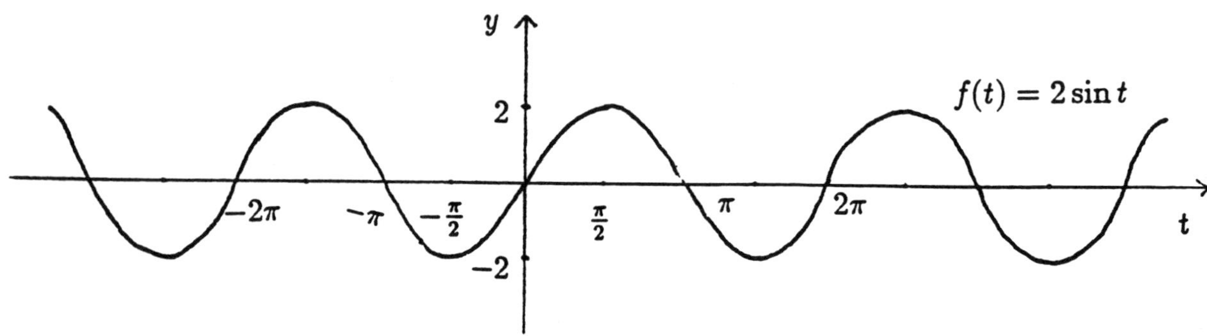

20. $f(t)$ has amplitude 1 and period $T = 2\pi$.

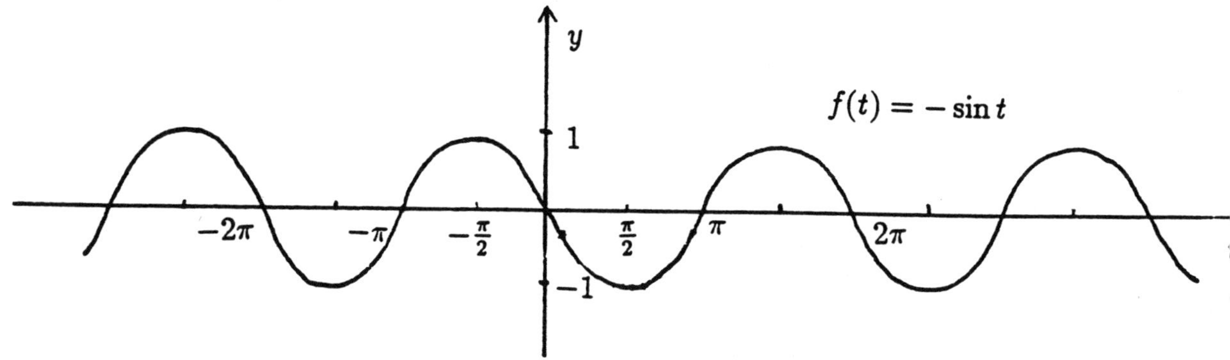

21. $f(t) = \sin 2t$ has amplitude 1 and period $T = \frac{2\pi}{2} = \pi$.

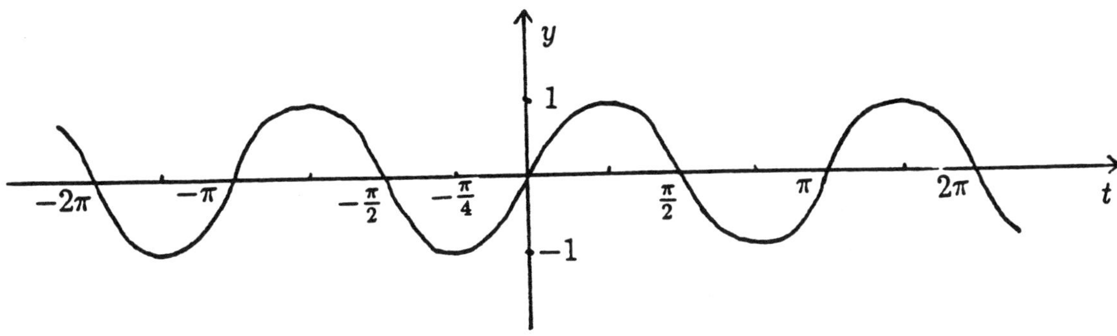

22. $f(t) = 3 \cdot \sin 2t$ has amplitude 3 and period $T = \frac{2\pi}{2} = \pi$.

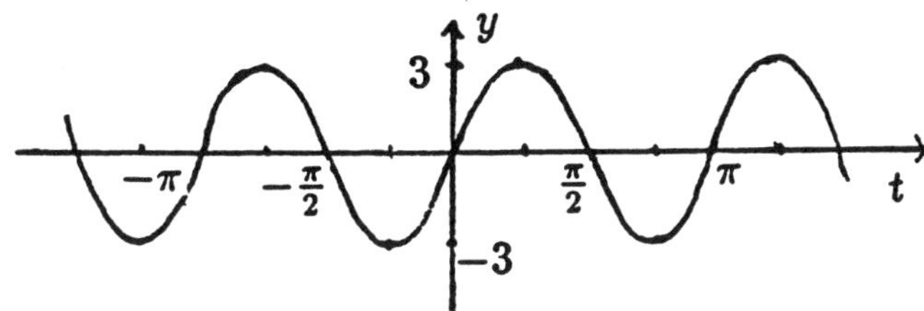

23. $f(t) = 4\cos(-t) = 4 \cdot \cos t$ has amplitude 4 and period $T = 2\pi$.

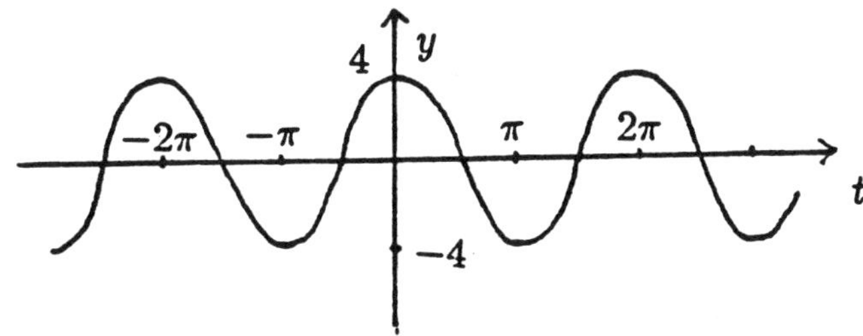

24. $f(t) = -\cos 2t$ has amplitude 1 and period $T = \frac{2\pi}{2} = \pi$.

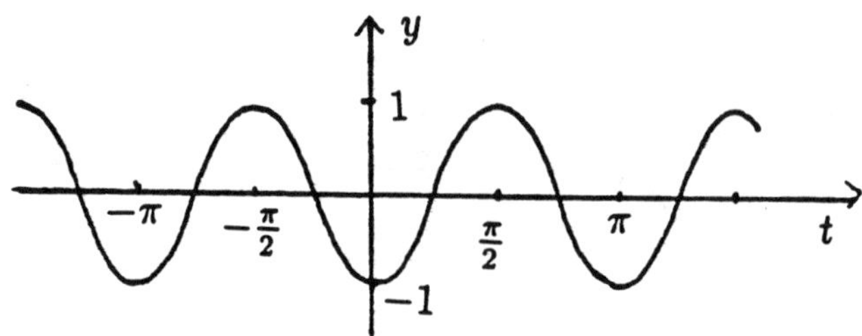

25. $\frac{y}{10} = \sin\frac{\pi}{4} = \frac{\sqrt{2}}{2} \implies y = \frac{\sqrt{2}}{2} \cdot 10 = 5\sqrt{2}$.
$\frac{x}{10} = \cos\frac{\pi}{4} = \frac{\sqrt{2}}{2} \implies x = \frac{\sqrt{2}}{2} \cdot 10 = 5\sqrt{2}$.

26. $\frac{y}{5} = \sin\frac{\pi}{3} = \frac{\sqrt{3}}{2} \implies y = \frac{5}{2}\sqrt{3}$.
$\frac{x}{5} = \cos\frac{\pi}{3} = \frac{1}{2} \implies x = \frac{5}{2}$.

27. $\frac{y}{5} = \sin\frac{\pi}{6} = \frac{1}{2} \implies y = \frac{5}{2}$.
$\frac{x}{5} = \cos\frac{\pi}{6} = \frac{\sqrt{3}}{2} \implies x = \frac{5}{2}\sqrt{3}$.

28. $\frac{y}{10} = \sin\frac{\pi}{12} = 0.2588 \implies y = 2.588$.
$\frac{x}{10} = \cos\frac{\pi}{12} = 0.9659 \implies x = 9.659$.

29. $\sin 15° = \sin\left(15 \cdot \frac{\pi}{180}\right) = \sin\left(\frac{\pi}{12}\right) = 0.2588$.

30. $\cos(68°) = \cos\left(\frac{68°}{180°} \cdot \pi\right) = \cos\left(\frac{17\pi}{45}\right) = .3746$.

31. $\cos\left(\frac{\pi}{7}\right) = .9010$.

32. $\sin 140° = \sin\left(140 \cdot \frac{\pi}{180}\right) = \sin\left(\frac{7}{9}\pi\right) = 0.6428$.

33. $\sin\left(\frac{2\pi}{9}\right) = .6428$.

34. $\cos\left(\frac{13\pi}{5}\right) = -.3090$.

35. $x = \sin\frac{\pi}{12} = 0.2588$ miles.

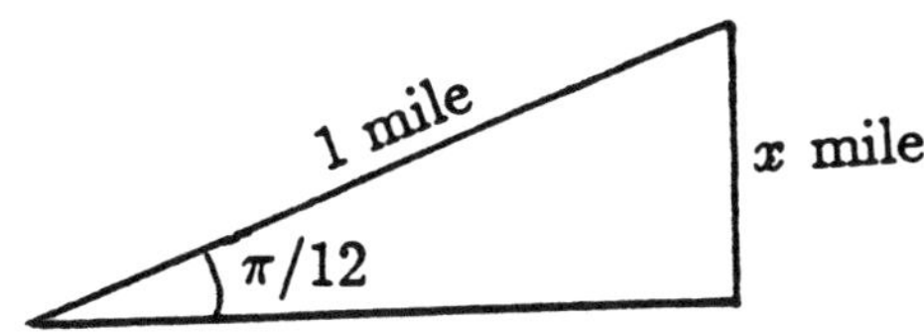

36.a. $T(6) = 60 - 15\cos\left(\frac{\pi}{12}\cdot 6\right) = 60 - 15\cos\left(\frac{\pi}{2}\right) = 60°$.

b. $T(12) = 60 - 15\cos\left(\frac{\pi}{12}\cdot 12\right) = 60 - 15\cdot\cos\pi = 75°$.

c. $T(16) = 60 - 15\cos\left(\frac{\pi}{12}\cdot 16\right) = 60 - 15\cdot\cos\frac{4}{3}\pi = 60 - 15\cdot\left(-\frac{1}{2}\right) = 67.5°$.

37.a. $D\left(\frac{5}{2}\right) = 12 + 3\sin\left(\frac{\pi}{6}\cdot\frac{5}{2} - \frac{5}{12}\pi\right)$

$= 12 + 3\sin 0 = 12$ hours.

b. $D\left(\frac{11}{2}\right) = 12 + 3\cdot\sin\left(\frac{\pi}{6}\cdot\frac{11}{2} - \frac{5}{12}\pi\right)$

$= 12 + 3\sin\left(\frac{\pi}{2}\right) = 15$ hours.

c. $D\left(\frac{23}{2}\right) = 12 + 3\sin\left(\frac{\pi}{6}\cdot\frac{23}{2} - \frac{5}{12}\pi\right)$

$= 12 + 3\cdot\sin\left(\frac{18}{12}\pi\right) = 12 + 3\sin\left(\frac{3}{2}\pi\right) = 9$ hours.

38.a. $h = 15 + 15\sin 0 = 15$ meters.

b. $h = 15 + 15\sin\frac{2}{3}\pi = 15 + 15\cdot\frac{\sqrt{3}}{2} = 15 + 7.5\sqrt{3}$ meters.

c. $h = 15 + 15\sin\frac{3}{2}\pi = 15 - 15 = 0$ meters.

39.a. $h(0) = 0 + \sin\left(\frac{\pi\cdot 0}{4}\right) + B = 0 + B = B$.

b. $h(4) = 4 + \sin\frac{4\pi}{4} + B = 4 + B$.

c. $h(6) = 6 + \sin\frac{6\pi}{4} + B = 5 + B$.

40.a. $P(0) = 10\cdot 0 + 4\sin\left(\frac{\pi\cdot 0}{2}\right) + 500 = \$500$.

b. $P(12) = 10(12) + 4\sin\frac{12\pi}{2} + 500 = \$620$.

c. $P(15) = 10(15) + 4\sin\frac{15\pi}{2} + 500 = \$646$.

41. $A = \frac{1}{2}xy$. $\frac{y}{h} = \sin\theta \implies y = h\sin\theta$. Therefore $A = \frac{1}{2}xh\sin\theta$.

42.a. maximum $P(t) = 105 + 12 \cdot (1) = 117$.

b. maximum $P(t) = 105 + 12 \cdot (-1) = 93$.

43. The period T is $\frac{2\pi}{5}$ seconds. So, the frequency is $\frac{1}{T} = \frac{5}{2\pi}$ beats per second. Hence the heart rate in beats per minute is $\frac{5}{2\pi} \cdot 60 \approx 48$.

44.a. minimum size F_0 of the fox population $= 200 + 60 \cdot (-1) = 140$.

b. minimum size R_0 of the rabbit population $= 100 + 80 \cdot (-1) = 20$.

c. ????

45.a. To find when high tide occurs, solve for t such that

$0 \le t \le 24$ and $\sin \frac{\pi(t-4)}{6} = 1$.

Case 1: $\frac{\pi(t-4)}{6} = \frac{\pi}{2} \implies \frac{t-4}{6} = \frac{1}{2} \implies t = 7$.

Case 2: $\frac{\pi(t-4)}{6} = \frac{5\pi}{2} \implies \frac{t-4}{6} = \frac{5}{2} \implies t = 19$.

The high tide occurs at 7am and 7pm.

b. To find when low tide occurs, solve for t such that

$0 \le t \le 24$ and $\sin \frac{\pi(t-4)}{6} = -1$.

Case 1: $\frac{\pi(t-4)}{6} = \frac{3\pi}{2} \implies \frac{t-4}{6} = \frac{3}{2} \implies t = 13$.

Case 2: $\frac{\pi(t-4)}{6} = -\frac{\pi}{2} \implies \frac{t-4}{6} = -\frac{1}{2} \implies t = 1$.

The low tide occurs at 1am and 1pm.

c.

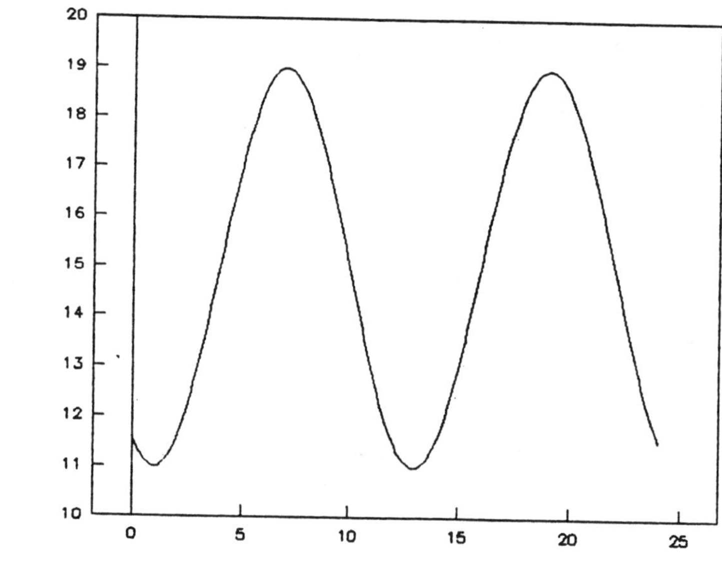

Solutions to Exercise Set 6.3 (7.3)

1. $f'(x) = 3\cos 3x$.
2. $f'(x) = -2\sin 2x$.
3. $y = x\sin(x+\pi)$.

   $y' = 1\cdot\sin(x+\pi) + x\cdot\cos(x+\pi) = \sin(x+\pi) + x\cdot\cos(x+\pi)$.
4. $f'(x) = \cos\left(\frac{\pi}{2} - x\right)\cdot(-1) = -\cos\left(\frac{\pi}{2} - x\right)$.
5. $f'(x) = 3x^2(\cos 2x) + x^3\cdot(-\sin 2x)(2) = 3x^2\cos 2x - 2x^3\sin 2x$.
6. $f(\theta) = e^{\sin^2\theta}$.

   $f'(\theta) = e^{\sin^2\theta}\cdot 2\sin\theta\cdot\cos\theta$.
7. $y = \sin^3 x\cos^2 x$.

   $y' = (3\sin^2 x\cos x)\cdot\cos^2 x + (\sin^3 x)\cdot(2\cos x\cdot(-\sin x))$

   $= 3\sin^2 x\cos^3 x - 2\sin^4 x\cos x$.
8. $f'(x) = \frac{1}{2}(x+\cos x)^{-1/2}\cdot(1-\sin x)$.
9. $f'(x) = \cos\sqrt{1+x^2}\{\frac{1}{2}(1+x^2)^{-1/2}\}\cdot\{2x\} = x(1+x^2)^{-1/2}\cdot\cos\sqrt{1+x^2}$.
10. $f(x) = \frac{\sin x}{x^2+1}$.

    $f'(x) = \frac{(x^2+1)\cdot\cos x - \sin x\cdot(2x)}{(x^2+1)^2}$.
11. $f(t) = \cos^2(\ln t)$.

    $f'(t) = 2\cos(\ln t)\cdot(-\sin(\ln t))\cdot\frac{1}{t} = -\frac{2\cos(\ln t)\cdot\sin(\ln t)}{t}$.
12. $y' = \frac{-\cos x(1+\cos x) - (1-\sin x)(-\sin x)}{(1+\cos x)^2} = \frac{-\cos x - \cos^2 x + \sin x - \sin^2 x}{(1+\cos x)^2}$.
13. $f(x) = 1$ for all $x \implies f'(x) = 0$ for all x.
14. $f(x) = x^3\cos\sqrt{1-x^2}$.

    $f'(x) = (3x^2)\cdot\cos\sqrt{1-x^2} + (x^3)\cdot\left(-\sin\sqrt{1-x^2}\right)\cdot\frac{1}{2}(1-x^2)^{-1/2}\cdot(-2x)$

    $= 3x^2\cdot\cos\sqrt{1-x^2} + \frac{x^4\sin\sqrt{1-x^2}}{\sqrt{1-x^2}}$.
15. $f(t) = \ln(t^4 + \sin t)$.

    $f'(t) = \frac{1}{t^4+\sin t}\cdot(4t^3 + \cos t) = \frac{4t^3+\cos t}{t^4+\sin t}$.
16. $y' = -\sin\theta(e^{\cos\theta}) + \cos\theta\cdot(e^{\cos\theta})\cdot(-\sin\theta) = -\sin\theta e^{\cos\theta}(1+\cos\theta)$.
17. $y' = \frac{\cos x(3+x^2) - (1+\sin x)(2x)}{(3+x^2)^2}$.

18. $f(x) = \sin^2\left(x^3 + \sqrt{x}\right)$.

$f'(x) = 2\sin\left(x^3 + \sqrt{x}\right) \cdot \cos\left(x^3 + \sqrt{x}\right) \cdot \left(3x^2 + \frac{1}{2\sqrt{x}}\right)$

$= \left(6x^2 + \frac{1}{\sqrt{x}}\right) \cdot \sin\left(x^3 + \sqrt{x}\right) \cdot \cos\left(x^3 + \sqrt{x}\right)$.

19. $f(x) = \cos x \cdot \sin \frac{1}{x}$.

$f'(x) = (-\sin x) \cdot \sin \frac{1}{x} + \cos x \cdot \left(\cos \frac{1}{x}\right) \cdot \left(-\frac{1}{x^2}\right)$

$= -\sin x \cdot \sin \frac{1}{x} - \frac{1}{x^2} \cdot \cos x \cdot \cos \frac{1}{x}$.

20. $y' = 3\sin^2 x(\cos x) \cdot \cos^5 x + \sin^3 x \cdot 5\cos^4 x \cdot (-\sin x)$.

$= 3\sin^2 x \cos^6 x - 5\sin^4 x \cos^4 x$.

21. Slope $m = y' = 2\cos 2x \implies y' = 2\cos \frac{2\pi}{6} \implies y' = 1$.

$y - \frac{\sqrt{3}}{2} = 1 \cdot \left(x - \frac{\pi}{6}\right) \implies y = x - \frac{\pi}{6} + \frac{\sqrt{3}}{2}$.

22. Slope $m = y' = e^{\cos t}(-\sin t) \implies y' = e^{\cos \frac{\pi}{2}}\left(-\sin \frac{\pi}{2}\right) = -1$.

$y = e^{\cos t} = e^{\cos \frac{\pi}{2}} = 1$.

$y - 1 = -1\left(x - \frac{\pi}{2}\right) \implies y = -x + \frac{\pi}{2} + 1$.

23. $s'(t) = \frac{\cos 2t(2)\cdot(3+\cos^2 t) - \sin 2t\cdot(2\cos t)\cdot(-\sin t)}{(3+\cos^2 t)^2}$

$s'\left(\frac{\pi}{4}\right) = \frac{2\cos\frac{2\pi}{4}\left(3+\cos^2\frac{\pi}{4}\right) - \sin\frac{2\pi}{4}\cdot\left(2\cos\frac{\pi}{4}\right)\left(-\sin\frac{\pi}{4}\right)}{\left(3+\cos^2\frac{\pi}{4}\right)^2}$

$= \frac{2\cdot 0\cdot\left(3+\frac{1}{2}\right) - 1\cdot\left(2\cdot\frac{1}{2}\sqrt{2}\right)\left(-\frac{1}{2}\sqrt{2}\right)}{\left(3+\frac{1}{2}\right)^2} = \frac{1}{\frac{49}{4}} = \frac{4}{49}$ m/s.

24. (a) $s'(t) = 18\cos 3t$.

$s'(t) = 0$ when $3t = \frac{\pi}{2}, \frac{3\pi}{2}, \frac{5\pi}{2}, \frac{7\pi}{2}, \cdots$

$\implies t = \frac{\pi}{6}, \frac{\pi}{2}, \frac{5\pi}{6}, \frac{7\pi}{6}, \frac{3\pi}{2}, \cdots$.

(b) $s'(t)$ is maximum when $\cos 3t = 1 \implies 3t = 0, 2\pi, 4\pi, \ldots$

$\implies t = 0, \frac{2\pi}{3}, \frac{4\pi}{3}, \cdots$. This maximum velocity is 6 m/s.

25. Because $\sin^2 x + \cos^2 x = 1$ for all x, $f(x) = g(x)$ for all x.

26. $f'(x) = \frac{1}{2}x^{-1/2} \cdot (\sin x) + \sqrt{x} \cdot (\cos x)$.

$f''(x) = -\frac{1}{4}x^{-3/2} \cdot (\sin x) + \frac{1}{2}x^{-1/2} \cdot (\cos x) + \frac{1}{2}x^{-1/2} \cdot (\cos x) + \sqrt{x} \cdot (-\sin x)$

$= -\frac{1}{4}x^{-3/2}(\sin x) + x^{-1/2}(\cos x) + \sqrt{x}(-\sin x)$.

27. $f'(t) = \cos(\ln t) \cdot \left(\frac{1}{t}\right)$.

$f''(t) = -\sin(\ln t) \cdot \left(\frac{1}{t}\right) \cdot \left(\frac{1}{t}\right) + \cos(\ln t) \cdot (-t^{-2}) = \frac{-\sin(\ln t) - \cos(\ln t)}{t^2}$

28. $x = \sin(x+y)$.

$1 = \cos(x+y)\cdot\left(1+\frac{dy}{dx}\right)$.

$\cos(x+y)\cdot\left(\frac{dy}{dx}\right) = 1-\cos(x+y)$.

$\frac{dy}{dx} = \frac{1-\cos(x+y)}{\cos(x+y)}$.

29. $\cos(xy) = x+y$.

$-\sin(xy)\cdot\left(1\cdot y + x\cdot\frac{dy}{dx}\right) = 1+\frac{dy}{dx}$.

$\frac{dy}{dx}\cdot(1+x\cdot\sin(xy)) = -y\cdot\sin(xy)-1$.

$\frac{dy}{dx} = -\frac{1+y\sin(xy)}{1+x\sin(xy)}$.

30. $\cos(x+y)\left(1+\frac{dy}{dx}\right)+(-\sin y)\left(\frac{dy}{dx}\right) = 0$.

$\cos(x+y)+\frac{dy}{dx}(\cos(x+y)-\sin y) = 0$,

$\frac{dy}{dx} = \frac{-\cos(x+y)}{\cos(x+y)-\sin y}$.

31. $\cos y\cdot\left(\frac{dy}{dx}\right) = 1\cdot\cos y + x\cdot(-\sin y)\left(\frac{dy}{dx}\right)$,

$\frac{dy}{dx}(\cos y + x\sin y) = \cos y$. $\qquad \frac{dy}{dx} = \frac{\cos y}{\cos y + x\sin y}$.

32. $f'(x) = 2\cos 2x$.

$f'(x) = 0$ when $\cos 2x = 0 \implies x = \frac{\pi}{4}$.

$f(0) = \sin 0 = 0, \quad f\left(\frac{\pi}{4}\right) = \sin\frac{2\pi}{4} = 1, \quad f\left(\frac{\pi}{2}\right) = \sin\frac{2\pi}{2} = 0$.

The maximum value is $f\left(\frac{\pi}{4}\right) = 1$.

The minimum value is $f(0) = f\left(\frac{\pi}{2}\right) = 0$.

33. $f'(x) = \cos x - \sin x$. $\quad f'(x) = 0$ when $x = \frac{\pi}{4}$ and $\frac{5\pi}{4}$.

$f(0) = 1, \quad f\left(\frac{\pi}{4}\right) = \sqrt{2}, \quad f\left(\frac{5\pi}{4}\right) = -\sqrt{2}, \quad f(2\pi) = 1$.

The maximum value is $f\left(\frac{\pi}{4}\right) = \sqrt{2}$.

The minimum value is $f\left(\frac{5\pi}{4}\right) = -\sqrt{2}$.

34. $f'(x) = \cos x + \sin x$. $\quad f'(x) = 0$ when $x = -\frac{\pi}{4}$.

$f\left(-\frac{\pi}{2}\right) = -1, \quad f\left(\frac{\pi}{2}\right) = 1, \quad f\left(-\frac{\pi}{4}\right) = -\sqrt{2}$.

The maximum value is $f\left(\frac{\pi}{2}\right) = 1$.

The minimum value is $f\left(-\frac{\pi}{4}\right) = -\sqrt{2}$.

35. $f'(x) = 1 - \cos x$.

$f'(x) = 0$ when $\cos x = 1 \implies x = 0$ and 2π.

$f\left(\frac{\pi}{2}\right) = \frac{\pi}{2} - 1, \qquad f\left(\frac{3\pi}{2}\right) = \frac{3\pi}{2} + 1.$

The maximum value is $f\left(\frac{3\pi}{2}\right) = \frac{3\pi}{2} + 1$.

The minimum value is $f\left(\frac{\pi}{2}\right) = \frac{\pi}{2} - 1$.

36. $f'(x) = 3\cos 3x$ for $0 \le x \le \frac{2\pi}{3}$. $f'(x) = 0$ for $x = \frac{\pi}{6}$ and $\frac{\pi}{2}$.

Intervals	Sign of $f'(x)$	Conclusion
$0 < x < \frac{\pi}{6}$	$f'(x) > 0$	increasing
$\frac{\pi}{6} < x < \frac{\pi}{2}$	$f'(x) < 0$	decreasing
$\frac{\pi}{2} < x < \frac{2\pi}{3}$	$f'(x) > 0$	increasing

$f\left(\frac{\pi}{6}\right)$ is the relative maximum. $f\left(\frac{\pi}{2}\right)$ is the relative minimum.

$f''(x) = -9\sin 3x$.

$f''(x) = 0$ for $x = 0$, $\frac{\pi}{3}$, and $\frac{2\pi}{3}$.

Intervals	Sign of $f''(x)$	Conclusion
$0 < x < \frac{\pi}{3}$	$f''(x) < 0$	concave down
$\frac{\pi}{3} < x < \frac{2\pi}{3}$	$f''(x) > 0$	concave up

The inflection point is when $x = \frac{\pi}{3}$.

x	0	$\frac{\pi}{6}$	$\frac{\pi}{3}$	$\frac{\pi}{2}$	$\frac{2\pi}{3}$
y	0	1	0	-1	0

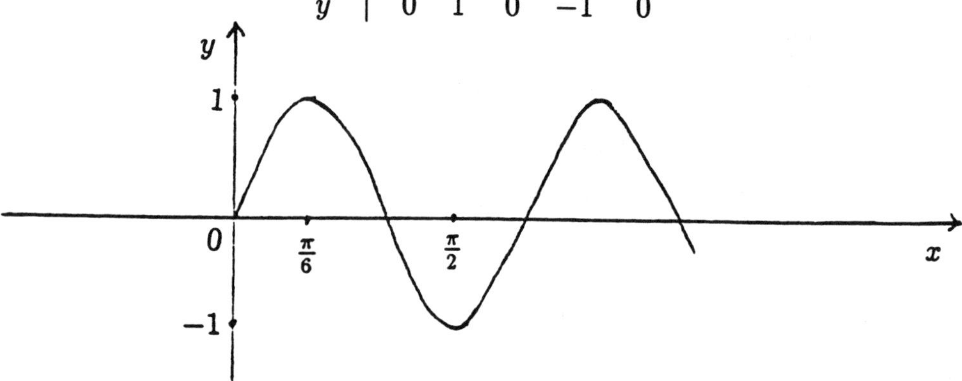

37. $f'(x) = -\sin\left(x - \frac{\pi}{2}\right)$ for $0 \le x \le 2\pi$.

$f'(x) = 0$ for $x = \frac{\pi}{2}$ and $\frac{3\pi}{2}$.

Interval	Sign of $f'(x)$	Conclusion
$0 < x < \frac{\pi}{2}$	$f'(x) > 0$	increasing
$\frac{\pi}{2} < x < \frac{3\pi}{2}$	$f'(x) < 0$	decreasing
$\frac{3\pi}{2} < x < 2\pi$	$f'(x) > 0$	increasing

$f\left(\frac{\pi}{2}\right)$ is the relative maximum. $f\left(\frac{3\pi}{2}\right)$ is the relative minimum.

$f''(x) = -\cos\left(x - \frac{\pi}{2}\right)$.

$f''(x) = 0$ for $x = \pi$ and 2π.

Intervals	Sign of $f''(x)$	Conclusion
$0 < x < \pi$	$f''(x) < 0$	concave down
$\pi < x < 2\pi$	$f''(x) > 0$	concave up

The inflection point is when $x = \pi$.

x	0	$\frac{\pi}{2}$	π	$\frac{3\pi}{2}$	2π
y	0	1	0	−1	0

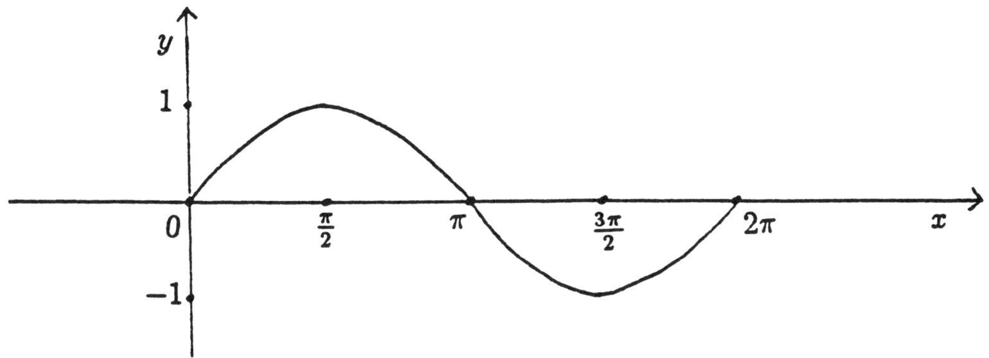

38. $f(x) = x \sin x$ in $[-\pi, \pi]$.

$f'(x) = \sin x + x \cos x$.

$f'(x) = 0$ for $x = -\frac{\sin x}{\cos x} \implies x = 0,\ x = \pm 2.028758$.

Interval I	Sign of $f'(x)$	Conclusion
$-\pi < x < -2.028758$	$+$	increasing
$-2.028758 < x < 0$	$-$	decreasing
$0 < x < 2.028758$	$+$	increasing
$2.028758 < x < \pi$	$-$	decreasing

$f(-2.028758)$ and $f(2.028758)$ are the relative maximum.

$f(0)$ is the relative minimum.

$f''(x) = \cos x + \cos x - x \sin x = 2\cos x - x \sin x$.

$f''(x) = 0$ for $x = \frac{2\cos x}{\sin x} \implies x = \pm 1.076874$.

Interval I	Sign of $f''(x)$	Conclusion
$-\pi < x < -1.076874$	$-$	concave down
$-1.076874 < x < 1.076874$	$+$	concave up
$1.076874 < x < \pi$	$-$	concave down

$f(x)$ has inflection points at $x = \pm 1.076874$.

x	0	$\pm\frac{\pi}{4}$	± 1.076874	$\pm\frac{\pi}{2}$	± 2.028758	$\pm\frac{2\pi}{3}$
$f(x)$	0	0.555	0.948	$\frac{\pi}{2}$	1.820	1.814

x	$\pm\frac{3\pi}{4}$	$\pm\frac{5\pi}{6}$	$\pm\frac{6\pi}{7}$	$\pm\frac{14\pi}{15}$
$f(x)$	1.666	1.309	1.168	0.61

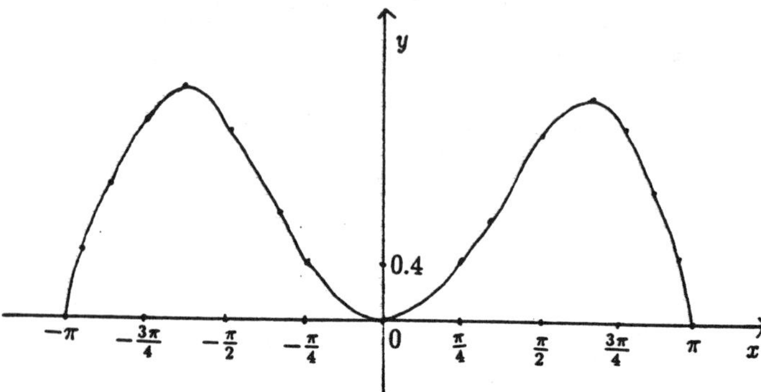

39. $f(x) = x + \sin x$ in $[0, 2\pi]$. $f'(x) = 1 + \cos x$.

$f'(x) = 0$ for $x = \pi$, and $f'(x) > 0$ for all $x \neq \pi$.

So $f(x)$ has no relative extremum.

$f''(x) = -\sin x = 0$ for $x = \pi$.

Interval	Sign of $f''(x)$	Conclusion
$0 < x < \pi$	$-$	concave down
$\pi < x < 2\pi$	$+$	concave up

The inflection point occurs when $x = \pi$

x	0	$\frac{\pi}{4}$	$\frac{\pi}{2}$	π	$\frac{3\pi}{2}$	$\frac{7\pi}{4}$	2π
$f(x)$	0	1.49	2.57	π	3.71	4.79	2π

40. $f(x) = \sin x + \cos x$ in $[0, 2\pi]$. $f'(x) = \cos x - \sin x$.

$f'(x) = 0$ for $\cos x = \sin x \implies x = \frac{\pi}{4}$ and $x = \frac{5\pi}{4}$.

Interval	Sign of $f'(x)$	Conclusion
$0 < x < \frac{\pi}{4}$	$+$	increasing
$\frac{\pi}{4} < x < \frac{5\pi}{4}$	$-$	decreasing
$\frac{5\pi}{4} < x < 2\pi$	$+$	increasing

$f\left(\frac{\pi}{4}\right)$ is the relative maximum and $f\left(\frac{5\pi}{4}\right)$ is the relative minimum.

$f''(x) = -\sin x - \cos x$.

$f''(x) = 0$ for $\sin x = -\cos x \implies x = \frac{3\pi}{4}$ and $x = \frac{7\pi}{4}$.

Interval	Sign of $f''(x)$	Conclusion
$0 < x < \frac{3\pi}{4}$	$-$	concave down
$\frac{3\pi}{4} < x < \frac{7\pi}{4}$	$+$	concave up
$\frac{7\pi}{4} < x < 2\pi$	$-$	concave down

$f(x)$ has inflection points when $x = \frac{3\pi}{4}$ and $x = \frac{7\pi}{4}$.

x	0	$\frac{\pi}{4}$	$\frac{\pi}{2}$	$\frac{3\pi}{4}$	π	$\frac{5\pi}{4}$	$\frac{3\pi}{2}$	$\frac{7\pi}{4}$	2π
$f(x)$	1	$\sqrt{2}$	1	0	-1	$-\sqrt{2}$	-1	0	1

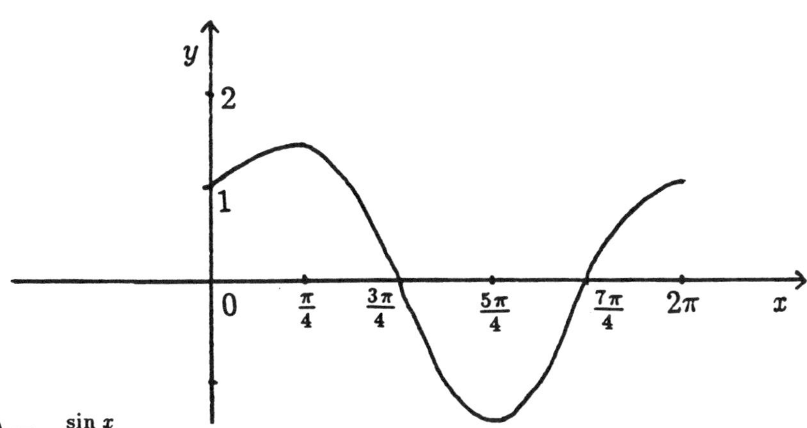

41. $f(x) = \frac{\sin x}{2+\cos x}$.

$f'(x) = \frac{\cos x(2+\cos x)-(-\sin x)\cdot\sin x}{(2+\cos x)^2} = \frac{2\cos x+\cos^2 x+\sin^2 x}{(2+\cos x)^2} = \frac{2\cos x+1}{(2+\cos x)^2}$.

$f'(x) = 0$ for $\cos x = -\frac{1}{2} \implies x = \frac{2\pi}{3}$ and $-\frac{2\pi}{3}$

Interval	Sign of $f'(x)$	Conclusion
$-\pi < x < -\frac{2\pi}{3}$	$-$	decreasing
$-\frac{2\pi}{3} < x < \frac{2\pi}{3}$	$+$	increasing
$\frac{2\pi}{3} < x < \pi$	$-$	decreasing

$f\left(-\frac{2\pi}{3}\right)$ is the relative minimum.

$f\left(\frac{2\pi}{3}\right)$ is the relative maximum.

$f''(x) = \frac{-2\sin x(2+\cos x)^2 - 2(2+\cos x)(-\sin x)\cdot(2\cos x+1)}{(2+\cos x)^4}.$

$f''(x) = \frac{-2\sin x(2+\cos x)+2\sin x(2\cos x+1)}{(2+\cos x)^3}.$

$f''(x) = \frac{-4\sin x - 2\sin x\cos x + 4\sin x\cos x + 2\sin x}{(2+\cos x)^3}.$

$f''(x) = \frac{-2\sin x(1-\cos x)}{(2+\cos x)^3}.$

$f''(x) = 0$ for $x = -\pi$, $x = 0$, and $x = \pi$.

Interval	Sign of $f''(x)$	Conclusion
$-\pi < x < 0$	+	concave up
$0 < x < \pi$	−	concave down

The inflection point is when $x = 0$.

x	$-\pi$	$-\frac{2\pi}{3}$	$-\frac{\pi}{2}$	$-\frac{\pi}{4}$	0	$\frac{\pi}{4}$	$\frac{\pi}{2}$	$\frac{2\pi}{3}$	π
$f(x)$	0	−0.577	−0.5	−0.261	0	0.261	0.5	0.577	0

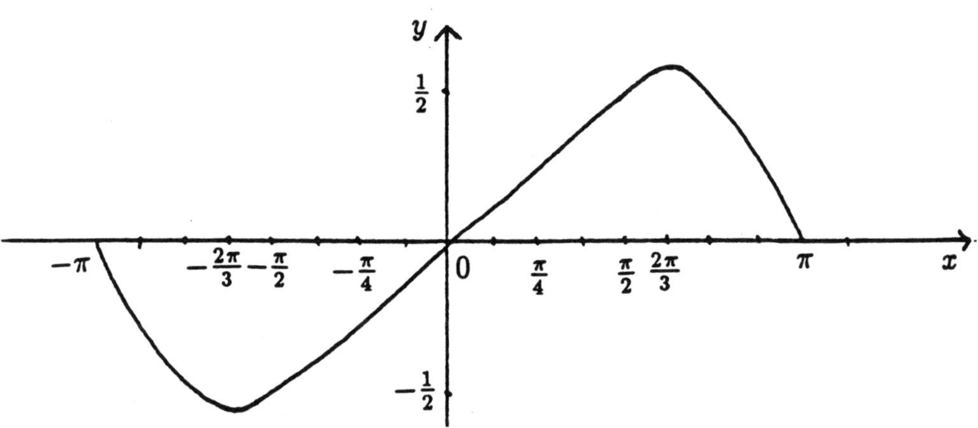

42. Let V denote the volume of the trough.

The area of each end panel $= 1h + h(\sin\theta)$. $h = \cos\theta$.

So, the area of each end panel $= \cos\theta + \cos\theta(\sin\theta)$.

$V = f(\theta) = 6\left(\cos\theta + \cos\theta(\sin\theta)\right)$.

$$\begin{aligned} f'(\theta) &= 6\left(-\sin\theta - \sin\theta(\sin\theta) + \cos\theta(\cos\theta)\right) \\ &= 6(-\sin\theta - \sin^2\theta + \cos^2\theta) \\ &= 6(-\sin\theta - \sin^2\theta + 1 - \sin^2\theta) \\ &= 6(-2\sin^2\theta - \sin\theta + 1) \end{aligned}$$

$= 6(-2\sin\theta + 1)(\sin\theta + 1).$

$f'(\theta) = 0$ for $-2\sin\theta + 1 = 0 \implies \sin\theta = \frac{1}{2} \implies \theta = \frac{\pi}{6}.$

$f(0) = 6, \quad f\left(\frac{\pi}{2}\right) = 0, f\left(\frac{\pi}{6}\right) = 6\left(\frac{\sqrt{3}}{2} + \frac{\sqrt{3}}{2} \cdot \frac{1}{2}\right).$

The maximum volume occurs when $\theta = \frac{\pi}{6}$.

43. $y = f(t) = t + \sin\left(\frac{\pi t}{4}\right) + B.$

$y' = f'(t) = 1 + \cos\left(\frac{\pi t}{4}\right)\left(\frac{\pi}{4}\right).$

The maximum value of $f'(t)$ occurs when $\cos\frac{\pi t}{4} = 1 \implies \frac{\pi t}{4} = 0$ and $2\pi \implies$ $t = 0$ and 8 months.

The minimum value of $f'(t)$ occurs when $\cos\frac{\pi t}{4} = -1 \implies \frac{\pi t}{4} = \pi$ and $3\pi \implies t = 4$ and 12 months.

44. $T(t) = 60 - 15\cos\left(\frac{\pi t}{12}\right).$

a. The maximum temperature occurs when

$\cos\left(\frac{\pi t}{12}\right) = -1 \implies \frac{\pi t}{12} = \pi \implies t = 12 =$ noon.

This maximum is $60 + 15 = 75$.

b. The minimum temperature occurs when

$\cos\left(\frac{\pi t}{12}\right) = 1 \implies \frac{\pi t}{12} = 0$ or $2\pi \implies t = 0$ or $24 =$midnight.

This minimum is $60 - 15 = 45°$.

c. $T'(t) = 15\sin\left(\frac{\pi t}{12}\right)\left(\frac{\pi}{12}\right).$

The maximum value of $T'(t)$ occurs when

$\sin\frac{\pi t}{12} = 1 \implies \frac{\pi t}{12} = \frac{\pi}{2} \implies t = 6 =$6:00 a.m..

45. $D(t) = 12 + 3\sin\left(\frac{\pi t}{6} - \frac{5\pi}{12}\right)$

(a) The days are longest when $\sin\left(\frac{\pi t}{6} - \frac{5\pi}{12}\right) = 1$

$\implies \left(\frac{\pi t}{6} - \frac{5\pi}{12}\right) = \frac{\pi}{2} \implies \frac{\pi t}{6} = \frac{11\pi}{12} \implies t = \frac{11}{2}$ months.

(b) The days are shortest when $\sin\left(\frac{\pi t}{6} - \frac{5\pi}{12}\right) = -1$

$\implies \left(\frac{\pi t}{6} - \frac{5\pi}{12}\right) = \frac{3\pi}{2} \implies \frac{\pi t}{6} = \frac{23\pi}{12} \implies t = \frac{23}{2}$ months.

46. $\lim\limits_{\theta \to 0} \frac{\cos\theta - 1}{\theta} = \lim\limits_{\theta \to 0} \left[\left(\frac{\cos\theta - 1}{\theta}\right) \left(\frac{\cos\theta + 1}{\cos\theta + 1}\right) \right]$
$= \lim\limits_{\theta \to 0} \frac{\cos^2\theta - 1}{\theta(1+\cos\theta)}$
$= \lim\limits_{\theta \to 0} \frac{-\sin^2\theta}{\theta(1+\cos\theta)}$ since $\sin^2\theta + \cos^2\theta = 1$
$= \lim\limits_{\theta \to 0} \left(\frac{\sin\theta}{\theta}\right) \left(\frac{-\sin\theta}{1+\cos\theta}\right)$
$= 1\left(\frac{0}{1+1}\right)$
since $\lim\limits_{\theta \to 0} \frac{\sin\theta}{\theta} = 1$ from formula (8), $\lim\limits_{\theta \to 0} \sin\theta = \sin 0 = 0$,
and $\lim\limits_{\theta \to 0} \cos\theta = \cos 0 = 1$.

47. $F(t) = 200 + 20\cos 2t + 20\sin 2t$.

$F'(t) = -40\sin 2t + 40\cos 2t$.

$F'(t) = 0 \implies \sin 2t = \cos 2t \implies 2t = \frac{\pi}{4}, \frac{5\pi}{4}, \frac{9\pi}{4}, \frac{13\pi}{4}, \ldots$.

$F''(t) = -80\cos 2t - 80\sin 2t$.

$F''\left(\frac{\pi}{8}\right) = -80 \cdot \cos\left(\frac{\pi}{4}\right) - 80\sin\left(\frac{\pi}{4}\right) < 0$.

$F''\left(\frac{5\pi}{8}\right) = -80 \cdot \cos\left(\frac{5\pi}{4}\right) - 80\sin\left(\frac{5\pi}{4}\right) > 0$.

Therefore, the function $F(t)$ has relative maxima at $t = \frac{\pi}{8} + n \cdot \pi$ and has relative minima at $t = \frac{5\pi}{8} + n \cdot \pi$ where $n = 0, 1, 2, 3, \ldots$.

48. $R(t) = 1000 + 200\sin t + 200\cos t$.

a. $R'(t) = 200\cos t - 200\sin t$.

$R'(t) = 0 \implies \cos t = \sin t \implies t = \frac{\pi}{4}, \frac{5\pi}{4}, \frac{9\pi}{4}, \frac{13\pi}{4}, \ldots$.

$R''(t) = -200\sin t - 200\cos t$.

$R''\left(\frac{\pi}{4}\right) = -200 \cdot \frac{\sqrt{2}}{2} - 200\frac{\sqrt{2}}{2} = -200 \cdot \sqrt{2} < 0$.

$R''\left(\frac{5\pi}{4}\right) = -200 \cdot \left(\frac{-\sqrt{2}}{2}\right) - 200 \cdot \left(\frac{-\sqrt{2}}{2}\right) = 200 \cdot \sqrt{2} > 0$.

The numbers t for which $F(t)$ has a relative maxima are $t = \frac{\pi}{4} + 2n\pi$ where $n = 0, 1, 2, 3, \ldots$.

b. The numbers t for which $F(t)$ has a relative minima are $t = \frac{5\pi}{4} + 2n\pi$ for $n = 0, 1, 2, 3, \ldots$.

c. These relative maxima are

$$1000 + 200 \cdot \frac{\sqrt{2}}{2} + 200 \cdot \frac{\sqrt{2}}{2} = 1000 + 200\sqrt{2}.$$

These relative minima are

$$1000 + 200 \cdot \left(\frac{-\sqrt{2}}{2}\right) + 200 \cdot \left(\frac{-\sqrt{2}}{2}\right) = 1000 - 200\sqrt{2}.$$

49. $F(t) = 500 - 50\cos t + 50\sin t$.

a. $F'(t) = 50\sin t + 50\cos t$.

$F'(t) = 0 \implies \sin t = -\cos t \implies t = \frac{3\pi}{4}, \frac{7\pi}{4}, \frac{11\pi}{4}, \frac{15\pi}{4}, \ldots$.

$F''(t) = 50\cos t - 50\sin t$.

$F''\left(\frac{3\pi}{4}\right) = 50 \cdot \left(\frac{-\sqrt{2}}{2}\right) - 50 \cdot \left(\frac{\sqrt{2}}{2}\right) = -50\sqrt{2} < 0.$

$F''\left(\frac{7\pi}{4}\right) = 50 \cdot \left(\frac{\sqrt{2}}{2}\right) - 50 \cdot \left(\frac{-\sqrt{2}}{2}\right) = 50\sqrt{2} > 0.$

The numbers t for which $F(t)$ has relative maxima are $t = \frac{3\pi}{4} + 2n\pi$ where $n = 0, 1, 2, 3, \ldots$.

b. The numbers t for which $F(t)$ has relative minima are $t = \frac{7\pi}{4} + 2n\pi$ where $n = 0, 1, 2, 3, \ldots$.

c. These relative maxima are

$$500 - 50 \cdot \left(\frac{-\sqrt{2}}{2}\right) + 50 \cdot \left(\frac{\sqrt{2}}{2}\right) = 500 + 50\sqrt{2}.$$

These relative minima are

$$500 - 50 \cdot \left(\frac{\sqrt{2}}{2}\right) + 50 \cdot \left(\frac{-\sqrt{2}}{2}\right) = 500 - 50\sqrt{2}.$$

50.

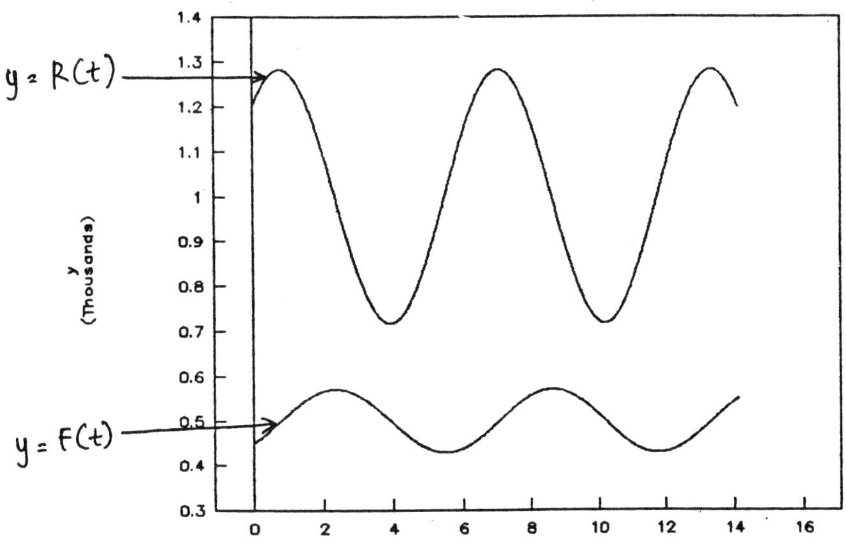

51. $f(t) = 55 + 35\sin(2\pi(t-90)/365)$.

a. $f'(t) = 35\cos(2\pi(t-90)/365)\cdot \frac{2\pi}{365}$.

$f'(t) = 0 \implies \frac{2\pi}{365}\cdot(t-90) = \frac{\pi}{2}, \frac{3\pi}{2}, \frac{5\pi}{2}, \frac{7\pi}{2}, \cdots \implies$

$t = 90 + \frac{365}{4},\ 90 + \frac{3\cdot 365}{4},\ 90 + \frac{5\cdot 365}{4},\ 90 + \frac{7\cdot 365}{4}, \cdots$

$\implies t = 181.25, \quad t = 363.75.$

$f''(t) = -35\sin(2\pi(t-90)/365)\cdot\left(\frac{2\pi}{365}\right)^2$.

$f''(181.25) = -35\cdot\sin\left(\frac{\pi}{2}\right)\cdot\left(\frac{2\pi}{365}\right)^2 < 0.$

$f''(363.75) = -35\cdot\sin\left(\frac{3\pi}{2}\right)\cdot\left(\frac{2\pi}{365}\right)^2 > 0.$

The maximum temperature occurs when $t = 181.25$.

b. The minimum temperature occurs when $t = 363.75$.

52. $f(t) = 100 + 40\cos\left(\frac{\pi t}{2}\right)$.

a. The population is the largest when $\cos\left(\frac{\pi t}{2}\right) = 1 \implies$

$\frac{\pi t}{2} = 0, 2\pi, 4\pi, \cdots \implies t = 0, 4.$

b. The population is the smallest when $\cos\left(\frac{\pi t}{2}\right) = -1 \implies$

$\frac{\pi t}{2} = \pi, 3\pi, 5\pi, \cdots \implies t = 2, 6.$

c. The length of time between the two maxima is 4 years.

Solutions to Exercise Set 6.4 (7.4)

1. $\int \cos 6x\,dx = \frac{1}{6}\sin 6x + C.$

2. $\int \sin 3x\,dx = -\frac{1}{3}\cos 3x + C.$

3. $\int_0^\pi 3\sin(\pi - x)dx = 3\cos(\pi - x)\Big]_0^\pi = 3\cos(\pi-\pi) - 3\cos(\pi - 0)$

$$= 3\cos 0 - 3\cos\pi = 3 + 3 = 6.$$

4. Let $u = x^2$, $du = 2x\ dx$, $x\ dx = \frac{1}{2}du$.

$$\int_0^{\sqrt{\pi}/2} x \cos x^2 dx = \int_0^{\sqrt{\pi}/2} \cos u \cdot \tfrac{1}{2}du = \tfrac{1}{2} \sin x^2\Big]_0^{\sqrt{\pi}/2}$$

$$= \tfrac{1}{2} \sin \tfrac{\pi}{4} - \tfrac{1}{2} \sin 0 = \tfrac{\sqrt{2}}{4}.$$

5. Let $u = 1 + x^3$, $du = 3x^2 dx$, $x^2 dx = \frac{1}{3}du$.

$\int x^2 \sin(1 + x^3)dx = \int \sin u \cdot \frac{1}{3}du = -\frac{1}{3} \cos(1 + x^3) + C.$

6. $\int_0^{\pi/4} sin^2 x cos x\, dx.$

Let $u = \sin x$.

$du = \cos x\ dx$.

$\int \sin^2 x \cos x\ dx = \int u^2\ du = \frac{1}{3}u^3 + C = \frac{1}{3} \sin^3 x + C.$

$\int_0^{\pi/4} \sin^2 x \cos x\ dx = \frac{1}{3} \sin^3 x\Big]_0^{\pi/4} = \frac{1}{3} \cdot \left(\frac{\sqrt{2}}{2}\right)^3 - \frac{1}{3} \cdot (0) = \frac{\sqrt{2}}{12}.$

7. $\int_0^{\sqrt{\pi}} t \cos(\pi - t^2)dt.$

Let $u = \pi - t^2$.

$du = -2t\, dt \implies t\, dt = -\frac{1}{2}\, du.$

$\int t \cos(\pi - t^2)dt = \int \cos u \left(-\frac{1}{2}\, du\right) = -\frac{1}{2} \sin u + C = -\frac{1}{2} \sin(\pi - t^2) + C.$

$\int_0^{\sqrt{\pi}} t \cos(\pi - t^2)dt = -\frac{1}{2} \sin(\pi - t^2)\Big]_0^{\sqrt{\pi}} = -\frac{1}{2} \cdot \sin(0) + \frac{1}{2} \cdot \sin(\pi) = 0.$

8. Let $u = \sqrt{t}$, $du = \frac{1}{2\sqrt{t}}\ dt$, $\frac{1}{\sqrt{t}}\ dt = 2\ du$.

$\int \frac{\cos \sqrt{t}}{\sqrt{t}}\ dt = \int (\cos u)2\ du = 2 \sin \sqrt{t} + C.$

9. Let $u = \cos t$, $du = -\sin t\ dt$.

$$\int_0^{\pi} \cos^3 t \sin t\ dt = \int_0^{\pi} u^3(-du) = -\tfrac{1}{4}u^4\Big]_0^{\pi} = -\tfrac{1}{4} \cos^4 t\Big]_0^{\pi}.$$

$$= -\tfrac{1}{4} \cos^4 \pi + \tfrac{1}{4} \cos^4 0 = 0.$$

10. Let $u = \cos x$, $du = -\sin x\ dx$.

$\int \frac{\sin x}{\cos^2 x}\ dx = \int \frac{1}{u^2}(-du) = \frac{1}{u} + C = \frac{1}{\cos x} + C.$

11. Let $u = 1 - \sin x$, $du = -\cos x dx$.

$\int \cos x\sqrt{1 - \sin x}\ dx = \int \sqrt{u}(-du) = -\frac{2}{3}u^{3/2} + C = -\frac{2}{3}(1 - \sin x)^{3/2} + C.$

12. $\int_0^1 \frac{\sin x}{\cos x}\,dx.$

Let $u = \cos x$.

$du = -\sin x\,dx \implies \sin x\,dx = -du.$

$\int \frac{\sin x}{\cos x}\,dx = \int \frac{1}{u}\cdot(-du) = -\ln(u) + C = -\ln(\cos x) + C.$

$\int_0^1 \frac{\sin x}{\cos x}\,dx = -\ln(\cos x)]_0^1 = -\ln(\cos 1) + \ln(1) = -\ln(\cos 1).$

13. $\int \frac{\sin(3+\ln x)}{x}\,dx.$

Let $u = 3 + \ln x$.

$du = \frac{1}{x}\,dx.$

$\int \frac{\sin(3+\ln x)}{x}\,dx = \int \sin(u)du = -\cos(u) + C = -\cos(3+\ln x) + C.$

14. Let $u = e^x$, $du = e^x\,dx$.

$\int_0^1 e^x \sin(e^x)dx = \int_{x=0}^{x=1} \sin u\,du = -\cos(e^x)]_{x=0}^{x=1} = -\cos e^1 + \cos e^0 = 1.452.$

15. $\int \frac{\cos(\ln^3 x)}{x}\,dx.$?

Suppose the author means $\int \frac{\cos(\ln(3x))}{x}\,dx.$

Let $u = \ln(3x)$.

$du = \frac{1}{3x}\cdot(3)dx = \frac{1}{x}\,dx.$

$\int \frac{\cos(\ln(3x))}{x}\,dx = \int \cos(u)du = \sin(u) + C = \sin(\ln(3x)) + C.$

16. Let $u = \sin x$, $du = \cos x\,dx$.

$\int \frac{\cos x}{\sin x}\,dx = \int \frac{1}{u}\,du = \ln|u| + C = \ln|\sin x| + C.$

17. Let $u = \cos t$, $du = -\sin t\,dt$.

$\int_0^{\pi/2} (\sin t)e^{\cos t}\,dt = \int_{t=0}^{t=\pi/2} e^u(-du) = -e^{\cos t}]_0^{\pi/2}$

$= -e^{\cos \pi/2} + e^{\cos 0} = e - 1 = 1.718.$

18. Let $u = \sin^2 t$, $du = 2\sin t\cos t\,dt$.

$\int \sin t(\cos t)e^{\sin^2 t}dt = \int e^u\left(\frac{1}{2}\,du\right) = \frac{1}{2}e^{\sin^2 t} + C.$

19. Identity: $\sin^2 x = \frac{1}{2} - \frac{1}{2}\cos 2x.$

Let $u = 2x$, $du = 2\,dx$.

$\int \sin^2 x\,dx = \int\left(\frac{1}{2} - \frac{1}{2}\cos 2x\right)dx = \int \frac{1}{2}\,dx - \int \frac{1}{4}\cos u\,du = \frac{1}{2}x - \frac{1}{4}\sin 2x + C.$

20. $\cos^2 x = \frac{1}{2} + \frac{1}{2}\cos 2x.$

Let $u = 2x, \qquad du = 2\ dx.$

$\int \cos^2 x dx = \int \frac{1}{2}\ dx + \int \frac{1}{4}\cos u\ du = \frac{1}{2}x + \frac{1}{4}\sin 2x + C.$

21. $\int\limits_0^{\sqrt{\pi}/2} x\sin^2 x^2 \cos x^2\ dx.$

Let $u = \sin x^2.$

$du = \cos x^2(2x)dx \implies x\cos x^2\ dx = \frac{1}{2}\ du.$

$\int x\sin^2 x^2 \cos x^2\ dx = \int u^2\left(\frac{1}{2}\ du\right) = \frac{1}{2}\cdot\frac{1}{3}u^3 + C = \frac{1}{6}\sin^3 x^2 + C.$

$$\int\limits_0^{\sqrt{\pi}/2} x\sin^2 x^2\cos x^2\ dx = \tfrac{1}{6}\sin^3 x^2\Big]_0^{\sqrt{\pi}/2} = \tfrac{1}{6}\cdot\sin^3\left(\tfrac{\pi}{4}\right) - \tfrac{1}{6}\cdot\sin^3(0)$$

$$= \tfrac{1}{6}\cdot\left(\tfrac{\sqrt{2}}{2}\right)^3 - 0 = \tfrac{\sqrt{2}}{24}.$$

22. Let $u = \sin\sqrt{x}, \qquad 2\ du = \frac{\cos\sqrt{x}}{\sqrt{x}}\ dx.$

$\int \frac{\sin\sqrt{x}\cos\sqrt{x}}{\sqrt{x}}\ dx = \int 2u\ du = u^2 + C = \sin^2\sqrt{x} + C.$

23. Let $u = \pi x, \qquad du = \pi\ dx.$

$$\int\limits_0^1 \sin\pi x\ dx = \int\limits_{x=0}^{x=1}\sin u\left(\tfrac{1}{\pi}\ du\right) = -\tfrac{1}{\pi}\cos u\Big]_{x=0}^{x=1} = -\tfrac{1}{\pi}\cos\pi x\Big]_0^1$$

$$= -\tfrac{1}{\pi}\cos\pi + \tfrac{1}{\pi}\cos 0 = \tfrac{2}{\pi}.$$

24. $$\int\limits_0^4(1000 + 400\cos 2\pi t)dt = \left[1000t + \tfrac{400\sin 2\pi t}{2\pi}\right]_0^4$$

$$= 4000 + \tfrac{200}{\pi}\sin 8\pi - 0 = \$4000.$$

25. Average value $= \int\limits_0^1 \cos\pi t\ dt = \frac{\sin\pi t}{\pi}\Big]_0^1 = 0.$

26. Area $= \int_0^{\pi}(x - \sin x)dx = \left[\frac{x^2}{2} + \cos x\right]_0^{\pi}$.
$= \frac{\pi^2}{2} - 1 - 1 = \frac{\pi^2}{2} - 2.$

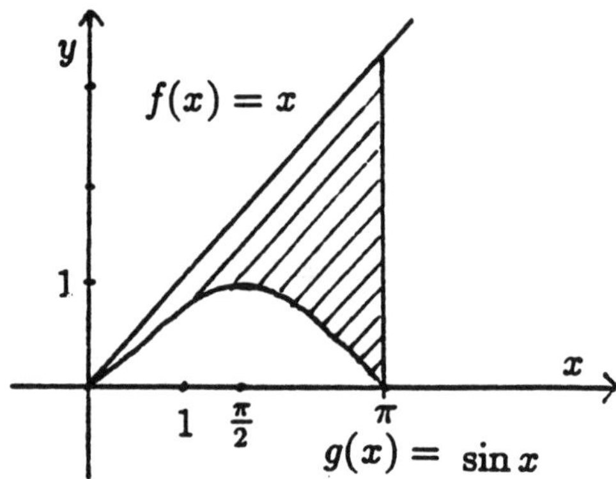

27.

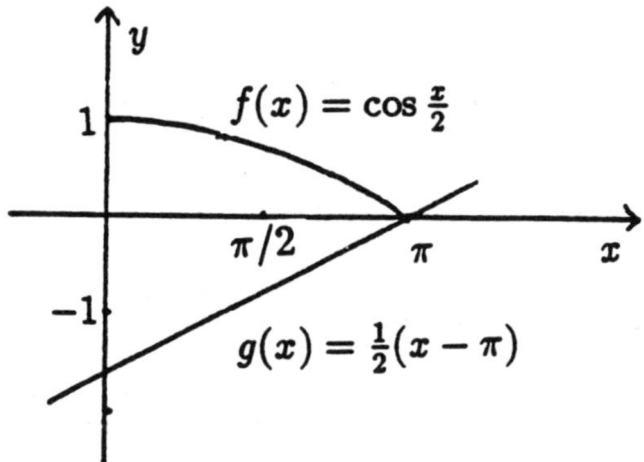

$$\begin{aligned} \text{Area} &= \int_0^{\pi}\left(\cos\tfrac{x}{2} - \tfrac{1}{2}x + \tfrac{1}{2}\pi\right)dx \\ &= \left(2\sin\tfrac{x}{2} - \tfrac{1}{4}x^2 + \tfrac{\pi}{2}x\right)\Big]_0^{\pi} \\ &= \left(2 - \tfrac{1}{4}\pi^2 + \tfrac{\pi^2}{2}\right) = 2 + \tfrac{1}{4}\pi^2. \end{aligned}$$

28. Volume $= \int_0^{\pi}\sin^2 x \cdot \pi\, dx = \pi \cdot \int_0^{\pi}\frac{1-\cos 2x}{2}\, dx$
$= \frac{\pi}{2}\cdot\left(x - \frac{\sin 2x}{2}\right)\Big]_0^{\pi} = \frac{\pi}{2}(\pi - 0) = \frac{\pi^2}{2}.$

29. Volume $= \int_0^{\pi/2} \cos^2 x \cdot \pi \, dx = \pi \int_0^{\pi/2} \frac{1+\cos 2x}{2} \, dx.$

$= \frac{\pi}{2} \cdot \left(x + \frac{\sin 2x}{2}\right)\Big]_0^{\pi/2} = \frac{\pi}{2} \cdot \left(\frac{\pi}{2} - 0\right) = \frac{\pi^2}{4}.$

30. Average value $= \int_0^6 (1000 + 500 \sin \pi t) dt/6$

$= \frac{1}{6}\left[1000t - \frac{500}{\pi}\cos \pi t\right]_0^6$

$= \frac{1}{6}\left(6000 - \frac{500}{\pi} + \frac{500}{\pi}\right) = \$1000.$

31. $P(t) = 40 + 60 \sin \pi t$ for $0 \le t \le 10$.

a. Total profit $= \int_0^{10} P(t)dt = \int_0^{10} (40 + 60 \sin \pi t)dt = \left(40t - \frac{60}{\pi}\cos \pi t\right)\Big]_0^{10}.$

$= \left(40(10) - \frac{60}{\pi}\cos(10\pi)\right) - \left(40(0) - \frac{60}{\pi}\cos(0)\right) = 400.$

b. Average profit $= \frac{\int_0^{10} P(t)dt}{10-0} = \frac{400}{10} = 40.$

32. $\int_0^5 P(t)dt = \left(40(5) - \frac{60}{\pi}\cos(5\pi)\right) - \left(40(0) - \frac{60}{\pi}\cos(0)\right) = 200 + \frac{120}{\pi}.$

Average profit for the first five years $= \frac{\int_0^5 P(t)dt}{5-0} = \frac{200+\frac{120}{\pi}}{5} = 40 + \frac{24}{\pi}.$

This is different than the average profit over 10 years because the annual profit function $P(t)$ is not a constant function.

33. $P(t) = 100 + 5t + 200 \cos(\pi(t-5)/5)$ for $0 \le t \le 10$.

a. $P(0) = 100 + 5(0) + 200 \cos\left(\frac{\pi}{5}(0-5)\right)$

$= 100 + 0 + 200\cos(-\pi) = 100 + 200(-1) = -100.$

At time $t = 0$ the company is losing money at the rate of 100,000 dollars per year.

b. Total earnings for the decade

$= \int_0^{10} P(t)dt = \int_0^{10} \left(100 + 5t + 200\cos\left(\frac{\pi}{5}(t-5)\right)\right) dt$

$= \left(100t + \frac{5}{2}t^2 + 200 \cdot \frac{5}{\pi} \cdot \sin\left(\frac{\pi}{5}(t-5)\right)\right)\Big]_0^{10}$

$= \left(100(10) + \frac{5}{2}(10)^2 + \frac{1000}{\pi}\sin(\pi)\right) - \left(0 + 0 + \frac{1000}{\pi}\sin(-\pi)\right)$

$= 1000 + 250 = 1250.$

Therefore the total earnings are predicted to be 1,250,000 dollars.

34. Suppose the author means $0 \le x \le \sqrt{\pi/c}$.

Let $u = cx^2$.

$du = 2cx\,dx \implies x\,dx = \frac{1}{2c}\,du.$

$\int x\sin(cx^2)dx = \int \sin u\left(\frac{1}{2c}\,du\right) = -\frac{1}{2c}\cos u + C = -\frac{1}{2c}\cos(cx^2) + C.$

$\text{Area} = \int_0^{\sqrt{\pi/c}} x\sin(cx^2)dx = -\frac{1}{2c}\cos(cx^2)\Big]_0^{\sqrt{\pi/c}}$

$= -\frac{1}{2c}\cos\left(c\cdot\frac{\pi}{c}\right) + \frac{1}{2c}\cdot\cos(0) = -\frac{1}{2c}\cdot(-1) + \frac{1}{2c}(1) = \frac{1}{c}.$

Therefore $\frac{1}{c} = 10 \implies c = \frac{1}{10}.$

Solutions to Exercise Set 6.5 (7.5)

1. a. $\tan\theta = \frac{1}{\sqrt{3}}.$ b. $\cot\theta = \frac{\sqrt{3}}{1} = \sqrt{3}.$

   c. $\sec\theta = \frac{2}{\sqrt{3}}.$ d. $\csc\theta = \frac{2}{1} = 2.$

2. a. $\tan\theta = \frac{1}{1} = 1.$ b. $\cot\theta = \frac{1}{1} = 1.$

   c. $\sec\theta = \frac{\sqrt{2}}{1} = \sqrt{2}.$ d. $\csc\theta = \frac{\sqrt{2}}{1} = \sqrt{2}.$

3. a. $\tan\theta = \frac{3}{4}.$ b. $\cot\theta = \frac{4}{3}.$

   c. $\sec\theta = \frac{5}{4}.$ d. $\csc\theta = \frac{5}{3}.$

4. a. $\tan\theta = \frac{\sqrt{3}}{1} = \sqrt{3}.$ b. $\cot\theta = \frac{1}{\sqrt{3}}.$

   c. $\sec\theta = \frac{2}{1} = 2$ d. $\csc\theta = \frac{2}{\sqrt{3}}.$

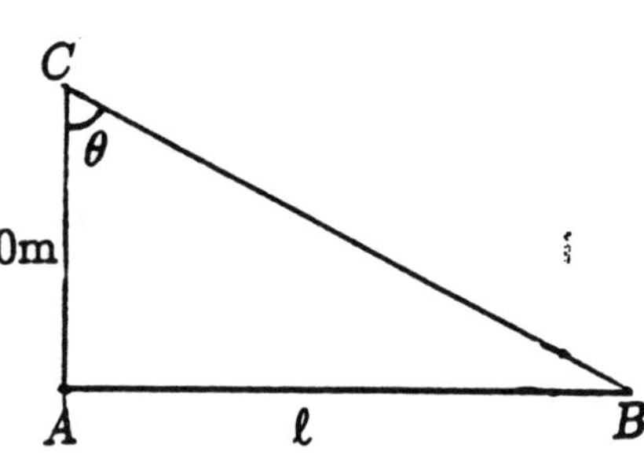

5. $\tan\theta = \frac{\ell}{50}.$

   $\ell = 50\cdot\tan\theta = 50\tan\frac{\pi}{6} = 50\cdot\frac{\sqrt{3}}{3} \approx 28.87$ m.

   The distance between A and B is 28.87 m.

6. $\ell = 50\cdot\tan\theta = 50\cdot\tan\frac{\pi}{3} = 50\cdot\sqrt{3} \approx 86.6$ m.

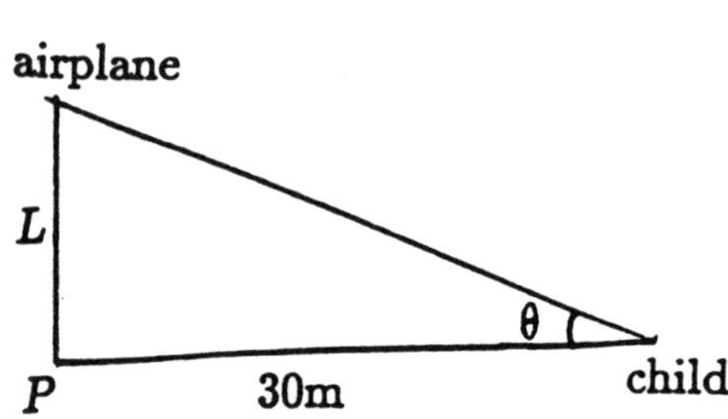

7. $\tan\theta = \frac{L}{30}.$

   $L = 30\cdot\tan\theta = 30\cdot\tan 30° = 30\cdot\frac{\sqrt{3}}{3} \approx 17.32$ m.

   The altitude of the airplane is 17.32 m.

8. $\frac{dy}{dx} = \sec^2(2\pi x) \cdot \frac{d}{dx}(2\pi x) = 2\pi \cdot \sec^2(2\pi x)$.

9. $f'(x) = (2x) \cdot (\sec x) + x^2 \cdot (\sec x \tan x)$.

10. $f(x) = \sec\left(x + \frac{\pi}{6}\right)$

$f'(x) = \sec\left(x + \frac{\pi}{6}\right) \tan\left(x + \frac{\pi}{6}\right)$.

11. $\frac{dy}{dx} = 3\cot^2(6x) \cdot \frac{d}{dx}\left(\cot(6x)\right) = 3\cot^2(6x) \cdot \left(-\csc^2(6x)\frac{d}{dx}(6)\right)$

$= -18\cot^2(6x) \cdot \csc^2(6x)$.

12. $\frac{dy}{dx} = (e^x) \cdot (\sec^3 x) + (e^x) \cdot (3\sec^2 x) \cdot (\sec x \tan x)$

$= e^x \sec^3 x \cdot (1 + 3\tan x)$.

13. $f(x) = \sec x \cdot \tan x$.

$f'(x) = \sec x \cdot \tan x \cdot \tan x + \sec x \cdot \sec^2 x = \sec x(\tan^2 x + \sec^2 x)$.

14. $f(x) = \ln \sec x$.

$f'(x) = \frac{1}{\sec x} \cdot \sec x \cdot \tan x = \tan x$.

15. $\frac{dy}{dx} = \frac{\sec x \cdot (1) - x \cdot (\sec x \tan x)}{(\sec x)^2} = \frac{1 - x\tan x}{\sec x} = \cos x - x\sin x$.

16. $f(x) = \sqrt{1 + \sec^2 x}$.

$f'(x) = \frac{1}{2}(1 + \sec^2 x)^{-1/2} \cdot 2\sec x \cdot \sec x \cdot \tan x = \sec^2 x \cdot \tan x(1 + \sec^2 x)^{-1/2}$.

17. $f(x) = \cot^2 3x$.

$f'(x) = 2\cot 3x \cdot (-\csc^2 3x)(3) = -6\cot 3x \cdot \csc^2 3x$.

18. $y = x^3 \csc(1 - x)$.

$\frac{dy}{dx} = 3x^2 \csc(1 - x) + x^3\left(-\csc(1 - x)\cot(1 - x)\right)(-1)$.

$= x^2 \csc(1 - x)\left[3 + x\cot(1 - x)\right]$.

19. $y = \tan x^3$.

$\frac{dy}{dx} = \sec^2 x^3 \cdot 3x^2 = 3x^2 \sec^2 x^3$.

20. $f(x) = \frac{\tan(x + \pi)}{x}$.

$f'(x) = \frac{x \cdot \sec^2(x + \pi) - \tan(x + \pi)}{x^2}$.

21. $f(x) = \sec\sqrt{1 + x^2}$.

$f'(x) = \sec\sqrt{1 + x^2} \cdot \tan\sqrt{1 + x^2} \cdot \frac{1}{2}(1 + x^2)^{-1/2} \cdot 2x$.

$= x(1 + x^2)^{-1/2} \sec\sqrt{1 + x^2} \cdot \tan\sqrt{1 + x^2}$.

22. $\frac{dy}{dx} = e^{\sec^2 \pi x} \cdot 2 \sec \pi x \cdot (\sec \pi x \tan \pi x) \cdot \pi$

$= 2\pi \sec^2 \pi x \tan \pi x \cdot e^{\sec^2 \pi x}.$

23. $f(x) = \cot \ln \sqrt{x}.$

$f'(x) = -\csc^2 \ln \sqrt{x} \cdot \frac{1}{\sqrt{x}} \cdot \frac{1}{2\sqrt{x}} = -\frac{\csc^2 \ln \sqrt{x}}{2x}.$

24. $\int \cot 2x \; dx = \frac{1}{2} \ln |\sin 2x| + C.$

25. $\int_0^{\pi/16} \sec^2 4x \, dx = \frac{1}{4} \tan 4x \Big]_0^{\pi/16} = \frac{1}{4} \cdot \left(\tan\left(4 \cdot \frac{\pi}{16}\right) - \tan(0)\right)$

$= \frac{1}{4} \cdot \tan \frac{\pi}{4} = \frac{1}{4} \cdot 1 = \frac{1}{4}.$

26. $\int_0^{\pi/4} 3 \sec^3 x \tan x \, dx.$

Let $u = \sec x.$

$du = \sec x \tan x \, dx.$

$\int 3 \sec^3 x \tan x \, dx = \int 3u^2 \, du = u^3 + C = \sec^3 x + C.$

$\int_0^{\pi/4} 3 \sec^3 x \tan x \, dx = \sec^3 x \Big]_0^{\pi/4} = \left(\sqrt{2}\right)^3 - (1)^3 = 2\sqrt{2} - 1.$

27. $\int x \sec^2 x^2 \; dx.$

Let $u(x) = x^2.$ $du = 2x \; dx.$

$\int x \sec^2 x^2 \; dx = \frac{1}{2} \int \sec^2 u \; du = \frac{1}{2} \tan u + C = \frac{1}{2} \tan x^2 + C.$

28. $\int \tan \pi x \; dx = \frac{1}{\pi} \ln |\sec \pi x| + C.$

29. $\int_0^{\pi/4} \sec^2(\pi + x) dx = \tan(\pi + x)]_0^{\pi/4} = \tan\left(\frac{5\pi}{4}\right) - \tan(\pi) = 1 - 0 = 1.$

30. $\int \frac{\sec^2 \sqrt{x}}{\sqrt{x}} \; dx.$

Let $u(x) = \sqrt{x}.$ $du = \frac{1}{2} \cdot \frac{1}{\sqrt{x}} \; dx.$

$\int \frac{\sec^2 \sqrt{x}}{\sqrt{x}} \; dx = 2 \int \sec^2 u \; du = 2 \tan u + C = 2 \tan \sqrt{x} + C.$

31. $\int (x - \csc^2 x) dx = \int x \; dx - \int \csc^2 x \; dx$

$= \frac{1}{2}x^2 + C_1 - (-\cot x + C_2) = \frac{1}{2}x^2 + \cot x + C.$

32. $\int_0^{\pi/\sqrt{2}} x\tan(\pi - x^2)dx.$

Let $u(x) = \pi - x^2.$ $\qquad du = -2x\ dx.$

$$\int x\tan(\pi - x^2)dx = -\tfrac{1}{2}\int \tan u\ du = -\tfrac{1}{2}\ln|\sec u| + C$$
$$= -\tfrac{1}{2}\ln|\sec(\pi - x^2)| + C.$$

$$\int_0^{\pi/\sqrt{2}} x\tan(\pi - x^2)dx = -\tfrac{1}{2}\ln|\sec(\pi - x^2)|\Big]_0^{\pi/\sqrt{2}}.$$
$$= -\tfrac{1}{2}\left[\ln\left|\sec\left(\pi - \left(\tfrac{\pi}{\sqrt{2}}\right)^2\right)\right| - \ln|\sec\pi|\right]$$
$$= -\tfrac{1}{2}\left(\ln\left|\sec\left(\pi - \tfrac{\pi^2}{2}\right)\right| - \ln 1\right)$$
$$= -\tfrac{1}{2}\ln\left|\sec\pi\left(1 - \tfrac{\pi}{2}\right)\right|.$$

33. $\int \csc^3(\pi x)\cot(\pi x)dx.$

Let $u = \csc(\pi x).$

$du = -\csc(\pi x)\cdot\cot(\pi x)\cdot\pi\ dx \implies \csc(\pi x)\cdot\cot(\pi x)dx = -\frac{1}{\pi}\,du.$

$\int \csc^3(\pi x)\cot(\pi x)dx = \int u^2\left(-\frac{1}{\pi}\,du\right) = -\frac{1}{\pi}\cdot\frac{1}{3}u^3 + C = -\frac{1}{3\pi}\csc^3(\pi x) + C.$

34. $\int \frac{\sec\sqrt{2x+1}}{\sqrt{2x+1}}\,dx.$

Let $u = \sqrt{2x+1}.$

$du = \frac{1}{2}\cdot(2x+1)^{-1/2}\cdot 2dx = \frac{1}{\sqrt{2x+1}}\,dx.$

$$\int \frac{\sec\sqrt{2x+1}}{\sqrt{2x+1}}\,dx = \int \sec u\,du = \ln|\sec u + \tan u| + C$$
$$= \ln|\sec\sqrt{2x+1} + \tan\sqrt{2x+1}| + C.$$

35. $\int\left(\csc^2\pi x - \cot\pi x\right)dx = \int \csc^2\pi x dx - \int \cot\pi x\ dx$

$= \left(-\frac{1}{\pi}\cot\pi x + C_1\right) - \left(\frac{1}{\pi}\ln|\sin\pi x| + C_2\right).$

$= -\frac{1}{\pi}(\cot\pi x + \ln|\sin\pi x|) + C.$

36. Area of $R = \int_0^{\pi/4} \tan x\ dx = \ln|\sec x|\Big]_0^{\pi/4}$

$= \ln\left|\sec\frac{\pi}{4}\right| - \ln|\sec 0| = \ln\sqrt{2} - \ln 1 = \ln\sqrt{2}.$

37. Average value $= \dfrac{\int_{-\pi/4}^{\pi/4}\sec^2 x\ dx}{\frac{\pi}{4} - \left(-\frac{\pi}{4}\right)} = \dfrac{\tan x\Big]_{-\pi/4}^{\pi/4}}{\frac{\pi}{2}} = \frac{2}{\pi}\cdot(1 - (-1)) = \frac{4}{\pi}.$

38. $\frac{d}{dx}(\tan x) = \sec^2 x > 0$ for all x on the interval $\left(-\frac{\pi}{2}, \frac{\pi}{2}\right)$.

Therefore $y = \tan x$ has neither a relative maximum nor a relative minimum on this interval.

39. $y = f(x) = \tan(\sin x)$ for $-\frac{\pi}{2} < x < \frac{\pi}{2}$.

$f'(x) = \sec^2(\sin x) \cdot \cos x$.

$$f''(x) = (2\sec(\sin x) \cdot \sec(\sin x) \cdot \tan(\sin x) \cdot \cos x) \cdot \cos x + \sec^2(\sin x) \cdot (-\sin x)$$
$$= 2\sec^2(\sin x) \cdot \tan(\sin x) \cdot \cos^2 x - \sec^2(\sin x) \cdot (\sin x).$$

Using Lotus, we find that $f''(x) = 0$ for $x \approx -.89$, $x = 0$, $x \approx .89$.

For $-\frac{\pi}{2} < x < -.89$, $f''(x) > 0$, so the graph of $y = f(x)$ is concave up.

For $-.89 < x < 0$, $f''(x) < 0$, so the graph of $y = f(x)$ is concave down.

For $0 < x < .89$, $f''(x) > 0$, so the graph of $y = f(x)$ is concave up.

For $.89 < x < \frac{\pi}{2}$, $f''(x) < 0$, so the graph of $y = f(x)$ is concave down.

40. From the sketch of the graphs of $y = \tan x$ and $y = \frac{4}{\pi}x$, there are two relevant points of intersection. They are $\left(-\frac{\pi}{4}, -1\right)$ and $\left(\frac{\pi}{4}, 1\right)$.

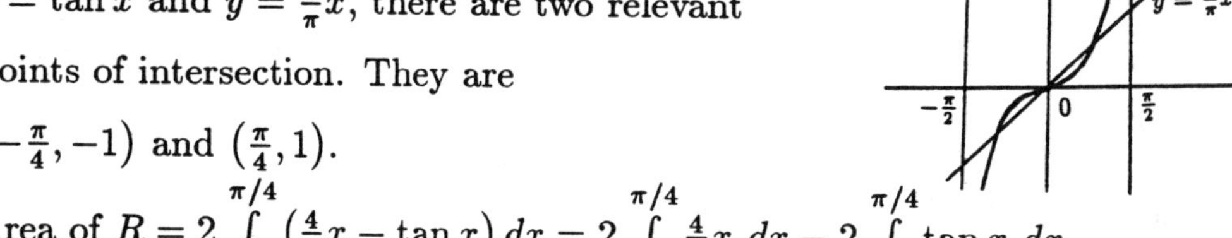

$$\text{Area of } R = 2\int_0^{\pi/4} \left(\tfrac{4}{\pi}x - \tan x\right) dx = 2\int_0^{\pi/4} \tfrac{4}{\pi}x\, dx - 2\int_0^{\pi/4} \tan x\, dx$$
$$= \left(\tfrac{4}{\pi}x^2 - 2\ln|\sec x|\right)\Big]_0^{\pi/4} = \tfrac{4}{\pi}\left(\tfrac{\pi}{4}\right)^2 - 2\ln\left(\sec\tfrac{\pi}{4}\right) + 2\ln|\sec 0|$$
$$= \tfrac{\pi}{4} - 2\ln\sqrt{2} + 2\ln 1 = \tfrac{\pi}{4} - 2\ln\sqrt{2} = \tfrac{\pi}{4} - \ln 2.$$

41. $$\text{Volume of } S = 2\int_0^{\pi/4} \pi\sec^2 x\, dx = 2\pi\tan x\Big]_0^{\pi/4}$$
$$= 2\pi\left(\tan\tfrac{\pi}{4} - \tan 0\right) = 2\pi.$$

42. $$\text{Volume of } S = \int_{\pi/4}^{3\pi/4} \pi\csc^2 x\, dx = -\pi\cot x\Big]_{\pi/4}^{3\pi/4}$$
$$= -\pi\left(\cot\tfrac{3\pi}{4} - \cot\tfrac{\pi}{4}\right) = -\pi(-1-1) = 2\pi.$$

43. $$\text{Average Value} = \frac{\int_0^{\pi/4} \sec x\, dx}{\pi/4} = \left[\ln|\sec x + \tan x|\Big]_0^{\pi/4}\right] \cdot \tfrac{4}{\pi}$$
$$= \left\{\ln\left|\sec\tfrac{\pi}{4} + \tan\tfrac{\pi}{4}\right| - \ln|\sec 0 + \tan 0|\right\} \cdot \tfrac{4}{\pi}$$
$$= \left\{\ln|\sqrt{2}+1| - \ln|1+0|\right\} \cdot \tfrac{4}{\pi} = \tfrac{4}{\pi} \cdot \ln\left(\sqrt{2}+1\right).$$

44. $x = \frac{\pi}{2} \pm n\pi$, $(n = 0, 1, 2, \dots)$, are the vertical asymptotes because

$\lim_{x \to \frac{\pi}{2} \pm n\pi} |\tan x| = \infty.$

45. (8) $\frac{d}{dx} \cot x = \frac{d}{dx}\left(\frac{\cos x}{\sin x}\right) = \frac{-\sin x \cdot \sin x - \cos x \cdot \cos x}{\sin^2 x}$

$= \frac{-(\sin^2 x + \cos^2 x)}{\sin^2 x} = -\frac{1}{\sin^2 x} = -\csc^2 x.$

(9) $\frac{d}{dx} \sec x = \frac{d}{dx}\left(\frac{1}{\cos x}\right) = \frac{-(-\sin x)}{\cos^2 x} = \frac{1}{\cos x} \cdot \frac{\sin x}{\cos x} = \sec x \cdot \tan x.$

(10) $\frac{d}{dx} \csc x = \frac{d}{dx}\left(\frac{1}{\sin x}\right) = \frac{d}{dx}(\sin x)^{-1} = -\sin^{-2} x \cdot \cos x = -\frac{1}{\sin x} \cdot \frac{\cos x}{\sin x}$

$= -\csc x \cdot \cot x.$

46. (16) $\int \cot x \, dx = \int \frac{\cos x}{\sin x} \, dx.$

Let $u(x) = \sin x.$ $\qquad du = \cos x \, dx.$

$\int \cot x \, dx = \int \frac{1}{u} \, du = \ln |u| + C = \ln |\sin x| + C.$

(17) $\int \sec x \, dx = \int \frac{\sec x \cdot (\sec x + \tan x)}{\sec x + \tan x} \, dx$

$= \int \frac{\sec x \tan x + \sec^2 x}{\sec x + \tan x} \, dx$

$= \int \frac{1}{\sec x + \tan x} d(\sec x + \tan x)$

$= \ln |\sec x + \tan x| + C.$

(18) $\int \csc x \, dx = \int \frac{\csc(\csc x - \cot x)}{\csc x - \cot x} \, dx$

$= \int \frac{-\csc x \cot x + \csc^2 x}{\csc x - \cot x} dx$

$= \int \frac{1}{\csc x - \cot x} d(\csc x - \cot x)$

$= \ln |\csc x - \cot x| + C.$

Solutions to the Review Exercises - Chapter 6 (Chapter 7)

1.

a. $\theta_r = 60 \cdot \frac{\pi}{180} = \frac{\pi}{3}.$

b. $\theta_r = 35 \cdot \frac{\pi}{180} = \frac{7}{36}\pi.$

c. $\theta_r = 120 \cdot \frac{\pi}{180} = \frac{2}{3}\pi.$

d. $\theta_r = 210 \cdot \frac{\pi}{180} = \frac{7}{6}\pi.$

e. $\theta_r = 10 \cdot \frac{\pi}{180} = \frac{1}{18}\pi.$

f. $\theta_r = 75 \cdot \frac{\pi}{180} = \frac{5}{12}\pi.$

2.

a. $\theta_d = \frac{\pi}{4} \cdot \frac{180}{\pi} = 45°.$ b. $\theta_d = \frac{3\pi}{4} \cdot \frac{180}{\pi} = 135°.$

c. $\theta_d = \frac{5\pi}{4} \cdot \frac{180}{\pi} = 225°.$ d. $\theta_d = \frac{3\pi}{2} \cdot \frac{180}{\pi} = 270°.$

e. $\theta_d = \frac{7\pi}{6} \cdot \frac{180}{\pi} = 210°.$ f. $\theta_d = \frac{17\pi}{12} \cdot \frac{180}{\pi} = 255°.$

3.

a. $\theta = -\frac{\pi}{2} + 2\pi = \frac{3}{2}\pi.$ b. $\theta = \frac{7\pi}{2} - 2\pi = \frac{3}{2}\pi.$

c. $\theta = \frac{11\pi}{4} - 2\pi = \frac{3}{4}\pi.$ d. $\theta = -\frac{5\pi}{3} + 2\pi = \frac{\pi}{3}.$

e. $\theta = -7\pi + 4 \cdot 2\pi = \pi.$ f. $\theta = 9\pi - 4 \cdot 2\pi = \pi.$

4.

a. $\sin \frac{5\pi}{6} = \frac{1}{2}.$ b. $\cos\left(-\frac{3\pi}{2}\right) = \cos\left(\frac{3\pi}{2}\right) = 0.$

c. $\sin \frac{9\pi}{2} = \sin\left(2 \cdot 2\pi + \frac{\pi}{2}\right) = \sin \frac{\pi}{2} = 1.$

d. $\cos \frac{11\pi}{3} = \cos\left(2\pi + \frac{5}{3}\pi\right) = \cos \frac{5}{3}\pi = \frac{1}{2}.$

e. $\sin\left(-\frac{5\pi}{3}\right) = -\sin\left(\frac{5\pi}{3}\right) = -\left(-\frac{\sqrt{3}}{2}\right) = \frac{\sqrt{3}}{2}.$

f. $\sin(8\pi) = \sin(4 \cdot 2\pi + 0) = \sin 0 = 0.$

g. $\cos \frac{7\pi}{2} = \cos\left(2\pi + \frac{3\pi}{2}\right) = \cos\left(\frac{3}{2}\pi\right) = 0.$

h. $\cos 3\pi = \cos(2\pi + \pi) = \cos \pi = -1.$

i. $\sin\left(-\frac{11}{6}\pi\right) = -\sin \frac{11}{6}\pi = -\left(-\frac{1}{2}\right) = \frac{1}{2}.$

5. $y' = -\cos(\pi - x).$

6. $f(x) = \pi x \cos(3x).$

$$\begin{aligned} f'(x) &= (\pi \cdot 1) \cdot (\cos(3x)) + (\pi x) \cdot (-3\sin(3x)) \\ &= \pi \cdot (\cos(3x) - 3x\sin(3x)). \end{aligned}$$

7. $f'(x) = 3 \cdot \sec^2 3x.$ 8. $y' = \cos\sqrt{x} \cdot \frac{1}{2\sqrt{x}} = \frac{\cos\sqrt{x}}{2\sqrt{x}}.$

9. $y = \cos(\pi \sin x).$

$\frac{dy}{dx} = -\sin(\pi \sin x) \cdot (\pi \cos x).$

10. $f'(x) = \frac{\cos x + \sin x \cdot x}{\cos^2 x}.$

11. $y = \ln(tan^2 x)$.

$\frac{dy}{dx} = \frac{1}{\tan^2 x} \cdot 2\tan x \cdot \sec^2 x = \frac{2\sec^2 x}{\tan x} = 2 \cdot \frac{1}{\cos^2 x} \cdot \frac{\cos x}{\sin x}$

$= \frac{2}{\cos x \cdot \sin x} = 2\sec x \cdot \csc x$.

12. $y = \sqrt{1+\cos^2 x}$.

$\frac{dy}{dx} = \frac{1}{2} \cdot (1+\cos^2 x)^{-1/2} \cdot (2\cos x \cdot (-\sin x)) = -\frac{\cos x \sin x}{\sqrt{1+\cos^2 x}}$.

13. $f'(x) = e^{\sec \pi x} \cdot \left(\frac{1}{\cos \pi x}\right)' = e^{\sec \pi x} \cdot \frac{\pi \cdot \sin \pi x}{\cos^2 \pi x} = e^{\sec \pi x} \cdot \pi \cdot (\sec \pi x)(\tan \pi x)$.

14. $f'(x) = \frac{-\sin x \cdot \tan x - \sec^2 x \cdot \cos x}{\tan^2 x} = \frac{-\sin x \cdot \tan x - \sec x}{\tan^2 x}$.

15. $y' = 3(x+\sec x)^2 \cdot \left(1 + \frac{\sin x}{\cos^2 x}\right) = 3(x+\sec x)^2 \, (1+\sec x \tan x)$.

16. $y' = 2(\cos x - \cot x) \cdot (-\sin x + \csc^2 x)$.

17. $f(x) = \frac{e^{\cos x}}{1+\sin^2 x}$.

$f'(x) = \frac{(1+\sin^2 x)\cdot(e^{\cos x}\cdot(-\sin x)) - (e^{\cos x})\cdot(2\sin x \cos x)}{(1+\sin^2 x)^2}$

$= \frac{-e^{\cos x}\cdot \sin x(1+\sin^2 x + 2\cos x)}{(1+\sin^2 x)^2}$

18. $f'(x) = (\sec^2 e^x) \cdot e^x$.

19. $y' = \frac{1}{\sqrt{4+\cos x}} \cdot \frac{-\sin x}{2\sqrt{4+\cos x}} = \frac{-sinx}{2(4+\cos x)}$.

20. $f'(x) = \frac{\sin x}{\cos^2 x} - 2 \cdot \sec^2 2x$.

21. $\int \sin 4x \; dx = -\frac{\cos 4x}{4} + C$.

22. $\int x \cos(\pi x^2) dx$.

Let $u = \pi x^2$.

$du = 2\pi x dx \implies x\,dx = \frac{1}{2\pi} du$.

$\int x\cos(\pi x^2)dx = \int \cos u \left(\frac{1}{2\pi}\right) du = \frac{1}{2\pi}\sin u + C = \frac{1}{2\pi}\sin(\pi x^2) + C$.

23. $\int \sec \pi x \cdot \tan \pi x \; dx = \int \frac{\sin \pi x}{\cos^2 \pi x} dx$.

Let $u = \cos \pi x \implies du = -\pi \sin \pi x \; dx$. Thus we have

$\int \sec \pi x \cdot \tan \pi x \; dx = \int \frac{1}{u^2} \cdot \frac{-1}{\pi} \; du = -\frac{1}{\pi} \int u^{-2} \; du$.

$= -\frac{1}{\pi}(-u^{-1}) + C = \frac{1}{\pi}(\cos \pi x)^{-1} + C = \frac{1}{\pi} \sec \pi x + C$.

24. Let $u = \tan x \implies du = \sec^2 x \; dx$. Thus $\int \sec^2 x \cdot \tan x \; dx = \int u \; du = \frac{u^2}{2} + C = \frac{1}{2}\tan^2 x + C$.

25. Let $u = \tan x \implies du = \sec^2 x \; dx$. Thus $\int \sqrt{\tan x} \cdot \sec^2 x \; dx = \int u^{1/2} \cdot du = \frac{2}{3}u^{3/2} + C = \frac{2}{3}(\tan x)^{3/2} + C$.

26. $\int \frac{dx}{\tan x \cos x} = \int \frac{1}{\sin x}\, dx = \int \csc x\, dx = \ln|\csc x - \cot x| + C.$

27. Let $u = \tan 3x \implies du = 3\sec^2 3x\ dx.$

$\int \sec^2(3x)\cdot e^{\tan 3x}\ dx = \int e^u \cdot \frac{1}{3}\ du = \frac{1}{3}e^u + C = \frac{1}{3}\cdot e^{\tan 3x} + C.$

28. Let $u = \sqrt{x} \implies du = \frac{1}{2\sqrt{x}}\ dx.$

$\int \frac{\sec^2\sqrt{x}}{\sqrt{x}}\ dx = \int \sec^2 u \cdot 2du = 2\cdot \tan u + C = 2\cdot \tan\sqrt{x} + C.$

29. $\int x\cos^{10}(x^2)\sin(x^2)dx.$

Let $u = \cos(x^2).$

$du = -\sin(x^2)\cdot(2x\,dx) \implies x\sin(x^2)dx = -\frac{1}{2}du.$

$$\begin{aligned}\int x\cos^{10}(x^2)\sin(x^2)dx &= \int u^{10}\left(-\tfrac{1}{2}\,du\right) = -\tfrac{1}{2}\cdot\tfrac{1}{11}u^{11} + C\\ &= -\tfrac{1}{22}\cos^{11}(x^2) + C.\end{aligned}$$

30. Let $u = 1 + \cos x \implies du = -\sin x\ dx.$

$\int \sin x\cdot\sqrt{1+\cos x}\ dx = \int \sqrt{u}(-du) = -\frac{2}{3}u^{3/2} + C = -\frac{2}{3}(1+\cos x)^{3/2} + C.$

31. $\int_0^{\pi/4} \cos\left(x + \frac{\pi}{2}\right) dx = \sin\left(x+\frac{\pi}{2}\right)\Big]_0^{\pi/4} = \sin\frac{3}{4}\pi - \sin\frac{\pi}{2} = \frac{\sqrt{2}}{2} - 1.$

32. $\int_{-\pi/4}^{\pi/4} \sec^2 x\ dx = \tan x\Big]_{-\pi/4}^{\pi/4} = \tan\frac{\pi}{4} - \tan\left(-\frac{\pi}{4}\right) = 1 - (-1) = 2.$

33. Let $u = \sec x \implies du = \sec x\tan x\ dx.$

$\int \sec x\tan x e^{\sec x}\ dx = \int e^u\ du = e^u + C = e^{\sec x} + C.$

Thus $\int_0^{\pi/4} \sec x\cdot\tan x\cdot e^{\sec x}\ dx = e^{\sec x}\Big]_0^{\pi/4} = e^{\sec\pi/4} - e^{\sec 0} = e^{\sqrt{2}} - e^1.$

34. Let $u = x^2 \implies du = 2x\ dx.$

$$\begin{aligned}\int x\cdot\csc^2 x^2\ dx &= \int \csc^2 u\cdot\tfrac{1}{2}\ du = \tfrac{1}{2}(-\cot u) + C\\ &= -\tfrac{1}{2}\cot x^2 + C.\end{aligned}$$

$\int_{\pi/4}^{\pi/2} x\cdot\csc^2 x^2\ dx = -\frac{1}{2}\cot x^2\Big]_{\pi/4}^{\pi/2} = -\frac{1}{2}\left(\cot\frac{\pi^2}{4} - \cot\frac{\pi^2}{16}\right).$

35. Let $u = \cos\pi x \implies du = -\pi\sin\pi x\ dx.$

$\int \sin\pi x\cdot\cos\pi x\ dx = \int u\cdot\left(-\frac{1}{\pi}\right)du = -\frac{1}{\pi}\cdot\frac{u^2}{2} + C = -\frac{1}{\pi}\cdot\frac{\cos^2\pi x}{2} + C.$

$\int_0^1 \sin\pi x\cdot\cos\pi x\ dx = -\frac{1}{\pi}\cdot\frac{\cos^2\pi x}{2}\Big]_0^1 = -\frac{1}{2\pi}(1-1) = 0.$

36. Let $u = \tan x \implies du = \sec^2 x\ dx$.

$\int \frac{\sec^2 x}{1+\tan x}\ dx = \int \frac{du}{1+u} = \ln(1+u) + C = \ln(1+\tan x) + C.$

$\int_0^{\pi/4} \frac{\sec^2 x}{1+\tan x}\ dx = \ln(1+\tan x)]_0^{\pi/4} = \ln\left(1+\tan\frac{\pi}{4}\right) - \ln 1$

$= \ln 2 - 0 = \ln 2.$

37. a.

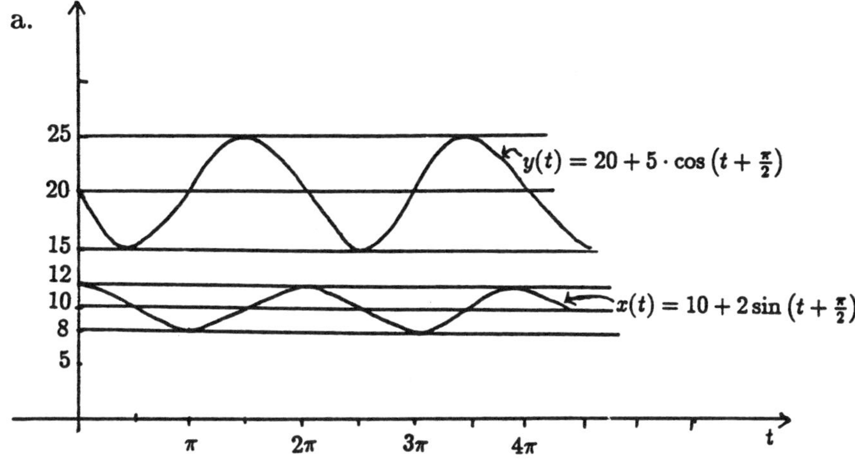

b. maximum of $x(t) = 12$.

minimum of $x(t) = 8$.

c. maximum of $y(t) = 25$.

minimum of $y(t) = 15$.

38. a. $x(0) = 10 \cdot \cos\left(-\frac{\pi}{4}\right) = 10 \cdot \frac{\sqrt{2}}{2} = 5\sqrt{2}.$

b. $v(t) = x'(t) = -10 \cdot \sin\left(\pi t - \frac{\pi}{4}\right) \cdot \pi = -10\pi \cdot \sin\left(\pi t - \frac{\pi}{4}\right).$

c. 10.

39. $y' = \sec^2 x \implies y'\left(\frac{\pi}{4}\right) = \sec^2\left(\frac{\pi}{4}\right) = 2.$

40. $y' = \sec x \cdot \tan x \implies y'\left(\frac{\pi}{3}\right) = \sec\frac{\pi}{3} \cdot \tan\frac{\pi}{3} = 2 \cdot \sqrt{3}.$

$y - 2 = y'\left(\frac{\pi}{3}\right) \cdot \left(x - \frac{\pi}{3}\right) = 2\sqrt{3}\left(x - \frac{\pi}{3}\right).$

$\therefore y = 2\sqrt{3}x - \frac{2\sqrt{3}}{3}\pi + 2.$

41. Area $= \int_0^{\sqrt{\pi/3}} x \cdot \sec x^2 \, dx.$

To find antiderivative $\int x \cdot \sec x^2 \, dx$, let $u = x^2 \implies du = 2x \, dx.$

$$\int x \cdot \sec x^2 \, dx = \int \sec u \cdot \tfrac{1}{2} \, du = \tfrac{1}{2} \ln|\sec u + \tan u| + C$$
$$= \tfrac{1}{2} \ln|\sec x^2 + \tan x^2| + C.$$

Thus
$$\text{Area} = \int_0^{\sqrt{\pi/3}} x \cdot \sec x^2 \, dx$$
$$= \tfrac{1}{2} \left(\ln|\sec x^2 + \tan x^2|\right)\Big]_0^{\sqrt{\pi/3}}$$
$$= \tfrac{1}{2} \left(\ln|\sec \tfrac{\pi}{3} + \tan \tfrac{\pi}{3}| - \ln 1\right)$$
$$= \tfrac{1}{2} \left(\ln(2 + \sqrt{3}) - \ln 1\right)$$
$$= \tfrac{1}{2} \ln(2 + \sqrt{3}).$$

42. Area $= 2 \cdot \int_0^{\pi/3} \sec x \tan x \, dx = 2 \sec x\Big]_0^{\pi/3} = 2 \cdot (2 - 1) = 2.$

43. Area $= \int_{\pi/6}^{\pi/3} \csc x \, dx = (\ln|\csc x - \cot x|)\Big]_{\pi/6}^{\pi/3}$

$= \ln|\csc \tfrac{\pi}{3} - \cot \tfrac{\pi}{3}| - \ln|\csc \tfrac{\pi}{6} - \cot \tfrac{\pi}{6}|$

$= \ln \tfrac{1}{\sqrt{3}} - \ln(2 - \sqrt{3}).$

Thus the average value of $f(x) = \csc x$ is (Area)/ $\left(\tfrac{\pi}{3} - \tfrac{\pi}{6}\right)$.

$= \tfrac{6}{\pi} \left(\ln \tfrac{1}{\sqrt{3}} - \ln(2 - \sqrt{3})\right).$

44. Volume $= \int_0^{\pi} \pi y^2 \, dx = \int_0^{\pi} \pi(1 + \sin x) dx$

$= \pi(x - \cos x)\Big]_0^{\pi} = \pi(\pi + 2).$

45. $f(x) = \sin x + \cos x.$

a. $f'(x) = \cos x - \sin x.$

$f''(x) = -\sin x - \cos x.$

$f''(x) = 0 \implies \sin x = -\cos x \implies \tan x = -1 \implies x = \tfrac{3\pi}{4}.$

For $0 < x < \tfrac{3\pi}{4}$, $f''(x) < 0$. For $\tfrac{3\pi}{4} < x < \pi$, $f''(x) > 0$.

Therefore the graph of $y = f(x)$ is concave up on the interval $\left(\tfrac{3\pi}{4}, \pi\right)$.

b. The graph of $y = f(x)$ is concave down on the interval $\left(0, \tfrac{3\pi}{4}\right)$.

46. a. $h(t) = 8 + 3\cos\left(\frac{\pi}{6} \cdot (t-2)\right)$.

$h(0) = 8 + 3\cos\left(\frac{\pi}{6} \cdot (0-2)\right) = 8 + 3 \cdot \frac{1}{2} = 9.5$ feet.

b. The times of high tide occur when $\cos\left(\frac{\pi}{6} \cdot (t-2)\right) = 1 \implies$

$\frac{\pi}{6} \cdot (t-2) = 0, 2\pi, 4\pi, \cdot \implies t = 2,\ t = 14.$

Thus the high tide occurs at 2 a.m. and 2 p.m.

c. The times of low tide occur when $\cos\left(\frac{\pi}{6} \cdot (t-2)\right) = -1 \implies$

$\frac{\pi}{6} \cdot (t-2) = \pi, 3\pi, 5\pi, \cdot \implies t = 8,\ t = 20.$

Thus the low tide occurs at 8 a.m. and 8 p.m.

47. $P(t) = 300 + 120\sin\left(\frac{\pi}{5} \cdot (2+t)\right)$.

a. The population size oscillates because of the term $120\sin\left(\frac{\pi}{5} \cdot (2+t)\right)$.

b. The population maximum occurs when $\sin\left(\frac{\pi}{5} \cdot (2+t)\right) = 1 \implies$

$\frac{\pi}{5} \cdot (2+t) = \frac{\pi}{2}, \frac{5\pi}{2}, \frac{9\pi}{2}, \cdots \implies t = \frac{1}{2},\ t = \frac{21}{2},\ t = \frac{41}{2}, \ldots$.

Therefore the time between population maxima is 10 years.

c. The minimum population size occurs when $\sin\left(\frac{\pi}{5} \cdot (2+t)\right) = -1$

$\implies \frac{\pi}{5} \cdot (2+t) = \frac{3\pi}{2}, \frac{7\pi}{2}, \frac{11\pi}{2}, \cdot \implies t = \frac{11}{2},\ t = \frac{31}{2},\ t = \frac{51}{2}, \cdots$.

Therefore the minimum population size occurs after 5.5 years.

CHAPTER 7

Solutions to Exercise Set 7.1

1. $\int \cos x\sqrt{4+\sin x}\; dx$.

   Let $u(x) = 4 + \sin x$. $\quad du = \cos x\; dx$.

   $\int \cos x\sqrt{4+\sin x}\; dx = \int \sqrt{u}\; du = \frac{2}{3}u^{3/2} + C = \frac{2}{3}(4+\sin x)^{3/2} + C.$

2. $\int_0^2 x\sqrt{1+x^2}\; dx$.

   Let $u(x) = 1 + x^2$. $\quad du = 2x\; dx$.

   $\int x\sqrt{1+x^2}\; dx = \frac{1}{2}\int \sqrt{u}\; du = \frac{1}{3}u^{3/2} + C = \frac{1}{3}(1+x^2)^{3/2} + C.$

   $\int_0^2 x\sqrt{1+x^2}\; dx = \frac{1}{3}(1+x^2)^{3/2}\Big]_0^2 = \frac{1}{3}(1+2^2)^{3/2} - \frac{1}{3}(1+0)^{3/2} = \frac{1}{3}(5^{3/2}-1).$

3. $\int_3^4 (x^2-3)\sqrt{x^3-9x}\; dx$.

   Let $u(x) = x^3 - 9x$. $\quad du = (3x^2-9)dx = 3(x^2-3)dx$.

   $\int (x^2-3)\sqrt{x^3-9x}\; dx = \frac{1}{3}\int \sqrt{u}\; du = \frac{1}{3}\cdot\frac{2}{3}u^{3/2} + C = \frac{2}{9}(x^3-9x)^{3/2} + C.$

   $\int_3^4 (x^2-3)\sqrt{x^3-9x}\; dx = \frac{2}{9}(x^3-9x)^{3/2}\Big]_3^4$

   $$= \frac{2}{9}(4^3-9\cdot 4)^{3/2} - \frac{2}{9}(3^3-9\cdot 3)^{3/2} = \frac{2}{9}\cdot 28^{3/2}.$$

4. $\int \frac{x}{1+x}\; dx = \int \left(1 - \frac{1}{1+x}\right) dx = x - \ln|1+x| + C.$

5. $\int_0^1 xe^{x^2}\; dx$

   Let $u(x) = x^2$. $\quad du = 2x\; dx, \quad x\; dx = \frac{1}{2}\; du.$

   $\int xe^{x^2}\; dx = \frac{1}{2}\int e^u\; du = \frac{1}{2}e^u + C = \frac{1}{2}e^{x^2} + C.$

   $\int_0^1 xe^{x^2}\; dx = \frac{1}{2}e^{x^2}\Big]_0^1 = \frac{1}{2}(e^1 - e^0) = \frac{1}{2}(e-1).$

6. $\int\limits_0^1 \sin \pi x \; dx = -\frac{1}{\pi} \cos \pi x\Big]_0^1$

$$= -\frac{1}{\pi}\left(\cos(\pi \cdot 1) - \cos(\pi \cdot 0)\right) = -\frac{1}{\pi}(-1-1) = \frac{2}{\pi}.$$

7. $\int \frac{x^4}{1+x^5} \; dx.$

   Let $u(x) = 1 + x^5$. $\qquad du = 5x^4 \; dx.$

   $\int \frac{x^4}{1+x^5} \; dx = \frac{1}{5}\int \frac{1}{u} \; du = \frac{1}{5}\ln|u| + C = \frac{1}{5}\ln|1+x^5| + C.$

8. $\int \frac{x^3}{\sqrt{1-x^4}} \; dx.$

   Let $u(x) = 1 - x^4$. $\qquad du = -4x^3 \; dx.$

   $\int \frac{x^3}{\sqrt{1-x^4}} \; dx = -\frac{1}{4}\int \frac{1}{\sqrt{u}} \; du = -\frac{1}{4}2 \cdot u^{1/2} + C = -\frac{1}{2}(1-x^4)^{1/2} + C.$

9. $\int \frac{e^x}{1+e^x} \; dx.$

   Let $u(x) = 1 + e^x$. $\qquad du = e^x \; dx.$

   $\int \frac{e^x}{1+e^x} \; dx = \int \frac{1}{u} \; du = \ln|u| + C = \ln(1+e^x) + C.$

10. $\int \tan x \cdot \sec^2 x \; dx.$

    Let $u(x) = \tan x$. $\qquad du = \sec^2 x \; dx.$

    $\int \tan x \cdot \sec^2 x \; dx = \int u \; du = \frac{1}{2}u^2 + C = \frac{1}{2}\tan^2 x + C.$

11. $\int\limits_0^{\pi/4} \tan^3 x \cdot \sec^2 x \; dx.$

    Let $u(x) = \tan x$. $\qquad du = \sec^2 x \; dx.$

    $\int \tan^3 x \cdot \sec^2 x \; dx = \int u^3 \; du = \frac{1}{4}u^4 + C = \frac{1}{4}\tan^4 x + C.$

    $\int\limits_0^{\pi/4} \tan^3 x \cdot \sec^2 x \; dx = \frac{1}{4}\tan^4 x\Big]_0^{\pi/4} = \frac{1}{4}\left(\tan^4 \frac{\pi}{4} - \tan^4 0\right) = \frac{1}{4}(1^4 - 0) = \frac{1}{4}.$

12. $\int \frac{x+2}{x^2+4x+7} \; dx.$

    Let $u(x) = x^2 + 4x + 7$. $\qquad du = (2x+4)dx = 2(x+2)dx.$

    $\int \frac{x+2}{x^2+4x+7} \; dx = \frac{1}{2}\int \frac{1}{u} \; du = \frac{1}{2}\ln|u| + C = \frac{1}{2}\ln|x^2+4x+7| + C.$

13. $\int (x+3)(x^2+6x+4)^3 \; dx.$

    Let $u(x) = x^2 + 6x + 4$. $\qquad du = (2x+6)dx = 2(x+3)dx.$

    $\int (x+3)(x^2+6x+4)^3 \; dx = \frac{1}{2}\int u^3 \; du = \frac{1}{8}u^4 + C = \frac{1}{8}(x^2+6x+4)^4 + C.$

14. $\int_0^1 \frac{\cos \pi x}{4+\sin \pi x}\, dx$

Let $u(x) = 4 + \sin \pi x$. $\quad du = \pi \cos \pi x\, dx$.

$\int \frac{\cos \pi x}{4+\sin \pi x}\, dx = \frac{1}{\pi}\int \frac{1}{u}\, du = \frac{1}{\pi}\ln|u| + C = \frac{1}{\pi}\ln|4+\sin \pi x| + C$.

$$\int_0^1 \frac{\cos \pi x}{4+\sin \pi x}\, dx = \frac{1}{\pi}\ln|4+\sin \pi x|\Big]_0^1 = \frac{1}{\pi}\left(\ln|4+\sin(\pi\cdot 1)| - \ln|4+\sin(\pi\cdot 0)|\right)$$

$$= \frac{1}{\pi}(\ln 4 - \ln 4) = 0.$$

15. $\int \sec x \cdot \tan x \cdot e^{\sec x}\, dx$.

Let $u(x) = \sec x$. $\quad du = \sec x \cdot \tan x\, dx$.

$\int \sec x \cdot \tan x \cdot e^{\sec x}\, dx = \int e^u\, du = e^u + C = e^{\sec x} + C$.

16. $\int x^{1/3}\sqrt{1+x^{4/3}}\, dx$.

Let $u(x) = 1 + x^{4/3}$. $\quad du = \frac{4}{3}x^{1/3}\, dx$.

$\int x^{1/3}\sqrt{1+x^{4/3}}\, dx = \frac{3}{4}\int \sqrt{u}\, du = \frac{3}{4}\cdot\frac{2}{3}u^{3/2} + C = \frac{1}{2}(1+x^{4/3})^{2/3} + C$.

17. $\int_1^4 \frac{e^{\sqrt{x}}}{\sqrt{x}}\, dx$.

Let $u(x) = e^{\sqrt{x}}$. $\quad du = \frac{1}{2\sqrt{x}}e^{\sqrt{x}}\, dx$.

$\int \frac{e^{\sqrt{x}}}{\sqrt{x}}\, dx = 2\int du = 2u + C = 2e^{\sqrt{x}} + C$.

$$\int_0^4 \frac{e^{\sqrt{x}}}{\sqrt{x}}\, dx = 2\, e^{\sqrt{x}}\Big]_1^4 = 2\left(e^{\sqrt{4}} - e^{\sqrt{1}}\right) = 2(e^2 - e).$$

18. $\int \frac{\sin x \cos x}{1+\sin^2 x}\, dx$.

Let $u(x) = 1 + \sin^2 x$. $\quad du = 2\sin x\cos x\, dx$.

$\int \frac{\sin x \cos x}{1+\sin^2 x}\, dx = \frac{1}{2}\int \frac{1}{u}\, du = \frac{1}{2}\ln|u| + C = \frac{1}{2}\ln(1+\sin^2 x) + C$.

19. $\int \sin^3 x \cos x\, dx$.

Let $u(x) = \sin x$. $\quad du = \cos x\, dx$.

$\int \sin^3 x \cos x\, dx = \int u^3\, du = \frac{1}{4}u^4 + C = \frac{1}{4}\sin^4 x + C$.

20. $\int_0^{\sqrt{\pi}/2} x\tan(x^2)\, dx$.

Let $u(x) = x^2$. $\quad du = 2x\, dx$.

$\int x\tan(x^2)dx = \frac{1}{2}\int \tan u\, du = \frac{1}{2}\ln|\sec u| + C = \frac{1}{2}\ln|\sec(x^2)| + C$.

$$\int_0^{\sqrt{\pi}/2} x\tan(x^2)dx = \tfrac{1}{2}\ln|\sec(x^2)|\Big]_0^{\sqrt{\pi}/2} = \tfrac{1}{2}\left(\ln\left|\sec\left(\left(\tfrac{\sqrt{\pi}}{2}\right)^2\right)\right| - \ln|\sec 0|\right)$$

$$= \tfrac{1}{2}\left(\ln\left(\sec\tfrac{\pi}{4}\right) - \ln(\sec 0)\right)$$

$$= \tfrac{1}{2}(\ln\sqrt{2} - \ln 1) = \tfrac{1}{2}\ln\sqrt{2} = \tfrac{1}{4}\ln 2.$$

21. $\int \frac{1}{\cos^2 x}\,dx = \int \sec^2 x\,dx = \tan x + C.$

22. $\int \frac{\sec^2 x}{\sqrt{1+2\tan x}}\,dx$

Let $u(x) = 1 + 2\tan x.$ $\quad du = 2\sec^2 x\,dx.$

$\int \frac{\sec^2 x}{\sqrt{1+2\tan x}}\,dx = \frac{1}{2}\int \frac{1}{\sqrt{u}}\,du = \frac{1}{2}2u^{1/2} + C = \sqrt{1+2\tan x} + C.$

23. $\int_1^2 \frac{dx}{x^2+2x+1} = \int_1^2 \frac{1}{(x+1)^2}dx = -\frac{1}{x+1}\Big]_1^2 = -\frac{1}{2+1} - \left(-\frac{1}{1+1}\right) = \frac{1}{6}.$

24. $\int_0^3 \frac{2^{\sqrt{x+1}}}{\sqrt{x+1}}\,dx$

Let $u(x) = \sqrt{x+1}.$ $\quad du = \frac{1}{2}\cdot\frac{dx}{\sqrt{x+1}}.$

$\int \frac{2^{\sqrt{x+1}}}{\sqrt{x+1}}\,dx = 2\int 2^u\,du = 2\cdot\frac{2^u}{\ln 2} + C = \frac{2^{\sqrt{x+1}+1}}{\ln 2} + C.$

$\int_0^3 \frac{2^{\sqrt{x+1}}}{\sqrt{x+1}}\,dx = \frac{2}{\ln 2}2^{\sqrt{x+1}}\Big]_0^3 = \frac{2}{\ln 2}\left(2^{\sqrt{3+1}} - 2^{\sqrt{0+1}}\right) = \frac{2}{\ln 2}(2^2 - 2^1) = \frac{4}{\ln 2}.$

25. $\int e^x\left(\tan(1+e^x)\right)dx$

Let $u(x) = 1 + e^x.$ $\quad du = e^x\,dx.$

$\int e^x\tan(1+e^x)dx = \int \tan u\,du = \ln|\sec u| + C = \ln|\sec(1+e^x)| + C.$

26. $\int x\sec(\pi + x^2)dx$

Let $u(x) = \pi + x^2.$ $\quad du = 2x\,dx.$

$$\int x\sec(\pi+x^2)dx = \tfrac{1}{2}\int \sec u\,du = \tfrac{1}{2}\ln|\sec u + \tan u| + C$$
$$= \tfrac{1}{2}\ln|\sec(\pi+x^2) + \tan(\pi+x^2)| + C.$$

27. $\int \frac{\cos x}{\sqrt{1-\sin^2 x}}\,dx = \int \frac{\cos x}{\sqrt{\cos^2 x}}\,dx = \int \frac{\cos x}{|\cos x|}\,dx$

$$= \begin{cases} \int 1\,dx & 2n\pi - \frac{\pi}{2} \le x \le 2n\pi + \frac{\pi}{2} \\ -\int 1\,dx & 2n\pi + \frac{\pi}{2} \le x \le 2n\pi + \frac{3\pi}{2} \end{cases}$$

$$= \begin{cases} x + C & 2n\pi - \frac{\pi}{2} \le x \le 2n\pi + \frac{\pi}{2} \\ -x + C & 2n\pi + \frac{\pi}{2} \le x \le 2n\pi + \frac{3\pi}{2} \end{cases},$$

where n is an integer.

28. $\int (e^x+7)^2\,dx = \int (e^{2x}+14e^x+49)dx = \frac{1}{2}e^{2x}+14e^x+49x+C.$

29. $\int_e^{e^2} \frac{1}{x\ln x}\,dx$

Let $u(x)=\ln x.\qquad du=\frac{1}{x}\,dx.$

$\int \frac{1}{x\ln x}\,dx = \int \frac{1}{u}\,du = \ln|u|+C = \ln|\ln x|+C.$

$\int_e^{e^2} \frac{1}{x\ln x}\,dx = \ln\ln x\Big]_e^{e^2} = \ln\ln e^2 - \ln\ln e = \ln 2 - \ln 1 = \ln 2.$

30. $\int \frac{\left(1+2\sqrt{x}\right)^3}{\sqrt{x}}\,dx$

Let $u(x)=1+2\sqrt{x}.\qquad du=\frac{1}{\sqrt{x}}\,dx.$

$\int \frac{\left(1+2\sqrt{x}\right)^3}{\sqrt{x}}\,dx = \int u^3\,du = \frac{1}{4}u^4+C = \frac{1}{4}\left(1+2\sqrt{x}\right)^4+C.$

31. $\int \frac{1}{\sqrt{x}\left(1+\sqrt{x}\right)}\,dx$

Let $u(x)=1+\sqrt{x}.\qquad du=\frac{1}{2}\cdot\frac{1}{\sqrt{x}}\,dx.$

$\int \frac{1}{\sqrt{x}\left(1+\sqrt{x}\right)}\,dx = 2\int \frac{1}{u}\,du = 2\ln|u|+C = 2\ln\left(1+\sqrt{x}\right)+C.$

32. $\int_1^2 \frac{x^2-3x-4}{x-4}\,dx = \int_1^2 \frac{(x-4)(x+1)}{(x-4)}\,dx$

$= \int_1^2 (x+1)dx = \left(\frac{1}{2}x^2+x\right)\Big]_1^2$

$= \left(\frac{1}{2}2^2+2\right) - \left(\frac{1}{2}\cdot 1+1\right) = \frac{5}{2}.$

33. $\int_3^5 x\sqrt{x^2-9}\,dx.$

Let $u(x)=x^2-9.\qquad du=2x\,dx \implies x\,dx=\frac{1}{2}\,du.$

$\int x\sqrt{x^2-9}\,dx = \int u^{1/2}\left(\frac{1}{2}\,du\right) = \frac{1}{2}\cdot\frac{2}{3}u^{3/2}+C = \frac{1}{3}(x^2-9)^{3/2}+C.$

$\int_3^5 x\sqrt{x^2-9}\,dx = \frac{1}{3}(x^2-9)^{3/2}\Big]_3^5 = \frac{1}{3}\cdot\left(16^{3/2}-0\right) = \frac{64}{3}.$

34. $\int_0^1 x(x^2+1)^4\,dx.$

Let $u(x)=x^2+1.\qquad du=2x\,dx \implies x\,dx=\frac{1}{2}\,du.$

$\int x(x^2+1)^4\,dx = \int u^4\left(\frac{1}{2}\,du\right) = \frac{1}{2}\cdot\frac{1}{5}u^5+C = \frac{1}{10}(x^2+1)^5+C.$

$\int_0^1 x(x^2+1)^4\,dx = \frac{1}{10}(x^2+1)^5\Big]_0^1 = \frac{1}{10}\cdot(2^5-1^5) = \frac{31}{10}.$

35. $\int_0^2 x\sqrt{1+x^2}\,dx.$

Let $u(x) = 1 + x^2$. $\quad du = 2x\,dx \implies x\,dx = \frac{1}{2}\,du.$

$\int x\sqrt{1+x^2}\,dx = \int u^{1/2}\left(\frac{1}{2}\,du\right) = \frac{1}{2}\cdot\frac{2}{3}u^{3/2} + C = \frac{1}{3}(1+x^2)^{3/2} + C.$

$\int_0^2 x\sqrt{1+x^2}\,dx = \frac{1}{3}(1+x^2)^{3/2}\Big]_0^2 = \frac{1}{3}\cdot(5^{3/2} - 1).$

36. $\int_0^2 xe^{2x^2}\,dx.$

Let $u(x) = 2x^2$. $\quad du = 4x\,dx \implies x\,dx = \frac{1}{4}\,du.$

$\int xe^{2x^2}\,dx = \int e^u\left(\frac{1}{4}\,du\right) = \frac{1}{4}e^u + C = \frac{1}{4}e^{2x^2} + C.$

$\int_0^2 xe^{2x^2}\,dx = \frac{1}{4}e^{2x^2}\Big]_0^2 = \frac{1}{4}(e^8 - 1).$

37. $\int_0^{\sqrt{\pi}/2} t\sin(\pi - t^2)dt.$

Let $u(t) = \pi - t^2$. $\quad du = -2t\,dt \implies t\,dt = -\frac{1}{2}\,du.$

$\int t\sin(\pi - t^2)dt = \int \sin u\left(-\frac{1}{2}\,du\right) = -\frac{1}{2}\cdot(-\cos u) + C = \frac{1}{2}\cos(\pi - t^2) + C.$

$\int_0^{\sqrt{\pi}/2} t\sin(\pi - t^2)dt = \frac{1}{2}\cos(\pi - t^2)\Big]_0^{\sqrt{\pi}/2} = \frac{1}{2}\cdot\left(\cos\frac{3\pi}{4} - \cos\pi\right) = \frac{1}{2}\cdot\left(-\frac{\sqrt{2}}{2} + 1\right).$

38. $\int_0^1 x^{1/2}(1 - x^{3/2})dx.$

Let $u(x) = 1 - x^{3/2}$. $\quad du = -\frac{3}{2}x^{1/2}\,dx \implies x^{1/2}\,dx = -\frac{2}{3}\,du.$

$\int x^{1/2}(1 - x^{3/2})dx = \int u\left(-\frac{2}{3}\,du\right) = -\frac{2}{3}\cdot\frac{1}{2}u^2 + C = -\frac{1}{3}(1 - x^{3/2})^2 + C.$

$\int_0^1 x^{1/2}\left(1 - x^{3/2}\right)dx = -\frac{1}{3}(1 - x^{3/2})^2\Big]_0^1 = \frac{1}{3}.$

39. $\int_0^{\pi/4} \tan^3 x\sec^2 x\,dx.$

Let $u(x) = \tan x$. $\quad du = \sec^2 x\,dx.$

$\int_0^{\pi/4} \tan^3 x\sec^2 x\,dx = \frac{1}{4}\tan^4 x\Big]_0^{\pi/4} = \frac{1}{4}.$

40. $\int_0^1 \frac{2x}{3x^2+1}\,dx.$

Let $u(x) = 3x^2 + 1$. $\quad du = 6x\,dx \implies 2x\,dx = \frac{1}{3}\,du.$

$\int \frac{2x}{3x^2+1}\,dx = \int \frac{1}{u}\cdot\left(\frac{1}{3}\,du\right) = \frac{1}{3}\ln|u| + C = \frac{1}{3}\ln(3x^2+1) + C.$

$\int_0^1 \frac{2x}{3x^2+1}\,dx = \frac{1}{3}\ln(3x^2+1)\Big]_0^1 = \frac{1}{3}\ln 4.$

41. $\int_1^4 \frac{(4+\sqrt{x})^2}{\sqrt{x}}\,dx.$

Let $u = 4+\sqrt{x}$. $\quad du = \frac{1}{2\sqrt{x}}\,dx \implies \frac{1}{\sqrt{x}}\,dx = 2\,du.$

$\int \frac{(4+\sqrt{x})^2}{\sqrt{x}}\,dx = \int u^2(2\,du) = \frac{2}{3}u^3 + C = \frac{2}{3}\cdot(4+\sqrt{x})^3 + C.$

$\int_1^4 \frac{(4+\sqrt{x})^2}{\sqrt{x}}\,dx = \frac{2}{3}(4+\sqrt{x})^3\Big]_1^4 = \frac{2}{3}\cdot(6^3-5^3) = \frac{182}{3}.$

42. $\int_0^{\ln 2} e^x \sin(\pi e^x)\cos(\pi e^x)dx.$

Let $u(x) = \sin(\pi e^x)$. $\quad du = \cos(\pi e^x)\cdot(\pi e^x\,dx) \implies e^x\cdot\cos(\pi e^x)dx = \frac{1}{\pi}\,du.$

$\int e^x \sin(\pi e^x)\cos(\pi e^x)dx = \int u\cdot\frac{1}{\pi}\,du = \frac{1}{\pi}\cdot\frac{1}{2}u^2 + C$

$= \frac{1}{2\pi}\sin^2(\pi e^x) + C.$

$\int_0^{\ln 2} e^x \sin(\pi e^x)\cos(\pi e^x)dx = \frac{1}{2\pi}\sin^2(\pi e^x)\Big]_0^{\ln 2}$

$= \frac{1}{2\pi}\cdot\left[\sin^2(\pi\cdot 2) - \sin^2(\pi\cdot 1)\right] = 0.$

43. Area of $R = \int_0^{\sqrt{\pi}/2} x\sin x^2\,dx.$

Let $u(x) = x^2$. $\quad du = 2x\,dx.$

$\int x\sin x^2\,dx = \frac{1}{2}\int \sin u\,du = -\frac{1}{2}\cos u + C = -\frac{1}{2}\cos x^2 + C.$

Area of $R = \int_0^{\sqrt{\pi}/2} x\sin x^2\,dx = -\frac{1}{2}\cos x^2\Big]_0^{\sqrt{\pi}/2} = -\frac{1}{2}\left(\cos\left(\frac{\sqrt{\pi}}{2}\right)^2 - \cos 0\right)$

$= -\frac{1}{2}\left(\frac{\sqrt{2}}{2} - 1\right) = -\frac{1}{4}(\sqrt{2}-2).$

44. Volume $V = \int_0^{\pi} \pi\sin^2 x\,dx = \pi\int_0^{\pi}\left(\frac{1}{2} - \frac{1}{2}\cos 2x\right)dx$

$= \pi\left(\frac{1}{2}x - \frac{1}{4}\sin 2x\right)\Big]_0^{\pi}$

$= \pi\left(\frac{1}{2}\pi - \frac{1}{4}\sin 2\pi - \frac{1}{2}\cdot 0 + \frac{1}{4}\sin 0\right) = \frac{\pi^2}{2}.$

45. Consumers' surplus $= \int_0^{16} \frac{20000}{20+5x}\,dx - (16)(200)$

$= 20000\cdot\frac{1}{5}\ln(20+5x)\Big]_0^{16} - 3200$

$= 4000\left(\ln(20+5\cdot 16) - \ln(20+5\cdot 0)\right) - 3200$

$= 4000(\ln 100 - \ln 20) - 3200$

$= 4000\ln\left(\frac{100}{20}\right) - 3200 = -3200 + 4000\ln 5.$

46. Let $u(x) = 9 + x^2$. $\quad du = 2x\,dx \implies x\,dx = \frac{1}{2}\,du$.

$\int MR(x)dx = \int u^{-1/2}\left(\frac{1}{2}\,du\right) = \frac{1}{2}\cdot 2u^{1/2} + C = \sqrt{9+x^2} + C$.

$\int_4^6 MR(x)dx = \sqrt{9+x^2}\,\Big]_4^6 = \sqrt{45} - 5$.

47. Let $u(t) = \sin \pi t$. $\quad du = \cos \pi t \cdot \pi\,dt \implies \cos \pi t\,dt = \frac{1}{\pi}\,du$.

$\int R(t)dt = 4,000t + 2,000 \cdot \int u \cdot \frac{1}{\pi}\,du = 4,000t + \frac{2,000}{\pi}\cdot \frac{1}{2}u^2 + C$

$= 4,000t + \frac{1,000}{\pi}\cdot \sin^2 \pi t + C$.

$\int_0^4 R(t)dt = \left(4,000t + \frac{1,000}{\pi}\cdot \sin^2 \pi t\right)\Big]_0^4 = \$16,000$.

48. Let $u(t) = 4 + t^2$. $\quad du = 2t\,dt \implies t\,dt = \frac{1}{2}\,du$.

$\int A(t)dt = 5,000t + 200 \cdot \int u^{1/2}\left(\frac{1}{2}\,du\right) = 5,000t + 100 \cdot \frac{2}{3}u^{3/2} + C$

$= 5,000t + \frac{200}{3}(4+t^2)^{3/2} + C$.

$\int_0^4 A(t)dt = \left(5,000t + \frac{200}{3}\cdot(4+t^2)^{3/2}\right)\Big]_0^4 = \left(20,000 + \frac{200}{3}\cdot(20)^{3/2}\right) - \left(0 + \frac{200}{3}\cdot 8\right)$

$= 20,000 + \frac{200}{3}\cdot\left((20)^{3/2} - 8\right) = \$25,429.52$.

49. Let $u(x) = 2 + x^4$. $\quad du = 4x^3\,dx \implies x^3\,dx = \frac{1}{4}\,du$.

$\int \frac{x^3}{2+x^4}\,dx = \int \frac{1}{u}\left(\frac{1}{4}\,du\right) = \frac{1}{4}\ln|u| + C = \frac{1}{4}\ln(2+x^4) + C$.

Average value of $f(x)$ on the interval $[0,2]$ is given by

$$\frac{\int_0^2 f(x)dx}{2-0} = \frac{1}{2}\cdot\frac{1}{4}\ln(2+x^4)\Big]_0^2 = \frac{1}{8}\cdot(\ln 18 - \ln 2) = \frac{1}{8}\ln 9.$$

50. Let $u(p) = 1 + p^2$. $\quad du = 2p\,dp \implies p\,dp = \frac{1}{2}\,du$.

$\int \frac{p}{1+p^2}\,dp = \int \frac{1}{u}\left(\frac{1}{2}\,du\right) = \frac{1}{2}\ln|u| + C = \frac{1}{2}\ln(1+p^2) + C$.

Average value of $q(p)$ on the interval $[0,3]$ is given by

$$\frac{\int_0^3 q(p)dp}{3-0} = \frac{1}{6}\ln(1+p^2)\Big]_0^3 = \frac{1}{6}\ln 10.$$

Solutions to Exercise Set 7.2

1. $\int xe^{-x}\,dx.$

   Let $u = x, \qquad dv = e^{-x}\,dx.$

   Then $du = dx, \qquad v = -e^{-x}.$

   $\int xe^{-x}\,dx = -xe^{-x} - \int(-e^{-x})dx = -xe^{-x} + (-1)e^{-x} + C = -e^{-x}(x+1) + C.$

2. $\int (x-4)e^{x}\,dx.$

   Let $u = x-4, \qquad dv = e^{x}\,dx.\ du = dx, \qquad v = e^{x}.$

   $\int (x-4)e^{x}\,dx = (x-4)e^{x} - \int e^{x}\,dx = (x-4)e^{x} - e^{x} + C = e^{x}(x-5) + C.$

3. $\int x\ln x\,dx.$

   Let $u = \ln x, \quad dv = x\,dx. \quad du = \frac{1}{x}\,dx, \quad v = \frac{1}{2}x^2.$

   $\int x\ln x\,dx = \frac{1}{2}x^2\ln x - \int \frac{1}{2}x^2\cdot\frac{1}{x}\,dx = \frac{1}{2}x^2\ln x - \frac{1}{4}x^2 + C = \frac{1}{4}x^2(2\ln x - 1) + C.$

4. $\int_1^e x^2\ln x\,dx.$

   Let $u = \ln x, \quad dv = x^2\,dx. \quad du = \frac{1}{x}\,dx, \quad v = \frac{1}{3}x^3.$

   $\int x^2\ln x\,dx = \frac{1}{3}x^3\ln x - \int \frac{1}{3}x^3\cdot\frac{1}{x}\,dx = \frac{1}{3}x^3\ln x - \frac{1}{9}x^3 + C = \frac{1}{9}x^3(3\ln x - 1) + C.$

   $\int_1^e x^2\ln x\,dx = \frac{1}{9}x^3(3\ln x - 1)\Big]_1^e = \frac{1}{9}\left[e^3(3\ln e - 1) - 1^3(3\ln 1 - 1)\right]$

   $= \frac{1}{9}\left(e^3\cdot 2 - (-1)\right) = \frac{1}{9}(2e^3 + 1).$

5. $\int_0^3 x\sqrt{x+1}\,dx.$

   Let $u = x, \quad dv = \sqrt{x+1}\,dx. \quad du = dx, \quad v = \frac{2}{3}(x+1)^{3/2}.$

   $\int x\sqrt{x+1}\,dx = \frac{2}{3}x(x+1)^{3/2} - \int \frac{2}{3}(x+1)^{3/2}dx = \frac{2}{3}x(x+1)^{3/2} - \frac{2}{3}\cdot\frac{2}{5}(x+1)^{5/2} + C$

   $= \frac{2}{15}(x+1)^{3/2}\left[5x - 2(x+1)\right] + C = \frac{2}{15}(3x-2)(x+1)^{3/2} + C.$

   $\int_0^3 x\sqrt{x+1}\,dx = \frac{2}{15}(3x-2)(x+1)^{3/2}\Big]_0^3$

   $= \frac{2}{15}\left[(3\cdot 3 - 2)(3+1)^{3/2} - (0-2)(0+1)^{3/2}\right] = \frac{2}{15}(7\cdot 8 + 2)$

   $= \frac{116}{15}.$

6. $\int \frac{x}{\sqrt{x+4}}\,dx$.

Let $u = x, \quad dv = \frac{1}{\sqrt{x+4}}\,dx. \quad du = dx, \quad v = 2\sqrt{x+4}.$

$$\int \frac{x}{\sqrt{x+4}}\,dx = 2x\sqrt{x+4} - \int 2\sqrt{x+4}\,dx = 2x\sqrt{x+4} - \tfrac{4}{3}(x+4)^{3/2} + C$$
$$= \tfrac{2}{3}\sqrt{x+4}\,[3x - 2(x+4)] + C = \tfrac{2}{3}(x-8)\sqrt{x+4} + C.$$

7. $\int x \sin x\,dx$.

Let $u = x, \quad dv = \sin x\,dx. \quad du = dx, \quad v = -\cos x.$

$\int x \sin x\,dx = -x\cos x - \int(-\cos x)dx = -x\cos x + \sin x + C.$

8. $\int x^2 \cos x\,dx$.

Let $u = x^2, \quad dv = \cos x\,dx. \quad du = 2x\,dx, \quad v = \sin x.$

$\int x^2 \cos x\,dx = x^2 \sin x - 2\int x \sin x\,dx.$

In $\int x \sin x\,dx$, let $u_1 = x, \quad dv_1 = \sin x\,dx. \quad du_1 = dx, \quad v_1 = -\cos x.$

Then $\int x \sin x\,dx = -x\cos x - \int(-\cos x)dx = -x\cos x + \sin x + C_1.$

Thus $\int x^2 \cos x\,dx = x^2 \sin x - 2(-x\cos x + \sin x + C_1)$

$$= (x^2 - 2)\sin x + 2x\cos x + C.$$

9. $\int x(2x+1)^5 dx$.

Let $u = x, \quad dv = (2x+1)^5 dx.$

$du = dx, \quad v = \frac{1}{2}\cdot\frac{1}{6}(2x+1)^6 = \frac{1}{12}(2x+1)^6.$

$$\int x(2x+1)^5 dx = \tfrac{1}{12}x(2x+1)^6 - \int \tfrac{1}{12}(2x+1)^6 dx$$
$$= \tfrac{1}{12}x(2x+1)^6 - \tfrac{1}{12}\cdot\tfrac{1}{2}\cdot\tfrac{1}{7}(2x+1)^7 + C.$$
$$= \tfrac{1}{12\cdot 14}(2x+1)^6(14x - 2x - 1) + C = \tfrac{1}{168}(12x-1)(2x+1)^6 + C.$$

10. $\int x \ln x^2\,dx = \int 2x \ln x\,dx$.

Let $u = \ln x, \quad dv = x\,dx. \quad du = \frac{1}{x}\,dx, \quad v = \frac{1}{2}x^2.$

$$\int x \ln x^2\,dx = 2\left[\tfrac{1}{2}x^2 \ln x - \int \tfrac{1}{2}x^2\cdot\tfrac{1}{x}\,dx\right] = 2\left(\tfrac{1}{2}x^2 \ln x - \tfrac{1}{4}x^2 + C_1\right)$$
$$= \tfrac{1}{2}x^2(2\ln x - 1) + C = \tfrac{1}{2}x^2(\ln x^2 - 1) + C.$$

11. $\int x^2 e^x\,dx$.

Let $u = x^2, \quad dv = e^x\,dx. \quad du = 2x\,dx, \quad v = e^x.$

$\int x^2 e^x\,dx = x^2 e^x - \int 2xe^x\,dx = x^2 e^x - 2\int xe^x\,dx.$

In $\int xe^x\,dx$, let $u_1 = x$, $dv_1 = e^x\,dx$. $du_1 = dx$, $v_1 = e^x$.

$\int xe^x\,dx = xe^x - \int e^x\,dx = xe^x - e^x + C_1 = e^x(x-1) + C_1$.

Thus, $\int x^2e^x\,dx = x^2e^x - 2\left[e^x(x-1) + C_1\right] = e^x(x^2 - 2x + 2) + C$.

12. $\int x^3 \ln x\,dx$.

Let $u = \ln x$, $dv = x^3\,dx$. $du = \frac{1}{x}\,dx$, $v = \frac{1}{4}x^4$.

$\int x^3 \ln x\,dx = \frac{1}{4}x^4 \ln x - \int \frac{1}{4}x^4 \cdot \frac{1}{x}\,dx = \frac{1}{4}x^4 \ln x - \frac{1}{16}x^4 + C = \frac{1}{16}x^4(4\ln x - 1) + C$.

13. $\int_0^{\pi/4} x \sec^2 x\,dx$.

Let $u = x$, $dv = \sec^2 x\,dx$. $du = dx$, $v = \tan x$.

$\int x \sec^2 x\,dx = x\tan x - \int \tan x\,dx = x\tan x - \ln|\sec x| + C$.

$$\int_0^{\pi/4} x \sec^2 x\,dx = (x\tan x - \ln\sec x)\Big]_0^{\pi/4} = \left(\tfrac{\pi}{4}\tan\tfrac{\pi}{4} - \ln\sec\tfrac{\pi}{4}\right) - (0 - \ln\sec 0)$$

$$= \left(\tfrac{\pi}{4}\cdot 1 - \ln\sqrt{2}\right) + \ln 1 = \tfrac{\pi}{4} - \tfrac{1}{2}\ln 2 + 0$$

$$= \tfrac{1}{4}(\pi - 2\ln 2).$$

14. $\int_0^{\pi/4} x \sec x \tan x\,dx$.

Let $u = x$, $dv = \sec x \tan x\,dx$. $du = dx$, $v = \sec x$.

$\int x\sec x\tan x\,dx = x\sec x - \int \sec x\,dx = x\sec x - \ln|\sec x + \tan x| + C$.

$$\int_0^{\pi/4} x\sec x\tan x\,dx = [x\sec x - \ln(\sec x + \tan x)]\Big]_0^{\pi/4}$$

$$= \left[\tfrac{\pi}{4}\sec\tfrac{\pi}{4} - \ln\left(\sec\tfrac{\pi}{4} + \tan\tfrac{\pi}{4}\right)\right] - [0 - \ln(\sec 0 + \tan 0)]$$

$$= \tfrac{\pi}{4}\cdot\sqrt{2} - \ln\left(\sqrt{2}+1\right) + \ln 1 = \tfrac{\sqrt{2}}{4}\pi - \ln\left(1+\sqrt{2}\right).$$

15. $\int_0^1 x^2e^{-x}\,dx$.

Let $u = x^2$, $dv = e^{-x}\,dx$. $du = 2x\,dx$, $v = -e^{-x}$.

$\int x^2e^{-x}\,dx = -x^2e^{-x} - \int(-2xe^{-x})dx = -x^2e^{-x} + 2\int xe^{-x}\,dx$.

In $\int xe^{-x}\,dx$, let $u_1 = x$, $dv_1 = e^{-x}\,dx$. $du_1 = dx$, $v_1 = -e^{-x}$.

$\int xe^{-x}dx = -xe^{-x} - \int(-e^{-x})dx = -xe^{-x} - e^{-x} + C_1$.

Thus $\int x^2e^{-x}\,dx = -x^2e^{-x} + 2(-xe^{-x} - e^{-x} + C_1) = -e^{-x}(x^2+2x+2) + C$.

$\int_0^1 x^2e^{-x}\,dx = -e^{-x}(x^2+2x+2)\Big]_0^1 = -e^{-1}(1+2+2) + e^0\cdot 2 = 2 - \frac{5}{e}$.

16. $\int_1^e (\ln x)^2\,dx$.

Let $u(x) = (\ln x)^2$, $dv = dx$. $du = \frac{2\ln x}{x}\,dx$, $v = x$.

$\int (\ln x)^2\,dx = x(\ln x)^2 - \int x\cdot\frac{2\ln x}{x}\,dx = x(\ln x)^2 - 2\int \ln x\,dx$.

In $\int \ln x\,dx$, let $u_1 = \ln x$, $dv_1 = dx$. $du_1 = \frac{1}{x}\,dx$, $v_1 = x$.

$\int \ln x\,dx = x\ln x - \int x\cdot\frac{1}{x}\,dx = x\ln x - x + C_1$.

$\int (\ln x)^2\,dx = x(\ln x)^2 - 2(x\ln x - x + C_1) = x\left[(\ln x)^2 - 2\ln x + 2\right] + C$.

$$\begin{aligned}\int_1^e (\ln x)^2\,dx &= x\left[(\ln x)^2 - 2\ln x + 2\right]\Big]_1^e\\ &= e\left[(\ln e)^2 - 2\ln e + 2\right] - 1\left[(\ln 1)^2 - 2\ln 1 + 2\right]\\ &= e(1-2+2) - 1(2) = e - 2.\end{aligned}$$

17. $\int_0^1 \frac{x^3}{\sqrt{1+x^2}}\,dx$.

Let $u = x^2$, $dv = \frac{x}{\sqrt{1+x^2}}\,dx$. $du = 2x\,dx$, $v = \sqrt{1+x^2}$.

$$\begin{aligned}\int \frac{x}{\sqrt{1+x^2}}\,dx &= x^2\cdot\sqrt{1+x^2} - \int 2x\cdot\sqrt{1+x^2}\,dx = x^2\cdot\sqrt{1+x^2} - \tfrac{2}{3}(1+x^2)^{3/2} + C\\ &= \tfrac{1}{3}\sqrt{1+x^2}\cdot(3x^2 - 2 - 2x^2) + C = \tfrac{1}{3}\sqrt{1+x^2}\cdot(x^2-2) + C.\end{aligned}$$

$\int_0^1 \frac{x^3}{\sqrt{1+x^2}}\,dx = \frac{1}{3}\sqrt{1+x^2}\,(x^2-2)\Big]_0^1 = \frac{1}{3}\left(-\sqrt{2}+2\right) = 0.1953$.

18. $\int x^3\sqrt{1+x^2}\,dx$.

Let $u(x) = x^2$, $dv = x\sqrt{1+x^2}\,dx$. $du = 2x\,dx$, $v = \frac{1}{3}(1+x^2)^{3/2}$.

$$\begin{aligned}\int x^3\sqrt{1+x^2}\,dx &= \tfrac{1}{3}x^2(1+x^2)^{3/2} - \tfrac{2}{3}\int x(1+x^2)^{3/2}\,dx\\ &= \tfrac{1}{3}x^2(1+x^2)^{3/2} - \tfrac{2}{15}(1+x^2)^{5/2} + C\\ &= \tfrac{1}{15}(1+x^2)^{3/2}(5x^2 - 2 - 2x^2) + C\\ &= \tfrac{1}{15}(1+x^2)^{3/2}(3x^2-2) + C.\end{aligned}$$

19. $\int e^{2x}\cos x\,dx$.

Let $u(x) = e^{2x}$, $dv = \cos x\,dx$. $du = 2e^{2x}\,dx$, $v = \sin x$.

$\int e^{2x}\cos x\,dx = e^{2x}\sin x - 2\int e^{2x}\sin x\,dx$.

In $\int e^{2x}\sin x\ dx$, let $u_1 = e^{2x}$, $dv_1 = \sin x\ dx$. $du_1 = 2e^{2x}\ dx$, $v_1 = -\cos x$.

$\int e^{2x}\sin x\ dx = -e^{2x}\cos x + 2\int e^{2x}\cos x\ dx$.

$$\int e^{2x}\cos x\ dx = e^{2x}\sin x - 2\left(-e^{2x}\cos x + 2\int e^{2x}\cos x\ dx\right)$$
$$= e^{2x}\sin x + 2e^{2x}\cos x - 4\int e^{2x}\cos x\ dx.$$

$5\int e^{2x}\cos x\ dx = e^{2x}(\sin x + 2\cos x)$.

$\int e^{2x}\cos x\ dx = \frac{1}{5}e^{2x}(\sin x + 2\cos x) + C$.

20. $\int \sin(\ln x)dx$.

Let $u(x) = \sin(\ln x)$, $dv = dx$. $du = (\cos(\ln x))\frac{1}{x}\ dx$, $v = x$.

$\int \sin(\ln x)dx = x\sin(\ln x) - \int \frac{1}{x}(\cos(\ln x))\cdot x\ dx$.

$\int \sin(\ln x)dx = x\sin(\ln x) - \int \cos(\ln x)dx$.

In $\int \cos(\ln x)dx$, let $u_1 = \cos(\ln x)$, $dv_1 = dx$.

$du_1 = -\sin(\ln x)\cdot\frac{1}{x}\ dx$, $v_1 = x$.

$\int \cos(\ln x)dx = x\cos(\ln x) + \int \sin(\ln x)dx$.

So $\int \sin(\ln x)dx = x\sin(\ln x) - \left(x\cos(\ln x) + \int \sin(\ln x)dx\right)$
$$= x\left[\sin(\ln x) - \cos(\ln x)\right] - \int \sin(\ln x)dx.$$

$\int \sin(\ln x)dx = \frac{1}{2}x\left[\sin(\ln x) - \cos(\ln x)\right] + C$.

21. $\int x^3e^{-x^2}\ dx$.

Let $u(x) = -x^2$. $du = -2x\ dx$, and $x^2 = -u$.

$\int x^3e^{-x^2}\ dx = \int x^2e^{-x^2}x\ dx = \frac{1}{2}\int ue^u du$.

In $\int ue^u\ du$, let $w = u$, $dv = e^u\ du$. $dw = du$, $v = e^u$.

$\int ue^u\ du = ue^u - \int e^u\ du = ue^u - e^u + C_1$.

So $\int x^3e^{-x^2}\ dx = \frac{1}{2}(ue^u - e^u + C_1) = \frac{1}{2}\left(-x^2e^{-x^2} - e^{-x^2} + C_1\right)$
$$= -\frac{1}{2}e^{-x^2}(x^2+1) + C.$$

22. $\int_0^4 e^{\sqrt{x}}\,dx$.

Let $u = e^{\sqrt{x}}$. Then $\sqrt{x} = \ln u$. $du = \frac{1}{2\sqrt{x}}e^{\sqrt{x}}\,dx$.

$dx = 2\sqrt{x}\cdot\frac{1}{e^{\sqrt{x}}}\,du = 2(\ln u)\cdot\frac{1}{u}\,du$.

$\int e^{\sqrt{x}}\,dx = \int u(2\ln u)\cdot\frac{1}{u}\,du = 2\int \ln u\,du$.

In $\int \ln u\,du$, let $w = \ln u$, $dv = du$. $dw = \frac{1}{u}\,du$, $v = u$.

$\int \ln u\,du = u\ln u - \int du = u\ln u - u + C_1$.

So $\int e^{\sqrt{x}}\,dx = 2(u\ln u - u + C_1) = 2u(\ln u - 1) + C$

$= 2e^{\sqrt{x}}(\sqrt{x} - 1) + C$.

$\int_0^4 e^{\sqrt{x}}\,dx = 2e^{\sqrt{x}}(\sqrt{x} - 1)\big]_0^4 = 2\left[e^2(2-1) - e^0(0-1)\right]$

$= 2(e^2 + 1) = 16.7781$.

23. Area of $R = \int_0^1 (x - \ln(1+x))\,dx = \int_0^1 x\ dx - \int_0^1 \ln(1+x)dx$

$= \frac{1}{2}x^2\big]_0^1 - \int_0^1 \ln(1+x)dx = \frac{1}{2}(1-0) - (1+x)\cdot(\ln(1+x) - 1)\big]_0^1$

$= \frac{1}{2} - (1+1)\left[\ln(1+1) - 1\right] + (1+0)\left[\ln(1+0) - 1\right]$

$= \frac{1}{2} - 2(\ln 2 - 1) + (0 - 1) = \frac{3}{2} - 2\ln 2$.

24. Volume of $S = \int_1^e \pi(\ln x)^2\ dx = \pi\int_1^e (\ln x)^2\ dx$

$= \pi(e-2)$ by the result of problem 16, Exercise Set 7.2.

25. a. Present value $= \int_0^5 (e^{-.10t})\cdot(1000t)dt$

$= 1000\cdot\left(-\frac{1}{.10}te^{-.10t} - \frac{1}{.01}e^{-.10t}\right)\big]_0^5$

$= 1000\cdot\left(-50e^{-.5} - 100e^{-.5} + 100\right) = \$9,020.40$.

b. Value of the total contributions $= \int_0^5 1000t\ dt$.

$= 500t^2\big]_0^5 = \$12,500$.

26. Amount of the gift $= \int_0^4 (e^{-.05t})\cdot(100t)dt$

$= 100\cdot\left(-\frac{1}{.05}te^{-.05t} - \frac{1}{.0025}e^{-.05t}\right)\big]_0^4$

$= 100\cdot\left(-80e^{-.2} - 400e^{-.2} + 400\right) = \$700.92$.

27. Average of $f = \frac{1}{e-1}\int_1^e \ln x\ dx = \frac{1}{e-1}\cdot x(\ln x - 1)\Big]_1^e$

$= \frac{1}{e-1}\left[e(\ln e - 1) - 1(\ln 1 - 1)\right] = \frac{1}{e-1}$.

28. Average value $= \dfrac{\int_0^{\ln 5} xe^{-2x}\,dx}{\ln 5 - 0} = \frac{1}{\ln 5}\cdot\left(-\frac{1}{2}xe^{-2x} - \frac{1}{4}e^{-2x}\right)\Big]_0^{\ln 5}$
$= \frac{1}{\ln 5}\cdot\left(-\frac{\ln 5}{2}e^{-2\ln 5} - \frac{1}{4}e^{-2\ln 5} + \frac{1}{4}\right)$
$= \frac{1}{\ln 5}\cdot\left(-\frac{\ln 5}{2}\cdot .04 - \frac{1}{4}\cdot(.04) + .25\right)$
$= \frac{1}{\ln 5}\cdot(-.02\ln 5 + .24) = 0.1291.$

29. Total earnings $= \int_0^5 10,000te^{-.2t}\,dt$
$= 10,000\cdot\left(-\frac{1}{.2}te^{-.2t} - \frac{1}{.04}e^{-.2t}\right)\Big]_0^5$
$= 10,000\cdot\left(-\frac{1}{.2}\cdot 5\cdot e^{-1} - \frac{1}{.04}e^{-1} + \frac{1}{.04}\right)$
$= 10,000\cdot\left(-50e^{-1} + 25\right) = \$66,060.28.$

30. Present value $= \int_0^5 (e^{-.05t})\cdot(10,000te^{-.2t})dt$
$= 10,000\cdot\left(-\frac{1}{.25}te^{-.25t} - \frac{1}{.0625}e^{-.25t}\right)\Big]_0^5$
$= 10,000\cdot\left(-36e^{-1.25} + 16\right) = \$56,858.27.$

31. Distance traveled $= \int_0^{10} v(t)dt = \int_0^{10} 2te^{0.5t}\,dt$
$= 2\cdot\left(\frac{1}{.5}te^{.5t} - \frac{1}{.25}e^{.5t}\right)\Big]_0^{10} = 2\cdot\left(16e^5 + 4\right) = 4,757.2211.$

32. Present value $= \int_0^5 e^{-.08t}(100,000 - 100t)dt$
$= \left\{100,000\cdot\frac{1}{-.08}e^{-.08t} - 100\cdot\left(-\frac{1}{.08}te^{-.08t} - \frac{1}{.0064}e^{-.08t}\right)\right\}\Big]_0^5$
$= 1,250,000\cdot\left(1 - e^{-.4}\right) + 100\cdot\left(62.5e^{-.4} + 156.25e^{-.4} - 156.25\right)$
$= -1,228,125e^{-.4} + 1,234,375 = \$411,138.19.$

Solutions to Exercise Set 7.3

1. $\int(4+3x)^5\,dx = \frac{1}{3}\cdot\frac{1}{6}(4+3x)^6 + C = \frac{1}{18}(4+3x)^6 + C.$
2. $\int\frac{1}{(3-2x)^2}\,dx = -\frac{1}{(-2)(3-2x)} + C = \frac{1}{2(3-2x)} + C.$
3. $\int\frac{x}{(5+x)^2}\,dx = \frac{1}{1^2}\left[\ln|5+x| + \frac{5}{5+x}\right] + C = \ln|5+x| + \frac{5}{5+x} + C.$
4. $\int\frac{dx}{x(6+3x)} = -\frac{1}{6}\ln\left|\frac{6+3x}{x}\right| + C.$
5. $\int x\sqrt{7+2x}\,dx = -\frac{2(2\cdot 7 - 3\cdot 2x)(7+2x)^{3/2}}{15\cdot 2^2} + C$
$= -\frac{(7-3x)(7+2x)^{3/2}}{15} + C.$

6. $\int x\sqrt{1+x}\;dx = -\frac{2(2\cdot 1-3\cdot 1x)(1+x)^{3/2}}{15\cdot 1^2} + C$
$= -\frac{2(2-3x)(1+x)^{3/2}}{15} + C.$

7. $\int \frac{x}{\sqrt{3+2x}}\;dx = -\frac{2(2\cdot 3-2x)}{3\cdot 2^2}\sqrt{3+2x} + C = -\frac{(3-x)\sqrt{3+2x}}{3} + C.$

8. $\int \frac{dx}{9-x^2} = \frac{1}{2\cdot 3}\ln\left|\frac{3+x}{3-x}\right| + C = \frac{1}{6}\ln\left|\frac{3+x}{3-x}\right| + C.$

9. $\int \frac{dx}{\sqrt{16+x^2}} = \ln\left|x+\sqrt{16+x^2}\right| + C.$

10. $\int xe^{6x}\;dx = \frac{e^{6x}}{6^2}(6x-1) + C = \frac{e^{6x}}{36}(6x-1) + C.$

11. $\int \frac{x\;dx}{(5+x^2)^2} = \int \frac{\frac{1}{2}d(x^2)}{(5+x^2)^2} = \frac{1}{2}\left(-\frac{1}{1(5+x^2)} + C_1\right) = \frac{-1}{2(5+x^2)} + C.$

12. $\int \frac{x\;dx}{x^2(3+5x^2)}$. Let $u = x^2$.
$\int \frac{x\;dx}{x^2(3+5x^2)} = \frac{1}{2}\int \frac{du}{u(3+5u)} = \frac{1}{2}\left(-\frac{1}{3}\ln\left|\frac{3+5u}{u}\right| + C_1\right) = -\frac{1}{6}\ln\left(\frac{3+5x^2}{x^2}\right) + C$

13. $\int \frac{dx}{9-16x^2} = \frac{1}{16}\int \frac{dx}{\frac{9}{16}-x^2} = \frac{1}{16}\left(\frac{1}{2\cdot\frac{3}{4}}\ln\left|\frac{\frac{3}{4}+x}{\frac{3}{4}-x}\right| + C_1\right)$
$= \frac{1}{24}\ln\left|\frac{3+4x}{3-4x}\right| + C.$

14. $\int x\sqrt{x^4+1}\;dx$. Let $u = x^2$.
$\int x\sqrt{x^4+1}\;dx = \frac{1}{2}\int \sqrt{u^2+1}\;du$
$= \frac{1}{2}\left\{\frac{1}{2}\left[u\sqrt{u^2+1} + 1^2\ln\left(u+\sqrt{u^2+1}\right)\right] + C_1\right\}$
$= \frac{1}{4}\left[x^2\sqrt{x^4+1} + \ln\left(x^2+\sqrt{x^4+1}\right)\right] + C.$

15. $\int \frac{dx}{x\sqrt{x^4+9}} = \int \frac{x}{x^2\sqrt{x^4+9}}\;dx$. Let $u = x^2$.
$\int \frac{dx}{x\sqrt{x^4+9}} = \frac{1}{2}\int \frac{du}{u\sqrt{u^2+9}} = \frac{1}{2}\left[-\frac{1}{3}\ln\left(\frac{3+\sqrt{u^2+9}}{u}\right) + C_1\right]$
$= -\frac{1}{6}\ln\left(\frac{3+\sqrt{x^4+9}}{x^2}\right) + C.$

16. $\int x^3e^{x^2}\;dx$. Let $u = x^2$.
$\int x^3e^{x^2}\;dx = \frac{1}{2}\int ue^u\;du = \frac{1}{2}\left[e^u(u-1) + C_1\right] = \frac{1}{2}e^{x^2}(x^2-1) + C.$

17. $\int \frac{4x}{1+e^{x^2}}\;dx$. Let $u = x^2$. $du = 2x\;dx$.
$\int \frac{4x}{1+e^{x^2}}\;dx = \int \frac{2}{1+e^u}\;du = 2\left[\ln\left(\frac{e^u}{1+e^u}\right) + C_1\right] = 2\ln\left(\frac{e^{x^2}}{1+e^{x^2}}\right) + C.$

18. $\int \frac{x\;dx}{7-2e^{x^2}}$. Let $u = x^2$.
$\int \frac{x\;dx}{7-2e^{x^2}} = \frac{1}{2}\int \frac{du}{7-2e^u} = \frac{1}{2}\left[\frac{u}{7} - \frac{1}{7\cdot 1}\ln|7-2e^u| + C_1\right]$
$= \frac{1}{14}\left(u - \ln|7-2e^u|\right) + C = \frac{1}{14}\left(x^2 - \ln\left|7-2e^{x^2}\right|\right) + C.$

19. $\int \frac{dx}{\sqrt{x}\left(1+e^{\sqrt{x}}\right)}$. Let $u = \sqrt{x}$. $du = \frac{1}{2}\cdot\frac{1}{\sqrt{x}}\;dx$.
$\int \frac{dx}{\sqrt{x}\left(1+e^{\sqrt{x}}\right)} = 2\int \frac{du}{1+e^u} = 2\left[\ln\left(\frac{e^u}{1+e^u}\right) + C_1\right] = 2\ln\left(\frac{e^{\sqrt{x}}}{1+e^{\sqrt{x}}}\right) + C.$

20. $\int \frac{dx}{x\ln x^2} = \int \frac{x\,dx}{x^2\ln x^2}.$ Let $u = x^2$.

$\int \frac{dx}{x\ln x^2} = \frac{1}{2}\int \frac{du}{u\ln u} = \frac{1}{2}\left[\ln|\ln u| + C_1\right] = \frac{1}{2}\ln|\ln x^2| + C.$

21. $\int x\sin^3(x^2)dx.$ Let $u = x^2$. $du = 2x\,dx.$

$$\int x\sin^3(x^2)dx = \frac{1}{2}\int \sin^3 u\,du = \frac{1}{2}\left[-\frac{1}{3}\cos u(\sin^2 u + 2) + C_1\right]$$
$$= -\frac{1}{6}\cos x^2\left(\sin^2(x^2) + 2\right) + C.$$

22. $\int e^x\cos^3(e^x)dx.$ Let $u = e^x$.

$$\int e^x\cos^3(e^x)dx = \int \cos^3 u\,du = \frac{1}{3}\sin u(\cos^2 u + 2) + C$$
$$= \frac{1}{3}\sin(e^x)\cdot\left(\cos^2(e^x) + 2\right) + C.$$

23. Let $u = \cos x$. $du = -\sin x\,dx.$

$$\int \frac{\sin x}{\cos x(5 - 2\cos x)}\,dx = \int \frac{-du}{u\cdot(5 - 2u)} = \frac{1}{5}\ln\left|\frac{5 - 2u}{u}\right| + C$$
$$= \frac{1}{5}\ln\left|\frac{5 - 2\cos x}{\cos x}\right| + C.$$

24. Let $u = \sqrt{x}$. $du = \frac{1}{2\sqrt{x}}\,dx \implies \frac{1}{\sqrt{x}}\,dx = 2\,du.$

$$\int \frac{1}{\sqrt{x}(4 - x)}\,dx = \int \frac{2\,du}{(4 - u^2)} = 2\cdot\frac{1}{4}\ln\left|\frac{u + 2}{u - 2}\right| + C$$
$$= \frac{1}{2}\ln\left|\frac{\sqrt{x} + 2}{\sqrt{x} - 2}\right| + C.$$

25. Let $u = \sec x$. $du = \sec x\tan x\,dx.$

$$\int \sqrt{\sec^2 x + 9}\sec x\tan x\,dx$$
$$= \int \sqrt{u^2 + 9}\,du$$
$$= \frac{1}{2}\left[u\sqrt{u^2 + 9} + 9\ln\left(u + \sqrt{u^2 + 9}\right)\right] + C$$
$$= \frac{1}{2}\left[\sec x\cdot\sqrt{\sec^2 x + 9} + 9\ln\left(\sec x + \sqrt{\sec^2 x + 9}\right)\right] + C.$$

26. Let $u = \sec x$. $du = \sec x \tan x\, dx \implies \tan x\, dx = \frac{1}{\sec x}\, du$
$\implies \tan x\, dx = \frac{1}{u}\, du.$

$$\int \sqrt{\sec^2 x + 25}\ \tan x\, dx = \int \frac{\sqrt{u^2+25}}{u}\, du$$
$$= \sqrt{u^2+25} - 5\ln\left(\frac{5+\sqrt{u^2+25}}{u}\right) + C$$
$$= \sqrt{\sec^2 x + 25} - 5\ln\left(\frac{5+\sqrt{\sec^2 x+25}}{\sec x}\right) + C.$$

27. Let $u = \sqrt{x}$. $du = \frac{1}{2\sqrt{x}}\, dx \implies \frac{1}{4\sqrt{x}}\, dx = \frac{1}{2}\, du.$

$$\int \frac{1}{4\sqrt{x}\left(1+e^{\sqrt{x}}\right)}\, dx = \frac{1}{2}\int \frac{1}{1+e^u}\, du = \frac{1}{2}\ln\left(\frac{e^u}{1+e^u}\right) + C$$
$$= \frac{1}{2}\ln\left(\frac{e^{\sqrt{x}}}{1+e^{\sqrt{x}}}\right) + C.$$

28. Let $u = \sin x$. $du = \cos x\, dx = \frac{1}{\sec x}\, dx.$

$$\int \frac{1}{\sec x\left(1+e^{\sin x}\right)}\, dx = \int \frac{1}{1+e^u}\, du = \ln\left(\frac{e^u}{1+e^u}\right) + C$$
$$= \ln\left(\frac{e^{\sin x}}{1+e^{\sin x}}\right) + C.$$

29. When $p_0 = 20$, $20 = \frac{100}{1+e^{2x_0}} \implies 20 + 20e^{2x_0} = 100 \implies e^{2x_0} = 4 \implies$
$2x_0 = \ln 4 \implies x_0 = \frac{1}{2}\ln 4 = \ln 4^{1/2} = \ln 2.$

$$\text{Consumer's surplus} = \int_0^{\ln 2} \frac{100}{1+e^{2x}}\, dx - 20\cdot\ln 2$$
$$= 50\ln\left(\frac{e^{2x}}{1+e^{2x}}\right)\Big]_0^{\ln 2} - 20\cdot\ln 2$$
$$= 50\left(\ln\left(\frac{4}{1+4}\right) - \ln\left(\frac{1}{1+1}\right)\right) - 20\cdot\ln 2$$
$$= 50\left(\ln\tfrac{4}{5} - \ln\tfrac{1}{2}\right) - 20\cdot\ln 2$$
$$= 50\ln 1.6 - 20\cdot\ln 2 = 9.6372.$$

30. Average value $= \dfrac{\int_0^{\pi/2} 4\sin^3 \pi x\,dx}{\frac{\pi}{2}-0} = \dfrac{\frac{4}{\pi}\cdot\left(-\frac{1}{3}\cos \pi x(\sin^2 \pi x+2)\right)\Big]_0^{\pi/2}}{\frac{\pi}{2}}$

$= \frac{8}{\pi^2}\cdot\left\{\left(-\frac{1}{3}\cos\frac{\pi^2}{2}\left(\sin^2\frac{\pi^2}{2}+2\right)\right)-\left(-\frac{1}{3}\cos 0(\sin^2 0+2)\right)\right\}$

$= \frac{8}{\pi^2}\{-.2170063+.6666667\} = 0.3645.$

31. Area $= \int_{-1}^{1}[f(x)-g(x)]\,dx = 2\cdot\int_{-1}^{1}\frac{1}{9-x^2}\,dx = 2\cdot\frac{1}{6}\ln\left|\frac{x+3}{x-3}\right|\Big]_{-1}^{1}$

$= \frac{1}{3}\left(\ln 2-\ln\frac{1}{2}\right) = \frac{1}{3}(\ln 2-(\ln 1-\ln 2)) = \frac{2}{3}\ln 2 = 0.4621.$

Solutions to Exercise Set 7.4

1. $\int_2^\infty \frac{1}{x^2}\,dx = \lim_{t\to\infty}\int_2^t \frac{1}{x^2}\,dx = \lim_{t\to\infty}\left(-\frac{1}{x}\right)\Big]_2^t$

$= \lim_{t\to\infty}\left(-\frac{1}{t}+\frac{1}{2}\right) = \frac{1}{2}$. It converges to $\frac{1}{2}$.

2. $\int_1^\infty \frac{1}{x^3}\,dx = \lim_{t\to\infty}\int_1^t \frac{1}{x^3}\,dx = \lim_{t\to\infty}\left(-\frac{1}{2x^2}\right)\Big]_1^t$

$= \lim_{t\to\infty}\left(-\frac{1}{2t^2}+\frac{1}{2}\right) = \frac{1}{2}$. It converges to $\frac{1}{2}$.

3. $\int_0^\infty e^{-x}\,dx = \lim_{t\to\infty}\int_0^t e^{-x}\,dx = \lim_{t\to\infty}(-e^{-x})\Big]_0^t$

$= \lim_{t\to\infty}(-e^{-t}+1) = (-0+1) = 1$. It converges to 1.

4. $\int_1^\infty xe^{x^2}\,dx = \lim_{t\to\infty}\int_1^t xe^{x^2}\,dx = \lim_{t\to\infty}\left(\frac{1}{2}e^{x^2}\Big]_1^t\right)$

$= \lim_{t\to\infty}\left(\frac{1}{2}e^{t^2}-\frac{1}{2}e\right) = +\infty$. Diverges.

5. $\int_1^\infty \frac{1}{\sqrt{x}}\,dx = \lim_{t\to\infty}\int_1^t \frac{1}{\sqrt{x}}\,dx = \lim_{t\to\infty}\left(2\sqrt{x}\Big]_1^t\right)$

$= \lim_{t\to\infty}\left(2\sqrt{t}-2\right) = +\infty$. Diverges.

6. $\int_0^\infty e^{\pi x}\,dx = \lim_{t\to\infty}\int_0^t e^{\pi x}\,dx = \lim_{t\to\infty}\frac{e^{\pi x}}{\pi}\Big]_0^t$

$= \lim_{t\to\infty}\frac{1}{\pi}(e^{\pi t}-1) = +\infty$. Diverges.

7. $\int_0^\infty \frac{x}{\sqrt{4+x^2}}\,dx = \lim_{t\to\infty}\int_0^t \frac{x}{\sqrt{4+x^2}}\,dx = \lim_{t\to\infty}\int_0^t \frac{\frac{1}{2}}{\sqrt{4+x^2}}\,dx^2 = \lim_{t\to\infty}\left(\sqrt{4+x^2}\Big]_0^t\right)$

$= \lim_{t\to\infty}\left(\sqrt{4+t^2}-2\right) = +\infty$. Diverges.

8. $\int_1^\infty \frac{e^{-\sqrt{x}}}{\sqrt{x}}\,dx = \lim_{t\to\infty}\int_1^t \frac{e^{-\sqrt{x}}}{\sqrt{x}}\,dx.$

Let $u = -\sqrt{x}$.

$\int \frac{e^{-\sqrt{x}}}{\sqrt{x}}\,dx = \int(-2e^u)du = -2e^u + C = -2e^{-\sqrt{x}} + C.$

$\int_1^\infty \frac{e^{-\sqrt{x}}}{\sqrt{x}}\,dx = \lim_{t\to\infty}\int_1^t \frac{e^{-\sqrt{x}}}{\sqrt{x}}\,dx = \lim_{t\to\infty}\left(-2e^{-\sqrt{x}}\Big]_1^t\right)$

$= \lim_{t\to\infty}\left(-2\left(e^{-\sqrt{t}}-\frac{1}{e}\right)\right) = -2\left(0-\frac{1}{e}\right) = \frac{2}{e}.$

It converges to $\frac{2}{e}$.

9. $\int_0^\infty \frac{1}{x+2}\,dx = \lim_{t\to\infty}\int_0^t \frac{1}{x+2}\,dx = \lim_{t\to\infty}\left(\ln|x+2|\right)]_0^t$

$= \lim_{t\to\infty}\left(\ln|t+2|-\ln 2\right) = +\infty$. Diverges.

10. $\int_0^\infty x\sqrt{1+x^2}\,dx = \lim_{t\to\infty}\int_0^t x\sqrt{1+x^2}\,dx = \lim_{t\to\infty}\int_0^t \frac{1}{2}\sqrt{1+x^2}\,dx^2$

$= \lim_{t\to\infty}\frac{1}{3}(1+x^2)^{3/2}\Big]_0^t = \lim_{t\to\infty}\frac{1}{3}\left[(1+t^2)^{3/2}-1\right]$

$= +\infty$. Diverges.

11. $\int_{-\infty}^{-2} \frac{1}{x^2}\,dx = \lim_{t\to-\infty}\int_t^{-2} \frac{1}{x^2}\,dx = \lim_{t\to-\infty}\left(-\frac{1}{x}\Big]_t^{-2}\right) = \lim_{t\to-\infty}\left(\frac{1}{2}+\frac{1}{t}\right) = \frac{1}{2}.$

It converges to $\frac{1}{2}$.

12. $\int_{-\infty}^{0} e^x\,dx = \lim_{t\to-\infty}\int_t^0 e^x\,dx = \lim_{t\to-\infty} e^x\Big]_t^0 = \lim_{t\to-\infty}(1-e^t) = 1.$

It converges to 1.

13. $\int_{-\infty}^{0} \frac{1}{\sqrt{1-x}}\,dx = \lim_{t\to-\infty}\int_t^0 \frac{1}{\sqrt{1-x}}\,dx = \lim_{t\to-\infty}\left(-2\sqrt{1-x}\right)\Big]_t^0$

$= \lim_{t\to-\infty}\left(-2\cdot\left(1-\sqrt{1-t}\right)\right) = +\infty$. Diverges.

14. $\int_0^\infty \frac{x}{1+x^2}\,dx = \lim_{t\to\infty}\int_0^t \frac{x}{1+x^2}\,dx = \lim_{t\to\infty}\int_0^t \frac{1}{2}\cdot\frac{1}{1+x^2}\,dx^2$

$= \lim_{t\to\infty}\frac{1}{2}\ln(1+x^2)\Big]_0^t = \lim_{t\to\infty}\frac{1}{2}\cdot\ln(1+t^2) = +\infty$. Diverges.

15. $\int_0^1 \frac{1}{\sqrt{1-x}}\,dx = \lim_{t\to 1^-} \int_0^t \frac{1}{\sqrt{1-x}}\,dx = \lim_{t\to 1^-} \left(-2\sqrt{1-x}\right)\Big]_0^t$

$= \lim_{t\to 1^-} \left(-2\left(\sqrt{1-t}-1\right)\right) = 2.$ It converges to 2.

16. $\int_0^1 \frac{1}{x}\,dx = \lim_{t\to 0+} \int_t^1 \frac{1}{x}\,dx = \lim_{t\to 0+} \ln|x|\Big]_t^1 = \lim_{t\to 0+} (\ln 1 - \ln|t|).$

$= +\infty.$ Diverges.

17. $\int_0^1 \frac{1}{x^2}\,dx = \lim_{t\to 0+} \int_t^1 \frac{1}{x^2}\,dx = \lim_{t\to 0+} \left(-\frac{1}{x}\right)\Big]_t^1 = \lim_{t\to 0+} \left(-1+\frac{1}{t}\right) = +\infty.$
Diverges.

18. $\int_1^3 \frac{x}{\sqrt{9-x^2}}\,dx = \lim_{t\to 3^-} \int_1^t \frac{x}{\sqrt{9-x^2}}\,dx = \lim_{t\to 3^-} \left(-\sqrt{9-x^2}\right)\Big]_1^t$

$= \lim_{t\to 3^-} \left(-\sqrt{9-t^2}+\sqrt{8}\right) = 2\sqrt{2}.$ It converges to $2\sqrt{2}$.

19. $\int_1^e \frac{1}{x\ln x}\,dx = \lim_{t\to 1+} \int_t^e \frac{1}{x\ln x}\,dx = \lim_{t\to 1+} (\ln|\ln x|)\Big]_t^e.$

$= \lim_{t\to 1+} (\ln\ln e - \ln|\ln t|) = +\infty.$ Diverges.

20. $\int_0^1 (1-x)^{-2/3}\,dx = \lim_{t\to 1^-} \int_0^t (1-x)^{-2/3}\,dx = \lim_{t\to 1^-} \left(-3(1-x)^{1/3}\right)\Big]_0^t$

$= \lim_{t\to 1^-} \left(-3\left[(1-t)^{1/3}-1\right]\right) = 3.$ It converges to 3.

21. $\int_1^3 \frac{1}{x^2-2x+1}\,dx = \lim_{t\to 1+} \int_t^3 \frac{dx}{(x-1)^2} = \lim_{t\to 1+} \left(-\frac{1}{(x-1)}\right)\Big]_t^3$

$= \lim_{t\to 1+} \left(-\frac{1}{2}+\frac{1}{t-1}\right) = +\infty.$ Diverges.

22. $\int_0^2 \frac{1}{(x-2)^2}\,dx = \lim_{t\to 2^-} \int_0^t \frac{dx}{(x-2)^2} = \lim_{t\to 2^-} \left(-\frac{1}{x-2}\right)\Big]_0^t.$

$= \lim_{t\to 2^-} \left(-\frac{1}{t-2}-\frac{1}{2}\right) = +\infty.$ Diverges.

23. $P_0 = \lim_{T\to\infty} \int_0^T 100e^{-0.10t}\,dt = \lim_{T\to\infty} \left(-\frac{100}{0.10}e^{-0.10t}\right)\Big]_0^T$
$= \lim_{T\to\infty} 1000\left(1-e^{-0.10T}\right) = \$1000.$

24. $P_0 = \lim_{T\to\infty} \int_0^T 100e^{-0.05t}\,dt = \lim_{T\to\infty} \left(-\frac{100}{0.05}e^{-0.05t}\right)\Big]_0^T$
$= \lim_{T\to\infty} 2000\left(1-e^{-0.05T}\right) = \$2000.$

25. a. $P_0 = \int_0^{10} 1e^{-0.10t}\, dt = -\frac{1}{0.10}e^{-0.10t}\Big]_0^{10} = 10 \cdot \left(1 - \frac{1}{e}\right)$ million dollars.

b. $P_0 = \lim_{T\to\infty} \int_0^{T} 1e^{-0.10t}\, dt = \lim_{T\to\infty} \left(-\frac{1}{0.10}e^{-0.10t}\right)\Big]_0^{T} = \lim_{T\to\infty} 10\left(1 - e^{-0.10T}\right)$
$= 10$ million dollars.

26. $P_0 = \lim_{T\to\infty} \int_0^{T} 1000e^{-0.05t}\, dt = \lim_{T\to\infty} \left(-\frac{1000}{0.05}e^{-0.05t}\right)\Big]_0^{T}$
$= \lim_{T\to\infty} 20,000\left(1 - e^{-0.05T}\right) = \$20,000.$

27. Volume $= \int_0^{\infty} \pi(e^{-x})^2 dx = \pi \int_0^{\infty} e^{-2x}\, dx = \pi \cdot \lim_{t\to\infty} \left(-\frac{1}{2}e^{-2x}\right)\Big]_0^{t}$
$= \frac{\pi}{2} \lim_{t\to\infty} \left(1 - e^{-2t}\right) = \frac{\pi}{2}.$

28. Case 1: $p = -1$.
$\int_1^{t} x^{-1}\, dx = \ln x\Big]_1^{t} = \ln t \to \infty$ as $t \to \infty$.
Case 2: $p \neq -1$.
$\int_1^{t} x^{p}\, dx = \frac{1}{p+1}x^{p+1}\Big]_1^{t} = \frac{t^{p+1}}{p+1} - \frac{1}{p+1}.$
$\lim_{t\to\infty} \frac{t^{p+1}}{p+1}$ is finite if and only if $p + 1 < 0 \implies p < -1$.
Therefore the improper integral $\int_1^{\infty} x^{p}\, dx$ converges if and only if $p < -1$.

29. Case 1: $p = 1$.
$\int_t^{1} \frac{1}{x}\, dx = \ln x\Big]_t^{1} = -\ln t \to \infty$ as $t \to 0^+$.
Case 2: $p \neq 1$.
$\int_t^{1} \frac{1}{x^p}\, dx = \int_t^{1} x^{-p}\, dx = \frac{1}{-p+1}x^{-p+1}\Big]_t^{1} = \frac{1}{-p+1} - \frac{1}{-p+1}t^{-p+1}.$
$\lim_{t\to 0^+} \frac{1}{-p+1}t^{-p+1}$ is finite if and only if $-p + 1 > 0 \implies p < 1$.
Therefore the improper integral $\int_0^{1} \frac{1}{x^p}\, dx$ converges if and only if $p < 1$.

30. a. $P_0 = \lim_{T\to\infty} \int_0^{T} 1000e^{-0.05t}\, dt = \lim_{T\to\infty} \left(-\frac{1000}{0.05}e^{-0.05t}\right)\Big]_0^{T}$
$= \lim_{T\to\infty} 20,000\left(1 - e^{-0.05T}\right) = \$20,000.$

b. $P_0 = \lim_{T\to\infty} \int_0^{T} 1000e^{-0.10t}\, dt = \lim_{T\to\infty} \left(-\frac{1000}{0.10}e^{-0.1t}\right)\Big]_0^{T}$
$= \lim_{T\to\infty} 10000\left(1 - e^{-0.1T}\right) = \$10,000.$

c. $P_0 = \lim_{T\to\infty} \int_0^T 1000e^{-0.15t}\,dt = \lim_{T\to\infty} \left(-\frac{1000}{0.15}e^{-0.15t}\right)\Big]_0^T$
$= \lim_{T\to\infty} 6666.67\left(1 - e^{-0.15T}\right) = \$6666.67.$

31. $P_0 = \lim_{T\to\infty} \int_0^T 5,000e^{-.10T}\,dt = \lim_{T\to\infty} 50,000(1 - e^{-.10T}) = \$50,000.$

32. Fair compensation $= \lim_{T\to\infty} \int_0^T e^{-.10t}(1,000,000e^{.05t})dt$
$= 1,000,000 \cdot \frac{1}{-.05} \cdot \lim_{T\to\infty} (e^{-.05T} - 1) = \$20,000,000.$

33. $\int_0^2 \frac{1}{x-2}\,dx = \lim_{t\to 2^-} \ln|x-2|\Big]_0^t = -\infty.$

Therefore $\int_0^6 \frac{1}{x-2}\,dx$ diverges.

34. $\int_0^2 \frac{1}{(x-2)^2}\,dx = \lim_{t\to 2^-} \left(\frac{-1}{x-2}\right)\Big]_0^t = \infty.$

Therefore $\int_0^6 \frac{1}{(x-2)^2}\,dx$ diverges.

35. $\int_0^2 \frac{1}{(x-2)^{2/3}}\,dx = 3 \cdot \lim_{t\to 2^-} (x-2)^{1/3}\Big]_0^t = 3 \cdot \left(0 - (-2)^{1/3}\right) = 3 \cdot \sqrt[3]{2}.$

$\int_2^4 \frac{1}{(x-2)^{2/3}}\,dx = 3 \cdot \lim_{t\to 2^+} (x-2)^{1/3}\Big]_t^4 = 3 \cdot \left((2)^{1/3} - 0\right) = 3 \cdot \sqrt[3]{2}.$

Therefore the improper integral $\int_0^4 \frac{1}{(x-2)^{2/3}}\,dx$ converges and has the value $3\left(\sqrt[3]{2} + \sqrt[3]{2}\right) = 6\sqrt[3]{2}.$

36. $\int_0^{\pi/2} \tan x\,dx = \lim_{t\to \frac{\pi}{2}^-} \ln\sec x\Big]_0^t = \infty.$

Therefore $\int_0^{\pi} \tan x\,dx$ diverges.

37. $\int_0^1 \frac{1}{x^2-2x+1}\,dx = \int_0^1 \frac{1}{(x-1)^2}\,dx = \lim_{t\to 1^-} \left(-\frac{1}{x-1}\right)\Big]_0^t = \infty.$

Therefore $\int_0^2 \frac{1}{x^2-2x+1}\,dx$ diverges.

38. $\int_{-5}^{-3} \frac{1}{\sqrt[3]{x+3}}\,dx = \lim_{t\to -3^-} \left(\frac{3}{2}(x+3)^{2/3}\right)\Big]_{-5}^{t} = -\frac{3}{2}\cdot 2^{2/3}.$

$\int_{-3}^{0} \frac{1}{\sqrt[3]{x+3}}\,dx = \lim_{t\to -3^+} \left(\frac{3}{2}(x+3)^{2/3}\right)\Big]_{t}^{0} = \frac{3}{2}\cdot 3^{2/3}.$

Therefore the improper integral $\int_{-5}^{0} \frac{1}{\sqrt[3]{x+3}}\,dx$ converges and has the value $\frac{3}{2}\cdot\left(-2^{2/3}+3^{2/3}\right)$.

Solutions to Exercise Set 7.5

1. $f(x) = (x^3 - 7x + 4)$. Since $n = 4$, one obtains

$\int_0^4 f(x)dx \approx \frac{4-0}{2\cdot 4}\left(f(0) + 2f(1) + 2f(2) + 2f(3) + f(4)\right)$

$= \frac{1}{2}\left(4 + 2(-2) + 2(-2) + 2(10) + 40\right) = \frac{1}{2}(56) = 28.$

2. Put $f(x) = \frac{4}{1+x^2}$. Since $n = 4$, one obtains

$$\int_0^2 f(x)dx \approx \frac{2-0}{2\cdot 4}\cdot\left(f(0) + 2f\left(\frac{1}{2}\right) + 2f(1) + 2f\left(\frac{3}{2}\right) + f(2)\right)$$

$$= \frac{1}{4}\cdot\left(4 + 2\cdot\frac{16}{5} + 2\cdot 2 + 2\cdot\frac{16}{13} + \frac{4}{5}\right)$$

$$= \frac{1}{4}\cdot\left(8 + \frac{36}{5} + \frac{32}{13}\right) = \frac{1}{4}\cdot\frac{1148}{65} \approx 4.4154.$$

3. Put $f(x) = \sin x$. Since $n = 4$, one obtains

$\int_0^{\pi} f(x)dx \approx \frac{\pi-0}{2\cdot 4}\left(f(0) + 2f\left(\frac{\pi}{4}\right) + 2f\left(\frac{\pi}{2}\right) + 2f\left(\frac{3\pi}{4}\right) + f(\pi)\right)$

$= \frac{\pi}{8}\left(0 + 2\cdot\frac{\sqrt{2}}{2} + 2\cdot 1 + 2\cdot\frac{\sqrt{2}}{2} + 0\right)$

$= \frac{\pi}{8}\left(2 + 2\sqrt{2}\right) = \frac{\pi}{4}\left(1 + \sqrt{2}\right) \approx 1.8961.$

4. Put $f(x) = \sin x$. Since $n = 6$, one obtains

$$\int_0^{\pi} f(x)dx \approx \tfrac{\pi-0}{2\cdot 6}\left(f(0)+2f\left(\tfrac{\pi}{6}\right)+2f\left(\tfrac{\pi}{3}\right)+2f\left(\tfrac{\pi}{2}\right)+2f\left(\tfrac{2\pi}{3}\right)\right.$$

$$\left.+2f\left(\tfrac{5\pi}{6}\right)+f(\pi)\right).$$

$$=\tfrac{\pi}{12}\left(0+1+\sqrt{3}+2+\sqrt{3}+1+0\right)$$

$$=\tfrac{\pi}{12}\left(4+2\sqrt{3}\right)=\tfrac{\pi}{6}\left(2+\sqrt{3}\right)\approx 1.9541.$$

5. Put $f(x) = \frac{4}{1+x^2}$. Since $n = 8$, one obtains

$$\int_0^4 f(x)dx \approx \frac{4-0}{2\cdot 8}\cdot\left(f(0)+2f\left(\frac{1}{2}\right)+2f(1)+2f\left(\frac{3}{2}\right)+2f(2)\right.$$

$$\left.+\,2f\left(\frac{5}{2}\right)+2f(3)+2f\left(\frac{7}{2}\right)+f(4)\right)$$

$$=\frac{1}{4}\cdot\left(4+2\cdot\frac{16}{5}+2\cdot 2+2\cdot\frac{16}{13}+2\cdot\frac{4}{5}+2\cdot\frac{16}{29}+2\cdot\frac{2}{5}+2\cdot\frac{16}{53}+\frac{4}{17}\right)$$

$$=\frac{1}{4}\cdot\left(8+\frac{44}{5}+\frac{32}{13}+\frac{32}{29}+\frac{32}{53}+\frac{4}{17}\right)$$

$$\approx 5.3010.$$

6. Put $f(x) = \sqrt{1+x^2}$. Since $n = 4$, one obtains

$$\int_0^2 f(x)dx \approx \tfrac{2-0}{2\cdot 4}\left(f(0)+2f\left(\tfrac{1}{2}\right)+2f(1)+2f\left(\tfrac{3}{2}\right)+f(2)\right)$$

$$=\tfrac{1}{4}\left(1+\sqrt{5}+2\sqrt{2}+\sqrt{13}+\sqrt{5}\right)$$

$$=\tfrac{1}{4}\left(1+2\sqrt{5}+2\sqrt{2}+\sqrt{13}\right)\approx 2.9765.$$

7. Put $f(x) = \cos\left(\frac{\pi x}{4}\right)$. Since $n = 4$, one obtains

$$\int_0^4 f(x)dx \approx \tfrac{4-0}{2\cdot 4}\left(f(0)+2f(1)+2f(2)+2f(3)+f(4)\right)$$

$$=\tfrac{1}{2}\left(1+2\cdot\tfrac{\sqrt{2}}{2}+2\cdot 0+2\cdot\left(-\tfrac{\sqrt{2}}{2}\right)+(-1)\right)$$

$$=\tfrac{1}{2}\left(1+\sqrt{2}-\sqrt{2}-1\right)=0.$$

8. Put $f(x)=\frac{x}{1+x^3}$. Since $n=4$, one obtains

$$\int_0^2 f(x)dx \approx \tfrac{2-0}{2\cdot 4}\left(f(0)+2f\left(\tfrac{1}{2}\right)+2f(1)+2f\left(\tfrac{3}{2}\right)+f(2)\right)$$

$$=\tfrac{1}{4}\left(0+2\cdot\tfrac{4}{9}+2\cdot\tfrac{1}{2}+2\cdot\tfrac{12}{35}+\tfrac{2}{9}\right)$$

$$=\tfrac{1}{4}\left(1+\tfrac{10}{9}+\tfrac{24}{35}\right)\approx 0.6992.$$

9. Put $f(x)=\frac{1}{1-x^2}$. Since $n=8$, one obtains

$$\int_2^4 f(x)dx \approx \tfrac{4-2}{2\cdot 8}\left(f(2)+2f\left(\tfrac{9}{4}\right)+2f\left(\tfrac{5}{2}\right)+2f\left(\tfrac{11}{4}\right)\right.$$

$$\left.+\,2f(3)+2f\left(\tfrac{13}{4}\right)+2f\left(\tfrac{7}{2}\right)+2f\left(\tfrac{15}{4}\right)+f(4)\right)$$

$$=\tfrac{1}{8}\left(-\tfrac{1}{3}+2\left(-\tfrac{16}{65}\right)+2\left(-\tfrac{4}{21}\right)+2\left(-\tfrac{16}{105}\right)+2\left(-\tfrac{1}{8}\right)\right.$$

$$\left.+2\left(-\tfrac{16}{153}\right)+2\left(-\tfrac{4}{45}\right)+2\left(-\tfrac{16}{209}\right)+\left(-\tfrac{1}{15}\right)\right)\approx -0.296.$$

10. Put $f(x)=x\cdot\sin\left(\frac{\pi x}{4}\right)$. Since $n=8$, one obtains

$$\int_{-4}^4 f(x)dx \approx \tfrac{4-(-4)}{2\cdot 8}\left(f(-4)+2f(-3)+2f(-2)+2f(-1)\right.$$

$$\left.+2f(0)+2f(1)+2f(2)+2f(3)+f(4)\right)$$

$$=\tfrac{1}{2}\left((-4)\cdot 0+2(-3)\left(-\tfrac{\sqrt{2}}{2}\right)+2(-2)(-1)+2(-1)\left(-\tfrac{\sqrt{2}}{2}\right)\right.$$

$$\left.+\,2\cdot 0+2\cdot\tfrac{\sqrt{2}}{2}+2\cdot 2\cdot 1+2\cdot 3\cdot\tfrac{\sqrt{2}}{2}+0\right)$$

$$=\tfrac{1}{2}\left(3\sqrt{2}+4+\sqrt{2}+\sqrt{2}+4+3\sqrt{2}\right)$$

$$=\tfrac{1}{2}\left(8\sqrt{2}+8\right)=4\sqrt{2}+4\approx 9.6569.$$

11. $\displaystyle\int_0^4 f(x)dx \approx \tfrac{4-0}{3\cdot 4}\left(f(0)+4f(1)+2f(2)+4f(3)+f(4)\right)$

$$=\tfrac{1}{3}\left(4+4(-2)+2(-2)+4\cdot 10+40\right)$$

$$=\tfrac{1}{3}(72)=24.$$

12. $\int_0^2 f(x)dx \approx \frac{2-0}{3\cdot 4}\left(f(0)+4f\left(\frac{1}{2}\right)+2f(1)+4f\left(\frac{3}{2}\right)+f(2)\right)$

$= \frac{1}{6}\left(4+4\cdot\frac{16}{5}+2\cdot 2+4\cdot\frac{16}{13}+\frac{4}{5}\right)$

$= \frac{1}{6}\left(8+\frac{68}{5}+\frac{64}{13}\right)\approx 4.4205.$

13. $\int_0^\pi f(x)dx \approx \frac{\pi-0}{3\cdot 4}\left(f(0)+4f\left(\frac{\pi}{4}\right)+2f\left(\frac{\pi}{2}\right)+4f\left(\frac{3\pi}{4}\right)+f(\pi)\right)$

$= \frac{\pi}{12}\left(0+4\cdot\frac{\sqrt{2}}{2}+2\cdot 1+4\cdot\frac{\sqrt{2}}{2}+0\right)$

$= \frac{\pi}{12}\left(2+4\sqrt{2}\right)=\frac{\pi}{6}\left(1+2\sqrt{2}\right)\approx 2.005.$

14. $\int_0^\pi f(x)dx \approx \frac{\pi-0}{3\cdot 6}\left(f(0)+4f\left(\frac{\pi}{6}\right)+2f\left(\frac{\pi}{3}\right)+4f\left(\frac{\pi}{2}\right)+2f\left(\frac{2\pi}{3}\right)\right.$

$\left.+4f\left(\frac{5\pi}{6}\right)+f(\pi)\right)$

$= \frac{\pi}{18}\left(0+4\left(\frac{1}{2}\right)+2\left(\frac{\sqrt{3}}{2}\right)+4\cdot 1+2\left(\frac{\sqrt{3}}{2}\right)+4\cdot\frac{1}{2}+0\right)$

$= \frac{\pi}{18}\left(2+\sqrt{3}+4+\sqrt{3}+2\right)=\frac{\pi}{18}\left(8+2\sqrt{3}\right)\approx 2.0009.$

15. $\int_0^4 f(x)dx \approx \frac{4-0}{3\cdot 8}\cdot\left(f(0)+4\cdot f\left(\frac{1}{2}\right)+2\cdot f(1)+4\cdot f\left(\frac{3}{2}\right)+2\cdot f(2)\right.$

$\left.+\,4\cdot f\left(\frac{5}{2}\right)+2\cdot f(3)+4\cdot f\left(\frac{7}{3}\right)+f(4)\right)$

$= \frac{1}{6}\cdot\left(4+4\cdot\frac{16}{5}+2\cdot 2+4\cdot\frac{16}{13}+2\cdot\frac{4}{5}+4\cdot\frac{16}{29}\right.$

$\left.+\,2\cdot\frac{2}{5}+4\cdot\frac{16}{53}+\frac{4}{17}\right)$

$= \frac{1}{6}\cdot\left(8+\frac{76}{5}+\frac{64}{13}+\frac{64}{29}+\frac{64}{53}+\frac{4}{17}\right)$

$\approx 5.2955.$

16. $\int_0^2 f(x)dx \approx \frac{2-0}{3\cdot 4}\left(f(0)+4f\left(\frac{1}{2}\right)+2f(1)+4f\left(\frac{3}{2}\right)+f(2)\right)$

$= \frac{1}{6}\left(\sqrt{1}+4\left(\frac{\sqrt{5}}{2}\right)+2\left(\sqrt{2}\right)+4\left(\frac{\sqrt{13}}{2}\right)+\sqrt{5}\right)$

$= \frac{1}{6}\left(1+3\sqrt{5}+2\sqrt{2}+2\sqrt{13}\right)\approx 2.9580.$

17. $\int\limits_0^4 f(x)dx \approx \frac{4-0}{3\cdot 4}\left(f(0)+4f(1)+2f(2)+4f(3)+f(4)\right)$

$$= \tfrac{1}{3}\left\{1+4\left(\tfrac{1}{2}\sqrt{2}\right)+2(0)+4\left(-\tfrac{1}{2}\sqrt{2}\right)+(-1)\right\}$$

$$= \tfrac{1}{3}(0) = 0.$$

18. $\int\limits_0^2 f(x)dx \approx \frac{2-0}{3\cdot 4}\left(f(0)+4f\left(\tfrac{1}{2}\right)+2f(1)+4f\left(\tfrac{3}{2}\right)+f(2)\right)$

$$= \tfrac{1}{6}\left(0+4\left(\tfrac{4}{9}\right)+2\left(\tfrac{1}{2}\right)+4\left(\tfrac{12}{35}\right)+\tfrac{2}{9}\right)$$

$$= \tfrac{1}{6}\left(3+\tfrac{48}{35}\right) \approx 0.7286.$$

19. $\int\limits_2^4 f(x)dx \approx \frac{4-2}{3\cdot 8}\left(f(2)+4f\left(\tfrac{9}{4}\right)+2f\left(\tfrac{10}{4}\right)+4f\left(\tfrac{11}{4}\right)+2f(3)+4f\left(\tfrac{13}{4}\right)\right.$

$$\left.+2f\left(\tfrac{14}{4}\right)+4f\left(\tfrac{15}{4}\right)+f(4)\right)$$

$$= \tfrac{1}{12}\left\{-\tfrac{1}{3}+4\left(-\tfrac{16}{65}\right)+2\left(-\tfrac{16}{84}\right)+4\left(-\tfrac{16}{105}\right)+2\left(-\tfrac{1}{8}\right)+4\left(-\tfrac{16}{153}\right)\right.$$

$$\left.+2\left(-\tfrac{16}{180}\right)+4\left(-\tfrac{16}{209}\right)+\left(-\tfrac{1}{15}\right)\right\}$$

$$= \tfrac{1}{12}(-3.5274) \approx -0.2939.$$

20. $\int\limits_{-4}^4 f(x)dx \approx \frac{4-(-4)}{3\cdot 8}\left(f(-4)+4f(-3)+2f(-2)+4f(-1)+2f(0)+4f(1)\right.$

$$\left.+2f(2)+4f(3)+f(4)\right)$$

$$= \tfrac{1}{3}\left\{-4(0)+4(-3)\left(-\tfrac{1}{2}\sqrt{2}\right)+2(-2)(-1)+4(-1)\left(-\tfrac{1}{2}\sqrt{2}\right)+2(0)(0)\right.$$

$$\left.+4(1)\left(\tfrac{1}{2}\sqrt{2}\right)+2(2)(1)+4(3)\left(\tfrac{1}{2}\sqrt{2}\right)+4(0)\right\}$$

$$= \tfrac{1}{3}\left(8+4\sqrt{2}+12\sqrt{2}\right) \approx 10.2091.$$

21. $\int_0^{\pi/4} \frac{2}{1+x^2}\,dx \approx \frac{\frac{\pi}{4}-0}{2\cdot 6}\left(f(0)+2\cdot f\left(\frac{\pi}{24}\right)+2\cdot f\left(\frac{\pi}{12}\right)+2\cdot f\left(\frac{\pi}{8}\right)\right.$

$\left.+2\cdot f\left(\frac{\pi}{6}\right)+2\cdot f\left(\frac{5\pi}{24}\right)+f\left(\frac{\pi}{4}\right)\right)$

$= \frac{\pi}{48}\cdot(2+2\cdot(1.966308)+2\cdot(1.871715)+2\cdot(1.732783)$

$+2\cdot(1.569667)+2\cdot(1.400200)+1.236973)$

$\approx 1.3298.$

22. Put $f(x)=\sqrt{1+x^2}$. Since $n=6$, one obtains

$$\int_0^3 f(x)dx \approx \frac{3-0}{2\cdot 6}\left(f(0)+2f\left(\frac{1}{2}\right)+2f(1)+2f\left(\frac{3}{2}\right)+2f(2)+2f\left(\frac{5}{2}\right)+f(3)\right)$$

$$= \frac{1}{4}\left(\sqrt{1}+2\left(\sqrt{\frac{5}{4}}\right)+2\left(\sqrt{2}\right)+2\left(\sqrt{\frac{13}{4}}\right)+2\left(\sqrt{5}\right)+2\left(\sqrt{\frac{29}{4}}\right)+\sqrt{10}\right)$$

$$\approx \frac{1}{4}(22.68962) \approx 5.6724.$$

23. Put $f(x)=\sqrt{x^2+2}$. Since $n=6$, one obtains

$$\int_1^3 f(x)dx \approx \frac{3-1}{2\cdot 6}\left(f(1)+2f\left(\frac{4}{3}\right)+2f\left(\frac{5}{3}\right)+2f(2)+2f\left(\frac{7}{3}\right)+2f\left(\frac{8}{3}\right)+f(3)\right)$$

$$= \frac{1}{6}\left\{\sqrt{3}+2\left(\sqrt{\frac{34}{9}}\right)+2\left(\sqrt{\frac{43}{9}}\right)+2\left(\sqrt{6}\right)+2\left(\sqrt{\frac{67}{9}}\right)+2\left(\sqrt{\frac{82}{9}}\right)+\left(\sqrt{11}\right)\right\}$$

$$= \frac{1}{6}(29.70041) \approx 4.9501.$$

24. $f(x)=\sin x^2$.

$$\int_0^1 \sin x^2 dx \approx \frac{1-0}{2\cdot 6}\left(f(0)+2f\left(\frac{1}{6}\right)+2f\left(\frac{1}{3}\right)+2f\left(\frac{1}{2}\right)+2f\left(\frac{2}{3}\right)+2f\left(\frac{5}{6}\right)+f(1)\right)$$

$$\approx \frac{1}{12}(0+2(0.02777)+2(0.11088)+2(0.24740)$$

$$+2(0.42996)+2(0.63996)+0.84147) \approx 0.3128.$$

25. $f(x) = x\sin\left(\frac{\pi x}{2}\right)$.

$$\int_0^2 x\sin\left(\tfrac{\pi x}{2}\right)dx \approx \tfrac{2-0}{2\cdot 6}\left(f(0)+2f\left(\tfrac{1}{3}\right)+2f\left(\tfrac{2}{3}\right)+2f(1)+2f\left(\tfrac{4}{3}\right)+2f\left(\tfrac{5}{3}\right)+f(2)\right)$$

$$= \tfrac{1}{6}\left(0+2\left(\tfrac{1}{3}\right)\left(\tfrac{1}{2}\right)+2\left(\tfrac{2}{3}\right)\left(\tfrac{\sqrt{3}}{2}\right)+2(1)(1)+2\left(\tfrac{4}{3}\right)\left(\tfrac{\sqrt{3}}{2}\right)\right.$$

$$\left.+2\left(\tfrac{5}{3}\right)\left(\tfrac{1}{2}\right)+0\right)$$

$$= \tfrac{1}{6}\left(\tfrac{1}{3}+\tfrac{2}{3}\sqrt{3}+2+\tfrac{4}{3}\sqrt{3}+\tfrac{5}{3}\right) = \tfrac{1}{6}\left(4+2\sqrt{3}\right) \approx 1.2440.$$

26. $f(x) = \tan\sqrt{x}$.

$$\int_0^{\pi/4} f(x)dx \approx \tfrac{\frac{\pi}{4}-0}{2\cdot 6}\left(f(0)+2f\left(\tfrac{\pi}{24}\right)+2f\left(\tfrac{\pi}{12}\right)+2f\left(\tfrac{\pi}{8}\right)+2f\left(\tfrac{\pi}{6}\right)+2f\left(\tfrac{5\pi}{24}\right)+f\left(\tfrac{\pi}{4}\right)\right)$$

$$= \tfrac{\pi}{48}(0+2(0.37846)+2(0.56154)+2(0.72401)$$

$$+\,2(0.88346)+2(1.04838)+1.22512) \approx 0.5509.$$

27. $$\int_0^{\pi/4} \tfrac{2}{1+x^2}\,dx \approx \tfrac{\frac{\pi}{4}-0}{3\cdot 6}\cdot\left(f(0)+4\cdot f\left(\tfrac{\pi}{24}\right)+2\cdot f\left(\tfrac{\pi}{12}\right)+4\cdot f\left(\tfrac{\pi}{8}\right)\right.$$

$$\left.+\,2\cdot f\left(\tfrac{\pi}{6}\right)+4\cdot f\left(\tfrac{5\pi}{24}\right)+f\left(\tfrac{\pi}{4}\right)\right)$$

$$= \tfrac{\pi}{72}(2+4\cdot(1.966308)+2\cdot(1.871715)+4\cdot(1.732833)$$

$$+\,2\cdot(1.569667)+4\cdot(1.400200)+1.236973)$$

$$\approx 1.3316.$$

28. $f(x) = \sqrt{1+x^2}$.

$$\int_0^3 f(x)dx \approx \tfrac{3-0}{3\cdot 6}\left(f(0)+4f\left(\tfrac{1}{2}\right)+2f(1)+4f\left(\tfrac{3}{2}\right)+2f(2)+4f\left(\tfrac{5}{2}\right)+f(3)\right)$$

$$= \tfrac{1}{6}\left(1+2\sqrt{5}+2\sqrt{2}+2\sqrt{13}+2\sqrt{5}+2\sqrt{29}+\sqrt{10}\right)$$

$$= \tfrac{1}{6}\left(1+4\sqrt{5}+2\sqrt{2}+2\sqrt{13}+2\sqrt{29}+\sqrt{10}\right) \approx 5.6527.$$

29. $f(x) = \sqrt{x^2+2}$.

$$\int_1^3 f(x)dx \approx \tfrac{3-1}{3\cdot 6}\left(f(1)+4f\left(\tfrac{4}{3}\right)+2f\left(\tfrac{5}{3}\right)+4f(2)+2f\left(\tfrac{7}{3}\right)+4f\left(\tfrac{8}{3}\right)+f(3)\right)$$

$$= \tfrac{1}{9}\left(\sqrt{3}+4\left(\tfrac{\sqrt{34}}{3}\right)+2\left(\tfrac{\sqrt{43}}{3}\right)+4\sqrt{6}+2\left(\tfrac{\sqrt{67}}{3}\right)+4\left(\tfrac{\sqrt{82}}{3}\right)+\sqrt{11}\right)$$

$$\approx 4.9471.$$

30. $f(x) = \sin x^2$.

$$\int_0^1 f(x)dx \approx \tfrac{1-0}{3\cdot 6}\left(f(0)+4f\left(\tfrac{1}{6}\right)+2f\left(\tfrac{1}{3}\right)+4f\left(\tfrac{1}{2}\right)+2f\left(\tfrac{2}{3}\right)+4f\left(\tfrac{5}{6}\right)+f(1)\right)$$

$$= \tfrac{1}{18}(0+4(0.02777)+2(0.11088)+4(0.24740)$$

$$+2(0.42996)+4(0.63996)+0.84147) \approx 0.3102.$$

31. $f(x) = x\sin\left(\frac{\pi x}{2}\right)$.

$$\int_0^2 f(x)dx \approx \tfrac{2-0}{3\cdot 6}\left(f(0)+4f\left(\tfrac{1}{3}\right)+2f\left(\tfrac{2}{3}\right)+4f(1)+2f\left(\tfrac{4}{3}\right)+4f\left(\tfrac{5}{3}\right)+f(2)\right)$$

$$= \tfrac{1}{9}\left(0+4\left(\tfrac{1}{3}\right)\left(\tfrac{1}{2}\right)+2\left(\tfrac{2}{3}\right)\left(\tfrac{\sqrt{3}}{2}\right)+4(1)(1)+2\left(\tfrac{4}{3}\right)\left(\tfrac{\sqrt{3}}{2}\right)+4\left(\tfrac{5}{3}\right)\left(\tfrac{1}{2}\right)+0\right)$$

$$= \tfrac{1}{9}\left(4+\tfrac{2+6\sqrt{3}+10}{3}\right) = \tfrac{1}{9}\left(8+2\sqrt{3}\right) \approx 1.27389.$$

32. $f(x) = \tan\sqrt{x}$.

$$\int_0^{\pi/4} f(x)dx \approx \tfrac{\frac{\pi}{4}-0}{3\cdot 6}\left(f(0)+4f\left(\tfrac{\pi}{24}\right)+2f\left(\tfrac{\pi}{12}\right)+4f\left(\tfrac{\pi}{8}\right)+2f\left(\tfrac{\pi}{6}\right)+4f\left(\tfrac{5\pi}{24}\right)+f\left(\tfrac{\pi}{4}\right)\right)$$

$$= \tfrac{\pi}{72}(0+4(0.37846)+2(0.56154)+4(0.72401)$$

$$+2(0.88346)+4(1.04838)+1.22512) \approx 0.5549.$$

33. $\int_0^{\pi/4} \frac{1}{1+x^2}\, dx \approx ?$

number of subintervals	approximation to the integral
$n = 5$	0.6649153
$n = 20$	0.6656965
$n = 100$	0.6657707
$n = 250$	0.6657731

34. $\int_0^{\pi/4} \frac{1}{1+x^2} dx \approx ?$

number of subintervals	approximation to the integral
$n = 5$	0.6657755
$n = 20$	0.6657739
$n = 100$	0.6657739
$n = 250$	0.665774

35. a. $f(x) = \sin\sqrt{x}$.

$$\int_0^3 f(x)dx \approx \frac{3-0}{2\cdot 6}\left(f(0) + 2f\left(\tfrac{1}{2}\right) + 2f(1) + 2f\left(\tfrac{3}{2}\right) + 2f(2) + 2f\left(\tfrac{5}{2}\right) + f(3)\right)$$

$$\approx \tfrac{1}{4}(0 + 2(0.64964) + 2(0.84147) + 2(0.94072) + 2(0.98777)$$

$$+2(0.99995) + 0.98703)$$

$$\approx 2.4565.$$

b. $f(x) = \frac{1}{\sqrt{\cos x}}$.

$$\int_0^{\pi/3} f(x)dx \approx \frac{\frac{\pi}{3}-0}{2\cdot 6}\left(f(0) + 2f\left(\tfrac{\pi}{18}\right) + 2f\left(\tfrac{\pi}{9}\right) + 2f\left(\tfrac{\pi}{6}\right) + 2f\left(\tfrac{2\pi}{9}\right)\right.$$

$$\left. +2f\left(\tfrac{5\pi}{18}\right) + f\left(\tfrac{\pi}{3}\right)\right)$$

$$\approx \tfrac{\pi}{36}(1 + 2(1.00768) + 2(1.03159) + 2(1.07457)$$

$$+2(1.14254) + 2(1.24729) + 1.41421) \approx 1.17125.$$

36. a. $f(x) = \sin\sqrt{x}$.

$$\int_0^3 f(x)dx \approx \tfrac{3-0}{3\cdot 6}\left(f(0) + 4f\left(\tfrac{1}{2}\right) + 2f(1) + 4f\left(\tfrac{3}{2}\right) + 2f(2) + 4f\left(\tfrac{5}{2}\right) + f(3)\right)$$

$$\approx \tfrac{1}{6}\left(0 + 4(0.64964) + 2(0.84147) + 4(0.94072)\right.$$

$$\left. + 2(0.98777) + 4(0.99995) + 0.98703\right) \approx 2.50118.$$

b. $f(x) = \frac{1}{\sqrt{\cos x}}$.

$$\int_0^{\pi/3} f(x)dx \approx \tfrac{\frac{\pi}{3}-0}{3\cdot 6}\left(f(0) + 4f\left(\tfrac{\pi}{18}\right) + 2f\left(\tfrac{\pi}{9}\right) + 4f\left(\tfrac{\pi}{6}\right) + 2f\left(\tfrac{2\pi}{9}\right)\right.$$

$$\left. +4f\left(\tfrac{5\pi}{18}\right) + f\left(\tfrac{\pi}{3}\right)\right)$$

$$\approx \tfrac{\pi}{54}\left(1 + 4(1.00768) + 2(1.03159) + 4(1.07457)\right.$$

$$\left. +2(1.14254) + 4(1.24729) + 1.41421\right) \approx 1.1682.$$

37. $\int_{-\Delta x}^{\Delta x} (ax^2 + bx + c)dx = \left(\tfrac{1}{3}ax^3 + \tfrac{1}{2}bx^2 + cx\right)\Big]_{-\Delta x}^{\Delta x}$

$= \left\{\tfrac{1}{3}a(\Delta x)^3 + \tfrac{1}{2}b(\Delta x)^2 + c(\Delta x)\right\} - \left\{\tfrac{1}{3}a(-\Delta x)^3 + \tfrac{1}{2}b(-\Delta x)^2 + c(-\Delta x)\right\}$

$= \tfrac{\Delta x}{6}\left\{2a(\Delta x)^2 + 3b(\Delta x) + 6c + 2a(\Delta x)^2 - 3b(\Delta x) + 6c\right\}$

$= \tfrac{\Delta x}{3}\left\{2a(\Delta x)^2 + 6c\right\}$

$= \tfrac{\Delta x}{3}\left\{\left[a(\Delta x)^2 - b(\Delta x) + c\right] + 4c + \left[a(\Delta x)^2 + b(\Delta x) + c\right]\right\}$

$= \tfrac{\Delta x}{3}\left[y_0 + 4y_1 + y_2\right]$.

38. a. $\ln 3 = \int_1^3 \tfrac{1}{x}dx \approx \tfrac{3-1}{2\cdot 8}\left[1 + 2\left(\tfrac{4}{5}\right) + 2\left(\tfrac{2}{3}\right) + 2\left(\tfrac{4}{7}\right) + 2\left(\tfrac{1}{2}\right) + 2\left(\tfrac{4}{9}\right) + 2\left(\tfrac{2}{5}\right)\right.$

$\left. +2\left(\tfrac{4}{11}\right) + \left(\tfrac{1}{3}\right)\right]$

$= \tfrac{1}{8}\left(2 + \tfrac{8}{5} + \tfrac{4}{3} + \tfrac{8}{7} + \tfrac{8}{9} + \tfrac{4}{5} + \tfrac{8}{11} + \tfrac{1}{3}\right)$

$= \tfrac{1}{8}(8.8254) \approx 1.10321.$

b. $f(x) = \frac{1}{x}$. $f'(x) = -\frac{1}{x^2}$. $f''(x) = \frac{2}{x^3}$.

The maximum value of $|f''(x)|$ for $1 \le x \le 3$ is 2. Therefore $|\text{Error}| \le \frac{(3-1)^3 \cdot 2}{12(8)^2} = \frac{16}{768} = 0.02083$.

c. From the calculator, we get $\ln 3 = 1.098612$.

$|1.098612 - 1.10321| = 0.0045977$.

This is less than the estimated error in part b.

39. a. $\ln 5 = \int_1^5 \frac{1}{x} dx \approx \frac{5-1}{3 \cdot 8} \left[1 + 4\left(\frac{2}{3}\right) + 2\left(\frac{1}{2}\right) + 4\left(\frac{2}{5}\right) + 2\left(\frac{1}{3}\right)\right.$

$$\left. + 4\left(\frac{2}{7}\right) + 2\left(\frac{1}{4}\right) + 4\left(\frac{2}{9}\right) + \frac{1}{5}\right]$$

$$= \frac{1}{6}\left[2 + \frac{8}{3} + \frac{8}{5} + \frac{2}{3} + \frac{8}{7} + \frac{1}{2} + \frac{8}{9} + \frac{1}{5}\right]$$

$$\approx \frac{1}{6}(9.6650794) \approx 1.6108466.$$

b. $f(x) = \frac{1}{x}$. $f'(x) = -\frac{1}{x^2}$. $f''(x) = \frac{2}{x^3}$. $f'''(x) = -\frac{6}{x^4}$. $f^{\text{iv}}(x) = \frac{24}{x^5}$.

The maximum value of $|f^{\text{iv}}(x)|$ for $1 \le x \le 5$ is 24. Therefore

$|\text{Error}| \le \frac{(5-1)^5 \cdot 24}{180 \cdot (8)^4} = \frac{24756}{737280} = 0.03333\ldots$.

c. From the calculator, we get $\ln 5 = 1.6094379$.

$|1.6094379 - 1.6108466| = 0.0014087$.

This is less than the estimated error in part b.

Solutions to the Chapter 7 Review Exercises

1. Let $u = x + 5$. $du = dx$.

   $\int \sqrt{x+5}\, dx = \int \sqrt{u}\, du = \frac{2}{3}u^{3/2} + C = \frac{2}{3}(x+5)^{3/2} + C$.

2. Let $u = x^2$. $du = 2x\ dx$, $x\ dx = \frac{1}{2}\ du$.

   $\int x \cos x^2\ dx = \int \cos u \cdot \frac{1}{2}\ du = \frac{1}{2}\sin u + C = \frac{1}{2}\sin x^2 + C$.

3. Let $u = 1 + x^2$. $du = 2x\ dx$, $x\ dx = \frac{1}{2}\ du$.

4. Let $u = \sqrt{x+1}$. $du = \frac{1}{2}(x+1)^{-1/2}\, dx$, $\frac{1}{\sqrt{x+1}}\, dx = 2\, du$.

$\int \frac{e^{\sqrt{x+1}}}{\sqrt{x+1}}\, dx = \int e^u \cdot 2\, du = 2e^u + C = 2e^{\sqrt{x+1}} + C$.

5. $\int_0^1 (x-2)(x^2+3)dx = \int_0^1 (x^3 - 2x^2 + 3x - 6)dx$

$$= \left(\tfrac{1}{4}x^4 - \tfrac{2}{3}x^3 + \tfrac{3}{2}x^2 - 6x\right)\Big]_0^1$$

$$= \tfrac{1}{4} - \tfrac{2}{3} + \tfrac{3}{2} - 6 = \tfrac{3-8+18-72}{12} = \tfrac{-59}{12}.$$

6. Let $u = 1 - x^2$. $du = -2x\, dx$, $x\, dx = -\frac{1}{2}\, du$.

$\int \frac{x}{\sqrt{1-x^2}}\, dx = \int u^{-1/2}\left(-\frac{1}{2}\, du\right) = \frac{-\frac{1}{2}u^{1/2}}{1/2} + C = -\sqrt{1-x^2} + C$.

7. Let $u = \sin x$. $du = \cos x\, dx$, $\cos x\, dx = du$.

$\int_0^{\pi/6} \sin^3 x \cos x\, dx = \int_{x=0}^{x=\pi/6} u^3\, du = \frac{1}{4}u^4\Big]_{x=0}^{x=\pi/6} = \frac{1}{4}\sin^4 x\Big]_0^{\pi/6}$

$$= \tfrac{1}{4}\left(\tfrac{1}{16} - 0\right) = \tfrac{1}{64}.$$

8. Let $u = e^{2x}$. $du = 2e^{2x}\, dx$, $e^{2x}\, dx = \frac{1}{2}\, du$.

$\int e^{2x} \ln e^{2x}\, dx = \int (\ln u)\frac{1}{2}\, du = \frac{1}{2}(u \ln u - u) + C$

$= \frac{1}{2}(e^{2x}\ln e^{2x} - e^{2x}) + C$.

9. Let $u = x^2 + 2$. $du = 2x\, dx \implies x\, dx = \frac{1}{2}\, du$.

$\int x\sqrt{x^2+2}\, dx = \int \sqrt{u}\left(\frac{1}{2}\, du\right) = \frac{1}{2} \cdot \frac{2}{3}u^{3/2} + C$

$= \frac{1}{3}(x^2+2)^{3/2} + C$.

10. Let $u = x$, $dv = e^{3x}\, dx$. $du = dx$, $v = \frac{1}{3}e^{3x}$.

$\int xe^{3x}\, dx = \int u\, dv = uv - \int v\, du = x\left(\frac{1}{3}e^{3x}\right) - \int \frac{1}{3}e^{3x}\, dx$

$= \frac{1}{3}xe^{3x} - \frac{1}{9}e^{3x} + C$.

$\int_0^1 xe^{3x}\, dx = \left(\frac{1}{3}xe^{3x} - \frac{1}{9}e^{3x}\right)\Big]_0^1 = \left(\frac{1}{3} \cdot 1 \cdot e^3 - \frac{1}{9}e^3\right) + \frac{1}{9} = \frac{2}{9}e^3 + \frac{1}{9}$.

11. Let $u = x - 3$. $du = dx$.

$\int_0^2 \frac{dx}{(x-3)^2} = \int_{x=0}^{x=2} u^{-2}\, du = \frac{u^{-1}}{-1}\Big]_{x=0}^{x=2} = -\frac{1}{x-3}\Big]_0^2 = 1 - \frac{1}{3} = \frac{2}{3}$.

12. Let $u = x$, $dv = e^{-3x}\, dx$. $du = dx$, $v = -\frac{1}{3}e^{-3x}$.

$\int xe^{-3x}\, dx = \int u\, dv = uv - \int v\, du = x\left(-\frac{1}{3}e^{-3x}\right) - \int \left(-\frac{1}{3}e^{-3x}\right) dx$

$= -\frac{1}{3}xe^{-3x} - \frac{1}{9}e^{-3x} + C$.

$= -\frac{1}{3}xe^{-3x} - \frac{1}{9}e^{-3x} + C.$

13. Let $u = \ln x$, $dv = x^2\,dx$. $du = \frac{1}{x}\,dx$, $v = \frac{1}{3}x^3$.

$\int x^2 \ln x\,dx = (\ln x)\left(\frac{1}{3}x^3\right) - \int \left(\frac{1}{3}x^3\right)\left(\frac{1}{x}\,dx\right)$

$= \frac{1}{3}x^3 \ln x - \frac{1}{3}\int x^2\,dx$

$= \frac{1}{3}x^3 \ln x - \frac{1}{9}x^3 + C.$

14. Let $u = x^2$, $dv = \sin x\,dx$. $du = 2x\,dx$, $v = -\cos x$.

$\int x^2 \sin x\,dx = x^2(-\cos x) + 2\int x\cos x\,dx$

$= -x^2\cos x + 2(x\sin x + \cos x) + C$ from Example 3 in Sec. 7.2.

15. $\int_0^3 \frac{1}{\sqrt[3]{x+1}}\,dx = \frac{3}{2}(x+1)^{2/3}\Big]_0^3 = \frac{3}{2}\left[(3+1)^{2/3} - 1\right] = \frac{3}{2}(4^{2/3} - 1).$

16. Let $u = x$, $dv = (x+3)^6\,dx$. $du = dx$, $v = \frac{1}{7}(x+3)^7$.

$\int x(x+3)^6\,dx = \frac{1}{7}x(x+3)^7 - \frac{1}{7}\int (x+3)^7 dx = \frac{1}{7}x(x+3)^7 - \frac{1}{56}(x+3)^8 + C$

$= \frac{1}{56}(x+3)^7(8x - x - 3) + C = \frac{1}{56}(x+3)^7(7x-3) + C.$

$\int_{-2}^{-1} x(x+3)^6 dx = \frac{1}{56}(x+3)^7(7x-3)\Big]_{-2}^{-1} = \frac{1}{56}\left[(-1+3)^7(-7-3) - (-2+3)^7(-14-3)\right]$

$= \frac{1}{56}(17 - 10\cdot 2^7) = -\frac{1263}{56}.$

17. $\int_0^{1/4} \sec^2 \pi x\,dx = \frac{1}{\pi}\tan \pi x\Big]_0^{1/4} = \frac{1}{\pi}\left(\tan\frac{\pi}{4} - \tan 0\right) = \frac{1}{\pi}(1-0) = \frac{1}{\pi}.$

18. Let $u = x$, $dv = e^{-x}\,dx$. $du = dx$, $v = -e^{-x}$.

$\int xe^{-x}\,dx = -xe^{-x} + \int e^{-x}\,dx = -xe^{-x} - e^{-x} + C = -e^{-x}(x+1) + C.$

19. Let $u = \ln x$, $dv = x^3\,dx$. $du = \frac{1}{x}\,dx$, $v = \frac{1}{4}x^4$.

$\int x^3 \ln x\,dx = (\ln x)\left(\frac{1}{4}x^4\right) - \int\left(\frac{1}{4}x^4\right)\left(\frac{1}{x}\,dx\right)$

$= \frac{1}{4}x^4 \ln x - \frac{1}{4}\int x^3\,dx$

$= \frac{1}{4}x^4 \ln x - \frac{1}{16}x^4 + C.$

20. $\int_1^\infty \frac{dx}{\sqrt[3]{x+2}} = \lim_{t\to\infty}\int_1^t \frac{1}{\sqrt[3]{x+2}}\,dx = \lim_{t\to\infty}\frac{3}{2}(x+2)^{2/3}\Big]_1^t$

$= \lim_{t\to\infty}\frac{3}{2}\left[(t+2)^{2/3} - 3^{2/3}\right] = +\infty.$ Diverges.

21. Let $u = x$, $dv = \csc^2 x\,dx$. $du = dx$, $v = -\cot x$.

$\int x\csc^2 x\,dx = -x\cot x + \int \cot x\,dx = -x\cot x + \ln|\sin x| + C.$

22. Let $u = \sin(\ln x)$, $dv = dx$. $du = \cos(\ln x) \cdot \frac{1}{x}\,dx$, $v = x$.

$\int \sin(\ln x)dx = x\sin(\ln x) - \int \cos(\ln x)dx.$

Let $u_1 = \cos(\ln x)$, $dv_1 = dx$. $du_1 = -\sin(\ln x) \cdot \frac{1}{x}\,dx$, $v_1 = x$.

$\int \cos(\ln x)dx = x\cos(\ln x) + \int \sin(\ln x)dx.$

$\int \sin(\ln x)dx = x\sin(\ln x) - \left(x\cos(\ln x) + \int \sin(\ln x)dx\right).$

$2\int \sin(\ln x)dx = x\sin(\ln x) - x\cos(\ln x) + C.$

$\int \sin(\ln x)dx = \frac{x}{2}\left(\sin(\ln x) - \cos(\ln x)\right) + C.$

23. $\int_{-\infty}^{0} e^x\,dx = \lim_{t\to-\infty} \int_t^0 e^x\,dx = \lim_{t\to-\infty} e^x\Big]_t^0 = \lim_{t\to-\infty}(1 - e^t) = 1.$

24. Let $u(x) = 9 + x^2$. $du = 2x\,dx$.

$\int \frac{x}{\sqrt{9+x^2}}\,dx = \frac{1}{2}\int \frac{1}{\sqrt{u}}\,du = \frac{1}{2} \cdot 2u^{1/2} + C = \sqrt{9+x^2} + C.$

25. Let $u = \ln x$. $du = \frac{1}{x}\,dx$.

$\int \frac{1}{x\ln x}\,dx = \int \frac{1}{u}\,du = \ln|u| + C = \ln|\ln x| + C.$

$$\int_{e^2}^{\infty} \frac{dx}{x\ln x} = \lim_{t\to\infty} \int_{e^2}^{t} \frac{1}{x\ln x}\,dx = \lim_{t\to\infty} \ln|\ln x|\Big]_{e^2}^{t}$$

$$= \lim_{t\to\infty}(\ln\ln t - \ln\ln e^2) = +\infty. \quad \text{Diverges.}$$

26. Let $u = 5 + x^2$. $du = 2x\,dx$.

$\int \frac{x}{5+x^2}\,dx = \frac{1}{2}\int \frac{1}{u}\,du = \frac{1}{2}\ln|u| + C = \frac{1}{2}\ln(5+x^2) + C.$

$$\int_2^{\infty} \frac{x}{5+x^2}\,dx = \lim_{t\to\infty} \int_2^t \frac{x}{5+x^2}\,dx = \lim_{t\to\infty} \frac{1}{2}\ln(5+x^2)\Big]_2^t$$

$$= \lim_{t\to\infty} \frac{1}{2}\left(\ln(5+t^2) - \ln 9\right) = +\infty. \quad \text{Diverges.}$$

27. Let $u = 16 - x^2$. $du = -2x\,dx$.

$\int \frac{x}{\sqrt{16-x^2}}\,dx = -\frac{1}{2}\int \frac{1}{\sqrt{u}}\,du = -\frac{1}{2} \cdot 2u^{1/2} + C = -\sqrt{16-x^2} + C.$

$$\int_1^4 \frac{x}{\sqrt{16-x^2}}\,dx = \lim_{b\to4^-} \int_1^b \frac{x}{\sqrt{16-x^2}}\,dx = \lim_{b\to4^-} \left(-\sqrt{16-x^2}\right)\Big]_1^b$$

$$= \lim_{b\to4^-} \left(-\sqrt{16-b^2} + \sqrt{15}\right) = \sqrt{15}.$$

28. $\int (x+2)\ln x\ dx = \int x \ln x\ dx + 2\int \ln x\ dx.$

From Ex. 13, $\int x\ln x\ dx = \frac{1}{2}x^2 \ln x - \frac{1}{4}x^2 + C_1.$

From Example 5, Set 7.2, $\int \ln x\ dx = x\ln x - x + C_2.$

$$\begin{aligned}\int (x+2)\ln x\ dx &= \tfrac{1}{2}x^2\ln x - \tfrac{1}{4}x^2 + C_1 + 2(x\ln x - x + C_2)\\ &= \tfrac{1}{4}x^2(2\ln x - 1) + 2x(\ln x - 1) + C.\end{aligned}$$

29. Let $u(x) = x^2 + 4x + 3$. $\quad du = (2x+4)dx.$

$\int \frac{2x+4}{x^2+4x+3}\ dx = \int \frac{1}{u}\ du = \ln|u| + C = \ln|x^2+4x+3| + C.$

30. $\int \tan 3x\ dx = \frac{1}{3}\ln|\sec 3x| + C.$

31. The increase $= \int_2^4 MC(x)dx = \int_2^4 \frac{40x^2}{\sqrt{4+x^3}}\ dx.$

Let $u = 4 + x^3$. $\quad du = 3x^2dx \implies x^2\ dx = \frac{1}{3}du.$

$\int \frac{40x^2}{\sqrt{4+x^3}}\ dx = \frac{40}{3}\int (u)^{-1/2}du = \frac{40}{3}\cdot 2u^{1/2} + C = \frac{80}{3}\sqrt{4+x^3} + C.$

$\int_2^4 \frac{40x^2}{\sqrt{4+x^3}}\ dx = \frac{80}{3}\sqrt{4+x^3}\ \Big]_2^4 = \frac{80}{3}\left(\sqrt{68} - \sqrt{12}\right) = 127.5229$ cents.

32. Anticipated royalties $= \int_0^5 R(t)dt$

$$\begin{aligned} &= \int_0^5 (500 + 300\sin^2 \pi t\cos \pi t)dt\\ &= \left(500t + \tfrac{100}{\pi}\sin^3 \pi t\right)\Big]_0^5\\ &= 2500 + \tfrac{100}{\pi}\sin^3 5\pi\\ &= 2500.\end{aligned}$$

33. Total amount deposited $= \int_0^4 A(t)dt = \int_0^4 400te^{2-t^2}\ dt.$

Let $u = 2 - t^2$. $\quad du = -2t\,dt \implies t\,dt = -\frac{1}{2}\,du.$

$\int 400te^{2-t^2}\ dt = -200\int e^u\ du = -200e^u + C = -200e^{2-t^2} + C.$

$$\begin{aligned}\int_0^4 400te^{2-t^2}\ dt &= -200e^{2-t^2}\Big]_0^4 = -200(e^{-14} - e^2)\\ &= 1477.8111 \text{ ounces.}\end{aligned}$$

34. Consumer's surplus $= \int_0^4 \frac{100}{x\sqrt{x^2+9}}\,dx - (5)(4).$

$$= 100 \cdot \lim_{a\to 0+} \int_a^4 \frac{dx}{x\sqrt{x^2+9}} - 20.$$

$$= 100 \cdot \lim_{a\to 0+} \left(-\tfrac{1}{3}\right) \ln\left(\frac{3+\sqrt{x^2+9}}{x}\right)\Big]_a^4 - 20.$$

$$= \frac{-100}{3}\left[\ln\left(\frac{3+5}{4}\right) - \lim_{a\to 0+} \ln\left(\frac{3+\sqrt{a^2+9}}{a}\right)\right] - 20$$

$$= \infty.$$

35. Average value $= \dfrac{\int_0^{\pi/2} 4\cos^3 \pi x\,dx}{\frac{\pi}{2}-0} = \dfrac{\frac{4}{3\pi}\sin(\pi x)\cdot\left(\cos^2(\pi x)+2\right)\Big]_0^{\pi/2}}{\frac{\pi}{2}}$

$$= \frac{\frac{4}{3\pi}\sin\left(\frac{\pi^2}{2}\right)\cdot\left(\cos^2\left(\frac{\pi^2}{2}\right)+2\right)}{\frac{\pi}{2}}$$

$$= -0.5399.$$

36. Area $= \int_{-1}^{1} \left(\frac{1}{4-x^2} - \frac{1}{x^2-4}\right) dx = 4 \cdot \int_0^1 \frac{1}{4-x^2}\,dx$

$$= 4 \cdot \tfrac{1}{4} \ln\left|\frac{2+x}{2-x}\right|\Big]_0^1 = \ln 3.$$

CHAPTER 8 (CHAPTER 6)

Solutions to Exercise Set 8.1 (6.1)

1. $f(x,y) = 3x + 7y$.
   a. $f(2,3) = 3(2) + 7(3) = 27$.
   b. $f(0,-6) = 3(0) + 7(-6) = -42$.
   c. $f(-1,5) = 3(-1) + 7(5) = 32$.
2. $f(x,y) = 4xy^2 - 3x^2y + 6$.
   a. $f(0,0) = 4(0)(0)^2 - 3(0)^2(0) + 6 = 6$.
   b. $f(1,-2) = 4(1)(-2)^2 - 3(1)^2(-2) + 6 = 28$.
   c. $f(3,5) = 4(3)(5)^2 - 3(3)^2(5) + 6 = 171$.
3. $f(x,y) = \sqrt{xy} - \frac{1}{xy}$.
   a. $f(1,1) = \sqrt{1 \cdot 1} - \frac{1}{1 \cdot 1} = 0$.
   b. $f(-2,-2) = \sqrt{(-2)(-2)} - \frac{1}{(-2)(-2)} = 2 - \frac{1}{4} = \frac{7}{4}$.
   c. $f(1,9) = \sqrt{(1)(9)} - \frac{1}{(1)(9)} = 3 - \frac{1}{9} = \frac{26}{9}$.
4. $f(x,y) = xe^y - ye^x + e^{xy}$
   a. $f(0,2) = 0 \cdot e^2 - 2e^0 + e^{0 \cdot 2} = 0 - 2 + 1 = -1$.
   b. $f(-2, \ln 2) = -2 \cdot e^{\ln 2} - (\ln 2)e^{-2} + e^{(-2)(\ln 2)} = -2 \cdot 2 - (\ln 2) \cdot e^{-2} + e^{\ln 2^{-2}}$

      $= -4 - e^{-2} \cdot \ln 2 + 2^{-2} = \frac{-15}{4} - e^{-2} \ln 2$.
   c. $f(4,1) = 4 \cdot e^1 - 1 \cdot e^4 + e^{4 \cdot 1} = 4e$.
5. $f(x,y,z) = x^2 - 2xy + xz^3$.
   a. $f(1,2,-1) = 1^2 - 2(1)(2) + (1)(-1)^3 = -4$.
   b. $f(3,-3,0) = 3^2 - 2(3)(-3) + 3(0)^3 = 27$.
   c. $f(-1,1,4) = (-1)^2 - 2(-1)(1) + (-1)4^3 = -61$.

6. $f(x,y,z) = xye^z + \ln(x+y) + 3$.

   a. $f(2,1,0) = (2)(1)e^0 + \ln(2+1) + 3 = 5 + \ln 3$.

   b. $f(2,2,-1) = (2)(2)e^{-1} + \ln(2+2) + 3 = 3 + \frac{4}{e} + \ln 4$.

   c. $f(0,1,3) = (0)(1)e^3 + \ln(0+1) + 3 = 3$.

7. $f(x,y,z) = \sqrt{x^2+y^2+z^2}$.

   a. $f(0,1,0) = \sqrt{0^2+1^2+0^2} = 1$.

   b. $f(-2,1,2) = \sqrt{(-2)^2+1^2+2^2} = \sqrt{9} = 3$.

   c. $f\left(3,-3,\sqrt{7}\right) = \sqrt{3^2+(-3)^2+\left(\sqrt{7}\right)^2} = \sqrt{25} = 5$.

8. $f(x,y,z) = x\sqrt{y} + \sqrt[3]{xz} - y^{2/3}z^{-1/3}$.

   a. $f(1,1,0) = 1\sqrt{1} + \sqrt[3]{1\cdot(0)} - 1^{2/3}\cdot 0^{-1/3}$

      $f(1,1,0)$ is undefined because $0^{-1/3}$ is undefined.

   b. $f(2,1,4) = 2\sqrt{1} + \sqrt[3]{2\cdot 4} - 1^{2/3}\cdot 4^{-1/3} = 2 + 2 - 4^{-1/3} = 4 - \frac{1}{4^{1/3}}$.

   c. $f(0,1,-8) = 0\cdot\sqrt{1} + \sqrt[3]{0\cdot(-8)} - 1^{2/3}\cdot(-8)^{-1/3} = \frac{1}{2}$.

9.

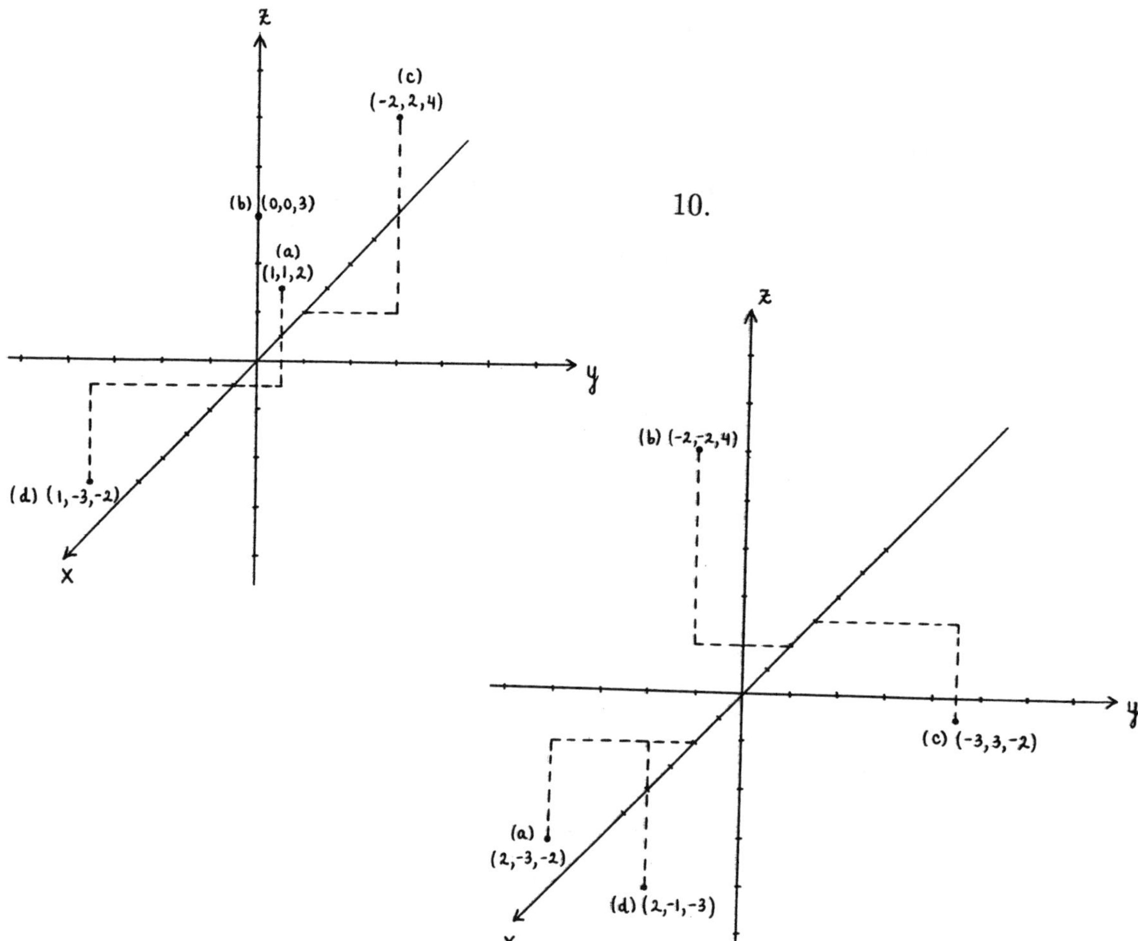

11. $f(x, y) = 3xy^2 - 5x^3y + 7$.

    The domain includes all ordered pairs of real numbers.

12. $f(x, y) = \sqrt{x^2 + y^2}$.

    The domain includes all ordered pairs of real numbers.

13. $f(x, y) = \frac{10x^2}{y^2 - x^2}$.

    The domain consists of all ordered pairs (x, y) of real numbers such that either $x \neq y$ or $x \neq -y$.

14. $f(x, y) = \ln(xy)$.

    The domain consists of all ordered pairs (x, y) of real numbers such that $x \cdot y > 0$.

15. $f(x, y) = 4x^{2/3}y^{1/4}$.

    The domain consists of all ordered pairs (x, y) of real numbers such that $y \geq 0$.

16. $f(x, y, z) = \sqrt{16 - (x^2 + y^2 + z^2)}$.

    The domain consists of all ordered triples (x, y, z) of all real numbers such that $x^2 + y^2 + z^2 \leq 16$.

17. Let x = number of plastic brooms sold

    y = number of fiber brooms sold.

    Then $z = R(x, y) = 10x + 15y$.

18. $C(x, y) = 4x + 6y + 20$.

19. $P(x, y) = R(x, y) - C(x, y) = (10x + 15y) - (4x + 6y + 20)$

    $= 6x + 9y - 20$.

20. a. $C(x, y) = 0.35x + 0.20y + 75$.

    b. $C(200, 150) = 0.35(200) + 0.20(150) + 75 = \$175$.

21. a. $R(x, y) = 0.75x + 0.60y$.

    b. $R(200, 150) = 0.75(200) + 0.60(150) = \$240$.

22. a. $P(x, y) = R(x, y) - C(x, y) = (0.75x + 0.60y) - (0.35x + 0.20y + 75)$

    $= 0.40x + 0.40y - 75$.

    b. $P(200, 150) = 0.40(200) + 0.40(150) - 75 = \$65$.

23. a. $P(r,t) = 500e^{rt}$.

    b. $P(0.03, 5) = 500e^{0.03(5)} \approx \$580.92$.

    c. $P(0.10, 7) = 500e^{0.1(7)} \approx \$1006.88$.

24. $P_0(r,T) = 10,000e^{-rT}$.

25. $f(K,L) = 100K^{1/4}L^{3/4}$.

    a. $f(81,16) = 100(81)^{1/4} \cdot (16)^{3/4} = 100(3)(8) = 2400$.

    b. $f(16,81) = 100(16)^{1/4}(81)^{3/4} = 100(2)(27) = 5400$.

26. a. $R(x,y) = xp + yq = x(200 - 2x) + y(150 - 3y) = -2x^2 - 3y^2 + 200x + 150y$.

    b. $$\begin{aligned} P(x,y) &= (-2x^2 - 3y^2 + 200x + 150y) - (500 + 4x + 5y) \\ &= -2x^2 - 3y^2 + 196x + 145y - 500. \end{aligned}$$

27. $E(x,t) = 20x^{3/2}e^{-0.05t}$.

    a. $E(16,20) = 20(16)^{3/2}e^{-0.05(20)} = 1280e^{-1}$.

    b. $E(64,10) = 20(64)^{3/2}e^{-0.05(10)} = 10240e^{-0.5}$.

28. $f(K,L) = \lambda K^{\alpha}L^{1-\alpha}$.

    $f(2K,2L) = \lambda(2K)^{\alpha}(2L)^{1-\alpha} = \lambda 2^{\alpha}K^{\alpha} \cdot 2^{1-\alpha}L^{1-\alpha} = 2\lambda K^{\alpha}L^{1-\alpha} = 2f(K,L)$.

Solutions to Exercise Set 8.2 (6.2)

1. $f(xy) = 3x - 6y + 5$.

    a. $\frac{\partial f}{\partial x}(x,y) = 3$. b. $\frac{\partial f}{\partial y}(x,y) = -6$.

2. $f(x,y) = x^2 - 2xy + y^3$.

    a. $\frac{\partial f}{\partial x}(x,y) = 2x - 2y$. b. $\frac{\partial f}{\partial y}(x,y) = -2x + 3y^2$.

3. $f(x,y) = 4xy^2 - 3x^3y + y^5$.

    a. $\frac{\partial f}{\partial x}(x,y) = 4y^2 - 9x^2y$. b. $\frac{\partial f}{\partial y}(x,y) = 8xy - 3x^3 + 5y^4$.

4. $f(x,y) = x\sin(xy)$.

    a. $\frac{\partial f}{\partial x}(x,y) = 1 \cdot \sin(xy) + x \cdot \cos(xy) \cdot y = \sin(xy) + xy \cdot \cos(xy)$.

    b. $\frac{\partial f}{\partial y}(x,y) = x \cdot \cos(xy) \cdot x = x^2 \cdot \cos(xy)$.

5. $f(x,y) = \sqrt{x+y} = (x+y)^{1/2}$.

a. $\frac{\partial f}{\partial x}(x,y) = \frac{1}{2}(x+y)^{-1/2} \cdot 1 = \frac{1}{2\sqrt{x+y}}$.

b. $\frac{\partial f}{\partial y}(x,y) = \frac{1}{2}(x+y)^{-1/2} \cdot 1 = \frac{1}{2\sqrt{x+y}}$.

6. $f(x,y) = x\sqrt{y} - y\sqrt{x}$.

a. $\frac{\partial f}{\partial x}(x,y) = \sqrt{y} - y \cdot \frac{1}{2\sqrt{x}} = \frac{2\sqrt{xy}-y}{2\sqrt{x}}$.

b. $\frac{\partial f}{\partial y}(x,y) = x \cdot \frac{1}{2\sqrt{y}} - \sqrt{x} = \frac{x-2\sqrt{xy}}{2\sqrt{y}}$.

7. $f(x,y) = \frac{x+y}{x-y}$.

a. $\frac{\partial f}{\partial x}(x,y) = \frac{(x-y)-(x+y)}{(x-y)^2} = \frac{-2y}{(x-y)^2}$.

b. $\frac{\partial f}{\partial y}(x,y) = \frac{(x-y)-(x+y)(-1)}{(x-y)^2} = \frac{2x}{(x-y)^2}$.

8. $f(x,y) = e^{x^2-2y}$.

a. $\frac{\partial f}{\partial x}(x,y) = e^{x^2-2y}(2x) = 2xe^{x^2-2y}$.

b. $\frac{\partial f}{\partial y}(x,y) = e^{x^2-2y}(-2) = -2e^{x^2-2y}$.

9. $f(x,y) = \ln\sqrt{x^2+y^2} = \frac{1}{2}\ln(x^2+y^2)$.

a. $\frac{\partial f}{\partial x}(x,y) = \frac{1}{2} \cdot \frac{2x}{x^2+y^2} = \frac{x}{x^2+y^2}$. b. $\frac{\partial f}{\partial y}(x,y) = \frac{1}{2} \cdot \frac{2y}{x^2+y^2} = \frac{y}{x^2+y^2}$.

10. $f(x,y) = xe^{\sqrt{y}}$.

a. $\frac{\partial f}{\partial x}(x,y) = e^{\sqrt{y}}$. b. $\frac{\partial f}{\partial y}(x,y) = xe^{\sqrt{y}} \cdot \frac{1}{2\sqrt{y}} = \frac{x}{2\sqrt{y}}e^{\sqrt{y}}$.

11. $f(x,y) = x^{2/3}y^{-1/3} + \sqrt{\frac{y}{x}}$.

a. $\frac{\partial f}{\partial x}(x,y) = \frac{2}{3}x^{-1/3}y^{-1/3} - \frac{1}{2}x^{-3/2}y^{1/2}$.

b. $\frac{\partial f}{\partial y}(x,y) = -\frac{1}{3}x^{2/3}y^{-4/3} + \frac{1}{2\sqrt{y}} \cdot \frac{1}{\sqrt{x}} = -\frac{1}{3}x^{2/3}y^{-4/3} + \frac{1}{2\sqrt{xy}}$.

12. $f(x,y) = \frac{x-e^y}{y+e^x}$.

a. $\frac{\partial f}{\partial x}(x,y) = \frac{1\cdot(y+e^x)-(x-e^y)\cdot e^x}{(y+e^x)^2} = \frac{e^x(1-x+e^y)+y}{(y+e^x)^2}$.

b. $\frac{\partial f}{\partial y}(x,y) = \frac{-e^y(y+e^x)-(x-e^y)\cdot 1}{(y+e^x)^2} = \frac{e^y(1-y-e^x)-x}{(y+e^x)^2}$.

13. $f(x,y) = \tan(x+y) - \sec(x-y)$.

a. $\frac{\partial f}{\partial x}(x,y) = \sec^2(x+y) - \sec(x-y)\tan(x-y)$.

b. $\frac{\partial f}{\partial y}(x,y) = \sec^2(x+y) - \sec(x-y)\tan(x-y)(-1)$

$$= \sec^2(x+y) + \sec(x-y)\tan(x-y).$$

14. $f(x,y) = x\sec(xy) - \cos(x-y)$.

a. $\frac{\partial f}{\partial x}(x,y) = 1\cdot\sec(xy) + x\cdot\sec(xy)\tan(xy)\cdot y + \sin(x-y)$

$= \sec(xy) + xy\cdot\sec(xy)\cdot\tan(xy) + \sin(x-y)$.

b. $\frac{\partial f}{\partial y}(x,y) = x\cdot\sec(xy)\cdot\tan(xy)\cdot x + \sin(x-y)(-1)$

$= x^2\sec(xy)\tan(xy) - \sin(x-y)$.

15. $z = \frac{x}{y^2} - \frac{y}{x^2} = xy^{-2} - yx^{-2}$.

$z_x = \frac{1}{y^2} + \frac{2y}{x^3}$. $\quad z_y = x(-2)y^{-3} - x^{-2} = -\frac{2x}{y^3} - \frac{1}{x^2}$.

16. $z = 4 - x^{2/3}\sqrt{y+1}$.

$z_x = -\frac{2}{3}x^{-1/3}\sqrt{y+1}$. $\quad z_y = -x^{2/3}\frac{1}{2\sqrt{y+1}} = -\frac{x^{2/3}}{2\sqrt{y+1}}$.

17. $z = ye^{\sqrt{x}-1}$. $\quad z_x = ye^{\sqrt{x}-1}\cdot\frac{1}{2\sqrt{x}} = \frac{y}{2\sqrt{x}}e^{\sqrt{x}-1}$. $\quad z_y = e^{\sqrt{x}-1}$.

18. $z = \ln(x+ye^x)$. $\quad z_x = \frac{1+ye^x}{x+ye^x}$. $\quad z_y = \frac{e^x}{x+ye^x}$.

19. $z = \sqrt{\cos^2(xy) + e^{\sin y}}$.

$z_x = \frac{1}{2}\cdot\left(\cos^2(xy) + e^{\sin y}\right)^{-1/2}\cdot(2\cos(xy)(-\sin(xy))\cdot y)$

$= -y\cos(xy)\sin(xy)\cdot\left(\cos^2(xy) + e^{\sin y}\right)^{-1/2}$.

$z_y = \frac{1}{2}\cdot\left(\cos^2(xy) + e^{\sin y}\right)^{-1/2}\cdot\left(2\cos(xy)(-\sin(xy))\cdot x + e^{\sin y}\cdot\cos y\right)$.

20. $z = \frac{x^{5/2}+y^{2/3}}{y-x}$.

$z_x = \frac{\frac{5}{2}x^{3/2}(y-x)-(x^{5/2}+y^{2/3})(-1)}{(y-x)^2} = \frac{-3x^{5/2}+5x^{3/2}y+2y^{2/3}}{2(y-x)^2}$.

$z_y = \frac{\frac{2}{3}y^{-1/3}(y-x)-(x^{5/2}+y^{2/3})\cdot 1}{(y-x)^2} = -\frac{y^{2/3}+2xy^{-1/3}+3x^{5/2}}{3(y-x)^2}$.

21. $f(x,y) = 2x(y-7)$. $\quad \frac{\partial f}{\partial x}(x,y) = 2(y-7)$. $\quad \frac{\partial f}{\partial x}(3,5) = 2(5-7) = -4$.

22. $f(x,y) = \frac{x}{y+x}$. $\quad \frac{\partial f}{\partial y}(x,y) = \frac{-x}{(y+x)^2}$. $\quad \frac{\partial f}{\partial y}(2,-3) = \frac{-2}{(-3+2)^2} = -2$.

23. $f(x,y) = \sqrt{y^2+2x}$.

$\frac{\partial f}{\partial y}(x,y) = \frac{2y}{2\sqrt{y^2+2x}} = \frac{y}{\sqrt{y^2+2x}}$. $\quad \frac{\partial f}{\partial y}(4,1) = \frac{1}{\sqrt{1^2+2(4)}} = \frac{1}{3}$.

24. $f(x,y) = ye^{-2x} + x\ln y$.

$\frac{\partial f}{\partial y}(x,y) = e^{-2x} + \frac{x}{y}$. $\quad \frac{\partial f}{\partial y}(1,4) = e^{-2(1)} + \frac{1}{4} = e^{-2} + \frac{1}{4}$.

25. $f(x,y,z) = x\sin(yz) + z\cos^2(xy)$.

$\frac{\partial f}{\partial z}(x,y,z) = x\cdot\cos(yz)\cdot y + \cos^2(xy)$.

$\frac{\partial f}{\partial z}\left(1,\frac{\pi}{4},-1\right) = 1\cdot\cos\left(\frac{\pi}{4}\cdot(-1)\right)\cdot\frac{\pi}{4} + \cos^2\left(1\cdot\frac{\pi}{4}\right) = \frac{\sqrt{2}}{2}\cdot\frac{\pi}{4} + \frac{1}{2}$.

26. $f(x,y,z) = xe^{yz}$.

$\frac{\partial f}{\partial y}(x,y,z) = xze^{yz}$. $\quad \frac{\partial f}{\partial y}(2,0,3) = (2)(3)e^{0\cdot 3} = 6$.

27. $f(x,y,z) = xe^{yz}$.

$\frac{\partial f}{\partial z}(x,y,z) = xye^{yz}$. $\quad \frac{\partial f}{\partial z}(2,0,3) = (2)(0)e^{0\cdot 3} = 0$.

28. $f(x,y,z) = \sqrt{xyz}$. $\quad \frac{\partial f}{\partial y}(x,y,z) = \frac{xz}{2\sqrt{xyz}}$. $\quad \frac{\partial f}{\partial y}(2,3,6) = \frac{(2)(6)}{2\sqrt{2\cdot 3\cdot 6}} = 1$.

29. $f(x,y) = x^2 - 2xy + y^2$.

a. $\frac{\partial^2 f}{\partial x^2} = \frac{\partial}{\partial x}\left(\frac{\partial f}{\partial x}\right) = \frac{\partial}{\partial x}(2x - 2y) = 2$.

b. $\frac{\partial^2 f}{\partial x \partial y} = \frac{\partial}{\partial x}\left(\frac{\partial f}{\partial y}\right) = \frac{\partial}{\partial x}(-2x + 2y) = -2$.

c. $\frac{\partial^2 f}{\partial y \partial x} = \frac{\partial}{\partial y}\left(\frac{\partial f}{\partial x}\right) = \frac{\partial}{\partial y}(2x - 2y) = -2$.

d. $\frac{\partial^2 f}{\partial y^2} = \frac{\partial}{\partial y}\left(\frac{\partial f}{\partial y}\right) = \frac{\partial}{\partial y}(-2x + 2y) = 2$.

30. $f(x,y) = x \cdot e^{y-x}$.

a. $$\begin{aligned}\frac{\partial^2 f}{\partial x^2} &= \frac{\partial}{\partial x}\left(e^{y-x} + x \cdot e^{y-x} \cdot (-1)\right)\\ &= -e^{y-x} + \left(-e^{y-x} - e^{y-x} \cdot x \cdot (-1)\right)\\ &= -2e^{y-x} + x \cdot e^{y-x}.\end{aligned}$$

b. $\frac{\partial^2 f}{\partial y \partial x} = \frac{\partial}{\partial y}(e^{y-x} - x \cdot e^{y-x}) = e^{y-x} - x \cdot e^{y-x}$.

c. $\frac{\partial^2 f}{\partial x \partial y} = \frac{\partial}{\partial x}(x \cdot e^{y-x}) = e^{y-x} - x \cdot e^{y-x}$.

d. $\frac{\partial^2 f}{\partial y^2} = \frac{\partial}{\partial y}(x \cdot e^{y-x}) = x \cdot e^{y-x}$.

31. $f(x,y) = \ln(x^2 + y^2)$.

a. $\frac{\partial^2 f}{\partial x^2} = \frac{\partial}{\partial x}\left(\frac{2x}{x^2+y^2}\right) = \frac{2(x^2+y^2) - 2x(2x)}{(x^2+y^2)^2} = \frac{-2x^2+2y^2}{(x^2+y^2)^2}$.

b. $\frac{\partial^2 f}{\partial y \partial x} = \frac{\partial}{\partial y}\left(\frac{2x}{x^2+y^2}\right) = \frac{-2x(2y)}{(x^2+y^2)^2} = \frac{-4xy}{(x^2+y^2)^2}$.

c. $\frac{\partial^2 f}{\partial x \partial y} = \frac{\partial}{\partial x}\left(\frac{2y}{x^2+y^2}\right) = \frac{-4xy}{(x^2+y^2)^2}$.

d. $\frac{\partial^2 f}{\partial y^2} = \frac{\partial}{\partial y}\left(\frac{2y}{x^2+y^2}\right) = \frac{2(x^2+y^2) - 2y\cdot 2y}{(x^2+y^2)^2} = \frac{2x^2-2y^2}{(x^2+y^2)^2}$.

32. $f(K,L) = 40K^{1/3} \cdot L^{2/3}$.

a. $\frac{\partial f}{\partial K} = 40 \cdot \left(\frac{1}{3}\right) \cdot K^{-2/3} \cdot L^{2/3}$

$\Longrightarrow \left(\frac{\partial f}{\partial K}\right)(27,8) = 40 \cdot \left(\frac{1}{3}\right) \cdot (27)^{-2/3} \cdot (8)^{2/3} = \frac{40}{3} \cdot 3^{-2} \cdot 2^2 = \frac{160}{27}$.

b. $\frac{\partial f}{\partial L} = 40K^{1/3} \cdot \left(\frac{2}{3}\right) L^{-1/3}$

$\Longrightarrow \left(\frac{\partial f}{\partial L}\right)(27,8) = 40 \cdot (27)^{1/3} \cdot \left(\frac{2}{3}\right) \cdot (8)^{-1/3} = 40 \cdot 3 \cdot \frac{2}{3} \cdot 2^{-1} = 40$.

33. $\left(\frac{\partial f}{\partial K}\right)(27,8) = \frac{160}{27}$.

34. $R(x,y) = 40x^2 + 80y^2 - 100x - 200y - 20\sqrt{xy}$.

a. $\frac{\partial R}{\partial x} = 80x - 100 - 10\frac{y}{\sqrt{xy}}$

$\implies \left(\frac{\partial R}{\partial x}\right)(4,9) = 80 \cdot 4 - 100 - 10 \cdot \frac{9}{\sqrt{36}} = 320 - 100 - 15 = 205.$

b. $\frac{\partial R}{\partial y} = 160y - 200 - 10\frac{x}{\sqrt{xy}}$

$\implies \left(\frac{\partial R}{\partial y}\right)(4,9) = 160 \cdot 9 - 200 - 10 \cdot \frac{4}{6} = 1233.33.$

35. $u(x,y) = 4x^{2/3} + 6y^{3/2} - xy^2$.

a. $\frac{\partial u}{\partial x} = 4 \cdot \frac{2}{3}x^{-1/3} - y^2 = \frac{8}{3}x^{-1/3} - y^2$

$\implies \frac{\partial u}{\partial x}(8,4) = \frac{8}{3} \cdot \frac{1}{2} - 16 = \frac{4}{3} - 16 = -\frac{44}{3}.$

b. $\frac{\partial u}{\partial y} = 6 \cdot \frac{3}{2} \cdot y^{1/2} - 2xy$

$\implies \frac{\partial u}{\partial y}(2,4) = 6 \cdot \frac{3}{2} \cdot 2 - 2 \cdot 2 \cdot 4 = 18 - 16 = 2.$

a. means that an increase in slices of pizza from $x = 8$ to $x = 9$ results in a decrease of $\frac{44}{3}$ units of a person's level of satisfaction at $y = 4$, approximately.

b. means that an increase in glasses of soda from $y = 4$ to $y = 5$ results in an increase of 2 units of a person's level of satisfaction at $x = 2$, approximately.

36. $P(r,t) = 100(e^{rt} - 1)$.

a. $\frac{\partial P}{\partial t} = 100 \cdot e^{rt} \cdot r \implies \left(\frac{\partial P}{\partial t}\right)(0.1,5) = 100 \cdot e^{0.5} \cdot (0.1) = 10 \cdot e^{0.5}$.

b. $\frac{\partial P}{\partial t}(0.1,5) = 10 \cdot e^{0.5}$.

37. $f(K,L) = 100\sqrt{KL}$.

a. $\frac{\partial f}{\partial K} = \frac{50 \cdot L}{\sqrt{KL}} \implies \frac{\partial f}{\partial K}(3,27) = \frac{50 \cdot 27}{9} = 150.$

b. $\frac{\partial f}{\partial L} = \frac{50 \cdot K}{\sqrt{KL}} \implies \frac{\partial f}{\partial L}(3,27) = \frac{50 \cdot 3}{9} = \frac{50}{3}.$

38. a. Negative since demand decreases as price increases.

b. Positive since demand increases as the money spent on advertising the automobile increases.

39. $W(x,y) = 150 - 30x^2 + 20y$.

$B(x,y) = 200 + 40x - 30y^2$.

a. $\frac{\partial W}{\partial x} = -60x \implies \left(\frac{\partial W}{\partial x}\right)(1,1) = -60 < 0$.

$\frac{\partial W}{\partial y} = 20 \implies \left(\frac{\partial W}{\partial y}\right)(1,1) = 20 > 0$.

If the price for white eggs increases from $x = 1$ to $x = 2$ for $y = 1$, then the demand for white eggs decreases approximately by 60 units.

If the price for brown eggs increases from $y = 1$ to $y = 2$ for $x = 1$, then the demand for white eggs increases approximately by 20 units.

b. $\frac{\partial B}{\partial x} = 40 \implies \left(\frac{\partial B}{\partial x}\right)(1,1) = 40 > 0$.

$\frac{\partial B}{\partial y} = -60y \implies \left(\frac{\partial B}{\partial y}\right)(1,1) = -60 < 0$.

If the price for white eggs increases from $x = 1$ to $x = 2$ for $y = 1$, then the demand for brown eggs increases approximately by 40 units.

If the price for brown eggs increases from $y = 1$ to $y = 2$ for $x = 1$, then the demand for brown eggs decreases approximately by 60 units.

40. $f(p,q) = 500 - 30p^2 + hq \implies \frac{\partial f}{\partial q}(p,q) = h > 0$.

$g(p,q) = 200 - 4q^2 + kp \implies \frac{\partial g}{\partial q}(p,q) = k > 0$.

41. $f(p,q) = 30 - 4p^2 + 16q \implies \frac{\partial f}{\partial q}(p,q) = 16 > 0$.

$g(p,q) = 80 - 12q^2 + 10p \implies \frac{\partial g}{\partial p}(p,q) = 10 > 0$.

Therefore the products are competitive.

42. $f(p,q) = 16 - 12p^2 - 4q \implies \frac{\partial f}{\partial q}(p,q) = -4 < 0$.

$g(p,q) = 40 - 6p - 8q^3 \implies \frac{\partial g}{\partial p}(p,q) = -6 < 0$.

Therefore the products are complementary.

43. $f(p,q) = 25 - 2p^{3/2} + 4\ln q \implies \frac{\partial f}{\partial q}(p,q) = 4 \cdot \frac{1}{q} > 0$.

$g(p,q) = 40 + e^{p/100} - q^{7/3} \implies \frac{\partial g}{\partial p}(p,q) = e^{p/100} \cdot \frac{1}{100} > 0$.

Therefore the products are competitive.

44. $f(p,q) = \frac{10p}{4p+2q} \implies \frac{\partial f}{\partial q}(p,q) = \frac{(4p+2q)(0)-(10p)(2)}{(4p+2q)^2} = \frac{-20p}{(4p+2q)^2} < 0.$

$g(p,q) = \frac{30q}{2p+5q} \implies \frac{\partial g}{\partial p}(p,q) = \frac{(2p+5q)(0)-(30q)(2)}{(2p+5q)^2} = \frac{-60q}{(2p+5q)^2} < 0.$

Therefore the products are complementary.

45. $f(p,q) = \frac{10q}{q-p} \implies \frac{\partial f}{\partial q}(p,q) = \frac{(q-p)(10)-(10q)(1)}{(q-p)^2} = \frac{-10p}{(q-p)^2} < 0.$

$q(p,q) = \frac{5p}{p-2q} \implies \frac{\partial g}{\partial p}(p,q) = \frac{(p-2q)(5)-(5p)(1)}{(p-2q)^2} = -\frac{10q}{(p-2q)^2} < 0.$

Therefore the products are complementary.

Solutions to Exercise Set 8.3 (6.3)

1. $f(x,y) = x^2 + y^2 + 4y + 4.$

$\frac{\partial f}{\partial x} = 2x = 0 \implies x = 0. \qquad \frac{\partial f}{\partial y} = 2y + 4 = 0 \implies y = -2.$

So (0,-2) is a critical point for f.

$\frac{\partial^2 f}{\partial x^2} = 2, \quad \frac{\partial^2 f}{\partial x \partial y} = 0$, and $\frac{\partial^2 f}{\partial y^2} = 2.$

Thus, at the critical point (0,-2) we have

$$A = 2, \quad B = 0, \quad C = 2, \quad \text{and } D = 0^2 - 2 \cdot 2 = -4.$$

Since $D < 0$ and $A > 0$, f has a relative minimum value at the critical point (0,-2), and the relative minimum value is $f(0,-2) = 0 + 4 - 8 + 4 = 0$.

2. $f(x,y) = x^2 - 3y^2 + 4x + 6y + 10.$

$\frac{\partial f}{\partial x} = 2x + 4 = 0 \implies x = -2, \quad \frac{\partial f}{\partial y} = -6y + 6 = 0 \implies y = 1.$

Thus the critical point is $(-2, 1)$.

$\frac{\partial^2 f}{\partial x^2} = 2, \quad \frac{\partial^2 f}{\partial y^2} = -6, \quad \frac{\partial^2 f}{\partial x \partial y} = 0.$

$D = 0^2 - 2 \cdot (-6) > 0.$

Therefore the point $(-2, 1, 9)$ is a saddle point.

3. $f(x,y) = x^2 + 2y^2 - 6x + 4y - 8.$

$\frac{\partial f}{\partial x} = 2x - 6 = 0 \implies x = 3, \quad \frac{\partial f}{\partial y} = 4y + 4 = 0 \implies y = -1.$

Thus the critical point is $(3, -1)$.

$\frac{\partial^2 f}{\partial x^2} = 2, \quad \frac{\partial^2 f}{\partial y^2} = 4, \quad \frac{\partial^2 f}{\partial x \partial y} = 0.$

$D = 0^2 - 2 \cdot (4) = -8 < 0$, and $\frac{\partial^2 f}{\partial x^2}(3,-1) > 0$.

Therefore $f(x,y)$ has a relative minimum value at $(3,-1)$.

4. $f(x,y) = 2x - 6y - x^2 - 2y^2 + 10$.

$\frac{\partial f}{\partial x} = 2 - 2x = 0 \implies x = 1, \quad \frac{\partial f}{\partial y} = -6 - 4y = 0 \implies y = -\frac{3}{2}$.

Thus $\left(1, -\frac{3}{2}\right)$ is a critical point for f.

$\frac{\partial^2 f}{\partial x^2} = -2, \quad \frac{\partial^2 f}{\partial x \partial y} = 0, \quad \frac{\partial^2 f}{\partial y^2} = -4$.

At the point $\left(1, -\frac{3}{2}\right)$, $A = -2, \quad B = 0, \quad C = -4$, and

$D = 0^2 - (-2)(-4) = -8$.

Since $A < 0$ and $D < 0$ at the critical point $\left(1, -\frac{3}{2}\right)$, f has a relative maximum value $f\left(1, -\frac{3}{2}\right) = 2 - 6\left(-\frac{3}{2}\right) - 1 - 2 \cdot \frac{9}{4} + 10 = \frac{31}{2}$.

5. $f(x,y) = x^2 - y^2 + 6x + 4y + 5$.

$\frac{\partial f}{\partial x} = 2x + 6 = 0 \implies x = -3, \quad \frac{\partial f}{\partial y} = -2y + 4 = 0 \implies y = 2$.

Thus (-3,2) is a critical point for f. $\quad \frac{\partial^2 f}{\partial x^2} = 2, \quad \frac{\partial^2 f}{\partial x \partial y} = 0, \quad \frac{\partial^2 f}{\partial y^2} = -2$.

At the point (-3,2), $A = 2, \quad B = 0, \quad C = -2$, and $D = 4$.

Since $D > 0$, $(-3, 2, f(-3,2)) = (-3, 2, 0)$ is a saddle point for f.

6. $f(x,y) = x^2 - 5xy + 6y^2 + 3x - 2y - 3$.

$\frac{\partial f}{\partial x} = 2x - 5y + 3, \quad \frac{\partial f}{\partial y} = -5x + 12y - 2$.

We now find the critical point of f.

$$\begin{cases} \frac{\partial f}{\partial x} = 0 \implies 2x - 5y + 3 = 0 \\ \frac{\partial f}{\partial y} = 0 \implies -5x + 12y - 2 = 0. \end{cases}$$

$$\begin{cases} 12 \cdot (2x - 5y + 3) = 24x - 60y + 36 = 0 \\ 5 \cdot (-5x + 12y - 2) = -25x + 60y - 10 = 0 \end{cases}$$

$$-x + 26 = 0 \implies x = 26.$$

$$2 \cdot 26 - 5y + 3 = 0 \implies y = 11.$$

Therefore the critical point is $(26, 11)$.

$\frac{\partial^2 f}{\partial x^2} = 2, \quad \frac{\partial^2 f}{\partial y^2} = 12, \quad \frac{\partial^2 f}{\partial x \partial y} = -5$.

At the critical point $(26, 11)$:

$$D = (-5)^2 - 2 \cdot 12 = 1 > 0.$$

Therefore the surface $z = f(x, y)$ has a saddle point at the point where $x = 26$, $y = 11$.

7. $f(x, y) = xy + 9$. $\frac{\partial f}{\partial x} = y = 0$, $\frac{\partial f}{\partial y} = x = 0$.

   Thus (0,0) is a critical point for f. $\frac{\partial^2 f}{\partial x^2} = 0$, $\frac{\partial^2 f}{\partial x \partial y} = 1$, $\frac{\partial^2 f}{\partial y^2} = 0$.

   Since $D = 1 > 0$ at the point (0,0), f has a saddle point at (0,0,9).

8. $f(x, y) = 3x^2 - 7xy + 4y^2 - x - y + 5$.

   $\frac{\partial f}{\partial x} = 6x - 7y - 1$, $\frac{\partial f}{\partial y} = -7x + 8y - 1$.

   We now find the critical point of f.

$$\begin{cases} \frac{\partial f}{\partial x} = 0 \implies 6x - 7y - 1 = 0 \\ \frac{\partial f}{\partial y} = 0 \implies -7x + 8y - 1 = 0. \end{cases}$$

$$\begin{cases} 8 \cdot (6x - 7y - 1) = 48x - 56y - 8 = 0 \\ 7 \cdot (-7x + 8y - 1) = -49x + 56y - 7 = 0. \end{cases}$$

$$-x - 15 = 0 \implies x = -15.$$

$$6 \cdot (-15) - 7y - 1 = 0 \implies 7y = -91 \implies y = -13.$$

   Thus the critical point is $(-15, -13)$.

   $\frac{\partial^2 f}{\partial x^2} = 6$, $\frac{\partial^2 f}{\partial y^2} = 8$, $\frac{\partial^2 f}{\partial x \partial y} = -7$.

   At the critical point $(-15, -13)$:

$$D = (-7)^2 - 6 \cdot 8 = 1 > 0.$$

   Therefore the surface $z = f(x, y)$ has a saddle point at the point where $x = -15$, $y = -13$.

9. $f(x, y) = 5x^2 + y^2 - 10x - 6y + 15$.

   $\frac{\partial f}{\partial x} = 10x - 10 = 0 \implies x = 1$, $\frac{\partial f}{\partial y} = 2y - 6 = 0 \implies y = 3$.

   $(1, 3)$ is a critical point for f. $\frac{\partial^2 f}{\partial x^2} = 10$, $\frac{\partial^2 f}{\partial x \partial y} = 0$, $\frac{\partial^2 f}{\partial y^2} = 2$.

   Hence, at the point (1,3), $A = 10$, $B = 0$, $C = 2$, and $D = -20$.

   Since $A > 0$ and $D < 0$, f has a relative minimum value $f(1, 3) = 1$ at the critical point (1,3).

10. $f(x,y) = x^2 + y^3 - 3y$.

$\frac{\partial f}{\partial x} = 2x = 0 \implies x = 0, \quad \frac{\partial f}{\partial y} = 3y^2 - 3 = 0 \implies y = \pm 1.$

Thus critical points for f are (0,1) and (0,-1).

$\frac{\partial f}{\partial x^2} = 2, \quad \frac{\partial^2 f}{\partial x \partial y} = 0, \quad \frac{\partial^2 f}{\partial y^2} = 6y.$

(i) At the point (0,1), $A = 2$, $B = 0$, $C = 6$, $D = -12$.

So f has a relative minimum value $f(0,1) = -2$ at the critical point (0,1).

(ii) At the point (0,-1), $A = 2$, $B = 0$, $C = -6$, and $D = 12$.

Since $D > 0$, f has a saddle point at (0,-1,2).

11. $f(x,y) = x^3 - y^3$.

$\frac{\partial f}{\partial x} = 3x^2 = 0 \implies x = 0, \quad \frac{\partial f}{\partial y} = -3y^2 = 0 \implies y = 0.$

So f has a critical point at (0,0). $\frac{\partial^2 f}{\partial x^2} = 6x, \quad \frac{\partial^2 f}{\partial x \partial y} = 0, \quad \frac{\partial^2 f}{\partial y^2} = -6y.$

Since $D = 0$ at the point (0,0), we can't conclude by this test. However, $f(x,0) = x^3$ has neither a relative minimum nor a relative maximum at (0,0); $f(0,y) = -y^3$ has neither a relative minimum nor a relative maximum at (0,0). Therefore, the point (0,0) provides neither a relative maximum nor a relative minimum for f.

12. $f(x,y) = e^{x^2-2x+y^2+4}$.

$\frac{\partial f}{\partial x} = e^{x^2-2x+y^2+4} \cdot (2x-2) = 0 \implies x = 1,$

$\frac{\partial f}{\partial y} = e^{x^2-2x+y^2+4} \cdot (2y) = 0 \implies y = 0.$

Thus a critical point for f is (1,0).

$$\begin{aligned}\frac{\partial^2 f}{\partial x^2} &= 2 \cdot e^{x^2-2x+y^2+4} + (2x-2)^2 e^{x^2-2x+y^2+4} \\ &= e^{x^2-2x+y^2+4}(4x^2 - 8x + 6),\end{aligned}$$

$$\frac{\partial^2 f}{\partial x \partial y} = (2x-2)(2y) \cdot e^{x^2-2x+y^2+4},$$

$$\begin{aligned}\frac{\partial^2 f}{\partial y^2} &= 2 \cdot e^{x^2-2x+y^2+4} + 2y \cdot e^{x^2-2x+y^2+4} \cdot 2y \\ &= (2+4y^2) \cdot e^{x^2-2x+y^2+4}.\end{aligned}$$

At the point (1,0), $A = e^3 \cdot 2$, $B = 0$, $C = 2 \cdot e^3$, and $D = -4e^6$.

Since $A > 0$ and $D < 0$, f has a relative minimum value $f(1,0) = e^3$ at the point (1,0).

13. $f(x,y) = x^3 + y^3 + 4xy.$

$$\frac{\partial f}{\partial x} = 3x^2 + 4y = 0 \quad (1)$$
$$\frac{\partial f}{\partial y} = 3y^2 + 4x = 0 \quad (2)$$

From (1) we obtain $y = -3x^2/4$. Thus (2) become $3\left(-\frac{3x^2}{4}\right)^2 + 4x = 0$
$\implies \frac{27}{16}x^4 + 4x = 0 \implies 27x^4 + 64x = 0$
$\implies x(27x^3 + 64) = 0 \implies x = 0,\ -\frac{4}{3}.$

Hence, critical points for f are (0,0) and $\left(-\frac{4}{3}, -\frac{4}{3}\right)$.

$\frac{\partial^2 f}{\partial x^2} = 6x, \qquad \frac{\partial^2 f}{\partial x \partial y} = 4, \qquad \frac{\partial^2 f}{\partial y^2} = 6y.$

(i) At the point (0,0), $A = 0$, $B = 4$, $C = 0$, and $D = 16$. f has a saddle point at (0,0).

(ii) At the point $\left(-\frac{4}{3}, -\frac{4}{3}\right)$, $A = -8$, $B = 4$, $C = -8$, and $D = -48$. f has a relative maximum value $f\left(-\frac{4}{3}, -\frac{4}{3}\right) = \frac{64}{27}$.

14. $f(x,y) = y^3 - x^2 - 5x - 12y + 5$

$\frac{\partial f}{\partial x} = -2x - 5 = 0 \implies x = -\frac{5}{2},$

$\frac{\partial f}{\partial y} = 3y^2 - 12 = 0 \implies y^2 = 4 \implies y = 2$ or $y = -2$.

Thus the critical points are $\left(-\frac{5}{2}, 2\right)$ and $\left(-\frac{5}{2}, -2\right)$.

$\frac{\partial^2 f}{\partial x^2} = -2, \quad \frac{\partial^2 f}{\partial y^2} = 6y, \quad \frac{\partial^2 f}{\partial x \partial y} = 0.$

At the critical point $\left(-\frac{5}{2}, 2\right)$:

$$D = 0^2 - (-2)(12) = 24 > 0.$$

Therefore the surface $z = f(x,y)$ has a saddle point where $x = -\frac{5}{2}$, $y = 2$.

At the critical point $\left(-\frac{5}{2}, -2\right)$:

$$D = 0^2 - (-2)(-12) = 24 > 0, \qquad \frac{\partial^2 f}{\partial x^2} = -2 < 0.$$

Therefore $f(x,y)$ has a relative maximum value at the critical point $\left(-\frac{5}{2}, 2\right)$.

15. $f(x,y) = x^2 + y^2 + x - 2y + xy + 5.$

$$\frac{\partial f}{\partial x} = 2x + 1 + y = 0 \qquad \textbf{(1)}$$
$$\frac{\partial f}{\partial y} = 2y - 2 + x = 0 \qquad \textbf{(2)}$$

$((\mathbf{1}) \times (-2)) + (\mathbf{2}) \implies -3x - 4 = 0 \implies x = -\frac{4}{3}.$

From $(\mathbf{1})$, $y = -2x - 1, \quad y = -2\left(-\frac{4}{3}\right) - 1 = \frac{5}{3}.$

The critical point is $\left(-\frac{4}{3}, \frac{5}{3}\right)$.

$\frac{\partial^2 f}{\partial x^2} = 2, \qquad \frac{\partial^2 f}{\partial x \partial y} = 1, \qquad \frac{\partial^2 f}{\partial y^2} = 2. \qquad D = 1^2 - 2 \cdot 2 = -3 < 0.$

Thus f has a relative minimum value at $\left(-\frac{4}{3}, \frac{5}{3}\right)$.

16. $f(x,y) = x^2 + 3y^2 - 2x + 3y + 2xy - 6.$

$$\frac{\partial f}{\partial x} = 2x - 2 + 2y = 0 \qquad \textbf{(1)}$$
$$\frac{\partial f}{\partial y} = 6y + 3 + 2x = 0 \qquad \textbf{(2)}$$

$((\mathbf{1}) \times (-1)) + (\mathbf{2}) \implies 4y + 5 = 0 \implies y = -\frac{5}{4}.$

From $(\mathbf{1})$, $x = 1 - y \implies x = 1 - \left(-\frac{5}{4}\right) = \frac{9}{4}.$

The critical point of f is $\left(\frac{9}{4}, -\frac{5}{4}\right)$.

$\frac{\partial^2 f}{\partial x^2} = 2, \qquad \frac{\partial^2 f}{\partial x \partial y} = 2, \qquad \frac{\partial^2 f}{\partial y^2} = 6. \qquad D = 2^2 - 2 \cdot 6 = -8 < 0.$

Thus f has a relative minimum at $\left(\frac{9}{4}, -\frac{5}{4}\right)$.

17. $P(x,y) = 30x + 90y - 0.5x^2 - 2y^2 - xy.$

$$\frac{\partial P}{\partial x} = 30 - x - y = 0 \qquad \textbf{(1)}$$
$$\frac{\partial P}{\partial y} = 90 - 4y - x = 0 \qquad \textbf{(2)}$$

$((\mathbf{1}) \times (-1)) + (\mathbf{2}) \implies -3y + 60 = 0 \implies y = 20.$

From $(\mathbf{1}) \implies x = 30 - y \implies x = 30 - 20 = 10.$

The critical point for P is (10,20).

$\frac{\partial^2 P}{\partial x^2} = -1, \qquad \frac{\partial^2 P}{\partial x \partial y} = -1, \qquad \frac{\partial^2 P}{\partial y^2} = -4.$

$D = (-1)^2 - (-1)(-4) = -3 < 0$, and $A < 0$.

Thus the profit P is maximum when $x = 10$ and $y = 20$.

18. $P(x, y) = 20x - x^2 + 40y - y^2$.

$\frac{\partial P}{\partial x} = 20 - 2x = 0 \implies x = 10$.

$\frac{\partial P}{\partial y} = 40 - 2y = 0 \implies y = 20$.

$\frac{\partial^2 P}{\partial x^2} = -2, \qquad \frac{\partial^2 P}{\partial x \partial y} = 0, \qquad \frac{\partial^2 P}{\partial y^2} = -2.$

$D = 0 - (-2)(-2) = -4 < 0$ and $A < 0$.

Thus profit is maximum when $(x, y) = (10, 20)$.

19. $P(x, y) = 40x + 80y - 2x^2 - 10y^2 - 4xy$.

$$\frac{\partial P}{\partial x} = 40 - 4x - 4y = 0 \qquad (1)$$

$$\frac{\partial P}{\partial y} = 80 - 20y - 4x = 0 \qquad (2)$$

$((\mathbf{1}) \times (-1)) + (\mathbf{2}) \implies 40 - 16y = 0 \implies y = \frac{40}{16} = \frac{5}{2} = 2.5.$

From $(\mathbf{1}) \implies x = 10 - y \implies x = 10 - \left(\frac{5}{2}\right) = \frac{15}{2} = 7.5.$

The critical point for P is (7.5,2.5).

$\frac{\partial^2 P}{\partial x^2} = -4, \qquad \frac{\partial^2 P}{\partial x \partial y} = -4, \qquad \frac{\partial^2 P}{\partial y^2} = -20.$

$D = (-4)^2 - (-4)(-20) = -64 < 0$, and $A < 0$.

Thus the maximum productivity occurs when $(x, y) = (7.5, 2.5)$.

20. $C(x, y) = 10,000 + 10x^2 + 15y^2 - 100x - 200y + 10xy$.

$$\frac{\partial C}{\partial x} = 20x - 100 + 10y = 0 \qquad (1)$$

$$\frac{\partial C}{\partial y} = 30y - 200 + 10x = 0 \qquad (2)$$

$(\mathbf{1}) + ((\mathbf{2}) \times (-2)) \implies -50y + 300 = 0 \implies y = 6.$

From $(\mathbf{2}) \implies x = 20 - 3y \implies x = 20 - (3 \cdot 6) = 2.$

The critical point for C is (2,6).

$\frac{\partial^2 C}{\partial x^2} = 20, \qquad \frac{\partial^2 C}{\partial x \partial y} = 10, \qquad \frac{\partial^2 C}{\partial y^2} = 30.$

$D = 10^2 - 20 \cdot 30 = -500 < 0$, and $A > 0$.

Thus costs are minimum when $(x, y) = (2, 6)$.

21. The revenue $R(x, y) = p \cdot x + q \cdot y$.

$$R(x, y) = (20 - x)x + \left(46 - \tfrac{5}{2}y\right)y = -x^2 - \tfrac{5}{2}y^2 + 20x + 46y.$$

The profit $P(x, y) = R(x, y) - C(x, y)$

$$= \left(-x^2 - \tfrac{5}{2}y^2 + 20x + 46y\right) - (100 + 4x + 2y + xy)$$

$$= -x^2 - \tfrac{5}{2}y^2 + 16x + 44y - xy - 100.$$

$$\frac{\partial P}{\partial x} = -2x + 16 - y = 0 \qquad (1)$$

$$\frac{\partial P}{\partial y} = -5y + 44 - x = 0 \qquad (2)$$

$\mathbf{(1)} + (\mathbf{(2)} \times (-2)) \implies -72 + 9y = 0 \implies y = 8.$

From $\mathbf{(1)} \implies x = 8 - \frac{y}{2} \implies x = 8 - \frac{1}{2}(8) = 4.$

The critical point for P is (4,8).

$\frac{\partial^2 P}{\partial x^2} = -2, \qquad \frac{\partial^2 P}{\partial x \partial y} = -1, \qquad \frac{\partial^2 P}{\partial y^2} = -5.$

$D = (-1)^2 - (-2)(-5) = -9 < 0$, and $A < 0$.

Thus the maximum profit occurs when $(x, y) = (4, 8)$.

22. a. $R(x, y) = xD_A(x, y) + yD_B(x, y) = x(30 - 5x + y) + y(40 - 4y + x)$

$$= -5x^2 - 4y^2 + 30x + 40y + 2xy.$$

b. $P(x, y) = R(x, y) - C(x, y) = (-5x^2 - 4y^2 + 30x + 40y + 2xy) - (2x + y).$

$P(x, y) = -5x^2 - 4y^2 + 28x + 39y + 2xy.$

(c)

$$\frac{\partial P}{\partial x} = -10x + 28 + 2y = 0 \qquad (1)$$

$$\frac{\partial P}{\partial y} = -8y + 39 + 2x = 0 \qquad (2)$$

$(\mathbf{(1)} \times 4) + \mathbf{(2)} \implies -38x + 151 = 0 \implies x = \frac{151}{38} \approx \$3.97.$

From $\mathbf{(1)} \implies y = 5x - 14 \implies y = 5\left(\frac{151}{38}\right) - 14 = \frac{223}{38} \approx \$5.87.$

$\frac{\partial^2 P}{\partial x^2} = -10, \qquad \frac{\partial^2 P}{\partial x \partial y} = 2, \qquad \frac{\partial^2 P}{\partial y^2} = -8.$

$D = 2^2 - (-10)(-8) = -76 < 0$, and $A < 0$.

Thus the profit is maximum when $(x, y) = (\$3.97, \$5.87)$.

23. Let x be the length of the box in meters, let y be the width of the box in meters, and let z be the height of the box in meters.

Let p be the price in dollars per square meter of the material for the side walls.

Since the volume of the box is to be 64 cubic meters, it follows that

$xyz = 64 \implies z = \frac{64}{xy}$.

Then let $f(x, y)$ be the cost in dollars of the material for the box. Note that

$$\begin{aligned} f(x,y) &= 2 \cdot (xz + yz) \cdot p + 2 \cdot xy \cdot 2p \\ &= 2 \cdot \left(x \cdot \frac{64}{xy} + y \cdot \frac{64}{xy} \right) \cdot p + 4xyp \\ &= \left(\frac{128}{y} + \frac{128}{x} + 4xy \right) \cdot p \\ &= 4 \cdot \left(\frac{32}{y} + \frac{32}{x} + xy \right) \cdot p. \end{aligned}$$

$$\frac{\partial f}{\partial x} = 4 \cdot \left(-\frac{32}{x^2} + y \right) \cdot p = 0 \implies y = \frac{32}{x^2}.$$

$$\frac{\partial f}{\partial y} = 4 \cdot \left(-\frac{32}{y^2} + x \right) \cdot p = 0 \implies x = \frac{32}{y^2}.$$

Substitute $y = \frac{32}{x^2}$ into the eqution $x = \frac{32}{y^2}$.

$$x = \frac{32}{\left(\frac{32}{x^2}\right)^2} \implies x = \frac{x^4}{32} \implies x^3 = 32 \implies x = \sqrt[3]{32}.$$

$$y = \frac{32}{\left(\sqrt[3]{32}\right)^2} = \sqrt[3]{32}.$$

$$z = \frac{64}{\sqrt[3]{32} \cdot \sqrt[3]{32}} = 2 \cdot \sqrt[3]{32}.$$

$$\frac{\partial^2 f}{\partial x^2} = \frac{256}{x^3} \cdot p, \qquad \frac{\partial^2 f}{\partial y^2} = \frac{256}{y^3} \cdot p, \qquad \frac{\partial^2 f}{\partial x \partial y} = 4p.$$

At the critical point $\left(\sqrt[3]{32}, \sqrt[3]{32}\right)$:

$$A = \frac{\partial^2 f}{\partial x^2}\left(\sqrt[3]{32}, \sqrt[3]{32}\right) = 8p, \qquad C = \frac{\partial^2 f}{\partial y^2}\left(\sqrt[3]{32}, \sqrt[3]{32}\right) = 8p,$$

$$B = \frac{\partial^2 f}{\partial x \partial y}\left(\sqrt[3]{32}, \sqrt[3]{32}\right) = 4p.$$

$$D = B^2 - AC = (4p)^2 - (8p)(8p) = -48p^2 < 0.$$

So therefore the minimum cost occurs when $x = \sqrt[3]{32}$, $y = \sqrt[3]{32}$, $z = 2 \cdot \sqrt[3]{32}$.

24. Let x be the length of the box in meters, let y be the width of the box in meters, let z be the height of the box in meters.

Since the volume of the box is to be 64 m^3, it follows that $xyz = 64 \implies z = \frac{64}{xy}$.

Let $f(x, y)$ be the surface area of the box in square meters.

$$f(x,y) = 2(xy + xz + yz) = 2 \cdot \left(xy + x \cdot \frac{64}{xy} + y \cdot \frac{64}{xy} \right)$$
$$= 2 \cdot \left(xy + \frac{64}{y} + \frac{64}{x} \right).$$
$$\frac{\partial f}{\partial x} = 2 \cdot \left(y - \frac{64}{x^2} \right) = 0 \implies y = \frac{64}{x^2}.$$
$$\frac{\partial f}{\partial y} = 2 \cdot \left(x - \frac{64}{y^2} \right) = 0 \implies x = \frac{64}{y^2}.$$

Substitute $y = \frac{64}{x^2}$ into the equation $x = \frac{64}{y^2}$.

$$x = \frac{64}{\left(\frac{64}{x^2}\right)^2} \implies x = \frac{x^4}{64} \implies x^3 = 64 \implies x = 4.$$
$$y = \frac{64}{(4)^2} = 4.$$
$$z = \frac{64}{(4 \cdot 4)} = 4.$$
$$\frac{\partial^2 f}{\partial x^2} = \frac{256}{x^3}, \qquad \frac{\partial^2 f}{\partial y^2} = \frac{256}{y^3}, \qquad \frac{\partial^2 f}{\partial x \partial y} = 2.$$

At the critical point $(4, 4)$:

$$A = \frac{\partial^2 f}{\partial x^2}(4,4) = 4, \qquad C = \frac{\partial^2 f}{\partial y^2}(4,4) = 4,$$
$$B = \frac{\partial^2 f}{\partial x \partial y}(4,4) = 2.$$
$$D = B^2 - AC = (2)^2 - (4)(4) = -12 < 0.$$

Therefore the minimum surface area occurs when $x = 4$, $y = 4$, $z = 4$.

25. Let $\ell =$ the length of the package

$w =$ the width of the package

$h =$ the height of the package

Then $\ell + 2w + 2h = 84 \implies \ell = 84 - 2w - 2h$.

The volume $V = hw\ell = hw(84 - 2w - 2h)$.

$$V = 84hw - 2hw^2 - 2h^2w \text{ for } h > 0, \quad w > 0.$$

$$\frac{\partial V}{\partial h} = 84w - 2w^2 - 4hw = 0 \implies 42 - w - 2h = 0 \qquad \textbf{(1)}$$

$$\frac{\partial V}{\partial w} = 84h - 4hw - 2h^2 = 0 \implies 42 - h - 2w = 0 \qquad \textbf{(2)}$$

From **(1)**: $w = 42 - 2h$. Then **(2)** becomes

$42 - h - 2(42 - 2h) = 0 \implies 3h - 42 = 0 \implies h = 14$.

Then $w = 42 - 2h = 42 - 2(14) = 14$.

$$\ell = 84 - 2(w + h) = 84 - 2(14 + 14) = 28.$$

$\frac{\partial^2 V}{\partial h^2} = -4w, \qquad \frac{\partial^2 V}{\partial h \partial w} = 84 - 4w - 4h, \qquad \frac{\partial^2 V}{\partial w^2} = -4h.$

At (14,14), $D = (-28)^2 - (-56)(-56) < 0, \quad A = -56 < 0$.

Thus the volume of the package is the largest when $h = w = 14$ inches, $\ell = 28$ inches.

26. (a) From the graph of $f(x, y) = \sqrt{x^2 + y^2}$, the graph is above or on the (x, y)- plane with the lowest point at (0,0,0).

Thus f has a relative minimum value of $f(0, 0) = 0$ at (0,0).

(b) $\frac{f(x,0) - f(0,0)}{x - 0} = \frac{|x|}{x}$ does not have a limit as $x \to 0$.

Thus $f_x(0, 0)$ does not exist.

Similarly $f_y(0, 0)$ does not exist.

Solutions to Exercise Set 8.4 (6.4)

1. $f(x,y) = xy. \qquad g(x,y) = x + 4y - 8 = 0.$

$L(x,y,\lambda) = xy + \lambda(x + 4y - 8).$

$$\frac{\partial L}{\partial x} = y + \lambda = 0 \qquad \textbf{(1)}$$
$$\frac{\partial L}{\partial y} = x + 4\lambda = 0 \qquad \textbf{(2)}$$
$$\frac{\partial L}{\partial \lambda} = x + 4y - 8 = 0 \qquad \textbf{(3)}$$
$$((\textbf{1}) \times (-4)) + (\textbf{2}) \to -4y + x = 0 \implies x = 4y.$$

Next we substitute $x = 4y$ into equation **(3)**.

$$4y + 4y - 8 = 0 \implies y = 1 \implies x = 4.$$

Note that $f(4,1) = 4 > f(12,-1) = -12.$

Therefore, $f(4,1) = 4$ is the maximum value of f, and f has no minimum value.

2. $f(x,y) = 2x^2 + 4y^2. \qquad g(x,y) = 2x - 4y + 3 = 0.$

$L(x,y,\lambda) = 2x^2 + 4y^2 + \lambda(2x - 4y + 3).$

$$\frac{\partial L}{\partial x} = 4x + 2\lambda = 0 \qquad \textbf{(1)}$$
$$\frac{\partial L}{\partial y} = 8y - 4\lambda = 0 \qquad \textbf{(2)}$$
$$\frac{\partial L}{\partial \lambda} = 2x - 4y + 3 = 0 \qquad \textbf{(3)}$$
$$((\textbf{1}) \cdot 2) + (\textbf{2}) \to 8x + 8y = 0 \implies x = -y.$$

Next we substitute $x = -y$ into equation **(3)**.

$$2(-y) - 4y + 3 = 0 \implies y = \frac{1}{2} \implies x = -\frac{1}{2}.$$

Note that $f\left(-\frac{1}{2},\frac{1}{2}\right)=\frac{3}{2}<f\left(-\frac{3}{2},0\right)=\frac{9}{2}$.

Therefore, $f\left(-\frac{1}{2},\frac{1}{2}\right)=\frac{3}{2}$ is the minimum value of f, and f has no maximum value.

3. $f(x,y)=x^2-8x+y^2+4y-6. \qquad g(x,y)=2x-y+5=0.$

$L(x,y,\lambda)=x^2-8x+y^2+4y-6+\lambda(2x-y+5).$

$$\frac{\partial L}{\partial x}=2x-8+2\lambda=0 \implies x-4+\lambda=0 \qquad (1)$$

$$\frac{\partial L}{\partial y}=2y+4-\lambda=0 \qquad (2)$$

$$\frac{\partial L}{\partial \lambda}=2x-y+5=0 \qquad (3)$$

$$\mathbf{(1)}+\mathbf{(2)} \implies x+2y=0 \implies x=-2y \qquad (4)$$

(3) and **(4)** give $2(-2y)-y+5=0 \implies y=1.$

$x=-2y=-2(1)=-2.$

Since $f(0,5)=39>f(-2,1)=19, \quad f(-2,1)=19$ is the minimum value of f, and there is no maximum value of f.

4. $f(x,y)=4y-2x. \qquad g(x,y)=x^2+y^2-2=0.$

$L(x,y,\lambda)=4y-2x+\lambda(x^2+y^2-2).$

$$\frac{\partial L}{\partial x}=-2+2x\lambda=0 \qquad (1)$$

$$\frac{\partial L}{\partial y}=4+2y\lambda=0 \qquad (2)$$

$$\frac{\partial L}{\partial \lambda}=x^2+y^2-2=0 \qquad (3)$$

$$((\mathbf{1})\cdot y)-((\mathbf{2})\cdot x)\to -2y-4x=0 \implies y=-2x.$$

Next substitute $y=-2x$ into equation **(3)**.

$x^2+4x^2-2=0 \implies 5x^2=2 \implies x^2=.4 \implies x=\sqrt{.4} \quad \text{or} \quad x=-\sqrt{.4}.$

The critical points are $\left(\sqrt{.4},-2\cdot\sqrt{.4}\right)$ and $\left(-\sqrt{.4},2\cdot\sqrt{.4}\right)$.

$$f\left(\sqrt{.4},-2\sqrt{.4}\right)=4\cdot\left(-2\sqrt{.4}\right)-2\cdot\sqrt{.4}=-10\sqrt{.4}.$$

$$f\left(-\sqrt{.4},2\sqrt{.4}\right)=4\cdot\left(2\sqrt{.4}\right)-2\cdot\left(-\sqrt{.4}\right)=10\sqrt{.4}.$$

The minimum value of f is $f\left(\sqrt{.4}, -2\sqrt{.4}\right) = -10\sqrt{.4}$, and the maximum value of f is $f\left(-\sqrt{.4}, 2\sqrt{.4}\right) = 10\sqrt{.4}$.

5. $f(x,y) = xy$. $\qquad g(x,y) = x^2 + y^2 - 32 = 0$.

$L(x,y,\lambda) = xy + \lambda(x^2 + y^2 - 32)$.

$$\frac{\partial L}{\partial x} = y + 2x\lambda = 0 \tag{1}$$

$$\frac{\partial L}{\partial y} = x + 2y\lambda = 0 \tag{2}$$

$$\frac{\partial L}{\partial \lambda} = x^2 + y^2 - 32 = 0 \tag{3}$$

$$((\mathbf{1}) \cdot (y)) + ((\mathbf{2}) \cdot (-x)) \implies y^2 - x^2 = 0 \implies y = \pm x.$$

For $y = x$: **(3)** gives $x^2 + x^2 - 32 = 0 \implies x = \pm 4$.

Then critical points are (4,4), (-4,-4).

For $y = -x$: **(3)** gives $x^2 + (-x)^2 - 32 = 0 \implies x = \pm 4$.

Then critical points are (4,-4), (-4,4).

$f(4,4) = f(-4,-4) = 16, \qquad f(4,-4) = f(-4,4) = -16.$

Thus the minimum value of f is -16 and the maximum value is 16.

6. $f(x,y) = x^2 + y, \qquad g(x,y) = x^2 + y^2 - 9 = 0.$

$L(x,y,\lambda) = x^2 + y + \lambda(x^2 + y^2 - 9)$.

$$\frac{\partial L}{\partial x} = 2x + 2\lambda x = 0 \implies 2x(1+\lambda) = 0 \implies \lambda = -1 \quad \text{or} \quad x = 0 \tag{1}$$

$$\frac{\partial L}{\partial y} = 1 + 2\lambda y = 0 \implies y = -\frac{1}{(2\lambda)} \tag{2}$$

$$\frac{\partial L}{\partial \lambda} = x^2 + y^2 - 9 = 0 \tag{3}$$

Substituting $x = 0$ and $y = -\frac{1}{2\lambda}$ into equation **(3)** gives

$$\frac{1}{4\lambda^2} - 9 = 0 \implies \lambda^2 = \frac{1}{36} \implies \lambda = \pm\frac{1}{6}.$$

Thus critical points for L are $\left(0, -3, \frac{1}{6}\right)$ and $\left(0, 3, -\frac{1}{6}\right)$.

Substituting $\lambda = -1$ and $y = -\frac{1}{2\lambda} = \frac{1}{2}$ into equation **(3)** gives

$$x^2 + \frac{1}{4} - 9 = 0 \implies x^2 = \frac{35}{4} \implies x = \pm\frac{\sqrt{35}}{2}.$$

Thus two other critical points for L are $\left(\frac{\sqrt{35}}{2}, \frac{1}{2}, -1\right)$ and $\left(-\frac{\sqrt{35}}{2}, \frac{1}{2}, -1\right)$.

$f(0,-3) = -3, \qquad f(0,3) = 3, \qquad f\left(\frac{\sqrt{35}}{2}, \frac{1}{2}\right) = \frac{37}{4}, \qquad f\left(-\frac{\sqrt{35}}{2}, \frac{1}{2}\right) = \frac{37}{4}.$

Hence, the maximum value of f is $\frac{37}{4}$ and the minimum value is -3.

7. $f(x,y) = x^3 - y^3, \qquad g(x,y) = x - y - 2 = 0.$

$L(x,y,\lambda) = x^3 - y^3 + \lambda(x - y - 2).$

$$\frac{\partial L}{\partial x} = 3x^2 + \lambda = 0. \tag{1}$$

$$\frac{\partial L}{\partial y} = -3y^2 - \lambda = 0. \tag{2}$$

$$\frac{\partial L}{\partial \lambda} = x - y - 2 = 0. \tag{3}$$

$$\textbf{(1)} + \textbf{(2)} \text{ gives } 3(x^2 - y^2) = 0 \implies x = \pm y \tag{4}$$

From **(3)** and **(4)** we obtain $x = 1$ and $y = -1$.

$f(1,-1) = 1+1 = 2$ is the minimum value of f since $f(2,0) = 8 > f(1,-1) = 2$, and there is no maximum value of f.

8. $f(x,y,z) = x + 2y - z, \qquad g(x,y,z) = x^2 + y^2 + z^2 - 24 = 0.$

$L(x,y,z,\lambda) = x + 2y - z + \lambda(x^2 + y^2 + z^2 - 24).$

$$\frac{\partial L}{\partial x} = 1 + 2\lambda x = 0 \implies x = -\frac{1}{2\lambda} \tag{1}$$

$$\frac{\partial L}{\partial y} = 2 + 2\lambda y = 0 \implies y = -\frac{1}{\lambda} \tag{2}$$

$$\frac{\partial L}{\partial z} = -1 + 2\lambda z = 0 \implies z = \frac{1}{2\lambda} \tag{3}$$

$$\frac{\partial L}{\partial \lambda} = x^2 + y^2 + z^2 - 24 = 0 \tag{4}$$

Substituting **(1)**, **(2)**, and **(3)** into **(4)**, we have

$$\frac{1}{4\lambda^2} + \frac{1}{\lambda^2} + \frac{1}{4\lambda^2} - 24 = 0 \implies \frac{6}{4\lambda^2} = 24 \implies \lambda = \pm\frac{1}{4}.$$

Therefore, two critical points to be checked are

$$(-2,-4,2) \quad \text{and} \quad (2,4,-2).$$

$f(-2,-4,2) = -2-8-2 = -12$ and $f(2,4,-2) = 12$.

Thus, the maximum value of f is 12 and the minimum value of f is -12.

9. $f(x,y,z) = xyz, \qquad g(x,y,z) = x^2+y^2+z^2-12 = 0.$

$L(x,y,z,\lambda) = xyz + \lambda(x^2+y^2+z^2-12).$

$$\frac{\partial L}{\partial x} = yz + 2\lambda x = 0 \tag{1}$$

$$\frac{\partial L}{\partial y} = xz + 2\lambda y = 0 \tag{2}$$

$$\frac{\partial L}{\partial z} = xy + 2\lambda z = 0 \tag{3}$$

$$\frac{\partial L}{\partial \lambda} = x^2+y^2+z^2-12 = 0 \tag{4}$$

From $((\mathbf{1})\cdot y) - ((\mathbf{2})\cdot x)$, $(y^2-x^2)z = 0 \implies y = x, \quad y = -x$, or $z = 0$.

(i) For $z = 0$:

(3) and **(4)** become

$$\begin{cases} xy = 0 \\ x^2+y^2 = 12 \end{cases} \implies \begin{cases} x = 0 \\ y = \pm 2\sqrt{3} \end{cases} \text{ or } \begin{cases} y = 0 \\ x = \pm 2\sqrt{3} \end{cases}$$

Critical points to be checked are $(0, 2\sqrt{3}, 0)$, $(0, -2\sqrt{3}, 0)$, $(2\sqrt{3}, 0, 0)$, and $(-2\sqrt{3}, 0, 0)$.

(ii) For $y = x$:

(2), **(3)**, and **(4)** become

$$x(z+2\lambda) = 0 \implies x = 0 \quad \text{or} \quad z = -2\lambda$$

$$x^2 + 2\lambda z = 0 \tag{5}$$

$$2x^2 + z^2 - 12 = 0. \tag{6}$$

Substituting $x = 0$ into **(6)**, we obtain $z = \pm 2\sqrt{3}$.

Substituting $z = -2\lambda$ into **(5)** and **(6)** gives

$$x^2 = 4\lambda^2 \tag{7}$$

$$2x^2 + 4\lambda^2 - 12 = 0. \tag{8}$$

From **(7)** and **(8)**, $\lambda = \pm 1$ and so $z = \mp 2$.

Critical points to be checked are $(0,0,2\sqrt{3})$, $(0,0,-2\sqrt{3})$, $(2,2,-2)$, $(-2,-2,-2)$, $(2,2,2)$ and $(-2,-2,2)$.

(iii) For $y = -x$:

(2), **(3)**, and **(4)** become

$$x(z - 2\lambda) = 0 \implies x = 0 \quad \text{or} \quad z = 2\lambda$$

$$-x^2 + 2\lambda z = 0 \tag{9}$$

$$2x^2 + z^2 - 12 = 0 \tag{10}$$

When $x = 0$, $\quad z = \pm 2\sqrt{3}$.

When $z = 2\lambda$, $8\lambda^2 + 4\lambda^2 - 12 = 0 \implies \lambda = \pm 1$ and so $z = \pm 2$.

Critical points to be checked are $(0,0,2\sqrt{3})$, $(0,0,-2\sqrt{3})$, $(2,-2,2)$, $(-2,2,2)$, $(2,-2,-2)$, and $(-2,2,-2)$.

$$
\begin{array}{ll}
f\left(0,2\sqrt{3},0\right) = 0 & f\left(0,-2\sqrt{3},0\right) = 0 \\
f\left(2\sqrt{3},0,0\right) = 0 & f\left(-2\sqrt{3},0,0\right) = 0 \\
f\left(0,0,2\sqrt{3}\right) = 0 & f\left(0,0,-2\sqrt{3}\right) = 0 \\
f(2,2,-2) = -8 & f(-2,-2,-2) = -8 \\
f(2,2,2) = 8 & f(-2,-2,2) = 8 \\
f(2,-2,2) = -8 & f(-2,2,2) = -8 \\
f(2,-2,-2) = 8 & f(-2,2,-2) = 8.
\end{array}
$$

Hence, the maximum value of f is 8 and the minimum value of f is -8.

10. $f(x,y,z) = x + y + z, \qquad g(x,y,z) = x^2 + y^2 + z^2 - 12 = 0.$

$L(x,y,z,\lambda) = x + y + z + \lambda(x^2 + y^2 + z^2 - 12).$

$$\frac{\partial L}{\partial x} = 1 + 2\lambda x = 0 \implies x = -\frac{1}{2\lambda} \quad (1)$$
$$\frac{\partial L}{\partial y} = 1 + 2\lambda y = 0 \implies y = -\frac{1}{2\lambda} \quad (2)$$
$$\frac{\partial L}{\partial z} = 1 + 2\lambda z = 0 \implies z = -\frac{1}{2\lambda} \quad (3)$$
$$\frac{\partial L}{\partial \lambda} = x^2 + y^2 + z^2 - 12 = 0 \quad (4)$$

Substituting equations **(1)**, **(2)**, and **(3)** into equation **(4)** gives

$$\frac{3}{4\lambda^2} = 12 \implies \lambda^2 = \frac{3}{48} = \frac{1}{16} \implies \lambda = \pm\frac{1}{4}.$$

Hence critical points to be checked are

$$(-2,-2,-2) \quad \text{and} \quad (2,2,2)$$

$f(-2,-2,-2) = -6$, and $f(2,2,2) = 6$.

The maximum value of f is 6 and the minimum value is -6.

11. $f(x,y,z) = x^2 + y^2 + z^2, \quad g(x,y,z) = x - y + z - 1 = 0.$

$L(x,y,z,\lambda) = x^2 + y^2 + z^2 + \lambda(x - y + z - 1).$

$$\frac{\partial L}{\partial x} = 2x + \lambda = 0 \implies x = -\frac{1}{2}\lambda. \quad (1)$$
$$\frac{\partial L}{\partial y} = 2y - \lambda = 0 \implies y = \frac{1}{2}\lambda. \quad (2)$$
$$\frac{\partial L}{\partial z} = 2z + \lambda = 0 \implies z = -\frac{1}{2}\lambda. \quad (3)$$
$$\frac{\partial L}{\partial \lambda} = x - y + z - 1 = 0. \quad (4)$$

Substituting equations **(1)**, **(2)**, and **(3)** into equation **(4)** gives

$$\left(-\frac{1}{2}\lambda\right) - \left(\frac{1}{2}\lambda\right) + \left(-\frac{1}{2}\lambda\right) - 1 = 0 \implies -\frac{3}{2}\lambda - 1 = 0 \implies \lambda = -\frac{2}{3}.$$

Then

$$x = -\frac{1}{2}\cdot\left(-\frac{2}{3}\right) = \frac{1}{3}, \quad y = \frac{1}{2}\cdot\left(-\frac{2}{3}\right) = -\frac{1}{3}, \quad z = -\frac{1}{2}\cdot\left(-\frac{2}{3}\right) = \frac{1}{3}.$$

Note

$$f\left(\frac{1}{3}, -\frac{1}{3}, \frac{1}{3}\right) = \frac{1}{3} < f(0,0,1) = 1.$$

Therefore, the minimum value of f is $f\left(\frac{1}{3}, -\frac{1}{3}, \frac{1}{3}\right) = \frac{1}{3}$, and f has no maximum value.

12. $f(x,y,z) = x + 2y + z. \qquad g(x,y,z) = x^2 + y^2 + z^2 - 4 = 0.$

$L(x,y,z,\lambda) = x + 2y + z + \lambda(x^2 + y^2 + z^2 - 4).$

$$\frac{\partial L}{\partial x} = 1 + 2\lambda x = 0 \implies x = -\frac{1}{2\lambda}. \qquad \textbf{(1)}$$
$$\frac{\partial L}{\partial y} = 2 + 2\lambda y = 0 \implies y = -\frac{1}{\lambda}. \qquad \textbf{(2)}$$
$$\frac{\partial L}{\partial z} = 1 + 2\lambda z = 0 \implies z = -\frac{1}{2\lambda}. \qquad \textbf{(3)}$$
$$\frac{\partial L}{\partial \lambda} = x^2 + y^2 + z^2 - 4 = 0. \qquad \textbf{(4)}$$

Substituting equations **(1)**, **(2)**, and **(3)** into equation **(4)** gives

$$\frac{1}{4\lambda^2} + \frac{1}{\lambda^2} + \frac{1}{4\lambda^2} - 4 = 0 \implies \frac{6}{4\lambda^2} = 4 \implies 16\lambda^2 = 6 \implies \lambda^2 = \frac{3}{8}$$
$$\implies \lambda = \sqrt{.375} \quad \text{or} \quad \lambda = -\sqrt{.375}.$$

The critical points are given by

$$x = -\frac{1}{2\cdot\sqrt{.375}} = -.8165, \qquad y = -\frac{1}{\sqrt{.375}} = -1.6330,$$
$$z = -\frac{1}{2\cdot\sqrt{.375}} = -.8165$$

and

$$x = .8165, \qquad y = 1.6330, \qquad z = .8165.$$

Note that

$$f(-.8165,-1.6330,-.8165) = -4.899 \quad \text{and} \quad f(.8165,1.6330,.8165) = 4.899.$$

Therefore, the minimum value of f is $f(-.8165,-1.6330,-8165) = -4.899$, and the maximum value of f is $f(.8165,1.6330,.8165) = 4.899$.

13. $P(x,y) = 60x^{1/4}y^{3/4}, \quad g(x,y) = 20x + 10y - 80 = 0.$

$L(x,y,\lambda) = 60x^{1/4}y^{3/4} + \lambda(20x + 10y - 80).$

$$\frac{\partial L}{\partial x} = 15x^{-3/4}y^{3/4} + 20\lambda = 0 \qquad \textbf{(1)}$$

$$\frac{\partial L}{\partial y} = 45y^{-1/4}x^{1/4} + 10\lambda = 0 \qquad \textbf{(2)}$$

$$\frac{\partial L}{\partial \lambda} = 20x + 10y - 80 = 0 \qquad \textbf{(3)}$$

From **(1)** – (**(2)** × 2) one obtains

$$15\left(\frac{y}{x}\right)^{3/4} - 90\left(\frac{y}{x}\right)^{-1/4} = 0 \implies \left(\frac{y}{x}\right) - 6 = 0 \implies y = 6x \qquad \textbf{(4)}$$

From **(3)** and **(4)**, $20x + 60x - 80 = 0 \implies x = 1$ and so $y = 6$.

$P(1,6) = 60 \cdot 6^{3/4}$. Since $P(4,0) = 0$, the maximum value of P is $60 \cdot 6^{3/4}$.

14. $L(x,y,\lambda) = 8x^2 + 2y^2 - 4xy + \lambda(4x + 2y - 20).$

$$\frac{\partial L}{\partial x} = 16x - 4y + 4\lambda = 0 \qquad \textbf{(1)}$$

$$\frac{\partial L}{\partial y} = 4y - 4x + 2\lambda = 0 \qquad \textbf{(2)}$$

$$\frac{\partial L}{\partial \lambda} = 4x + 2y - 20 = 0 \qquad \textbf{(3)}$$

From **(1)** – (**(2)** × 2), one obtains

$$24x - 12y = 0 \implies y = 2x \qquad \textbf{(4)}$$

From **(3)** and **(4)**, $4x + 4x - 20 = 0 \implies 8x = 20 \implies x = \frac{5}{2}$, and so $y = 5$.

Thus, the critical point for L is $\left(\frac{5}{2}, 5, -5\right)$.

$f\left(\frac{5}{2},5\right) = 8 \cdot \frac{25}{4} + 2 \cdot 25 - 4 \cdot \frac{5}{2} \cdot 5 = 50.$

Since $f(5,0) = 8 \cdot 25 = 200$, f has the minimum value 50 at the point $\left(\frac{5}{2},5\right)$.

15. $P(x,y) = 200x + 100y - 4x^2 - 2y^2.$

$g(x,y) = 20x + 10y - 600 = 0.$

$L(x,y,\lambda) = 200x + 100y - 4x^2 - 2y^2 + \lambda(20x + 10y - 600).$

$$\frac{\partial L}{\partial x} = 200 - 8x + 20\lambda = 0 \qquad (1)$$

$$\frac{\partial L}{\partial y} = 100 - 4y + 10\lambda = 0 \qquad (2)$$

$$\frac{\partial L}{\partial \lambda} = 20x + 10y - 600 = 0 \qquad (3)$$

From **(1)** $-$ (**(2)** $\times$ 2), $-8x + 8y = 0 \implies y = x.$ **(4)**

From **(3)** and **(4)**, $30x = 600 \implies x = 20.$

Thus, the critical point for L is $(20, 20, -2)$.

$P(20,20) = 4000 + 2000 - 1600 - 800 = 3600.$

Since $P(30,0) = 6000 - 3600 = 2400 < 3600$, P has the maximum value 3600 at (20,20).

Therefore, 20 color sets and 20 black and white sets should be produced in order to maximize profit.

16. $P(x,y) = 10x + 25y - 5xy, \quad g(x,y) = 4x + 2y - 40 = 0.$

$L(x,y,\lambda) = 10x + 25y - 5xy + \lambda(4x + 2y - 40).$

$$\frac{\partial L}{\partial x} = 10 - 5y + 4\lambda = 0 \qquad (1)$$

$$\frac{\partial L}{\partial y} = 25 - 5x + 2\lambda = 0 \qquad (2)$$

$$\frac{\partial L}{\partial \lambda} = 4x + 2y - 40 = 0 \qquad (3)$$

From **(1)** $-$ (**(2)** $\times$ 2), $\quad -40 - 5y + 10x = 0,$

$$2x - y - 8 = 0. \qquad (4)$$

From **(3)** and **(4)**, $x = 7$ and $y = 6$.

$$P(7,6) = 70 + 150 - 210 = 10.$$

Since $P(0,20) = 500 > 10$, P(7,6) is not the maximum value. By solving the constraint equation $4x + 2y = 20$ for y in terms of x and substituting this expression into the production function, we can show that the maximum value of the production function occurs when $x = 0$, $y = 20$.

17. $P(x,y) = 20x + 40y - x^2 - y^2$. $\quad g(x,y) = x + y - 40 = 0$.

$L(x,y,\lambda) = 20x + 40y - x^2 - y^2 + \lambda(x + y - 40)$.

$$\frac{\partial L}{\partial x} = 20 - 2x + \lambda = 0 \qquad \textbf{(1)}$$
$$\frac{\partial L}{\partial y} = 40 - 2y + \lambda = 0 \qquad \textbf{(2)}$$
$$\frac{\partial L}{\partial \lambda} = x + y - 40 = 0 \qquad \textbf{(3)}$$

From **(1)** – **(2)**,

$$-2x + 2y - 20 = 0. \qquad (4)$$

From **(3)** and **(4)**, $4y = 100 \implies y = 25$ and $x = 15$.

$$P(15,25) = 300 + 1000 - 225 - 625 = 1300 - 850 = 450.$$

Since $P(20,20) = 400 + 800 - 400 - 400 = 400 < 450$, P has the maximum value 450 when $x = 15$ and $y = 25$.

18. Let x be the length of the box in meters, let y be the width of the box in meters, and let z be the height of the box in meters.

Let p be the price in dollars per square meter of the material for the sidewalls.

Let $f(x,y,z)$ be the cost in dollars of the material for the box.

Note that

$$\begin{aligned} f(x,y,z) &= 2 \cdot (xz + yz) \cdot p + 2 \cdot xy \cdot 2p \\ &= 2p(xz + yz + 2xy). \end{aligned}$$

We want to minimize $f(x, y, z)$ subject to the constraint that the volume of the box is 16 cubic meters, so the constraint equation is

$$g(x, y, z) = xyz - 16 = 0.$$

Let

$$\begin{aligned} L(x, y, z, \lambda) &= f(x, y, z) + \lambda \cdot g(x, y, z) \\ &= 2p(xz + yz + 2xy) + \lambda(xyz - 16). \end{aligned}$$

$$\frac{\partial L}{\partial x} = 2p(z + 2y) + \lambda yz = 0 \quad (1)$$

$$\frac{\partial L}{\partial y} = 2p(z + 2x) + \lambda xz = 0 \quad (2)$$

$$\frac{\partial L}{\partial z} = 2p(x + y) + \lambda xy = 0 \quad (3)$$

$$\frac{\partial L}{\partial \lambda} = xyz - 16 = 0. \quad (4)$$

$$((\mathbf{1}) \cdot x) - ((\mathbf{2}) \cdot y) \rightarrow 2pz(x - y) = 0 \implies x = y$$

$$((\mathbf{1}) \cdot x) - ((\mathbf{3}) \cdot z) \rightarrow 2py(2x - z) = 0 \implies z = 2x.$$

Substituting $x = y$ and $z = 2x$ into equation $(\mathbf{4})$ gives

$$x \cdot x \cdot 2x - 16 = 0 \implies 2x^3 = 16 \implies x^3 = 8 \implies x = 2.$$

Then $y = x = 2$ and $z = 2x = 4$.

Observe next that

$$f(2, 2, 4) = 2p(8 + 8 + 8) = 48p.$$

Also observe that

$$f(1, 2, 8) = 2p(8 + 16 + 4) = 56p.$$

Since $f(2,2,4) < f(1,2,8)$, the minimum cost occurs when $x = 2$ meters, $y = 2$ meters, $z = 4$ meters.

19. Let x, y, and z be the length, width, and height in inches, respectively, of the rectangular package.

Let $f(x,y,z)$ be the volume in cubic inches of the box.

Note that

$$f(x,y,z) = xyz.$$

We want to maximize $f(x,y,z)$ subject to the constraint

$$x + 2y + 2z = 84.$$

Let

$$L(x,y,z,\lambda) = xyz + \lambda(x + 2y + 2z - 84).$$

$$\frac{\partial L}{\partial x} = yz + \lambda = 0 \qquad \mathbf{(1)}$$

$$\frac{\partial L}{\partial y} = xz + 2\lambda = 0 \qquad \mathbf{(2)}$$

$$\frac{\partial L}{\partial z} = xy + 2\lambda = 0 \qquad \mathbf{(3)}$$

$$\frac{\partial L}{\partial \lambda} = x + 2y + 2z - 84 = 0. \qquad \mathbf{(4)}$$

$$((\mathbf{1}) \cdot 2) - (\mathbf{2}) \to 2yz - xz = 0 \implies z(2y - x) = 0 \implies y = \frac{1}{2}x.$$

$$((\mathbf{1}) \cdot 2) - (\mathbf{3}) \to 2yz - xy = 0 \implies y(2z - x) = 0 \implies z = \frac{1}{2}x.$$

Substituting $y = \frac{1}{2}x$ and $z = \frac{1}{2}x$ into equation **(4)** gives

$$x + 2 \cdot \left(\frac{1}{2}x\right) + 2 \cdot \left(\frac{1}{2}x\right) - 84 = 0 \implies 3x - 84 = 0 \implies x = 28.$$

Then $y = 14$ and $z = 14$.

Observe next that

$$f(28, 14, 14) = 5488.$$

Also observe that

$$f(40, 20, 2) = 1600.$$

Since $f(28, 14, 14) > f(40, 20, 2)$, the maximum volume occurs when $x = 28$ inches, $y = 14$ inches, $z = 14$ inches.

20. Let x, y, and z be the length, width, and height in feet, respectively, of the open rectangular container.

Let $f(x, y, z)$ be the volume in cubic feet of the container.

Note that

$$f(x, y, z) = xyz.$$

The cost of materials for the container is given by

$$(2xz + 2yz) \cdot 4 + (xy) \cdot 5 = 8xz + 8yz + 5xy.$$

We want to maximize $f(x, y, z)$ subject to the constraint

$$8xz + 8yz + 5xy = 960.$$

Let

$$L(x, y, z, \lambda) = xyz + \lambda(8xz + 8yz + 5xy - 960).$$

$$\frac{\partial L}{\partial x} = yz + 8z\lambda + 5y\lambda = 0 \qquad \textbf{(1)}$$

$$\frac{\partial L}{\partial y} = xz + 8z\lambda + 5x\lambda = 0 \qquad \textbf{(2)}$$

$$\frac{\partial L}{\partial z} = xy + 8x\lambda + 8y\lambda = 0 \qquad \textbf{(3)}$$

$$\frac{\partial L}{\partial \lambda} = 8xz + 8yz + 5xy - 960 = 0 \qquad \textbf{(4)}$$

$$((\mathbf{1}) \cdot x) - ((\mathbf{2}) \cdot y) \to 8xz\lambda - 8yz\lambda = 0 \implies 8z\lambda(x - y) = 0 \implies y = x.$$

$$((\mathbf{1}) \cdot x) - ((\mathbf{3}) \cdot z) \to 5xy\lambda - 8yz\lambda = 0 \implies y\lambda(5x - 8z) = 0 \implies z = \frac{5}{8}x.$$

Substituting $y = x$ and $z = \frac{5}{8}x$ into equation **(4)** gives

$$8x \cdot \left(\frac{5}{8}x\right) + 8x \cdot \left(\frac{5}{8}x\right) + 5x \cdot x - 960 = 0 \implies 15x^2 = 960 \implies x^2 = 64 \implies x = 8.$$

Then $y = 8$ and $z = 5$.

Observe next that

$$f(8,8,5) = 320.$$

Also observe that

$$f(4,4,13.75) = 220.$$

Since $f(8,8,5) > f(4,4,13.75)$, the maximum volume occurs when $x = 8$ feet, $y = 8$ feet, $z = 5$ feet.

21. Let r be the radius of the cylindrical can in cm, and h be the height of the cylindrical can in cm.

Let $f(r,h)$ denote the volume in cubic centimeters of the can.

Note that

$$f(r,h) = \pi r^2 h.$$

The amount of material used to make the can in square cm is given by

$$2\pi rh + 2(\pi r^2).$$

We want to maximize $f(r,h)$ subject to the constraint

$$2\pi rh + 2\pi r^2 = 100.$$

Let

$$L(r,h,\lambda) = \pi r^2 h + \lambda(2\pi rh + 2\pi r^2 - 100).$$

$$\frac{\partial L}{\partial r} = 2\pi rh + 2\pi h\lambda + 4\pi r\lambda = 0 \qquad (1)$$

$$\frac{\partial L}{\partial h} = \pi r^2 + 2\pi r\lambda = 0 \qquad (2)$$

$$\frac{\partial L}{\partial \lambda} = 2\pi rh + 2\pi r^2 - 100 = 0 \qquad (3)$$

$$\textbf{(2)} \implies \lambda = -\frac{r}{2}.$$

Substituting $\lambda = -\frac{r}{2}$ into equation **(1)** gives

$$2\pi rh + 2\pi h \cdot \left(-\frac{r}{2}\right) + 4\pi r \cdot \left(-\frac{r}{2}\right) = 0 \implies \pi rh - 2\pi r^2 = 0 \implies$$
$$\pi r(h - 2r) = 0 \implies h = 2r.$$

Substituting $h = 2r$ into equation **(3)** gives

$$2\pi r \cdot 2r + 2\pi r^2 - 100 = 0 \implies 6\pi r^2 = 100 \implies r^2 = \frac{100}{6\pi} \implies r = \frac{10}{\sqrt{6\pi}}.$$

Then $h = \frac{20}{\sqrt{6\pi}}$.

Observe that

$$f\left(\frac{10}{\sqrt{6\pi}}, \frac{20}{\sqrt{6\pi}}\right) = \pi \cdot \frac{100}{6\pi} \cdot \frac{20}{\sqrt{6\pi}} = \frac{1000}{3\sqrt{6\pi}} > f\left(2, \frac{25 - 2\pi}{\pi}\right).$$

Therefore, the maximum volume occurs when $r = \frac{10}{\sqrt{6\pi}}$ cm and $h = \frac{20}{\sqrt{6\pi}}$ cm.

22. $g(x, y, z) = xyz - V = 0.$

$$C(x, y, z) = 4xy + 3(2yz + xz) + 2xz$$
$$= 4xy + 6yz + 5xz.$$

$$L(x, y, z) = 4xy + 6yz + 5xz + \lambda(xyz - V).$$

$$\frac{\partial L}{\partial x} = 4y + 5z + \lambda yz = 0 \qquad \textbf{(1)}$$
$$\frac{\partial L}{\partial y} = 4x + 6z + \lambda xz = 0 \qquad \textbf{(2)}$$
$$\frac{\partial L}{\partial z} = 6y + 5x + \lambda xy = 0 \qquad \textbf{(3)}$$
$$\frac{\partial L}{\partial \lambda} = xyz - V = 0 \qquad \textbf{(4)}$$

From $((\mathbf{1}) \cdot x) - ((\mathbf{2}) \cdot y)$, $z(5x - 6y) = 0 \implies y = \frac{5}{6}x.$ **(5)**

From $((\mathbf{1}) \cdot x) - ((\mathbf{3}) \cdot z)$, $y(4x - 6z) = 0 \implies z = \frac{2}{3}x.$ **(6)**

From **(4)**, **(5)**, and **(6)**,

$$\tfrac{5}{9}x^3 = V \implies x^3 = \tfrac{9}{5}V \implies x = \left(\tfrac{9}{5}V\right)^{1/3}.$$

Thus $y = \frac{5}{6} \cdot \left(\frac{9}{5}V\right)^{1/3}$ and $z = \frac{2}{3} \cdot \left(\frac{9}{5}V\right)^{1/3}$.

$$\begin{aligned} &C\left(\left(\tfrac{9}{5}V\right)^{1/3}, \tfrac{5}{6} \cdot \left(\tfrac{9}{5}V\right)^{1/3}, \tfrac{2}{3} \cdot \left(\tfrac{9}{5}V\right)^{1/3}\right) \\ &= \tfrac{10}{3}\left(\tfrac{9}{5}V\right)^{2/3} + \tfrac{10}{3}\left(\tfrac{9}{5}V\right)^{2/3} + \tfrac{10}{3}\left(\tfrac{9}{5}V\right)^{2/3} \\ &= 10\left(\tfrac{9}{5}V\right)^{2/3}. \end{aligned}$$

Since $C\left(\sqrt[3]{V}, \sqrt[3]{V}, \sqrt[3]{V}\right) = 15V^{2/3} > 10\left(\frac{9}{5}V\right)^{2/3}$, the minimum cost occurs at

$x = \left(\frac{9}{5}V\right)^{1/3}$ feet, $y = \frac{5}{6} \cdot \left(\frac{9}{5}V\right)^{1/3}$ feet, and $z = \frac{2}{3} \cdot \left(\frac{9}{5}V\right)^{1/3}$ feet.

23. Let r and h be the radius and height in centimeters, respectively, of the cylindrical jar.

Let $f(r, h)$ denote the exterior surface area of the jar in square centimeters.

Note that

$$f(r, h) = 2\pi rh + 2\pi r^2.$$

The volume of the jar is given by

$$\pi r^2 h = 2000.$$

We want to minimize $f(r, h)$ subject to the constraint

$$\pi r^2 h - 2000 = 0.$$

Let

$$L(r, h, \lambda) = 2\pi rh + 2\pi r^2 + \lambda(\pi r^2 h - 2000).$$

$$\frac{\partial L}{\partial r} = 2\pi h + 4\pi r + 2\pi rh\lambda = 0 \quad \textbf{(1)}$$

$$\frac{\partial L}{\partial h} = 2\pi r + \pi r^2 \lambda = 0 \quad \textbf{(2)}$$

$$\frac{\partial L}{\partial \lambda} = \pi r^2 h - 2000 = 0 \quad \textbf{(3)}$$

$$((\textbf{1}) \cdot r) - ((\textbf{2}) \cdot 2h) \to (2\pi rh + 4\pi r^2 + 2\pi r^2 h\lambda) - (4\pi rh + 2\pi r^2 h\lambda) = 0$$

$$\implies -2\pi rh + 4\pi r^2 = 0 \implies 2\pi r(-h + 2r) = 0 \implies h = 2r.$$

Substituting $h = 2r$ into equation **(3)** gives

$$\pi r^2 \cdot 2r - 2000 = 0 \implies r^3 = \frac{1000}{\pi} \implies r = \frac{10}{\pi^{1/3}}.$$

Then $h = \frac{20}{\pi^{1/3}}$.

Observe that

$$f\left(\frac{10}{\pi^{1/3}}, \frac{20}{\pi^{1/3}}\right) = 2\pi \cdot \frac{10}{\pi^{1/3}} \cdot \frac{20}{\pi^{1/3}} + 2\pi\left(\frac{10}{\pi^{1/3}}\right)^2 = 400\pi^{1/3} + 200\pi^{1/3}$$
$$= 600\pi^{1/3}.$$

Also observe that

$$f\left(10, \frac{20}{\pi}\right) = 2\pi \cdot 10 \cdot \frac{20}{\pi} + 2\pi(10)^2 = 400 + 200\pi.$$

Since $f\left(\frac{10}{\pi^{1/3}}, \frac{20}{\pi^{1/3}}\right) < f\left(10, \frac{20}{\pi}\right)$, the minimum exterior surface area of the jar is obtained when $r = \frac{10}{\pi^{1/3}}$ cm and $h = \frac{20}{\pi^{1/3}}$ cm.

24. $P(x, y) = 100x^{1/4}y^{3/4}$.

a. $\frac{\partial p}{\partial x} = 25x^{-3/4} \cdot y^{3/4} \implies \frac{\partial p}{\partial x}(1, 6) = 25 \cdot 6^{3/4}$.

b. $\frac{\partial p}{\partial y} = 75x^{1/4}y^{-1/4} \implies \frac{\partial p}{\partial y}(1, 6) = 75 \cdot 6^{-1/4}$.

c. $\frac{\frac{\partial p}{\partial x}(1,6)}{\frac{\partial p}{\partial y}(1,6)} = \frac{1}{3} \cdot 6 = 2$.

d. In Exercise 13, $P(x, y) = 60x^{1/4}y^{3/4}$.

$$\frac{\frac{\partial p}{\partial x}}{\frac{\partial p}{\partial y}} = \frac{15x^{-3/4}y^{3/4}}{45x^{1/4}y^{-1/4}} = \frac{y}{3x}.$$

Thus, for $x = 1$ and $y = 6$, $\frac{\frac{\partial p}{\partial x}}{\frac{\partial p}{\partial y}} = \frac{6}{3} = 2$.

Solutions to Exercise Set 8.5 (6.5)

1. $f(x,y) = 6x^2 + 4y^3$.

   $\frac{\partial f}{\partial x} = 12x, \qquad \frac{\partial f}{\partial y} = 12y^2.$

   $\frac{\partial f}{\partial x}(1,4) = 12, \qquad \frac{\partial f}{\partial y}(1,4) = 192.$

   $$\begin{aligned} f(1.2, 3.9) &\approx f(1,4) + \tfrac{\partial f}{\partial x}(1,4)(1.2-1) + \tfrac{\partial f}{\partial y}(1,4)(3.9-4) \\ &= 262 + 12 \cdot (0.2) + 192 \cdot (-0.1) \\ &= 245.2. \end{aligned}$$

2. $f(x,y) = 4x^3y^6$.

   $\frac{\partial f}{\partial x} = 12x^2y^6, \qquad \frac{\partial f}{\partial y} = 24x^3y^5.$

   $\frac{\partial f}{\partial x}(0,2) = 0, \qquad \frac{\partial f}{\partial y}(0,2) = 0.$

   $$\begin{aligned} f(0.2, 1.9) &\approx f(0,2) + \tfrac{\partial f}{\partial x}(0,2) \cdot (0.2-0) + \tfrac{\partial f}{\partial y}(0,2) \cdot (1.9-2) \\ &= 0. \end{aligned}$$

3. $f(x,y) = 3x^2e^{-y}$.

   $\frac{\partial f}{\partial x} = 6xe^{-y}, \qquad \frac{\partial f}{\partial y} = -3x^2e^{-y}.$

   $\frac{\partial f}{\partial x}(4,-2) = 24e^2, \qquad \frac{\partial f}{\partial y}(4,-2) = -48e^2.$

   $$\begin{aligned} f(4.1, -2) &\approx f(4,-2) + \tfrac{\partial f}{\partial x}(4,-2) \cdot (4.1-4) + \tfrac{\partial f}{\partial y}(4,-2) \cdot (-2-(-2)) \\ &= 48e^2 + 24e^2(0.1) + (-48e^2)(0) \\ &= 50.4e^2 = 372.4084. \end{aligned}$$

4. $f(x,y) = 4e^{xy^2}$.

   $\frac{\partial f}{\partial x} = 4e^{xy^2} \cdot y^2, \qquad \frac{\partial f}{\partial y} = 4e^{xy^2} \cdot 2xy.$

   $\frac{\partial f}{\partial x}(1,1) = 4e, \qquad \frac{\partial f}{\partial y}(1,1) = 4e \cdot 2 = 8e.$

   $$\begin{aligned} f(1.1, 0.9) &\approx f(1,1) + \tfrac{\partial f}{\partial x}(1,1) \cdot (1.1-1) + \tfrac{\partial f}{\partial y}(1,1) \cdot (0.9-1) \\ &= 4e + 4e \cdot (0.1) + 8e \cdot (-0.1) \\ &= 3.6e. \end{aligned}$$

5. $f(x,y) = \sqrt{x}\sin y$.

$\frac{\partial f}{\partial x} = \frac{1}{2\sqrt{x}}\sin y, \qquad \frac{\partial f}{\partial y} = \sqrt{x}\cdot\cos y.$

$\frac{\partial f}{\partial x}\left(4, \frac{\pi}{2}\right) = \frac{1}{4}, \qquad \frac{\partial f}{\partial y}\left(4, \frac{\pi}{2}\right) = 0.$

$$\begin{aligned} f\left(4.15, \tfrac{13\pi}{24}\right) &\approx f\left(4, \tfrac{\pi}{2}\right) + \tfrac{\partial f}{\partial x}\left(4, \tfrac{\pi}{2}\right)\cdot(4.15-4) + \tfrac{\partial f}{\partial y}\left(4, \tfrac{\pi}{2}\right)\cdot\left(\tfrac{13\pi}{24} - \tfrac{\pi}{2}\right) \\ &= 2 + \tfrac{1}{4}\cdot 0.15 + 0\cdot\tfrac{\pi}{24} \\ &= 2.0375. \end{aligned}$$

6. $f(x,y) = x^{4/3}y^{2/3}$.

$\frac{\partial f}{\partial x} = \frac{4}{3}x^{1/3}y^{2/3}, \qquad \frac{\partial f}{\partial y} = \frac{2}{3}x^{4/3}y^{-1/3}.$

$\frac{\partial f}{\partial x}(8,27) = 24, \qquad \frac{\partial f}{\partial y}(8,27) = \frac{32}{9}.$

$$\begin{aligned} f(8.1, 26.5) &\approx f(8,27) + \tfrac{\partial f}{\partial x}(8,27)\cdot(8.1-8) + \tfrac{\partial f}{\partial y}(8,27)\cdot(26.5-27) \\ &= 144 + 24\cdot(0.1) + \left(\tfrac{32}{9}\right)\cdot(-0.5) \\ &= 144.6222. \end{aligned}$$

7. $f(x,y) = \frac{x}{x+y}$.

$\frac{\partial f}{\partial x} = \frac{(x+y)\cdot(1)-(x)\cdot(1)}{(x+y)^2} = \frac{y}{(x+y)^2}, \qquad \frac{\partial f}{\partial y} = \frac{(x+y)\cdot(0)-(x)\cdot(1)}{(x+y)^2} = \frac{-x}{(x+y)^2}.$

$\frac{\partial f}{\partial x}(6,2) = \frac{2}{64} = \frac{1}{32}, \qquad \frac{\partial f}{\partial y}(6,2) = \frac{-6}{64} = -\frac{3}{32}.$

$$\begin{aligned} f(6.1, 2.05) &\approx f(6,2) + \tfrac{\partial f}{\partial x}(6,2)\cdot(6.1-6) + \tfrac{\partial f}{\partial y}(6,2)\cdot(2.05-2) \\ &= \tfrac{3}{4} + \tfrac{1}{32}\cdot(0.1) + \left(-\tfrac{3}{32}\right)\cdot(0.05) \\ &= 0.7484. \end{aligned}$$

8. $f(x,y) = x\cos y^2$.

$\frac{\partial f}{\partial x} = \cos y^2, \qquad \frac{\partial f}{\partial y} = x\cdot(-\sin y^2)\cdot(2y) = -2xy\sin y^2.$

$\frac{\partial f}{\partial x}(7,0) = 1, \qquad \frac{\partial f}{\partial y}(7,0) = 0.$

$$\begin{aligned} f(7.15, -0.1) &\approx f(7,0) + \tfrac{\partial f}{\partial x}(7,0)\cdot(7.15-7) + \tfrac{\partial f}{\partial y}(7,0)(-0.1-0) \\ &= 7 + 1\cdot(0.15) + 0\cdot(-0.1) \\ &= 7.15. \end{aligned}$$

9. $f(x,y,z) = xy + yz + xz$.

$\frac{\partial f}{\partial x} = y + z, \qquad \frac{\partial f}{\partial y} = x + z, \qquad \frac{\partial f}{\partial z} = y + x.$

$\frac{\partial f}{\partial x}(1,3,7) = 10, \qquad \frac{\partial f}{\partial y}(1,3,7) = 8, \qquad \frac{\partial f}{\partial z}(1,3,7) = 4.$

$$\begin{aligned} f(0.95, 3.10, 7.05) &\approx f(1,3,7) + \tfrac{\partial f}{\partial x}(1,3,7)\cdot(0.95-1) + \tfrac{\partial f}{\partial y}(1,3,7)\cdot(3.10-3) \\ &\quad + \tfrac{\partial f}{\partial z}(1,3,7)\cdot(7.05-7) \\ &= 31 + 10\cdot(-0.05) + 8\cdot(0.10) + 4\cdot(0.05) \\ &= 31.5. \end{aligned}$$

10. $f(x,y,z) = e^{xyz}$.

$\frac{\partial f}{\partial x} = e^{xyz}\cdot(yz), \qquad \frac{\partial f}{\partial y} = e^{xyz}\cdot(xz), \qquad \frac{\partial f}{\partial z} = e^{xyz}(xy).$

$\frac{\partial f}{\partial x}(1,1,1) = e, \qquad \frac{\partial f}{\partial y}(1,1,1) = e, \qquad \frac{\partial f}{\partial z}(1,1,1) = e.$

$$\begin{aligned} f(0.9, 1.10, 1.05) &\approx f(1,1,1) + \tfrac{\partial f}{\partial x}(1,1,1)\cdot(0.9-1) + \tfrac{\partial f}{\partial y}(1,1,1)(1.10-1) \\ &\quad + \tfrac{\partial f}{\partial z}(1,1,1)\cdot(1.05-1) \\ &= e + e\cdot(-0.1) + e\cdot(0.10) + e\cdot(0.05) \\ &= 1.05e. \end{aligned}$$

11. $f(x,y) = 3xy^2 + 4x^3y$.

$\frac{\partial f}{\partial x} = 3y^2 + 12x^2y, \qquad \frac{\partial f}{\partial y} = 6xy + 4x^3.$

$\frac{\partial f}{\partial x}(1,-1) = -9, \qquad \frac{\partial f}{\partial y}(1,-1) = -2.$

$df = \frac{\partial f}{\partial x}(1,-1)dx + \frac{\partial f}{\partial y}(1,-1)dy = -9\,dx - 2\,dy.$

12. $f(x,y) = \sqrt{x^2+y^2}$.

$\frac{\partial f}{\partial x} = \frac{x}{\sqrt{x^2+y^2}}, \qquad \frac{\partial f}{\partial y} = \frac{y}{\sqrt{x^2+y^2}}.$

$\frac{\partial f}{\partial y}(1,2) = \frac{1}{\sqrt{5}}, \qquad \frac{\partial f}{\partial y}(1,2) = \frac{2}{\sqrt{5}}.$

$df = \frac{\partial f}{\partial x}(1,2)dx + \frac{\partial f}{\partial y}(1,2)dy = \frac{1}{\sqrt{5}}\,dx + \frac{2}{\sqrt{5}}\,dy.$

13. $f(x,y) = \sin(xy)$.

$\frac{\partial f}{\partial x} = \cos(xy)\cdot(y), \qquad \frac{\partial f}{\partial y} = \cos(xy)\cdot x.$

$\frac{\partial f}{\partial x}\left(\frac{\pi}{4},1\right) = \frac{\sqrt{2}}{2}, \qquad \frac{\partial f}{\partial y}\left(\frac{\pi}{4},1\right) = \frac{\pi}{4}\cdot\frac{\sqrt{2}}{2}.$

$df = \frac{\partial f}{\partial x}\left(\frac{\pi}{4},1\right)dx + \frac{\partial f}{\partial y}\left(\frac{\pi}{4},1\right)dy = \frac{\sqrt{2}}{2}\,dx + \frac{\pi}{4}\cdot\frac{\sqrt{2}}{2}\,dy = \frac{\sqrt{2}}{2}\,dx + \frac{\pi\sqrt{2}}{8}\,dy.$

14. $f(x,y) = \frac{x}{x-y}$.

$\frac{\partial f}{\partial x} = \frac{(x-y)\cdot(1)-x(1)}{(x-y)^2} = \frac{-y}{(x-y)^2}$, $\quad \frac{\partial f}{\partial y} = \frac{(x-y)\cdot(0)-x(-1)}{(x-y)^2} = \frac{x}{(x-y)^2}$.

$\frac{\partial f}{\partial x}(2,1) = -1$, $\quad \frac{\partial f}{\partial y}(2,1) = 2$.

$df = \frac{\partial f}{\partial x}(2,1)dx + \frac{\partial f}{\partial y}(2,1)dy = -1\,dx + 2\,dy$.

15. $f(x,y) = \ln(x^2+y^2)$.

$\frac{\partial f}{\partial x} = \frac{2x}{x^2+y^2}$, $\quad \frac{\partial f}{\partial y} = \frac{2y}{x^2+y^2}$.

$\frac{\partial f}{\partial x}(0,e) = 0$, $\quad \frac{\partial f}{\partial y}(0,e) = \frac{2e}{e^2} = 2e^{-1}$.

$df = \frac{\partial f}{\partial x}(0,e)dx + \frac{\partial f}{\partial y}(0,e)dy = 0\,dx + 2e^{-1}\,dy = \frac{2}{e}\,dy$.

16. $f(x,y) = x^3y - ye^{-2x}$.

$\frac{\partial f}{\partial x} = 3x^2y + 2ye^{-2x}$, $\quad \frac{\partial f}{\partial y} = x^3 - e^{-2x}$.

$\frac{\partial f}{\partial x}(0,2) = 4$, $\quad \frac{\partial f}{\partial y}(0,2) = -1$.

$df = \frac{\partial f}{\partial x}(0,2)dx + \frac{\partial f}{\partial y}(0,2)dy = 4\,dx - 1\,dy$.

17. $f(x,y,z) = xyz + xe^{-z}$.

$\frac{\partial f}{\partial x} = yz + e^{-z}$, $\quad \frac{\partial f}{\partial y} = xz$, $\quad \frac{\partial f}{\partial z} = xy - xe^{-z}$.

$\frac{\partial f}{\partial x}(1,1,0) = 1$, $\quad \frac{\partial f}{\partial y}(1,1,0) = 0$, $\quad \frac{\partial f}{\partial z}(1,1,0) = 0$.

$$\begin{aligned} df &= \frac{\partial f}{\partial x}(1,1,0)dx + \frac{\partial f}{\partial y}(1,1,0)dy + \frac{\partial f}{\partial z}(1,1,0)dz \\ &= dx + 0 + 0 = dx. \end{aligned}$$

18. $f(x,y,z) = x\sin y - z\cos x$.

$\frac{\partial f}{\partial x} = \sin y + z\sin x$, $\quad \frac{\partial f}{\partial y} = x\cos y$, $\quad \frac{\partial f}{\partial z} = -\cos x$.

$\frac{\partial f}{\partial x}\left(\frac{\pi}{4},\frac{\pi}{2},1\right) = 1 + \frac{\sqrt{2}}{2}$, $\quad \frac{\partial f}{\partial y}\left(\frac{\pi}{4},\frac{\pi}{2},1\right) = 0$, $\quad \frac{\partial f}{\partial z}\left(\frac{\pi}{4},\frac{\pi}{2},1\right) = -\frac{\sqrt{2}}{2}$.

$$\begin{aligned} df &= \frac{\partial f}{\partial x}\left(\frac{\pi}{4},\frac{\pi}{2},1\right)dx + \frac{\partial f}{\partial y}\left(\frac{\pi}{4},\frac{\pi}{2},1\right)dy + \frac{\partial f}{\partial z}\left(\frac{\pi}{4},\frac{\pi}{2},1\right)dz \\ &= \left(1+\frac{\sqrt{2}}{2}\right)dx + 0\,dy + \left(-\frac{\sqrt{2}}{2}\right)dz \\ &= \left(1+\frac{\sqrt{2}}{2}\right)dx + \left(-\frac{\sqrt{2}}{2}\right)dz. \end{aligned}$$

19. $M = C + D$.

$dM = \frac{\partial M}{\partial C}\cdot dC + \frac{\partial M}{\partial D}\cdot dD = 1\cdot(0.10C) + 1\cdot(-0.05D)$.

$\frac{dM}{M} = \frac{0.10C - 0.05D}{C+D}$.

(percentage change in the money supply) $\approx \frac{0.10C-0.05D}{C+D}\cdot 100$.

20. Volume $V = \pi \cdot \left(\frac{d}{2}\right)^2 \cdot \ell = \frac{\pi}{4} d^2 \ell$.

$\frac{\partial V}{\partial d} = \frac{\pi}{2} d\ell, \qquad \frac{\partial V}{\partial \ell} = \frac{\pi}{4} d^2.$

$\frac{\partial V}{\partial d}(4,12) = 24\pi, \qquad \frac{\partial V}{\partial \ell}(4,12) = 4\pi.$ $dV = \frac{\partial V}{\partial d}(4,12) \cdot (dd) + \frac{\partial V}{\partial \ell}(4.12) \cdot (d\ell) =$ $24\pi \cdot (dd) + 4\pi \cdot (d\ell)$.

(Maximum relative error in the volume) $\approx \frac{24\pi \cdot (0.2) + 4\pi \cdot (0.2)}{48\pi} = \frac{5.6\pi}{48\pi} = 0.1167$.

(Maximum percentage error in the volume) $\approx 11.67\%$.

21. $\frac{\partial f}{\partial x} = 300 x^{-1/4} y^{1/4}, \qquad \frac{\partial f}{\partial y} = 100 x^{3/4} y^{-3/4}.$

$\frac{\partial f}{\partial x}(16,256) = 600, \qquad \frac{\partial f}{\partial y}(15,256) = 12.5.$

$$\begin{aligned}\text{(Approximate change in output)} = df &= \frac{\partial f}{\partial x}(16,256) \cdot dx + \frac{\partial f}{\partial y}(16,256) \cdot dy \\ &= 600 \cdot (-2) + 12.5 \cdot 24 \\ &= -900.\end{aligned}$$

22. $\frac{1}{R} = \frac{1}{R_1} + \frac{1}{R_2} + \frac{1}{R_3}$.

$R = \frac{1}{\frac{1}{R_1} + \frac{1}{R_2} + \frac{1}{R_3}} = \frac{R_1 R_2 R_3}{R_2 R_3 + R_1 R_3 + R_1 R_2}$.

$$\begin{aligned}\frac{\partial R}{\partial R_1} &= \frac{(R_2 R_3 + R_1 R_3 + R_1 R_2) \cdot (R_2 R_3) - (R_1 R_2 R_3) \cdot (R_3 + R_2)}{(R_2 R_3 + R_1 R_3 + R_1 R_2)^2} \\ &= \frac{(R_2 R_3)^2}{(R_2 R_3 + R_1 R_3 + R_1 R_2)^2}.\end{aligned}$$

$\frac{\partial R}{\partial R_2} = \frac{(R_1 R_3)^2}{(R_2 R_3 + R_1 R_3 + R_1 R_2)^2}, \qquad \frac{\partial R}{\partial R_3} = \frac{(R_1 R_2)^2}{(R_2 R_3 + R_1 R_3 + R_1 R_2)^2}.$

$$\begin{aligned}dR &= \frac{\partial R}{\partial R_1} \cdot dR_1 + \frac{\partial R}{\partial R_2} \cdot dR_2 + \frac{\partial R}{\partial R_3} \cdot dR_3 \\ &= \frac{1}{(R_2 R_3 + R_1 R_3 + R_1 R_2)^2} \cdot (0.1) \cdot \left((R_2 R_3)^2 \cdot R_1 + (R_1 R_3)^2 \cdot R_2 + (R_1 R_2)^2 \cdot R_3\right).\end{aligned}$$

$\frac{dR}{R} = (0.1) \cdot \frac{1}{R_2 R_3 + R_1 R_3 + R_1 R_2} \cdot R_1 R_2 R_3 \cdot (R_2 R_3 + R_1 R_3 + R_1 R_2) \cdot \frac{1}{R_1 R_2 R_3} = 0.1.$

The approximate percentage change in R is 10 percent.

23. $V(T,P) = \frac{2600T}{P}$.

$V(300,780) = 1000, \qquad V(310,775) = 1040.$

The actual change in the volume is

$$V(310,775) - V(300,780) = 40 \text{ cm}^3.$$

24. The relative error in the approximation to the change in the volume is

$$\left|\frac{39.74 - 40}{40}\right| = \frac{.26}{40} = 0.0065.$$

25. a. $S(K, L) = 4KL - L^2$.

$S(16, 20) = 4(16)(20) - (20)^2 = 880$.

b. $\frac{\partial S}{\partial K} = 4L, \qquad \frac{\partial S}{\partial L} = 4K - 2L.$

$\frac{\partial S}{\partial K}(16, 20) = 80, \qquad \frac{\partial S}{\partial L}(16, 20) = 24.$

The approximate change in sales is

$$\begin{aligned} dS &= \frac{\partial S}{\partial K}(16, 20) \cdot dK + \frac{\partial S}{\partial L}(16, 20) \cdot dL \\ &= 80 \cdot dK + 24 \cdot dL \\ &= 80 \cdot 2 + 24 \cdot 0 \\ &= 160. \end{aligned}$$

26. $0 = dS = 64 \cdot dK + 48 \cdot dL \implies 0 = 64 \cdot (-3) + 48dL \implies dL = 4.$

Solutions to Exercise Set 8.6 (6.6)

1.

j	x_j	y_j	x_j^2	$x_j y_j$
1	4	9	16	36
2	3	6	9	18
3	5	7	25	35
4	5	8	25	40
Σ	17	30	75	129

$m = \frac{4(129)-(17)(30)}{4(75)-(17)^2} = \frac{6}{11} \approx 0.545.$

$b = \frac{(75)(30)-(17)(129)}{4(75)-(17)^2} = \frac{57}{11} \approx 5.182.$

The least squares regression line is $y = 0.545x + 5.182$.

2.

j	x_j	y_j	x_j^2	x_jy_j
1	12	4	144	48
2	6	14	36	84
3	10	9	100	90
4	18	2	324	36
5	3	16	9	48
6	5	16	25	80
Σ	54	61	638	386

$m = \frac{6(386)-54(61)}{6(638)-(54)^2} = -\frac{978}{912} \approx -1.072.$

$b = \frac{(638)(61)-(54)(386)}{6(638)-(54)^2} = \frac{18074}{912} \approx 19.818.$

The least squares regression line is $y = -1.072x + 19.818$.

3.

j	x_j	y_j	x_j^2	x_jy_j
1	52	74	2704	3848
2	46	66	2116	3036
3	69	94	4761	6486
4	54	91	2916	4914
5	61	84	3721	5124
6	48	80	2304	3840
Σ	330	489	18522	27248

$m = \frac{6(27248)-330(489)}{6(18522)-(330)^2} = \frac{2118}{2232} \approx 0.949.$

$b = \frac{18522(489)-330(27248)}{6(18522)-(330)^2} = \frac{65418}{2232} \approx 29.309.$

The least squares regression line is $y = 0.949x + 29.309$.

4.

j	x_j	y_j	x_j^2	x_jy_j
1	6	2	36	12
2	8	5	64	40
3	9	5	81	45
4	10	7	100	70
5	12	9	144	108
6	6	4	36	24
7	11	8	121	88
Σ	62	40	582	387

$m = \frac{7(387)-62(40)}{7(582)-(62)^2} = \frac{229}{230} \approx 0.9957.$

$b = \frac{582(40)-62(387)}{7(582)-(62)^2} = -\frac{714}{230} \approx -3.1043.$

The least squares regression line is $y = 0.9957x - 3.1043$.

5.

j	x_j	y_j	x_j^2	x_jy_j
1	0	1	0	0
2	1	0	1	0
3	1	2	1	2
4	2	1	4	2
5	2	2	4	4
6	3	2	9	6
7	3	1	9	3
8	3	3	9	9
9	4	2	16	8
10	4	3	16	12
Σ	23	17	69	46

$m = \frac{10(46)-23(17)}{10(69)-(23)^2} = \frac{69}{161} \approx 0.429.$

$b = \frac{69(17)-23(46)}{10(69)-(23)^2} = \frac{115}{161} \approx 0.714.$

The least squares regression line is $y = 0.429x + 0.714$.

6.

j	x_j	y_j	x_j^2	x_jy_j
1	0	5.6	0	0.0
2	1	6.2	1	6.2
3	2	5.3	4	10.6
Σ	3	17.1	5	16.8

Let x =the number of years after 1991, and let y =the percentage of disposable income.

a. $y = mx + b$.

$m = \frac{(3)(16.8)-3(17.1)}{3(5)-3^2} = -\frac{0.9}{6} = -0.15.$

$b = \frac{5(17.1)-3(16.8)}{3(5)-3^2} = \frac{35.1}{6} = 5.85.$

The least squares regression line is $y = -0.15x + 5.85$.

b. For the year 1995, $x = 4$. The predicted personal savings rate for 1995 is $y = -0.15(4) + 5.85 = 5.25\%$.

7.

j	x_j	y_j	x_j^2	x_jy_j
1	0	10	0	0
2	1	14	1	14
3	2	8	4	16
4	3	4	9	12
5	4	2	16	8
6	5	3	25	15
Σ	15	41	55	65

Let x = the number of years after 1988, and let y = the number of orders for new railroad cars received by the manufacturer.

a. $y = mx + b$.

$m = \frac{6(65)-15(41)}{6(55)-(15)^2} = -\frac{225}{105} = -2.143.$

$b = \frac{55(41)-15(65)}{6(55)-(15)^2} = \frac{1280}{105} = 12.19.$

The least squares regression line is $y = -2.143x + 12.19$.

b. For the year 1994, $x = 6$. The predicted number of orders the company would receive in 1994 is

$$y = -2.143(6) + 12.19 = -0.668 \text{ (668 cars returned)}.$$

8.

j	x_j	y_j	x_j^2	x_jy_j
1	25	52	625	1300
2	35	30	1225	1050
3	45	26	2025	1170
4	55	21	3025	1155
Σ	160	129	6900	4675

$m = \frac{4(4675)-160(129)}{4(6900)-(160)^2} = \frac{-1940}{2000} = -0.97.$

$b = \frac{6900(129)-160(4675)}{4(6900)-(160)^2} = \frac{142100}{2000} = 71.05.$

a. The least squares regression line is $y = -0.97x + 71.05$.

b. At $x = 65$, $y = -0.97(65) + 71.05 = 8$ deaths.

9.

j	x_j	y_j	x_j^2	x_jy_j
1	0	3.21	0	0
2	1	3.29	1	3.29
3	2	3.08	4	6.16
4	3	3.15	9	9.45
5	4	3.23	16	12.92
Σ	10	15.96	30	31.82

Let $x =$ the number of years after 1989, and let $y =$ the corresponding overall grade point average for undergraduates.

a. $y = mx + b$.

$m = \frac{5(31.82)-10(15.96)}{5(30)-(10)^2} = -\frac{0.5}{50} = -.01.$

$b = \frac{30(15.96)-10(31.82)}{5(30)-(10)^2} = \frac{160.6}{50} = 3.212.$

The least squares regression line is $y = -0.01x + 3.212$.

b. For the year 1995, $x = 6$. The predicted overall grade point average for undergraduates for the year 1995 is

$$y = -0.01(6) + 3.212 = 3.152.$$

c. This least squares regression line does not suggest grade inflation since the slope is negative.

10.

j	x_j	y_j	x_j^2	x_jy_j
1	0	7	0	0
2	1	11	1	11
3	2	16	4	32
4	3	23	9	69
5	4	25	16	100
Σ	10	82	30	212

Let $x =$ the number of decades after 1950, and let $y =$the corresponding percentage of the United States adults aged 25-29 who had graduated from college.

a. $y = mx + b$.

$m = \frac{5(212)-10(82)}{5(30)-(10)^2} = \frac{240}{50} = 4.8.$

$b = \frac{30(82)-10(212)}{5(30)-(10)^2} = \frac{340}{50} = 6.8.$

The least squares regression line is $y = 4.8x + 6.8$.

b. For the year 2000, $x = 5$. The predicted percentage described above for the year 2000 is

$$y = 4.8(5) + 6.8 = 30.8\% \ .$$

11.

j	x_j	y_j	x_j^2	x_jy_j
1	0	20	0	0
2	1	24	1	24
3	2	25	4	50
4	3	32	9	96
5	4	34	16	136
Σ	10	135	30	306

$$m = \frac{5(306) - 10(135)}{5(30) - (10)^2} = \frac{180}{50} = 3.6.$$

$$b = \frac{30(135) - 10(306)}{5(30) - (10)^2} = \frac{990}{50} = 19.8.$$

a. For the year 1994, $x = 5$. The predicted $y = 3.6(5) + 19.8 = 37.8\%$.

b. For the year 2000, $x = 8$. The predicted $y = 3.6(8) + 19.8 = 48.6\%$.

12.

j	x_j	y_j	x_j^2	x_jy_j
1	0	160	0	0
2	1	164	1	164
3	2	168	4	336
4	3	171	9	513
5	4	175	16	700
Σ	10	838	30	1713

$$m = \frac{5(1713) - 10(838)}{5(30) - (10)^2} = \frac{185}{50} = 3.7.$$

$$b = \frac{30(838) - 10(1713)}{5(30) - (10)^2} = \frac{8010}{50} = 160.2.$$

a. For the year 1995, $x = 5$. The predicted average weight $y = 3.7(5) + 160.2 = 178.7$ lbs.

b. For the year 1998, $x = 5.6$. The predicted average weight $y = 3.7(5.6) + 160.2 = 180.92$ lbs.

13.

j	x_j	y_j	x_j^2	$x_j y_j$
1	2	5	4	10
2	0	7	0	0
3	4	2	16	8
4	6	4	36	24
5	8	3	64	24
Σ	20	21	120	66

$$m = \frac{5(66) - 20(21)}{5(120) - (20)^2} = -\frac{90}{200} = -0.45.$$

$$b = \frac{120(21) - 20(66)}{5(120) - (20)^2} = \frac{1200}{200} = 6.$$

The regression line is $y = -0.45x + 6$.

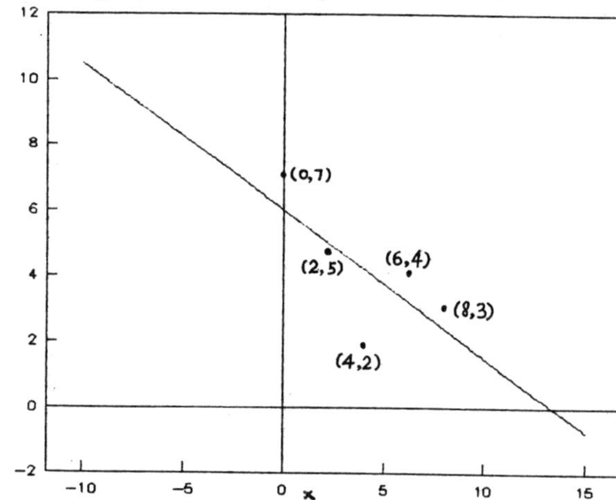

14.

j	x_j	y_j	x_j^2	$x_j y_j$
1	0	3	0	0
2	4	4	16	16
3	3	2	9	6
4	6	9	36	54
Σ	13	18	61	76

$$m = \frac{4(76) - 13(18)}{4(61) - (13)^2} = \frac{70}{75} = 0.9333.$$
$$b = \frac{61(18) - 13(76)}{4(61) - (13)^2} = \frac{110}{75} = 1.4667.$$

The regression line is $y = 0.9333x + 1.4667$.

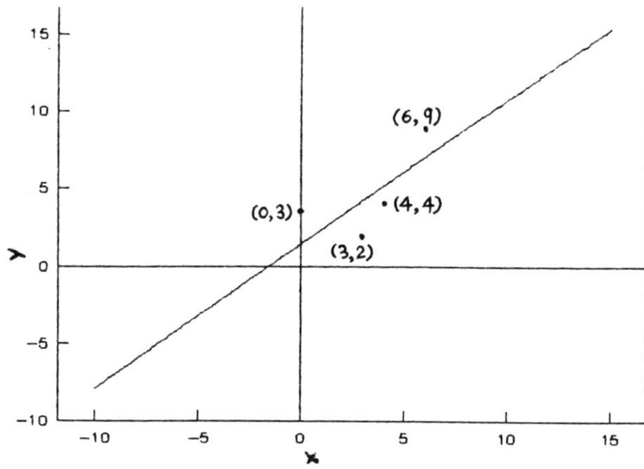

15.

j	x_j	y_j	x_j^2	x_jy_j
1	4	−2	16	−8
2	2	6	4	12
3	0	3	0	0
4	−2	8	4	−16
Σ	4	15	24	−12

$$m = \frac{4(-12) - 4(15)}{4(24) - (4)^2} = -\frac{108}{80} = -1.35.$$
$$b = \frac{24(15) - 4(-12)}{4(24) - (4)^2} = \frac{408}{80} = 5.1.$$

The regression line is $y = -1.35x + 5.1$.

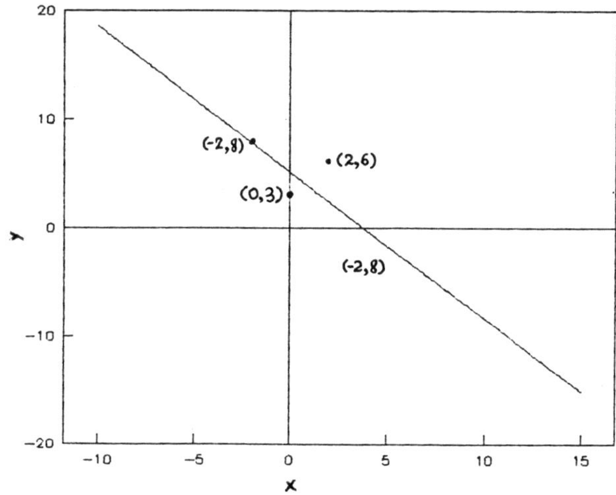

16.

j	x_j	y_j	x_j^2	x_jy_j
1	0	0	0	0
2	1	4	1	4
3	2	3	4	6
4	4	−1	16	−4
5	5	5	25	25
Σ	12	11	46	31

$$m = \frac{5(31) - 12(11)}{5(46) - (12)^2} = \frac{23}{86} = 0.267.$$

$$b = \frac{46(11) - 12(31)}{5(46) - (12)^2} = \frac{134}{86} = 1.558.$$

The regression line is $y = 0.267x + 1.558$.

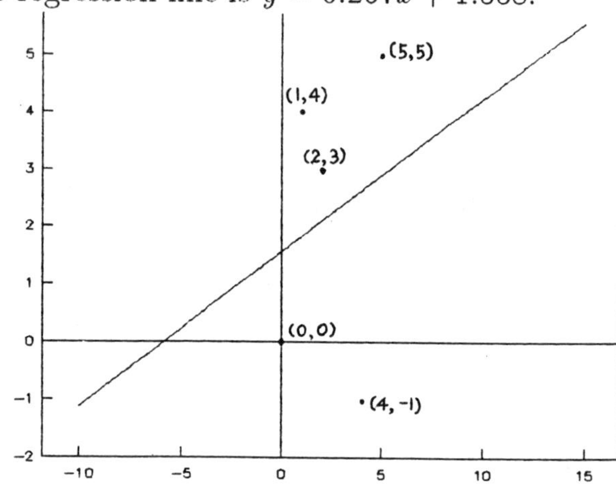

17.

j	x_j	y_j	x_j^2	x_jy_j
1	2	4	4	8
2	3	5	9	15
3	3.5	3	12.25	10.5
4	4.2	4	17.64	16.8
5	5	6	25	30
Σ	17.7	22	67.89	80.3

$$m = \frac{5(80.3) - 17.7(22)}{5(67.89) - (17.7)^2} = \frac{12.1}{26.16} = 0.463.$$

$$b = \frac{(67.89)(22) - 17.7(80.3)}{5(67.89) - (17.7)^2} = \frac{72.27}{26.16} = 2.763.$$

a. The least squares regression line is $y = 0.463x + 2.763$.

b. The size of the litter increases with age since the slope of the regression line is positive.

18.

j	x_j	y_j	x_j^2	x_jy_j
1	40	65	1600	2600
2	50	70	2500	3500
3	60	55	3600	3300
4	70	52	4900	3640
5	80	46	6400	3680
Σ	300	288	19000	16720

$$m = \frac{5(16720) - 300(288)}{5(19000) - (300)^2} = -\frac{2800}{5000} = -0.56.$$

$$b = \frac{19000(288) - 300(16720)}{5(19000) - (300)^2} = \frac{456000}{5000} = 91.2.$$

a. Here x denotes the price of an ice cream cone in cents, and y denotes the number of ice cream cones sold per day. The least squares regression line for sales y as a function of price x is given by

$$y = mx + b = -0.56x + 91.2.$$

b. According to this regression model, the sales decrease as price increases since the slope of the regression line is negative.

c. If the price is increased to 1 dollar per ice cream cone, the predicted daily sales are given by

$$y = -0.56(100) + 91.2 = 35.2 \text{ ice creams cones per day.}$$

19. Here let p denote the price of an ice cream cone in cents, and let r denote the corresponding daily revenue in cents.

j	p_j	r_j	p_j^2	$p_j r_j$
1	40	2600	1600	104000
2	50	3500	2500	175000
3	60	3300	3600	198000
4	70	3640	4900	254800
5	80	3680	6400	294400
Σ	300	16720	19000	1,026,200

$$m = \frac{5(1,025,200) - 300(16,720)}{5(19,000) - (300)^2} = \frac{115,000}{5,000} = 23.$$

$$b = \frac{19,000(16,720) - 300(1,026,200)}{5(19,000) - (300)^2} = \frac{9,820,000}{5,000} = 1964.$$

a. The least squares regression line for revenue as a function of price is given by

$$r = mp + b = 23p + 1964.$$

b. According to this regression model, the daily revenue increases with increases in price since the slope of the regression line is positive.

Solutions to Exercise Set 8.7 (6.7)

1. $\int_0^2 \int_0^1 (x-y)dxdy = \int_0^2 \left[\left(\frac{x^2}{2} - xy \right) \Big]_0^1 \right] dy$

$$= \int_0^2 \left(\tfrac{1}{2} - y\right) dy = \left(\tfrac{1}{2}y - \tfrac{1}{2}y^2\right)\Big]_0^2 = \tfrac{1}{2}(2-4) = -1.$$

2. $\int_0^1 \int_1^3 (2x+4y)dydx = \int_0^1 \left[(2xy + 2y^2)\Big]_1^3 \right] dx = \int_0^1 (2x \cdot 3 + 2 \cdot 3^2 - 2x - 2)dx$

$$= \int_0^1 (4x+16)dx = (2x^2 + 16x)\Big]_0^1 = 2 + 16 = 18.$$

3. $\int_{-1}^1 \int_0^2 x^2 y\, dydx = \int_{-1}^1 \left[\tfrac{1}{2}x^2y^2\Big]_0^2 \right] dx$

$$= \int_{-1}^1 \tfrac{1}{2}x^2 \cdot 4\, dx = \tfrac{2}{3}x^3\Big]_{-1}^1 = \tfrac{4}{3}.$$

4. $\int_{-2}^{1}\int_{0}^{1} x\sqrt[3]{y}\,dydx = \int_{-2}^{1}\left[x\cdot\frac{3}{4}y^{4/3}\Big]_0^1\right]dx = \int_{-2}^{1} x\cdot\frac{3}{4}\,dx = \frac{3}{8}x^2\Big]_{-2}^{1}$

$= \frac{3}{8}\cdot(1-4) = -\frac{9}{8}.$

5. $\int_{0}^{2}\int_{0}^{x^2}(x-2y)dydx = \int_{0}^{2}\left[(xy-y^2)\Big]_0^{x^2}\right]dx = \int_{0}^{2}(x^3-x^4)dx$

$= \left(\frac{1}{4}x^4 - \frac{1}{5}x^5\right)\Big]_0^2 = \frac{1}{4}\cdot 16 - \frac{1}{5}\cdot 32 = -\frac{48}{20} = -\frac{12}{5}.$

6. $\int_{-1}^{2}\int_{0}^{x} y\sqrt{x^3+1}dydx = \int_{-1}^{2}\left[\left(\frac{1}{2}y^2\sqrt{x^3+1}\right)\Big]_0^x\right]dx = \int_{-1}^{2}\frac{1}{2}x^2\sqrt{x^3+1}\,dx.$

Let $u(x) = x^3+1, \quad du = 3x^2\,dx.$

$\int \frac{1}{2}x^2\sqrt{x^3+1}\,dx = \frac{1}{2}\cdot\frac{1}{3}\int\sqrt{u}\,du = \frac{1}{6}\cdot\frac{2}{3}u^{3/2}+C = \frac{1}{9}(x^3+1)^{3/2}+C.$

$\int_{-1}^{2}\int_{0}^{x} y\sqrt{x^3+1}\,dydx = \frac{1}{9}(x^3+1)^{3/2}\Big]_{-1}^{2} = \frac{1}{9}\left((8+1)^{3/2}-(-1+1)^{3/2}\right)$

$= \frac{1}{9}\cdot 27 = 3.$

7. $\int_{0}^{1}\int_{0}^{1} ye^{x-y^2}\,dydx.$

Let $u(y) = x-y^2. \quad du = -2y\,dy.$

$\int_{0}^{1} ye^{x-y^2}\,dy = -\frac{1}{2}\int_{y=0}^{y=1} e^u\,du = -\frac{1}{2}e^u\Big]_{y=0}^{y=1} = -\frac{1}{2}e^{x-y^2}\Big]_0^1 = -\frac{1}{2}\left(e^{x-1}-e^x\right).$

$\int_{0}^{1}\int_{0}^{1} ye^{x-y^2}\,dydx = \int_{0}^{1}\left[\int_{0}^{1} ye^{x-y^2}\,dy\right]dx = \int_{0}^{1}\left(-\frac{1}{2}(e^{x-1}-e^x)\right)dx$

$= -\frac{1}{2}(e^{x-1}-e^x)\Big]_0^1 = -\frac{1}{2}(e^0-e^1-e^{-1}+e^0) = \frac{1}{2}\left(e+\frac{1}{e}-2\right).$

8. $\int_{-1}^{0}\int_{-1}^{y+1}(xy-x)dxdy = \int_{-1}^{0}\left[\left(\frac{1}{2}x^2y-\frac{1}{2}x^2\right)\Big]_{-1}^{y+1}\right]dy$

$= \int_{-1}^{0}\left[\frac{1}{2}(y+1)^2y - \frac{1}{2}(y+1)^2 - \frac{1}{2}y + \frac{1}{2}\right]dy = \int_{-1}^{0}\frac{1}{2}(y^3+y^2-2y)dy$

$= \frac{1}{2}\left(\frac{1}{4}y^4+\frac{1}{3}y^3-y^2\right)\Big]_{-1}^{0} = -\frac{1}{2}\left(\frac{1}{4}-\frac{1}{3}-1\right) = \frac{13}{24}.$

9. $\int_0^1 \int_{-x}^{\sqrt{x}} \frac{y}{1+x}\, dy dx = \int_0^1 \left[\frac{y^2}{2(1+x)} \right]_{-x}^{\sqrt{x}} dx = \int_0^1 \frac{x-x^2}{2(1+x)}\, dx = \int_0^1 \frac{(x-x^2+2)-2}{2(1+x)}\, dx$

$= \frac{1}{2} \int_0^1 \left(2 - x - \frac{2}{1+x}\right) dx = \frac{1}{2}\, \left[2x - \frac{1}{2}x^2 - 2\ln(1+x)\right]\Big]_0^1$

$= \frac{1}{2}\left(2 - \frac{1}{2} - 2\ln 2\right) = \frac{3}{4} - \ln 2.$

10. $\int_1^2 \int_0^{y^3} e^{x/y} dx dy = \int_1^2 \left[y e^{x/y} \Big]_0^{y^3} \right] dy = \int_1^2 y(e^{y^2} - 1) dy = \int_1^2 (e^{y^2} - 1)\frac{1}{2} d(y^2)$

$= \frac{1}{2}(e^{y^2} - y^2)\Big]_1^2 = \frac{1}{2}(e^4 - 4 - e^1 + 1) = \frac{1}{2}(e^4 - e - 3).$

11. $\int_0^2 \int_0^{\pi/2} x^2 \cos y\, dy dx = \int_0^2 \left[x^2 \sin y \Big]_0^{\pi/2} \right] dx$

$= \int_0^2 x^2\, dx = \frac{1}{3}x^3\Big]_0^2 = \frac{8}{3}.$

12. $\int_0^{\sqrt{2}/2} \int_0^{\sqrt{y}} x \sin y^2\, dx dy = \int_0^{\sqrt{2}/2} \left[\frac{1}{2}x^2 \sin y^2 \Big]_0^{\sqrt{y}} \right] dy = \int_0^{\sqrt{2}/2} \frac{1}{2} y \sin y^2\, dy$

$= \frac{1}{2} \int_0^{\sqrt{2}/2} \sin y^2 \left(\frac{1}{2} d(y^2)\right) = -\frac{1}{4} \cos y^2\Big]_0^{\sqrt{2}/2}$

$= -\frac{1}{4} \cos\left(\frac{1}{2}\right) + \frac{1}{4} = 0.0306.$

13. $\iint_R (x + y^2) dA = \int_0^2 \int_0^1 (x + y^2) dx dy = \int_0^2 \left[\left(\frac{1}{2}x^2 + xy^2\right)\Big]_0^1 \right] dy$

$= \int_0^2 \left(\frac{1}{2} + y^2\right) dy = \left(\frac{1}{2}y + \frac{1}{3}y^3\right)\Big]_0^2 = \frac{1}{2} \cdot 2 + \frac{1}{3} \cdot 8 = \frac{11}{3}.$

14. $\iint_R (x^2 + y^2) dA = \int_0^b \int_0^a (x^2 + y^2) dx dy = \int_0^b \left[\left(\frac{1}{3}x^3 + xy^2\right)\Big]_0^a \right] dy$

$= \int_0^b \left(\frac{1}{3}a^3 + ay^2\right) dy = \left(\frac{1}{3}a^3 y + \frac{a}{3}y^3\right)\Big]_0^b = \frac{1}{3}(a^3 b + ab^3) = \frac{1}{3}ab(a^2 + b^2).$

15. $\iint_R \frac{xy}{\sqrt{x^2+y^2}}\,dA = \int_1^2\int_1^2 \frac{xy}{\sqrt{x^2+y^2}}\,dxdy = \int_1^2 \left[\int_1^2 \frac{y}{\sqrt{x^2+y^2}}\frac{1}{2}d(x^2+y^2)\right] dy$

$= \int_1^2 \left[\frac{1}{2}y\cdot 2\sqrt{x^2+y^2}\right]_1^2 dy = \int_1^2 y\left(\sqrt{4+y^2}-\sqrt{1+y^2}\right) dy$

$= \int_1^2 \left(\sqrt{4+y^2}-\sqrt{1+y^2}\right)\frac{1}{2}d(y^2) = \frac{1}{2}\left(\frac{2}{3}(4+y^2)^{3/2}-\frac{2}{3}(1+y^2)^{3/2}\right)\Big]_1^2$

$= \frac{1}{3}\left[(4+4)^{3/2}-(1+4)^{3/2}\right] - \frac{1}{3}\left[(4+1)^{3/2}-(1+1)^{3/2}\right]$

$= \frac{1}{3}8^{3/2} - \frac{2}{3}5^{3/2} + \frac{1}{3}2^{3/2}.$

16. $\iint_R \frac{x^2}{1+y}dA = \int_0^1\int_0^{e^x-1} \frac{x^2}{1+y}dydx = \int_0^1 \left[x^2\ln(1+y)\Big]_0^{e^x-1}\right] dx$

$= \int_0^1 x^2(\ln e^x - \ln 1)dx = \int_0^1 x^3\,dx = \frac{1}{4}x^4\Big]_0^1 = \frac{1}{4}.$

17. $\iint_R xy\,dA = \int_0^1\int_y^{\sqrt{y}} xy\,dxdy = \int_0^1 \left[\frac{1}{2}x^2y\Big]_y^{\sqrt{y}}\right] dy$

$= \int_0^1 \left[\frac{1}{2}\left(\sqrt{y}\right)^2 y - \frac{1}{2}y^2y\right] dy = \frac{1}{2}\int_0^1 (y^2-y^3)dy = \frac{1}{2}\left(\frac{1}{3}y^3-\frac{1}{4}y^4\right)\Big]_0^1$

$= \frac{1}{2}\left(\frac{1}{3}-\frac{1}{4}\right) = \frac{1}{24}.$

18. $\iint_R (2x+y)dA = \int_0^1\int_0^{\sqrt{1-y^2}} (2x+y)dxdy = \int_0^1 \left[(x^2+xy)\Big]_0^{\sqrt{1-y^2}}\right] dy$

$= \int_0^1 \left[(1-y^2)+\sqrt{1-y^2}\cdot y\right] dy$

$= \left[y-\frac{1}{3}y^3-\frac{1}{3}(1-y^2)^{3/2}\right]_0^1$

$= \left[1-\frac{1}{3}-0\right]-\left[0-0-\frac{1}{3}\right] = 1.$

19. $\int_{-1}^1\int_0^{x+1} (x+y)dydx = \int_0^2\int_{y-1}^1 (x+y)dxdy$

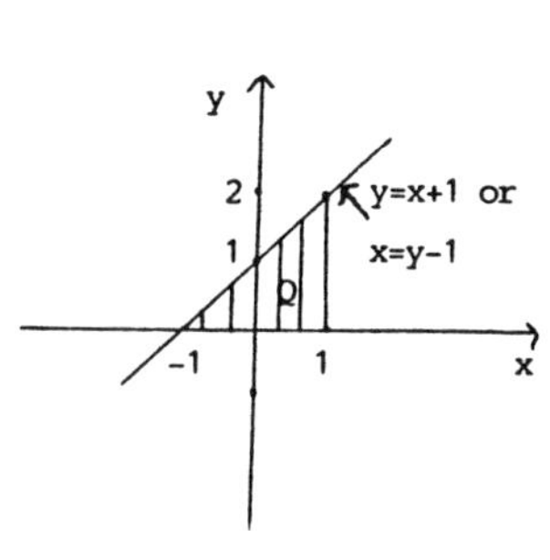

$= \int_0^2 \left[\left(\frac{1}{2}x^2+xy\right)\Big]_{y-1}^1\right] dy$

$= \int_0^2 \left\{\left(\frac{1}{2}+y\right)-\left[\frac{1}{2}(y-1)^2+(y-1)y\right]\right\} dy$

$= \int_0^2 \left(-\frac{3}{2}y^2+3y\right) dy = \left(-\frac{1}{2}y^3+\frac{3}{2}y^2\right)\Big]_0^2$

$= -\frac{1}{2}\cdot 8 + \frac{3}{2}\cdot 4 = 2.$

20. $\int_0^1 \int_{x^2}^1 xe^{y^2}\, dydx = \int_0^1 \int_0^{\sqrt{y}} xe^{y^2}\, dxdy$

$= \int_0^1 \left[\left[\frac{1}{2}x^2 e^{y^2} \right]_0^{\sqrt{y}} \right] dy = \int_0^1 \frac{1}{2} y e^{y^2}\, dy$

$= \frac{1}{4} \int_0^1 e^{y^2} d(y^2) = \frac{1}{4} e^{y^2} \Big]_0^1 = \frac{1}{4}(e^1 - e^0) = \frac{1}{4}(e-1).$

y=x2 or x=$\sqrt{y}$

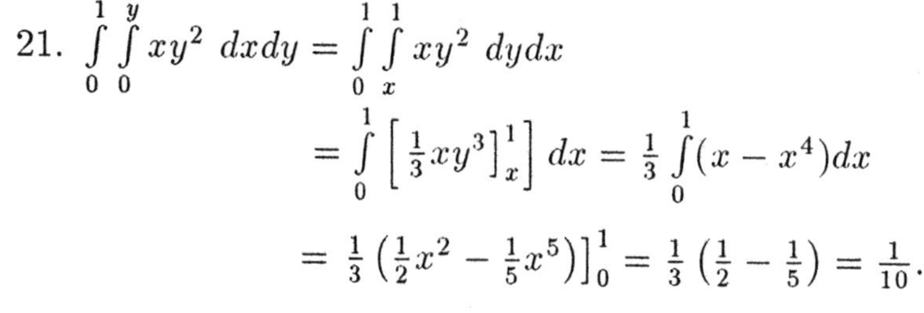

21. $\int_0^1 \int_0^y xy^2\, dxdy = \int_0^1 \int_x^1 xy^2\, dydx$

$= \int_0^1 \left[\left[\frac{1}{3}xy^3 \right]_x^1 \right] dx = \frac{1}{3} \int_0^1 (x - x^4) dx$

$= \frac{1}{3}\left(\frac{1}{2}x^2 - \frac{1}{5}x^5\right)\Big]_0^1 = \frac{1}{3}\left(\frac{1}{2} - \frac{1}{5}\right) = \frac{1}{10}.$

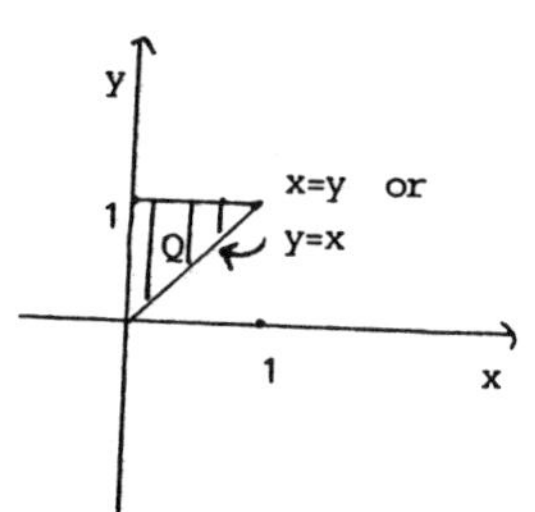

22. $\int_{-2}^0 \int_{x^2}^4 xe^{y^2}\, dydx = \int_0^4 \int_{-\sqrt{y}}^0 xe^{y^2}\, dxdy$

$= \int_0^4 \left[\left[\frac{1}{2}x^2 e^{y^2} \right]_{-\sqrt{y}}^0 \right] dy$

$= \frac{1}{2} \int_0^4 (0 - ye^{y^2}) dy = -\frac{1}{4} \int_0^4 e^{y^2} d(y^2)$

$= -\frac{1}{4} e^{y^2} \Big]_0^4 = -\frac{1}{4}(e^{16} - e^0) = \frac{1}{4}(1 - e^{16}).$

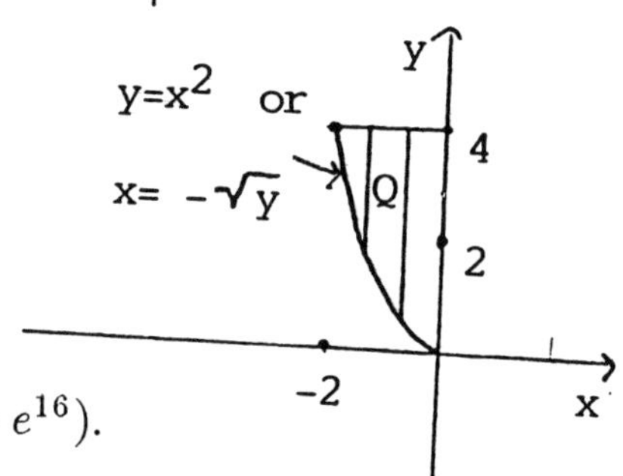

23. Volume $= \int_0^3 \int_0^3 (6 - x - y) dxdy = \int_0^3 \left[\left(6x - \frac{1}{2}x^2 - xy\right)\Big]_0^3 \right] dy$

$= \int_0^3 \left(18 - \frac{9}{2} - 3y\right) dy = \left(18y - \frac{9}{2}y - \frac{3}{2}y^2\right)\Big]_0^3 = 54 - \frac{27}{2} - \frac{3}{2} \cdot 9 = 27.$

24. Volume $= \int_0^2 \int_0^2 (9 - x^2 - y^2) dydx = \int_0^2 \left\{ \left[(9 - x^2)y - \frac{1}{3}y^3 \right] \Big]_0^2 \right\} dx$

$= \int_0^2 \left((9 - x^2) \cdot 2 - \frac{1}{3} \cdot 8\right) dx = \int_0^2 \left(\frac{46}{3} - 2x^2\right) dx$

$= \left(\frac{46}{3}x - \frac{2}{3}x^3\right)\Big]_0^2 = \frac{46}{3} \cdot 2 - \frac{2}{3} \cdot 8 = \frac{76}{3}.$

25. Volume $= \int_0^2 \int_0^{4-x^2} 4\, dydx = 4 \int_0^2 \left(y]_0^{4-x^2} \right) dx = 4 \int_0^2 (4 - x^2) dx$

$= 4\left(4x - \frac{1}{3}x^3\right)\Big]_0^2 = 4\left(8 - \frac{8}{3}\right) = \frac{64}{3}.$

26. $\text{Volume} = \int_0^2 \int_0^{3-\frac{3}{2}x} (9 - x^2 - y^2)dydx = \int_0^2 \left\{ \left[(9 - x^2)y - \frac{1}{3}y^3\right]\Big]_0^{3-\frac{3}{2}x} \right\} dx$

$= \int_0^2 \left[(9 - x^2)\left(3 - \frac{3}{2}x\right) - \frac{1}{3}\left(3 - \frac{3}{2}x\right)^3\right] dx$

$= \frac{3}{8}\int_0^2 (7x^3 - 26x^2 + 48)dx = \frac{3}{8}\left(\frac{7}{4}x^4 - \frac{26}{3}x^3 + 48x\right)\Big]_0^2$

$= \frac{3}{8}\left(\frac{7}{4}\cdot 16 - \frac{26}{3}\cdot 8 + 48\cdot 2\right) = \frac{3}{8}\left(\frac{164}{3}\right) = \frac{164}{8} = \frac{41}{2}.$

27. $\text{Volume} = \iint_R (12 - 4x - 2y)dA = \int_0^2 \int_0^1 (12 - 4x - 2y)dydx$

$= \int_0^2 \left\{ \left[(12 - 4x)y - y^2\right]\Big]_0^1 \right\} dx = \int_0^2 (12 - 4x - 1)dx$

$= \int_0^2 (11 - 4x)dx = (11x - 2x^2)\Big]_0^2 = 11\cdot(2) - 2\cdot(4) = 14.$

28. $\text{Volume} = \iint_R f(x,y)dA = \int_1^4 \int_1^9 \frac{e^{\sqrt{x}}}{\sqrt{xy}}\, dydx$

$= \int_1^4 \left[\frac{e^{\sqrt{x}}}{\sqrt{x}}\left(2\sqrt{y}\Big]_1^9\right)\right] dx = \int_1^4 \frac{e^{\sqrt{x}}}{\sqrt{x}} 2\left(\sqrt{9} - 1\right) dx = 4\int_1^4 \frac{e^{\sqrt{x}}}{\sqrt{x}}\, dx$

$= 8\int_1^4 e^{\sqrt{x}} d\left(\sqrt{x}\right) = 8\cdot e^{\sqrt{x}}\Big]_1^4 = 8\left(e^{\sqrt{4}} - e^1\right) = 8(e^2 - e).$

29. $\text{Volume} = \iint_R f(x,y)dA = \int_0^2 \int_0^{3-\frac{3}{2}x} (4 + x + y)dydx$

$= \int_0^2 \left\{ \left[(4 + x)y + \frac{1}{2}y^2\right]\Big]_0^{3-\frac{3}{2}x} \right\} dx$

$= \int_0^2 \left[(4 + x)\left(3 - \frac{3}{2}x\right) + \frac{1}{2}\left(3 - \frac{3}{2}x\right)^2\right] dx$

$= \int_0^2 \frac{3}{8}(44 - 20x - x^2)dx = \frac{3}{8}\left(44x - 10x^2 - \frac{1}{3}x^3\right)\Big]_0^2$

$= \frac{3}{8}\left(88 - 40 - \frac{1}{3}\cdot 8\right) = 17.$

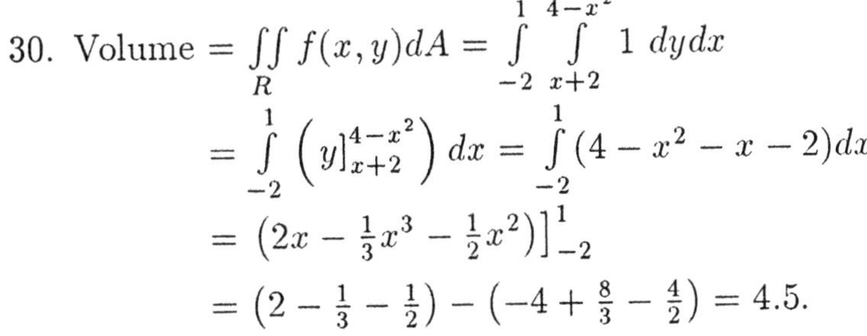

30. $\text{Volume} = \iint_R f(x,y)dA = \int_{-2}^1 \int_{x+2}^{4-x^2} 1\, dydx$

$= \int_{-2}^1 \left(y\Big]_{x+2}^{4-x^2}\right) dx = \int_{-2}^1 (4 - x^2 - x - 2)dx$

$= \left(2x - \frac{1}{3}x^3 - \frac{1}{2}x^2\right)\Big]_{-2}^1$

$= \left(2 - \frac{1}{3} - \frac{1}{2}\right) - \left(-4 + \frac{8}{3} - \frac{4}{2}\right) = 4.5.$

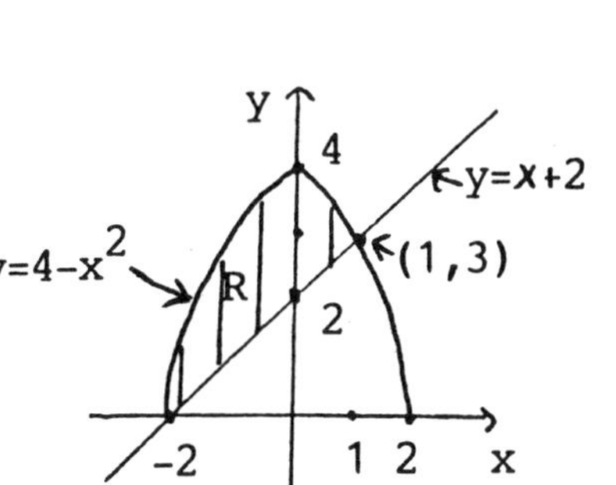

31. Area $= \iint\limits_R dA = \int\limits_0^1 \int\limits_{x^3}^{x^2} dy dx$
$= \int\limits_0^1 \left(y]_{x^3}^{x^2} \right) dx = \int\limits_0^1 (x^2 - x^3) dx$
$= \left(\frac{1}{3}x^3 - \frac{1}{4}x^4\right)\Big]_0^1$
$= \frac{1}{3} - \frac{1}{4} = \frac{1}{12}.$

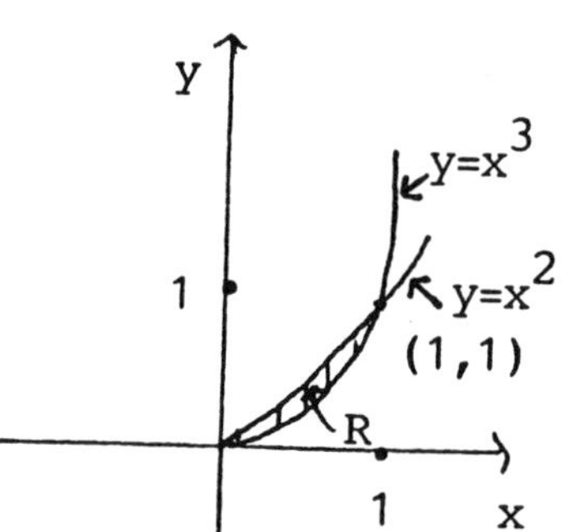

32. Area of $R = \frac{\pi}{2} \cdot \frac{\pi}{2} = \frac{\pi^2}{4}.$

$\iint\limits_R f(x,y) dA = \int\limits_0^{\pi/2} \int\limits_0^{\pi/2} (xy + x\cos y + y \sin x) dy dx$
$= \int\limits_0^{\pi/2} \left(\frac{1}{2}xy^2 + x\sin y + \frac{1}{2}y^2 \sin x\right)\Big]_0^{\pi/2} dx$
$= \int\limits_0^{\pi/2} \left(\frac{1}{2}x \cdot \left(\frac{\pi}{2}\right)^2 + x \sin\frac{\pi}{2} + \frac{1}{2}\left(\frac{\pi}{2}\right)^2 \sin x\right) dx$
$= \left(\frac{\pi^2}{8} \cdot \frac{1}{2}x^2 + \frac{1}{2}x^2 + \frac{\pi^2}{8}(-\cos x)\right)\Big]_0^{\pi/2}$
$= \left(\frac{\pi^2}{8} \cdot \frac{\pi^2}{8} + \frac{\pi^2}{8}\right) - \left(\frac{\pi^2}{8} \cdot (-1)\right) = \frac{\pi^4}{64} + \frac{\pi^2}{8} + \frac{\pi^2}{8} = \frac{\pi^4}{64} + \frac{\pi^2}{4}.$

Average value $= \frac{\frac{\pi^4}{64} + \frac{\pi^2}{4}}{\frac{\pi^2}{4}} = \frac{\pi^2}{16} + 1.$

33. This formula for the average value of f over R follows because

$$\text{Area of } R = \iint\limits_R 1 \cdot dA.$$

34. The area of the region R is $(\ln 2) \cdot (\ln 3)$.

$\iint\limits_R f(x,y) dA = \int\limits_0^{\ln 3} \int\limits_0^{\ln 2} (6x + e^y) dy dx = \int\limits_0^{\ln 3} (6xy + e^y)]_0^{\ln 2} dx$
$= \int\limits_0^{\ln 3} \left((6 \ln 2) \cdot x + e^{\ln 2} - 1\right) dx = \int\limits_0^{\ln 3} \left((\ln 64) \cdot x + 1\right) dx$
$= \left((\ln 64) \cdot \frac{1}{2}x^2 + x\right)\Big]_0^{\ln 3} = (\ln 8)(\ln 3)^2 + \ln 3.$

The average value $= \frac{(\ln 8)(\ln 3)^2 + \ln 3}{(\ln 2) \cdot (\ln 3)} = \frac{(\ln 8)(\ln 3) + 1}{\ln 2}.$

35. The area of the region R is 6.

$$\iint_R f(x,y)dA = \int_0^4 \int_0^{-\frac{3}{4}x+3} 12x\,dydx = \int_0^4 12xy]_0^{-\frac{3}{4}x+3}\,dx$$

$$= \int_0^4 (-9x^2+36x)dx = (-3x^3+18x^2)]_0^4 = 96.$$

The average value $= \frac{96}{6} = 16.$

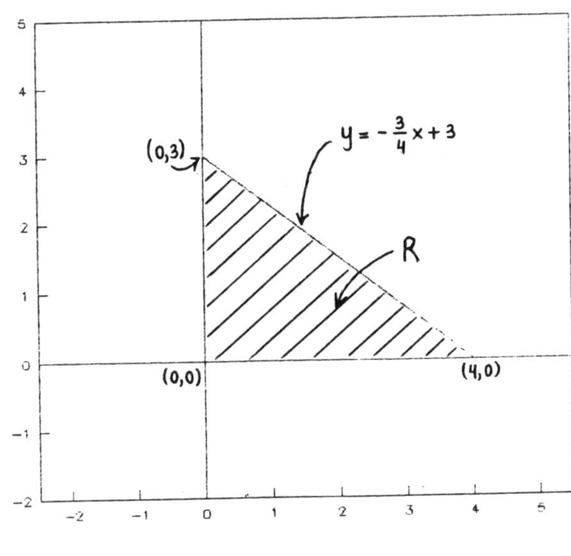

Solutions to Chapter 8 (Chapter 6) - Review Exercises

1. $f(x,y) = \sqrt{25-x^2-y^2}.$

   The domain of f is the set of ordered pairs (x,y) of real numbers such that $x^2+y^2 \le 25.$

2. $f(x,y) = x^3 - y\ln(3+x).$

   a. $f(0,2) = 0^3 - 2\ln(3+0) = -2\ln 3.$

   b. $f(1,-2) = 1^3 - (-2)\ln(3+1) = 1+2\ln 4.$

3. $f(x,y) = \frac{y-x}{y+x}.$

   a. $f(2,-1) = \frac{-1-2}{-1+2} = -3.$

   b. $f(1,-2) = \frac{-2-1}{-2+1} = 3.$

4. 

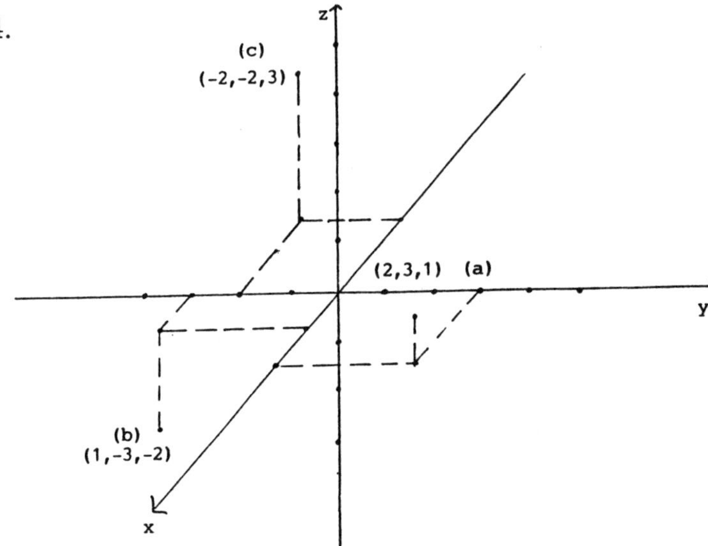

5. $f(x,y) = \frac{1+x^2}{1-xy}$.

   $f(x,y)$ is discontinuous on the set of points $\{(x,y) \mid xy = 1\}$.

6. $f(x,y) = x^4 + 2xy - 3y^3$.

   $\frac{\partial f}{\partial x} = 4x^3 + 2y$. $\qquad \frac{\partial f}{\partial y} = 2x - 9y^2$.

7. $f(x,y) = (x-y)^3 + \ln xy$.

   $\frac{\partial f}{\partial x} = 3(x-y)^2 + \frac{y}{xy} = 3(x-y)^2 + \frac{1}{x}$.

   $\frac{\partial f}{\partial y} = 3(x-y)^2(-1) + \frac{x}{xy} = -3(x-y)^2 + \frac{1}{y}$.

8. $f(x,y) = (4-xy)^5 + \sqrt{xy}$.

   $\frac{\partial f}{\partial x} = 5(4-xy)^4(-y) + \frac{y}{2\sqrt{xy}} = -5y(4-xy)^4 + \frac{1}{2}\sqrt{\frac{y}{x}}$.

   $\frac{\partial f}{\partial y} = 5(4-xy)^4(-x) + \frac{x}{2\sqrt{xy}} = -5x(4-xy)^4 + \frac{1}{2}\sqrt{\frac{x}{y}}$.

9. $f(x,y) = \frac{xy}{x+y}$.

   $\frac{\partial f}{\partial x} = \frac{y(x+y)-xy}{(x+y)^2} = \left(\frac{y}{x+y}\right)^2$.

   $\frac{\partial f}{\partial y} = \frac{x(x+y)-xy}{(x+y)^2} = \left(\frac{x}{x+y}\right)^2$.

10. $f(x,y) = xye^{xy}$.

    $\frac{\partial f}{\partial x} = ye^{xy} + xye^{xy} \cdot y = ye^{xy}(1+xy)$.

    $\frac{\partial f}{\partial y} = xe^{xy} + xye^{xy} \cdot x = xe^{xy}(1+xy)$.

11. $f(x,y)=x^{2/3}y^{1/3}-\frac{x^{3/4}}{y^{1/4}}$.

$\frac{\partial f}{\partial x}=\frac{2}{3}x^{-1/3}y^{1/3}-\frac{3x^{-1/4}}{4y^{1/4}}=\frac{2y^{1/3}}{3x^{1/3}}-\frac{3}{4x^{1/4}y^{1/4}}$.

$\frac{\partial f}{\partial y}=\frac{1}{3}x^{2/3}y^{-2/3}-\left(-\frac{1}{4}\right)y^{-5/4}x^{3/4}=\frac{x^{2/3}}{3y^{2/3}}+\frac{x^{3/4}}{4y^{5/4}}$.

12. $f(x,y)=\ln\sqrt{x^2+y^2}$.

$\frac{\partial f}{\partial x}=\frac{1}{\sqrt{x^2+y^2}}\cdot\frac{2x}{2\sqrt{x^2+y^2}}=\frac{x}{x^2+y^2}$.

$\frac{\partial f}{\partial y}=\frac{1}{\sqrt{x^2+y^2}}\cdot\frac{2y}{2\sqrt{x^2+y^2}}=\frac{y}{x^2+y^2}$.

13. $f(x,y)=xy\sqrt{y^2-x^2}$.

$\frac{\partial f}{\partial x}=y\sqrt{y^2-x^2}+xy\cdot\frac{-2x}{2\sqrt{y^2-x^2}}=y\sqrt{y^2-x^2}-\frac{x^2y}{\sqrt{y^2-x^2}}$.

$\frac{\partial f}{\partial y}=x\sqrt{y^2-x^2}+xy\cdot\frac{2y}{2\sqrt{y^2-x^2}}=x\sqrt{y^2-x^2}+\frac{xy^2}{\sqrt{y^2-x^2}}$.

14. $f(x,y)=\frac{e^{x+y}}{1+xy}$.

$\frac{\partial f}{\partial x}=\frac{(1+xy)\cdot e^{x+y}-e^{x+y}\cdot y}{(1+xy)^2}=\frac{e^{x+y}(1+xy-y)}{(1+xy)^2}$.

$\frac{\partial f}{\partial y}=\frac{(1+xy)\cdot e^{x+y}-e^{x+y}\cdot x}{(1+xy)^2}=\frac{e^{x+y}(1+xy-x)}{(1+xy)^2}$.

15. $f(x,y)=(xy^2-x^2y)^{2/3}$.

$\frac{\partial f}{\partial x}=\frac{2}{3}(xy^2-x^2y)^{-1/3}(y^2-2xy)=\frac{2}{3}y(y-2x)(xy^2-x^2y)^{-1/3}$.

$\frac{\partial f}{\partial y}=\frac{2}{3}(xy^2-x^2y)^{-1/3}(2xy-x^2)=\frac{2}{3}x(2y-x)(xy^2-x^2y)^{-1/3}$.

16. $f(x,y,z)=\frac{xyz^2}{x+y+z}$.

$\frac{\partial f}{\partial x}=\frac{(x+y+z)\cdot yz^2-xyz^2\cdot 1}{(x+y+z)^2}=\frac{(y+z)yz^2}{(x+y+z)^2}$.

$\frac{\partial f}{\partial y}=\frac{(x+y+z)\cdot xz^2-xyz^2\cdot 1}{(x+y+z)^2}=\frac{(x+z)xz^2}{(x+y+z)^2}$.

$\frac{\partial f}{\partial z}=\frac{(x+y+z)zxyz-xyz^2\cdot 1}{(x+y+z)^2}=\frac{(x+y)2xyz+xyz^2}{(x+y+z)^2}$.

17. $f(x,y,z)=\ln\sqrt{x^2+4y^2+2z}$.

$\frac{\partial f}{\partial x}=\frac{1}{\sqrt{x^2+4y^2+2z}}\cdot\frac{1}{2}\left(x^2+4y^2+2z\right)^{-1/2}\cdot 2x=\frac{x}{x^2+4y^2+2z}$.

$\frac{\partial f}{\partial y}=\frac{1}{\sqrt{x^2+4y^2+2z}}\cdot\frac{1}{2}\left(x^2+4y^2+2z\right)^{-1/2}\cdot 8y=\frac{4y}{x^2+4y^2+2z}$.

$\frac{\partial f}{\partial z}=\frac{1}{\sqrt{x^2+4y^2+2z}}\cdot\frac{1}{2}\left(x^2+4y^2+2z\right)^{-1/2}\cdot 2=\frac{1}{x^2+4y^2+2z}$.

18. $f(x,y) = ye^{x+y}$.

$\frac{\partial f}{\partial x} = ye^{x+y}$. $\quad \frac{\partial f}{\partial y} = e^{x+y} + ye^{x+y} = e^{x+y}(y+1)$.

$\frac{\partial^2 f}{\partial x^2} = \frac{\partial}{\partial x}(ye^{x+y}) = ye^{x+y}$.

$\frac{\partial^2 f}{\partial x \partial y} = e^{x+y} + ye^{x+y} = e^{x+y}(y+1)$.

$\frac{\partial^2 f}{\partial y^2} = e^{x+y} + (y+1)e^{x+y} = e^{x+y}(y+2)$.

19. $f(x,y) = \sqrt{y^2 - x^2}$.

$\frac{\partial f}{\partial x} = \frac{-2x}{2\sqrt{y^2-x^2}} = -x(y^2-x^2)^{-1/2}$. $\quad \frac{\partial f}{\partial y} = \frac{2y}{2\sqrt{y^2-x^2}} = y(y^2-x^2)^{-1/2}$.

$\frac{\partial^2 f}{\partial x^2} = -(y^2-x^2)^{-1/2} - x\left(-\frac{1}{2}\right)(y^2-x^2)^{-3/2}(-2x)$

$= -(y^2-x^2)^{-1/2} - x^2(y^2-x^2)^{-3/2} = \frac{-(y^2-x^2+x^2)}{(y^2-x^2)^{3/2}} = \frac{-y^2}{(y^2-x^2)^{3/2}}$.

$\frac{\partial^2 f}{\partial x \partial y} = -x\left(-\frac{1}{2}\right)(y^2-x^2)^{-3/2}(2y) = \frac{xy}{(y^2-x^2)^{3/2}}$.

$\frac{\partial^2 f}{\partial y^2} = (y^2-x^2)^{-1/2} + y\left(-\frac{1}{2}\right)(y^2-x^2)^{-3/2}(2y)$

$= \frac{1}{\sqrt{y^2-x^2}} - \frac{y^2}{(y^2-x^2)^{3/2}} = \frac{y^2-x^2-y^2}{(y^2-x^2)^{3/2}} = \frac{-x^2}{(y^2-x^2)^{3/2}}$.

20. $f(x,y) = 4y^2 - 2x^2$.

$\frac{\partial f}{\partial x} = -4x = 0 \implies x = 0$.

$\frac{\partial f}{\partial y} = 8y = 0 \implies y = 0$.

So the critical point is (0,0).

$\frac{\partial^2 f}{\partial x^2} = -4, \quad \frac{\partial^2 f}{\partial x \partial y} = 0, \quad \frac{\partial^2 f}{\partial y^2} = 8$.

At (0,0), $D = 0 - (-4)\cdot 8 = 32 > 0 \implies f(x,y)$ has a saddle point at (0,0,0).

21. $f(x,y) = 3x^2 + xy - 6y^2$.

$$\frac{\partial f}{\partial x} = 6x + y = 0 \qquad \textbf{(1)}$$

$$\frac{\partial f}{\partial y} = x - 12y = 0 \qquad \textbf{(2)}$$

$$\textbf{(1)} + (\textbf{(2)} \times (-6)) \implies 73y = 0 \implies y = 0 \qquad \textbf{(3)}$$

From **(2)** and **(3)**: $x = 12y = 12 \cdot (0) = 0$.

So the critical point is (0,0).

$\frac{\partial^2 f}{\partial x^2} = 6, \quad \frac{\partial^2 f}{\partial x \partial y} = 1, \quad \frac{\partial^2 f}{\partial y^2} = -12$.

At (0,0), $D = 1 - 6(-12) = 73 > 0$.

$f(x,y)$ has a saddle point at (0,0,0).

22. $f(x,y) = x^2y + xy^2 + 4x + 4y$.

$$\frac{\partial f}{\partial x} = 2xy + y^2 + 4 = 0 \qquad (1)$$
$$\frac{\partial f}{\partial y} = x^2 + 2xy + 4 = 0 \qquad (2)$$

$(1) - (2) \implies y^2 - x^2 = 0 \implies (y+x)(y-x) = 0 \implies y = x$ or $y = -x$.

For $y = -x$: From **(1)** $\implies 2x \cdot (-x) + (-x)^2 + 4 = 0 \implies x = \pm 2$.

So (2,-2) and (-2,2) are critical points.

For $y = x$: From **(1)** $\implies 2x \cdot (x) + (x^2) + 4 = 0 \implies$ no solution.

$\frac{\partial^2 f}{\partial x^2} = 2y, \quad \frac{\partial^2 f}{\partial x \partial y} = 2x + 2y, \quad \frac{\partial^2 f}{\partial y^2} = 2x.$

At (2,-2): $A = -4, \quad B = 4 - 4 = 0, \quad C = 4.$

$D = 0 - (-4)(4) = 16 > 0.$ (2,-2) $\rightarrow$ a saddle point.

At (-2,2): $A = 4, \quad B = 0, \quad C = -4.$

$D = 16 > 0.$ (-2,2) $\rightarrow$ a saddle point.

23. $f(x,y) = e^{x^2 - 4xy}$.

$$\frac{\partial f}{\partial x} = e^{x^2-4xy}(2x - 4y) = 0 \implies 2x - 4y = 0 \qquad (1)$$
$$\frac{\partial f}{\partial y} = e^{x^2-4xy}(-4x) = 0 \implies 4x = 0 \qquad (2)$$

From **(2)**: $4x = 0 \implies x = 0$. **(3)**

From **(1)** and **(3)**: $2 \cdot (0) - 4y = 0 \implies y = 0$.

So (0,0) is the critical point.

$$\frac{\partial^2 f}{\partial x^2} = e^{x^2-4xy}(2x-4y)^2 + 2e^{x^2-4xy},$$
$$\frac{\partial^2 f}{\partial x \partial y} = e^{x^2-4xy}(2x-4y)(-4x) + (-4)e^{x^2-4xy},$$
$$\frac{\partial^2 f}{\partial y^2} = e^{x^2-4xy} \cdot 16x^2.$$

At (0,0): $A = 2, \quad B = -4, \quad C = 0. \qquad D = 16 > 0.$

Thus (0,0,1) is a saddle point of f.

24. $f(x, y) = \ln(1 + x^2 + y^2)$.

$$\frac{\partial f}{\partial x} = \frac{2x}{1 + x^2 + y^2} = 0 \implies x = 0.$$
$$\frac{\partial f}{\partial y} = \frac{2y}{1 + x^2 + y^2} = 0 \implies y = 0.$$

So (0,0) is a critical point.

$$\frac{\partial^2 f}{\partial x^2} = \frac{2(1 + x^2 + y^2) - 2x \cdot 2x}{(1 + x^2 + y^2)^2} = \frac{2(1 - x^2 + y^2)}{(1 + x^2 + y^2)^2},$$
$$\frac{\partial^2 f}{\partial x \partial y} = \frac{-2x \cdot 2y}{(1 + x^2 + y^2)^2} = \frac{-4xy}{(1 + x^2 + y^2)^2},$$
$$\frac{\partial^2 f}{\partial y^2} = \frac{2(1 + x^2 - y^2)}{(1 + x^2 + y^2)^2}.$$

At (0,0): $A = 2, \quad B = 0, \quad C = 2. \quad D = -4 < 0, \quad A > 0.$

Thus $f(x, y)$ has a relative minimum at (0,0).

25. $f(x, y) = e^{1+x^2-y^2}$.

$$\frac{\partial f}{\partial x} = e^{1+x^2-y^2}(2x) = 0 \implies x = 0.$$
$$\frac{\partial f}{\partial y} = e^{1+x^2-y^2}(-2y) = 0 \implies y = 0.$$

So (0,0) is the critical point.

$$\frac{\partial^2 f}{\partial x^2} = e^{1+x^2-y^2}(4x^2) + 2e^{1+x^2-y^2},$$
$$\frac{\partial^2 f}{\partial x \partial y} = e^{1+x^2-y^2}(-4xy),$$
$$\frac{\partial^2 f}{\partial y^2} = e^{1+x^2-y^2}(4y^2) - 2e^{1+x^2-y^2}.$$

At (0,0): $A = 2, \quad B = 0, \quad C = -2.$ Then $D = 4 > 0.$

So (0,0) $\rightarrow$ a saddle point.

26. $f(x,y) = 6x^2 - 2x - 3xy + y^2 + 5y + 5.$

$$\frac{\partial f}{\partial x} = 12x - 2 - 3y = 0 \qquad (1)$$

$$\frac{\partial f}{\partial y} = -3x + 2y + 5 = 0 \qquad (2)$$

$$(1) + ((2) \times 4) \implies 5y + 18 = 0 \implies y = -\frac{18}{5}.$$

From (2): $-3x + 2\left(-\frac{18}{5}\right) + 5 = 0 \implies x = \frac{-11}{15}$. So the critical point is $\left(-\frac{11}{15}, -\frac{18}{5}\right)$.

$$\frac{\partial^2 f}{\partial x^2} = 12, \qquad \frac{\partial^2 f}{\partial x \partial y} = -3, \qquad \frac{\partial^2 f}{\partial y^2} = 2.$$

At $\left(-\frac{11}{15}, -\frac{18}{5}\right)$: $D = 9 - (12)\cdot 2 = -15 < 0$ and $A > 0$.

Thus $f(x,y)$ has a relative minimum at $\left(-\frac{11}{15}, -\frac{18}{5}\right)$.

27. $\int\limits_{-1}^{1}\int\limits_{0}^{2}(2x+3y)dxdy = \int\limits_{-1}^{1}\left[(x^2+3xy)\Big]_0^2\right] dy$
$= \int\limits_{-1}^{1}(4+6y)dy = (4y+3y^2)\Big]_{-1}^{1} = 4+3-(-4+3) = 8.$

28. $\int\limits_{0}^{3}\int\limits_{0}^{1}\frac{xy}{\sqrt{x^2+y^2}}\,dxdy = \int\limits_{0}^{3}\left[y\sqrt{x^2+y^2}\Big]_0^1\right] dy$
$= \int\limits_{0}^{3}\left(y\sqrt{1+y^2} - y^2\right) dy = \frac{1}{2}\int\limits_{0}^{3}\sqrt{1+y^2}\; d(y^2) - \int\limits_{0}^{3} y^2\,dy$
$= \left[\frac{1}{2}\cdot\frac{2}{3}(1+y^2)^{3/2} - \frac{1}{3}y^3\right]\Big]_0^3 = \frac{1}{3}\left[(1+9)^{3/2} - 27 - 1\right] = \frac{1}{3}(10^{3/2} - 28).$

29. $\int\limits_{0}^{1}\int\limits_{0}^{x} xy\sqrt{x^2+y^2}\;dydx = \int\limits_{0}^{1}\left[\int\limits_{0}^{x} x\sqrt{x^2+y^2}\;\frac{1}{2}\;d(y^2)\right] dx$
$= \int\limits_{0}^{1}\left[\frac{1}{2}x\cdot\frac{2}{3}(x^2+y^2)^{3/2}\Big]_0^x\right] dx = \int\limits_{0}^{1}\frac{1}{3}x\left[(x^2+x^2)^{3/2} - (x^2+0)^{3/2}\right] dx$
$= \frac{1}{3}(2^{3/2}-1)\int\limits_{0}^{1} x^4\;dx = \frac{1}{3}(2^{3/2}-1)\cdot\frac{1}{5}x^5\Big]_0^1 = \frac{1}{15}(2^{3/2}-1).$

30. $\int\limits_{0}^{1}\int\limits_{0}^{y} y^4 e^{xy^2}\;dxdy = \int\limits_{0}^{1}\left[y^4\cdot\frac{1}{y^2}e^{xy^2}\Big]_0^y\right] dy = \int\limits_{0}^{1} y^2(e^{y^3}-1)dy$
$= \frac{1}{3}\int\limits_{0}^{1}(e^{y^3}-1)d(y^3) = \frac{1}{3}(e^{y^3}-y^3)\Big]_0^1 = \frac{1}{3}(e-1-e^0) = \frac{1}{3}(e-2).$

31. $\int_0^1 \int_0^x xy(7+y^2)dydx = \int_0^1 \left[\int_0^x x(7+y^2)\frac{1}{2}d(y^2)\right] dx = \int_0^1 \left[\frac{1}{4}x(7+y^2)^2\right]_0^x dx$

$= \frac{1}{4}\int_0^1 x\left[(7+x^2)^2 - 7^2\right] dx = \frac{1}{4}\int_0^1 \left[(7+x^2)^2 - 49\right] \cdot \frac{1}{2}d(x^2)$

$= \frac{1}{8}\left[\frac{1}{3}(7+x^2)^3 - 49x^2\right]\Big|_0^1 = \frac{1}{8}\left(\frac{1}{3}\cdot 8^3 - 49 - \frac{1}{3}\cdot 7^3\right) = \frac{1}{8}\cdot\frac{22}{3} = \frac{11}{12}.$

32. $\int_{-1}^1 \int_{-y}^y (xy^3 + x^3y)dxdy = \int_{-1}^1 \left[\left(\frac{1}{2}x^2y^3 + \frac{1}{4}x^4y\right)\right]_{-y}^y dy$

$= \int_{-1}^1 \left(\frac{1}{2}y^2y^3 + \frac{1}{4}y^4y - \frac{1}{2}y^2y^3 - \frac{1}{4}y^4y\right) dy = 0.$

33. a. $R(x,y) = xp + yq$

$= x(80-2x) + y(120-4y)$

$= -2x^2 - 4y^2 + 80x + 120y.$

b. $P(x,y) = R(x,y) - C(x,y)$

$= (-2x^2 - 4y^2 + 80x + 120y) - (1000 + 20x + 20y)$

$= -2x^2 - 4y^2 + 60x + 100y - 1000.$

c. $\frac{\partial R}{\partial x} = -4x + 80 = 0 \implies x = 20$

$\frac{\partial R}{\partial y} = -8y + 120 = 0 \implies y = 15$

$\frac{\partial^2 R}{\partial x^2} = -4, \quad \frac{\partial^2 R}{\partial x \partial y} = 0, \quad \frac{\partial^2 R}{\partial y^2} = -8.$

At (20,15), $D = -32 < 0$ and $A < 0$,

thus the revenue is a maximum when $(x,y) = (20,15)$.

d. $\frac{\partial P}{\partial x} = -4x + 60 = 0 \implies x = 15$

$\frac{\partial P}{\partial y} = -8y + 100 = 0 \implies y = 12.5$

$\frac{\partial^2 P}{\partial x^2} = -4, \quad \frac{\partial^2 P}{\partial x \partial y} = 0, \quad \frac{\partial^2 P}{\partial y^2} = -8.$

At (15,12.5), $D = -32 < 0$ and $A < 0$.

Thus the profit is a maximum when $(x,y) = (15,12.5)$.

34. $P(x,y) = -2x^2 - 4y^2 + 60x + 100y - 1000$ subject to $x + y = 20$.

$L(x,y,\lambda) = P(x,y) + \lambda \cdot (x + y - 20)$.

$$\frac{\partial L}{\partial x} = -4x + 60 + \lambda = 0 \qquad (1)$$
$$\frac{\partial L}{\partial y} = -8y + 100 + \lambda = 0 \qquad (2)$$
$$\frac{\partial L}{\partial \lambda} = x + y - 20 = 0 \qquad (3)$$
$$\mathbf{(1)} - \mathbf{(2)} \implies -4x + 8y - 40 = 0 \implies x = 2y - 10 \qquad (4)$$

(3) and **(4)** give $(2y - 10) + y - 20 = 0 \implies y = 10 \implies x = 2(10) - 10 = 10$.

Since $P(20,0) = -2(400) + 1200 - 1000 = -600$

$$P(10,10) = -200 - 400 + 600 + 1000 - 1000 = 0,$$

thus the profit is a maximum subject to the given constraint when

$x = y = 10$ items.

35. a. $P(4,16) = 200\sqrt{4 \cdot 16} = 1600$.

b. $\frac{\partial P}{\partial x} = 200 \cdot \frac{y}{2\sqrt{xy}} = 100\sqrt{\frac{y}{x}}$.

When $y = 16$, the increase in $P \approx \frac{\partial P}{\partial x}(4,16) \cdot (5-4) = 100\sqrt{\frac{16}{4}} = 200$.

36. $V(r,t) = 10{,}000e^{-(0.05+r)t}$.

a. $V(0.10,2) = 10{,}000e^{-(0.05+0.1)\cdot 2} \approx \$7408.18$.

b. $V(0.10,3) = 10{,}000e^{-(0.05+0.1)\cdot 3} \approx \$6376.28$.

So the decrease between year 2 and 3 while $r = 0.10$ is

$V(0.10,2) - V(0.10,3) \approx \$7408.18 - \$6376.28 = \$1031.90$.

37. $P(x,y) = 25x + 5y - x^2 - xy$.

$$\frac{\partial P}{\partial x} = 25 - 2x - y = 0 \qquad (1)$$
$$\frac{\partial P}{\partial y} = 5 - x = 0 \qquad (2)$$

From **(2)** $\implies x = 5$.

From **(1)** $\implies y = 25 - 2x \implies y = 25 - 2(5) = 15$.

$$\frac{\partial^2 P}{\partial x^2} = -2, \quad \frac{\partial^2 P}{\partial x \partial y} = -1, \quad \frac{\partial^2 P}{\partial y^2} = 0.$$

At (5,15): $D = 1 > 0$. $(5,15) \to$ a saddle point of P.

Note that $P(2.5, 20) > P(5, 15)$. Thus there is no maximum profit.

38. Let x, y, z be the dimensions of the box where

$x =$ length, $y =$ width, $z =$ height.

Then the cost $C = 2(2xy) + 3(2yz + 2xz) = 4xy + 6(yz + xz)$,

and $xyz = 1000$.

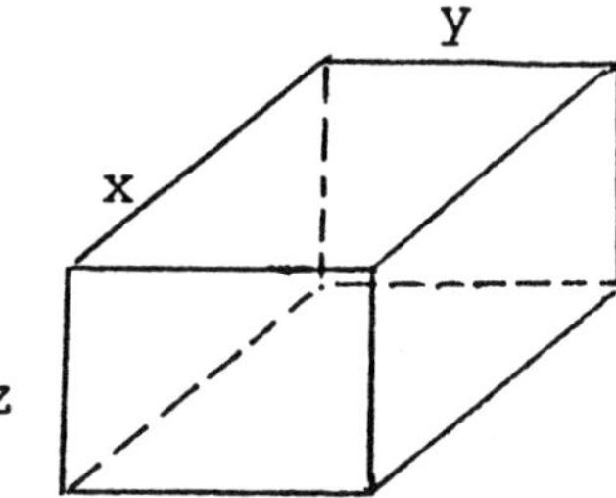

Then, $C = 4xy + 6000\left(\frac{1}{x} + \frac{1}{y}\right)$.

$$\frac{\partial C}{\partial x} = 4y - \frac{6000}{x^2} = 0 \implies yx^2 = 1500 \qquad (1)$$

$$\frac{\partial C}{\partial y} = 4x - \frac{6000}{y^2} = 0 \implies xy^2 = 1500 \qquad (2)$$

$$((\mathbf{1}) \times y) - ((\mathbf{2}) \times x) = 0 \implies 1500(y - x) = 0 \implies y = x. \qquad (3)$$

(**1**) and (**3**) give $x^3 = 1500 \implies x = \sqrt[3]{1500}$.

Then $y = \sqrt[3]{1500}$ and $z = 1000/(xy) = 1000/1500^{2/3} = \frac{\frac{2}{3}\cdot 1500}{1500^{2/3}} = \frac{2}{3} \cdot \sqrt[3]{1500}$.

$$\frac{\partial^2 C}{\partial x^2} = \frac{12000}{x^3}, \quad \frac{\partial^2 C}{\partial x \partial y} = 4, \quad \frac{\partial^2 C}{\partial y^2} = \frac{12000}{y^3}.$$

At $\left(\sqrt[3]{1500}, \sqrt[3]{1500}\right)$, we have that

$$D = 16 - 64 < 0, \quad \text{and} \quad A > 0.$$

Thus, the cost C is minimum when $x = y = \sqrt[3]{1500}$ and $z = \frac{2}{3}\sqrt[3]{1500}$.

39. $f(x,y,z) = x + 2y - 3z + 1, \qquad g(x,y,z) = x^2 + y^2 + z^2 - 14 = 0.$

$L(x,y,z,\lambda) = f(x,y,z) + \lambda g(x,y,z).$

$$\frac{\partial L}{\partial x} = 1 + 2x\lambda = 0 \tag{1}$$

$$\frac{\partial L}{\partial y} = 2 + 2y\lambda = 0 \tag{2}$$

$$\frac{\partial L}{\partial z} = -3 + 2z\lambda = 0 \tag{3}$$

$$\frac{\partial L}{\partial \lambda} = x^2 + y^2 + z^2 - 14 = 0 \tag{4}$$

From **(1)**, **(2)**, and **(3)** $\implies x = -\frac{1}{2\lambda}, \quad y = -\frac{1}{\lambda}, \quad z = \frac{3}{2\lambda}.$ **(5)**

(4) and **(5)** give $\frac{1}{4\lambda^2} + \frac{1}{\lambda^2} + \frac{9}{4\lambda^2} = 14 \implies 1 + 4 + 9 = 14 \cdot (4\lambda^2)$

$\implies \lambda^2 = \frac{1}{4} \implies \lambda = \pm\frac{1}{2}.$

For $\lambda = \frac{1}{2}$: $\quad x = -1, \quad y = -2, \quad z = 3.$

For $\lambda = -\frac{1}{2}$: $\quad x = 1, \quad y = 2, \quad z = -3.$

Since $f(-1,-2,3) = -13, \quad f(1,2,-3) = 15,$

thus the maximum value of f is $f(1,2,-3) = 15.$

40. $V(x,y,z) = 8xyz, \qquad g(x,y,z) = \frac{x^2}{4} + \frac{y^2}{9} + \frac{z^2}{4} - 1 = 0.$

$L(x,y,z,\lambda) = 8xyz + \lambda \cdot \left(\frac{x^2}{4} + \frac{y^2}{9} + \frac{z^2}{4} - 1\right).$

$$\frac{\partial L}{\partial x} = 8yz + \frac{x}{2}\lambda = 0 \tag{1}$$

$$\frac{\partial L}{\partial y} = 8xz + \frac{2y}{9}\lambda = 0 \tag{2}$$

$$\frac{\partial L}{\partial z} = 8xy + \frac{z}{2}\lambda = 0 \tag{3}$$

$$\frac{\partial L}{\partial \lambda} = \frac{x^2}{4} + \frac{y^2}{9} + \frac{z^2}{4} - 1 = 0 \tag{4}$$

$\left(\frac{2y}{9} \cdot \mathbf{(1)}\right) - \left(\frac{x}{2} \cdot \mathbf{(2)}\right) \implies \frac{2y}{9}\left(8yz + \frac{x}{2}\lambda\right) - \frac{x}{2} \cdot \left(8xz + \frac{2y}{9}\lambda\right)$

$= \frac{16y^2z}{9} + \frac{2xy\lambda}{18} - \frac{8x^2z}{2} - \frac{2xy\lambda}{18} = \frac{16y^2z}{9} - 4x^2z$

$= 4z \cdot \left(\frac{4y^2}{9} - x^2\right) = 0 \implies z = 0 \text{ or } \frac{4y^2}{9} = x^2 \implies y^2 = \frac{9}{4}x^2.$

$(z \cdot \mathbf{(1)}) - (x \cdot \mathbf{(3)}) \implies z \cdot \left(8yz + \frac{x}{2}\lambda\right) - x \cdot \left(8xy + \frac{z}{2}\lambda\right)$

$= 8yz^2 + \frac{xz\lambda}{2} - 8x^2y - \frac{xz\lambda}{2} = 8y \cdot (z^2 - x^2) = 0$

$\Longrightarrow\ y = 0$ or $z^2 = x^2$.

Substituting $y^2 = \frac{9}{4}x^2$ and $z^2 = x^2$ into (4) gives

$\frac{x^2}{4} + \frac{1}{9} \cdot \frac{9}{4}x^2 + \frac{x^2}{4} - 1 = \frac{3}{4}x^2 - 1 = 0, \quad x^2 = \frac{4}{3} \Longrightarrow x = \frac{2}{\sqrt{3}}.$

$y^2 = \frac{9}{4}\left(\frac{4}{3}\right) \Longrightarrow y = \sqrt{3}. \quad z = \frac{2}{\sqrt{3}}.$

The maximum volume $V = 8 \cdot \frac{2}{\sqrt{3}} \cdot \sqrt{3} \cdot \frac{2}{\sqrt{3}} = \frac{32}{\sqrt{3}}$.

41. We are to maximize $S = 10xy + 2x + 4y$ subject to the constraint $x + y = 10$.

Let $L(x, y, \lambda) = (10xy + 2x + 4y) + \lambda \cdot (x + y - 10)$.

$\frac{\partial L}{\partial x} = 10y + 2 + \lambda(1) = 0.$

$\frac{\partial L}{\partial y} = 10x + 4 + \lambda(1) = 0.$

$\frac{\partial L}{\partial \lambda} = x + y - 10 = 0.$

$\frac{\partial L}{\partial x} = 0 \Longrightarrow \lambda = -10y - 2.$

$\frac{\partial L}{\partial y} = 0 \Longrightarrow \lambda = -10x - 4.$

$$-10y - 2 = -10x - 4.$$

$$10x - 10y = -2.$$

$$x + y = 10.$$

$$10x - 10(10 - x) = -2 \Longrightarrow 20x = 98 \Longrightarrow x = 4.9 \Longrightarrow y = 5.1.$$

42.

j	x_j	y_j	x_j^2	x_jy_j
1	1	34	1	34
2	2	72	4	144
3	3	220	9	660
4	4	290	16	1160
5	5	345	25	1725
Σ	15	961	55	3723

Let x denote the year, and let y denote the corresponding sales in thousands.

The least squares regression line is given by

$$y = mx + b$$

where

$$m = \frac{5\cdot(3723)-(15)\cdot(961)}{5\cdot(55)-(15)^2} = \frac{4200}{50} = 84.$$
$$b = \frac{(55)\cdot(961)-(15)\cdot(3723)}{5\cdot(55)-(15)^2} = \frac{-2990}{50} = -59.80.$$

a. The least squares regression line is given by

$$y = 84x - 59.80.$$

b. The sales predicted for the year seven is

$$y = 84\cdot(7) - 59.80$$
$$y = 528.20.$$

43. The area of the region is given by

$$\int_0^1 \left\{ \int_{x^2}^{\sqrt{x}} 1\, dy \right\} dx = \int_0^1 \left\{ \sqrt{x} - x^2 \right\} dx$$
$$= \left(\frac{2}{3}x^{3/2} - \frac{1}{3}x^3 \right)\Big]_0^1 = \left(\frac{2}{3} - \frac{1}{3} \right) = \frac{1}{3}.$$

44. a. $A_1v_1 = A_2v_2 \implies v_2 = \frac{A_1}{A_2}v_1.$

b. $\frac{\partial v_2}{\partial A_2} = -\frac{A_1v_1}{A_2^2}.$

c. $\frac{\partial v_2}{\partial A_2}\Big]_{A_1=5,\ A_2=3,\ v_1=20} = -\frac{5\cdot 20}{3^2} = -\frac{100}{9}$ cm/s.

45. Revenue $R = (1200 - 40x + 20y)\cdot x + (2650 + 10x - 30y)\cdot y$

$$= -40x^2 + 30xy - 30y^2 + 1200x + 2650y.$$

Cost $C = (1200 - 40x + 20y)\cdot 65 + (2650 + 10x - 30y)\cdot 80$

$$= 78000 - 2600x + 1300y + 212000 + 800x - 2400y$$
$$= -1800x - 1100y + 290,000.$$

Profit $P = R - C$

$$= -40x^2 + 30xy - 30y^2 + 3000x + 3750y - 290,000.$$

$\frac{\partial P}{\partial x} = -80x + 30y + 3000.$

$\frac{\partial P}{\partial y} = 30x - 60y + 3750.$

$\frac{\partial P}{\partial x} = 0$ and $\frac{\partial P}{\partial y} = 0$:

$$\begin{cases} -80x + 30y + 3000 = 0 \\ 30x - 60y + 3750 = 0 \end{cases}$$

$$\begin{cases} -160x + 60y + 6000 = 0 \\ 30x - 60y + 3750 = 0 \end{cases}$$

$$-130x + 9750 = 0 \implies x = 75.$$

$$-80 \cdot 75 + 30y + 3000 = 0 \implies 30y = 3000 \implies y = 100.$$

$\frac{\partial^2 P}{\partial x^2} = -80, \quad \frac{\partial^2 P}{\partial y^2} = -60, \quad \frac{\partial P}{\partial x \partial y} = 30.$

$D = (30^2) - (-80)(-60) = -3900 < 0.$

$A = -80 < 0.$

Therefore the maximum profit P occurs when $x = 75$, $y = 100$.

CHAPTER 9

Solutions to Exercise Set 9.1

1. $\{2n+3\}$.

   $a_1 = 2(1)+3 = 5, \quad a_2 = 2(2)+3 = 7, \quad a_3 = 2(3)+3 = 9,$

   $a_4 = 2(4)+3 = 11, \quad a_5 = 2(5)+3 = 13.$

2. $\left\{\frac{n}{2n+1}\right\}$.

   $a_1 = \frac{1}{2\cdot1+1} = \frac{1}{3}, \quad a_2 = \frac{2}{2\cdot2+1} = \frac{2}{5}, \quad a_3 = \frac{3}{2\cdot3+1} = \frac{3}{7},$

   $a_4 = \frac{4}{2\cdot4+1} = \frac{4}{9}, \quad a_5 = \frac{5}{2\cdot5+1} = \frac{5}{11}.$

3. $\left\{\frac{2n+1}{n+5}\right\}$.

   $a_1 = \frac{2\cdot1+1}{1+5} = \frac{1}{2}, \quad a_2 = \frac{2\cdot2+1}{2+5} = \frac{5}{7}, \quad a_3 = \frac{2\cdot3+1}{3+5} = \frac{7}{8},$

   $a_4 = \frac{2\cdot4+1}{4+5} = 1, \quad a_5 = \frac{2\cdot5+1}{5+5} = \frac{11}{10}.$

4. $\left\{\frac{(-1)^n}{1+n^2}\right\}$.

   $a_1 = \frac{(-1)}{1+1^2} = -\frac{1}{2}, \quad a_2 = \frac{(-1)^2}{1+2^2} = \frac{1}{5}, \quad a_3 = \frac{(-1)^3}{1+3^2} = -\frac{1}{10}.$

   $a_4 = \frac{(-1)^4}{1+4^2} = \frac{1}{17}, \quad a_5 = \frac{(-1)^5}{1+5^2} = -\frac{1}{26}.$

5. $\left\{\frac{(-1)^n}{e^n}\right\}$.

   $a_1 = \frac{(-1)}{e^1} = -\frac{1}{e}, \quad a_2 = \frac{(-1)^2}{e^2} = \frac{1}{e^2}, \quad a_3 = \frac{(-1)^3}{e^3} = -\frac{1}{e^3},$

   $a_4 = \frac{(-1)^4}{e^4} = \frac{1}{e^4}, \quad a_5 = \frac{(-1)^5}{e^5} = -\frac{1}{e^5}.$

6. $\left\{\frac{20n}{1+n^3}\right\}$.

$a_1 = \frac{20\cdot 1}{1+1^3} = 10, \quad a_2 = \frac{20\cdot 2}{1+2^3} = \frac{40}{9}, \quad a_3 = \frac{20\cdot 3}{1+3^3} = \frac{15}{7},$

$a_4 = \frac{20\cdot 4}{1+4^3} = \frac{80}{65} = \frac{16}{13}, \quad a_5 = \frac{20\cdot 5}{1+5^3} = \frac{100}{126} = \frac{50}{63}.$

7. $\left\{\frac{(n-1)(n+1)}{2n^2+2n+2}\right\} = \left\{\frac{n^2-1}{2(n^2+n+1)}\right\}$.

$a_1 = \frac{1^2-1}{2(1^2+1+1)} = 0, \quad a_2 = \frac{2^2-1}{2(2^2+2+1)} = \frac{3}{14}, \quad a_3 = \frac{3^2-1}{2(3^2+3+1)} = \frac{8}{26} = \frac{4}{13},$

$a_4 = \frac{4^2-1}{2(4^2+4+1)} = \frac{15}{42} = \frac{5}{14}, \quad a_5 = \frac{5^2-1}{2(5^2+5+1)} = \frac{24}{62} = \frac{12}{31}.$

8. $\{\ln(n+2)\}$.

$a_1 = \ln(1+2) = \ln 3, \quad a_2 = \ln(2+2) = \ln 4, \quad a_3 = \ln(3+2) = \ln 5,$

$a_4 = \ln(4+2) = \ln 6, \quad a_5 = \ln(5+2) = \ln 7.$

9. $\left\{(-1)^{n+1}\sin\left(\frac{n\pi}{4}\right)\right\}$.

$a_1 = (-1)^{1+1}\sin\left(\frac{1\cdot\pi}{4}\right) = \frac{\sqrt{2}}{2}, \quad a_2 = (-1)^{2+1}\sin\left(\frac{2\cdot\pi}{4}\right) = -1,$

$a_3 = (-1)^{3+1}\sin\left(\frac{3\cdot\pi}{4}\right) = \frac{\sqrt{2}}{2},$

$a_4 = (-1)^{4+1}\sin\left(\frac{4\cdot\pi}{4}\right) = 0,$

$a_5 = (-1)^{5+1}\sin\left(\frac{5\cdot\pi}{4}\right) = -\frac{\sqrt{2}}{2}.$

10. $\left\{\frac{1+n^2}{5+n^3}\right\}$.

$a_1 = \frac{1+1^1}{5+1^3} = \frac{1}{3}, \quad a_2 = \frac{1+2^2}{5+2^3} = \frac{5}{13}, \quad a_3 = \frac{1+3^2}{5+3^3} = \frac{10}{32} = \frac{5}{16},$

$a_4 = \frac{1+4^2}{5+4^3} = \frac{17}{69}, \quad a_5 = \frac{1+5^2}{5+5^3} = \frac{26}{130} = \frac{13}{65} = \frac{1}{5}.$

11. $\left\{\frac{n}{3n+6}\right\}$.

$\lim_{n\to\infty}\frac{n}{3n+6} = \lim_{n\to\infty}\frac{\frac{n}{n}}{\frac{3n}{n}+\frac{6}{n}} = \lim_{n\to\infty}\frac{1}{3+\frac{6}{n}} = \frac{1}{3}.$

12. $\left\{\frac{2n-\sqrt{n}}{n+3}\right\}$.

$$\lim_{n\to\infty}\frac{2n-\sqrt{n}}{n+3} = \lim_{n\to\infty}\frac{2n-\sqrt{n}}{n+3}\cdot\frac{\frac{1}{n}}{\frac{1}{n}}.$$
$$= \lim_{n\to\infty}\frac{2-\frac{1}{\sqrt{n}}}{1+\frac{3}{n}} = \frac{2-0}{1+0} = 2.$$

13. $\{\sqrt{5+n}\}$.

$$\lim_{n\to\infty}\sqrt{5+n} = \infty \implies \text{the sequence diverges.}$$

14. $\left\{\frac{n-1}{n^3+3}\right\}$.

$$\lim_{n\to\infty}\frac{n-1}{n^3+3} = \lim_{n\to\infty}\frac{\frac{n}{n^3}-\frac{1}{n^3}}{\frac{n^3}{n^3}+\frac{3}{n^3}} = \lim_{n\to\infty}\frac{\frac{1}{n^2}-\frac{1}{n^3}}{1+\frac{3}{n^3}} = 0.$$

15. $\left\{\frac{e^n+3}{e^n}\right\}$.

$$\lim_{n\to\infty}\frac{e^n+3}{e^n} = \lim_{n\to\infty}\left(1+\frac{3}{e^n}\right) = 1 \quad \text{since } \lim_{n\to\infty}\frac{3}{e^n} = 0.$$

16. $\left\{\frac{2n-n^2}{2n^2+1}\right\}$.

$$\lim_{n\to\infty}\frac{2n-n^2}{2n^2+1}\cdot\frac{\frac{1}{n^2}}{\frac{1}{n^2}} = \lim_{n\to\infty}\frac{\frac{2}{n}-1}{2+\frac{1}{n^2}} = \frac{0-1}{2+0} = -2.$$

17. $\left\{\frac{n^2-2n+1}{3+n^2}\right\}$.

$$\lim_{n\to\infty}\frac{n^2-2n+1}{3+n^2} = \lim_{n\to\infty}\frac{\frac{(n^2-2n+1)}{n^2}}{\frac{(3+n^2)}{n^2}} = \lim_{n\to\infty}\frac{1-\frac{2}{n}+\frac{1}{n^2}}{\frac{3}{n^2}+1} = 1.$$

18. $\left\{\frac{n^2+3n+5}{n^5+2}\right\}$.

$$\lim_{n\to\infty}\frac{n^2+3n+5}{n^5+2} = \lim_{n\to\infty}\frac{\frac{1}{n^3}+\frac{3}{n^4}+\frac{5}{n^5}}{1+\frac{2}{n^5}} = 0.$$

19. $\{\ln(n+1)-\ln(n)\}$.

$$\lim_{n\to\infty}[\ln(n+1)-\ln(n)] = \lim_{n\to\infty}\ln\left(\frac{n+1}{n}\right) = \lim_{n\to\infty}\ln\left(1+\frac{1}{n}\right) = \ln 1 = 0.$$

20. $\lim\limits_{n\to\infty} e^{\frac{n-1}{n}} = \lim\limits_{n\to\infty} e^{\left(1-\frac{1}{n}\right)} = e^1 = e.$

21. Since $\sin n\pi = 0$ for all $n \ge 1$, $\lim\limits_{n\to\infty}\sin n\pi = 0$.

22. $\left\{\frac{(-1)^n n^2}{9-n^2}\right\}$.

For even n, $\lim_{n\to\infty}(-1)^n\frac{n^2}{9-n^2} = \lim_{n\to\infty} 1\cdot\frac{n^2}{9-n^2} = -1$.

For odd n, $\lim_{n\to\infty}(-1)^n\frac{n^2}{9-n^2} = \lim_{n\to\infty}(-1)\cdot\frac{n^2}{9-n^2} = 1$.

Thus the sequence diverges.

23. $\lim_{n\to\infty}\cos\frac{n-1}{n^2} = \lim_{n\to\infty}\cos\frac{\frac{1}{n}-\frac{1}{n^2}}{1} = \cos 0 = 1$.

24. $\lim_{n\to\infty}\frac{e^n-e^{-n}}{e^n+e^{-n}} = \lim_{n\to\infty}\frac{\frac{(e^n-e^{-n})}{e^n}}{\frac{(e^n+e^{-n})}{e^n}} = \lim_{n\to\infty}\frac{1-e^{-2n}}{1+e^{-2n}} = 1$

since $\lim_{n\to\infty} e^{-2n} = 0$.

25. $\lim_{n\to\infty}\left\{\frac{1}{n}-\frac{1}{n+1}\right\} = \lim_{n\to\infty}\frac{(n+1)-n}{n(n+1)} = \lim_{n\to\infty}\frac{1}{n(n+1)} = 0$.

26. $\lim_{n\to\infty}\frac{2^n}{3^n+4^n} = \lim_{n\to\infty}\frac{\frac{2^n}{4^n}}{\frac{(3^n+4^n)}{4^n}} = \lim_{n\to\infty}\frac{\left(\frac{1}{2}\right)^n}{\left(\frac{3}{4}\right)^n+1} = 0$.

27. $\lim_{n\to\infty}\left(1+\frac{1}{n}\right)^n = e$ as developed in Chapter 4.

28. $\lim_{n\to\infty}\left(1-\frac{1}{n}\right)^n = \lim_{n\to\infty}\frac{1}{\left(1+\frac{1}{-n}\right)^{-n}} = \frac{1}{e}$.

29. $\lim_{n\to\infty}\frac{\sqrt{n}}{1+\sqrt{n}} = \lim_{n\to\infty}\frac{\frac{\sqrt{n}}{\sqrt{n}}}{\frac{(1+\sqrt{n})}{\sqrt{n}}} = \lim_{n\to\infty}\frac{1}{\frac{1}{\sqrt{n}}+1} = 1$.

30. $\lim_{n\to\infty}\ln\left(\frac{n^2+6}{9+n^2}\right) = \lim_{n\to\infty}\ln\left(\frac{1+\frac{6}{n^2}}{\frac{9}{n^2}+1}\right) = \ln 1 = 0$.

31. $\left\{\cos\left(\frac{n\pi}{4}\right)\right\}$.

The terms of this sequence oscillate as $n\to\infty$.

Therefore this sequence diverges.

32. $\left\{\sin\left(\frac{(n+1)\cdot\pi}{n^2}\right)\right\}$.

$\lim_{n\to\infty}\frac{(n+1)\cdot\pi}{n^2} = 0 \Longrightarrow \lim_{n\to\infty}\sin\frac{(n+1)\cdot\pi}{n^2} = 0$.

33. $N(n) = \frac{A}{B+Ce^{-n}}$.

$\lim_{n\to\infty} N(n) = \frac{A}{B}$ since $\lim_{n\to\infty} e^{-n} = 0$.

34. $N(n) = \frac{A}{B+Ce^n}$.

$\lim_{n\to\infty} N(n) = 0$ since $\lim_{n\to\infty} e^n = \infty$.

35. $a_n = 1000 \cdot (1.08)^n$.

a. $a_1 = 1000 \cdot 1.08 = \$1080$.

b. $a_2 = 1000 \cdot (1.08)^2 = \$1166.40$.

c. $a_5 = 1000 \cdot (1.08)^5 = \$1469.33$.

36. a. $a_0 = 1000 \cdot e^{-.05 \cdot 7} = \$704.69$.

b. $a_1 = 1000 \cdot e^{-.05 \cdot 6} = \$740.82$.

c. $a_5 = 1000 \cdot e^{-.05 \cdot 2} = \$904.84$.

37. Assume that the fleet average fuel consumption in miles per gallon increases by 10 percent at the end of each year.

a. $a_1 = 22 + 22 \cdot .10 = 22 \cdot (1.10) = 24.2$.

b. $a_2 = 22 \cdot (1.10) + 22 \cdot (1.10) \cdot .10 = 22 \cdot (1.10) \cdot (1.10) = 22 \cdot (1.10)^2 = 26.62$.

c. $a_n = 22 \cdot (1.10)^n$.

38. We treat the problem as if the population is increasing continuously at the rate of 6 percent per year. Then this problem is analogous to continuous compounding of interest.

a. $P_1 = 3 \cdot e^{.06 \cdot 1} = 3.1855$ million.

b. $P_2 = 3 \cdot e^{.06 \cdot 2} = 3.3825$ million.

c. $P_n = 3 \cdot e^{.06 \cdot n}$.

39. $a_1 = 50.$

$a_2 = 50 \cdot \frac{1}{2} + 50 = 50 \cdot \left(1 + \frac{1}{2}\right).$

$a_3 = 50 \cdot \left(1 + \frac{1}{2}\right) \cdot \frac{1}{2} + 50 = 50 \cdot \left(1 + \frac{1}{2} + \frac{1}{4}\right).$

$a_4 = 50 \cdot \left(1 + \frac{1}{2} + \frac{1}{4}\right) \cdot \frac{1}{2} + 50 = 50 \cdot \left(1 + \frac{1}{2} + \frac{1}{4} + \frac{1}{8}\right).$

etc.

$a_n = 50 \cdot \left(1 + \frac{1}{2} + \frac{1}{4} + \frac{1}{8} + \cdots + \frac{1}{2^{n-1}}\right).$

40. $a_1 = 10 \cdot (1.01).$

$a_2 = (10 \cdot (1.01) + 10) \cdot (1.01) = 10 \cdot ((1.01) + (1.01)^2).$

$a_3 = \left[10 \cdot ((1.01) + (1.01)^2) + 10\right](1.01) = 10 \cdot \left((1.01) + (1.01)^2 + (1.01)^3\right).$

etc.

$a_n = 10 \cdot \left((1.01) + (1.01)^2 + (1.01)^3 \cdots + (1.01)^n\right).$

41. $P(n) = 1 - \left(\frac{100-10}{100}\right)^n = 1 - \left(\frac{9}{10}\right)^n$ or $100 \cdot \left[1 - \left(\frac{9}{10}\right)^n\right]$ percent.

$P(1) = 1 - \frac{9}{10} = .1$ or 10 percent.

$P(2) = 1 - \left(\frac{9}{10}\right)^2 = 1 - \frac{81}{100} = .19$ or 19 percent.

$P(3) = 1 - \left(\frac{9}{10}\right)^3 = 1 - \frac{729}{1000} = .271$ or 27.1 percent.

42. Let a_n be the number of pairs of rabbits at the beginning of the nth month.
At the beginning of the 1st month, we have 1 pair of rabbits, so $a_1 = 1$.
At the beginning of the 2nd month, we still have 1 pair of rabbits, so $a_2 = 1$.
At the beginning of the 3rd month, this original pair of rabbits gives birth to a pair of rabbits, so we then have 2 pairs of rabbits. Thus $a_3 = 2 = a_1 + a_2$.
At the beginning of the 4th month, the original pair of rabbits again gives birth to a pair of rabbits, but the pair born at the beginning of the 3rd month is now just becoming fertile, so we then have 3 pairs of rabbits. Thus $a_4 = 3 = a_2 + a_3$.
At the beginning of the 5th month, the pairs born at the beginning of the 1st month and the 3rd month each give birth to a pair of rabbits, but the pair born at the beginning of the 4th month is now just becoming fertile. So we then have 5 pairs of rabbits. Thus $a_5 = 5 = a_3 + a_4$. This process continues.

Solutions to Exercise Set 9.2

1. $\sum_{k=1}^{\infty}(2k+3)$.

   $a_1 = 2(1)+3 = 5, \quad a_2 = 2(2)+3 = 7, \quad a_3 = 2(3)+3 = 9, \quad a_4 = 2(4)+3 = 11,$

   $S_4 = \sum_{k=1}^{4} a_k = 5+7+9+11.$

2. $\sum_{k=1}^{\infty} \sqrt{1+k^2}$.

   $S_4 = \sum_{k=1}^{4} \sqrt{1+k^2} = \sqrt{1+1^2}+\sqrt{1+2^2}+\sqrt{1+3^2}+\sqrt{1+4^2}$

   $= \sqrt{2}+\sqrt{5}+\sqrt{10}+\sqrt{17}.$

3. $\sum_{k=0}^{\infty} \frac{1}{5^k}$.

   $S_4 = \frac{1}{5^0}+\frac{1}{5^1}+\frac{1}{5^2}+\frac{1}{5^3} = 1+\frac{1}{5}+\frac{1}{25}+\frac{1}{125}.$

4. $\sum_{k=0}^{\infty} \frac{1+3^k}{5^k}$.

   $S_4 = \sum_{k=0}^{3} \frac{1+3^k}{5^k} = \frac{1+3^0}{5^0}+\frac{1+3^1}{5^1}+\frac{1+3^2}{5^2}+\frac{1+3^3}{5^3} = 2+\frac{4}{5}+\frac{2}{5}+\frac{28}{125}.$

5. $\sum_{k=1}^{\infty} \frac{(-1)^k}{1+k^2}$.

   $S_4 = \sum_{k=1}^{4} \frac{(-1)^k}{1+k^2} = \frac{(-1)^1}{1+1^2}+\frac{(-1)^2}{1+2^2}+\frac{(-1)^3}{1+3^2}+\frac{(-1)^4}{1+4^2} = -\frac{1}{2}+\frac{1}{5}-\frac{1}{10}+\frac{1}{17}.$

6. $\sum_{k=0}^{\infty} \frac{2^k}{7^k}$.

   $S_4 = \frac{2^0}{7^0}+\frac{2^1}{7^1}+\frac{2^2}{7^2}+\frac{2^3}{7^3} = 1+\frac{2}{7}+\frac{4}{49}+\frac{8}{343}.$

7. $\sum_{k=1}^{\infty} \frac{\cos(\pi k)}{1+k}$.

   $S_4 = \sum_{k=1}^{4} \frac{\cos(\pi k)}{1+k} = \frac{\cos(\pi\cdot 1)}{1+1}+\frac{\cos(\pi\cdot 2)}{1+2}+\frac{\cos(\pi\cdot 3)}{1+3}+\frac{\cos(\pi\cdot 4)}{1+4}$

   $= -\frac{1}{2}+\frac{1}{3}-\frac{1}{4}+\frac{1}{5}.$

8. $\sum_{k=1}^{\infty} \sqrt{k}^k$.

   $S_4 = \sqrt{1}^1+\sqrt{2}^2+\sqrt{3}^3+\sqrt{4}^4 = 1+2+3^{3/2}+16.$

9. $\sum\limits_{k=0}^{\infty} \frac{1}{4^k} = \sum\limits_{k=0}^{\infty} \left(\frac{1}{4}\right)^k = \frac{1}{1-\frac{1}{4}} = \frac{4}{3}$ since $\left|\frac{1}{4}\right| < 1$.

10. $\sum\limits_{k=0}^{\infty} \frac{3^k}{2^k} = \sum\limits_{k=0}^{\infty} \left(\frac{3}{2}\right)^k$ diverges since $\frac{3}{2} > 1$.

11. $\sum\limits_{k=0}^{\infty} \frac{3^k}{7^k} = \sum\limits_{k=0}^{\infty} \left(\frac{3}{7}\right)^k = \frac{1}{1-\frac{3}{7}} = \frac{7}{4}$ since $\left|\frac{3}{7}\right| < 1$.

12. $\sum\limits_{k=0}^{\infty} \left(\frac{\sqrt{2}}{2}\right)^k$.

Since $\left|\frac{\sqrt{2}}{2}\right| < 1$ this geometric series converges and has sum $\frac{1}{1-\frac{\sqrt{2}}{2}}$.

13. $\sum\limits_{k=1}^{\infty} \frac{1}{e^{-k}} = \sum\limits_{k=1}^{\infty} e^k$.

This is a divergent geometric series since $e > 1$.

14. $\sum\limits_{k=2}^{\infty} \frac{1}{2^k} = \sum\limits_{k=0}^{\infty} \left(\frac{1}{2}\right)^k - \left(\frac{1}{2}\right)^0 - \left(\frac{1}{2}\right)^1 = \frac{1}{1-\frac{1}{2}} - 1 - \frac{1}{2} = 2 - 1 - \frac{1}{2} = \frac{1}{2}$.

15. $\sum\limits_{k=1}^{\infty} \frac{7^k}{5^k} = \sum\limits_{k=0}^{\infty} \left(\frac{7}{5}\right)^k - 1$ diverges since $\frac{7}{5} > 1$.

16. $\sum\limits_{k=2}^{\infty} \frac{9^k}{7^k} = \sum\limits_{k=0}^{\infty} \left(\frac{9}{7}\right)^k - 1 - \frac{9}{7}$ diverges since $\frac{9}{7} > 1$.

17. $\sum\limits_{k=0}^{\infty} \frac{1+3^k}{8^k} = \sum\limits_{k=0}^{\infty} \left(\frac{1}{8}\right)^k + \sum\limits_{k=0}^{\infty} \left(\frac{3}{8}\right)^k = \frac{1}{1-\frac{1}{8}} + \frac{1}{1-\frac{3}{8}} = \frac{1}{7/8} + \frac{1}{5/8}$

$= \frac{8}{7} + \frac{8}{5} = \frac{40}{35} + \frac{56}{35} = \frac{96}{35}$.

18. $\sum\limits_{k=0}^{\infty} \frac{2 \cdot 3^k}{2^k} = 2 \cdot \sum\limits_{k=0}^{\infty} \left(\frac{3}{2}\right)^k$.

The infinite series $\sum\limits_{k=0}^{\infty} \left(\frac{3}{2}\right)^k$ is a divergent geometric series.

Therefore the given infinite series diverges.

19. $\sum\limits_{k=1}^{\infty} \frac{2^k - 3^k}{4^k} = \sum\limits_{k=0}^{\infty} \left[\left(\frac{2}{4}\right)^k - \left(\frac{3}{4}\right)^k\right] - \frac{2^0 - 3^0}{4^0} = \sum\limits_{k=0}^{\infty} \left(\left(\frac{1}{2}\right)^k + (-1)\left(\frac{3}{4}\right)^k\right)$

$= \sum\limits_{k=0}^{\infty} \left(\frac{1}{2}\right)^k + \sum\limits_{k=0}^{\infty} (-1)\left(\frac{3}{4}\right)^k = \sum\limits_{k=0}^{\infty} \left(\frac{1}{2}\right)^k + (-1)\sum\limits_{k=0}^{\infty} \left(\frac{3}{4}\right)^k = \frac{1}{1-\frac{1}{2}} - \frac{1}{1-\frac{3}{4}} = -2$.

20. $\sum\limits_{k=0}^{\infty} \frac{2^k + 5^k}{4^k} = \sum\limits_{k=0}^{\infty} \left(\left(\frac{1}{2}\right)^k + \left(\frac{5}{4}\right)^k\right)$ diverges since $\sum\limits_{k=0}^{\infty} \left(\frac{5}{4}\right)^k$ diverges.

21. $\sum_{k=1}^{\infty} \frac{5\cdot 3^k - 3\cdot 7^k}{5^k} = \sum_{k=0}^{\infty}\left[5\left(\frac{3}{5}\right)^k - 3\cdot\left(\frac{7}{5}\right)^k\right] - \frac{5\cdot 3^0 - 3\cdot 7^0}{5^0}$

The infinite series $\sum_{k=0}^{\infty}\left(\frac{3}{5}\right)^k$ is a convergent geometric series, but the infinite series $\sum_{k=0}^{\infty}\left(\frac{7}{5}\right)^k$ is a divergent geometric series. Therefore the given infinite series diverges.

22. $\sum_{k=0}^{\infty}\frac{7\cdot 2^{k+1} - 3\cdot 4^{k+1}}{5^{k+1}} = \sum_{k=0}^{\infty}\frac{7\cdot 2\cdot 2^k - 3\cdot 4\cdot 4^k}{5\cdot 5^k} = \sum_{k=0}^{\infty}\left[\frac{14}{5}\left(\frac{2}{5}\right)^k - \frac{12}{5}\left(\frac{4}{5}\right)^k\right]$

$= \frac{14}{5}\sum_{k=0}^{\infty}\left(\frac{2}{5}\right)^k - \frac{12}{5}\sum_{k=0}^{\infty}\left(\frac{4}{5}\right)^k = \frac{14}{5}\cdot\frac{1}{1-\frac{2}{5}} - \frac{12}{5}\cdot\frac{1}{1-\frac{4}{5}} = -\frac{22}{3}.$

23. $0.333\bar{3} = 3\sum_{k=1}^{\infty}\left(\frac{1}{10}\right)^k = 3\left[\sum_{k=0}^{\infty}\left(\frac{1}{10}\right)^k - \left(\frac{1}{10}\right)^0\right]$

$= 3\left(\frac{1}{1-\frac{1}{10}} - 1\right) = 3\left(\frac{10}{9} - 1\right) = 3\cdot\frac{1}{9} = \frac{1}{3}.$

24. $0.222\bar{2} = 2\sum_{k=1}^{\infty}\left(\frac{1}{10}\right)^k = \frac{2}{10}\sum_{k=0}^{\infty}\left(\frac{1}{10}\right)^k = \frac{2}{10}\cdot\frac{1}{1-\frac{1}{10}} = \frac{2}{10}\cdot\frac{10}{9} = \frac{2}{9}.$

25. a. Total amount in the first five years

$= 1000\left(1 + \frac{1}{2} + \frac{1}{2^2} + \frac{1}{2^3} + \frac{1}{2^4}\right) = 1000\left(\frac{1-\frac{1}{2^5}}{1-\frac{1}{2}}\right) = 1000\cdot\frac{31}{16} = \$1937.50.$

b. Total amount received in perpetuity $= 1000\left(\frac{1}{1-\frac{1}{2}}\right) = \$2000.$

26. a. $h + 2\cdot\frac{2}{3}h + 2\cdot\left(\frac{2}{3}\right)^2 h + 2\cdot\left(\frac{2}{3}\right)^3 h + \ldots.$

b. $h + \frac{2\cdot\frac{2}{3}h}{1-\frac{2}{3}} = h + 4h = 5h.$

27. $1 + 0.75 + (0.75)^2 + (0.75)^3 + \cdots = \frac{1}{1-0.75} = 4.$ Total is \$4 million.

28. $500 + (0.6)\cdot 500 + (0.6)^2\cdot 500 + (0.6)^3\cdot 500 + \cdots = \frac{500}{1-0.6} = 500\cdot\frac{5}{2} = 1250.$

29. a. $S(n) = 1000\cdot e^{-r} + 1000\cdot e^{-2r} + \cdots + 1000\cdot e^{-nr} = \sum_{k=1}^{n} 1000\cdot e^{-kr}.$

b. $S(n) = \frac{1000\cdot e^{-r}(1-e^{-nr})}{1-e^{-r}}.$

c. $S = \lim_{n\to\infty} S(n) = \sum_{k=1}^{\infty} 1000\cdot e^{-rk}.$

d. $e^{-r} < 1$ for $r > 0$, so the infinite series in (c) converges.

e. $S = \frac{1000 \cdot e^{-r}}{1-e^{-r}} = \frac{1000}{e^r - 1}$.

30. For $x = 1$, $\sum_{k=1}^{\infty} x^k = \sum_{k=1}^{\infty} 1 = \lim_{n\to\infty} \sum_{k=1}^{n} 1 = \lim_{n\to\infty} n = \infty$.

31. $20 + (0.75) \cdot 20 + (0.75)^2 \cdot 20 + \cdots = \frac{20}{1-0.75} = \$80$.

32. a. $\sum_{k=1}^{\infty} x^k = x \cdot \sum_{k=1}^{\infty} x^{k-1} = x \cdot (1 + x + x^2 + \ldots) = x \cdot \frac{1}{1-x} = \frac{x}{1-x}$.

b. $\sum_{k=2}^{\infty} x^k = x^2 \cdot \sum_{k=2}^{\infty} x^{k-2} = x^2 \cdot (1 + x + x^2 + \ldots) = x^2 \frac{1}{1-x} = \frac{x^2}{1-x}$.

33. Let a_n denote the number of people who have heard the rumor by the end of the nth day.

$a_1 = 3$.

$a_2 = 3 + 3 \cdot 2 = 9$.

$a_3 = 9 + 9 \cdot 2 = 27$.

$a_4 = 27 + 27 \cdot 2 = 81$.

In general, $a_n = 3^n$ which is analogous to the compound interest formula.

Note that $a_n \to \infty$ as $n \to \infty$.

34. Let a_n denote the number of people who have heard the rumor by the end of the nth day.

$a_1 = 1 + 2$.

$a_2 = 1 + 2 + 2 \cdot 2 = 1 + 2 + 2^2$.

$a_3 = 1 + 2 + 2^2 + 2^2 \cdot 2 = 1 + 2 + 2^2 + 2^3$.

In general, $a_n = 1 + 2 + 2^2 + 2^3 + \cdots + 2^n$.

a. $a_4 = 1 + 2 + 2^2 + 2^3 + 2^4 = 1 + 2 + 4 + 8 + 16 = 31$.

b. The series of rumor-knowers is a divergent geometric series.

35. Let a_0 be the initial amount of the pollutant.

Let a_n be the amount of the pollutant that remains after n days.

$a_1 = .40 \cdot a_0$.

$a_2 = .40 \cdot a_0 - (.40 \cdot a_0) \cdot (.60) = .16a_0$

$a_3 = .16a_0 - (.16a_0) \cdot (.60) = .064a_0$.

$a_4 = .064a_0 - (.064a_0) \cdot (.60) = .0256a_0$.

In general, $a_n = .4^n a_0$.

$\lim_{n\to\infty} a_n = 0$.

Solutions to Exercise Set 9.3

1. $\int_1^\infty \frac{1}{x+2}\,dx = \lim_{t\to\infty}\int_1^t \frac{1}{x+2}\,dx = \lim_{t\to\infty} \ln(x+2)\Big]_1^t$
$= \lim_{t\to\infty}[\ln(t+2)-\ln 3] = \infty$. So $\sum_{k=1}^\infty \frac{1}{k+2}$ diverges.

2. $\int_1^\infty \frac{1}{2x+5}\,dx = \lim_{t\to\infty}\int_1^t \frac{1}{2x+5}\,dx = \lim_{t\to\infty}\frac{1}{2}\ln(2x+5)\Big]_1^t$
$= \lim_{t\to\infty}\left[\frac{1}{2}\ln(2t+5)-\frac{1}{2}\ln(7)\right] = \infty$. So $\sum \frac{1}{2k+5}$ diverges.

3. Put $f(x) = \frac{x}{(x^2+1)^{3/2}}$. $f'(x) = \frac{(x^2+1)^{3/2} - x\cdot\frac{3}{2}(x^2+1)^{1/2}2x}{(x^2+1)^3} = \frac{1-2x^2}{(x^2+1)^{5/2}} < 0$
for $x \ge 1$. Thus $f(x)$ is decreasing.
$\int_1^\infty \frac{x}{(x^2+1)^{3/2}}\,dx = \lim_{t\to\infty}\int_1^t \frac{x}{(x^2+1)^{3/2}}\,dx = \lim_{t\to\infty}\int_1^t \frac{1}{(x^2+1)^{3/2}}\cdot\frac{1}{2}\,d(x^2+1)$
$= \lim_{t\to\infty}\frac{1}{2}(-2)(x^2+1)^{-1/2}\Big]_1^t = \lim_{t\to\infty}\left[-(t^2+1)^{-1/2}+2^{-1/2}\right] = \frac{1}{\sqrt{2}}$.
So $\sum \frac{k}{(k^2+1)^{3/2}}$ converges.

4. $\int_1^\infty \frac{x}{x^2+5}\,dx = \lim_{t\to\infty}\int_1^t \frac{x}{x^2+5}\,dx = \lim_{t\to\infty}\frac{1}{2}\ln(x^2+5)\Big]_1^t$
$= \lim_{t\to\infty}\frac{1}{2}\left[\ln(t^2+5)-\frac{1}{2}\ln(6)\right] = \infty$. So $\sum \frac{k}{k^2+5}$ diverges.

5. $\int_2^\infty \frac{1}{x\ln x}\,dx = \lim_{t\to\infty}\int_2^t \frac{1}{x\ln x}\,dx = \lim_{t\to\infty}\int_2^t \frac{1}{\ln x}\,d(\ln x)$
$= \lim_{t\to\infty}(\ln\ln x)\Big]_2^t = \lim_{t\to\infty}(\ln\ln t - \ln\ln 2) = \infty$. So $\sum \frac{1}{k\ln k}$ diverges.

6. Put $f(x) = xe^{-x^2}$.
$f'(x) = e^{-x^2} + xe^{-x^2}(-2x) = e^{-x^2}(1-2x^2) < 0$ for $x \ge 1$.
Thus $f(x)$ is decreasing.
$\int_1^\infty f(x)dx = \lim_{t\to\infty}\int_1^t xe^{-x^2}\,dx = \lim_{t\to\infty}\int_1^t e^{-x^2}\cdot\frac{1}{2}\,d(x^2)$
$= \lim_{t\to\infty}\left(-\frac{1}{2}e^{-x^2}\right)\Big]_1^t = \lim_{t\to\infty}\left(-\frac{1}{2}e^{-t^2}+\frac{1}{2}e^{-1}\right) = \frac{1}{2e}$.
So $\sum ke^{-k^2}$ converges.

7. Put $f(x) = x^2 e^{-x^3}$.
$f'(x) = 2xe^{-x^3} + x^2 e^{-x^3}(-3x^2) = xe^{-x^3}(2 - 3x^3) < 0$ for $x \geq 1$.
Thus $f(x)$ is decreasing.
$\int_1^\infty f(x)\,dx = \lim_{t\to\infty} \int_1^t x^2 e^{-x^3}\,dx = \lim_{t\to\infty} \int_1^t e^{-x^3}\left(\frac{1}{3}\right) d(x^3)$
$= \lim_{t\to\infty} \left(-\frac{1}{3}e^{-x^3}\right)\Big]_1^t = \lim_{t\to\infty}\left(-\frac{1}{3}e^{-t^3} + \frac{1}{3}e^{-1}\right) = \frac{1}{3e}$.
So $\sum k^2 e^{-k^3}$ converges.

8. Put $f(x) = xe^x$. $f'(x) = e^x + xe^x > 0$ for $x > 0$.
Since $f(x)$ is increasing for $x > 0$, we cannot apply the integral test to this problem.

9. Since $\frac{3}{k} > \frac{1}{k}$ and $\sum \frac{1}{k}$ diverges, $\sum \frac{3}{k}$ diverges.

10. Since $\frac{1}{1+k^2} < \frac{1}{k^2}$ and $\sum \frac{1}{k^2}$ converges, $\sum \frac{1}{1+k^2}$ converges, and hence $\sum \frac{3}{1+k^2}$ converges.

11. Since $\frac{1}{2+k^{3/2}} < \frac{1}{k^{3/2}}$ and $\sum \frac{1}{k^{3/2}}$ converges, $\sum \frac{1}{2+k^{3/2}}$ converges.

12. Since $\frac{\sqrt{k}}{1+k^3} < \frac{\sqrt{k}}{k^3} = \frac{1}{k^{5/2}}$ and $\sum \frac{1}{k^{5/2}}$ converges, $\sum \frac{\sqrt{k}}{1+k^3}$ converges.

13. Since $\frac{3+k}{k^2} > \frac{k}{k^2} = \frac{1}{k}$ and $\sum \frac{1}{k}$ diverges, $\sum \frac{3+k}{k^2}$ diverges.

14. Since $\frac{\sqrt{k}}{1+k} \geq \frac{1}{1+k} > \frac{1}{2+k}$ and $\sum \frac{1}{2+k}$ diverges by problem 1,
$\sum \frac{\sqrt{k}}{1+k}$ diverges.

15. Since $\frac{2k}{1+k^3} < \frac{2k}{k^3} = \frac{2}{k^2}$ and $\sum \frac{2}{k^2}$ converges, $\sum \frac{2k}{1+k^3}$ converges.

16. Since $\frac{3}{\sqrt{k}+2} \geq \frac{3}{\sqrt{k}+2\sqrt{k}} = \frac{1}{\sqrt{k}}$ and $\sum \frac{1}{\sqrt{k}}$ diverges, $\sum \frac{3}{\sqrt{k}+2}$ diverges.

17. $\lim_{k\to 0} \frac{\frac{k+3}{2k^2+1}}{\frac{1}{k}} = \lim_{k\to\infty} \frac{k^2+3k}{2k^2+1} = \frac{1}{2}$. Since $\sum \frac{1}{k}$ diverges, $\sum \frac{k+3}{2k^2+1}$ diverges.

18. $\lim_{k\to\infty} \frac{\frac{k^2-4}{k^3+k+5}}{\frac{1}{k}} = 1$. Since $\sum \frac{1}{k}$ diverges, $\sum \frac{k^2-4}{k^3+k+5}$ diverges.

19. $\lim_{k\to\infty} \frac{\frac{k+\sqrt{k}}{k+k^3}}{\frac{1}{k^2}} = \lim_{k\to\infty} \frac{k^3+k^2\sqrt{k}}{k+k^3} = 1$. Since $\sum \frac{1}{k^2}$ converges, $\sum \frac{k+\sqrt{k}}{k+k^3}$ converges.

20. $\lim_{k\to\infty} \frac{\frac{2k+3}{\sqrt{k^3+2}}}{\frac{1}{\sqrt{k}}} = \lim_{k\to\infty} \frac{2\sqrt{k^3}+3\sqrt{k}}{\sqrt{k^3+2}} = 2$.
Since $\sum \frac{1}{\sqrt{k}}$ diverges, $\sum \frac{2k+3}{\sqrt{k^3+2}}$ diverges.

21. Comparison Test.
Since $\frac{k^2+5}{3k^2} > \frac{k^2}{3k^2} = \frac{1}{3}$ and $\sum \frac{1}{3}$ diverges, $\sum \frac{k^2+5}{3k^2}$ diverges.

22. Comparison Test.
Since $\frac{3^k}{3k+5^k} < \frac{3^k}{5^k} = \left(\frac{3}{5}\right)^k$ and $\sum \left(\frac{3}{5}\right)^k$ converges, $\sum \frac{3^k}{3k+5^k}$ converges.

23. Comparison Test.
Since $\frac{k}{4+k^3} < \frac{k}{k^3} = \frac{1}{k^2}$ and $\sum \frac{1}{k^2}$ converges, $\sum \frac{k}{4+k^3}$ converges.

24. Integral Test.
$\int_1^\infty \frac{1}{\sqrt{x+1}}\,dx = \lim_{t\to\infty} \int_1^t (x+1)^{-1/2}\,dx = \lim_{t\to\infty} 2(x+1)^{1/2}\Big]_1^t$
$= \lim_{t\to\infty} \left(2(t+1)^{1/2} - 2\sqrt{2}\right) = \infty.$ So $\sum \frac{1}{\sqrt{k+1}}$ diverges.

25. Comparison Test.
Since $\frac{\cos^2 \pi k}{k^2} \le \frac{1}{k^2}$ and $\sum \frac{1}{k^2}$ converges, $\sum \frac{\cos^2 \pi k}{k^2}$ converges.

26. $\frac{k-3}{k} = 1 - \frac{3}{k} \to 1$ as $k \to \infty$.
Therefore, the series $\sum \frac{k-3}{k}$ diverges by Theorem 5.

27. Comparison Test.
Note that $e^{\sqrt{k}} > 1$ for $k > 0$.
$\frac{e^{\sqrt{k}}}{\sqrt{k}} > \frac{1}{\sqrt{k}}$ and $\sum \frac{1}{\sqrt{k}}$ diverges. So $\sum \frac{e^{\sqrt{k}}}{\sqrt{k}}$ diverges.

28. No test can be applied since $\cos\left(\frac{\pi k}{4}\right)$ is not always positive. Since $\lim_{k\to\infty} \cos\left(\frac{\pi k}{4}\right)$ does not exist, by Theorem 5 in this section, $\sum \cos\left(\frac{\pi k}{4}\right)$ diverges.

29. Use the Limit Comparison Test.
Since $\lim_{k\to\infty} \frac{\frac{2k^2-2k}{k^4+3k}}{\frac{1}{k^2}} = \lim_{k\to\infty} \frac{2k^4-2k^3}{k^4+3k} = 2$ and $\sum \frac{1}{k^2}$ converges, $\sum \frac{2k^2-2k}{k^4+3k}$ converges.

30. Theorem 5.
Since $\lim_{k\to\infty} \frac{k^2-2}{k^2+2} = 1 \ne 0$, $\sum \frac{k^2-2}{k^2+2}$ diverges.

31. Limit Comparison Test.
Since $\lim_{k\to\infty} \frac{\frac{k}{\sqrt{1+k^2}}}{1} = 1$ and $\sum 1$ diverges, $\sum \frac{k}{\sqrt{1+k^2}}$ diverges.

32. Use the Comparison Test.
Since $\frac{4}{1+k^3} < \frac{4}{k^3}$ and $\sum \frac{4}{k^3} = 4\sum \frac{1}{k^3}$ converges, $\sum \frac{4}{1+k^3}$ converges.

33. Comparison Test.
$\frac{1}{\sqrt{4k(k+1)}} \ge \frac{1}{\sqrt{4k(k+k)}} = \frac{1}{\sqrt{8}\,k}$ for $k \ge 1$.
Since $\sum \frac{1}{\sqrt{8}\,k} = \frac{1}{\sqrt{8}} \sum \frac{1}{k}$ diverges, $\sum \frac{1}{\sqrt{4k(k+1)}}$ diverges.

34. Limit Comparison Test.
Since $\lim_{k\to\infty} \frac{\frac{2k}{\sqrt{k^3-1}}}{\frac{1}{\sqrt{k}}} = 2$ and $\sum \frac{1}{\sqrt{k}}$ diverges, $\sum \frac{2k}{\sqrt{k^3-1}}$ diverges.

35. Comparison Test.
$\frac{2^k}{5+3^{k+1}} < \frac{2^k}{3^{k+1}} = \frac{1}{3}\left(\frac{2}{3}\right)^k$ and $\sum \frac{1}{3}\left(\frac{2}{3}\right)^k = \frac{1}{3}\sum \left(\frac{2}{3}\right)^k$ converges.
So $\sum \frac{2^k}{5+3^{k+1}}$ converges.

36. Comparison Test.

Since $\frac{2}{(k+3)^{3/2}} < \frac{2}{k^{3/2}}$ and $\sum \frac{2}{k^{3/2}} = 2\sum \frac{1}{k^{3/2}}$ converges, $\sum \frac{2}{(k+3)^{3/2}}$ converges.

37. Integral Test.

Put $f(x) = \frac{\ln x}{x}$. $\quad f'(x) = \frac{\frac{1}{x}\cdot x - \ln x}{x^2} = \frac{1-\ln x}{x^2} < 0$ for $x > e$.

Thus $f(x)$ is decreasing for $x > e$.

$$\int_3^\infty f(x)dx = \lim_{t\to\infty}\int_3^t \frac{\ln x}{x}\,dx = \lim_{t\to\infty}\int_3^t \ln x\; d(\ln x) = \lim_{t\to\infty} \tfrac{1}{2}(\ln x)^2\Big]_3^t$$

$= \lim_{t\to\infty}\left[\frac{1}{2}(\ln t)^2 - \frac{1}{2}(\ln 3)^2\right] = \infty$. So $\sum \frac{\ln k}{k}$ diverges.

38. $\lim_{k\to\infty} \frac{\sqrt{k}\cdot\ln k}{1+\ln k} = \lim_{k\to\infty} \frac{\sqrt{k}}{\frac{1}{\ln k}+1} = \infty$.

Therefore $\sum \frac{\sqrt{k}\cdot\ln k}{1+\ln k}$ diverges by Theorem 5.

39. Use the Limit Comparison Test.

$\lim_{k\to\infty} \frac{\frac{7+k}{1+k^2}}{\frac{1}{k}} = \lim_{k\to\infty} \frac{7k+k^2}{1+k^2} = \lim_{k\to\infty} \frac{\frac{7}{k}+1}{\frac{1}{k}+1} = \frac{0+1}{0+1} = 1$.

$\sum \frac{1}{k}$ diverges. Therefore $\sum \frac{7+k}{1+k^2}$ diverges.

40. Limit Comparison Test.

Since $\lim_{k\to\infty} \frac{\frac{k^2}{k^4+2k+1}}{\frac{1}{k^2}} = 1$ and $\sum \frac{1}{k^2}$ converges, $\sum \frac{k^2}{k^4+2k+1}$ converges.

41. Comparison Test.

Since $\frac{2^k}{k+5} > \frac{1}{k+5}$ for $k > 0$ and $\sum \frac{1}{k+5}$ diverges, $\sum \frac{2^k}{k+5}$ diverges.

42. $\lim_{k\to\infty} \frac{2k^2}{\sqrt{k^3+6}} = \lim_{k\to\infty} \frac{2k^2}{\sqrt{k^3+6}} \cdot \frac{\frac{1}{k^{3/2}}}{\sqrt{\frac{1}{k^3}}} = \lim_{k\to\infty} \frac{2k^{1/2}}{\sqrt{1+\frac{6}{k^3}}} = \infty$.

Therefore $\sum \frac{2k^2}{\sqrt{k^3+6}}$ diverges by Theorem 5.

43. Limit Comparison Test.

Since $\lim_{k\to\infty} \frac{\frac{k}{(k+2)2^k}}{\frac{1}{2^k}} = 1$ and $\sum \frac{1}{2^k}$ converges, $\sum \frac{k}{(k+2)2^k}$ converges.

44. $\frac{\sin(\pi k)}{\sqrt{1+k^2}} = 0$ for all integers k. Therefore the infinite series $\sum \frac{\sin(\pi k)}{\sqrt{1+k^2}}$ converges and has sum 0.

45. Limit Comparison Test.

Since $\lim_{k\to\infty} \frac{\frac{2k+2}{\sqrt{k^3+2}}}{\frac{1}{\sqrt{k}}} = 2$ and $\sum \frac{1}{\sqrt{k}}$ diverges, $\sum \frac{2k+2}{\sqrt{k^3+2}}$ diverges.

46. Comparison Test.

Since $\ln k < k$ for $k > 1$, $\frac{\ln k}{k^3} < \frac{k}{k^3} = \frac{1}{k^2}$; and $\sum \frac{1}{k^2}$ converges.

Thus $\sum \frac{\ln k}{k^3}$ converges.

47. Comparison Test.
$\frac{k}{2+\ln k} > \frac{k}{2+k}$, so $\sum \frac{k}{2+\ln k}$ diverges.
Here $\sum \frac{k}{2+k}$ diverges by Theorem 5.

48. Limit Comparison Test.
Since $\lim\limits_{k\to\infty} \frac{\frac{k+3}{(k+2)2^k}}{\frac{1}{2^k}} = 1$ and $\sum \frac{1}{2^k}$ converges, $\sum \frac{k+3}{(k+2)2^k}$ converges.

Solutions to Exercise Set 9.4

1. $a_k = \frac{2^k}{k+2}$.

   $\rho = \lim\limits_{k\to\infty} \frac{a_{k+1}}{a_k} = \lim\limits_{k\to\infty} \frac{2^{k+1}}{(k+1)+2} \cdot \frac{k+2}{2^k} = \lim\limits_{k\to\infty} \frac{2(k+2)}{k+3} = 2 > 1.$

   Thus $\sum \frac{2^k}{k+2}$ diverges.

2. $a_k = \frac{3^k}{k!}$.

   $\rho = \lim\limits_{k\to\infty} \frac{a_{k+1}}{a_k} = \lim\limits_{k\to\infty} \frac{3^{k+1}}{(k+1)!} \cdot \frac{k!}{3^k} = \lim\limits_{k\to\infty} \frac{3}{k+1} = 0 < 1.$

   Therefore $\sum \frac{3^k}{k!}$ converges.

3. $a_k = k^2 e^{-k}$.

   $\rho = \lim\limits_{k\to\infty} \frac{a_{k+1}}{a_k} = \lim\limits_{k\to\infty} \frac{(k+1)^2 e^{-k-1}}{k^2 e^{-k}} = \lim\limits_{k\to\infty} \left(\frac{k+1}{k}\right)^2 \cdot e^{-1}$

   $= \lim\limits_{k\to\infty} \frac{1}{e}\left(1+\frac{1}{k}\right)^2 = \frac{1}{e} < 1.$ So $\sum k^2 e^{-k}$ converges.

4. $a_k = \frac{k!}{5^k}$.

   $\rho = \lim\limits_{k\to\infty} \frac{a_{k+1}}{a_k} = \lim\limits_{k\to\infty} \frac{(k+1)!}{5^{k+1}} \cdot \frac{5^k}{k!} = \lim\limits_{k\to\infty} \frac{k+1}{5} = \infty > 1.$

   Therefore $\sum \frac{k!}{5^k}$ diverges.

5. $a_k = \frac{\ln k}{ke^k}$.

   $a_k < \frac{k}{ke^k} = \frac{1}{e^k}$ for integers $k \geq 1$.

   The infinite series $\sum \frac{1}{e^k}$ converges. Therefore $\sum \frac{\ln k}{ke^k}$ converges.

6. $a_k = \frac{(3k)!}{(k!)^2}$.

$$a_{k+1} = \frac{(3(k+1))!}{((k+1)!)^2} = \frac{(3k+3)!}{((k+1)k!)^2} = \frac{3(k+1)(3k+2)(3k+1)\cdot(3k)!}{(k+1)^2(k!)^2}.$$

$$\rho = \lim_{k\to\infty} \frac{a_{k+1}}{a_k} = \lim_{k\to\infty} \frac{3(k+1)(3k+2)(3k+1)\cdot(3k)!}{(k+1)^2\cdot(k!)^2} \cdot \frac{(k!)^2}{(3k)!}$$

$$= \lim_{k\to\infty} \frac{3(k+1)(3k+2)(3k+1)}{(k+1)^2} = \infty > 1. \quad \text{So } \sum \frac{(3k)!}{(k!)^2} \text{ diverges.}$$

7. $a_k = \frac{k+2}{1+k^3}$.

$$\rho = \lim_{k\to\infty} \frac{a_{k+1}}{a_k} = \lim_{k\to\infty} \frac{k+3}{1+(k+1)^3} \cdot \frac{1+k^3}{k+2} = \lim_{k\to\infty} \frac{k+3}{k+2} \cdot \frac{k^3+1}{k^3+3k^2+3k+2} = 1.$$

Therefore there is no conclusion from the Ratio Test.

We use the Limit Comparison Test.

$\lim_{k\to\infty} \frac{\frac{k+2}{1+k^3}}{\frac{1}{k^2}} = \lim_{k\to\infty} \frac{k^3+2k^2}{1+k^3} = 1$. Since $\sum \frac{1}{k^2}$ converges, the infinite series $\sum \frac{k+2}{1+k^3}$ converges.

8. $a_k = \frac{\sqrt{k}}{k!}$.

$$\rho = \lim_{k\to\infty} \frac{a_{k+1}}{a_k} = \lim_{k\to\infty} \frac{\sqrt{k+1}}{(k+1)!} \cdot \frac{k!}{\sqrt{k}} = \lim_{k\to\infty} \frac{1}{k+1} \cdot \sqrt{\frac{k+1}{k}} = 0 < 1.$$

So $\sum \frac{\sqrt{k}}{k!}$ converges.

9. $a_k = \frac{k!}{ke^k}$.

$$\rho = \lim_{k\to\infty} \frac{a_{k+1}}{a_k} = \lim_{k\to\infty} \frac{(k+1)!}{(k+1)e^{k+1}} \cdot \frac{ke^k}{k!} = \lim_{k\to\infty} \frac{k}{k+1} \cdot \frac{k+1}{e} = \infty > 1.$$

So $\sum \frac{k!}{ke^k}$ diverges.

10. $a_k = \frac{k^2e^k}{k!}$.

$$\rho = \lim_{k\to\infty} \frac{a_{k+1}}{a_k} = \lim_{k\to\infty} \frac{(k+1)^2e^{k+1}}{(k+1)!} \cdot \frac{k!}{k^2e^k} = \lim_{k\to\infty} \left(\frac{k+1}{k}\right)^2 \cdot \frac{e}{k+1} = 0 < 1.$$

So $\sum \frac{k^2e^k}{k!}$ converges.

11. $a_k = \frac{e^{5k}}{k!}$.

$$\rho = \lim_{k\to\infty} \frac{a_{k+1}}{a_k} = \lim_{k\to\infty} \frac{e^{5(k+1)}}{(k+1)!} \cdot \frac{k!}{e^{5k}} = \lim_{k\to\infty} \frac{e^5}{k+1} = 0 < 1.$$

Therefore $\sum \frac{e^{5k}}{k!}$ converges.

12. $a_k = \frac{\pi k}{(k+1)!}$.

$$\rho = \lim_{k\to\infty} \frac{a_{k+1}}{a_k} = \lim_{k\to\infty} \frac{\pi(k+1)}{(k+2)!} \cdot \frac{(k+1)!}{\pi k}$$

$$= \lim_{k\to\infty} \frac{k+1}{k} \cdot \frac{1}{k+2} = 1 \cdot 0 = 0 < 1.$$

Therefore $\sum \frac{\pi k}{(k+1)!}$ converges.

13. $a_k = \frac{1}{k+2}$.

Since $a_k = \frac{1}{k+2} > \frac{1}{k+3} = a_{k+1}$ and $\lim_{k\to\infty} a_k = \lim_{k\to\infty} \frac{1}{k+2} = 0$,

$\sum_{k=0}^{\infty} \frac{(-1)^k}{k+2}$ converges.

14. Since $a_k = \frac{1}{k^2} > \frac{1}{(k+1)^2} = a_{k+1}$ and $\lim_{k\to\infty} a_k = \lim_{k\to\infty} \frac{1}{k^2} = 0$,

the alternating series $\sum \frac{(-1)^{k+1}}{k^2}$ converges.

15. $\sum_{k=1}^{\infty} \frac{\cos \pi k}{k} = -1 + \frac{1}{2} - \frac{1}{3} + \frac{1}{4} - \cdots = \sum_{k=1}^{\infty} \frac{(-1)^k}{k}$ where $a_k = \frac{1}{k}$.

Since $a_k = \frac{1}{k} > \frac{1}{k+1} = a_{k+1}$ and $\lim_{k\to\infty} a_k = \lim_{k\to\infty} \frac{1}{k} = 0$,

$\sum_{k=1}^{\infty} \frac{\cos \pi k}{k}$ converges.

16. $a_k = \frac{k}{k+5}$.

Since $\lim_{k\to\infty} a_k = \lim_{k\to\infty} \frac{k}{k+5} = 1 \neq 0$, the condition (ii) of the alternating series test fails. $\sum_{k=1}^{\infty} \frac{(-1)^k k}{k+5}$ diverges by Theorem 5.

17. $a_k = \frac{k^2}{\sqrt{k+1}}$.

Since $\lim_{k\to\infty} a_k = \lim_{k\to\infty} \frac{k^2}{\sqrt{k+1}} = \infty \neq 0$, the condition (ii) of the alternating series test fails. $\sum_{k=1}^{\infty} \frac{(-1)^k k^2}{\sqrt{k+1}}$ diverges by Theorem 5.

18. $a_k = \frac{k}{\sqrt{k+1}}$.

Since $\lim_{k\to\infty} a_k = \lim_{k\to\infty} \frac{k}{\sqrt{k+1}} = \infty$, the condition (ii) of the alternating series test fails. $\sum_{k=1}^{\infty} \frac{(-1)^k k}{\sqrt{k+1}}$ diverges by Theorem 5.

19. $\sum_{k=1}^{\infty} \frac{(-1)^k}{k^2}$.

$\sum_{k=1}^{\infty} \left|\frac{(-1)^k}{k^2}\right| = \sum_{k=1}^{\infty} \frac{1}{k^2}$ converges, so $\sum \frac{(-1)^k}{k^2}$ converges absolutely.

20. $\sum_{k=1}^{\infty} \frac{(-1)^k}{2k+1}$.

Since $\sum_{k=1}^{\infty} \left|\frac{(-1)^k}{2k+1}\right| = \sum_{k=1}^{\infty} \frac{1}{2k+1}$ diverges, but $\sum_{k=1}^{\infty} \frac{(-1)^k}{2k+1}$ converges by the alternating series test, $\sum_{k=1}^{\infty} \frac{(-1)^k}{2k+1}$ converges conditionally.

21. $\sum_{k=1}^{\infty} \frac{(-1)^k \cdot k}{(k+1)!}$.

$a_k = \left|\frac{(-1)^k \cdot k}{(k+1)!}\right| = \frac{k}{(k+1)!}$

$$\lim_{k\to\infty} \frac{a_{k+1}}{a_k} = \lim_{k\to\infty} \frac{\frac{k+1}{(k+2)!}}{\frac{k}{(k+1)!}} = \lim_{k\to\infty} \frac{(k+1)(k+1)!}{k(k+2)!}$$
$$= \lim_{k\to\infty} \frac{k+1}{k(k+2)} = 0.$$

Therefore $\sum_{k=1}^{\infty} \frac{(-1)^k \cdot k}{(k+1)!}$ converges absolutely by the ratio test.

22. $\sum_{k=1}^{\infty} \frac{(-1)^k \sqrt{k}}{k!}$

$a_k = \left|\frac{(-1)^k \sqrt{k}}{k!}\right| = \frac{\sqrt{k}}{k!}$.

$$\lim_{k\to\infty} \frac{a_{k+1}}{a_k} = \lim_{k\to\infty} \frac{\frac{\sqrt{k+1}}{(k+1)!}}{\frac{\sqrt{k}}{k!}} = \lim_{k\to\infty} \frac{\sqrt{k+1}\cdot k!}{(k+1)!\sqrt{k}} = \lim_{k\to\infty} \frac{1}{\sqrt{k(k+1)}} = 0.$$

Therefore $\sum_{k=1}^{\infty} \frac{(-1)^k \sqrt{k}}{k!}$ converges absolutely by the ratio test.

23. $\sum_{k=1}^{\infty} \frac{(-1)^k}{1+\sqrt{k}}$.

$$\lim_{k\to\infty} \frac{\frac{1}{1+\sqrt{k}}}{\frac{1}{\sqrt{k}}} = \lim_{k\to\infty} \frac{\sqrt{k}}{1+\sqrt{k}} = \lim_{k\to\infty} \frac{1}{\frac{1}{\sqrt{k}}+1} = 1.$$

$\sum_{k=1}^{\infty} \frac{1}{\sqrt{k}}$ diverges. Therefore $\sum_{k=1}^{\infty} \frac{(-1)^k}{1+\sqrt{k}}$ does not converge absolutely. But this series does converge by the alternating series test.

24. $\sum_{k=1}^{\infty} \frac{\cos \pi k}{\sqrt{k}} = \sum_{k=1}^{\infty} \left((-1)^k \cdot \frac{1}{\sqrt{k}}\right)$ converges by the alternating series test. The series of absolute values is $\sum_{k=1}^{\infty} \frac{1}{\sqrt{k}}$ which diverges.

25. $\sum_{k=1}^{\infty} \frac{1}{k} \sin \frac{k\pi}{2} = \sum_{k=1}^{\infty} (-1)^{k+1} \cdot \frac{1}{2k-1}$ converges by the alternating series test, but does not converge absolutely since $\sum_{k=1}^{\infty} \frac{1}{2k-1} \geq \sum_{k=1}^{\infty} \frac{1}{2k} = \frac{1}{2} \sum_{k=1}^{\infty} \frac{1}{k}$.

26. $\sum_{k=1}^{\infty} \frac{(-1)^k k^3}{2^k}$.

$a_k = \left| \frac{(-1)^k k^3}{2^k} \right| = \frac{k^3}{2^k}$.

$$\lim_{k \to \infty} \frac{a_{k+1}}{a_k} = \lim_{k \to \infty} \frac{\frac{(k+1)^3}{2^{k+1}}}{\frac{k^3}{2^k}} = \lim_{k \to \infty} \frac{(k+1)^3 \cdot 2^k}{k^3 \cdot 2^{k+1}}$$

$$= \lim_{k \to \infty} \frac{1}{2} \left(\frac{k+1}{k}\right)^3 = \lim_{k \to \infty} \frac{1}{2} \left(1 + \frac{1}{k}\right)^3 = \frac{1}{2}.$$

Therefore $\sum_{k=1}^{\infty} \frac{(-1)^k k^3}{2^k}$ converges absolutely by the ratio test.

27. $\sum_{k=1}^{\infty} \frac{(-1)^k \sqrt{k}}{(1+k^2)}$.

$a_k = \left| \frac{(-1)^k \sqrt{k}}{(1+k^2)} \right| = \frac{\sqrt{k}}{(1+k^2)} < \frac{\sqrt{k}}{k^2} = \frac{1}{k^{3/2}}$.

Therefore, $\sum_{k=1}^{\infty} \frac{(-1)^k \sqrt{k}}{(1+k^2)}$ converges absolutely by the comparison test.

28. $\sum_{k=1}^{\infty} \frac{(-1)^k}{1+k^2}$.

$a_k = \left| \frac{(-1)^k}{1+k^2} \right| = \frac{1}{1+k^2} < \frac{1}{k^2}$.

By the comparison test, $\sum_{k=1}^{\infty} \frac{(-1)^k}{1+k^2}$ converges absolutely.

29. Note that $3^{20} < 20!$.

Therefore $3^k < k!$ for $k \geq 20$, so that $\frac{k!}{3^k} > 1$ for $k \geq 20$.

Hence $\frac{(-1)^{k+1} k!}{3^k}$ does not approach 0 as k approaches ∞.

Thus the given infinite series does not converge.

30. $\sum_{k=2}^{\infty} \frac{(-1)^k \cdot k}{\ln \sqrt{k}}$.

$a_k = \left|\frac{(-1)^k \cdot k}{\ln \sqrt{k}}\right| = \frac{k}{\ln \sqrt{k}} > \frac{k}{\sqrt{k}} = \sqrt{k} \to \infty$ as $k \to \infty$.

Therefore $a_k \to \infty$ as $k \to \infty$.

$\sum_{k=2}^{\infty} \frac{(-1)^k \cdot k}{\ln \sqrt{k}}$ diverges by Theorem 5.

31. $\sum(-1)^k a_k$ with $a_k > a_{k+1}$ for all k.

Since $\lim_{k\to\infty} a_k = L \neq 0$, by Theorem 5, it does not converge.

32. a. Assume that $\rho < 1$, where $\rho = \lim_{k\to\infty} \frac{a_{k+1}}{a_k}$.

b. Let $\alpha = \frac{\rho+1}{2}$. Since $(\rho+1) < (1+1) = 2$, $\quad \alpha = \frac{\rho+1}{2} < 1$.

c. Since $\alpha = \frac{\rho+1}{2} > \frac{\rho+\rho}{2} = \rho$, there exists an integer N such that $\frac{a_{k+1}}{a_k} < \alpha$ for all integers $k \geq N$.

Otherwise, $\lim_{k\to\infty} \frac{a_{k+1}}{a_k} \geq \alpha > \rho$ which is a contradiction.

d. Since $\frac{a_{k+1}}{a_k} < \alpha$ for all $k \geq N$ from part (c),

$$\begin{aligned} a_{N+1} < \alpha \cdot a_N &\implies \alpha \cdot a_{N+1} < \alpha^2 \cdot a_N \\ &\implies a_{N+2} < \alpha \cdot a_{N+1} < \alpha^2 \cdot a_N \\ &\implies a_{N+3} < \alpha \cdot a_{N+2} < \alpha^2 \cdot a_{N+1} < a^3 \cdot a_N. \end{aligned}$$

By repeating this process, in general we obtain $a_{N+P} < \alpha^P a_N$.

e. Since $|\alpha| < 1$, $\sum_{P=1}^{\infty} \alpha^P$ is a geometric series which converges.

f. From part e, $\sum \alpha^P a_N = a_N \sum \alpha^P$ converges.

Therefore $\sum_{P=1}^{\infty} a_{N+P}$ converges by the Comparison Test.

$\therefore \sum_{k=1}^{\infty} a_k = \sum_{k=1}^{N} a_k + \sum_{P=1}^{\infty} a_{N+P}$ converges since $\sum_{k=1}^{N} a_k$ is a finite sum and $\sum_{P=1}^{\infty} a_{N+P}$ converges.

g. Assume that $\rho > 1$ where $\rho = \lim_{k\to\infty} \frac{a_{k+1}}{a_k}$.

Let $\alpha = \frac{\rho+1}{2}$. Then $1 < \alpha < \rho$ since $\alpha = \frac{\rho+1}{2} > \frac{1+1}{2} = 1$ and $\alpha = \frac{\rho+1}{2} < \frac{\rho+\rho}{2} = \rho$.

Hence, there exists an integer N such that $\frac{a_{k+1}}{a_k} > \alpha$

for all integers $k \geq N$.

Thus we obtain $a_{N+P} > \alpha^P a_N$.

Since $\alpha > 1$, $\lim_{P\to\infty} \alpha^P \cdot a_N = a_N \cdot \lim_{P\to\infty} \alpha^P = \infty$.

Hence $\lim_{P\to\infty} a_{N+P} = \infty \implies \lim_{k\to\infty} a_k = \infty$.

Therefore, $\sum a_k$ does not converge.

Solutions to Exercise Set 9.5

1. $f(x) = e^{-x}$. $\quad f(0) = 1$

   $f'(x) = -e^{-x}$, $\quad f'(0) = -1$

   $f''(x) = e^{-x}$, $\quad f''(0) = 1$

   $f'''(x) = -e^{-x}$, $\quad f'''(0) = -1$

   $P_3(x) = 1 - x + \frac{x^2}{2} - \frac{x^3}{3!}$.

2. $f(x) = \cos x$. $\quad f(0) = 1$

   $f'(x) = -\sin x$, $\quad f'(0) = 0$

   $f''(x) = -\cos x$, $\quad f''(0) = -1$

   $P_2(x) = 1 - \frac{x^2}{2}$.

3. $f(x) = \cos x$. $\quad f(0) = 1$

   $f'(x) = -\sin x$, $\quad f'(0) = 0$

   $f''(x) = -\cos x$, $\quad f''(0) = -1$

   $f'''(x) = \sin x$, $\quad f'''(0) = 0$

   $f^{(4)}(x) = \cos x$, $\quad f^{(4)}(0) = 1$

   $P_4(x) = 1 - \frac{x^2}{2} + \frac{x^4}{4!}$.

4. $f(x) = \ln(2 + x)$. $\quad f(0) = \ln 2$

   $f'(x) = \frac{1}{2+x}$, $\quad f'(0) = \frac{1}{2}$.

   $f''(x) = \frac{-1}{(x+2)^2}$, $\quad f''(0) = -\frac{1}{4}$

   $f'''(x) = \frac{2}{(x+2)^3}$, $\quad f'''(0) = \frac{1}{4}$

$f^{(4)}(x) = \frac{-6}{(x+2)^4}, \qquad f^{(4)}(0) = -\frac{3}{8}$

$P_4(x) = \ln 2 + \frac{1}{2}x - \frac{1}{8}x^2 + \frac{1}{24}x^3 - \frac{1}{64}x^4.$

5. $f(x) = \frac{1}{1+x}. \qquad f(0) = 1.$

$f'(x) = \frac{-1}{(1+x)^2}, \qquad f'(0) = -1$

$f''(x) = \frac{2}{(1+x)^3}, \qquad f''(0) = 2$

$P_2(x) = 1 - x + x^2.$

6. $f(x) = \tan x. \qquad f(0) = 0$

$f'(x) = \sec^2 x, \qquad f'(0) = 1$

$f''(x) = 2 \cdot \sec^2 x \cdot \tan x, \qquad f''(0) = 0$

$P_2(x) = x.$

7. $f(x) = (x+2)^{1/2}. \qquad f(0) = \sqrt{2}.$

$f'(x) = \frac{1}{2}(x+2)^{-1/2}, \qquad f'(0) = \frac{1}{2\sqrt{2}}$

$f''(x) = -\frac{1}{4}(x+2)^{-3/2}, \qquad f''(0) = -\frac{1}{4 \cdot 2^{3/2}} = -\frac{1}{8\sqrt{2}}$

$f'''(x) = \frac{3}{8}(x+2)^{-5/2}, \qquad f'''(0) = \frac{3}{8 \cdot 2^{5/2}} = \frac{3}{32\sqrt{2}}.$

$P_3(x) = \sqrt{2} + \frac{1}{2\sqrt{2}}x - \frac{1}{16\sqrt{2}}x^2 + \frac{1}{64\sqrt{2}}x^3.$

8. $f(x) = \sqrt{1+x}. \qquad f(0) = 1$

$f'(x) = \frac{1}{2\sqrt{1+x}}, \qquad f'(0) = \frac{1}{2}$

$f''(x) = \frac{-1}{4(1+x)\sqrt{1+x}}, \qquad f''(0) = -\frac{1}{4}$

$P_2(x) = 1 + \frac{1}{2}x - \frac{1}{8}x^2.$

9. $f(x) = \sin x. \qquad f\left(\frac{\pi}{4}\right) = \frac{\sqrt{2}}{2}$

$f'(x) = \cos x, \qquad f'\left(\frac{\pi}{4}\right) = \frac{\sqrt{2}}{2}$

$f''(x) = -\sin x, \qquad f''\left(\frac{\pi}{4}\right) = -\frac{\sqrt{2}}{2}$

$f'''(x) = -\cos x, \qquad f'''\left(\frac{\pi}{4}\right) = -\frac{\sqrt{2}}{2}.$

$f^{(4)}(x) = \sin x, \qquad f^{(4)}\left(\frac{\pi}{4}\right) = \frac{\sqrt{2}}{2}.$

$P_4(x) = \frac{\sqrt{2}}{2} + \frac{\sqrt{2}}{2}\left(x - \frac{\pi}{4}\right) - \frac{\sqrt{2}}{4}\left(x - \frac{\pi}{4}\right)^2 - \frac{\sqrt{2}}{12}\left(x - \frac{\pi}{4}\right)^3 + \frac{\sqrt{2}}{48}\left(x - \frac{\pi}{4}\right)^4.$

10. $f(x) = \cos x.$ $f\left(\frac{\pi}{3}\right) = \frac{1}{2}$

$f'(x) = -\sin x,$ $f'\left(\frac{\pi}{3}\right) = -\frac{\sqrt{3}}{2}$

$f''(x) = -\cos x,$ $f''\left(\frac{\pi}{3}\right) = -\frac{1}{2}.$

$f'''(x) = \sin x,$ $f'''\left(\frac{\pi}{3}\right) = \frac{\sqrt{3}}{2}.$ $f^{(4)}(x) = \cos x,$ $f^{(4)}\left(\frac{\pi}{3}\right) = \frac{1}{2}.$

$P_4(x) = \frac{1}{2} - \frac{\sqrt{3}}{2}\left(x - \frac{\pi}{3}\right) - \frac{1}{4}\left(x - \frac{\pi}{3}\right)^2 + \frac{\sqrt{3}}{12}\left(x - \frac{\pi}{3}\right)^3 + \frac{1}{48}\left(x - \frac{\pi}{3}\right)^4.$

11. $f(x) = e^x.$ $f(1) = e$

$f'(x) = e^x,$ $f'(1) = e$

$f''(x) = e^x,$ $f''(1) = e$

$f'''(x) = e^x,$ $f'''(1) = e$

$f^{(4)}(x) = e^x,$ $f^{(4)}(1) = e$

$$P_4(x) = e + e(x-1) + \tfrac{e}{2}(x-1)^2 + \tfrac{e}{3!}(x-1)^3 + \tfrac{e}{4!}(x-1)^4$$
$$= e\left[1 + (x-1) + \tfrac{1}{2}(x-1)^2 + \tfrac{1}{3!}(x-1)^3 + \tfrac{1}{4!}(x-1)^4\right].$$

12. $f(x) = \ln x.$ $f(1) = 0.$

$f'(x) = \frac{1}{x},$ $f'(1) = 1$

$f''(x) = -\frac{1}{x^2},$ $f''(1) = -1$

$f'''(x) = \frac{2}{x^3},$ $f'''(1) = 2$

$f^{(4)}(x) = \frac{-6}{x^4},$ $f^{(4)}(1) = -6$

$f^{(5)}(x) = \frac{24}{x^5},$ $f^{(5)}(1) = 24.$

$P_5(x) = (x-1) - \frac{1}{2}(x-1)^2 + \frac{1}{3}(x-1)^3 - \frac{1}{4}(x-1)^4 + \frac{1}{5}(x-1)^5.$

13. $f(x) = \sqrt{x}.$ $f(9) = 3$

$f'(x) = \frac{1}{2\sqrt{x}},$ $f'(9) = \frac{1}{6}$

$f''(x) = \frac{-1}{4x\sqrt{x}},$ $f''(9) = \frac{-1}{108}$

$f'''(x) = \frac{3}{8x^2\sqrt{x}},$ $f'''(9) = \frac{1}{648}$

$P_3(x) = 3 + \frac{1}{6}(x-9) - \frac{1}{216}(x-9)^2 + \frac{1}{3888}(x-9)^3.$

14. $f(x) = \sqrt{1+x}$. $\quad f(3) = 2$

$f'(x) = \frac{1}{2\sqrt{1+x}}$, $\quad f'(3) = \frac{1}{4}$

$f''(x) = \frac{-1}{4(1+x)\sqrt{1+x}}$, $\quad f''(3) = \frac{-1}{32}$

$f'''(x) = \frac{3}{8(1+x)^2\sqrt{1+x}}$, $\quad f'''(3) = \frac{3}{256}$.

$P_3(x) = 2 + \frac{1}{4}(x-3) - \frac{1}{64}(x-3)^2 + \frac{1}{512}(x-3)^3$.

15. $f(x) = e^{x^2}$. $\quad f(0) = 1$

$f'(x) = e^{x^2} \cdot 2x$, $\quad f'(0) = 0$

$f''(x) = 2 \cdot e^{x^2} + 4x^2 \cdot e^{x^2}$, $\quad f''(0) = 2$

$P_2(x) = 1 + x^2$.

16. $f(x) = \sin x$. $\quad f\left(\frac{\pi}{2}\right) = 1$.

$f'(x) = \cos x$, $\quad f'\left(\frac{\pi}{2}\right) = 0$

$f''(x) = -\sin x$, $\quad f''\left(\frac{\pi}{2}\right) = -1$

$f'''(x) = -\cos x$, $\quad f'''\left(\frac{\pi}{2}\right) = 0$

$f^{(4)}(x) = \sin x$, $\quad f^{(4)}\left(\frac{\pi}{2}\right) = 1$

$f^{(5)}(x) = \cos x$, $\quad f^{(5)}\left(\frac{\pi}{2}\right) = 0$

$f^{(6)}(x) = -\sin x$, $\quad f^{(6)}\left(\frac{\pi}{2}\right) = -1$

$P_6(x) = 1 - \frac{1}{2}\left(x - \frac{\pi}{2}\right)^2 + \frac{1}{4!}\left(x - \frac{\pi}{2}\right)^4 - \frac{1}{6!}\left(x - \frac{\pi}{2}\right)^6$.

17. $f(x) = x \cdot \cos x$. $\quad f\left(\frac{\pi}{4}\right) = \frac{\sqrt{2}}{8}\pi$

$f'(x) = \cos x - x \sin x$, $\quad f'\left(\frac{\pi}{4}\right) = \frac{\sqrt{2}}{2} - \frac{\sqrt{2}}{8}\pi$

$f''(x) = -2\sin x - x \cdot \cos x$, $\quad f''\left(\frac{\pi}{4}\right) = -\sqrt{2} - \frac{\sqrt{2}}{8}\pi$

$P_2(x) = \frac{\sqrt{2}}{8}\pi + \left(\frac{\sqrt{2}}{2} - \frac{\sqrt{2}}{8}\pi\right)\left(x - \frac{\pi}{4}\right) + \left(-\frac{\sqrt{2}}{2} - \frac{\sqrt{2}}{16}\pi\right)\left(x - \frac{\pi}{4}\right)^2$.

18. $f(x) = \frac{1}{1+x}$. $\quad f(1) = \frac{1}{2}$

$f'(x) = \frac{-1}{(1+x)^2}$, $\quad f'(1) = -\frac{1}{4}$

$f''(x) = \frac{2}{(1+x)^3}$, $\quad f''(1) = \frac{1}{4}$

$f'''(x) = -\frac{6}{(1+x)^4}$, $\quad f'''(1) = -\frac{6}{16} = -\frac{3}{8}$

$P_3(x) = \frac{1}{2} - \frac{1}{4}(x-1) + \frac{1}{8}(x-1)^2 - \frac{1}{16}(x-1)^3$.

19. $f(x) = e^{-x}$. $\quad f(0) = 1$

$f'(x) = -e^{-x}, \quad f'(0) = -1$

$f''(x) = e^{-x}, \quad f''(0) = 1$

$f'''(x) = -e^{-x}, \quad f'''(0) = -1$

$P_3(x) = 1 - x + \frac{1}{2}x^2 - \frac{1}{3!}x^3$.

$e^{-1} = f(1) \approx P_3(1) = 1 - 1 + \frac{1}{2} - \frac{1}{6} = \frac{1}{3} = 0.3333.$

20. $f(x) = \cos x$.

$P_2(x) = 1 - \frac{1}{2}x^2$.

$f\left(\frac{\pi}{12}\right) = \cos\left(\frac{\pi}{12}\right) \approx P_2\left(\frac{\pi}{12}\right) = 1 - \frac{1}{2} \cdot \frac{\pi^2}{144} = 1 - \frac{\pi^2}{288}.$

21. $f(x) = \sqrt{x}$.

$P_3(x) = 3 + \frac{1}{6}(x-9) - \frac{1}{216}(x-9)^2 + \frac{1}{3888}(x-9)^3$ by Problem 13.

$\sqrt{10} = f(10) \approx P_3(10) = 3 + \frac{1}{6} - \frac{1}{216} + \frac{1}{3888} = \frac{12295}{3888}.$

22. $f(x) = e^x$.

$P_n(x) = 1 + x + \frac{x^2}{2!} + \frac{x^3}{3!} + \cdots + \frac{x^n}{n!}$.

$P_n'(x) = 1 + x + \frac{x^2}{2!} + \cdots + \frac{x^{n-1}}{(x-1)!} = P_{n-1}(x).$

23. $f(x) = e^{-x}$.

$P_n(x) = 1 - x + \frac{x^2}{2!} - \frac{x^3}{3!} + \cdots + (-1)^n \frac{x^n}{n!}$.

$$\begin{aligned} P_n'(x) &= -1 + x - \frac{x^2}{2} + \cdots + (-1)^n \frac{x^{n-1}}{(n-1)!} \\ &= -\left(1 - x + \frac{x^2}{2} + \cdots + (-1)^{n-1}\frac{x^{n-1}}{(n-1)!}\right) \\ &= -P_{n-1}(x). \end{aligned}$$

24. $f(x) = \sin x, \quad g(x) = \cos x$.

$P_5(x) = x - \frac{x^3}{3!} + \frac{x^5}{5!}$.

$Q_5(x) = 1 - \frac{x^2}{2!} + \frac{x^4}{4!}$.

$P_5'(x) = 1 - \frac{x^2}{2!} + \frac{x^4}{4!} = Q_5(x).$

25. $P_4(x) = x - \frac{x^3}{3!}$.

$Q_4(x) = 1 - \frac{x^2}{2!} + \frac{x^4}{4!}$.

$Q_4'(x) = -x + \frac{x^3}{3!} = -\left(x - \frac{x^3}{3!}\right) = -P_4(x).$

Solutions to Exercise Set 9.6

1. Put $f(x) = \ln x$. $\quad f(1) = 0$.

$f'(x) = \frac{1}{x}$, $\quad f''(x) = -\frac{1}{x^2}$, $\quad f'''(x) = \frac{2}{x^3}$.

$f'(1) = 1$, $\quad f''(1) = -1$, $\quad f'''(1) = 2$.

$P_3(x) = f(1) + (x-1)f'(1) + \frac{1}{2}f''(1)(x-1)^2 + \frac{1}{3!}f'''(1)(x-1)^3$.

$$\ln(1.4) \approx P_3(1.4) = 0 + (1.4-1) + \tfrac{1}{2}(-1)(1.4-1)^2 + \tfrac{1}{6}\cdot 2(1.4-1)^3$$
$$= 0 + 0.4 - \tfrac{1}{2}\cdot(0.16) + \tfrac{1}{3}(0.064) = 0.3413.$$

$|f^{(4)}(x)| = \left|-\frac{6}{x^4}\right| = \frac{6}{x^4} \le 6$ in [1,1.4]. So $M = 6$.

$|R_3(1.4)| \le \frac{M}{4!}(1.4-1)^4 = \frac{6}{4\cdot 3\cdot 2}0.4^4 = 0.0064$.

2. Put $f(x) = \cos x$.

$f\left(\frac{\pi}{4}\right) = \frac{\sqrt{2}}{2}$. $\quad f'(x) = -\sin x$, $\quad f''(x) = -\cos x$.

$f'\left(\frac{\pi}{4}\right) = -\frac{\sqrt{2}}{2}$, $\quad f''\left(\frac{\pi}{4}\right) = -\frac{\sqrt{2}}{2}$.

$$\cos 36^\circ = \cos\left(\tfrac{36\cdot\pi}{180}\right) = \cos\left(\tfrac{\pi}{5}\right)$$
$$\approx P_2\left(\tfrac{\pi}{5}\right) = f\left(\tfrac{\pi}{4}\right) + f'\left(\tfrac{\pi}{4}\right)\left(\tfrac{\pi}{5} - \tfrac{\pi}{4}\right) + \frac{f''\left(\frac{\pi}{4}\right)}{2}\left(\tfrac{\pi}{5} - \tfrac{\pi}{4}\right)^2$$
$$= \tfrac{\sqrt{2}}{2} + \tfrac{\sqrt{2}}{2}\left(\tfrac{\pi}{20}\right) - \tfrac{\sqrt{2}}{4}\left(\tfrac{\pi}{20}\right)^2 = \tfrac{\sqrt{2}}{2}\left(1 + \tfrac{\pi}{20} - \tfrac{1}{2}\left(\tfrac{\pi}{20}\right)^2\right) \approx 0.8095.$$

$|f'''(x)| = |\sin x| \le \sin\frac{\pi}{4} = \frac{\sqrt{2}}{2}$ in $\left[\frac{\pi}{5}, \frac{\pi}{4}\right]$.

$|R_2(36^\circ)| \le \frac{M}{3!}\left|\frac{\pi}{5} - \frac{\pi}{4}\right|^3 = \frac{\sqrt{2}}{6\cdot 2}\left(\frac{\pi}{20}\right)^3 \approx 0.00046$.

3. Put $f(x) = \sqrt{x}$.

$f(4) = 2$. $\quad f'(x) = \frac{1}{2\sqrt{x}}$, $\quad f''(x) = -\frac{1}{4}x^{-3/2}$.

$f'(4) = \frac{1}{4}$, $\quad f''(4) = -\frac{1}{4}2^{-3} = -\frac{1}{32}$.

$$\sqrt{3.91} \approx P_2(3.91) = f(4) + f'(4)(3.91-4) + \tfrac{1}{2}f''(4)(3.91-4)^2$$
$$= 2 - \tfrac{1}{4}(0.09) - \tfrac{1}{64}(0.09)^2 = 1.9774.$$

$|f'''(x)| = \left|\frac{3}{8}x^{-5/2}\right| \le \frac{3}{8}(3.91)^{-5/2} \approx 0.0124$ for x in [3.91,4], so $M = 0.0124$.

$|R_2(3.91)| \le \frac{M}{3!}|3.91-4|^3 = \frac{0.0124}{6}\cdot(0.09)^3 \approx 1.506\cdot 10^{-6}$.

4. Put $f(x) = e^x$.

$f'(x) = f''(x) = f'''(x) = f^{(4)}(x) = e^x$.

$f(0) = 1, \qquad f'(0) = f''(0) = f'''(0) = 1$.

$$e^{0.2} \approx P_3(0.2) = f(0) + f'(0)(0.2) + \tfrac{f''(0)}{2}(0.2)^2 + \tfrac{f'''(0)}{3!}(0.2)^3$$
$$= 1 + 0.2 + \tfrac{1}{2}(0.2)^2 + \tfrac{1}{6}(0.2)^3 = 1.2213.$$

$|f^{(4)}(x)| = e^x \leq e^1 < 3$ in [0,1], so $M = 3$.

$|R_3(0.2)| \leq \frac{M}{4!}(0.2)^4 = \frac{3}{4!}0.2^4 = 0.0002$.

5. Put $f(x) = \sin x$. $\qquad f\left(\frac{\pi}{4}\right) = \frac{\sqrt{2}}{2}$.

$f'(x) = \cos x, \quad f''(x) = -\sin x, \quad f'''(x) = -\cos x, \quad f^{(4)}(x) = \sin x$.

$48° = \frac{48 \cdot \pi}{180} = \frac{4\pi}{15}$.

$$\sin 48° = \sin\left(\tfrac{4\pi}{15}\right) \approx P_4\left(\tfrac{4\pi}{15}\right) = f\left(\tfrac{\pi}{4}\right) + f'\left(\tfrac{\pi}{4}\right)\left(\tfrac{4\pi}{15} - \tfrac{\pi}{4}\right)$$
$$+ \tfrac{1}{2}f''\left(\tfrac{\pi}{4}\right)\left(\tfrac{4\pi}{15} - \tfrac{\pi}{4}\right)^2 + \tfrac{1}{3!}f'''\left(\tfrac{\pi}{4}\right)\left(\tfrac{4\pi}{15} - \tfrac{\pi}{4}\right)^3 + \tfrac{1}{4!}f^{(4)}\left(\tfrac{\pi}{4}\right)\left(\tfrac{4\pi}{15} - \tfrac{\pi}{4}\right)^4$$
$$= \tfrac{\sqrt{2}}{2} + \tfrac{\sqrt{2}}{2}\left(\tfrac{\pi}{60}\right) - \tfrac{\sqrt{2}}{4}\left(\tfrac{\pi}{60}\right)^2 - \tfrac{\sqrt{2}}{12}\left(\tfrac{\pi}{60}\right)^3 + \tfrac{\sqrt{2}}{48}\left(\tfrac{\pi}{60}\right)^4$$
$$= \tfrac{\sqrt{2}}{2}\left[1 + \tfrac{\pi}{60} - \tfrac{1}{2}\left(\tfrac{\pi}{60}\right)^2 - \tfrac{1}{6}\left(\tfrac{\pi}{60}\right)^3 + \tfrac{1}{24}\left(\tfrac{\pi}{60}\right)^4\right]$$
$$= \tfrac{\sqrt{2}}{2}\left(1 + 0.05236 - 0.00137 - 0.00002 + 3 \times 10^{-7}\right) \approx 0.7431.$$

$|f^{(5)}(x)| = |\cos x| \leq 1$ in $\left[\frac{\pi}{4}, \frac{4\pi}{15}\right]$, so $M = 1$.

$|R_4(48°)| \leq \frac{M}{5!}\left(\frac{4\pi}{15} - \frac{\pi}{4}\right)^5 = \frac{1}{5!}\left(\frac{\pi}{60}\right)^5 = 3 \times 10^{-9}$.

6. Put $f(x) = \sqrt[3]{x}$.

$f'(x) = \frac{1}{3}x^{-2/3}, \qquad f''(x) = -\frac{2}{9}x^{-5/3}$.

$f(8) = 2, \qquad f'(8) = \frac{1}{3} \cdot 2^{-2} = \frac{1}{12}, \qquad f''(8) = -\frac{2}{9}2^{-5} = -\frac{1}{144}$.

$$\sqrt[3]{10} \approx P_2(10) = f(8) + f'(8)(10 - 8) + \tfrac{1}{2}f''(8)(10 - 8)^2$$
$$= 2 + \tfrac{1}{12} \cdot 2 + \tfrac{1}{2}\left(-\tfrac{1}{144}\right) \cdot 2^2 = 2 + \tfrac{1}{6} - \tfrac{1}{72} \approx 2.1528.$$

$|f'''(x)| = \left|-\frac{2}{9} \cdot \left(-\frac{5}{3}x^{-8/3}\right)\right| = \frac{10}{27}x^{-8/3} \leq \frac{10}{27}8^{-8/3} = \frac{10}{27 \cdot 2^8} = M$.

$|R_2(10)| \leq \frac{M}{3!}(10 - 8)^3 = \frac{1}{6} \cdot \frac{10}{27 \cdot 2^8} \cdot 2^3 \approx 0.00193$.

7. Put $f(x) = \tan x$.

$f'(x) = \sec^2 x, \quad f''(x) = 2\sec x \cdot \sec x \tan x = 2\sec^2 x \tan x,$

$f'''(x) = 4\sec x \cdot \sec x \cdot \tan x \cdot \tan x + 2\sec^2 x \cdot \sec^2 x = 4\sec^2 x \tan^2 x + 2\sec^4 x.$

$f(0) = 0, \quad f'(0) = 1, \quad f''(0) = 0, \quad f'''(0) = 2.$

$$\tan\tfrac{\pi}{12} \approx P_3\left(\tfrac{\pi}{12}\right) = f(0) + f'(0)\cdot\tfrac{\pi}{12} + \tfrac{1}{2}f''(0)\left(\tfrac{\pi}{12}\right)^2 + \tfrac{1}{3!}f'''(0)\left(\tfrac{\pi}{12}\right)^3$$
$$= \tfrac{\pi}{12} + \tfrac{1}{3}\left(\tfrac{\pi}{12}\right)^3 \approx 0.2618 + 0.00598 = 0.2678.$$

$$f^{(4)}(x) = 8\sec x \cdot \sec x \tan x \cdot \tan^2 x + 4\sec^2 x \cdot 2\tan x \cdot \sec^2 x$$
$$+ 8\sec^3 x \sec x \tan x$$
$$= 8\sec^2 x \tan^3 x + 16\sec^4 x \cdot \tan x.$$

$$|f^{(4)}(x)| \le 8\sec^2\left(\tfrac{\pi}{4}\right)\tan^3\left(\tfrac{\pi}{4}\right) + 16\sec^4\left(\tfrac{\pi}{4}\right)\cdot\tan\left(\tfrac{\pi}{4}\right)$$
$$= 8\left(\sqrt{2}\right)^2\cdot 1 + 16\cdot\left(\sqrt{2}\right)^4 = 16 + 64 = 80 = M \text{ in } \left[0, \tfrac{\pi}{12}\right].$$

$$\left|R_3\left(\tfrac{\pi}{12}\right)\right| \le \tfrac{M}{4!}\left(\tfrac{\pi}{12}\right)^4 = \tfrac{80}{24}\left(\tfrac{\pi}{12}\right)^4 \approx 0.0157.$$

8. Put $f(x) = \sqrt{x}$.

$f'(x) = \frac{1}{2}x^{-1/2}, \quad f''(x) = -\frac{1}{4}x^{-3/2}, \quad f'''(x) = \frac{3}{8}x^{-5/2}.$

$f(49) = 7, \quad f'(49) = \frac{1}{14}, \quad f''(49) = -\frac{1}{4}\cdot\frac{1}{7^3} = -\frac{1}{1372},$

$f'''(49) = \frac{3}{8}\cdot\frac{1}{7^5} = \frac{3}{134456}.$

$$\sqrt{50} \approx f(49) + f'(49)\cdot 1 + \tfrac{1}{2}f''(49)\cdot 1^2 + \tfrac{1}{3!}f'''(49)\cdot 1^3$$
$$= 7 + \tfrac{1}{14} - \tfrac{1}{2(1372)} + \tfrac{3}{3!(134456)} \approx 7.0711.$$

$$|f^{(4)}(x)| = \left|-\tfrac{15}{16}x^{-7/2}\right| \le \tfrac{15}{16}(49)^{-7/2} = \tfrac{15}{16\cdot 7^7} = M.$$

$$|R_4(50)| \le \tfrac{M}{4!}\cdot 1^4 = \tfrac{1}{4!}\left(\tfrac{15}{16\cdot 7^7}\right) = 4.7\times 10^{-8}.$$

9. $e^x \approx 1 + x + \frac{x^2}{2} = P_2(x).$

$|f'''(x)| = e^x \le e < 3$ in [0,1].

$|R_2(x)| \le \frac{M}{3!}x^3 \le \frac{3}{3!}\cdot 1^3 = \frac{1}{2}.$

10. $\sin x \approx x - \frac{x^3}{3!} = P_3(x).$

$|f^{(4)}(x)| = |\sin x| \le \frac{\sqrt{2}}{2}$ for $|x| < \frac{\pi}{4}$.

$|R_3(x)| \le \frac{M}{4!}|x|^4 \le \frac{M}{4!}\left(\frac{\pi}{4}\right)^4 = \frac{\sqrt{2}}{2\cdot 4!}\left(\frac{\pi}{4}\right)^4.$

A bound on error $|R_3(x)| = \frac{\sqrt{2}}{2\cdot 4!}\left(\frac{\pi}{4}\right)^4 \approx 0.0112.$

11. The error associated with the approximation $\sin x \approx P_n(x)$ satisfies

$$|R_n(x)| \leq \frac{1}{(n+1)!}|x|^{n+1} \text{ since } \left|\frac{d^n}{dx^n}\sin x\right| \leq 1 \text{ for all } n > 0$$
$$\leq \frac{1}{(n+1)!}(0.2)^{n+1} \text{ since } |x| < 0.2.$$

Therefore to get the accuracy $|R_n(x)| \leq 0.001$, it is enough to get $\frac{1}{(n+1)!}(0.2)^{n+1} \leq 0.001$.

If $n = 2$: $\frac{1}{3!}(0.2)^3 \leq 0.001$?

$0.00133 \not\leq 0.001$

If $n = 3$: $\frac{1}{4!}(0.2)^4 \leq 0.001$?

$0.000067 < 0.001$.

Thus the integer n can be ≥ 3 in order to get the desired accuracy.

12. $\cos x \approx P_n(x)$. $\left|\frac{d^n}{dx^n}\cos x\right| \leq 1$ for all $n > 0$.

$|R_n(x)| \leq \frac{1}{(n+1)!}|x|^{n+1} \leq \frac{1}{(n+1)!}(0.2)^{n+1}$ since $|x| < 0.2$.

To get $|R_n(x)| \leq 0.0001$, let $\frac{1}{(n+1)!}(0.2)^{n+1} \leq 0.0001$.

If $n = 2$: $\frac{1}{3!}(0.2)^3 \leq 0.0001$?

$0.00133 \not\leq 0.0001$

If $n = 3$: $\frac{1}{4!}(0.2)^4 \leq 0.0001$?

$0.000067 < 0.0001$.

Thus $n \geq 3$.

13. a. Define $h(x) = g(x) - f(x)$.

b. Since $g(x) \geq f(x)$ for $a \leq x \leq b$,

$h(x) = g(x) - f(x) \geq 0$ for $a \leq x \leq b$.

c. Then $g(x) = f(x) + h(x)$ and

$\int_a^b g(x)dx = \int_a^b [f(x) + h(x)]\,dx$.

Since $f(x)$ and $g(x)$ are continuous on $[a, b]$, $h(x)$ is continuous on $[a, b]$. Hence $h(x)$ is also integrable on $[a, b]$, and by the Property (I1) in Section 5.5

$$\int_a^b [f(x) + h(x)]\,dx = \int_a^b f(x)dx + \int_a^b h(x)dx.$$

So

$$\int_a^b g(x)dx = \int_a^b f(x)dx + \int_a^b h(x)dx.$$

d. Since $h(x) \geq 0$ for $a \leq x \leq b$, the graph of $h(x)$ is above the x-axis. Geometrically $\int_a^b h(x)dx$ is the area of the region bounded by the graph of $y = h(x)$ and the x-axis for $a \leq x \leq b$. Therefore $\int_a^b h(x)dx \geq 0$.

e. From (c) and (d) we have

$$\int_a^b g(x)dx = \int_a^b f(x)dx + \int_a^b h(x) \geq \int_a^b f(x)dx$$

since $\int_a^b h(x)dx \geq 0$. That is,

$$\int_a^b f(x)dx \leq \int_a^b g(x)dx.$$

14. a. Since $-|f(x)| \leq f(x)$ for $f(x) \geq 0$
$-|f(x)| = f(x)$ for $f(x) \leq 0$, and since obviously $f(x) \leq |f(x)|$ for all $a \leq x \leq b$, then $-|f(x)| \leq f(x) \leq |f(x)|$.

b. Using the results of Exercise 13 and (a), we get

$$\int_a^b (-|f(x)|)\,dx \leq \int_a^b f(x)dx \leq \int_a^b |f(x)|dx$$

so $$-\int_a^b |f(x)|dx \leq \int_a^b f(x)dx \leq \int_a^b |f(x)|dx.$$

c. Thus

$$\left|\int_a^b f(x)dx\right| \leq \int_a^b |f(x)|dx.$$

Solutions to Exercise Set 9.7

1. $f(x) = e^{2x}, \qquad a = 0.$

   $f(0) = 1, \quad f'(x) = 2e^{2x}, \quad f''(x) = 4e^{2x}, \cdots, f^{(k)}(x) = 2^k e^{2x}$ for all k.

   $f'(0) = 2, \quad f''(0) = 4, \cdots, f^{(k)}(0) = 2^k$ for all k.

   The Taylor series is

   $\sum_{k=0}^{\infty} \frac{f^{(k)}(0)}{k!} x^k = \sum_{k=0}^{\infty} \frac{2^k}{k!} x^k = 1 + 2x + \frac{4}{2}x^2 + \frac{8}{3!}x^3 + \cdots + \frac{2^k}{k!}x^k + \cdots.$

2. $f(x) = e^{-x}, \quad a = 0.$

   $f(0) = 1, \quad f'(x) = -e^{-x}, \quad f''(x) = e^{-x}, \cdots, f^{(k)}(x) = (-1)^k e^{-x}$

   for $k = 0, 1, 2, \cdots$.

   $f'(0) = -1, \quad f''(0) = 1, \cdots, f^{(k)}(0) = (-1)^k$ for $k = 0, 1, 2, \cdots$.

   The Taylor series is

   $\sum_{k=0}^{\infty} \frac{f^{(k)}(0)}{k!} x^k = 1 - x + \frac{1}{2}x^2 - \frac{1}{3!}x^3 + \cdots + \frac{(-1)^k}{k!}x^k + \cdots.$

3. $f(x) = \frac{1}{1+x}, \quad a = 0.$

   $f(0) = 1, \quad f'(x) = \frac{-1}{(1+x)^2}, \quad f''(x) = 2(1+x)^{-3}, \quad f'''(x) = -6(1+x)^{-4}, \cdots.$

   $f^{(k)}(x) = (-1)^k k!(1+x)^{-k-1}$ for all $k = 0, 1, 2, \cdots$.

   $f'(0) = -1, \quad f''(0) = 2, \cdots, f^{(k)}(0) = (-1)^k k!$ for all $k = 0, 1, 2, \cdots$.

   The Taylor series is

   $\sum_{k=0}^{\infty} \frac{f^{(k)}(0)}{k!} x^k = \sum_{k=0}^{\infty} \frac{(-1)^k k!}{k!} x^k = \sum_{k=0}^{\infty} (-1)^k x^k = 1 - x + x^2 - x^3 + \cdots + (-1)^k x^k + \cdots.$

4. $f(x) = \sqrt{1+x}, \quad a = 0.$

   $f(0) = 1, \quad f'(x) = \frac{1}{2}(1+x)^{-1/2}, \quad f''(x) = -\frac{1}{4}(1+x)^{-3/2},$

   $f'''(x) = \frac{3}{8}(1+x)^{-5/2}, \cdots,$

   $f^{(k)}(x) = (-1)^{k-1} \frac{(2k-3)(2k-5)\cdots 5\cdot 3\cdot 1}{2^k}(1+x)^{-(2k-1)/2}$ for $k = 1, 2, 3, \cdots$.

   Thus $f'(0) = \frac{1}{2}, \quad f''(0) = -\frac{1}{4}, \quad f'''(0) = \frac{3}{8}, \quad f^{(4)}(0) = -\frac{3\cdot 5}{16}, \cdots.$

   The Taylor series is $\sum_{k=0}^{\infty} \frac{f^{(k)}(0)}{k!} x^k = 1 + \frac{1}{2}x + \frac{1}{2}\left(-\frac{1}{4}\right)x^2 + \frac{1}{3!}\left(\frac{3}{2^3}\right)x^3 - \frac{1}{4!}\left(\frac{3\cdot 5}{2^4}\right)x^4 + \cdots.$

5. $f(x) = \ln x$, $a = 1$.

$f'(x) = \frac{1}{x}$, $f''(x) = -x^{-2}$, $f'''(x) = 2x^{-3}$, $\cdots$,

$f^{(k)}(x) = (-1)^{k-1}(k-1)!x^{-k}$ for $k = 1, 2, 3, \cdots$.

$f(1) = 0$, $f'(1) = 1$, $f''(1) = -1$, $f'''(1) = 2$, $\cdots$,

$f^{(k)}(1) = (-1)^{k-1}(k-1)!$ for $k = 1, 2, 3, \cdots$.

The Taylor series is

$$\sum_{k=0}^{\infty} \frac{f^{(k)}(1)}{k!}(x-1)^k = (x-1) - \frac{1}{2}(x-1)^2 + \frac{2}{3!}(x-1)^3 + \cdots + \frac{(-1)^{k-1}(k-1)!}{k!}(x-1)^k + \cdots$$

$$= \sum_{k=1}^{\infty} \frac{(-1)^{k-1}}{k}(x-1)^k.$$

6. $f(x) = \frac{1}{x}$, $a = 1$.

$f'(x) = -x^{-2}$, $f''(x) = 2x^{-3}$, $f'''(x) = -3!x^{-4}$, $\cdots$, $f^{(k)}(x) = (-1)^k k! x^{-(k+1)}$ for $k = 0, 1, 2, \cdots$.

$f(1) = 1$, $f'(1) = -1$, $f''(1) = 2$, $f'''(1) = -3!$, $\cdots$, $f^{(k)}(1) = (-1)^k k!$ for $k = 0, 1, 2, \cdots$.

The Taylor series is $\sum\limits_{k=0}^{\infty} \frac{f^{(k)}(1)}{k!}(x-1)^k = 1 - (x-1) + (x-1)^2 - (x-1)^3 + \cdots + (-1)^k(x-1)^k + \cdots$.

7. $f(x) = \sin x$, $a = \frac{\pi}{2}$.

$f\left(\frac{\pi}{2}\right) = 1$, $f'(x) = \cos x$, $f''(x) = -\sin x$, $f'''(x) = -\cos x$,

$f^{(4)}(x) = \sin x, \cdots$.

$$f'\left(\tfrac{\pi}{2}\right) = 0, \quad f''\left(\tfrac{\pi}{2}\right) = -1, \quad f'''\left(\tfrac{\pi}{2}\right) = 0, \quad f^{(4)}\left(\tfrac{\pi}{2}\right) = 1, \cdots.$$

$\Longrightarrow f^{(2k)}\left(\frac{\pi}{2}\right) = (-1)^k$ and $f^{(2k+1)}\left(\frac{\pi}{2}\right) = 0$ for $k = 0, 1, 2, \cdots$.

The Taylor series is

$$\sum_{k=0}^{\infty} \frac{f^{(k)}\left(\frac{\pi}{2}\right)}{k!}\left(x - \frac{\pi}{2}\right)^k = 1 - \frac{1}{2!}\left(x - \frac{\pi}{2}\right)^2 + \frac{1}{4!}\left(x - \frac{\pi}{2}\right)^4 + \cdots$$

$$+ \frac{(-1)^k}{(2k)!}\left(x - \frac{\pi}{2}\right)^{2k} + \cdots.$$

8. $f(x) = \cos x, \quad a = \frac{\pi}{4}.$

$f\left(\frac{\pi}{4}\right) = \frac{\sqrt{2}}{2}, \quad f'(x) = -\sin x, \quad f''(x) = -\cos x, \quad f'''(x) = \sin x,$

$f^{(4)}(x) = \cos x, \cdots.$

$f'\left(\frac{\pi}{4}\right) = -\frac{\sqrt{2}}{2}, \quad f''\left(\frac{\pi}{4}\right) = \frac{-\sqrt{2}}{2}, \quad f'''(x) = \frac{\sqrt{2}}{2}, \quad f^{(4)}(x) = \frac{\sqrt{2}}{2}, \cdots.$

The Taylor series is

$$\sum_{k=0}^{\infty} \frac{f^{(k)}\left(\frac{\pi}{4}\right)}{k!}\left(x - \frac{\pi}{4}\right)^k = \frac{\sqrt{2}}{2}\left(1 - \left(x - \frac{\pi}{4}\right) - \frac{1}{2!}\left(x - \frac{\pi}{4}\right)^2 + \frac{1}{3!}\left(x - \frac{\pi}{4}\right)^3 + \cdots\right).$$

9. $f(x) = \sin x, \quad a = \frac{\pi}{3}.$

$f\left(\frac{\pi}{3}\right) = \frac{\sqrt{3}}{2}, \quad f'(x) = \cos x, \quad f''(x) = -\sin x, \quad f'''(x) = -\cos x,$

$f^{(4)}(x) = \sin x, \cdots.$

$f'\left(\frac{\pi}{3}\right) = \frac{1}{2}, \quad f''\left(\frac{\pi}{3}\right) = -\frac{\sqrt{3}}{2}, \quad f'''\left(\frac{\pi}{3}\right) = -\frac{1}{2}, \quad f^{(4)}\left(\frac{\pi}{3}\right) = \frac{\sqrt{3}}{2}, \cdots.$

The Taylor series is

$$\sum_{k=0}^{\infty} \frac{f^{(k)}\left(\frac{\pi}{3}\right)}{k!}\left(x - \frac{\pi}{3}\right)^k = \frac{\sqrt{3}}{2} + \frac{1}{2}\left(x - \frac{\pi}{3}\right) - \frac{\sqrt{3}}{2}\,\frac{\left(x - \frac{\pi}{3}\right)^2}{2!} - \frac{1}{2}\,\frac{\left(x - \frac{\pi}{3}\right)^3}{3!} + \cdots.$$

10. $f(x) = x \cdot \sin x, \quad a = 0.$

$f(0) = 0, \quad f'(x) = \sin x + x \cdot \cos x, \quad f''(x) = 2\cos x - x \cdot \sin x,$

$f'''(x) = -3\sin x - x \cdot \cos x, \quad f^{(4)}(x) = -4\cos x + x \cdot \sin x, \cdots.$

$f'(0) = 0, \quad f''(0) = 2, \quad f'''(0) = 0, \quad f^{(4)}(0) = -4, \cdots.$

The Taylor series is

$$\sum_{k=0}^{\infty} \frac{f^{(k)}(0)}{k!}x^k = x^2 - \frac{1}{4!} \cdot 4x^4 + \cdots + \frac{(-1)^{k-1}}{(2k)!} \cdot (2k)x^{2k} + \cdots$$

$$= \sum_{k=1}^{\infty} \frac{(-1)^{k-1}}{(2k-1)!}x^{2k}.$$

11. $a_k = \frac{1}{k!}(2x)^k$.

$\lim_{k\to\infty} \frac{|a_{k+1}|}{|a_k|} = \lim_{k\to\infty} \frac{\frac{|2x|^{k+1}}{(k+1)!}}{\frac{|2x|^k}{k!}} = \lim_{k\to\infty} \frac{2|x|}{k+1} = 0$ for all real numbers $x \neq 0$.

By the ratio test, the Taylor series $\sum_{k=0}^{\infty} \frac{(2x)^k}{k!}$ converges absolutely for all real numbers x. Therefore, the interval of convergence is $(-\infty, \infty)$.

12. $\sum_{k=0}^{\infty} (-1)^k x^k$.

Let $a_k = (-1)^k x^k$.

$\rho = \lim_{k\to\infty} \frac{|a_{k+1}|}{|a_k|} = \lim_{k\to\infty} \frac{|x|^{k+1}}{|x|^k} = |x|$. By the ratio test, $\sum_{k=0}^{\infty} (-1)^k x^k$ converges absolutely when $|x| < 1$.

Case $x = 1$: $\sum_{k=0}^{\infty} (-1)^k x^k = 1 - 1 + 1 - 1 + \cdots$ diverges since $\lim_{k\to\infty} a_k = \lim_{k\to\infty} (-1)^k$ does not exist.

Case $x = -1$: $\sum_{k=0}^{\infty} (-1)^k x^k = \sum_{k=0}^{\infty} (-1)^{2k} = \sum_{k=0}^{\infty} 1$ diverges.

Case $|x| > 1$: $\sum_{k=0}^{\infty} (-1)^k x^k$ diverges since $\lim_{k\to\infty} |a_k| = \lim_{k\to\infty} |x|^k = \infty$.

Therefore, the interval of convergence is (-1,1).

13. $\sum_{k=1}^{\infty} \frac{(-1)^{k-1}}{k}(x-1)^k$. $a_k = \frac{(-1)^{k-1}}{k}(x-1)^k$.

$\lim_{k\to\infty} \frac{|a_{k+1}|}{|a_k|} = \lim_{k\to\infty} \frac{\frac{|x-1|^{k+1}}{k+1}}{\frac{|x-1|^k}{k}} = \lim_{k\to\infty} \frac{k}{k+1}|x-1| = |x-1|$.

By the Ratio Test, the Taylor series converges absolutely when $|x-1| < 1$, i.e., $0 < x < 2$.

Case $|x-1| > 1$: The Taylor series diverges since

$$\lim_{k\to\infty} \frac{|x-1|^k}{k} = \infty \text{ if } |x-1| > 1.$$

Case $x = 2$: $\sum_{k=1}^{\infty} \frac{(-1)^{k-1}}{k} = 1 - \frac{1}{2} + \frac{1}{3} - \frac{1}{4} + \cdots$ converges by the Alternating Series Test.

Case $x = 0$: $\sum_{k=1}^{\infty} \frac{(-1)^{k-1}}{k}(-1)^k = \sum_{k=1}^{\infty} \frac{(-1)^{2k-1}}{k} = -\sum_{k=1}^{\infty} \frac{1}{k}$ diverges.

Hence, the interval of convergence for the Taylor series is $(0, 2]$.

14. $\frac{1}{1+x} = \sum_{k=0}^{\infty}(-1)^k x^k$

$\Longrightarrow \left(\frac{1}{1+x}\right)' = \left(\sum_{k=0}^{\infty}(-1)^k x^k\right)'$

$\Longrightarrow -\frac{1}{(1+x)^2} = \sum_{k=1}^{\infty}(-1)^k \cdot k \cdot x^{k-1}$

$\Longrightarrow \frac{1}{(1+x)^2} = \sum_{k=1}^{\infty}(-1)^{k-1} \cdot k \cdot x^{k-1}.$

Hence, the Taylor series for $\frac{1}{(1+x)^2}$ is

$$\sum_{k=1}^{\infty}(-1)^{k-1} \cdot k \cdot x^{k-1} = 1 - 2x + 3x^2 - 4x^3 + \cdots .$$

15. $\sin x = \sum_{k=0}^{\infty} \frac{(-1)^k}{(2k)!}\left(x - \frac{\pi}{2}\right)^{2k}$

$\Longrightarrow (\sin x)' = \left(\sum_{k=0}^{\infty} \frac{(-1)^k}{(2k)!}\left(x - \frac{\pi}{2}\right)^{2k}\right)'$

$\Longrightarrow \cos x = \sum_{k=1}^{\infty} \frac{(-1)^k}{(2k-1)!}\left(x - \frac{\pi}{2}\right)^{2k-1}$

$= -\left(x - \frac{\pi}{2}\right) + \frac{\left(x-\frac{\pi}{2}\right)^3}{3!} - \frac{\left(x-\frac{\pi}{2}\right)^5}{5!} + \cdots .$

16. $f(x) = \frac{2}{(1+x)^2}.$

By Problem 14, a power series is the Taylor series

$$2 - 4x + 6x^2 - 8x^2 + \cdots .$$

17. $\frac{1}{1-x} = 1 + x + x^2 + x^3 + x^4 + \cdots$ for $|x| < 1$.

$\frac{1}{(1-x)^2} = \frac{d}{dx}\,\frac{1}{1-x} = 1 + 2x + 3x^2 + 4x^3 + 5x^4 + 6x^5 + \cdots .$

$\frac{1}{(1-x)^3} = \frac{d}{dx}\,\frac{1}{(1-x)^2} = 2 + 6x + 12x^2 + 20x^3 + 30x^4 + \cdots .$

18. $\frac{d}{dx}\left(\frac{1}{2}\sin(x^2)\right) = x\cos(x^2).$

$\sin(x^2) = (x^2) - \frac{1}{3!}(x^2)^3 + \frac{1}{5!}(x^2)^5 - \frac{1}{7!}(x^2)^7 + \cdots .$

$\frac{1}{2}\sin(x^2) = \frac{1}{2}x^2 - \frac{1}{2\cdot 3!}x^6 + \frac{1}{2\cdot 5!}x^{10} - \frac{1}{2\cdot 7!}x^{14} + \cdots .$

$x\cos(x^2) = x - \frac{1}{2}x^5 + \frac{1}{24}x^9 - \frac{1}{720}x^{13} + \cdots$

$= x - \frac{1}{2!}x^5 + \frac{1}{4!}x^9 - \frac{1}{6!}x^{13} + \cdots .$

19. $\int \frac{1}{1-x}\,dx = -\ln(1-x) + C.$

$\frac{1}{1-x} = 1 + x + x^2 + x^3 + x^4 + \cdots$ for $|x| < 1$.

$x + \frac{1}{2}x^2 + \frac{1}{3}x^3 + \frac{1}{4}x^4 + \frac{1}{5}x^5 \cdots = -\ln(1-x) + C.$

$C = 0$, by considering $x = 0$ in the preceding formula.

$$\ln(1-x) = -x - \frac{1}{2}x^2 - \frac{1}{3}x^3 - \frac{1}{4}x^4 - \frac{1}{5}x^5 + \cdots .$$

20. $\int \frac{1}{4+x}\,dx = \ln(4+x) + C.$

$\frac{1}{4+x} = \frac{1}{4} \cdot \frac{1}{1+\frac{x}{4}} = \frac{1}{4} \cdot \left(1 - \frac{x}{4} + \left(\frac{x}{4}\right)^2 - \left(\frac{x}{4}\right)^3 + \left(\frac{x}{4}\right)^4 - \cdots\right).$

$\frac{1}{4} \cdot \left(x - \frac{x^2}{8} + \frac{x^3}{48} - \frac{x^4}{256} + \frac{x^5}{1280} - \cdots\right) = \ln(4+x) + C.$

$C = -\ln 4$, by considering $x = 0$ in the preceding formula.

$$\begin{aligned}\ln(4+x) &= \ln 4 + \frac{1}{4}x - \frac{1}{32}x^2 + \frac{1}{192}x^3 - \frac{1}{1024}x^4 + \frac{1}{5120}x^5 - \cdots \\ &= \ln 4 + \frac{1}{4}x - \frac{1}{2 \cdot 4^2}x^2 + \frac{1}{3 \cdot 4^3}x^3 - \frac{1}{4 \cdot 4^4}x^4 + \frac{1}{5 \cdot 4^5}x^5 - \cdots .\end{aligned}$$

21.

$$\lim_{k\to\infty} \frac{\left|\frac{x^{k+1}}{k+3}\right|}{\left|\frac{x^k}{k+2}\right|} = |x| \cdot \lim_{k\to\infty} \frac{k+2}{k+3} = |x| \cdot 1 = |x|.$$

The series converges absolutely for $|x| < 1 \implies -1 < x < 1$.

Test $x = -1$: $\frac{1}{2} - \frac{1}{3} + \frac{1}{4} - \frac{1}{5} + \cdots$, which converges by the alternating series test.

Test $x = 1$: $\frac{1}{2} + \frac{1}{3} + \frac{1}{4} + \frac{1}{5} + \cdots$, which diverges by the integral test.

The interval of convergence is $[-1, 1)$.

The radius of convergence is 1.

22.

$$\lim_{k\to\infty} \frac{\left|\frac{x^{k+1}}{2(k+1)}\right|}{\left|\frac{x^k}{2k}\right|} = \lim_{k\to\infty} |x| \cdot \frac{k}{k+1} = |x| \cdot 1 = |x|.$$

The series converges absolutely for $|x| < 1 \implies -1 < x < 1$.

Test $x = -1$: $-\frac{1}{2} + \frac{1}{4} - \frac{1}{6} + \frac{1}{8} - \frac{1}{10} + \cdots$, which converges by the alternating series test.

Test $x = 1$: $\frac{1}{2} + \frac{1}{4} + \frac{1}{6} + \frac{1}{8} + \cdots$, which diverges by the integral test.

The interval of convergence is $[-1, 1)$.

The radius of convergence is 1.

23.
$$\lim_{k\to\infty} \left| \frac{\frac{(-1)^{k+2}}{(k+1)!} x^{k+1}}{\frac{(-1)^{k+1}}{k!} x^k} \right| = |x| \cdot \lim_{k\to\infty} \frac{1}{k+1} = 0.$$

The series converges absolutely for all x.

24.
$$\lim_{k\to\infty} \frac{\left|\frac{2^{k+1}x^{k+1}}{(k+2)!}\right|}{\left|\frac{2^k x^k}{(k+1)!}\right|} = 2 \cdot |x| \cdot \lim \frac{1}{k+2} = 0.$$

The series converges absolutely for all x.

25.
$$\lim_{k\to\infty} \frac{\left|\frac{(k+1)^2+1}{(k+1)!} \cdot x^{k+1}\right|}{\left|\frac{k^2+1}{k!} \cdot x^k\right|} = \lim_{k\to\infty} |x| \cdot \frac{k^2+2k+2}{k^2+1} \cdot \frac{1}{k+1}$$
$$= |x| \cdot 1 \cdot 0 = 0.$$

The series converges absolutely for all x.

26.
$$\lim_{k\to\infty} \frac{\left|\frac{(-1)^{k+1}}{(k+1)(k+2)} \cdot x^{k+1}\right|}{\left|\frac{(-1)^k}{k(k+1)} \cdot x^k\right|} = \lim_{k\to\infty} |x| \cdot \frac{k}{k+2} = |x| \cdot 1 = |x|.$$

The series converges absolutely for $|x| < 1 \implies -1 < x < 1$.

Test $x = -1$: $\sum\limits_{k=1}^{\infty} \frac{1}{k^2+k}$ which converges by comparison to the known convergent series $\sum\limits_{k=1}^{\infty} \frac{1}{k^2}$.

Test $x = 1$: $\sum\limits_{k=1}^{\infty} \frac{(-1)^k}{k(k+1)}$ which converges by the alternating series test.

The interval of convergence is $[-1, 1]$.

The radius of convergence is 1.

27. Since $\lim\limits_{k\to\infty} \frac{x^k}{k} = +\infty$ for $x > 1$, $\lim\limits_{k\to\infty} |a_k| = \lim\limits_{k\to\infty} \frac{x^k}{k} = \infty$ for all $x > 1$. Hence $\sum\limits_{k=1}^{\infty} a_k = \sum\limits_{k=1}^{\infty} (-1)^{k+1} \frac{x^k}{k}$ does not converge by Theorem 5.

28. For $x < -1$, $\frac{(-1)^{k+1}}{k} x^k = -\frac{(-x)^k}{k}$ for $k = 1, 2, 3, \cdots$.
$\frac{(-x)^k}{k} > \frac{1}{k}$ for $k = 1, 2, 3, \cdots$ if $x < -1$, and $\sum_{k=1}^{\infty} \frac{1}{k}$ diverges. Thus for $x < -1$, $\sum_{k=1}^{\infty} \frac{(-x)^k}{k}$ diverges, so $\sum_{k=1}^{\infty} \frac{(-1)^{k+1}}{k} x^k$ diverges.

29. $f(x) = \frac{1}{1-x} \implies f(0) = 1$.

$f'(x) = \frac{1}{(1-x)^2} \implies f'(0) = 1$

$f''(x) = \frac{2!}{(1-x)^3} \implies f''(0) = 2!$

$f'''(x) = \frac{3!}{(1-x)^4} \implies f'''(0) = 3!$

$\vdots$

$f^{(k)}(x) = \frac{k!}{(1-x)^{k+1}} \implies f^{(k)}(0) = k!$

Hence $f(x) = \sum_{k=0}^{\infty} \frac{f^{(k)}(0)}{k!} x^k = \sum_{k=0}^{\infty} \frac{k!}{k!} x^k = \sum_{k=0}^{\infty} x^k$.

30. $f(x) = \sin x \implies f(0) = 0$

$f'(x) = \cos x \implies f'(0) = 1$

$f''(x) = -\sin x \implies f''(0) = 0$

$f'''(x) = -\cos x \implies f'''(0) = -1$

$f^{(4)}(x) = \sin x \implies f^{(4)}(0) = 0$, etc.

Hence $f^{(2k)}(0) = 0$ and $f^{(2k+1)}(0) = (-1)^k$ for $k = 0, 1, 2, \cdots$. The Taylor series for $\sin x$ is

$$\sum_{k=0}^{\infty} \frac{f^{(k)}(0)}{k!} x^k = \sum_{k=0}^{\infty} \frac{f^{(2k+1)}(0)}{(2k+1)!} x^{2k+1} = \sum_{k=0}^{\infty} \frac{(-1)^k}{(2k+1)!} x^{2k+1}.$$

31. $f(x) = \cos x \implies f(0) = 1$

$f'(x) = -\sin x \implies f'(0) = 0$

$f''(x) = -\cos x \implies f''(0) = -1$

$f'''(x) = \sin x \implies f'''(0) = 0$

$f^{(4)}(x) = \cos x \implies f^{(4)}(0) = 1$, etc.

Hence $f^{(2k+1)}(0) = 0$ and $f^{(2k)}(0) = (-1)^k$ for $k = 0, 1, 2, \cdots$. The Taylor series for $\cos x$ is

$$\sum_{k=0}^{\infty} \frac{f^{(k)}(0)}{k!} x^k = \sum_{k=0}^{\infty} \frac{f^{(2k)}(0)}{(2k)!} x^{2k} = \sum_{k=0}^{\infty} \frac{(-1)^k}{(2k)!} x^{2k}.$$

32. Case 1: Suppose x is a positive real number.

For all t between 0 and x, $|f^{(n+1)}(t)| = e^t \leq e^x$. Hence $|R_n(x)| \leq \frac{e^x}{(n+1)!}x^{n+1}$.

In view of Example 2, the series $\sum_{k=0}^{\infty} \frac{x^k}{k!}$ converges, so $\lim_{k\to\infty} \frac{x^k}{k!} = 0$. Thus $\lim_{n\to\infty} R_n(x) = 0$.

Case 2: Suppose x is a negative real number. For all t between 0 and x, $|f^{(n+1)}(t)| = e^t \leq 1$.

Hence $|R_n(x)| \leq \frac{1}{(n+1)!} \cdot |x|^{n+1}$.

Proceeding as in Case 1, we can show that $\lim_{n\to\infty} R_n(x) = 0$.

Solutions to Exercise Set 9.8

1. $f(x) = x^2 - \frac{7}{2}x + \frac{3}{4}$, $a = 0, b = 2$.

Since $f(0) = \frac{3}{4}$ and $f(2) = -\frac{9}{4}$, a zero of $f(x)$ must lie within [0,2].

The approximation scheme for Newton's Method is

$$x_{n+1} = x_n - \frac{f(x_n)}{f'(x_n)} = x_n - \frac{x_n^2 - \frac{7}{2}x_n + \frac{3}{4}}{2x_n - \frac{7}{2}}.$$

Using $x_1 = 0$ as a first approximation, we get results in the table:

n	x_n	$f(x_n)$	$f'(x_n)$	x_{n+1}
1	0.0	0.75	−3.5	0.214286
2	0.214286	0.045917	−3.071428	0.229236
3	0.229236	0.000223	−3.041528	0.229309
4	0.229309	0.000001	−3.041382	0.229309

Thus $x_4 = 0.229309$ is an approximation to the zero.

2. $f(x) = 1 - x - x^2$, $a = -2$, $b = -1$.

Since $f(-2) = -1$ and $f(-1) = 1$, there must exist a zero in [-2,-1]. The approximation scheme for Newton's Method is

$$x_{n+1} = x_n - \frac{1 - x_n - x_n^2}{-1 - 2x_n} = x_n + \frac{1 - x_n - x_n^2}{1 + 2x_n}.$$

Using a first aproximation $x_1 = -2$, we obtain the results as follows.

n	x_n	$f(x_n)$	$f'(x_n)$	x_{n+1}
1	−2.0	−1.0	3.0	−1.666667
2	−1.666667	−0.111112	2.333334	−1.619048
3	−1.619048	−0.002268	2.238096	−1.618035
4	−1.618035	−0.000002	2.236070	−1.618034

Thus the zero is approximately $x_5 = -1.618034$.

3. $f(x) = 1 - x - x^2, \quad a = 0, \quad b = 1.$

Since $f(0) = 1$ and $f(1) = -1$ there must exist a zero in [0,1]. The approximation scheme for Newton's Method is

$$x_{n+1} = x_n - \frac{1 - x_n - x_n^2}{-1 - 2x_n} = x_n + \frac{1 - x_n - x_n^2}{1 + 2x_n}.$$

Using $x_1 = 1$ as a first approximation, we obtain the results:

n	x_n	$f(x_n)$	$f'(x_n)$	x_{n+1}
1	1.0	−1.0	−3.0	0.666667
2	0.666667	−0.111112	−2.333334	0.619048
3	0.619048	−0.002268	−2.238096	0.618035
4	0.618035	−0.000002	−2.236070	0.618034

Thus $x_5 = 0.618034$ is a good approximation to the zero.

4. $f(x) = x^3 + x - 3, \quad a = 1, \quad b = 2.$

Since $f(1) = -1$ and $f(2) = 7$ there must exist a zero in [1,2]. The approximation scheme for Newton's Method is

$$x_{n+1} = x_n - \frac{x_n^3 + x_n - 3}{3x_n^2 + 1} = \frac{2x_n^3 + 3}{3x_n^2 + 1}.$$

Using $x_1 = 1$, we obtain the following results:

n	x_n	$f(x_n)$	$f'(x_n)$	x_{n+1}
1	1.0	−1.0	4.0	1.25
2	1.25	0.203125	5.6875	1.214286
3	1.214286	0.004739	5.423471	1.213412
4	1.213412	0.000002	5.417106	1.213412

Thus $x_4 = 1.213412$ is a good approximation to the zero.

5. $f(x) = \sqrt{x+3} - x, \quad a = 1, \quad b = 3.$

Since $f(1) = 1$ and $f(3) = \sqrt{6} - 3 < 0$ there must exist a zero in [1,3]. The approximation scheme for Newton's method is

$$x_{n+1} = x_n - \frac{\sqrt{x_n+3} - x_n}{\frac{1}{2\sqrt{x_n+3}} - 1} = x_n + \frac{2(x_n+3) - 2x_n\sqrt{x_n+3}}{2\sqrt{x_n+3} - 1}$$

$$x_{n+1} = \frac{2x_n\sqrt{x_n+3} - x_n + 2(x_n+3) - 2x_n\sqrt{x_n+3}}{2\sqrt{x_n+3} - 1}$$

$$= \frac{x_n + 6}{2\sqrt{x_n+3} - 1}.$$

Using $x_1 = 1$, we obtain the following results.

n	x_n	$f(x_n)$	$f'(x_n)$	x_{n+1}
1	1.0	1.0	−0.75	2.333333
2	2.333333	−0.023932	−0.783494	2.302788
3	2.302788	−0.000010	−0.782871	2.302775
4	2.302775	0.0000005	−0.782871	2.302775

Thus $x_4 = 2.302775$ is a good approximation to the zero.

6. $f(x) = x^3 + x^2 + 3, \quad a = -3, \quad b = -1.$

Since $f(-3) = -15$ and $f(-1) = 3$ there must exist a zero in [-3,-1]. The approximation scheme for Newton's Method is

$$x_{n+1} = x_n - \frac{x_n^3 + x_n^2 + 3}{3x_n^2 + 2x_n}.$$

Using $x_1 = -2$, we obtain the following results.

n	x_n	$f(x_n)$	$f'(x_n)$	x_{n+1}
1	−2.0	−1.0	8.0	−1.875
2	−1.875	−0.076172	6.796875	−1.863793
3	−1.863793	−0.000579	6.693587	−1.863706
4	−1.863706	0.000003	6.692788	−1.863707

Thus $x_5 = -1.863707$ is a good approximation to the zero.

7. $f(x) = x^4 - 5, \quad a = -2, \quad b = -1.$

Since $f(-2) = 11$ and $f(-1) = -4$ there must be a zero in [-2,-1]. The approximation scheme for Newton's method is

$$x_{n+1} = x_n - \frac{x_n^4 - 5}{4x_n^3}.$$

Using $x_1 = -2$, we have the following results.

n	x_n	$f(x_n)$	$f'(x_n)$	x_{n+1}
1	-2.0	11.0	-32.0	-1.65625
2	-1.65625	2.524949	-18.173462	-1.517314
3	-1.517314	0.300317	-13.972895	-1.495821
4	-1.495821	0.006319	-13.387481	-1.495349
5	-1.495349	0.000003	-13.374812	-1.495349

Thus $x_5 = -1.495349$ is a good approximation to the zero.

8. $f(x) = 5 - x^4, \quad a = 1, \quad b = 2.$

Since $f(1) = 4$ and $f(2) = -11$ there must be a zero in [1,2]. The approximation scheme for Newton's Method is

$$x_{n+1} = x_n - \frac{5 - x_n^4}{-4x_n^3}.$$

Using $x_1 = 2$, we have the following information.

n	x_n	$f(x_n)$	$f'(x_n)$	x_{n+1}
1	2.0	-11.0	-32.0	1.65625
2	1.65625	-2.524949	-18.173462	1.517314
3	1.517314	-0.300317	-13.972895	1.495821
4	1.495821	-0.006319	-13.387481	1.495349
5	1.495349	-0.000003	-13.374812	1.495349

Thus $x_5 = 1.495349$ is an approximation to the zero.

<u>Another method</u>:

Since $f(x) = 5 - x^4 = -(x^4 - 5)$, all zeros of $g(x) = x^4 - 5$ are also the zeros of $f(x) = 5 - x^4$. For Exercise 7, $x_5 = -1.495349$ is an approximation to the zero of $g(x)$ in [-2,-1]; obviously 1.495349 is the approximation to another zero of $g(x)$ in [1,2]. Therefore 1.495349 is also the approximation to the zero of $f(x) = 5 - x^4$ in [1,2].

9. Finding a number for which $\sqrt{x+3} = x$ is equivalent to finding a zero of the function $h(x) = \sqrt{x+3} - x$. From the results of Exercise 5, $h(x) = \sqrt{x+3} - x$ has a zero in [1,3] approximately equal to 2.302776. Thus a point of intersection between the graphs of $f(x)$ and $g(x)$ is approximately (2.302776, 2.302776).

10. a. Let $\sqrt{40} = x \implies 40 = x^2 \implies x^2 - 40 = 0$.

Now we will find a positive zero for the function $f(x) = x^2 - 40$. Since $f(6) = -4$ and $f(7) = 9$ there must exist a zero in [6,7]. The approximation scheme for Newton's Method is

$$x_{n+1} = x_n - \frac{x_n^2 - 40}{2x_n}.$$

Using $x_1 = 7$, we obtain the following information.

n	x_n	$f(x_n)$	$f'(x_n)$	x_{n+1}
1	7.0	9.0	14.0	6.357143
2	6.357143	0.413267	12.714286	6.324639
3	6.324639	0.001058	12.649278	6.324555
4	6.324555	−0.000004	12.64911	6.324555

So the desired zero is approximately $x_4 = 6.324555$. Thus $\sqrt{40} \approx 6.324555$.

b. Let $\sqrt[3]{49} = x \implies 49 = x^3 \implies x^3 - 49 = 0$.

Now we need to find a zero for the function $f(x) = x^3 - 49$. Since $f(3) = -22$ and $f(4) = 15$ there must exist a zero in [3,4]. The approximation scheme for Newton's Method is

$$x_{n+1} = x_n - \frac{x_n^3 - 49}{3x_n^2}.$$

Using $x_1 = 4$, we obtain the following information.

n	x_n	$f(x_n)$	$f'(x_n)$	x_{n+1}
1	4.0	15.0	48.0	3.6875
2	3.6875	1.141357	40.792969	3.659521
3	3.659521	0.008649	40.176282	3.659306
4	3.659306	0.000012	40.171561	3.659306

The zero is approximately $x_4 = 3.659306$. Thus $\sqrt[3]{49} \approx 3.659306$.

c. Let $\sqrt[5]{18} = x \implies x^5 - 18 = 0$.

Now we shall find a zero for the function $f(x) = x^5 - 18$. Since $f(1) = -17$ and $f(2) = 14$ there must exist a zero in [1,2]. The approximation scheme for Newton's Method is

$$x_{n+1} = x_n - \frac{x_n^5 - 18}{5x_n^4}.$$

Using $x_1 = 2$, we obtain the following information:

n	x_n	$f(x_n)$	$f'(x_n)$	x_{n+1}
1	2.0	14.0	80.0	1.825
2	1.825	2.244840	55.465314	1.784527
3	1.784527	0.097376	50.706367	1.782607
4	1.782607	0.000229	50.488496	1.782602
5	1.782602	−0.000023	50.487930	1.782602

The zero is approximately $x_5 = 1.782602$. Thus $\sqrt[5]{18} \approx 1.782602$.

d. Let $\sqrt{37} = x \implies x^2 - 37 = 0$.

We now will find a positive zero for the function $f(x) = x^2 - 37$. Since $f(6) = -1$ and $f(7) = 12$ there must exist a zero in [6,7]. The approximation scheme for Newton's Method is

$$x_{n+1} = x_n - \frac{x_n^2 - 37}{2x_n}.$$

Using $x_1 = 7$, we obtain the following information.

n	x_n	$f(x_n)$	$f'(x_n)$	x_{n+1}
1	7.0	12.0	14.0	6.142857
2	6.142857	0.734692	12.285714	6.083056
3	6.083056	0.003570	12.166112	6.082763
4	6.082763	0.000006	12.165526	6.082763

The desired zero is approximately $x_4 = 6.082763$. Thus $\sqrt{37} \approx 6.082763$.

11. If the first guess x_1 is precisely the desired zero the results of every iteration will be the same number as the first guess. In this case $f(x_1) = 0$, hence $x_2 = x_1 - \frac{f(x_1)}{f'(x_1)} = x_1$, and, in consequence of the Newton's Method, $x_{n+1} = x_n = \cdots = x_1$ for all $n \geq 1$.

12. $f(x) = 2x = \tan x = g(x) \implies 2x - \tan x = 0.$

    We now will find a zero for the function $h(x) = 2x - \tan x$ in $\left(0, \frac{\pi}{2}\right)$. Since $h\left(\frac{\pi}{4}\right) \approx 0.571$ and $h(1.3) \approx -1.002$ there must exist a zero in $\left(0, \frac{\pi}{2}\right)$.

    The approximation scheme for Newton's Method is

$$x_{n+1} = x_n - \frac{2x_n - \tan x_n}{2 - \sec^2 x_n}.$$

Using $x_1 = 1.3$, we obtain the following information.

n	x_n	$h(x_n)$	$h'(x_n)$	x_{n+1}
1	1.3	−1.002102	−11.975142	1.216318
2	1.216318	−0.269250	−6.300186	1.173581
3	1.173581	−0.036545	−4.682060	1.165776
4	1.165776	−0.000953	−4.440581	1.165561
5	1.165561	0.000001	−4.434126	1.165561

Thus the desired point of intersection is approximately (1.165561, 2.331121).

13. $f(x) = (x - 2)^{1/3}$.

    Use Newton's Method.

$$x_{n+1} = x_n - \frac{(x_n - 2)^{1/3}}{\frac{1}{3}(x_n - 2)^{-2/3}} = x_n - 3(x_n - 2) = 6 - 2x_n, \quad x_1 = 3.$$

n	x_n	$f(x_n)$	$f'(x_n)$	x_{n+1}
1	3.0	1.0	$\frac{1}{3}$	0.0
2	0.0	−1.259921	0.209987	6.0
3	6.0	1.587401	0.132283	−6.0
4	−6.0	−2.0	0.083333	18.0

The results show that this approximation does not converge to the zero of $f(x)$ but moves away.

14. Let $2x^3 = 5x + 1 \implies h(x) = 2x^3 - (5x + 1) = 0$.

Since $h(0) = -1$, $h(-1) = 2$, and $h(-2) = -7$, there must exist one zero in each interval [-2,-1] and [-1,0].

The approximation iteration is

$$x_{n+1} = x_n - \frac{2x_n^3 - 5x_n - 1}{6x_n^2 - 5}.$$

For the zero in [-2,-1], use $x_1 = -2$.

n	x_n	$h(x_n)$	$h'(x_n)$	x_{n+1}
1	−2.0	−7.0	19.0	−1.631579
2	−1.631579	−1.528795	10.972300	−1.492247
3	−1.492247	−0.184640	8.360807	−1.470163
4	−1.470163	−0.004345	7.968275	−1.469618
5	−1.469618	−0.000005	7.958662	−1.469617

The point of intersection in [-2,-1] is approximately (-1.469617, -6.348082).

For the zero in [-1,0], use $x_1 = 0$.

n	x_n	$h(x_n)$	$h'(x_n)$	x_{n+1}
1	0.0	−1.0	−5.0	−0.2
2	−0.2	−0.016	−4.76	−0.203361
3	−0.203361	−0.000015	−4.751866	−0.203364

The point of intersection in [-1,0] is approximately (-0.203364,-0.016821).

Solutions to Chapter 9 - Review Exercises

1. $\{\sin\frac{n\pi}{2}\}$ does not converge since
$\{\sin\frac{n\pi}{2}\} = \{\sin\frac{\pi}{2},\ \sin\pi,\ \sin\frac{3\pi}{2},\ \sin 2\pi,\ \sin\frac{5\pi}{2},\cdots\}$
$= \{1,0,-1,0,1,0,-1,\cdots\}.$
These terms oscillate among 1, 0 and -1 and do not approach any single number as $n\to\infty$.

2. $\lim\limits_{n\to\infty}\frac{1}{\sqrt{n+5}} = 0.$

3. $\lim\limits_{n\to\infty}(\ln\sqrt{n} - \ln n) = \lim\limits_{n\to\infty}\ln\left(\frac{\sqrt{n}}{n}\right) = \lim\limits_{n\to\infty}\ln\left(\frac{1}{\sqrt{n}}\right) = -\infty.$
Thus $\{\ln\sqrt{n} - \ln n\}$ diverges.

4. $\lim\limits_{n\to\infty}\left(1+\frac{1}{n}\right)^{2n} = \left[\lim\limits_{n\to\infty}\left(1+\frac{1}{n}\right)^n\right]^2 = e^2.$

5. $\lim\limits_{n\to\infty}\frac{n^2+n-5}{4+2n^2} = \lim\limits_{n\to\infty}\frac{1+\frac{1}{n}-\frac{5}{n^2}}{\frac{4}{n^2}+2} = \frac{1+0-0}{0+2} = \frac{1}{2}.$

6. $\lim\limits_{n\to\infty}\frac{4+\sqrt{n}}{n^{2/3}+3} = \lim\limits_{n\to\infty}\frac{4+n^{1/2}}{n^{2/3}+3}\cdot\frac{\frac{1}{n^{2/3}}}{\frac{1}{n^{2/3}}} = \lim\limits_{n\to\infty}\frac{\frac{4}{n^{2/3}}+\frac{1}{n^{1/6}}}{1+\frac{3}{n^{2/3}}}$
$= \frac{0+0}{1+0} = \frac{0}{1} = 0.$

7. $\lim\limits_{n\to\infty}\frac{3^n+4^n}{5^n} = \lim\limits_{n\to\infty}\left[\left(\frac{3}{5}\right)^n + \left(\frac{4}{5}\right)^n\right] = \lim\limits_{n\to\infty}\left(\frac{3}{5}\right)^n + \lim\limits_{n\to\infty}\left(\frac{4}{5}\right)^n = 0+0 = 0.$

8. $\lim\limits_{n\to\infty}[\ln n - \ln(n+1)] = \lim\limits_{n\to\infty}\ln\left(\frac{n}{n+1}\right) = \lim\limits_{n\to\infty}\ln\left(1-\frac{1}{n+1}\right) = \ln 1 = 0.$

9. $\sum\limits_{k=0}^{\infty}\frac{1}{3^k} = \sum\limits_{k=0}^{\infty}\left(\frac{1}{3}\right)^k = \frac{1}{1-\frac{1}{3}} = \frac{3}{2}.$

10. $\sum\limits_{k=0}^{\infty}\frac{1+3^k}{5^k} = \sum\limits_{k=0}^{\infty}\left(\frac{1}{5}\right)^k + \sum\limits_{k=0}^{\infty}\left(\frac{3}{5}\right)^k = \frac{1}{1-\frac{1}{5}} + \frac{1}{1-\frac{3}{5}}$
$= \frac{1}{\frac{4}{5}} + \frac{1}{\frac{2}{5}} = \frac{5}{4} + \frac{5}{2} = \frac{15}{4}.$

11. $\sum\limits_{k=1}^{\infty}\frac{3^k}{2^k} = \sum\limits_{k=1}^{\infty}\left(\frac{3}{2}\right)^k$ is a divergent geometric series.

12. $\sum\limits_{k=1}^{\infty}\frac{7}{4^k} = 7\sum\limits_{k=1}^{\infty}\frac{1}{4^k} = 7\left(\sum\limits_{k=0}^{\infty}\frac{1}{4^k} - \frac{1}{4^0}\right)$
$= 7\left(\frac{1}{1-\frac{1}{4}} - 1\right) = 7\left(\frac{4}{3}-1\right) = \frac{7}{3}.$

13. $\sum_{k=1}^{\infty} \frac{1}{1+k^2}$ converges by comparison to the known convergent series $\sum_{k=1}^{\infty} \frac{1}{k^2}$.

14. $\sum_{k=1}^{\infty} \frac{1}{k(k+1)}$.

Since $\frac{1}{k(k+1)} < \frac{1}{k^2}$ for $k \geq 1$ and $\sum_{k=1}^{\infty} \frac{1}{k^2}$ converges, $\sum_{k=1}^{\infty} \frac{1}{k(k+1)}$ converges by the Comparison Test.

15. $\sum_{k=1}^{\infty} \frac{3^k}{k^2}$.

By the Ratio Test, since

$$\lim_{k\to\infty} \frac{a_{k+1}}{a_k} = \lim_{k\to\infty} \frac{3^{k+1}}{(k+1)^2} \cdot \frac{k^2}{3^k} = \lim_{k\to\infty} \left(\frac{k}{k+1}\right)^2 \cdot 3 = 3 > 1, \sum_{k=1}^{\infty} \frac{3^k}{k^2} \text{ diverges.}$$

16. $\sum_{k=1}^{\infty} \frac{3^{2k}}{k^3}$.

$$\lim_{k\to\infty} \frac{a_{k+1}}{a_k} = \lim_{k\to\infty} \frac{3^{2(k+1)}}{(k+1)^3} \cdot \frac{k^3}{3^{2k}} = \lim_{k\to\infty} \left(\frac{k}{k+1}\right)^3 \cdot 9 = 9 > 1.$$

Thus $\sum_{k=1}^{\infty} \frac{3^{2k}}{k^3}$ diverges by the Ratio Test.

17. $\sum_{k=1}^{\infty} \frac{k^2}{k!}$.

$$\lim_{k\to\infty} \frac{a_{k+1}}{a_k} = \lim_{k\to\infty} \frac{(k+1)^2}{(k+1)!} \cdot \frac{k!}{k^2} = \lim_{k\to\infty} \frac{1}{k+1} \cdot \left(\frac{k+1}{k}\right)^2 = 0.$$

Thus $\sum_{k=1}^{\infty} \frac{k^2}{k!}$ converges by the Ratio Test.

18. $\lim_{k\to\infty} \frac{k^2}{k^2+3} = 1 \neq 0$. Therefore $\sum_{k=0}^{\infty} \frac{k^2}{k^2+3}$ diverges by Theorem 5 in Section 9.3.

19. $\sum_{k=0}^{\infty} \frac{k-1}{k+1}$. Since $\lim_{k\to\infty} \frac{k-1}{k+1} = 1 \neq 0$, by Theorem 5 in Section 9.3 $\sum_{k=0}^{\infty} \frac{k-1}{k+1}$ diverges.

20. $\sum_{k=0}^{\infty} \frac{3}{k+2}$.

By the Limit Comparison Test:

$$\lim_{k\to\infty} \frac{3}{k+2} \Big/ \frac{1}{k} = \lim_{k\to\infty} \frac{3k}{k+2} = 3 > 0, \text{ and } \sum_{k=1}^{\infty} \frac{1}{k} \text{ diverges.}$$

Thus $\sum_{k=0}^{\infty} \frac{3}{k+2}$ diverges.

21. $\sum_{k=1}^{\infty} \frac{1}{\sqrt{k+1}}$.

By the Limit Comparison Test:

$\lim_{k\to\infty} \frac{1}{\sqrt{k+1}} \Big/ \frac{1}{\sqrt{k}} = \lim_{k\to\infty} \sqrt{\frac{k}{k+1}} = 1 > 0$, and $\sum_{k=1}^{\infty} \frac{1}{\sqrt{k}}$, a p-series with $p = \frac{1}{2} < 1$, diverges. Thus $\sum_{k=1}^{\infty} \frac{1}{\sqrt{k+1}}$ diverges.

22. $\sum_{k=1}^{\infty} \frac{1}{k \ln k}$.

By the Integral Test:

$$\int_1^{\infty} \frac{1}{x \ln x}\, dx = \lim_{t\to\infty} \int_1^t \frac{1}{x \ln x}\, dx = \lim_{t\to\infty} \int_1^t \frac{1}{\ln x} d(\ln x)$$
$$= \lim_{t\to\infty} (\ln \ln x)\Big]_1^t = \infty.$$

Thus $\sum_{k=1}^{\infty} \frac{1}{k \ln k}$ diverges.

23. $\sum_{k=1}^{\infty} \frac{1}{k^2+1}$.

Since $\frac{1}{k^2+1} < \frac{1}{k^2}$ and $\sum_{k=1}^{\infty} \frac{1}{k^2}$ converges, by the Comparison Test $\sum_{k=1}^{\infty} \frac{1}{k^2+1}$ converges.

24. $\sum_{k=1}^{\infty} \frac{(-1)^k}{2k+1}$. $\qquad a_k = \frac{1}{2k+1}$.

Since $a_{k+1} < a_k$ for all $k \geq 0$ and $\lim_{k\to\infty} a_k = \lim_{k\to\infty} \frac{1}{2k+1} = 0$, by the Alternating Series Test $\sum_{k=1}^{\infty} \frac{(-1)^k}{2k+1}$ converges.

25. $\sum_{k=0}^{\infty} \frac{(-1)^k k}{\sqrt{k+1}}$. $\qquad a_k = \frac{k}{\sqrt{k+1}}$.

Since $\lim_{k\to\infty} a_k = \lim_{k\to\infty} \frac{k}{\sqrt{k+1}} = \infty$ the condition (ii) in the Alternating Series Test fails. By the necessary condition for convergence of series, Theorem 5 in Section 9.3, $\sum_{k=0}^{\infty} \frac{(-1)^k k}{\sqrt{k+1}}$ diverges.

26. $\sum_{k=0}^{\infty} \frac{\cos(\pi k^2)}{k+1} = 1 - \frac{1}{2} + \frac{1}{3} - \frac{1}{4} + \frac{1}{5} - \frac{1}{6} + \cdots$ which converges by the alternating series test.

27. $\sum_{k=1}^{\infty} \frac{k \cdot 2^k}{k!}$.

$\lim_{k\to\infty} \frac{a_{k+1}}{a_k} = \lim_{k\to\infty} \frac{(k+1)2^{k+1}}{(k+1)!} \cdot \frac{k!}{k \cdot 2^k} = \lim_{k\to\infty} \frac{2(k+1)}{(k+1)\cdot k} = 0 < 1.$

By the Ratio Test $\sum_{k=1}^{\infty} \frac{k \cdot 2^k}{k!}$ converges.

28. $\sum_{k=1}^{\infty} (-1)^{k+1} \frac{\sqrt{k}}{k+2}$. $\qquad a_k = \frac{\sqrt{k}}{k+2}$.

Let $f(x) = \frac{\sqrt{x}}{x+2}$. $\quad f'(x) = \frac{1}{2\sqrt{x}(x+2)} - \frac{\sqrt{x}}{(x+2)^2} = \frac{2-x}{2\sqrt{x}(x+2)^2} < 0$ for $x > 2$.

So $a_k = \frac{\sqrt{k}}{k+2} > \frac{\sqrt{k+1}}{(k+1)+2} = a_{k+1}$ for all $k \geq 2$.

And $\lim_{k\to\infty} a_k = \lim_{k\to\infty} \frac{\sqrt{k}}{k+2} = 0$.

By the Alternating Series Test, $\sum_{k=1}^{\infty} (-1)^{k+1} \cdot \frac{\sqrt{k}}{k+2}$ converges.

29. $\sum_{k=0}^{\infty} \frac{1}{(k+2)^{3/2}}$.

Since $\frac{1}{(k+2)^{3/2}} < \frac{1}{k^{3/2}}$ for all $k > 0$ and $\sum_{k=1}^{\infty} \frac{1}{k^{3/2}}$, a p-series with $p = \frac{3}{2} > 1$, converges, $\sum_{k=0}^{\infty} \frac{1}{(k+2)^{3/2}}$ converges.

30. $\sum_{k=1}^{\infty} \frac{(-1)^k k}{(k+1)!}$. $\qquad a_k = \frac{k}{(k+1)!} = \frac{1}{(k+1)(k-1)!}$.

$a_k > a_{k+1}$ for all $k \geq 1$ and $\lim_{k\to\infty} a_k = 0$.

By the Alternating Series Test $\sum_{k=1}^{\infty} \frac{(-1)^k k}{(k+1)!}$ converges.

31. $f(x) = e^{-3x}, \quad a = 0, \quad n = 3.$

$f'(x) = -3e^{-3x}, \quad f''(x) = 9e^{-3x}, \quad f'''(x) = -27e^{-3x}.$

$$P_3(x) = f(0) + f'(0)x + \tfrac{1}{2!}f''(0)x^2 + \tfrac{1}{3!}f'''(0)x^3$$

$$= 1 - 3x + \tfrac{9}{2}x^2 - \tfrac{27}{3!}x^3 = 1 - 3x + \tfrac{9}{2}x^2 - \tfrac{9}{2}x^3.$$

32. $f(x) = \cos x, \quad a = \frac{\pi}{6}, \quad n = 3.$

$f'(x) = -\sin x, \quad f''(x) = -\cos x, \quad f'''(x) = \sin x.$

$f\left(\frac{\pi}{6}\right) = \frac{\sqrt{3}}{2}, \quad f'\left(\frac{\pi}{6}\right) = -\frac{1}{2}, \quad f''\left(\frac{\pi}{6}\right) = -\frac{\sqrt{3}}{2}, \quad f'''\left(\frac{\pi}{6}\right) = \frac{1}{2}.$

$$P_3(x) = f\left(\frac{\pi}{6}\right) + f'\left(\frac{\pi}{6}\right)\left(x - \frac{\pi}{6}\right) + \frac{1}{2!}f''\left(\frac{\pi}{6}\right)\left(x - \frac{\pi}{6}\right)^2 + \frac{1}{3!}f'''\left(\frac{\pi}{6}\right)\left(x - \frac{\pi}{6}\right)^3$$

$$= \frac{\sqrt{3}}{2} - \frac{1}{2}\left(x - \frac{\pi}{6}\right) - \frac{\sqrt{3}}{4}\left(x - \frac{\pi}{6}\right)^2 + \frac{1}{12}\left(x - \frac{\pi}{6}\right)^3.$$

33. $f(x) = \tan x, \quad a = 0, \quad n = 2.$

$f'(x) = \sec^2 x, \quad f''(x) = 2\sec x \cdot \sec x \tan x = 2\sec^2 x \cdot \tan x.$

$f(0) = 0, \quad f'(0) = 1, \quad f''(0) = 0.$

$P_2(x) = f(0) + f'(0)x + \frac{1}{2!}f''(0)x^2 = x.$

34. $f(x) = e^{x^2}, \quad a = 0, \quad n = 3.$

$f'(x) = 2xe^{x^2}, \quad f''(x) = 2(e^{x^2} + x \cdot 2xe^{x^2}) = 2e^{x^2}(1 + 2x^2),$

$f'''(x) = 2 \cdot 4xe^{x^2} + 2 \cdot (1 + 2x^2) \cdot 2xe^{x^2} = 4xe^{x^2}(3 + 2x^2).$

$f(0) = 1, \quad f'(0) = 0, \quad f''(0) = 2, \quad f'''(0) = 0.$

$P_3(x) = f(0) + f'(0)x + \frac{1}{2!}f''(0)x^2 + \frac{1}{3!}f'''(0)x^3 = 1 + x^2.$

35. $f(x) = \ln(1 + x^2), \quad a = 0, \quad n = 2.$

$f'(x) = \frac{2x}{1+x^2}, \quad f''(x) = \frac{2(1+x^2) - 2x \cdot 2x}{(1+x^2)^2} = \frac{2(1-x^2)}{(1+x^2)^2}.$

$f(0) = 0, \quad f'(0) = 0, \quad f''(0) = 2.$

$P_2(x) = f(0) + f'(0)x + \frac{1}{2!}f''(0)x^2 = x^2.$

36. $f(x) = \sin x, \quad a = \frac{\pi}{6}, \quad n = 3.$

$f'(x) = \cos x, \quad f''(x) = -\sin x, \quad f'''(x) = -\cos x.$

$f\left(\frac{\pi}{6}\right) = \frac{1}{2}, \quad f'\left(\frac{\pi}{6}\right) = \frac{\sqrt{3}}{2}, \quad f''\left(\frac{\pi}{6}\right) = -\frac{1}{2}, \quad f'''\left(\frac{\pi}{6}\right) = -\frac{\sqrt{3}}{2}.$

$$P_3(x) = f\left(\tfrac{\pi}{6}\right) + f'\left(\tfrac{\pi}{6}\right)\left(x - \tfrac{\pi}{6}\right) + \tfrac{1}{2!}f''\left(\tfrac{\pi}{6}\right)\left(x - \tfrac{\pi}{6}\right)^2 + \tfrac{1}{3!}f'''\left(\tfrac{\pi}{6}\right)\left(x - \tfrac{\pi}{6}\right)^3$$
$$= \tfrac{1}{2} + \tfrac{\sqrt{3}}{2}\left(x - \tfrac{\pi}{6}\right) - \tfrac{1}{4}\left(x - \tfrac{\pi}{6}\right)^2 - \tfrac{\sqrt{3}}{12}\left(x - \tfrac{\pi}{6}\right)^3.$$

37. $f(x) = \sqrt{x}, \quad a = 9, \quad n = 3.$

$f'(x) = \frac{1}{2\sqrt{x}}, \quad f''(x) = -\frac{1}{4}x^{-3/2}, \quad f'''(x) = \frac{3}{8}x^{-5/2}.$

$f(9) = 3, \quad f'(9) = \frac{1}{6}, \quad f''(9) = -\frac{1}{4\cdot 27}, \quad f'''(9) = \frac{1}{8\cdot 81}.$

$$P_3(x) = f(9) + f'(9)(x-9) + \tfrac{1}{2!}f''(9)(x-9)^2 + \tfrac{1}{3!}f'''(9)(x-9)^3$$
$$= 3 + \tfrac{1}{6}(x-9) - \tfrac{1}{216}(x-9)^2 + \tfrac{1}{3888}(x-9)^3.$$

38. $f(x) = \sec x, \quad a = \frac{\pi}{4}, \quad n = 3.$

$f'(x) = \sec x \tan x, \quad f''(x) = \sec x \tan^2 x + \sec^3 x = \sec x(1 + 2\tan^2 x),$

$f'''(x) = \sec x \tan^3 x + 2\sec^3 x \tan x + 3\sec^3 x \tan x.$

$f\left(\frac{\pi}{4}\right) = \sqrt{2}, \quad f'\left(\frac{\pi}{4}\right) = \sqrt{2}, \quad f''\left(\frac{\pi}{4}\right) = 3\sqrt{2}, \quad f'''\left(\frac{\pi}{4}\right) = 11\sqrt{2}.$

$$P_3(x) = f\left(\tfrac{\pi}{4}\right) + f'\left(\tfrac{\pi}{4}\right)\left(x - \tfrac{\pi}{4}\right) + \tfrac{1}{2!}f''\left(\tfrac{\pi}{4}\right)\left(x - \tfrac{\pi}{4}\right)^2 + \tfrac{1}{3!}f'''\left(\tfrac{\pi}{4}\right)\left(x - \tfrac{\pi}{4}\right)^3$$
$$= \sqrt{2} + \sqrt{2}\left(x - \tfrac{\pi}{4}\right) + \tfrac{3\sqrt{2}}{2}\left(x - \tfrac{\pi}{4}\right)^2 + \tfrac{11\sqrt{2}}{6}\left(x - \tfrac{\pi}{4}\right)^3.$$

39. $f(x) = xe^x, \quad a = 0.$

$f'(x) = e^x(1+x), \quad f''(x) = e^x(2+x), \quad \cdots, \quad f^{(k)}(x) = e^x(k+x)$ for all $k \geq 0$.

$f^{(k)}(0) = k, \quad k \geq 0.$

The Taylor series for $f(x)$ is

$$\sum_{k=0}^{\infty} \frac{f^{(k)}(0)}{k!}x^k = x + \tfrac{2}{2!}x^2 + \tfrac{3}{3!}x^3 + \cdots + \tfrac{k}{k!}x^k + \cdots$$
$$= x + x^2 + \tfrac{1}{2!}x^3 + \cdots + \tfrac{1}{(k-1)!}x^k + \cdots.$$

40. $\sum_{k=1}^{\infty} \frac{k}{k!} x^k = \sum_{k=1}^{\infty} \frac{1}{(k-1)!} x^k$.

$\lim_{k\to\infty} \left|\frac{a_{k+1}}{a_k}\right| = \lim_{k\to\infty} \left|\frac{x^{k+1}}{k!} \cdot \frac{(k-1)!}{x^k}\right| = \lim_{k\to\infty} \frac{1}{k}|x| = 0$ for any $x \neq 0$. Thus the Taylor series for $f(x) = xe^x$ converges absolutely for all x.

41. $f(x) = \frac{1}{1-x} - 1, \quad a = 0.$

$f'(x) = \frac{1}{(1-x)^2}, \quad f''(x) = \frac{2!}{(1-x)^3}, \quad f'''(x) = \frac{3!}{(1-x)^4}, \cdots,$

$f^{(k)}(x) = \frac{k!}{(1-x)^{k+1}}$ for all $k \geq 1$.

$f^{(k)}(0) = k!$ for all $k \geq 1$.

The Taylor series for $f(x)$ is

$\sum_{k=0}^{\infty} \frac{f^{(k)}(0)}{k!} x^k = \sum_{k=1}^{\infty} \frac{k!}{k!} x^k = \sum_{k=1}^{\infty} x^k.$

42. $\sum_{k=1}^{\infty} x^k$. $\quad \lim_{k\to\infty} \left|\frac{a_{k+1}}{a_k}\right| = \lim_{k\to\infty} |x| = |x|$.

By the Ratio test, $\sum_{k=1}^{\infty} x^k$ converges absolutely for all x such that $|x| < 1$.

At $x = 1$: $\sum_{k=1}^{\infty} 1^k = 1 + 1 + 1 + \cdots$ diverges.

At $x = -1$: $\sum_{k=1}^{\infty} (-1)^k = -1 + 1 - 1 + \cdots$ diverges since the k-th term does not converge to 0.

For $|x| > 1$, $\sum_{k=1}^{\infty} x^k$ diverges since the k-th term does not converge to 0.

Thus, this series converges only for all x such that $|x| < 1$.

43. $f(x) = x^3 + 2x^2 - x + 3.$

$f'(x) = 3x^2 + 4x - 1, \quad f''(x) = 6x + 4, \quad f'''(x) = 6, \quad f^{(k)}(x) = 0$ for all $k \geq 4$.

a. $f'(0) = -1, \quad f''(0) = 4, \quad f'''(0) = 6, \quad f^{(k)}(0) = 0$ for $k \geq 4$.

The Taylor Series for $f(x)$ is

$$\sum_{k=0}^{\infty} \frac{f^{(k)}(0)}{k!} x^k = 3 - x + \frac{4}{2!}x^2 + \frac{6}{3!}x^3 = 3 - x + 2x^2 + x^3$$
$$= x^3 + 2x^2 - x + 3.$$

b. $f(1) = 5, \quad f'(1) = 6, \quad f''(1) = 10, \quad f'''(1) = 6, \quad f^{(k)}(1) = 0$ for $k \geq 4$.

The Taylor Series for $f(x)$ is

$$\sum_{k=0}^{\infty} \frac{f^{(k)}(1)}{k!}(x-1)^k = 5 + 6(x-1) + \frac{10}{2!}(x-1)^2 + \frac{6}{3!}(x-1)^3$$
$$= 5 + 6(x-1) + 5(x-1)^2 + (x-1)^3$$
$$= x^3 + 2x^2 - x + 3.$$

44. (a) $0.7777\cdots = \frac{7}{10} + \frac{7}{10^2} + \cdots + \frac{7}{10^k} + \cdots = \sum_{k=1}^{\infty} \frac{7}{10^k}$.

(b) $\sum_{k=1}^{\infty} \frac{7}{10^k} = \left(\sum_{k=0}^{\infty} \frac{7}{10^k}\right) - \frac{7}{10^0} = 7\left(\sum_{k=0}^{\infty} \frac{1}{10^k}\right) - 7$
$= 7 \cdot \frac{1}{1-\frac{1}{10}} - 7 = 7\left(\frac{10}{9} - 1\right) = \frac{7}{9}$.

45. Let $S(n) =$ the total amount paid by the annuity in n years. Then

$$S(1) = 500$$
$$S(2) = 500 + 500\left(\frac{1}{3}\right) = 500\left(1 + \frac{1}{3}\right)$$
$$S(3) = 500 + 500\left(\frac{1}{3}\right) + 500\left(\frac{1}{3}\right)^2 = 500\left(1 + \frac{1}{3} + \frac{1}{3^2}\right)$$
$$\vdots$$
$$S(n) = 500 + 500\left(\frac{1}{3}\right) + \cdots + 500\left(\frac{1}{3}\right)^{n-1} = 500\sum_{k=0}^{n-1} \frac{1}{3^k}.$$

Thus the total amount paid by the annuity in perpetuity is

$$S = \lim_{n\to\infty} S_n = \lim_{n\to\infty} 500 \sum_{k=0}^{n-1} \frac{1}{3^k} = 500 \sum_{k=0}^{\infty} \frac{1}{3^k}$$

$$= 500 \left(\frac{1}{1-\frac{1}{3}} \right) = \$750.$$

46. $f(x) = x^2 - x - 1.$

Since $f(0) = -1$ and $f(2) = 1$, there must exist a zero in [0,2].

The approximation scheme for Newton's Method is

$$x_{n+1} = x_n - \frac{x_n^2 - x_n - 1}{2x_n - 1}.$$

Using $x_1 = 2$, we obtain the following information:

n	x_n	$f(x_n)$	$f'(x_n)$	x_{n+1}
1	2.0	1.0	3.0	1.666667
2	1.666667	0.111112	2.333334	1.619048
3	1.619048	0.002268	2.238096	1.618035
4	1.618035	0.000002	2.23607	1.618034
5	1.618034	0.000000025	2.236068	1.618034

The desired solution of the equation is approximately $x_6 = 1.618034$.

47. $\cot x = x \implies f(x) = \cot x - x = 0.$

Since $f(\frac{\pi}{4}) = 1 - \frac{\pi}{4} > 0$ and $f\left(\frac{\pi}{2}\right) = -\frac{\pi}{2} < 0$ there must exist a zero of $f(x)$ in $\left[0, \frac{\pi}{2}\right]$. The approximation scheme is

$$x_{n+1} = x_n - \frac{\cot x_n - x_n}{-\csc^2 x_n - 1}.$$

Using $x_1 = \frac{\pi}{2}$, we obtain the following information:

n	x_n	$f(x_n)$	$f'(x_n)$	x_{n+1}
1	$\frac{\pi}{2}$	$-\frac{\pi}{2}$	−2.0	0.785398
2	0.785398	0.214602	−3.000001	0.856932
3	0.856932	0.009338	−2.750424	0.860327
4	0.860327	0.000018	−2.740194	0.860334
5	0.860334	0.000001	−2.740173	0.860334

The solution of the equation in $\left[0, \frac{\pi}{2}\right]$ is approximately $x_5 = 0.860334$.

48. Let $\sqrt[3]{175} = x \implies x^3 - 175 = 0$.

We now need to find a zero for the function $f(x) = x^3 - 175$. Since $f(5) = -50$ and $f(6) = 41$ there must exist a zero in [5,6]. The approximation scheme is

$$x_{n+1} = x_n - \frac{x_n^3 - 175}{3x_n^2}.$$

Using $x_1 = 6$, we obtain the following information:

n	x_n	$f(x_n)$	$f'(x_n)$	x_{n+1}
1	6.0	41.0	108.0	5.620370
2	5.620370	2.539389	94.765677	5.593573
3	5.593573	0.012042	93.864177	5.593445
4	5.593445	0.000027	93.859881	5.593445

Thus $x = \sqrt[3]{175} \approx 5.593445$.

49. Note that at a relative maximum point x of f, $f'(x) = 0$, i.e., $\sin x + x\cos x = 0$.

Let $F(x) = \sin x + x\cos x$. We then need to find a zero of $F(x)$ in [0,3]. Since $F(1) = 1.381773$ and $F(3) = -2.828857$ there must exist a zero of $F(x)$ in [1,3].

By Newton's Method

$$x_{n+1} = x_n - \frac{\sin x_n + x_n \cos x_n}{2\cos x_n - x_n \sin x_n},$$

and using $x_1 = 3$ we have

n	x_n	$F(x_n)$	$F'(x_n)$	x_{n+1}
1	3.0	−2.828857	−2.403345	1.822950
2	1.822950	0.513569	−2.264284	2.049763
3	2.049763	−0.057187	−2.740831	2.028898
4	2.028898	−0.000379	−2.704198	2.028758
5	2.028758	−0.0000004377	−2.703947	2.028758

Note that $F(x) > 0$ for all $0 < x \leq \frac{\pi}{2}$. Also note that $F'(x) < 0$ for all $\frac{\pi}{2} < x < 3$. Therefore there is exactly one point x in the open interval (0,3) such that $F(x) = 0$. Hence $f(x)$ has the unique critical value 2.028758 in the open interval (0,3). Since $f(0) = 0$, $f(3) = 3\sin 3 = 0.42336$, and $f(2.028758) = 1.8197$, the maximum value of $f(x)$ for x in [0,3] is $f(2.028758) = 1.8197$.

50. The present value of the entire revenue stream is

$$\sum_{n=1}^{\infty} P(n) = \sum_{n=1}^{\infty} 100 \cdot (e^{-.06})^n = 100 \cdot e^{-.06} \cdot \sum_{n=0}^{\infty} (e^{-.06})^n$$
$$= 100e^{-.06} \cdot \frac{1}{1 - e^{-.06}} = \$1,617.17.$$

51. a. The total distance traveled by the ball when it has struck the ground 3 times is

$$10 + 2 \cdot (10) \cdot \frac{5}{6} + 2 \cdot (10) \left(\frac{5}{6}\right)^2 = 10 + \frac{50}{3} + \frac{125}{9} = \frac{365}{9} \text{ feet.}$$

b. The total distance traveled by the ball when it has come to rest is

$$10 + 2 \cdot 10 \cdot \frac{5}{6} \cdot \sum_{n=0}^{\infty} \left(\frac{5}{6}\right)^n = 10 + \frac{50}{3} \cdot \frac{1}{1 - \frac{5}{6}} = 110 \text{ feet.}$$

CHAPTER 10

Solutions to Exercise Set 10.1

1. $y = Ce^{\pi x} \implies \frac{dy}{dx} = \pi C e^{\pi x} = \pi y.$

2. $y = \sqrt{x^2 + C} \implies \frac{dy}{dx} = \frac{x}{\sqrt{x^2+C}} = \frac{x}{y}.$

3. $y = A\sin 3x + B\cos 3x.$

   $\frac{dy}{dx} = 3A\cos 3x - 3B\sin 3x.$

   $\frac{d^2y}{dx^2} = -9A\sin 3x - 9B\cos 3x = -9y.$

   Therefore $\frac{d^2y}{dx^2} + 9y = 0.$

4. $f(t) = Ae^{4t} + Be^{-4t}.$

   $f'(t) = 4Ae^{4t} - 4Be^{-4t}.$

   $f''(t) = 16Ae^{4t} + 16Be^{-4t} = 16(Ae^{4t} + Be^{-4t}) = 16f(t).$

   Therefore, $f''(t) - 16f(t) = 0.$

5. $y = \frac{1}{2} + C\ e^{-4t}.$

   $\frac{dy}{dt} = -4C\ e^{-4t} = -4\left(y - \frac{1}{2}\right) = 2 - 4y.$

6. $f(t) = 100 + C\ e^{-0.02t}.$

   $f'(t) = -0.02C\ e^{-0.02t} = -0.02\left(f(t) - 100\right) = 2 - 0.02f(t).$

7. $y = 22 - C\ e^{-10x}$.

$\frac{dy}{dx} = 10C\ e^{-10x} = 10 \cdot (22 - y)$.

8. $y = Ae^{-2x} + Be^{3x}$.

$\frac{dy}{dx} = -2Ae^{-2x} + 3Be^{3x}$.

$\frac{d^2y}{dx^2} = 4Ae^{-2x} + 9Be^{3x}$.

$$\frac{d^2y}{dx^2} - \frac{dy}{dx} - 6y = \left(4Ae^{-2x} + 9Be^{3x}\right) - \left(-2Ae^{-2x} + 3Be^{3x}\right)$$
$$-6 \cdot \left(Ae^{-2x} + Be^{3x}\right) = 0.$$

9. $f(t) = Ae^{-2t} + Bte^{-2t}$.

$f'(t) = -2Ae^{-2t} - 2Bte^{-2t} + Be^{-2t}$.

$$f''(t) = 4Ae^{-2t} - 2Be^{-2t} + 4Bte^{-2t} - 2Be^{-2t}$$
$$= (4A - 4B)e^{-2t} + 4Bte^{-2t}.$$

$$f''(t) + 4f'(t) + 4f(t) = (4A - 4B)e^{-2t} + 4Bte^{-2t}$$
$$+4(-2Ae^{-2t} + Be^{-2t} - 2Bte^{-2t}) + 4f(t)$$
$$= -4Ae^{-2t} - 4Bte^{-2t} + 4f(t)$$
$$= -4(Ae^{-2t} + Bte^{-2t}) + 4f(t)$$
$$= -4f(t) + 4f(t) = 0.$$

10. $y = C\ e^{-t^2} + \frac{1}{2}$.

$\frac{dy}{dt} = (-2te^{-t^2}) \cdot C = -2t(C\ e^{-t^2}) = -2t \cdot \left(y - \frac{1}{2}\right) = -2ty + t$.

Thus, $\frac{dy}{dt} + 2ty = t$.

11. $\frac{dy}{dx} = 9 - x$.

$\int dy = \int (9 - x)dx \implies y = 9x - \frac{x^2}{2} + C$.

12. $f'(t) = e^{2t} + 2\sqrt{t} + 5.$

$$\int f'(t)dt = \int \left(e^{2t} + 2\sqrt{t} + 5\right) dt = \tfrac{1}{2}e^{2t} + 2 \cdot \tfrac{2}{3}t^{3/2} + 5t + C$$

$$= \tfrac{1}{2}e^{2t} + \tfrac{4}{3}t^{3/2} + 5t + C.$$

13. $\frac{dy}{dt} = t\ e^t.$

$$\int dy = \int te^t\ dt \implies y = t\ e^t - \int e^t\ dt$$

$$\implies y = t \cdot e^t - e^t + C.$$

14. $\frac{dy}{dx} = \sec x \cdot \tan x.$

$$\int dy = \int \sec x\ \ \tan x\ dx \implies y = \sec x + C.$$

15. $\frac{d^2y}{dt^2} = \frac{t+1}{\sqrt{t}}, \quad t > 0.$

$$\int \tfrac{d^2y}{dt^2} \cdot dt = \int \tfrac{t+1}{\sqrt{t}}\ dt \implies \int d\left(\tfrac{dy}{dt}\right) = \int \tfrac{t+1}{\sqrt{t}}\ dt$$

$$\implies \tfrac{dy}{dt} = \int \left(t^{1/2} + t^{-1/2}\right) dt = \tfrac{2}{3}t^{3/2} + 2t^{1/2} + C_1.$$

$$\therefore\ \tfrac{dy}{dt} = \tfrac{2}{3}t^{3/2} + 2t^{1/2} + C_1.$$

$$\int dy = \int \left(\tfrac{2}{3}t^{3/2} + 2t^{1/2} + C_1\right) dt$$

$$\implies y = \tfrac{4}{15}t^{5/2} + \tfrac{4}{3}t^{3/2} + C_1t + C_2.$$

16. $\frac{d^2y}{dt^2} = \cos t - \sin t.$

$$\tfrac{dy}{dt} = \int (\cos t - \sin t)\, dt = \sin t + \cos t + C_1.$$

$$y = \int(\sin t + \cos t + C_1)dt = -\cos t + \sin t + C_1t + C_2.$$

17. $\frac{dy}{dx} = x\sqrt{x^2+1}$.

$y = \int x\sqrt{x^2+1}\ dx = \int \frac{1}{2}\sqrt{t}\ dt$, where $t = x^2+1$.

$\therefore\ y = \frac{1}{3}t^{3/2} + C = \frac{1}{3}\left(x^2+1\right)^{3/2} + C$.

Since $y(0) = 1$, $\quad y(0) = \frac{1}{3} + C = 1 \implies C = \frac{2}{3}$.

Thus, $y = \frac{1}{3}\left(x^2+1\right)^{3/2} + \frac{2}{3}$.

18. $\frac{dy}{dx} = \frac{x}{y}$.

By Exercise 2, $y = \sqrt{x^2+C}$.

From $y(0) = 1$, $\quad y(0) = \sqrt{C} = 1 \implies C = 1$.

Thus, $y = \sqrt{x^2+1}$.

19. $\frac{dy}{dx} = 2 - 4y$.

By Exercise 5, $y = \frac{1}{2} + C\ e^{-4t}$.

From $y(0) = 3$, $\quad y(0) = \frac{1}{2} + C = 3 \implies C = \frac{5}{2}$.

Thus, $y = \frac{1}{2} + \frac{5}{2}e^{-4t}$.

20. $\frac{dy}{dt} = \sec^2 t$.

$\int dy = \int \sec^2 t\ dt \implies y = \tan t + C$.

From $y(0) = \pi$, $\quad y(0) = C = \pi$.

Thus, $y = \tan t + \pi$.

21. $\frac{dy}{dx} = -y$.

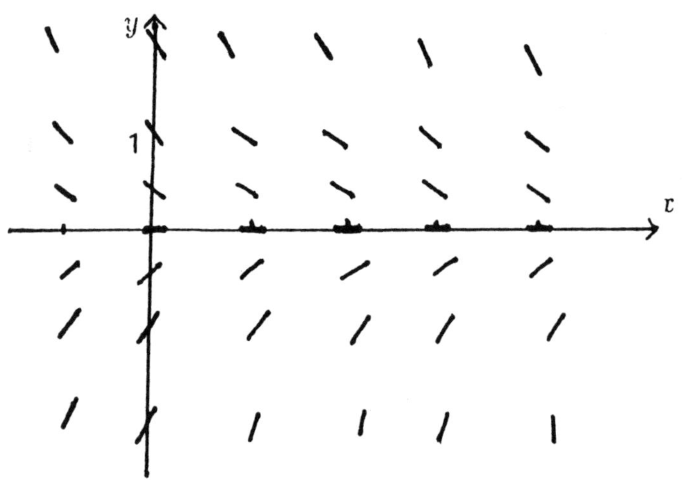

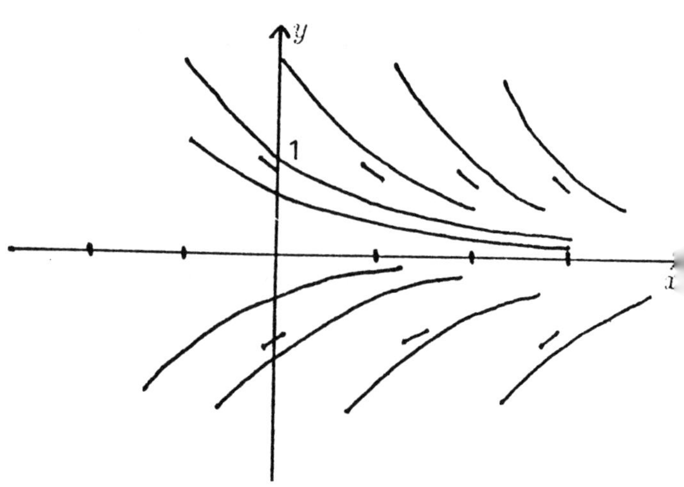

Direction field for

$$\frac{dy}{dx} = -y.$$

Solution curves associated

with direction field.

22. $\frac{dy}{dx} = y(4-y)$.

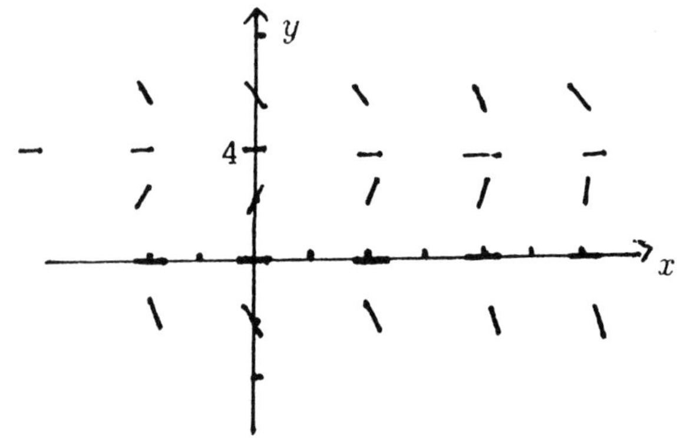

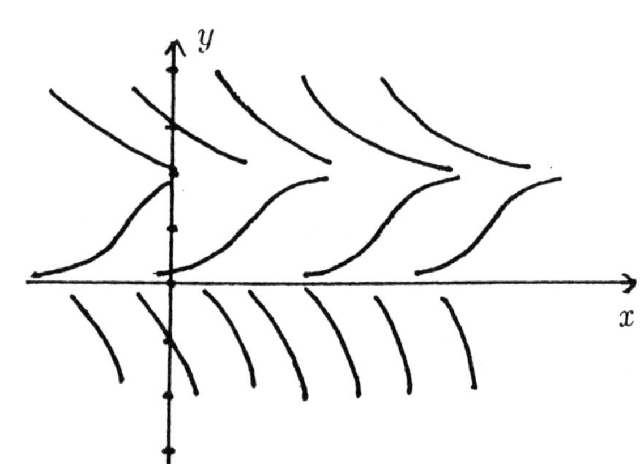

Direction field for

$$\frac{dy}{dx} = y(4-y).$$

Solution curves associated

with direction field.

23. $\frac{dy}{dx} = y(y-3)$.

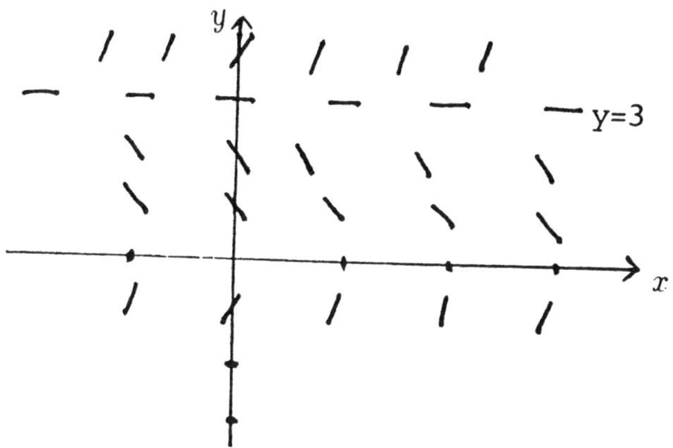

Direction field for

$$\frac{dy}{dx} = y(y-3).$$

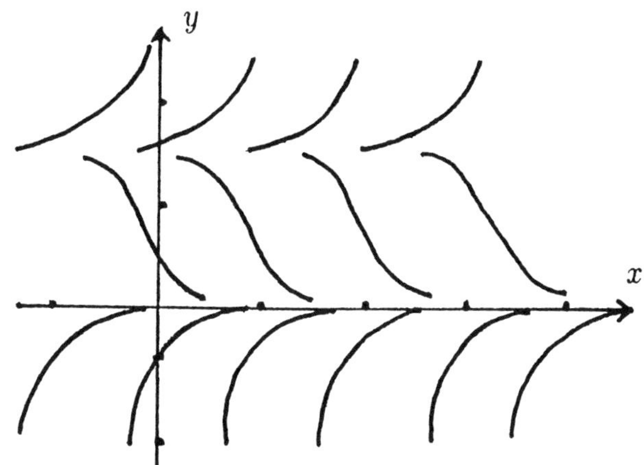

Solution curves associated

with direction field.

24. $f'(t) = 1 - f(t)$.

If we put $y = f(t)$, $\quad \frac{dy}{dt} = 1 - y$.

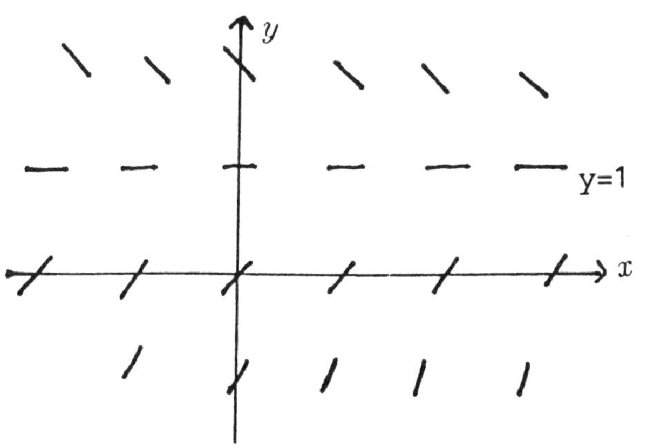

Direction field

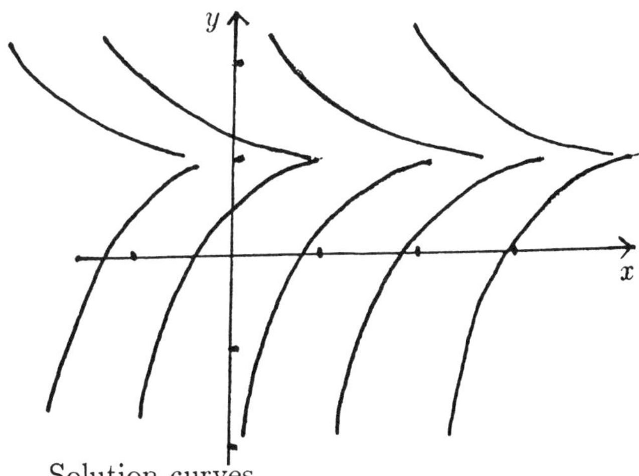

Solution curves

25. a. $\begin{cases} \frac{dV}{dt} = \sqrt{t+1} \\ V(0) = 5. \end{cases}$

b. $V = \int \sqrt{t+1}\ dt = \frac{2}{3}(t+1)^{3/2} + C.$

Since $V(0) = \frac{2}{3} + C = 5, \quad C = 5 - \frac{2}{3} = \frac{13}{3}.$

$\therefore\ V(t) = \frac{2}{3}(t+1)^{3/2} + \frac{13}{3}.$

26. a. $\begin{cases} C'(x) = 2x \\ C(0) = 400. \end{cases}$

b. $C(x) = \int (2x)dx = x^2 + K.$

Since $C(0) = K = 400, \quad C(x) = x^2 + 400.$

27. a. $\begin{cases} U'(x) = 6 - 2x \\ U(0) = 0. \end{cases}$

b. $U(x) = 6x - x^2 + C.$

$U(0) = 0 = C.$

$U(x) = 6x - x^2.$

c. $U'(x) < 0$ for $x > 3$.

Therefore $U(x)$ decreases as x increases for $x > 3$.

Hence, the answer is no.

28. a. $f(t) = Ce^{6t}$. $f'(t) = 6Ce^{6t} = 6f(t)$.

$\therefore\ f'(t) - 6f(t) = 0$.

b. $f(t) = A\sin 3t + B\cos 3t$. $f'(t) = 3(A\cos 3t - B\sin 3t)$.

$f''(t) = 9(-A\sin 3t - B\cos 3t) = -9f(t)$.

$\therefore f''(t) + 9f(t) = 0$.

Solutions to Exercise Set 10.2

1. $\frac{dy}{dx} = k\cdot(M-y)$ where $k=1$, $M=1$.

$y = M\cdot\left(1 - Ce^{-kx}\right) \implies y = 1 - Ce^{-x}$.

2. $\frac{dy}{dt} = 4 - 2y = 2\cdot(2-y)$. Here $\frac{dy}{dt} = k\cdot(M-y)$ where $k=2$, $M=2$.

$y = M\cdot\left(1 - Ce^{-kt}\right) \implies y = 2\cdot\left(1 - Ce^{-2t}\right)$.

3. $\frac{dy}{dt} = 10 - 2y = 2\cdot(5-y)$. Here $\frac{dy}{dt} = k\cdot(M-y)$ where $k=2$, $M=5$.

$y = M\cdot\left(1 - Ce^{-kt}\right) \implies 5\cdot\left(1 - Ce^{-2t}\right)$.

$y(0) = 1 \implies 1 = 5\cdot\left(1 - Ce^{-2\cdot(0)}\right) \implies .2 = 1 - C \implies C = .8$.

$y = 5\cdot\left(1 - 0.8\cdot e^{-2t}\right)$.

4. $\frac{dy}{dt} = 10\cdot(2-4y) = 40\cdot\left(\frac{1}{2} - y\right)$. Here $\frac{dy}{dt} = k\cdot(M-y)$ where $k=40$, $M=\frac{1}{2}$.

$y = M\cdot\left(1 - Ce^{-kt}\right) \implies y = \frac{1}{2}\cdot\left(1 - Ce^{-40t}\right)$.

$y(0) = 0 \implies 0 = \frac{1}{2}\cdot\left(1 - Ce^{-40\cdot(0)}\right) \implies C = 1$.

$y = \frac{1}{2}\cdot\left(1 - e^{-40t}\right)$.

5. $\frac{dy}{dt} = ky \cdot \left(\frac{M-y}{M}\right)$ where $k = 10$, $M = 50$.

$y = \frac{M}{1+Ce^{-kt}} \implies y = \frac{50}{1+Ce^{-10t}}.$

6. $\frac{dN}{dt} = kN \cdot \left(\frac{M-N}{M}\right)$ where $k = 5$, $M = 100$.

$N = \frac{M}{1+Ce^{-kt}} \implies N = \frac{100}{1+Ce^{-5t}}.$

$N(0) = 10 \implies 10 = \frac{100}{1+Ce^{-5(0)}} \implies 1 + C = 10 \implies C = 9.$

$N = \frac{100}{1+9e^{-5t}}.$

7. $P(t) = 100\left(1 - e^{\frac{\ln 0.6}{2}t}\right).$

$$P(5) = 100\left(1 - e^{\frac{\ln 0.6}{2}\cdot 5}\right) = 100\left(1 - e^{2.5\cdot\ln 0.6}\right)$$

$$\simeq 100(1 - 0.279) = 72.1\%.$$

8. $P(t) = \frac{600}{1+59\cdot e^{\left(\frac{1}{5}\ln\frac{19}{59}\right)t}}.$

$$P(10) = \frac{600}{1+59\cdot e^{\left(2\cdot\ln\frac{19}{59}\right)}} \doteqdot 84.3.$$

$\therefore$ 84 fish are present after 10 weeks.

9. $300 = \frac{600}{1+59\cdot e^{\left(\frac{1}{5}\ln\frac{19}{59}\right)t}}$

$\implies 1 + 59 \cdot e^{\left(\frac{1}{5}\ln\frac{19}{59}\right)t} = 2$

$\implies 59 \cdot e^{\left(\frac{1}{5}\ln\frac{19}{59}\right)t} = 1$

$\implies e^{\left(\frac{1}{5}\ln\frac{19}{59}\right)t} = \frac{1}{59}$

$\implies \left(\frac{1}{5}\ln\frac{19}{59}\right)t = -\ln 59$

$\implies t = -\frac{5\ln 59}{\ln\frac{19}{59}} \doteqdot 17.993.$

$\therefore$ After 18 weeks the pond will contain 300 fish.

10. $W(t) = 140\left(1 - e^{-0.5t}\right)$.

a. $W(6) = 140\left(1 - e^{-0.5\cdot 6}\right) = 140(1 - e^{-3}) \doteqdot 133.03$.

$\therefore$ 133 employees used word processing equipment.

b. $\lim\limits_{t\to\infty} W(t) = \lim\limits_{t\to\infty} 140(1 - e^{-0.5t}) = 140$ employees.

11. $N(x) = 10,000(1 - e^{-0.002x})$.

$N'(x) = -10,000 \cdot e^{-0.002x} \cdot (-0.002) = 20 \cdot e^{-0.002x}$.

At $x = 0$, sales increase most rapidly.

12. $N(t) = \frac{500}{1+10\cdot e^{-kt}}$.

a. $N(10) = 100$, so $\frac{500}{1+10\cdot e^{-10k}} = 100$.

$1 + 10 \cdot e^{-10k} = 5 \implies 10 \cdot e^{-10k} = 4$

$\implies e^{-10k} = 0.4 \implies -10k = \ln 0.4$.

$\therefore\ k = -\frac{\ln 0.4}{10}$.

b. $N(t) = \frac{500}{1+10\cdot e^{\frac{\ln 0.4}{10}t}}$.

$200 = \frac{500}{1+10\cdot e^{\frac{\ln 0.4}{10}t}}$.

$1 + 10 \cdot e^{\frac{\ln 0.4}{10}t} = \frac{5}{2}$. $10 \cdot e^{\frac{\ln 0.4}{10}t} = \frac{3}{2}$. $e^{\frac{\ln 0.4}{10}t} = \frac{3}{20}$. $\frac{\ln 0.4}{10}t = \ln\frac{3}{20}$.

$t = \ln\frac{3}{20} \cdot \frac{10}{\ln 0.4} = (\ln 0.15) \cdot \frac{10}{\ln 0.4} \doteqdot 20.7$.

$\therefore$ After 21 days, 200 of the residents will have caught a cold.

13. $N(t) = \frac{80,000}{1+200\cdot e^{-0.2t}}$.

a. $N(6) = \frac{80,000}{1+200\cdot e^{-1.2}} \doteqdot 1306.4$.

$\therefore$ 1306 subscribers after 6 months.

b. $\lim\limits_{t\to\infty} N(t) = \lim\limits_{t\to\infty} \frac{80,000}{1+200\cdot e^{-0.2t}} = 80,000$ subscribers.

14. $f(t) = \frac{1}{1+e^{-t}}$. $\quad f'(t) = \frac{e^{-t}}{(1+e^{-t})^2}$.

$$f''(t) = \frac{-e^{-t}\left(1+e^{-t}\right)^2 - e^{-t}\cdot\left(2(1+e^{-t})(-e^{-t})\right)}{(1+e^{-t})^4}$$

$$= \frac{-e^{-t}(1+e^{-t})+2e^{-t}\cdot e^{-t}}{(1+e^{-t})^3}$$

$$= \frac{e^{-t}(-1-e^{-t}+2\cdot e^{-t})}{(1+e^{-t})^3} = \frac{e^{-t}(-1+e^{-t})}{(1+e^{-t})^3}.$$

$f''(t) = 0$ for $-1 + e^{-t} = 0 \implies e^{-t} = 1 \implies t = 0.$

$\therefore$ $\left(0, \frac{1}{2}\right)$ is an inflection point for $f(t)$.

15. $N(t) = \frac{400}{1+100e^{-t}}$.

$$N'(t) = \frac{-400(-100e^{-t})}{(1+100e^{-t})^2} = \frac{40000e^{-t}}{(1+100e^{-t})^2}, \quad t > 0.$$

$$N''(t) = \frac{-40000e^{-t}\left(1+100e^{-t}\right)^2 - 40000e^{-t}\left(2\left(1+100e^{-t}\right)\cdot\left(-100\cdot e^{-t}\right)\right)}{(1+100e^{-t})^4}$$

$$= \frac{-40000e^{-t}(1+100e^{-t})+40000\cdot 200\cdot e^{-2t}}{(1+100e^{-t})^3}$$

$$= \frac{40000e^{-t}(-1-100e^{-t}+200e^{-t})}{(1+100e^{-t})^3}$$

$$= \frac{40000e^{-t}(-1+100\cdot e^{-t})}{(1+100e^{-t})^3}.$$

$N''(t) = 0$ for $100 \cdot e^{-t} = 1 \implies e^{-t} = \frac{1}{100} \implies t = -\ln\frac{1}{100} = \ln 100 \doteqdot 4.6.$

Since $N''(t) > 0$ on $(0, \ln 100)$ and $N''(t) < 0$ on $(\ln 100, \infty)$, $\quad N'(t)$ has a maximum at $t = \ln 100$.

Thus the influenza is spreading the greatest after 4.6 days.

16. $N(4) = \frac{50}{1+49e^{-4}} \doteqdot 26.35.$

$\therefore$ 26 people have heard the rumor after 4 hours.

17. $N(t) = 30(1 - e^{-0.05t}).$

a. $N(20) = 30(1 - e^{-1}) \doteqdot 18.96 \approx 19.$

$\therefore$ 19 words can be mastered in 20 minutes.

b. $20 = 30\left(1 - e^{-0.05t}\right)$. $\quad 1 - e^{-0.05t} = \frac{2}{3}$.

$e^{-0.05t} = \frac{1}{3} \implies -0.05t = -\ln 3 \implies t = 20 \cdot \ln 3 \doteqdot 21.97$.

$\therefore$ A student has mastered 20 words after 22 minutes.

18. $\frac{dy}{dt} = k(M - y)$ has a solution $y(t) = M(1 - c \cdot e^{-kt})$.

Since $y(0) = 6$ and $M = 22$, $\quad 6 = 22(1 - c)$

$\implies \frac{6}{22} = 1 - c \implies c = \frac{16}{22}$.

$\therefore\ y(t) = 22\left(1 - \frac{16}{22}e^{-kt}\right)$.

Since $y(10) = 14$, $\quad 14 = 22\left(1 - \frac{16}{22}e^{-10k}\right) \implies e^{-10k} = \frac{1}{2}$.

$y(20) = 22\left(1 - \frac{16}{22}e^{-20k}\right) = 22 - 16\left(\frac{1}{2}\right)^2 = 18°\text{C}$.

19. a. $N(t) = \frac{2000}{1+Ce^{-kt}}$.

$N(0) = 500 \implies 500 = \frac{2000}{1+Ce^{-k(0)}} \implies 1 + C = 4 \implies C = 3$.

$N(t) = \frac{2000}{1+3e^{-kt}}$.

$N(2) = 800 \implies 800 = \frac{2000}{1+3e^{-k(2)}} \implies 1 + 3e^{-2k} = 2.5 \implies 3e^{-2k} = 1.5$

$\implies e^{-2k} = .5 \implies -2k = \ln .5 \implies 2k = \ln 2 \implies k = \frac{\ln 2}{2}$.

$N(t) = \frac{2000}{1+3e^{-\left(\frac{\ln 2}{2}\right)t}}$.

b. $N(6) = \frac{2000}{1+3e^{-\left(\frac{\ln 2}{2}\right)\cdot 6}} = \frac{2000}{1+3e^{-3\ln 2}} \approx 1455$.

20. $N(t) = \frac{M}{1+Ce^{-kt}}$.

$N(0) = 5,000 \implies 5,000 = \frac{300,000}{1+Ce^{-k(0)}} \implies 1 + C = 60 \implies C = 59$.

$N(t) = \frac{300,000}{1+59e^{-kt}}$.

$N(2) = 15,000 \implies 15,000 = \frac{300,000}{1+59e^{-k(2)}} \implies 1 + 59e^{-2k} = 20 \implies$

$59e^{-2k} = 19 \implies e^{-2k} = \frac{19}{59} \implies -2k = \ln\frac{19}{59} \implies k = -\frac{1}{2}\ln\frac{19}{59}$

$N(t) = \frac{300{,}000}{1+59e^{\left(\frac{1}{2}\ln\frac{19}{59}\right)t}}.$

$N(5) = \frac{300{,}000}{1+59e^{\left(\frac{1}{2}\ln\frac{19}{59}\right)5}} = \frac{300{,}000}{1+59e^{\ln\left(\frac{19}{59}\right)^{5/2}}} = \frac{300{,}000}{1+59\cdot\left(\frac{19}{59}\right)^{5/2}}$

$\approx 67{,}081.$

21. $\frac{dN}{dt} = kN \cdot \left(\frac{M-N}{M}\right) \implies N = \frac{M}{1+Ce^{-kt}}.$

Here $M = 10{,}000$, $C = 999$, $k = 5$.

$\frac{dN}{dt} = 5N \cdot \left(\frac{10{,}000-N}{10{,}000}\right).$

$N(0) = \frac{10{,}000}{1+999} = 10.$

22. $y = M\left(1 - Ce^{-kt}\right).$

$\frac{dy}{dt} = -M \cdot Ce^{-kt} \cdot (-k) = k \cdot CMe^{-kt} = k \cdot (M - y).$

23. $P(t) = \frac{M}{1+C\cdot e^{-kt}}.$

$\frac{dP}{dt} = \frac{-M\cdot C\cdot e^{-kt}\cdot(-k)}{(1+C\cdot e^{-kt})^2} = \frac{M\cdot k\cdot C\cdot e^{-kt}}{(1+C\cdot e^{-kt})^2}$

$= P(t) \cdot \frac{k\cdot C\cdot e^{-kt}}{(1+C\cdot e^{-kt})} = k \cdot P \cdot \frac{1+C\cdot e^{-kt}-1}{1+C\cdot e^{-kt}}$

$= k \cdot P \cdot \left(1 - \frac{1}{1+C\cdot e^{-kt}}\right) = k \cdot P \cdot \left(1 - \frac{P}{M}\right) = k \cdot P \cdot \left(\frac{M-P}{M}\right).$

Solutions to Exercise Set 10.3

1. $\frac{dy}{dx} = \frac{y}{x}, \quad x \neq 0.$

$\frac{dy}{y} = \frac{dx}{x} \implies \int \frac{dy}{y} = \int \frac{dx}{x} \implies \ln|y| = \ln|x| + C_1$

$\ln|y| = \ln|x| + \ln e^{C_1} = \ln\left(e^{C_1}|x|\right).$

$\therefore\ y = e^{\hat{C}}x = Cx.$

2. $\frac{dy}{dx} = \frac{x}{y^2}, \quad y \neq 0. \qquad y^2 dy = x\ dx.$

$\int y^2\ dy = \int x\ dx \implies \frac{y^3}{3} = \frac{x^2}{2} + C_1.$

$\therefore\ y = \left(\frac{3}{2}x^2 + C\right)^{1/3}.$

3. $\frac{dy}{dx} = -4xy^2. \qquad \frac{dy}{y^2} = -4x\ dx$

$\int \frac{1}{y^2}\ dy = -\int 4x\ dx \implies -\frac{1}{y} = -2x^2 + C_1.$

$\therefore\ y = \frac{1}{2x^2+C}.$

4. $\frac{dy}{dx} = \frac{\sqrt{x}}{\sqrt{y}}, \qquad y > 0.$

$\sqrt{y}\ dy = \sqrt{x}\ dx \implies \int \sqrt{y}\ dy = \int \sqrt{x}\ dx.$

$\frac{2}{3}y^{3/2} = \frac{2}{3}x^{3/2} + C_1. \qquad y^{3/2} = x^{3/2} + C.$

$\therefore\ y = \left(x^{3/2} + C\right)^{2/3}.$

5. $\frac{dy}{dx} = e^{x-y}. \qquad \frac{dy}{dx} = \frac{e^x}{e^y} \implies e^y\ dy = e^x\ dx.$

$\int e^y\ dy = \int e^x\ dx \implies e^y = e^x + C.$

$\therefore\ y = \ln(e^x + C).$

6. $\frac{dy}{dx} = \frac{y \ln^2 x}{x} \implies \int \frac{1}{y}\, dy = \int \frac{\ln^2 x}{x}\, dx.$

$\ln|y| = \frac{1}{3}(\ln x)^3 + C_1.$

$y = Ce^{\frac{1}{3}(\ln x)^3}.$

7. $e^y \frac{dy}{dt} - t^2 = 0. \qquad e^y\ dy = t^2\ dt \implies \int e^y\ dy = \int t^2\ dt.$

$e^y = \frac{1}{3}t^3 + C.$

$\therefore\ y = \ln\left(\frac{1}{3}t^3 + C\right).$

8. $\frac{dy}{dt} + y\cos t = 0 \implies \frac{1}{y}\,dy = -\cos t\,dt.$

$\ln|y| = -\sin t + C_1.$

$y = Ce^{-\sin t}.$

9. $\frac{dy}{dx} = (1+x)(1+y).$

$\frac{dy}{1+y} = (1+x)dx \implies \int \frac{dy}{1+y} = \int (1+x)dx.$

$\ln|1+y| = x + \frac{x^2}{2} + C_1 \implies (1+y) = C\cdot e^{x+\frac{x^2}{2}}.$

$\therefore\ y = C\cdot e^{x+\frac{x^2}{2}} - 1.$

10. $2t\cdot\frac{dy}{dt} = \frac{t^2+1}{y^2}.$

$y^2\,dy = \frac{t^2+1}{2t}\,dt = \left(\frac{t}{2}+\frac{1}{2t}\right)dt.$

$\int y^2\,dy = \frac{1}{2}\int \left(t+\frac{1}{t}\right)dt.$

$\frac{1}{3}y^3 = \frac{1}{2}\left(\frac{t^2}{2}+\ln|t|\right) + C_1 \implies y^3 = \frac{3}{2}\left(\frac{t^2}{2}+\ln|t|\right) + C.$

$\therefore\ y = \sqrt[3]{\frac{3}{4}t^2 + \frac{3}{2}\ln|t| + C}.$

11. $y' = 5y \implies \frac{1}{y}\,dy = 5\,dx.$

$\ln|y| = 5x + C_1 \implies y = Ce^{5x}.$

$y(1) = \pi e^5 \implies \pi e^5 = Ce^5 \implies C = \pi.$

$y = \pi e^{5x}.$

12. $yy' = x+2, \quad y(0) = 3.$

$y\,dy = (x+2)dx \implies \int y\,dy = \int (x+2)dx.$

$\frac{1}{2}y^2 = \frac{x^2}{2} + 2x + C_1$

$\implies y^2 = x^2 + 4x + C \implies y = \sqrt{x^2+4x+C}.$

From $y(0) = 3$, $\sqrt{C} = 3 \implies C = 9$. Thus, $y = \sqrt{x^2+4x+9}$.

13. $\frac{dy}{dx} = xy \implies \frac{1}{y}\,dy = x\,dx \implies \ln|y| = \frac{1}{2}x^2 + C_1 \implies y = Ce^{\frac{1}{2}x^2}$.

$y(2) = 7 \implies 7 = Ce^{\frac{1}{2}\cdot(2)^2} \implies Ce^2 = 7 \implies C = 7e^{-2}$.

$y = 7e^{-2} \cdot e^{\frac{1}{2}x^2} = 7e^{\frac{1}{2}x^2-2}$.

14. $\frac{dy}{dx} = -\frac{y}{x^2}, \quad y(1) = 2$.

$\frac{dy}{y} = -\frac{1}{x^2}\,dx \implies \ln|y| = \frac{1}{x} + C_1 \implies y = Ce^{1/x}$.

From $y(1) = 2, \quad C \cdot e = 2 \implies C = 2 \cdot e^{-1}$. Thus, $y = 2 \cdot e^{\frac{1}{x}-1}$.

15. $\frac{dy}{dx} = y\cos x, \quad y(0) = \pi$.

$\frac{1}{y}\,dy = \cos x\ dx \implies \int \frac{1}{y}\,dy = \int \cos x\ dx$.

$\ln|y| = \sin x + C_1 \implies y = C \cdot e^{\sin x}$.

From $y(0) = \pi, \quad \pi = C$. Thus, $y = \pi \cdot e^{\sin x}$.

16. $y' = \sqrt{y}, \quad y(0) = 1$.

$\frac{1}{\sqrt{y}}\,dy = dx \implies \int y^{-1/2}\,dy = \int dx$.

$2y^{1/2} = x + C_1$. From $y(0) = 1, \quad 2 = C_1$.

Thus, $2y^{1/2} = x + 2 \implies \sqrt{y} = \frac{x}{2} + 1 \implies y = \left(\frac{x}{2} + 1\right)^2$.

17. $y' = 2(y-1), \quad y(0) = 1$.

$\frac{dy}{y-1} = 2\ dx \implies \int \frac{1}{y-1}\,dy = \int 2\ dx \implies \ln|y-1| = 2x + C_1$

$\implies y - 1 = C \cdot e^{2x} \implies y = 1 + Ce^{2x}$.

From $y(0) = 1, \quad 1 = 1 + C \implies C = 0$. Thus, $y \equiv 1$.

18. $\frac{dy}{dt} = 5(10-y), \quad y(0) = 5$.

$\frac{dy}{y-10} = -5\ dt \implies \ln|y-10| = -5t + C_1$

$\implies y - 10 = Ce^{-5t} \implies y = 10 + Ce^{-5t}$.

From $y(0) = 5, \quad 5 = 10 + C \implies C = -5$. Thus, $y = 10 - 5e^{-5t}$.

19. $y' - 3y = 6, \quad y(0) = 2.$

$y' = 3(y+2) \implies \frac{dy}{y+2} = 3 \cdot dx \implies \ln|y+2| = 3x + C_1.$

$y + 2 = C \cdot e^{3x} \implies y = -2 + C \cdot e^{3x}.$

From $y(0) = 2, \quad 2 = -2 + C \implies C = 4$. Thus, $y = -2 + 4e^{3x}$.

20. $\frac{dy}{dt} + 2y = 2, \quad y(0) = 4.$

$\frac{dy}{dt} = 2(1-y) \implies \frac{dy}{y-1} = -2dt \implies \ln|y-1| = -2t + C_1.$

$y - 1 = C\ e^{-2t} \implies y = 1 + C\ e^{-2t}.$

From $y(0) = 4, \quad 4 = 1 + C \implies C = 3$. Thus $y = 1 + 3 \cdot e^{-2t}$.

21. From Example 4, $k = 2$, and $K = 100$, the solution $P(t) = \frac{100}{1+M \cdot e^{-2t}}$. The carrying capacity is 100.

22. $\frac{dP}{dt} = P(1-P)$ with $P(0) > 0$.

$\frac{dP}{P(1-P)} = dt \implies \left(\frac{1}{P} + \frac{1}{1-P}\right) dP = dt.$

$\int \left(\frac{1}{P} + \frac{1}{1-P}\right) dP = \int dt \implies \ln|P| - \ln|P-1| = t + C_1$

$\implies \ln\left|\frac{P}{P-1}\right| = t + C_1 \implies \frac{P}{P-1} = Ce^t.$

$\implies P = Ce^t(P-1) = PCe^t - Ce^t$

$\implies P(1 - Ce^t) = -Ce^t. \therefore P = \frac{Ce^t}{Ce^t - 1}.$

$\lim_{t\to\infty} P(t) = \lim_{t\to\infty} \frac{C \cdot e^t}{C \cdot e^t - 1} = \lim_{t\to\infty} \frac{C}{C - e^{-t}} = \frac{C}{C} = 1.$

23. $\frac{dU}{dt} = kU \quad (k < 0).$

$\frac{dU}{U} = k\ dt \implies \ln|U| = kt + C_1 \implies U = Ce^{kt}.$

Here $U(0) = C \implies U = U(0) \cdot e^{kt}$.

Since $U(4.5) = \frac{1}{2}U(0)$,

$\frac{1}{2} = e^{4.5k} \implies 4.5k = \ln\frac{1}{2}$

$\therefore\ k = -\frac{\ln 2}{4.5} \approx -0.154.$

24. $\frac{dP}{dt} = kP \implies P = C \cdot e^{kt}$.

$$P(1970) = 0.2 \implies 0.2 = C \cdot e^{1970 \cdot k}. \tag{1}$$

$$P(1980) = 0.5 \implies 0.5 = C \cdot e^{1980 \cdot k}. \tag{2}$$

From **(1)** and **(2)**, $\frac{0.5}{0.2} = \frac{C \cdot e^{1980 \cdot k}}{C \cdot e^{1970 \cdot k}}$. Hence

$$2.5 = e^{10k}. \tag{3}$$

$P(1990) = C \cdot e^{1990 \cdot k} = C \cdot e^{1980k} \cdot e^{10k} = 0.5 \cdot (2.5) = 1.25.$

$\therefore$ Population in 1990 will be 1.25 million.

25. a. $\frac{dp}{dt} = -\frac{3}{200}p, \quad p(0) = 200(0.5) = 100.$

b. $\frac{dp}{p} = -\frac{3}{200}dt.$

$\ln|p| = -\frac{3}{200}t + C_1 \implies p = C \cdot e^{-0.015t}.$

Since $p(0) = 100, \quad C = 100.$ $\therefore p(t) = 100 \cdot e^{-0.015t}$

So $p(20) = 100 \cdot e^{-0.3}$.

c. $\lim\limits_{t\to\infty} p(t) = \lim\limits_{t\to\infty}(100 \cdot e^{-0.015t}) = 0.$

26. $\frac{dy}{dt} = k(y - 12) \implies y = 12 + C \cdot e^{kt}$.

Since $y(0) = 26$ and $y(30) = 20, \quad 14 = C, \quad$ and $8 = C \cdot e^{k \cdot 30}$

$\implies e^{30k} = \frac{8}{14} = \frac{4}{7}.$

$y(120) = 12 + 14 \cdot e^{k \cdot 120} = 12 + 14(e^{30k})^4 = 12 + 14\left(\frac{4}{7}\right)^4$

$= 12 + 14 \cdot \frac{16}{49} \cdot \frac{16}{49} \doteqdot 13.49.$

27. a. $\frac{dP}{dt} = 0.1P + 100, \quad P(0) = 100.$

b. Solving the above differential equation for P,

$$\ln|0.1P + 100| = \frac{1}{10}t + C_1$$

$$0.1P + 100 = C_2 \cdot e^{0.1t}$$

$$0.1P = C_2 \cdot e^{0.1t} - 100$$

$$\therefore P(t) = C \cdot e^{0.1t} - 1000.$$

Since $P(0) = 100, \quad 100 = C - 1000 \implies C = 1100.$

Hence $P(t) = 1100 \cdot e^{0.1t} - 1000.$

$P(7) = 1100 \cdot e^{0.7} - 1000 = \$1215.13.$

28. $\frac{dy}{dt} = y(1-y) \implies \frac{d^2y}{dt^2} = \frac{dy}{dt}(1-y) + y\left(-\frac{dy}{dt}\right)$

$\implies \frac{d^2y}{dt^2} = \frac{dy}{dt}(1-2y) = y(1-y)(1-2y).$

Thus $\frac{d^2y}{dt^2} = 0$ when $y = \frac{1}{2}$, and $\frac{d^2y}{dt^2}$ changes sign at $y = \frac{1}{2}$.

Hence y has an inflection point where $y = \frac{1}{2}$.

29. $P(t) = \frac{1}{1+M\cdot e^{-kt}}.$

Since $P(0) = \frac{100}{100{,}000} = \frac{1}{1000}, \quad \frac{1}{1+M} = \frac{1}{1000} \implies M = 999.$

Since $P(2) = \frac{500}{100{,}000} = \frac{1}{200}, \quad \frac{1}{200} = \frac{1}{1+999\cdot e^{-2k}}$

$\implies e^{-2k} = \frac{199}{999}.$

$P(5) = \frac{1}{1+999\cdot e^{-5k}} = \frac{1}{1+999\cdot(e^{-2k})^{5/2}}$

$= \frac{1}{1+999\cdot\left(\frac{199}{999}\right)^{5/2}} \doteqdot 0.053498.$

Hence the number of people who know the rumor is $100{,}000 \cdot 0.053498 \doteqdot 5350.$

30. $\frac{dA}{dt} = k \cdot A$, where $A(t)$ is the amount of raw sugar remaining after t hours.

So $A = A(0) \cdot e^{kt}$. Since $A(0) = 100$, $A(t) = 100 \cdot e^{kt}$.

Since $A(6) = 75$, $75 = 100 \cdot e^{6k} \implies e^{6k} = \frac{3}{4} \implies k = \frac{1}{6}\ln 0.75$.

Hence $A(t) = 100 \cdot e^{\frac{\ln 0.75}{6}t}$.

a. $50 = 100 \cdot e^{\frac{\ln 0.75}{6}t}$.

$0.5 = e^{\frac{\ln 0.75}{6}t} \implies \frac{\ln 0.75}{6}t = \ln 0.5$.

$\therefore\ t = \ln 0.5 \cdot \frac{6}{\ln 0.75} \doteqdot 14.5$ hours.

b. $t = \ln 0.1 \cdot \frac{6}{\ln 0.75} \doteqdot 48.02$ hours.

31. $\frac{dy}{dt} = k(y - 30)$, where $y(t)$ is the temperature in t minutes.

So $y(t) = 30 + C \cdot e^{kt}$.

Since $y(0) = 100$, $C = 70$.

$\therefore\ y(t) = 30 + 70 \cdot e^{kt}$.

Since $y(3) = 80$, $y(3) = 30 + 70 \cdot e^{3k} = 80$,

$70 \cdot e^{3k} = 50 \implies e^{3k} = \frac{5}{7}$.

Then, $y(10) = 30 + 70 \cdot e^{10k} = 30 + 70 \cdot (e^{3k})^{10/3}$

$= 30 + 70 \cdot \left(\frac{5}{7}\right)^{10/3} = 52.8°$ C.

32. $\frac{dy}{dt} = 5000 - \frac{y}{1000} \cdot 50$, where $y(t)$ is the number of parts of pollutants after t minutes.

Then $y(t) = 100,000 + Ce^{-0.05t}$.

Since $y(0) = 0$, $C = -100,000$.

$\therefore\ y(t) = 100,000 \cdot \left(1 - e^{-0.05t}\right)$.

Thus $P(t) = \frac{y(t)}{1000} = 100 \cdot \left(1 - e^{-0.05t}\right)$.

33. $\frac{dy}{dt} = 60 - \frac{y}{500} \cdot 10$, where $y(t)$ is the number of pounds of salt after t minutes.

$\frac{dy}{dt} = 60 - \frac{y}{50} = 60 - 0.02y = -0.02 \cdot (y - 3000)$

$\implies y(t) = 3000 + C \cdot e^{-0.02t}$.

Since $y(0) = 1000$, $-2000 = C$.

$\therefore\ y(t) = 3000 - 2000 \cdot e^{-0.02t}$.

Hence the concentration of salt after t minutes is $\frac{y(t)}{500} = 6 - 4 \cdot e^{-0.02t}$.

34. $\frac{dP}{dt} = k \cdot P \cdot (1 - P) \implies P(t) = \frac{1}{1+C \cdot e^{-kt}}$.
Since $P(0) = 0.1$, $0.1 = \frac{1}{1+C} \implies C = 9$.
So $P(t) = \frac{1}{1+9e^{-kt}}$.
Since $P(3) = 0.3$, $0.3 = \frac{1}{1+9e^{-3k}}$
$\implies 1 + 9e^{-3k} = \frac{10}{3} \implies 9e^{-3k} = \frac{7}{3}$
$\implies e^{-3k} = \frac{7}{27} \therefore k = -\frac{1}{3} \ln \frac{7}{27}$.
Therefore, $P(t) = \frac{1}{1+9 \cdot e^{\frac{1}{3}\left(\ln \frac{7}{27}\right)t}}$.

35. Let A be the value of the annuity after t years.
$\frac{dA}{dt} = 0.1 \cdot A - 2000 \implies \frac{dA}{dt} = 0.1(A - 20000)$.
$\implies A(t) = 20000 + C \cdot e^{0.1t}$.
Since $A(0) = 10000$, $C = -10000$.
$\therefore A(t) = 10000\left(2 - e^{0.1t}\right)$.
From $A(t) = 0$, $0 = 2 - e^{0.1t} \implies e^{0.1t} = 2$
$\implies 0.1t = \ln 2 \therefore t = 10 \cdot \ln 2 \doteqdot 6.9315$ years.

36. $\frac{dy}{dt} = -ky \cdot \ln y$. $\frac{dy}{y \cdot \ln y} = -k\ dt$.
$\ln|\ln y| = -kt + C_1$. $\ln y = C \cdot e^{-kt} \therefore y = e^{C \cdot e^{-kt}}$.

37. $\frac{dy}{ds} = k \cdot \frac{y}{s}$. $\frac{dy}{y} = \frac{k}{s}\ ds$.
$\ln|y| = k \ln|s| + C_1 \therefore y = C \cdot e^{k \ln s} = C \cdot s^k$.

Solutions to Exercise Set 10.4

1. $\frac{d^2y}{dt^2} + 5\frac{dy}{dt} + 6y = 0$.
$r^2 + 5r + 6 = 0 \implies (r+2)(r+3) = 0 \implies r_1 = -2, \quad r_2 = -3$.
The solution is $y = Ae^{-2t} + Be^{-3t}$.

2. $y'' + 4y' - 5y = 0$.
$r^2 + 4r - 5 = 0 \implies (r+5)(r-1) = 0 \implies r_1 = -5, \quad r_2 = 1$.
The general solution is $y = Ae^{-5t} + Be^{t}$.

3. $y'' - y = 0.$

   $r^2 - 1 = 0 \implies r_1 = 1, \quad r_2 = -1.$

   General solution is $y = Ae^{-t} + Be^{t}.$

4. $\frac{d^2y}{dt^2} + \frac{dy}{dt} - 12y = 0.$

   $r^2 + r - 12 = 0 \implies (r+4)(r-3) = 0 \implies r_1 = -4, \quad r_2 = 3.$

   General solution is $y = Ae^{-4t} + Be^{3t}.$

5. $\frac{d^2y}{dt^2} + 3\frac{dy}{dt} - 10y = 0.$

   $r^2 + 3r - 10 = 0 \implies (r+5)(r-2) = 0 \implies r_1 = -5, \quad r_2 = 2.$

   General solution is $y = Ae^{2t} + Be^{-5t}.$

6. $y'' - 4y = 0.$

   $r^2 - 4 = 0 \implies r_1 = -2, \quad r_2 = 2.$

   General solution is $y = Ae^{-2t} + Be^{2t}.$

7. $y'' - \pi y = 0.$

   $r^2 - \pi = 0 \implies r_1 = \sqrt{\pi}, \quad r_2 = -\sqrt{\pi}.$

   General solution is $y = Ae^{\sqrt{\pi}\,t} + Be^{-\sqrt{\pi}\,t}.$

8. $y'' - 4y' - 21y = 0.$

   $r^2 - 4r - 21 = 0 \implies (r-7)(r+3) = 0 \implies r_1 = 7, \quad r_2 = -3.$

   General solution is $y = Ae^{7t} + Be^{-3t}.$

9. $y'' + 2y' + y = 0.$

   $r^2 + 2r + 1 = 0 \implies (r+1)^2 = 0 \implies r_1 = r_2 = -1.$

   General solution is $y = Ae^{-t} + Bte^{-t}.$

10. $y'' - 4y' + 4y = 0.$

    $r^2 - 4r + 4 = 0 \implies (r-2)^2 = 0 \implies r_1 = r_2 = 2.$

    General solution is $y = Ae^{2t} + Bte^{2t}.$

11. $\frac{d^2y}{dt^2} + 6\frac{dy}{dt} + 9y = 0.$

    $r^2 + 6r + 9 = 0 \implies (r+3)^2 = 0 \implies r_1 = r_2 = -3.$

    General solution is $y = Ae^{-3t} + Bte^{-3t}.$

12. $y'' - 2y' + y = 0.$

$r^2 - 2r + 1 = 0 \implies (r-1)^2 = 0 \implies r_1 = r_2 = 1.$

General solution is $y = Ae^t + Bte^t$.

13. $y'' - 2Cy' + C^2y = 0.$

$r^2 - 2Cr + C^2 = 0 \implies (r-C)^2 = 0 \implies r_1 = r_2 = C.$

The general solution is $y = Ae^{Ct} + Bte^{Ct}$.

14. $y'' - 8y' + 64y = 0.$

$$r^2 - 8r + 64 = 0 \implies r = \frac{8 \pm \sqrt{(-8)^2 - 4(1)(64)}}{2\cdot(1)} = \frac{8 \pm \sqrt{-192}}{2}.$$
$$= \frac{8 \pm 8\sqrt{3}\,i}{2} = 4 \pm 4\sqrt{3}\,i.$$

The general solution is

$$y = Ae^{4t}\cos\left(4\sqrt{3}\,t\right) + Ce^{4t}\sin\left(4\sqrt{3}\,t\right).$$

15. $\frac{d^2y}{dt^2} + 4y = 0.$

$r^2 + 4 = 0 \implies r = \pm 2i.$

General solution is $y = A\sin 2t + B\cos 2t$.

16. $y'' + 16y = 0$

$r^2 + 16 = 0 \implies r = \pm 4i.$

General solution is $y = A\sin 4t + B\cos 4t$.

17. $y'' = -7y \implies y'' + 7y = 0.$

$r^2 + 7 = 0 \implies r = \pm\sqrt{7}\,i.$

The general solution is $y = A\cos\left(\sqrt{7}\,t\right) + B\sin\left(\sqrt{7}\,t\right)$.

18. $\frac{d^2y}{dt^2} = -36y \implies \frac{d^2y}{dt^2} + 36y = 0.$

$r^2 + 36 = 0 \implies r = \pm 6i.$

The general solution is $y = A\cos(6t) + B\sin(6t)$.

19. $y'' - 4y' + 13y = 0.$

$r^2 - 4r + 13 = 0 \implies r = \frac{4 \pm \sqrt{4^2 - 4(13)}}{2} = 2 \pm \frac{\sqrt{-36}}{2} = 2 \pm 3i.$

General solution is $y = Ae^{2t}\sin 3t + Be^{2t}\cos 3t$.

20. $y'' - 6y' + 25y = 0.$

$r^2 - 6r + 25 = 0 \implies r = \frac{6 \pm \sqrt{6^2 - 100}}{2} = 3 \pm 4i.$

General solution is $y = Ae^{3t}\sin 4t + Be^{3t}\cos 4t$.

21. $y'' + y' + \frac{y}{2} = 0.$

$r^2 + r + \frac{1}{2} = 0 \implies r = \frac{-1 \pm \sqrt{1^2 - 4(\frac{1}{2})}}{2} = -\frac{1}{2} \pm \frac{i}{2}.$

General solution is $y = Ae^{-t/2} \sin \frac{1}{2}t + Be^{-t/2} \cos \frac{1}{2}t.$

22. $y'' + y' + y = 0.$

$r^2 + r + 1 = 0 \implies r = \frac{-1 \pm \sqrt{1-4}}{2} = -\frac{1}{2} \pm \frac{\sqrt{3}\,i}{2}.$

General solution is $y = Ae^{-t/2} \sin \frac{\sqrt{3}}{2}t + Be^{-t/2} \cos \frac{\sqrt{3}}{2}t.$

23. $y'' - y' - 6y = 0.$

$r^2 - r - 6 = 0 \implies (r-3)(r+2) = 0 \implies r_1 = 3, \quad r_2 = -2.$ Then

$y = Ae^{3t} + Be^{-2t} \implies y' = 3Ae^{3t} - 2Be^{-2t}.$

$y(0) = 0, \quad y'(0) = 5 \implies \begin{cases} A + B = 0 & \textbf{(1)} \\ 3A - 2B = 5 & \textbf{(2)} \end{cases}$

From **(1)** $A = -B$; then **(2)** becomes $3A - 2(-A) = 5 \implies A = 1, \quad B = -1.$

Then the solution is $y = e^{3t} - e^{-2t}.$

24. $y'' - 5y' + 6y = 0.$

$r^2 - 5r + 6 = 0 \implies (r-2)(r-3) = 0 \implies r_1 = 2, \quad r_2 = 3.$

$y = Ae^{2t} + Be^{3t}. \quad y' = 2Ae^{2t} + 3Be^{3t}.$

$y(0) = 2, \quad y'(0) = 0 \implies \begin{cases} A + B = 2 \\ 2A + 3B = 0 \end{cases} \implies A = 6, \quad B = -4.$

The solution is $y = 6e^{2t} - 4e^{3t}.$

25. $y'' + 2y' + y = 0.$

$r^2 + 2r + 1 = 0 \implies r = -1.$

$y = Ae^{-t} + Bte^{-t}.$

$y' = -Ae^{-t} + Be^{-t} - Bte^{-t} = -Ae^{-t} + Be^{-t}(1-t).$

$y(0) = 3, \quad y'(0) = 1 \implies \begin{cases} A = 3 \\ -A + B = 1 \end{cases} \implies A = 3, \quad B = 4.$

The solution is $y = 3e^{-t} + 4te^{-t}.$

26. $y'' - 4y' + 4y = 0.$

$r^2 - 4r + 4 = 0 \implies r = 2.$

$y = Ae^{2t} + Bte^{2t}.$

$y' = 2Ae^{2t} + Be^{2t}(1+2t).$

$$y(0)=2,\quad y'(0)=5 \implies \begin{cases} A=2 \\ 2A+B=5 \end{cases} \implies A=2, \quad B=1.$$

Thus, the solution is $y = 2e^{2t} + te^{2t}$.

27. $y'' + 9y = 0. \quad r^2 + 9 = 0 \implies r = \pm 3i.$

$y = A\sin 3t + B\cos 3t.$

$y' = 3A\cos 3t - 3B\sin 3t.$

$$y(0)=3,\quad y'(0)=-3 \implies \begin{cases} B=3 \\ 3A=-3 \end{cases} \implies A=-1, \quad B=3.$$

Thus, the solution is $y = -\sin 3t + 3\cos 3t$.

28. $y'' + y = 0. \quad r^2 + 1 = 0 \implies r = \pm i.$

$y = A\sin t + B\cos t. \quad y' = A\cos t - B\sin t.$

$$y(0)=0,\quad y'(0)=2 \implies \begin{cases} B=0 \\ A=2. \end{cases}$$

Thus, the solution is $y = 2\sin t$.

29. $\frac{dy}{dt} = 4x, \qquad a = 4.$

$\frac{dx}{dt} = -y, \qquad b = 1.$

General solution, from (24) and (25) in Example 8, is

$x(t) = A\sin\sqrt{ab}\,t + B\cos\sqrt{ab}\,t$

$y(t) = -\sqrt{\frac{a}{b}}\left[A\cos\sqrt{ab}\,t - B\sin\sqrt{ab}\,t\right]$

$\implies x(t) = A\sin 2t + B\cos 2t.$

$\qquad y(t) = -2A\cos 2t + 2B\sin 2t.$

30. $\frac{dy}{dt} = 2(x-3). \qquad \frac{dx}{dt} = 8(y+5).$

Let $X = x - 3, \quad Y = y + 5.$

$\frac{dY}{dt} = 2X, \qquad \frac{dX}{dt} = 8Y.$

$\frac{d^2X}{dt^2} = 8\frac{dY}{dt} = 8(2X) = 16X.$

$\frac{d^2X}{dt^2} - 16X = 0.$

$X = Ae^{4t} + Be^{-4t}.$

$Y = \frac{1}{8}\frac{dX}{dt} = \frac{1}{2}Ae^{4t} - \frac{1}{2}Be^{-4t}.$

The general solution is

$x(t) = 3 + Ae^{4t} + Be^{-4t}.$

$y(t) = -5 + \frac{1}{2}Ae^{4t} - \frac{1}{2}Be^{-4t}.$

31. From Example 9, the general solution for the predator-prey system is

$x(t) = 20 + A\sin\sqrt{2}\,t + B\cos\sqrt{2}\,t$

$y(t) = 10 - \sqrt{2}\left(A\cos\sqrt{2}\,t - B\sin\sqrt{2}\,t\right).$

Now $x(0) = 100, \quad y(0) = 40.$

$$\Longrightarrow \begin{cases} 20 + B = 100 \\ 10 - \sqrt{2}\,A = 40 \end{cases} \Longrightarrow A = -\tfrac{30}{\sqrt{2}} = -15\sqrt{2}, \quad B = 80.$$

Thus the solution is

$x(t) = 20 - 15\sqrt{2}\sin\sqrt{2}\,t + 80\cos\sqrt{2}\,t$

$y(t) = 10 + 30\cos\sqrt{2}\,t + 80\sqrt{2}\sin\sqrt{2}\,t.$

32. $a = 1, \quad b = 2, \quad \bar{x} = 25, \quad \bar{y} = 16.$

The general solution is

$x(t) = 25 + A\sin\sqrt{2}\,t + B\cos\sqrt{2}\,t$

$y(t) = 16 - \tfrac{1}{\sqrt{2}}\left(A\cos\sqrt{2}\,t - B\sin\sqrt{2}\,t\right).$

$$x(0) = y(0) = 20 \Longrightarrow \begin{cases} 25 + B = 20 \\ 16 - \tfrac{1}{\sqrt{2}}\,A = 20 \end{cases} \Longrightarrow A = -4\sqrt{2}, \quad B = -5.$$

So the solution is

$x(t) = 25 - 4\sqrt{2}\sin\sqrt{2}\,t - 5\cos\sqrt{2}\,t$

$y(t) = 16 + 4\cos\sqrt{2}\,t - \tfrac{5}{2}\sqrt{2}\sin\sqrt{2}\,t.$

33. In Exercise 32, if $x(0) = 25, \quad y(0) = 16$:

$$\begin{cases} 25 + B = 25 \\ 16 - \tfrac{1}{\sqrt{2}}\,A = 16 \end{cases} \Longrightarrow A = B = 0.$$

The solution then is $x(t) = 25, \quad y(t) = 16.$

34. For both $a > 0$ and $b > 0$, the nature of the solutions of the given predator-prey system is the same as that given on p. 568. Consider this same system with $a > 0$ and $b < 0$. Then

$\frac{d^2x}{dt^2} = -b\frac{dy}{dt} = -abx.$

$\frac{d^2x}{dt^2} + abx = 0.$

$x = Ae^{\sqrt{-ab}\,t} + Be^{-\sqrt{-ab}\,t}.$

$$y = -\tfrac{1}{b}\,\tfrac{dx}{dt} = -\tfrac{1}{b}\sqrt{-ab}\left[Ae^{\sqrt{-ab}\,t} - Be^{-\sqrt{-ab}\,t}\right]$$

$$= -\sqrt{\tfrac{-a}{b}}\left[Ae^{\sqrt{-ab}\,t} - Be^{-\sqrt{-ab}\,t}\right].$$

It would not make sense to refer to the system when $b < 0$ as a predator-prey system since the number of rabbits is increasing as the number of foxes increases.

35. $y = te^{-2t}$. $y' = e^{-2t} - 2te^{-2t}$.

$$y'' = -2e^{-2t} - 2e^{-2t} + 4te^{-2t} = -4e^{-2t} + 4te^{-2t}.$$

$$\begin{aligned} y'' + 4y' + 4y &= (-4e^{-2t} + 4te^{-2t}) + 4(e^{-2t} - 2te^{-2t}) + 4te^{-2t} \\ &= -4e^{-2t} + 8te^{-2t} + 4e^{-2t} - 8te^{-2t} = 0. \end{aligned}$$

Thus $y = te^{-2t}$ is a solution to $y'' + 4y' + 4y = 0$.

36. From $y = Ay_1 + By_2$, $y' = Ay_1' + By_2'$ and $y'' = Ay_1'' + By_2''$

Then
$$\begin{aligned} y'' + ay' + by &= Ay_1'' + By_2'' + a(Ay_1' + By_2') + b(Ay_1 + By_2) \\ &= A(y_1'' + ay_1' + by_1) + B(y_2'' + ay_2' + by_2). \qquad (1) \end{aligned}$$

Since y_1 and y_2 are solutions to $y'' + ay' + by = 0$,

we obtain $y_1'' + ay_1' + by_1 = 0$ and $y_2'' + ay_2' + by_2 = 0$. **(2)**

Therefore, from **(1)** and **(2)**, $y'' + ay' + by = 0$ for $y = Ay_1 + By_2$.

Solutions to Exercise Set 10.5

1. $y' = y(3 - y)$.

a. From $y(3 - y) = 0$, there are two constant solutions $y = 0$ and $y = 3$.

b.

Range of y	Sign of y'	Behavior of Solution
$y < 0$	Negative	decreasing
$0 < y < 3$	Positive	increasing
$y > 3$	Negative	decreasing

c. $$\begin{aligned} y'' &= \tfrac{d}{dt}(y') = \tfrac{d}{dt}(y(3 - y)) = \tfrac{d}{dt}(3y - y^2) = (3 - 2y) \cdot y' \\ &= (3 - 2y) \cdot y \cdot (3 - y). \end{aligned}$$

Range of y	Sign of y''	Concavity
$y < 0$	Negative	Down
$0 < y < \frac{3}{2}$	Positive	Up
$\frac{3}{2} < y < 3$	Negative	Down
$y > 3$	Positive	Up

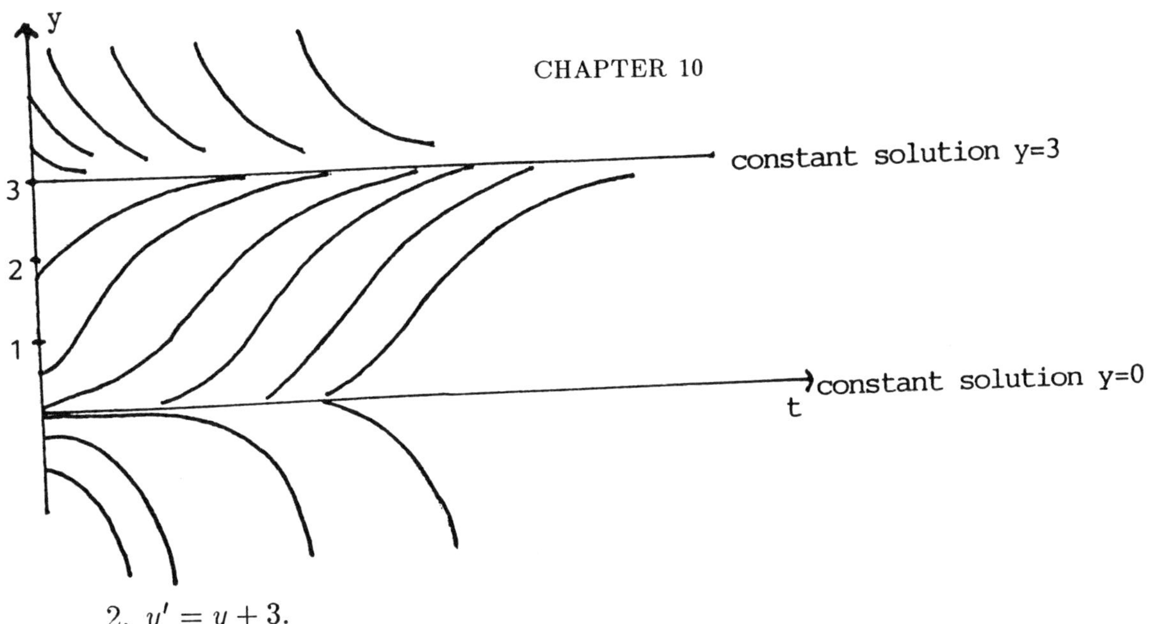

2. $y' = y + 3$.

a. $y = -3$ is a constant solution.

b.

Range of y	Sign of y'	Behavior of Solution
$y < -3$	Negative	decreasing
$y > -3$	Positive	increasing

c. $y'' = \frac{d}{dt}(y') = \frac{d}{dt}(y+3) = y' = y + 3$.

Range of y	Sign of y''	Concavity
$y < -3$	Negative	Down
$y > -3$	Positive	Up

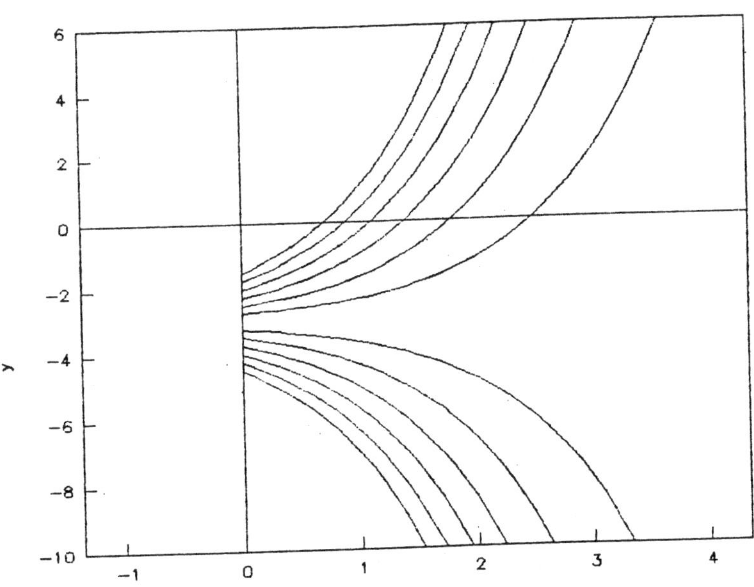

3. $y' = y^2$.

   a. $y = 0$ is a constant solution.

   b. For all $y \neq 0$, y is increasing.

   c. $y'' = 2y \cdot y' = 2y^3$.

Range of y	Sign of y''	Concavity
$y < 0$	Negative	Down
$y > 0$	Positive	Up

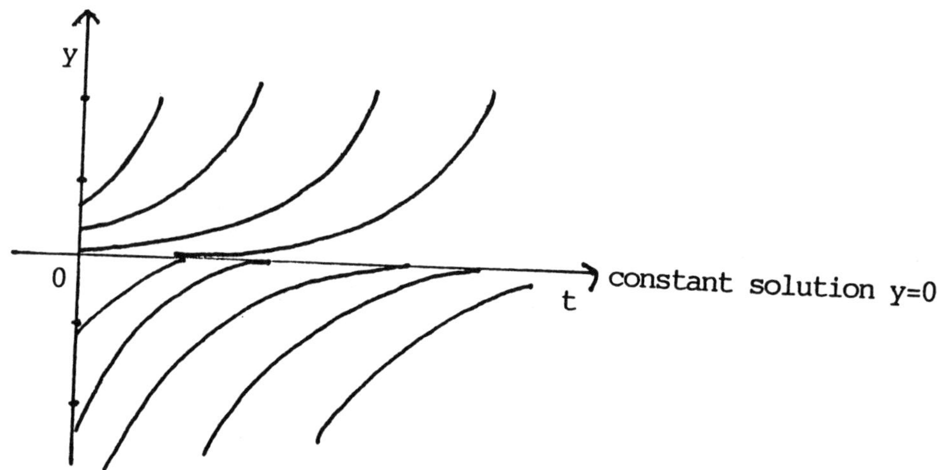

4. $y' = y(2 + y)$.

   a. $y = 0$ and $y = -2$ are constant solutions.

   b. y is increasing on $(-\infty, -2)$, decreasing on (-2,0), and increasing on $(0, \infty)$.

   c. $y'' = y'(2 + y) + y \cdot y' = 2yy' + 2y'$
   $= 2(y + 1)y' = 2(y + 1)y(2 + y)$.

Range of y	Sign of y''	Concavity
$y < -2$	Negative	Down
$-2 < y < -1$	Positive	Up
$-1 < y < 0$	Negative	Down
$y > 0$	Positive	Up

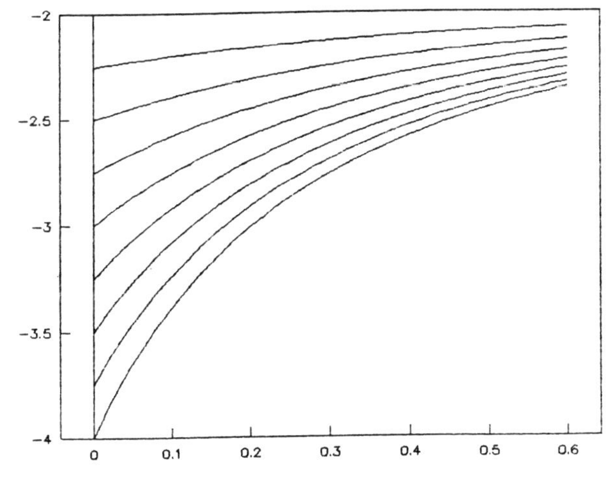

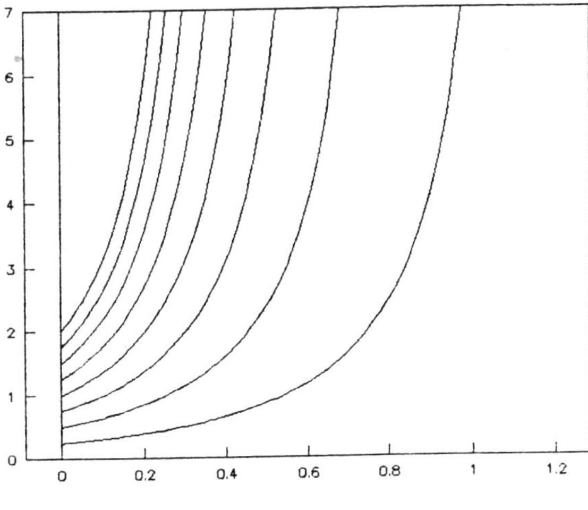

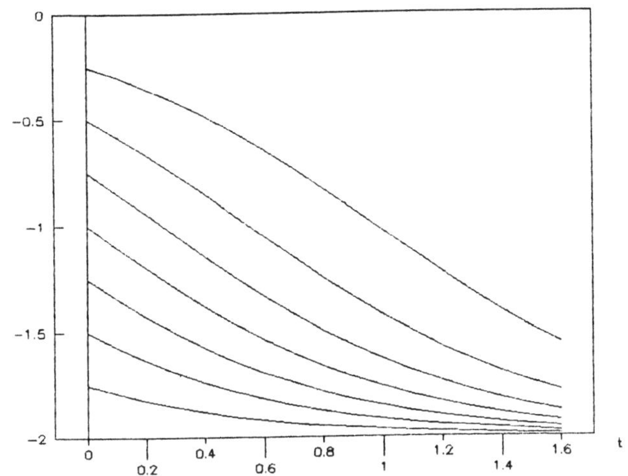

5. $y' = e^y$.

a. Since $e^y > 0$, there is no constant solution.

b¿ y is increasing for all y.

c. $y'' = e^y \cdot y' = e^{2y} > 0$, so y is concave up for all y.

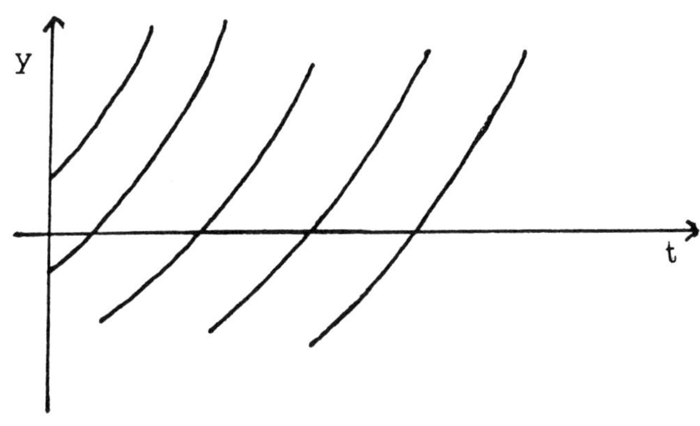

6. $y' = -\frac{1}{1+y^2}$.

a. Since $-\frac{1}{1+y^2} < 0$, there is no constant solution.

b. y is decreasing for all y.

c. $y'' = -\frac{-2y}{(1+y^2)^2} \cdot y' = \frac{-2y}{(1+y^2)^3}$.

$y'' > 0$ on $(-\infty, 0)$ and $y'' < 0$ on $(0, \infty)$.

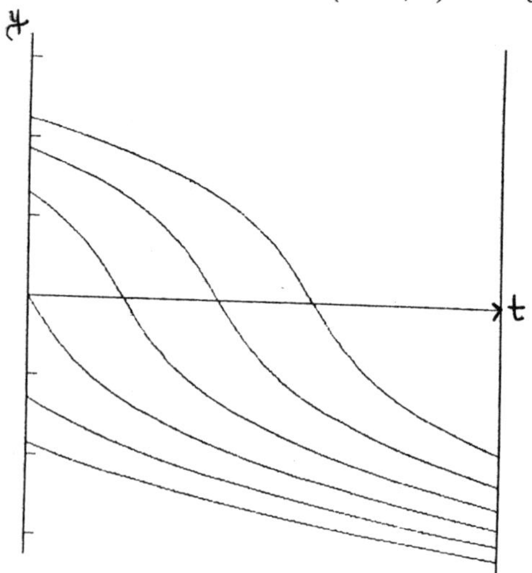

7. $y' = \cos y$.

a. $\cos y = 0$ for $y = \frac{(2n-1)\pi}{2}, \quad n = 0, \pm 1, \pm 2, \ldots$.

Thus $y = \left(n - \frac{1}{2}\right)\pi$, $(n = 0, \pm 1, \pm 2, \ldots)$, are constant solutions.

b.

Range of y	Sign of y'	Behavior of Solution
$-\frac{\pi}{2} < y < \frac{\pi}{2}$	Positive	increasing
$\frac{\pi}{2} < y < \frac{3\pi}{2}$	Negative	decreasing
$\left(2n - \frac{1}{2}\right)\pi < y < \left(2n + \frac{1}{2}\right)\pi$	Positive	increasing
$\left(2n - \frac{3}{2}\right)\pi < y < \left(2n - \frac{1}{2}\right)\pi$	Negative	decreasing

c. $y'' = (-\sin y) \cdot y' = -(\sin y)(\cos y) = -\frac{1}{2}\sin 2y$.

$y'' = 0$ for $2y = n\pi \implies y = \frac{n}{2}\pi, \quad n = 0, \pm 1, \pm 2, \ldots$.

Range of y	Sign of y''	Concavity
$0 < y < \frac{\pi}{2}$	Negative	Down
$\frac{\pi}{2} < y < \pi$	Positive	Up
$n\pi < y < \frac{2n+1}{2}\pi$	Negative	Down
$\frac{2n-1}{2}\pi < y < n\pi$	Positive	Up

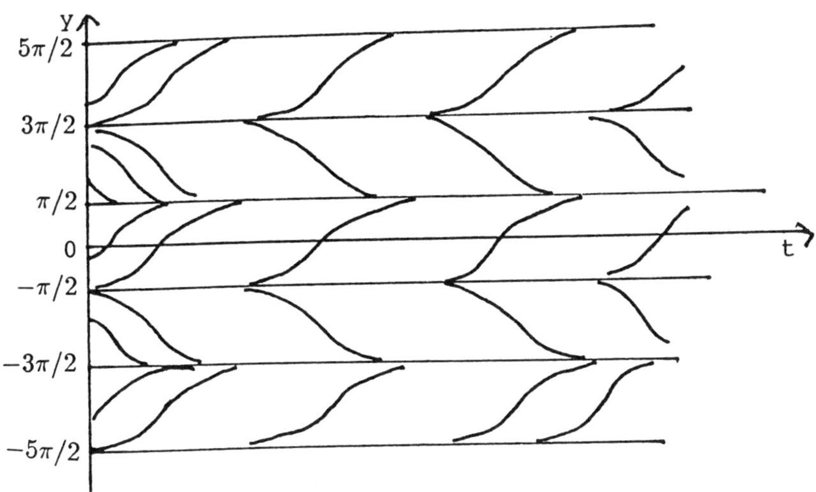

8. $y' = y \cdot (2 - y^2)$.

a. $y = 0$, $y = -\sqrt{2}$, and $y = \sqrt{2}$ are the constant solutions.

b.

Range of y	Sign of y'	Behavior of y
$y < -\sqrt{2}$	Positive	Increasing
$-\sqrt{2} < y < 0$	Negative	Decreasing
$0 < y < \sqrt{2}$	Positive	Increasing
$y > \sqrt{2}$	Negative	Decreasing

c. $y'' = y' \cdot (2 - y^2) + y \cdot (-2yy') = (2 - 3y^2) \cdot y' = (2 - 3y^2) \cdot y \cdot (2 - y^2)$.
$y'' = 0$ for $y = 0$, $\quad y = \pm\sqrt{\frac{2}{3}}$, $\quad y = \pm\sqrt{2}$.

Range of y	Sign of y''	Concavity
$y < -\sqrt{2}$	Negative	Down
$-\sqrt{2} < y < -\sqrt{\frac{2}{3}}$	Positive	Up
$-\sqrt{\frac{2}{3}} < y < 0$	Negative	Down
$0 < y < \sqrt{\frac{2}{3}}$	Positive	Up
$\sqrt{\frac{2}{3}} < y < \sqrt{2}$	Negative	Down
$y > \sqrt{2}$	Positive	Up

d.

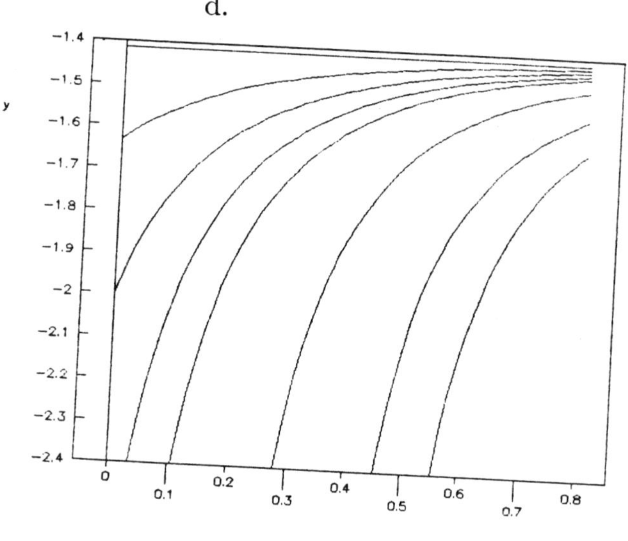

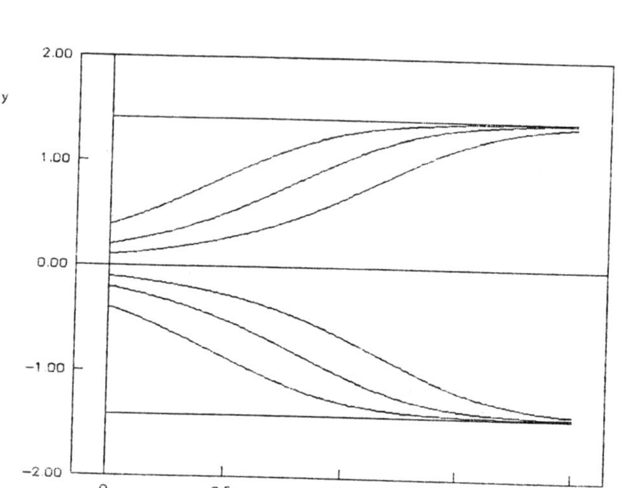

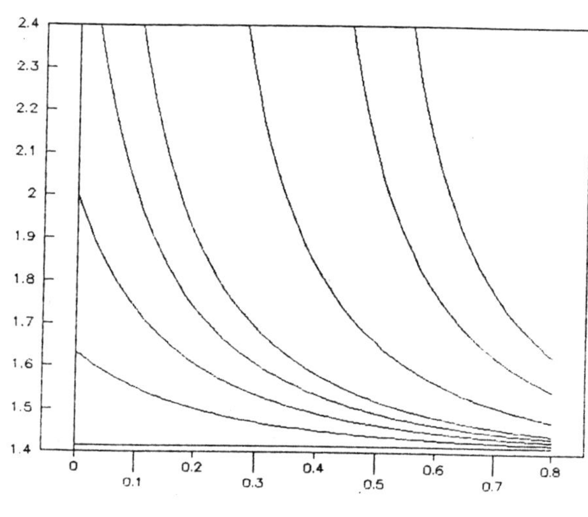

9. $y' = y(1-y)(2-y)$.

a. The constant solutions are $y = 0$, $y = 1$, and $y = 2$.

b.

Range of y	Sign of y'	Behavior of y
$y < 0$	Negative	decreasing
$0 < y < 1$	Positive	increasing
$1 < y < 2$	Negative	decreasing
$y > 2$	Positive	increasing

c. $y'' = [(1-y)(2-y) - y(1-y) - y(2-y)]\, y'.$

$= (3y^2 - 6y + 2)(1-y)(2-y)y.$

$y'' = 0$ for $y = 0, 1, 2$ and $1 \pm \frac{\sqrt{3}}{3}$.

Range of y	Sign of y''	Concavity
$y < 0$	Negative	Down
$0 < y < 1 - \frac{\sqrt{3}}{3}$	Positive	Up
$1 - \frac{\sqrt{3}}{3} < y < 1$	Negative	Down
$1 < y < 1 + \frac{\sqrt{3}}{3}$	Positive	Up
$1 + \frac{\sqrt{3}}{3} < y < 2$	Negative	Down
$y > 2$	Positive	Up

d.

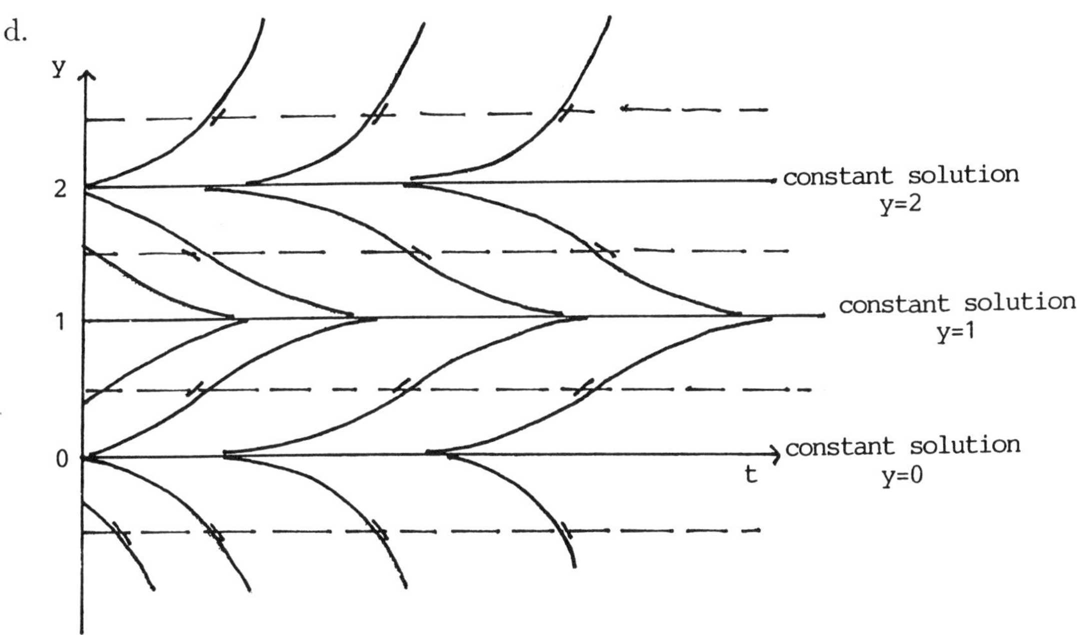

10. $y' = \frac{4}{(1+y)}$.

a. There is no constant solution.

b. $y' > 0$ for $y > -1$, so y increases as t increases when $y > -1$.

$y' < 0$ for $y < -1$, so y decreases as t increases when $y < -1$.

c. $y'' = -\frac{4}{(1+y)^2} \cdot y' = -\frac{16}{(1+y)^3}$.

$y'' > 0$ for $y < -1$, so the graph is concave up when $y < -1$.

$y'' < 0$ for $y > -1$, so the graph is concave down when $y > -1$.

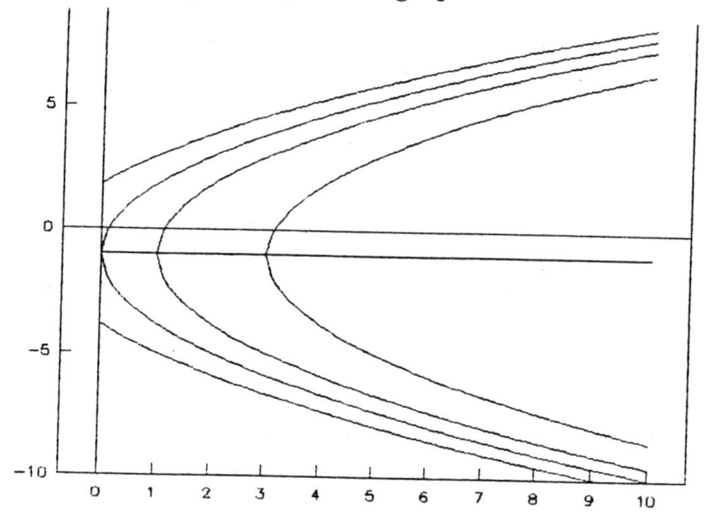

11. $y' = y(10 - y), \quad y(0) = 5.$

a. Because the initial condition $y(0) = 5$ is between $y = 0$ and $y = 10$, the solution $y(t)$ of the initial value problem is such that $0 < y(t) < 10$ for all $t \geq 0$.

b. Because $y'(t) > 0$ for all $t > 0$, it follows that $y(t)$ increases as t increases for all $t \geq 0$, and $5 < y(t) < 10$ for all $t > 0$.

c. $y'' = y' \cdot (10 - y) + y \cdot (-y') = y' \cdot (10 - 2y) = y(10 - y)(10 - 2y)$.
Because $y'' < 0$ for all $5 < y < 10$, the graph of $y(t)$ is concave down for all $t > 0$.

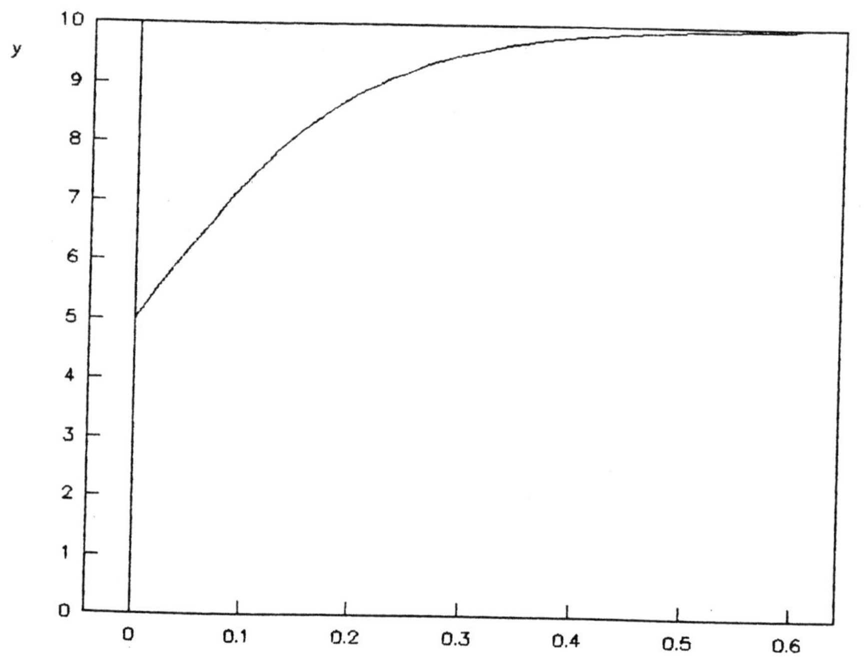

12. $y' = y(10-y), \quad y(0) = 20.$

a. Because the initial condition $y(0) = 20$ is greater than $y = 10$, the solution $y(t)$ of the initial value problem is such that $y(t) > 10$ for all $t \geq 0$.

b. Because $y'(t) < 0$ for all $t > 0$, it follows that $y(t)$ decreases as t increases for all $t \geq 0$, and $10 < y(t) < 20$ for all $t > 0$.

c. Because $y''(t) > 0$ for all $t > 0$, the graph of $y(t)$ is concave up for all $t > 0$.

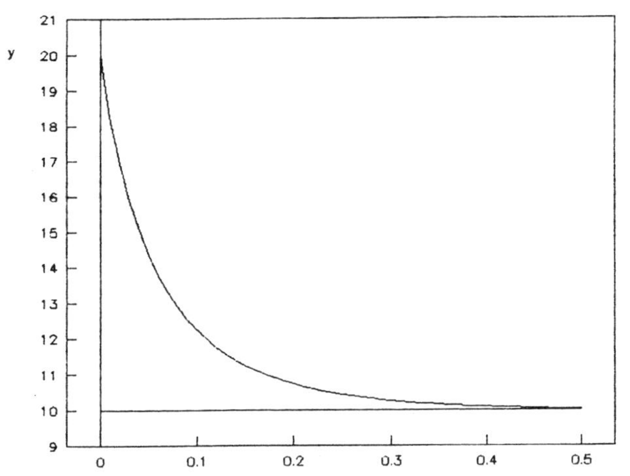

13. $y' = \sin y, \quad y(0) = -\frac{\pi}{4}.$

Since $\sin y = 0$ for $y = n\pi$ for n any integer, and since $-\pi < y(0) < 0$, the solution $y(t)$ to the initial value problem is such that $-\pi < y(t) < 0$ for all $t \geq 0$. Since $y'(t) < 0$ for all $t > 0$, $y(t)$ decreases as t increases for all $t \geq 0$, and therefore $-\pi < y(t) \leq -\frac{\pi}{4}$ for all $t \geq 0$. Since $y''(t) = \cos y \cdot y' = \cos y \cdot \sin y = \frac{1}{2}\sin 2y$, $y'' = 0$ for $y = -\frac{\pi}{2}$, and $y'' > 0$ for $-\pi < y < -\frac{\pi}{2}$ and $y'' < 0$ for $-\frac{\pi}{2} < y < -\frac{\pi}{4}$.

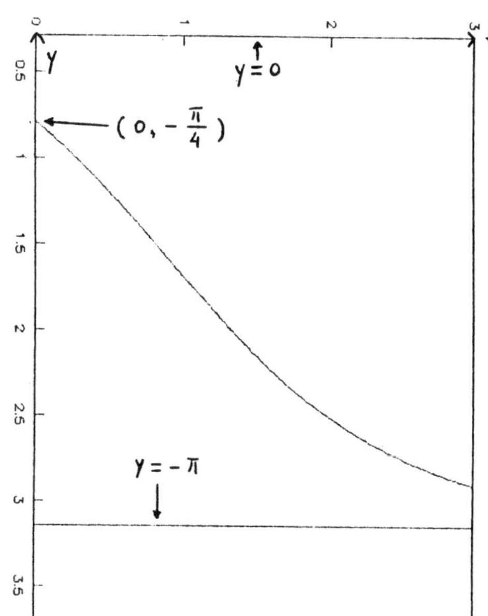

14. $y' = \cos y, \quad y(0) = \pi.$

Since $\cos y = 0$ for $y = \left(n - \frac{1}{2}\right)\pi$, $(n = 0, \pm 1, \pm 2, \dots)$, and $\frac{\pi}{2} < y(0) = \pi < \frac{3\pi}{2}$, the initial value problem has the solution $y(t)$ such that $\frac{\pi}{2} < y(t) < \frac{3\pi}{2}$ for all $t > 0$. Since $y'(t) < 0$ for all $t > 0$, $y(t)$ is decreasing for all $t > 0$. Thus $\frac{\pi}{2} < y(t) < \pi$ for all $t > 0$. Since $y'' = -\sin y \cdot \cos y = -\frac{1}{2}\sin 2y > 0$ for all $t > 0$, the graph of $y(t)$ is concave up for all $t > 0$.

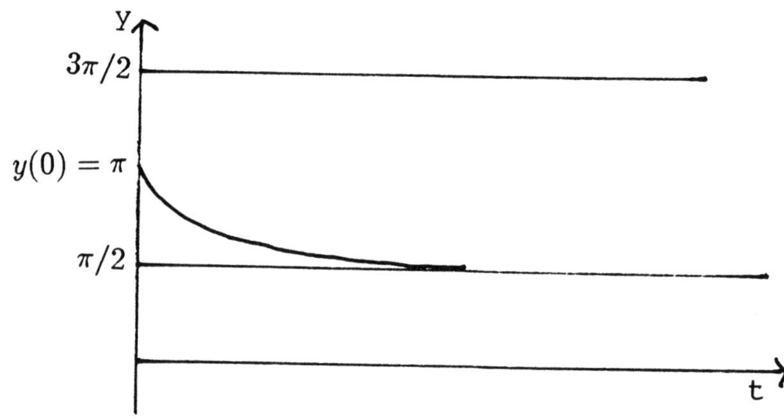

15. $y' = y(1 - y^2), \quad y(0) = -2.$

Since $y = 0, -1$ and 1 are the constant solutions of $y' = y(1 - y^2)$, and $y(0) = -2 < -1$, the initial value problem has the solution $y(t)$ such that $y(t) < -1$ for all $t > 0$. Since $y'(t) > 0$ for all $t > 0$, $y(t)$ increases for all $t > 0$. Thus $-2 < y(t) < -1$ for all $t > 0$. Since $y'' = (1 - y^2 - 2y^2)y' = y(1 - y^2)(1 - 3y^2)$, $y'' < 0$ for all $t > 0$. So the graph of $y(t)$ is concave down for all $t > 0$.

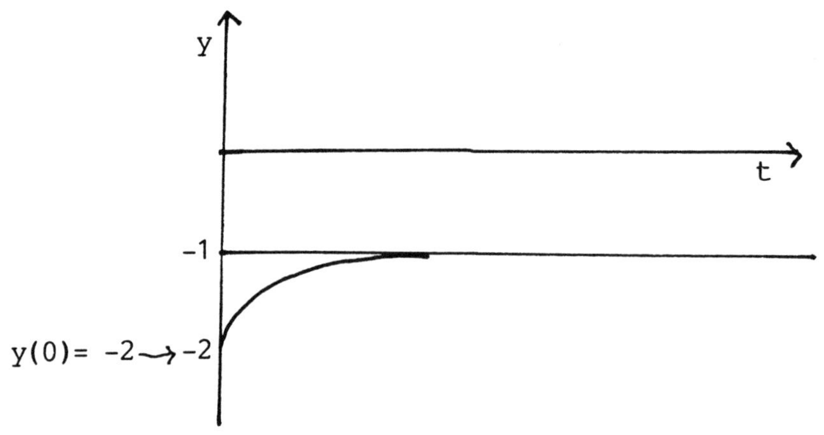

16. $y' = e^y, \quad y(0) = -2.$

Since $e^y > 0$ the solution $y(t)$ of this initial value problem is increasing for all $t > 0$. Thus $y(t) > -2$ for all $t > 0$. Since $y'' = e^y \cdot y' = e^{2y} > 0$, $y(t)$ is concave up for all $t > 0$.

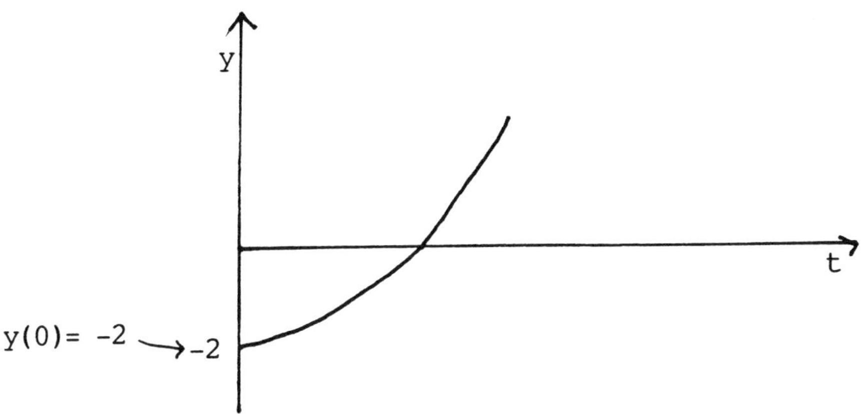

17. $y' = y^2, \quad y(0) = -1.$

Since $y^2 = 0$ for $y = 0$ and $y(0) = -1 < 0$, the solution $y(t)$ of this initial value problem has values with $y(t) < 0$ for all $t > 0$. Here, $y(t)$ is increasing for all $t > 0$. Since $y'' = 2yy' = 2y^3 < 0$ for all $t > 0$, the graph of $y(t)$ is concave down for all $t > 0$.

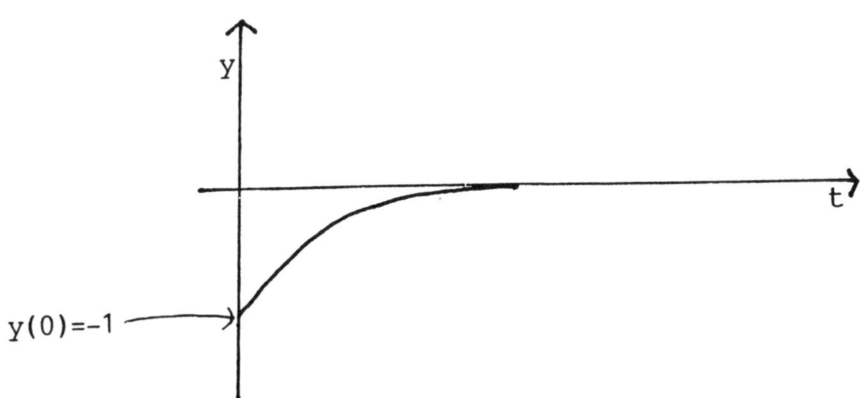

18. $y' = y^2, \quad y(0) = 3.$

As shown in Problem 17, the solution $y(t)$ of this initial value problem increases from $y(0) = 3$ with $y'(0) = y^2(0) = 9$, and the graph of $y(t)$ is concave up for all $t > 0$.

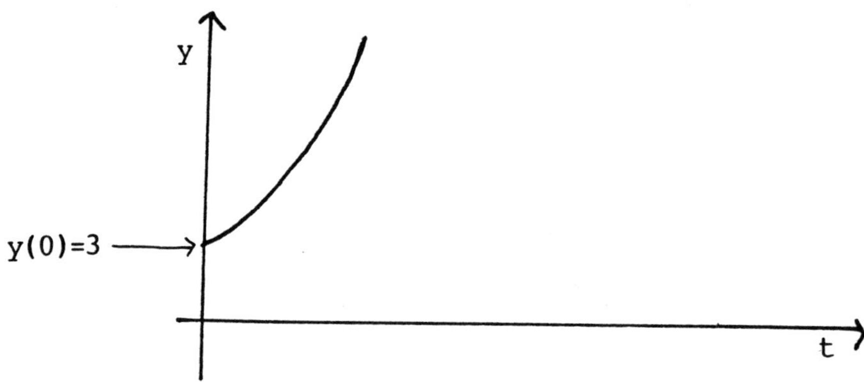

19. $y' = (y + 3)(y - 2), \quad y(0) = 1.$

Since $(y+3)(y-2) = 0$ for $y = -3$ and 2, and $-3 < y(0) = 1 < 2$, the solution $y(t)$ of this problem has values with $-3 < y(t) < 2$ for all $t > 0$. Since $y'(t) < 0$ for all $t > 0$, $y(t)$ is decreasing for all $t > 0$.

$$y'' = [(y-2) + (y+3)]\, y' = (2y+1)(y+3)(y-2).$$

Range of y	Sign of y''	Concavity
$-3 < y < -\frac{1}{2}$	Positive	Up
$-\frac{1}{2} < y < 1$	Negative	Down

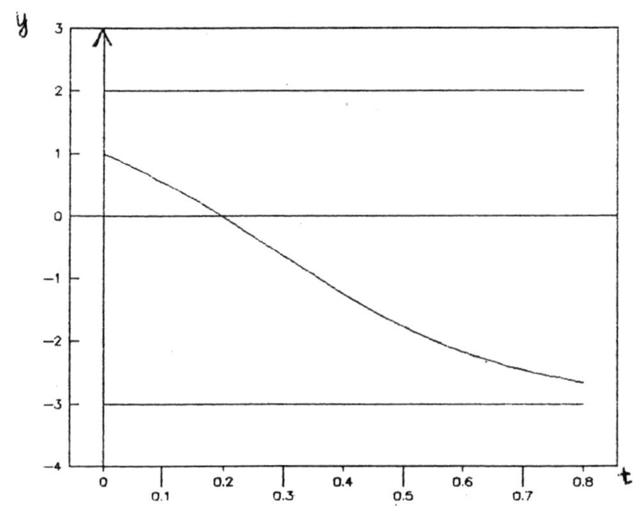

20. $y' = (y-1)(y-4), \quad y(0) = 3.$

Since $(y-1)(y-4) = 0$ for $y = 1$ and 4, and $1 < y(0) = 3 < 4$, the solution $y(t)$ of this problem has values with $1 < y(t) < 4$ for all $t \geq 0$. $y'(t) < 0$ for all $t > 0$, so $y(t)$ decreases from $y(0) = 3$.

Here $y'' = [(y-4) + (y-1)]\, y' = (2y-5)(y-1)(y-4)$.

Range of y	Sign of y''	Concavity
$1 < y < 2.5$	Positive	Up
$2.5 < y < 3$	Negative	Down

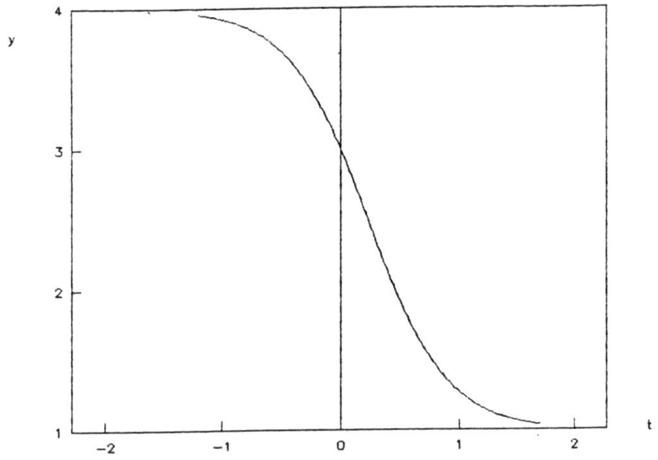

21. The values of y_1 and y_2 become closer together as t increases. Note that $y' = y(1-y)$ has two constant solutions $y = 0$ and $y = 1$, and $0 < y_1(0) = \frac{1}{2}$, $y_2(0) = \frac{3}{4} < 1$. Thus solutions y_1 and y_2 are in the same strip $0 < y < 1$ in which, by property A-111, any non-constant solution approaches the constant solution $y = 1$ as t increases. Hence the values of y_1 and y_2 become closer together as t increases.

22. The solution y_1 with initial value $y_1(0) = \frac{1}{2}$ is in the strip $0 < y < 1$, and solution y_2 with $y_2(0) = -\frac{1}{2}$ is in another strip $y < 0$. $y_2'(t) < 0$ shows that y_2 will decrease and becomes, as t increases, negatively infinitely large, while, as shown in problem 21, y_1 approaches $y = 1$ as t increases. Thus the values of y_1 and y_2 do not become closer together as t increases.

23. The solution y with $y(0) = \frac{3}{2}$ decreases for all $t > 0$ because $1 < y(t) < 2$ for all $t > 0$ so that $y'(t) < 0$ for all $t > 0$.

24. The solution y with $y(0) = 3$ increases for all appropriate $t > 0$ because $y(t) > 2$ for all such $t > 0$ so that $y'(t) > 0$ for all such $t > 0$.

25. The differential equation has three constant solutions $y = 0$, $y = 1$, and $y = 2$. We now determine whether all solutions in each strip decrease or increase.

Range of y	Sign of y'	Conclusion
$(-\infty, 0)$	Negative	decreasing
$(0, 1)$	Positive	increasing
$(1, 2)$	Negative	decreasing
$(2, \infty)$	Positive	increasing

From the above information, the only constant solution $y = c$ which has the described property is $y = 1$. Thus $c = 1$.

26. The differential equation has two constant solutions $y = -3$ and $y = 2$. Now determine whether the solutions in each strip decrease or increase.

Range of y	Sign of y'	Conclusion
$(-\infty, -3)$	Positive	increasing
$(-3, 2)$	Negative	decreasing
$(2, \infty)$	Positive	increasing

Thus the constant solution $y = c$ which has the described property is $y = 2$. That is, $c = 2$.

27. $y' = 30y(60 - 4y)$ has constant solutions $y = 0$ and $y = 15$.

Range of y	Sign of y'	Conclusion
$(-\infty, 0)$	Negative	decreasing
$(0, 15)$	Positive	increasing
$(15, \infty)$	Negative	decreasing

The constant solution to which all nearby solutions are asymptotic is $y = 15$.

28. $y' = 3y(20 - 3y)$, $\quad y(0) = 8$.

The solution for this model with $y(0) = 8$ decreases for $t > 0$ because $y'(0) = 3(8)(20 - 24) = -96$ implies that $y'(t) < 0$ for all $t > 0$.

29. Since the solution with $y(0) = 10$ is decreasing for all $t > 0$, we must have $y'(0) < 0 \implies y'(0) = k(10)(20 - 10) = 100k < 0 \implies k < 0$.

30. Since the solution with $y(0) = -2$ decreases for $t > 0$, we must have $y'(0) < 0 \implies y'(0) = k(-2)(1 - 4) = 6k < 0 \implies k < 0$.

Solutions to Exercise Set 10.6

1. $y' = 2y, \quad y(0) = 1.$
   $h = 0.25$ on $[0, 1]$: $\quad t_0 = 0,\ t_1 = \frac{1}{4},\ t_2 = \frac{1}{2},\ t_3 = \frac{3}{4},\ t_4 = 1.$
   By Euler's Method the approximation iteration scheme is

$$y_{i+1} = y_i + f(t_i, y_i)h \qquad (i = 0, 1, 2, 3) \quad \text{with } y_0 = y(0) = 1$$

   and $f(t, y) = 2y$. Thus

$$y_1 = y_0 + 2y_0h = 1 + 2(1)\left(\frac{1}{4}\right) = 1.5$$

$$y_2 = y_1 + 2y_1h = 1.5 + 2(1.5)\left(\frac{1}{4}\right) = 2.25$$

$$y_3 = y_2 + 2y_2h = 2.25 + 2(2.25)\left(\frac{1}{4}\right) = 3.375$$

$$y_4 = y_3 + 2y_3h = 3.375 + 2(3.375)\left(\frac{1}{4}\right) = 5.0625$$

2. $y' = 9y(1 - y), \quad y(0) = \frac{1}{2}.$
   Here $f(t, y) = 9y(1 - y)$ and $y_0 = y(0) = \frac{1}{2}$. Then

$$\begin{aligned} y_1 &= y_0 + f(t_0, y_0)h \\ &= \frac{1}{2} + 9\left(\frac{1}{2}\right)\left(1 - \frac{1}{2}\right)\left(\frac{1}{4}\right) = \frac{1}{2} + \frac{9}{16} = 1.0625 \\ y_2 &= y_1 + 9y_1(1 - y_1)h \\ &= 1.0625 + 9(1.0625)(1 - 1.0625)\left(\frac{1}{4}\right) = 0.913086 \end{aligned}$$

$$\begin{aligned} y_3 &= y_2 + 9y_2(1 - y_2)h \\ &= 0.913086 + 9(0.913086)(1 - 0.913086)\left(\frac{1}{4}\right) = 1.091646 \\ y_4 &= y_3 + 9y_3(1 - y_3)h \\ &= 1.091646 + 9(1.091646)(1 - 1.091646)\left(\frac{1}{4}\right) = 0.866545. \end{aligned}$$

3. $\begin{array}{l} y' + 2y = 4 \\ y(0) = 1 \end{array} \implies \begin{array}{l} y' = 4 - 2y \\ y(0) = 1 \end{array}.$

The approximations with $f(t, y) = 4 - 2y$ and $y_0 = y(0) = 1$ are

$$y_1 = y_0 + (4 - 2y_0)h = 1 + (4 - 2(1))\left(\frac{1}{4}\right) = 1.5$$

$$y_2 = y_1 + (4 - 2y_1)h = 1.5 + (4 - 2(1.5))\left(\frac{1}{4}\right) = 1.75$$

$$y_3 = y_2 + (4 - 2y_2)h = 1.75 + (4 - 2(1.75))\left(\frac{1}{4}\right) = 1.875$$

$$y_4 = y_3 + (4 - 2y_3)h = 1.875 + (4 - 2(1.875))\left(\frac{1}{4}\right) = 1.9375.$$

4. $\frac{dy}{dx} = \frac{y+1}{x+1}, \quad y(0) = 0.$

The approximations with $f(x, y) = \frac{y+1}{x+1}$ and $y_0 = y(0) = 0$ are

$$y_1 = y_0 + \frac{y_0 + 1}{x_0 + 1}h = 0 + \frac{0 + 1}{0 + 1}\left(\frac{1}{4}\right) = 0.25$$

$$y_2 = y_1 + \frac{y_1 + 1}{x_1 + 1}h = 0.25 + \frac{0.25 + 1}{0.25 + 1}\left(\frac{1}{4}\right) = 0.5$$

$$y_3 = y_2 + \frac{y_2 + 1}{x_2 + 1}h = 0.5 + \frac{0.5 + 1}{0.5 + 1}\left(\frac{1}{4}\right) = 0.75$$

$$y_4 = y_3 + \frac{y_3 + 1}{x_3 + 1}h = 0.75 + \frac{0.75 + 1}{0.75 + 1}\left(\frac{1}{4}\right) = 1.0.$$

5. $\frac{dy}{dx} = x(y + 1), \quad y(0) = 1.$

The approximations on $[0, 1]$ with $f(x, y) = x(y + 1)$, $h = \frac{1}{4}$, and $y_0 = y(0) = 1$ are:

$$y_1 = y_0 + x_0(y_0 + 1)h = 1 + 0(1 + 1)\left(\frac{1}{4}\right) = 1$$

$$y_2 = y_1 + x_1(y_1 + 1)h = 1 + \frac{1}{4}(1 + 1)\frac{1}{4} = 1.125$$

$$y_3 = y_2 + x_2(y_2 + 1)h = 1.125 + \left(\frac{1}{2}\right)(1.125 + 1)\left(\frac{1}{4}\right) = 1.390625$$

$$y_4 = y_3 + x_3(y_3 + 1)h = 1.390625 + \left(\frac{3}{4}\right)(1.390625 + 1)\left(\frac{1}{4}\right) \approx 1.838867.$$

6. $\frac{dy}{dx} = 2 - 4y, \quad y(0) = 1.$

The approximations on [0,1] with $f(x,y) = 2 - 4y, \quad h = \frac{1}{4}$, and $y_0 = y(0) = 1$ are:

$$y_1 = y_0 + (2 - 4y_0)h = 1 + (2 - 4(1))\left(\frac{1}{4}\right) = 0.5$$

$$y_2 = y_1 + (2 - 4y_1)h = 0.5 + (2 - 4(0.5))\left(\frac{1}{4}\right) = 0.5$$

$$y_3 = y_4 = 0.5.$$

7. $\begin{array}{l} \frac{dy}{dt} + 2ty = t \\ y(0) = \frac{3}{2} \end{array} \Longrightarrow \begin{array}{l} \frac{dy}{dt} = t(1 - 2y) \\ y(0) = \frac{3}{2} \end{array}.$

The approximations on [0,1] with $f(t,y) = t(1-2y)$, $h = \frac{1}{4}$, and $y_0 = y(0) = \frac{3}{2}$ are:

$$y_1 = y_0 + t_0(1 - 2y_0)h = \frac{3}{2} + 0\left(1 - 2\left(\frac{3}{2}\right)\right)\left(\frac{1}{4}\right) = 1.5$$

$$y_2 = y_1 + t_1(1 - 2y_1)h = 1.5 + \left(\frac{1}{4}\right)(1 - 2(1.5))\left(\frac{1}{4}\right) = 1.375$$

$$y_3 = y_2 + t_2(1 - 2y_2)h = 1.375 + \left(\frac{1}{2}\right)(1 - 2(1.375))\left(\frac{1}{4}\right) = 1.15625$$

$$y_4 = y_3 + t_3(1 - 2y_3)h = 1.15625 + \left(\frac{3}{4}\right)(1 - 2(1.15625))\left(\frac{1}{4}\right) = 0.910156.$$

8. $\begin{array}{l} e^y \cdot \frac{dy}{dt} - t^2 = 0 \\ y(0) = 0 \end{array} \Longrightarrow \begin{array}{l} \frac{dy}{dt} = t^2 e^{-y} \\ y(0) = 0 \end{array}.$

The approximations on [0,1] with $f(t,y) = t^2 e^{-y}$, $h = \frac{1}{4}$, and $y_0 = y(0) = 0$ are:

$$y_1 = y_0 + t_0^2 \cdot e^{-y_0} h = 0 + 0^2 \cdot e^{-0}\left(\frac{1}{4}\right) = 0$$

$$y_2 = y_1 + t_1^2 e^{-y_1} h = 0 + \left(\frac{1}{4}\right)^2 e^{-0}\left(\frac{1}{4}\right) = 0.015625$$

$$y_3 = y_2 + t_2^2 e^{-y_2} h = 0.015625 + \left(\frac{1}{2}\right)^2 e^{-0.015625}\left(\frac{1}{4}\right) = 0.077156$$

$$y_4 = y_3 + t_3^2 e^{-y_3} h = 0.077156 + \left(\frac{3}{4}\right)^2 e^{-0.077156}\left(\frac{1}{4}\right) = 0.207339.$$

9. $\frac{dy}{dt} = 2y - 2t, \quad y(0) = \frac{3}{2}.$

The approximations on [0,1] with $f(t,y) = 2(y-t)$, $h = \frac{1}{4}$ and $y_0 = y(0) = \frac{3}{2}$ are:

$$y_1 = y_0 + 2(y_0 - t_0)h = \frac{3}{2} + 2\left(\frac{3}{2} - 0\right)\left(\frac{1}{4}\right) = 2.25$$

$$y_2 = y_1 + 2(y_1 - t_1)h = 2.25 + 2(2.25 - 0.25)\left(\frac{1}{4}\right) = 3.25$$

$$y_3 = y_2 + 2(y_2 - t_2)h = 3.25 + 2(3.25 - 0.5)\left(\frac{1}{4}\right) = 4.625$$

$$y_4 = y_3 + 2(y_3 - t_3)h = 4.625 + 2(4.625 - 0.75)\left(\frac{1}{4}\right) = 6.5625.$$

10. $\begin{array}{l} \frac{dy}{dt} - 2y = e^{2t}\sin t \\ y(0) = 0 \end{array} \implies \begin{array}{l} \frac{dy}{dt} = 2y + e^{2t}\sin t \\ y(0) = 0 \end{array}.$

The approximations on [0,1] with $f(t,y) = 2y + e^{2t}\sin t$, $h = \frac{1}{4}$, and $y_0 = y(0) = 0$ are:

$$y_1 = y_0 + (2y_0 + e^{2t_0}\sin t_0)h = 0 + \left(2(0) + e^{2(0)}\sin 0\right)\left(\frac{1}{4}\right) = 0$$

$$y_2 = y_1 + (2y_1 + e^{2t_1}\sin t_1)h = 0 + \left(2(0) + e^{2\left(\frac{1}{4}\right)}\sin\frac{1}{4}\right)\left(\frac{1}{4}\right) = 0.101975$$

$$y_3 = y_2 + (2y_2 + e^{2t_2}\sin t_2)h$$
$$= 0.101975 + \left(2(0.101975) + e^{2\left(\frac{1}{2}\right)}\sin\left(\frac{1}{2}\right)\right)\left(\frac{1}{4}\right) = 0.478766$$

$$y_4 = y_3 + (2y_3 + e^{2t_3}\sin t_3)h$$
$$= 0.478766 + \left(2(0.478766) + e^{2\left(\frac{3}{4}\right)}\sin\left(\frac{3}{4}\right)\right)\left(\frac{1}{4}\right) = 1.481872.$$

11. The initial value problem in Exercise 3 is

$$\begin{array}{l} y' + 2y = 4 \\ y(0) = 1 \end{array} \implies \begin{array}{l} y' = -2(y-2) \\ y(0) = 1 \end{array}.$$

Using equations (14) and (15) of Section 10.3, we have the general solution for the differential equation with $k = -2$ and $a = 2$.

$$y = 2 + Ce^{-2t}$$

Now $y(0) = 1 \implies 1 = 2 + Ce^{-2(0)} \implies C = -1$.

Thus the actual solution of this initial value problem is

$$y = 2 - e^{-2t}.$$

Now compare the actual values of $y(t)$ with the values from Euler's method. The results are as follows. (See Exercise 3.)

t_j	0	0.25	0.5	0.75	1.0
y_j	1.0	1.5	1.75	1.875	1.9375
$y(t_j)$	1.0	1.393469	1.632121	1.776870	1.864665

12. (a) The 1000 in the differential equation comes from the fact that the investor is making deposits of \$1000 per year in a continuous manner. The $0.10P(t)$ comes from the fact that the bank is paying 10% annual interest rate compounded continuously.

(b) With $h = 1$, $f(t, P) = 0.1P + 1000$, and $P_0 = P(0) = 1000$, we have:

$$\begin{aligned} P(1) \approx P_1 &= P_0 + (0.1P_0 + 1000)h \\ &= 1000 + (0.1(1000) + (1000)) \cdot 1 = \$2100 \\ P(2) \approx P_2 &= P_1 + (0.1P_1 + 1000)h \\ &= 2100 + (0.1(2100) + 1000) \cdot 1 = \$3310 \\ P(3) \approx P_3 &= P_2 + (0.1P_2 + 1000)h \\ &= 3310 + (0.1(3310) + 1000) \cdot 1 = \$4641 \\ P(4) \approx P_4 &= P_3 + (0.1P_3 + 1000)h \\ &= 4641 + (0.1(4641) + 1000) \cdot 1 = \$6105.10. \end{aligned}$$

Solutions to the Review Exercises - Chapter 10

1. $\frac{dx}{dt} = x + 3.$

   $\frac{dx}{x+3} = dt \implies \int \frac{1}{x+3}\,dx = \int dt \implies \ln|x+3| = t + C_1 \implies$

   $x + 3 = Ce^t \implies x = -3 + Ce^t.$

2. $\frac{dy}{dx} = 4 + \sqrt{x}.$

   $\int dy = \int (4 + \sqrt{x})\,dx \implies y = 4x + \frac{2}{3}x^{3/2} + C.$

3. $\frac{dy}{dx} = \frac{\sec^2\sqrt{x}\tan\sqrt{x}}{\sqrt{x}}.$

   Let $u = \tan\sqrt{x}$. Then $du = \sec^2\sqrt{x}\cdot\frac{1}{2\sqrt{x}}\,dx \implies \frac{\sec^2\sqrt{x}}{\sqrt{x}}\,dx = 2\,du.$

   $\int \frac{\sec^2\sqrt{x}\tan\sqrt{x}}{\sqrt{x}}\,dx = \int u(2\,du) = u^2 + C = \tan^2\sqrt{x} + C.$

   $y = \tan^2\sqrt{x} + C.$

4. $\frac{d^2y}{dt^2} = \sqrt{1+t}.$

   Let $u = \frac{dy}{dt}$; then the differential equation becomes $\frac{du}{dt} = \sqrt{1+t}$

   $\implies \int du = \int \sqrt{1+t}\,dt \implies u = \frac{2}{3}(1+t)^{3/2} + C_1$

   $\implies u = \frac{dy}{dt} = \frac{2}{3}(1+t)^{3/2} + C_1$

   $\implies \int dy = \int \left[\frac{2}{3}(1+t)^{3/2} + C_1\right] dt$

   $\implies y = \frac{2}{3}\cdot\frac{2}{5}(1+t)^{5/2} + C_1 t + C_2$

   $y = \frac{4}{15}(1+t)^{5/2} + C_1 t + C_2$

5. $\frac{dy}{dt} = 3ty^2.$

   $\int \frac{1}{y^2}dy = \int 3t\,dt \implies -\frac{1}{y} = \frac{3}{2}t^2 + C_1$

   $\implies y = \frac{-1}{\frac{3}{2}t^2+C_1} = -\frac{2}{3t^2+C}$ where $C = 2C_1$.

6. $\frac{dy}{dx} = \frac{\sqrt{x+1}}{y}.$

   $\int y\,dy = \int \sqrt{x+1}\,dx \implies \frac{1}{2}y^2 = \frac{2}{3}(1+x)^{3/2} + C_1$

   $\implies y = \pm\frac{2}{3}\sqrt{3(1+x)^{3/2} + C}.$

7. $\frac{dy}{dt} = (1+t)(2+y).$

   $\int \frac{1}{2+y}\,dy = \int (1+t)dt \implies \ln|2+y| = t + \frac{1}{2}t^2 + C_1$

   $\implies 2 + y = C\cdot e^{\frac{1}{2}t^2+t} \implies y = Ce^{\frac{1}{2}t^2+t} - 2.$

8. $ty' = \frac{t+3}{y} \implies y\,dy = \frac{t+3}{t}\,dt \implies y\,dy = \left(1+\frac{3}{t}\right)dt.$

$\frac{1}{2}y^2 = t + 3\ln|t| + C_1 \implies y^2 = 2t + 6\ln|t| + C.$

$y = \pm\sqrt{2t + 6\ln|t| + C}.$

9. $y' + 2y = 4.$

$\frac{dy}{dt} = -2(y-2) \implies \int \frac{1}{y-2}\,dy = \int(-2)dt \implies \ln|y-2| = -2t + C_1$

$\implies y - 2 = Ce^{-2t} \implies y = 2 + Ce^{-2t}.$

10. $y' = y(3-y).$

$\int \frac{1}{(y-3)y}\,dy = \int(-1)dt \implies \int \frac{1}{3}\left(\frac{1}{y-3} - \frac{1}{y}\right)dy = -t + C_1$

$\implies \frac{1}{3}\left(\ln|y-3| - \ln|y|\right) = -t + C_1$

$\implies \ln\left|\frac{y-3}{y}\right| = -3t + C_2$

$\implies \frac{y-3}{y} = Ce^{-3t}$

$\implies y - 3 = yCe^{-3t} \implies y = \frac{3}{1-Ce^{-3t}}.$

11. $y' + y\cos t = 0.$

$y' = -y\cos t \implies \int \frac{1}{y}\,dy = \int(-\cos t)dt$

$\implies \ln|y| = -\sin t + C_1 \implies y = Ce^{-\sin t}.$

12. $y' = 4y\ln y.$

$\int \frac{1}{y\ln y}\,dy = \int 4\;dt \implies \ln|\ln y| = 4t + C_1$

$\implies \ln y = Ce^{4t} \implies y = e^{Ce^{4t}}.$

13. $(t-1)y' = t+1 \implies dy = \frac{t+1}{t-1}\,dt \implies dy = \frac{(t-1)+2}{t-1}\,dt \implies dy = \left(1 + \frac{2}{t-1}\right)dt.$

$y = t + 2\ln|t-1| + C.$

14. $\frac{dy}{dt} + \frac{1}{t^2}y = 0.$

$\frac{dy}{dt} = -\frac{y}{t^2} \implies \int \frac{1}{y}\,dy = \int\left(-\frac{1}{t^2}\right)dt \implies \ln|y| = \frac{1}{t} + C_1$

$\implies y = Ce^{1/t}.$

15. $\frac{dy}{dx} = y(2-y).$

$\int \frac{1}{y(2-y)}\,dy = \int\,dx \implies \frac{1}{2}\int\left(\frac{1}{y} + \frac{1}{2-y}\right)dy = x + C_1$

$\implies \ln|y| - \ln|2-y| = 2x + C_2 \implies \ln\left|\frac{y}{2-y}\right| = 2x + C_2$

$\implies \frac{y}{2-y} = C_3e^{2x} \implies y = 2C_3e^{2x} - yC_3e^{2x} \implies y = \frac{2C_3e^{2x}}{1+C_3e^{2x}}$

$= \frac{2}{\frac{1}{C_3}e^{-2x}+1} = \frac{2}{1+Ce^{-2x}}.$

16. $ty' + 2y = 0.$

$y' = -\frac{2y}{t} \implies \int \frac{1}{y}\,dy = \int \left(-\frac{2}{t}\right) dt \implies \ln|y| = -\ln t^2 + C_1$

$\implies |y| = e^{-\ln t^2 + C_1} \implies y = Ce^{\ln(1/t^2)} \implies y = \frac{C}{t^2}.$

17. $\frac{dy}{dt} = \frac{t+3}{y} \implies y\,dy = (t+3)dt.$

$\frac{1}{2}y^2 = \frac{1}{2}t^2 + 3t + C_1 \implies y^2 = t^2 + 6t + C.$

$y = \pm\sqrt{t^2 + 6t + C}.$

18. $\frac{dy}{dt} = \frac{y}{t+1}.$

$\int \frac{dy}{y} = \int \frac{dt}{t+1} \implies \ln|y| = \ln|t+1| + C_1$

$\implies y = C_2 e^{\ln|t+1|} \implies y = C(t+1).$

19. $y'' - 5y' - 6y = 0.$

The characteristic equation is

$(r^2 - 5r - 6) = 0 \implies (r+1)(r-6) = 0 \implies r_1 = -1,\ r_2 = 6.$

General solution is $y = Ae^{-t} + Be^{6t}.$

20. $\frac{d^2y}{dx^2} - 9y = 0.$

$r^2 - 9 = 0 \implies r = \pm 3.$

General solution is $y = Ae^{-3x} + Be^{3x}.$

21. $\frac{d^2y}{dx^2} - 5\frac{dy}{dx} - 14y = 0.$

$r^2 - 5r - 14 = 0 \implies (r-7)(r+2) = 0 \implies r_1 = 7,\ r_2 = -2.$

General solution is $y = Ae^{7x} + Be^{-2x}.$

22. $y'' - y' - 6y = 0.$

The characteristic equation is

$(r^2 - r - 6) = 0 \implies (r+2)(r-3) = 0 \implies r_1 = -2,\quad r_2 = 3.$

The general solution is $y = Ae^{-2t} + Be^{3t}.$

23. $y'' - 8y' + 16y = 0.$

The characteristic equation is

$(r^2 - 8r + 16) = 0 \implies (r-4)^2 = 0 \implies r = 4$ is a double root.

General solution is $y = Ae^{4t} + Bte^{4t}.$

24. $\frac{d^2y}{dt^2} = 3\frac{dy}{dt} + 10y$.

$r^2 - 3r - 10 = 0 \implies (r-5)(r+2) = 0 \implies r_1 = 5,\ r_2 = -2$.

General solution is $y = Ae^{5t} + Be^{-2t}$.

25. $\frac{dy}{dx} = \frac{x}{\sqrt{1+x^2}}, \quad y(0) = 3$.

$\int dy = \int \frac{x}{\sqrt{1+x^2}}\,dx \implies y = \frac{1}{2}\int \frac{1}{\sqrt{u}}\,du$ where $u = 1 + x^2$

$\implies y = \sqrt{u} + C = \sqrt{1+x^2} + C$.

$y(0) = 3 \implies 3 = 1 + C \implies C = 2$.

The solution is $y = 2 + \sqrt{1+x^2}$.

26. $\frac{d^2y}{dt^2} = 1 + t, \quad y(0) = 1, \quad y'(0) = 0$.

$\frac{dy}{dt} = \int (1+t)dt = t + \frac{1}{2}t^2 + C_1$.

$y = \int \left(t + \frac{1}{2}t^2 + C_1\right) dt = \frac{1}{2}t^2 + \frac{1}{6}t^3 + C_1 t + C_2$.

$y(0) = 1 \implies C_2 = 1$.

$y'(0) = 0 \implies C_1 = 0$.

So, the solution is $y = \frac{1}{6}t^3 + \frac{1}{2}t^2 + 1$.

27. $y' = 4y, \quad y(0) = 3$.

$\frac{dy}{y} = 4\,dt \implies \ln|y| = 4t + C_1 \implies y = Ce^{4t}$.

$y(0) = 3 \implies 3 = Ce^{4(0)} \implies C = 3$.

The solution is $y = 3e^{4t}$.

28. $\frac{dy}{dt} = ty, \quad y(1) = 3$.

$\int \frac{dy}{y} = \int t\ dt \implies \ln|y| = \frac{1}{2}t^2 + C_1 \implies y = Ce^{t^2/2}$.

$y(1) = 3 \implies 3 = Ce^{1/2} \implies C = 3e^{-1/2}$.

So the solution is $y = 3e^{-1/2} \cdot e^{t^2/2} = 3e^{\frac{1}{2}(t^2-1)}$.

29. $\frac{dy}{dx} = y \sin x, \quad y(0) = \pi$.

$\int \frac{dy}{y} = \int \sin x\ dx \implies \ln|y| = -\cos x + C_1 \implies y = Ce^{-\cos x}$.

$y(0) = \pi \implies \pi = Ce^{-\cos 0} \implies \pi = C \cdot e^{-1} \implies C = \pi e$.

So the solution is $y = \pi e \cdot e^{-\cos x} = \pi e^{1-\cos x}$.

30. $y' = 4(1-y), \quad y(0) = 1$.

$\int \frac{1}{y-1}\ dy = \int (-4)dt \implies \ln|y-1| = -4t + C_1 \implies y = 1 + Ce^{-4t}$ where $C \neq 0$.

$y(t) = 1$ is also a solution for $C = 0$.

$y(0) = 1 \implies 1 = 1 + C \implies C = 0$.

So the solution is $y = 1$.

31. $y' = 10(5 - t), \quad y(0) = 2$.

$y = 10\int(5-t)dt = 10\left(5t - \frac{1}{2}t^2\right) + C. \ \ y(0) = 2 \implies 2 = C$.

So the solution is $y = 50t - 5t^2 + 2$.

32. $\frac{d^2y}{dx^2} = -9y, \quad y(0) = 3, \quad y'(0) = 9$.

$r^2 + 9 = 0 \implies r = \pm 3i \implies y = A\sin 3x + B\cos 3x$.

$y(0) = 3 \implies B = 3; \quad y'(0) = 9 \implies 3A = 9 \implies A = 3$

So the solution is $y = 3(\sin 3x + \cos 3x)$.

33. $\frac{d^2y}{dx^2} + 16y = 0, \quad y(0) = 5, \quad y'(0) = 0$.

$r^2 + 16 = 0 \implies r = \pm 4i \implies y = A\sin 4x + B\cos 4x$.

$y(0) = 5 \implies B = 5; \quad y'(0) = 0 \implies 4A = 0 \implies A = 0$.

So the solution is $y = 5\cos 4x$.

34. $\frac{dy}{dt} = 5y(10 - y), \quad y(0) = 5$.

$\int \frac{1}{y(10-y)}\,dy = \int 5\,dt \implies \frac{1}{10}(\ln|y| - \ln|10-y|) = 5t + C_1$.

$\implies \ln\left|\frac{y}{10-y}\right| = 50t + C_2 \implies \frac{y}{10-y} = C_3e^{50t} \implies y = (10-y)C_3e^{50t}$.

$y = \frac{10C_3e^{50t}}{1+C_3e^{50t}} = \frac{10}{1+Ce^{-50t}}$.

$y(0) = 5 \implies 5 = \frac{10}{1+C} \implies C = 1$.

So the solution is $y = \frac{10}{1+e^{-50t}}$.

35. $y'' - 6y' + 9y = 0, \quad y(0) = 0, \quad y'(0) = 6$.

$r^2 - 6r + 9 = 0 \implies (r-3)^2 = 0 \implies r = 3$ is a double root.

So $y = Ae^{3t} + Bte^{3t}$. $\qquad y(0) = 0 \implies A = 0$.

$y' = (Bte^{3t})' = B(e^{3t} + 3te^{3t})$. $\qquad y'(0) = 6 \implies B = 6$.

So the solution is $y = 6te^{3t}$.

36. $y'' - 16y = 0, \quad y(0) = 2, \quad y'(0) = 8$.

$r^2 - 16 = 0 \implies r = \pm 4 \implies y = Ae^{4t} + Be^{-4t}$.

$$y(0) = 2, \quad y'(0) = 8 \implies \begin{cases} A + B = 2 \\ 4A - 4B = 8 \end{cases} \implies A = 2, \quad B = 0.$$

So, the solution is $y = 2e^{4t}$.

37. $\frac{dy}{dt} = 2y(4-y)$.

(1) $y = 0$ and $y = 4$ are the constant solutions

(2)

Range of y	Sign of y'	Conclusion
$y < 0$	negative	decreasing
$0 < y < 4$	positive	increasing
$y > 4$	negative	decreasing

(3) $y'' = 2[4 - y - y]\,y' = 8(2-y)y(4-y)$.

Range of y	Sign of y''	Concavity
$y < 0$	negative	down
$0 < y < 2$	positive	up
$2 < y < 4$	negative	down
$y > 4$	positive	up

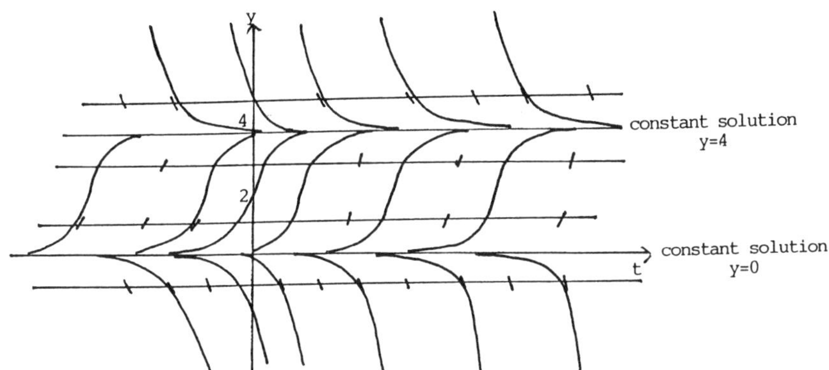

38. $\frac{dy}{dx} = xy$.

(1) $y = 0$ is the constant solution.

(2) Consider $x \geq 0$ only.

For $y > 0$, $x > 0$, $y' > 0$, so y increases.

For $y < 0$, $x > 0$, $y' < 0$, so y decreases.

Note that $y' = 0$ for $x = 0$.

(3) $y'' = y + xy' = (1 + x^2)y$.

$y'' > 0$ for $y > 0$ and $y'' < 0$ for $y < 0$.

So the graph of y is concave up for $y > 0$ and concave down for $y < 0$.

For any fixed y, y' increases with values of $x > 0$ if $y > 0$ and decreases with values of $x > 0$ if $y < 0$.

(4)

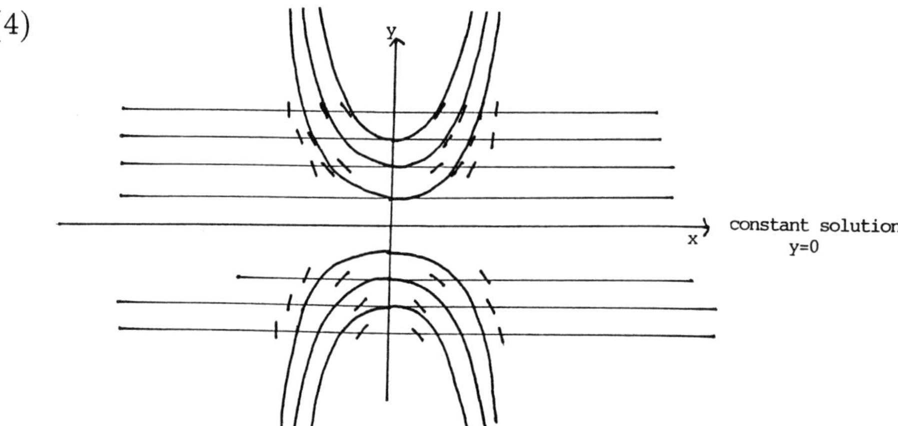

39. $\frac{dy}{dt} = y + 3$.

(1) $y = -3$ is the constant solution.

(2) For $y > -3$, $y' > 0$, so y increases.

For $y < -3$, $y' < 0$, so y decreases.

(3) $y'' = y' = y + 3$.

For $y > -3$, $y'' > 0$, so the graph of y is concave up.

For $y < -3$, $y'' < 0$, so the graph of y is concave down.

(4)

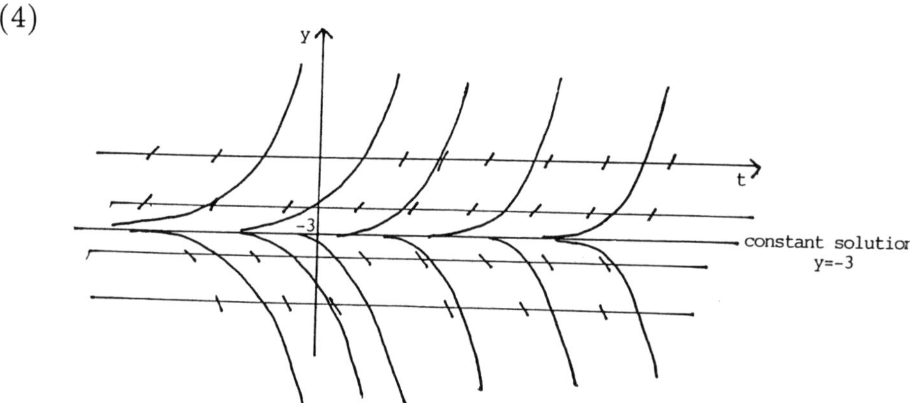

40. $\frac{dy}{dx} = \frac{x}{y}$.

(1) There is no constant solution.

(2) Consider $x \geq 0$ only.

Note when $x = 0$ and $y(0) \neq 0$, $y' = 0$.

For $y > 0$, $x > 0$, $y' > 0$, so y increases.

For $y < 0$, $x > 0$, $y' < 0$, so y decreases.

Note that $y = x$ $(y \neq 0)$ and $y = -x$ $(y \neq 0)$ are solutions.

For any fixed y, y increases with x if $y > 0$, and y decreases with x if $y < 0$.

(3) $y'' = \frac{y - xy'}{y^2} = \frac{y^2 - x^2}{y^3}$.

In the region R_1 such that

$R_1 = \{(x, y) \,|\, 0 < x < y \text{ or } -x < y < 0\}, \quad y'' > 0$.

In the region $R_2 = \{(x, y) \,|\, 0 < y < x \text{ or } y < -x < 0\}, \quad y'' < 0$.

So the graph of y is concave up in region R_1 and concave down in region R_2.

(4) For $x < 0$, we have a similar analysis as in parts (2) and (3) for $x > 0$.

(5)

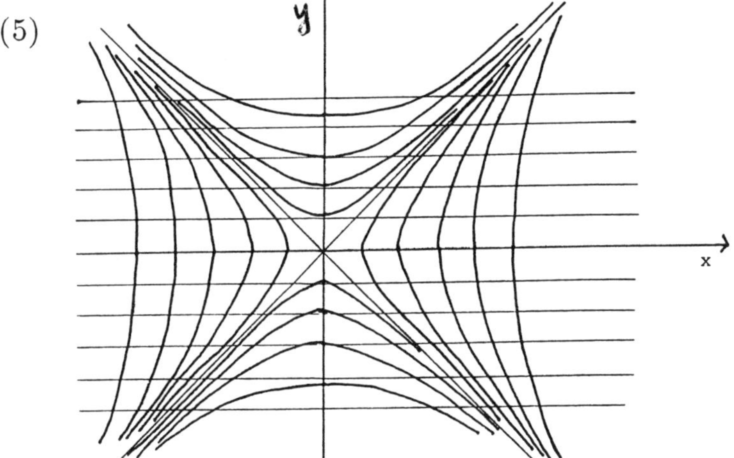

41. $V'(t) = \sqrt{36 + t}$. $\qquad V(0) = 200{,}000$.

$V(t) = \int \sqrt{36 + t}\, dt = \frac{2}{3}(36 + t)^{3/2} + C$.

$V(0) = 200{,}000 \implies \frac{2}{3}36^{3/2} + C = 200{,}000$

$\implies 144 + C = 200{,}000 \implies C = 199{,}856$.

So $V(t) = \frac{2}{3}(t + 36)^{3/2} + 199{,}856$ dollars.

42. $R'(x) = 60 + 2x - 0.2x^2$. $\qquad R(0) = 0$.

$R(x) = \int (60 + 2x - 0.2x^2)dx = 60x + x^2 - \frac{0.2}{3}x^3 + C$.

$R(0) = 0 \implies C = 0$, so $R(x) = 60x + x^2 - \frac{1}{15}x^3$.

43. (a) $y = Ce^{2t}$. $y' = 2Ce^{2t} = 2y$, so $y' = 2y$.

(b) $y = C_1e^t + C_2e^{-t}$.

$y' = C_1e^t - C_2e^{-t}$, $y'' = C_1e^t + C_2e^{-t} = y$. So $y'' - y = 0$.

44. Let y be the amount of salt in the tank after t minutes.

Then $\frac{dy}{dt} = 2(50) - \frac{y}{500}(50)$ with $y(0) = 500(1) = 500$.

$y' = 100 - \frac{1}{10}y$, $y(0) = 500$.

$\int \frac{dy}{1000-y} = \frac{1}{10}t + C_1$.

$-\ln|1000 - y| = \frac{t}{10} + C_1$

$1000 - y = Ce^{-t/10} \implies y = 1000 - Ce^{-t/10}$.

$y(0) = 500 \implies 1000 - C = 500 \implies C = 500$.

So $y = 500(2 - e^{-t/10})$.

45. $\frac{du}{dt} = -\frac{u}{5}(1)$, $u(0) = 5$. Here $u =$ the number of gallons of antifreeze.

$\int \frac{du}{u} = \int \left(-\frac{dt}{5}\right) \implies \ln u = -\frac{1}{5}t + C_1$.

$u = Ce^{-t/5}$.

$u(0) = 5 \implies C = 5$. So $u = 5e^{-t/5}$.

The concentration of antifreeze after t minutes is $e^{-t/5}$.

46. (a) $P'(t) = 0.1P(t) + 2000$.

(b) Using Euler's Method with $f(t, P) = 0.1P + 2000$, $h = 1$, and $P_0 = P(0) = 5000$, we have

$$\begin{aligned}
P(1) \approx P_1 &= P_0 + f(0, P_0)h \\
&= 5000 + (0.1(5000) + 2000) \cdot 1 = \$7500. \\
P(2) \approx P_2 &= P_1 + f(1, P_1)h \\
&= 7500 + (0.1(7500) + 2000) = \$10,250. \\
P(3) \approx P_3 &= P_2 + f(2, P_2)h \\
&= 10250 + (0.1(10250) + 2000) = \$13275. \\
P(4) \approx P_4 &= P_3 + f(3, P_3)h \\
&= 13275 + (0.1(13275) + 2000) = \$16602.50.
\end{aligned}$$

47. Denote by $y(t)$ the temperature of the drink t minutes after removing it. Thus

$y'(t) = K(80 - y(t))$ with $y(0) = 40°\text{F}$.

$\int \frac{dy}{80-y} = Kt + C_1 \implies -\ln(80-y) = Kt + C_1$.

$80 - y = Ce^{-Kt} \implies y = 80 - Ce^{-Kt}$.

$y(0) = 40 \implies 40 = 80 - C \implies C = 40$.

So $y(t) = 40(2 - e^{-Kt})$.

From $y(5) = 50 \implies 50 = 40(2 - e^{-5K})$

$\implies e^{-5K} = 2 - \frac{5}{4} = \frac{3}{4} \implies e^{-K} = \left(\frac{3}{4}\right)^{1/5}$.

Thus $y(t) = 40(2 - e^{-Kt}) = 40\left[2 - \left(\frac{3}{4}\right)^{(1/5)t}\right]$.

So $y(10) = 40\left[2 - \left(\frac{3}{4}\right)^{\frac{1}{5}(10)}\right] = 40\left[2 - \frac{9}{16}\right] = 57.5°\text{F}$.

48. $\frac{dv}{dt} = (t+5)^{2/3}, \quad v(0) = 5,000$.

$v = \frac{3}{5}(t+5)^{5/3} + C$.

$v(0) = \frac{3}{5} \cdot 5^{5/3} + C \implies 5000 = 3 \cdot 5^{2/3} + C \implies C = 5000 - 3 \cdot 5^{2/3}$.

$v(t) = \frac{3}{5}(t+5)^{5/3} + (5000 - 3 \cdot 5^{2/3})$.

49. $\frac{du}{ds} = \frac{1}{s+1}, \quad u(0) = 5$.

$u(s) = \ln(s+1) + C$.

$u(0) = 5 \implies 5 = \ln 1 + C \implies C = 5$.

$u(s) = \ln(s+1) + 5$.

CHAPTER 11

Solutions to Exercise Set 11.1

1. a. $Pr(E_1) = Pr(1) + Pr(3) = \frac{1}{3} + \frac{1}{6} = \frac{1}{2}$.

   b. $Pr(E_2) = Pr(1) + Pr(4) + Pr(5) = \frac{1}{3} + \frac{1}{12} + \frac{1}{12} = \frac{1}{2}$.

   c. $Pr(E_3) = Pr(2) + Pr(3) + Pr(5) = \frac{1}{3} + \frac{1}{6} + \frac{1}{12} = \frac{7}{12}$.

2. a. $Pr(E_1) = Pr(10) + Pr(20) = 0.1 + 0.3 = 0.4$.

   b. $Pr(E_2) = Pr(15) + Pr(20) + Pr(30) = 0.1 + 0.3 + 0.3 = 0.7$.

   c. $Pr(E_3) = Pr(10) + Pr(15) + Pr(20) + Pr(30)$

   $= 0.1 + 0.1 + 0.3 + 0.3$

   $= 0.8$.

3.

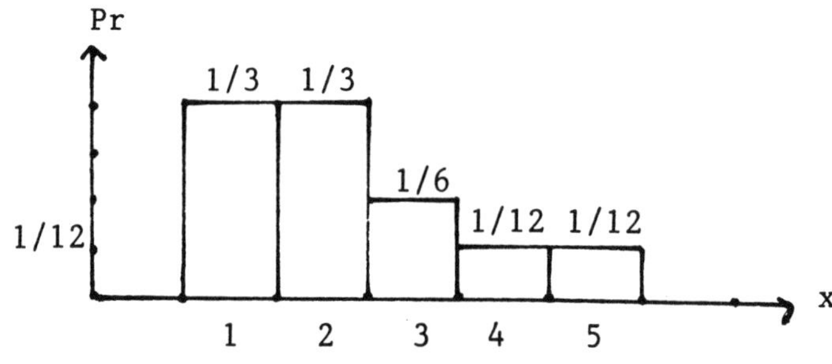

4.

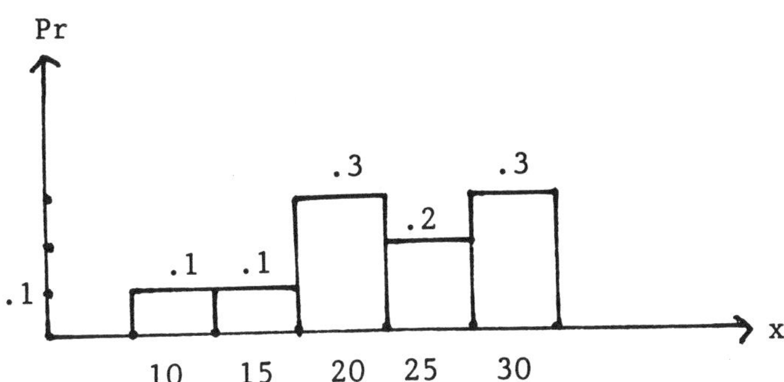

5. a. $Pr(X = 3) + Pr(X = 5) + Pr(X = 9)$

$= \frac{2}{36} + \frac{4}{36} + \frac{4}{36} = \frac{10}{36} = \frac{5}{18}.$

b. $Pr(X = 8) + Pr(X = 9) + Pr(X = 10) + Pr(X = 11) + Pr(X = 12)$

$= \frac{5}{36} + \frac{4}{36} + \frac{3}{36} + \frac{2}{36} + \frac{1}{36} = \frac{15}{36} = \frac{5}{12}.$

c. $1 - Pr(X = 6) = 1 - \frac{5}{36} = \frac{31}{36}.$

6. a. $Pr(X = 5) = \frac{1}{6}.$

b. $Pr(X = 2) + Pr(X = 4) = \frac{1}{6} + \frac{1}{6} = \frac{1}{3}.$

c. $1 - Pr(X = 3) = 1 - \frac{1}{6} = \frac{5}{6}.$

7. a. $Pr(\text{red}) = \frac{3}{5}$, (b) $Pr(\text{black}) = \frac{2}{5}.$

8. $Pr(\text{red}) = \frac{3}{3+x} = \frac{1}{4}$, $12 = 3 + x$, $\therefore x = 9.$

9. a. $\frac{8}{15}$ b. $\frac{7}{15}$ c. $\frac{3}{15} = \frac{1}{5}.$

10. a. The number of hours that both John and Mary are at their desks is 2. So the probability is $\frac{2}{9}$.

b. $\frac{7}{9}$. c. 0.

11. The area of the target is $12^2\pi = 144\pi$ and the area of the bull's eye is $3^2\pi = 9\pi$.

So the probability that it hits the bull's eye is $\frac{9\pi}{144\pi} = \frac{1}{16}$.

12. a. There are 12 face cards, and 3 of these cards are hearts.

Therefore the probability the card is a heart is $\frac{3}{12} = \frac{1}{4}$.

b. There are 4 queens. Therefore the probability the card is a queen is $\frac{4}{12} = \frac{1}{3}$.

c. The probability that the card is not a jack is $1 - \frac{1}{3} = \frac{2}{3}$.

d. The probability that the card is red and odd-numbered is 0.

13. a. $E(X) = x_0p_0 + x_1p_1 + x_2p_2 + x_3p_3$

$= 0 \cdot 0.2 + 1 \cdot 0.2 + 2 \cdot 0.3 + 3 \cdot 0.3 = 0.2 + 0.6 + 0.9 = 1.7.$

b. $\text{Var}(X) = 0.2(0 - 1.7)^2 + 0.2(1 - 1.7)^2 + 0.3(2 - 1.7)^2 + 0.3(3 - 1.7)^2$

$= 0.2 \cdot 2.89 + 0.2 \cdot 0.49 + 0.3 \cdot 0.09 + 0.3 \cdot 1.69$

$= 0.578 + 0.098 + 0.027 + 0.507 = 1.21.$

$\sigma = \sqrt{\text{Var}(X)} = 1.1.$

14. a. $E(X) = 0.1 \cdot 1 + 0.3 \cdot 2 + 0.3 \cdot 3 + 0.3 \cdot 4 = 0.1 + 0.6 + 0.9 + 1.2 = 2.8.$

b. $\text{Var}(X) = 0.1(1 - 2.8)^2 + 0.3(2 - 2.8)^2 + 0.3(3 - 2.8)^2 + 0.3(4 - 2.8)^2$

$= 0.1 \cdot 3.24 + 0.3 \cdot 0.64 + 0.3 \cdot 0.04 + 0.3 \cdot 1.44$

$= 0.324 + 0.192 + 0.012 + 0.432 = 0.960.$

$\sigma = \sqrt{\text{Var}(X)} \approx 0.9798.$

15. $E(X) = 0.2 \cdot 10 + 0.1 \cdot 100 + 0.7 \cdot 0 = 2 + 10 = \$12.$

16. $E(X) = 0.3 \cdot 5 + 0.7 \cdot 40 = 1.5 + 28 = \$29.50.$

17. $E(X) = 0.005 \cdot 100,000 + 0.025 \cdot 50,000 + 0.97 \cdot 0 = 500 + 1250 = \$1750.$

Solutions to Exercise Set 11.2

1. a. continuous b. discrete c. continuous d. continuous

   e. continuous f. discrete g. discrete

2. a. $\int_0^2 f(x)dx = \int_0^2 \frac{1}{6}\,dx = \frac{x}{6}\Big]_0^2 = \frac{1}{3}.$

   b. $\int_1^4 f(x)dx = \int_1^3 \frac{1}{6}\,dx + \int_3^4 \frac{1}{4}\,dx = \frac{x}{6}\Big]_1^3 + \frac{x}{4}\Big]_3^4$

   $= \frac{1}{3} + \frac{1}{4} = \frac{7}{12}.$

   c. $\int_0^5 f(x)dx = \int_0^3 \frac{1}{6}\,dx + \int_3^5 \frac{1}{4}\,dx$

   $= \frac{x}{6}\Big]_0^3 + \frac{x}{4}\Big]_3^5 = \frac{1}{2} + \frac{1}{2} = 1.$

3. a. $\int_0^2 f(x)\,dx = \int_0^2 \frac{3}{16}\sqrt{x}\,dx = \frac{3}{16} \cdot \frac{2}{3}x^{3/2}\Big]_0^2 = \frac{1}{8} \cdot 2^{3/2} = \frac{\sqrt{2}}{4}.$

   b. $\int_2^4 f(x)dx = \int_2^4 \frac{3}{16}\sqrt{x}\,dx = \frac{1}{8}x^{3/2}\Big]_2^4$

   $= \frac{1}{8}\left(4^{3/2} - 2^{3/2}\right) = \frac{1}{8}\left(\sqrt{64} - \sqrt{8}\right) = \frac{1}{8}\left(8 - 2\sqrt{2}\right) \approx 0.6464.$

   c. $\int_0^4 f(x)dx = \int_0^4 \frac{3}{16}\sqrt{x}\,dx = \frac{1}{8}x^{3/2}\Big]_0^4 = 1.$

4. a. $\int_0^2 \frac{3}{32}x(4-x)dx = \frac{3}{32}\int_0^2 x(4-x)dx = \frac{3}{32}\left(2x^2 - \frac{1}{3}x^3\right)\Big]_0^2$

   $= \frac{3}{32}\left(8 - \frac{8}{3}\right) = \frac{3}{32} \cdot \frac{16}{3} = \frac{1}{2}.$

b. $\int_3^4 f(x)\,dx = \int_3^4 \frac{3}{32}x(4-x)\,dx = \frac{3}{32}\cdot\left(\int_3^4 (4x-x^2)dx\right)$

$= \frac{3}{32}\cdot\left(2x^2 - \frac{1}{3}x^3\right)\Big]_3^4 = \frac{3}{32}\left[\left(32-\frac{64}{3}\right)-\left(18-\frac{27}{3}\right)\right]$

$= \frac{3}{32}\cdot\left[\frac{32}{3}-\frac{27}{3}\right] = \frac{3}{32}\cdot\frac{5}{3} = \frac{5}{32}.$

c. $\int_0^4 \frac{3}{32}x(4-x)dx = \frac{3}{32}\left(2x^2-\frac{1}{3}x^3\right)\Big]_0^4$

$= \frac{3}{32}\left(32-\frac{64}{3}\right) = \frac{3}{32}\cdot\frac{32}{3} = 1.$

5. a. $\int_0^3 f(x)\,dx = \int_0^3 \frac{x}{9}\,dx = \frac{x^2}{18}\Big]_0^3 = \frac{9}{18} = \frac{1}{2}.$

b. $\int_1^5 f(x)dx = \int_1^3 \frac{x}{9}\,dx + \int_3^5 \left(\frac{2}{3}-\frac{1}{9}x\right)dx$

$= \frac{x^2}{18}\Big]_1^3 + \left(\frac{2}{3}x - \frac{1}{18}x^2\right)\Big]_3^5$

$= \frac{1}{2} - \frac{1}{18} + \frac{10}{3} - \frac{25}{18} - 2 + \frac{1}{2}$

$= -1 + \frac{10}{3} - \frac{26}{18} = \frac{7}{3} - \frac{26}{18} = \frac{16}{18} = \frac{8}{9}.$

c. $\int_2^6 f(x)dx = \int_2^3 \frac{x}{9}\,dx + \int_3^6 \left(\frac{2}{3}-\frac{1}{9}x\right)dx$

$= \frac{x^2}{18}\Big]_2^3 + \left(\frac{2}{3}x - \frac{1}{18}x^2\right)\Big]_3^6$

$= \frac{5}{18} + 2 - \frac{27}{18} = 2 - \frac{22}{18} = \frac{14}{18} = \frac{7}{9}.$

6. $\int_0^1 ax(1-x)dx = 1 \implies a\int_0^1 (x-x^2)dx = 1$

$\implies a\left(\frac{1}{2}-\frac{1}{3}\right) = 1 \implies a = 6.$

7. $\int_0^{\ln 2} a\cdot e^{-x}\,dx = 1 \implies a\left(-e^{-x}\right)\Big]_0^{\ln 2} = 1$

$\implies a\left(-\frac{1}{2}+1\right) = 1 \implies a = 2.$

8. $E(X) = \int_0^5 x\cdot f(x)dx = \int_0^5 \frac{x}{5}\,dx = \frac{x^2}{10}\Big]_0^5 = 2.5.$

9. $E(X) = \int_0^4 x \cdot f(x)dx = \int_0^4 \frac{x^2}{8}\, dx = \left.\frac{x^3}{24}\right]_0^4 = \frac{64}{24} = \frac{8}{3}.$

10. $E(X) = \int_0^4 x \cdot f(x)dx = \int_0^4 x \cdot \frac{3}{32}x(4-x)dx$

$$= \frac{3}{32}\int_0^4 (4x^2 - x^3)dx = \frac{3}{32}\left(\frac{4}{3}x^3 - \frac{1}{4}x^4\right)\Big]_0^4$$

$$= \frac{3}{32}\left(\frac{256}{3} - 64\right) = \frac{3}{32}\left(\frac{64}{3}\right) = 2.$$

11. $\mathrm{Var}(X) = \int_0^5 (x - E(X))^2 f(x)dx$

$$= \int_0^5 (x-2.5)^2 \cdot \frac{1}{5}\, dx = \frac{25}{12}.$$

12. $\mathrm{Var}(X) = \int_0^4 \left(x - \frac{8}{3}\right)^2 \cdot \frac{x}{8}\, dx$

$$= \int_0^4 \left(x^2 - \frac{16}{3}x + \frac{64}{9}\right) \cdot \frac{x}{8}\, dx$$

$$= \frac{1}{8}\left(\frac{x^4}{4} - \frac{16}{9}x^3 + \frac{32}{9}x^2\right)\Bigg]_0^4$$

$$= \frac{1}{8}\left(64 - \frac{1024}{9} + \frac{512}{9}\right) = \frac{1}{8}\left(64 - \frac{512}{9}\right) = \frac{8}{9}.$$

13. $\mathrm{Var}(X) = \int_0^4 (x-2)^2 \cdot \frac{3}{32}x(4-x)dx$

$$= \frac{3}{32}\int_0^4 (x-2)^2(4x - x^2)dx$$

$$= \frac{3}{32}\left(-\frac{x^5}{5} + 2x^4 - \frac{20}{3}x^3 + 8x^2\right)\Bigg]_0^4$$

$$= \frac{3}{32}\left(-\frac{1024}{5} + 512 - \frac{1280}{3} + 128\right)$$

$$= \frac{3}{32}\left(640 - \frac{9472}{15}\right) = \frac{3}{32} \cdot \frac{128}{15} = \frac{4}{5}.$$

14. (a) $Pr\{0 \le X \le 2\} = F(2) - F(0) = \frac{1}{4} - 0 = \frac{1}{4}.$

(b) $Pr\{1 \le X \le 2\} = F(2) - F(1) = \frac{1}{4} - \frac{1}{16} = \frac{3}{16}.$

(c) $f(x) = F'(x) = \frac{x}{8}.$

15. a. $Pr\{0 \le X \le \frac{1}{2}\} = F\left(\frac{1}{2}\right) - F(0) = 3 \cdot \frac{1}{4} - 2 \cdot \frac{1}{8} - 0 = \frac{1}{2}.$

b. $Pr\{\frac{1}{2} \le X \le 1\} = F(1) - F\left(\frac{1}{2}\right) = 1 - \frac{1}{2} = \frac{1}{2}.$

c. $f(x) = F'(x) = 6x - 6x^2$.

16. $F(x) = \int f(x)dx = \int \frac{3}{16}\sqrt{x}\, dx = \frac{1}{8}x^{3/2} + C$.

Since $F(0) = 0$, $C = 0$. $\therefore F(x) = \frac{1}{8}x^{3/2}$.

17. $F(x) = \int f(x)dx = \int \frac{3}{32}x(4-x)dx$

$$= \frac{3}{32}\left(2x^2 - \frac{1}{3}x^3\right) + C = \frac{3}{16}x^2 - \frac{1}{32}x^3 + C.$$

Since $F(0) = 0$, $C = 0$. $\therefore F(x) = \frac{3}{16}x^2 - \frac{1}{32}x^3$.

18. $F(x) = \int f(x)dx = \int \frac{1}{5}\, dx = \frac{x}{5} + C$.

Since $F(0) = 0$, $C = 0$. $\therefore F(x) = \frac{x}{5}$.

19. a. $\int_0^2 \frac{5}{4}\left(\frac{1}{1+x}\right)^2 dx = \frac{5}{4}\left(-\frac{1}{1+x}\right)\Big]_0^2 = \frac{5}{4}\left(-\frac{1}{3}+1\right) = \frac{5}{6}$.

b. $\int_1^4 \frac{5}{4}\left(\frac{1}{1+x}\right)^2 dx = \frac{5}{4}\left(-\frac{1}{1+x}\right)\Big]_1^4 = \frac{3}{8}$.

20. a. $Pr\{0 \le X \le 15\} = \int_0^{15} \frac{\pi}{60}\sin\left(\frac{\pi x}{30}\right) dx$

$$= \frac{\pi}{60}\cdot\frac{30}{\pi}\left[\left(-\cos\frac{\pi x}{30}\right)\right]_0^{15}$$

$$= \frac{1}{2}\left(-\cos\frac{\pi}{2} + \cos 0\right) = \frac{1}{2}.$$

b. $Pr\{15 \le X \le 30\} = \int_{15}^{30} \frac{\pi}{60}\sin\left(\frac{\pi x}{30}\right) dx$

$$= \frac{1}{2}\left(-\cos\frac{\pi}{30}x\right)\Big]_{15}^{30}$$

$$= \frac{1}{2}\left(-\cos\pi + \cos\frac{\pi}{2}\right) = \frac{1}{2}.$$

21. a. $\int_0^{\infty} \frac{k}{(1+t)^2}dt = 1 \implies k\int_0^{\infty}(1+t)^{-2}dt = 1 \implies k\left[-(1+t)^{-1}\right]_0^{\infty} = 1$

$\implies k(0+1) = 1$. $\therefore k = 1$.

b. $Pr\{0 \le X \le 9\} = \int_0^9 \frac{1}{(1+t)^2}\, dt = -(1+t)^{-1}\Big]_0^9 = -\frac{1}{10} + 1 = \frac{9}{10}$.

Solutions to Exercise Set 11.3

1. 0.9772

2. 0.9744

3. $Pr\{1.2 \leq Z \leq 2.8\} = Pr\{Z \leq 2.8\} - Pr\{Z \leq 1.2\} = 0.9974 - 0.8849 = 0.1125$

4. $Pr\{0 \leq Z \leq 1.45\} = Pr\{Z \leq 1.45\} - Pr\{Z \leq 0\} = 0.9265 - 0.5 = 0.4265$

5. $Pr\{-1 \leq Z \leq 2\} = Pr\{Z \leq 2\} - Pr\{Z \leq -1\} = Pr\{Z \leq 2\} - Pr\{Z \geq 1\}$

$$= Pr\{Z \leq 2\} - (1 - Pr\{Z \leq 1\}) = 0.9772 - (1 - 0.8413)$$

$$= 0.9772 - 0.1587 = 0.8185$$

6. $Pr\{-2.35 \leq Z \leq 0\} = Pr\{0 \leq Z \leq 2.35\} = Pr\{Z \leq 2.35\} - Pr\{Z \leq 0\}$

$$= 0.9906 - 0.5 = 0.4906$$

7. $Pr\{-1.6 \leq Z \leq 1.55\} = Pr\{Z \leq 1.55\} - Pr\{Z \leq -1.6\}$

$$= Pr\{Z \leq 1.55\} - Pr\{Z \geq 1.6\}$$

$$= Pr\{Z \leq 1.55\} - (1 - Pr\{Z \leq 1.6\})$$

$$= 0.9394 - (1 - 0.9452) = 0.9394 - 0.0548 = 0.8846.$$

8. $Pr\{-1 \leq Z \leq -0.25\} = Pr\{0.25 \leq Z \leq 1\} = Pr\{Z \leq 1\} - Pr\{Z \leq 0.25\}$

$$= 0.8413 - 0.5987 = 0.2426.$$

9. $Z = \frac{X-\mu}{\sigma} = \frac{X-4}{2}.$

When $X = 7$, $\quad Z = \frac{7-4}{2} = \frac{3}{2}.$

$Pr\{X \leq 7\} = Pr\{Z \leq 1.5\} = 0.9332.$

10. $Z = \frac{X-\mu}{\sigma} = \frac{X-20}{20}$.

When $X = 0$, $Z = -1$.

When $X = 35$, $Z = \frac{35-20}{20} = \frac{15}{20} = \frac{3}{4}$.

$$Pr\{0 \leq X \leq 35\} = \{-1 \leq Z \leq 0.75\}$$

$$= Pr\{Z \leq 0.75\} - Pr\{Z \geq 1\}$$

$$= Pr\{Z \leq 0.75\} - (1 - Pr\{Z \leq 1\})$$

$$= 0.7734 - (1 - 0.8413) = 0.7734 - 0.1587 = 0.6147.$$

11. $A = \frac{1-2}{2} = -0.5$, $B = \frac{7-2}{2} = 2.5$.

So $Pr\{1 \leq X \leq 7\} = Pr\{-0.5 \leq Z \leq 2.5\} = Pr\{Z \leq 2.5\} - Pr\{Z \geq 0.5\}$

$$= 0.9938 - (1 - 0.6915) = 0.9938 - 0.3085 = 0.6853.$$

12. $A = \frac{-3-1}{2} = -2$, $B = \frac{5-1}{2} = 2$.

So $Pr\{-3 \leq X \leq 5\} = Pr\{-2 \leq Z \leq 2\} = Pr\{Z \leq 2\} - Pr\{Z \geq 2\}$

$$= 0.9772 - (1 - 0.9772) = 0.9772 - 0.0228 = 0.9544.$$

13. $A = \frac{35-50}{10} = -1.5$, $B = \frac{65-50}{10} = 1.5$.

So $Pr\{35 \leq X \leq 65\} = Pr\{-1.5 \leq Z \leq 1.5\}$

$$= Pr\{Z \leq 1.5\} - (1 - Pr\{Z \leq 1.5\})$$

$$= 0.9332 - (1 - 0.9332) = 0.9332 - 0.0668 = 0.8664.$$

14. $A = \frac{2-1}{1} = 1$, $B = \frac{3-1}{1} = 2$.

So $Pr\{2 \leq X \leq 3\} = Pr\{1 \leq Z \leq 2\} = Pr\{Z \leq 2\} - Pr\{Z \leq 1\}$

$$= 0.9772 - 0.8413 = 0.1359.$$

15. $A = \frac{-5-(-1)}{4} = -1, \quad B = \frac{1-(-1)}{4} = 0.5.$

So $Pr\{-5 \leq X \leq 1\} = Pr\{-1 \leq Z \leq 0.5\}$

$= Pr\{Z \leq 0.5\} - (1 - Pr\{Z \leq 1\})$

$= 0.6915 - (1 - 0.8413) = 0.6915 - 0.1587 = 0.5328.$

16. $A = \frac{17-22}{5} = -1, \quad B = \frac{25-22}{5} = 0.6.$

$Pr\{17 \leq X \leq 25\} = Pr\{-1 \leq Z \leq 0.6\} = Pr\{Z \leq 0.6\} - (1 - Pr\{Z \leq 1\})$

$= 0.7257 - (1 - 0.8413) = 0.7257 - 0.1587 = 0.5670.$

17. a. $A = \frac{72-70}{2} = 1$, so $Pr\{Z \geq 1\} = 1 - Pr\{Z \leq 1\} = 1 - 0.8413 = 0.1587.$

b. $A = \frac{66-70}{2} = -2, \quad B = \frac{74-70}{2} = 2,$

so $Pr\{-2 \leq Z \leq 2\} = 0.9772 - (1 - 0.9772) = 0.9544$

18. a. $A = \frac{16{,}000-18{,}000}{2500} = \frac{-2000}{2500} = -0.8, \quad B = \frac{20{,}000-18{,}000}{2500} = 0.8.$

So $Pr\{16,000 \leq X \leq 20,000\} = Pr\{-0.8 \leq Z \leq 0.8\}$

$= 0.7881 - (1 - 0.7881) = 0.5762.$

b. $A = \frac{24{,}000-18{,}000}{2500} = \frac{6{,}000}{2{,}500} = 2.4.$

So $Pr\{X \geq 24,000\} = Pr\{Z \geq 2.4\} = 1 - Pr\{Z \leq 2.4\}$

$= 1 - 0.9918 = 0.0082.$

19. $A = \frac{15-10}{3} = \frac{5}{3} = 1.67.$

$Pr\{Z \geq 1.67\} = 1 - Pr\{Z \leq 1.67\} = 1 - 0.9525 = 0.0475.$

20. $B = \frac{115-20}{10} = 9.5.$ $\quad Pr\{X \leq 115\} = Pr\{Z \leq 9.5\} \approx 1$ but is not in the Table 4.

21. $A = \frac{5-6}{1.5} = -0.67, \quad B = \frac{7-6}{1.5} = 0.67.$

$$Pr\{-0.67 \le Z \le 0.67\} = 0.7486 - (1 - 0.7486) = 0.4972.$$

22. $A = \frac{1020-1080}{40} = -1.5. \quad Pr\{Z \ge -1.5\} = Pr\{Z \le 1.5\} = 0.9332.$

23. $A = \frac{42000-30000}{8000} = \frac{12000}{8000} = 1.5.$

$$Pr\{Z > 1.5\} = 1 - Pr\{Z \le 1.5\} = 1 - 0.9332 = 0.0668.$$

24. $A = \frac{3.24-3}{0.1} = \frac{0.24}{0.1} = 2.4.$

$$Pr\{Z \ge 2.4\} = 1 - Pr\{Z \le 2.4\} = 1 - 0.9918 = 0.0082.$$

25. $$\begin{aligned} E(Z) &= \int_{-\infty}^{\infty} \frac{1}{\sqrt{2\pi}} z e^{-\frac{1}{2}z^2}\, dz \\ &= \lim_{t\to-\infty} \int_t^0 \frac{1}{\sqrt{2\pi}} z e^{-\frac{1}{2}z^2}\, dz + \lim_{t\to\infty} \int_0^t \frac{1}{\sqrt{2\pi}} z e^{-\frac{1}{2}z^2}\, dz \\ &= \lim_{t\to-\infty} \int_{\frac{1}{2}t^2}^0 \frac{1}{\sqrt{2\pi}} e^{-u}\, du + \lim_{t\to\infty} \int_0^{\frac{1}{2}t^2} \frac{1}{\sqrt{2\pi}} e^{-u}\, du \\ &= \lim_{t\to-\infty} \left(-\frac{1}{\sqrt{2\pi}} e^{-u}\right)\Big]_{\frac{1}{2}t^2}^0 + \lim_{t\to\infty} \left(-\frac{1}{\sqrt{2\pi}} e^{-u}\right)\Big]_0^{\frac{1}{2}t^2} \\ &= \lim_{t\to-\infty} \left(-\frac{1}{\sqrt{2\pi}} + \frac{1}{\sqrt{2\pi}} e^{-\frac{1}{2}t^2}\right) + \lim_{t\to\infty} \left(-\frac{1}{\sqrt{2\pi}} e^{-\frac{1}{2}t^2} + \frac{1}{\sqrt{2\pi}}\right) \\ &= -\frac{1}{\sqrt{2\pi}} + \frac{1}{\sqrt{2\pi}} = 0. \end{aligned}$$

26. a. It is sufficient to show that $f(\mu - a) = f(\mu + a)$ for all a.

$$f(\mu - a) = \frac{1}{\sqrt{2\pi}\sigma} e^{-\frac{1}{2}\left(\frac{-a}{\sigma}\right)^2} = \frac{1}{\sqrt{2\pi}\sigma} e^{-\frac{1}{2}\left(\frac{a}{\sigma}\right)^2} = f(\mu + a).$$

$\therefore f(x)$ is symmetric with respect to $x = \mu$.

b. $f'(x) = \frac{1}{\sqrt{2\pi}\,\sigma} e^{-\frac{1}{2}\left(\frac{x-\mu}{\sigma}\right)^2} \cdot \left(-\frac{x-\mu}{\sigma^2}\right)$ and

$$\begin{aligned} f''(x) &= \frac{1}{\sqrt{2\pi}\,\sigma} \left(-\frac{1}{\sigma^2} e^{-\frac{1}{2}\left(\frac{x-\mu}{\sigma}\right)^2} + \left(\frac{x-\mu}{\sigma^2}\right)^2 e^{-\frac{1}{2}\left(\frac{x-\mu}{\sigma}\right)^2}\right) \\ &= \frac{1}{\sqrt{2\pi}\,\sigma} e^{-\frac{1}{2}\left(\frac{x-\mu}{\sigma}\right)^2} \left(-\frac{1}{\sigma^2} + \left(\frac{x-\mu}{\sigma^2}\right)^2\right) \\ &= \frac{1}{\sqrt{2\pi}\,\sigma} e^{-\frac{1}{2}\left(\frac{x-\mu}{\sigma}\right)^2} \left(\frac{x-\mu}{\sigma^2} + \frac{1}{\sigma}\right)\left(\frac{x-\mu}{\sigma^2} - \frac{1}{\sigma}\right) \end{aligned}$$

$$= \frac{1}{\sqrt{2\pi}\,\sigma^5} e^{-\frac{1}{2}\left(\frac{x-\mu}{\sigma}\right)^2} (x-\mu+\sigma)(x-\mu-\sigma) = 0$$

for $x = \mu - \sigma$ and $\mu + \sigma$.

Interval	Sign of $f''(x)$	$\bigcup$ or $\bigcap$
$x < \mu - \sigma$	$+$	$\bigcup$
$\mu - \sigma < x < \mu + \sigma$	$-$	$\bigcap$
$x > \mu + \sigma$	$+$	$\bigcup$

$\therefore f(x)$ has inflection points $(\mu - \sigma,\ f(\mu - \sigma))$ and

$(\mu + \sigma,\ f(\mu + \sigma))$.

Solutions to Exercise Set 11.4

1. $\int_0^6 \alpha e^{-\alpha x}\,dx = (-e^{-\alpha x})]_0^6 = -e^{-(0.05)(6)} + e^0 = -e^{-0.3} + 1$

   $= 0.2592.$

2. $\int_0^{15} \alpha e^{-\alpha x}\,dx = (-e^{-\alpha x})]_0^{15} = -e^{-(0.10)(15)} + e^0 = -e^{-1.5} + 1$

   $= 0.7769.$

3. $\int_2^6 \alpha e^{-\alpha x}\,dx = (-e^{-ax})]_2^6 = -e^{-(0.25)(6)} + e^{-(0.25)(2)}$

   $= -e^{-1.5} + e^{-0.5} = 0.3834.$

4. $\int_1^6 \alpha e^{-\alpha x}\,dx = (-e^{-\alpha x})]_1^6 = -e^{-(0.10)(6)} + e^{-(0.10)(1)}$

   $= -e^{-0.6} + e^{-0.1} = 0.3560.$

5. $\int_2^5 \alpha e^{-\alpha x}\,dx = (-e^{-\alpha x})]_2^5 = -e^{-(0.50)(5)} + e^{-(0.50)(2)}$

   $= -e^{-2.5} + e^{-1} = 0.2858.$

6. $Pr\{0 \le X \le 3\} = 1 - e^{-0.2\cdot 3} = 1 - e^{-0.6} = 1 - 0.5488 = 0.4512.$

7. $Pr\{0 \leq X \leq 6\} = 1 - e^{-0.1 \cdot 6} = 1 - e^{-0.6} = 0.4512$

8. $Pr\{2 \leq X \leq 5\} = Pr\{X \leq 5\} - Pr\{X \leq 2\} = (1 - e^{-0.4 \cdot 5}) - (1 - e^{-0.4 \cdot 2})$

$$= e^{-0.8} - e^{-2.0} = 0.4493 - 0.1353 = 0.3140.$$

9. $Pr\{5 \leq X \leq 20\} = Pr\{X \leq 20\} - Pr\{X \leq 5\} = e^{-0.02 \cdot 5} - e^{-0.02 \cdot 20}$

$$= e^{-0.1} - e^{-0.4}$$

$$= 0.9048 - 0.6703 = 0.2345.$$

10. $Pr\{10 \leq X \leq 50\} = Pr\{X \leq 50\} - Pr\{X \leq 10\} = e^{-0.01 \cdot 10} - e^{-0.01 \cdot 50}$

$$= e^{-0.1} - e^{-0.5}$$

$$= 0.9048 - 0.6065 = 0.2983.$$

11. $\frac{1}{\alpha} = 20 \implies \alpha = \frac{1}{20} = 0.05.$

$Pr\{5 \leq X \leq 30\} = Pr\{X \leq 30\} - Pr\{X \leq 5\} = e^{-0.05 \cdot 5} - e^{-0.05 \cdot 30}$

$$= e^{-0.25} - e^{-1.5} = 0.7788 - 0.2231 = 0.5557.$$

12. $\frac{1}{\alpha} = 10 \implies \alpha = 0.1.$

$Pr\{5 \leq X \leq 15\} = e^{-0.1 \cdot 5} - e^{-0.1 \cdot 15} = e^{-0.5} - e^{-1.5}$

$$= 0.6065 - 0.2231 = 0.3834.$$

13. $\frac{1}{\alpha} = 1 \implies \alpha = 1.$

14. $\frac{1}{\alpha} = 40,000 \implies \alpha = \frac{1}{40,000}.$

$Pr\{X \geq 50,000\} = 1 - Pr\{X \leq 50,000\} = 1 - \left(1 - e^{-5/4}\right) = e^{-5/4} = 0.2865.$

15. $\frac{1}{\alpha} = 4 \implies \alpha = \frac{1}{4}.$

$Pr\{X \leq 2\} = 1 - e^{-\frac{1}{4} \cdot 2} = 1 - e^{-0.5} = 1 - 0.6065 = 0.3935.$

16. $\frac{1}{\alpha} = 10 \implies \alpha = 0.1.$

$$Pr\{X \geq 20\} = 1 - Pr\{X \leq 20\} = 1 - \left(1 - e^{-0.1 \cdot 20}\right) = e^{-2} = 0.1353.$$

17. $E(X) = \frac{1}{\alpha}.$

$$Pr\{X > E(X)\} = Pr\{X > \tfrac{1}{\alpha}\} = 1 - Pr\{X \leq \tfrac{1}{\alpha}\}$$
$$= 1 - \left(1 - e^{-\alpha \cdot \frac{1}{\alpha}}\right) = e^{-1} = 0.3679.$$

The answer is independent of α since $Pr\{X > E(X)\} = e^{-1}$ for any α.

18. $Pr\{0 \leq X \leq 5\} = 1 - e^{-\alpha \cdot 5} = 0.6320.$

$$\therefore e^{-5\alpha} = 0.3680 \implies -5\alpha = \ln(0.3680)$$
$$\implies \alpha = -\tfrac{\ln(0.3680)}{5} = 0.1999.$$

19. $\text{Var}(X) = \int\limits_0^\infty (x - E(X))^2 \cdot \alpha \cdot e^{-\alpha x}\, dx = \int\limits_0^\infty \left(x - \frac{1}{\alpha}\right)^2 \cdot \alpha \cdot e^{-\alpha x}\, dx$

$$= \int\limits_0^\infty \left(x^2 - \tfrac{2}{\alpha}x + \tfrac{1}{\alpha^2}\right) \cdot \alpha \cdot e^{-\alpha x}\, dx$$
$$= \alpha \int\limits_0^\infty x^2 e^{-\alpha x}\, dx - 2\int\limits_0^\infty x e^{-\alpha x}\, dx + \int\limits_0^\infty \tfrac{1}{\alpha} e^{-\alpha x}\, dx$$
$$= \alpha \lim_{t \to \infty} \left(-\tfrac{x^2}{\alpha} e^{-\alpha x} - \tfrac{2x}{\alpha^2} e^{-\alpha x} - \tfrac{2}{\alpha^3} e^{-\alpha x}\right)\Big]_0^t$$
$$-2 \lim_{t \to \infty} \left(-\tfrac{x}{\alpha} e^{-\alpha x} - \tfrac{1}{\alpha^2} e^{-\alpha x}\right)\Big]_0^t + \lim_{t \to \infty} \left(-\tfrac{1}{\alpha^2} e^{-\alpha x}\right)\Big]_0^t$$
$$= \tfrac{2}{\alpha^2} - 2 \cdot \tfrac{1}{\alpha^2} + \tfrac{1}{\alpha^2} = \tfrac{1}{\alpha^2}.$$

20. a. $f(x) = e^{\alpha x} \implies f^{(n)}(x) = \alpha^n e^{\alpha x} \implies f^{(n)}(0) = \alpha^n.$

$$\therefore f(x) = \sum_{n=0}^{\infty} \tfrac{f^{(n)}(0)}{n!} x^n = \sum_{n=0}^{\infty} \tfrac{\alpha^n}{n!} x^n = 1 + \alpha x + \tfrac{\alpha^2 x^2}{2} + \cdots + \tfrac{\alpha^n x^n}{n!} + \ldots$$

b. For $\alpha > 0$ and $x > 0$, each term of the Taylor series of $f(x)$ is positive.

Thus $f(x) = e^{\alpha x} > \frac{\alpha^2 x^2}{2}.$

c. $x \cdot e^{-\alpha x} = \frac{x}{e^{\alpha x}} < \frac{x}{\left(\frac{\alpha^2 x^2}{2}\right)} = \frac{2}{\alpha^2 x}.$

d. $\lim_{x \to \infty} x \cdot e^{-\alpha x} \leq \lim_{x \to \infty} \frac{2}{\alpha^2 x} = 0$ for $\alpha > 0$.

Since $\lim_{x \to \infty} x \cdot e^{-\alpha x} \geq 0$, $\lim_{x \to \infty} x \cdot e^{-\alpha x}$ should be zero.

Solutions to the Review Exercises for Chapter 11

1. a. $0.1 + 0.1 + 0.3 = 0.5$

   b. $0.1 + 0.2 + 0.3 = 0.6$

   c. $0.1 + 0.1 + 0.2 = 0.4$

2. $E(X) = 1 \cdot 0.1 + 2 \cdot 0.1 + 3 \cdot 0.2 + 4 \cdot 0.3 + 5 \cdot 0.3 = 0.1 + 0.2 + 0.6 + 1.2 + 1.5 = 3.6$

3. $\text{Var}(X) = E\left([X - E(X)]^2\right)$

   $= (1-3.6)^2 \cdot 0.1 + (2-3.6)^2 \cdot 0.1 + (3-3.6)^2 \cdot 0.2 + (4-3.6)^2 \cdot 0.3 + (5-3.6)^2 \cdot 0.3$

   $= 0.676 + 0.256 + 0.072 + 0.048 + 0.588 = 1.640.$

4. The number of red cards numbered less than 8 is 12.

   Therefore the probability of drawing at random a red card numbered less than 8 is $\frac{12}{52} = \frac{3}{13}$.

5.

x_i	\$490	−\$10
p_i	$\frac{1}{2000}$	$\frac{1999}{2000}$

$E(X) = 490 \cdot \frac{1}{2000} - \frac{1999}{2000} \cdot 10 = \frac{-19500}{2000} = -\$9.75.$

6. a. $\int_0^2 f(x)dx = \int_0^2 \frac{x}{6}\, dx = \left.\frac{x^2}{12}\right]_0^2 = \frac{1}{3}$.

   b. $\int_1^2 f(x)dx = \int_1^2 \frac{x}{6}\, dx = \left.\frac{x^2}{12}\right]_1^2 = \frac{4}{12} - \frac{1}{12} = \frac{3}{12} = \frac{1}{4}$.

c. $\int_1^4 f(x)dx = \int_1^2 \frac{x}{6}\, dx + \int_2^4 \frac{1}{3}\, dx$

$= \frac{x^2}{12}\Big]_1^2 + \frac{1}{3}\cdot 2 = \frac{1}{4} + \frac{2}{3} = \frac{11}{12}.$

7. $\int_0^5 kx\, dx = 1 \implies \frac{kx^2}{2}\Big]_0^5 = 1 \implies \frac{25}{2}k = 1 \therefore k = \frac{2}{25}.$

8. a. $\int_0^2 \frac{2}{25}x\, dx = \frac{2}{25}\,\frac{x^2}{2}\Big]_0^2 = \frac{4}{25}.$

b. $\int_1^4 \frac{2}{25}x\, dx = \frac{2}{25}\cdot\frac{x^2}{2}\Big]_1^4 = \frac{2}{25}\left(\frac{16}{2} - \frac{1}{2}\right) = \frac{2}{25}\cdot\frac{15}{2} = \frac{3}{5}.$

9. a. $\int_0^2 k(4 - x^2)dx = 1 \implies k\left(4x - \frac{1}{3}x^3\right)\Big]_0^2 = 1$

$\implies k\left(8 - \frac{8}{3}\right) = 1 \therefore k = \frac{3}{16}.$

b. $\int_0^1 \frac{3}{16}(4 - x^2)dx = \frac{3}{16}\left(4x - \frac{1}{3}x^3\right)\Big]_0^1 = \frac{3}{16}\left(\frac{11}{3}\right) = \frac{11}{16}.$

c. $\int_1^2 \frac{3}{16}(4 - x^2)dx = \frac{3}{16}\left(4x - \frac{1}{3}x^3\right)\Big]_1^2$

$= \frac{3}{16}\left(4 - \frac{7}{3}\right) = \frac{3}{16}\cdot\frac{5}{3} = \frac{5}{16}.$

10. $E(X) = \int_0^5 x\cdot f(x)dx = \int_0^5 x\cdot\frac{2}{25}\cdot x\, dx = \frac{2}{25}\cdot\frac{x^3}{3}\Big]_0^5$

$= \frac{2}{25}\cdot\frac{125}{3} = \frac{10}{3}.$

11. $\mathrm{Var}(X) = \int_0^5 \left(x - \frac{10}{3}\right)^2\cdot\frac{2}{25}\cdot x\, dx$

$= \frac{2}{25}\int_0^5 \left(x^3 - \frac{20}{3}x^2 + \frac{100}{9}x\right)dx$

$= \frac{2}{25}\left(\frac{x^4}{4} - \frac{20}{9}x^3 + \frac{50}{9}x^2\right)\Big]_0^5$

$= \frac{2}{25}\left(\frac{625}{4} - \frac{20\cdot 125}{9} + \frac{50\cdot 25}{9}\right) = \frac{25}{2} - \frac{200}{9} + \frac{100}{9}$

$= \frac{25}{2} - \frac{100}{9} = \frac{225-200}{18} = \frac{25}{18}.$

12. $E(X) = \int_0^2 \frac{3}{16}(4 - x^2)\cdot x\, dx = \frac{3}{16}\int_0^2 (4x - x^3)dx$

$= \frac{3}{16}\left(2x^2 - \frac{x^4}{4}\right)\Big]_0^2 = \frac{3}{16}(8 - 4) = \frac{3}{4}.$

13. $Pr\{Z \le -0.5\} = Pr\{Z \ge 0.5\} = 1 - Pr\{Z \le 0.5\} = 1 - 0.6915 = 0.3085.$

14. $Pr\{-2.1 \le Z \le 2.0\} = Pr\{Z \le 2.0\} - Pr\{Z \le -2.1\}$

$$= Pr\{Z \le 2.0\} - (1 - Pr\{Z \le 2.1\})$$

$$= 0.9772 - 1 + 0.9778 = 0.995.$$

15. $Pr\{Z \ge -0.6\} = Pr\{Z \le 0.6\} = 0.7257.$

16. $Pr\{Z \le 1.7\} = 0.9554.$

17. $A = \frac{7-10}{2} = -1.5.$

$Pr\{X \ge 7\} = Pr\{Z \ge -1.5\} = Pr\{Z \le 1.5\} = 0.9332.$

18. $A = \frac{6.5-8.6}{1.4} = \frac{-2.1}{1.4} = -1.5,$

$B = \frac{10-8.6}{1.4} = \frac{1.4}{1.4} = 1.$

$Pr\{6.5 \le X \le 10\} = Pr\{-1.5 \le Z \le 1\} = Pr\{Z \le 1\} - Pr\{Z \le -1.5\}$

$$= Pr\{Z \le 1\} - 1 + Pr\{Z \le 1.5\} = 0.8413 - 1 + 0.9332$$

$$= 0.7745.$$

19. $A = \frac{-11-2}{6} = -2.17, \qquad B = \frac{13-2}{6} = 1.83.$

$Pr\{-11 \le X \le 13\} = Pr\{-2.17 \le Z \le 1.83\}$

$$= Pr\{Z \le 1.83\} - Pr\{Z \le -2.17\}$$

$$= Pr\{Z \le 1.83\} - (1 - Pr\{Z \le 2.17\})$$

$$= 0.9664 - (1 - 0.9850) = 0.9514.$$

20. $A = \frac{136-140}{8} = -0.5, \qquad B = \frac{152-140}{8} = 1.5.$

$$Pr\{136 \leq X \leq 152\} = Pr\{-0.5 \leq Z \leq 1.5\} = Pr\{Z \leq 1.5\} - Pr\{Z \leq -0.5\}$$
$$= Pr\{Z \leq 1.5\} - (1 - Pr\{Z \leq 0.5\})$$
$$= 0.9332 - (1 - 0.6915) = 0.6247.$$

21. a. $A = \frac{110-100}{10} = 1.$

$$Pr\{X \geq 110\} = Pr\{Z \geq 1\} = 1 - Pr\{Z \leq 1\} = 1 - 0.8413 = 0.1587.$$

b. $A = \frac{120-100}{10} = 2.$

$$Pr\{X \geq 120\} = Pr\{Z \geq 2\} = 1 - Pr\{Z \leq 2\} = 1 - 0.9772 = 0.0228.$$

22. $\mu = 15$ and $\sigma = 4$.

$$\Longrightarrow Pr\{X \geq 9\} = Pr\{Z \geq \tfrac{-6}{4}\} = Pr\{Z \geq -1.5\} = Pr\{Z \leq 1.5\} = 0.9332.$$

23. $\mu = 16$ and $\sigma = 3$.

$$\Longrightarrow Pr\{X < 10\} = Pr\{Z < \tfrac{10-16}{3}\} = Pr\{Z < -2\}$$
$$= 1 - Pr\{Z < 2\} = 1 - 0.9772 = 0.0228.$$

24. $Pr\{4 \leq X \leq 8\} = Pr\{\frac{-2}{\sigma} \leq Z \leq \frac{2}{\sigma}\}$

$$= Pr\{Z \leq \tfrac{2}{\sigma}\} - 1 + Pr\{Z \leq \tfrac{2}{\sigma}\}$$
$$= 2 \cdot Pr\{Z \leq \tfrac{2}{\sigma}\} - 1 = 0.8664.$$

$$\Longrightarrow Pr\{Z \leq \tfrac{2}{\sigma}\} = 0.9332 \Longrightarrow \tfrac{2}{\sigma} = 1.5 \therefore \sigma = \tfrac{4}{3}.$$

25. $\int_4^\infty 0.2e^{-0.2x}dx = 0.2\left(\frac{e^{-0.2x}}{-0.2}\right)\Big]_4^\infty = e^{-0.8} = 0.4493.$

26. $\int_0^{40} 0.2e^{-0.2x}dx = -e^{-0.2x}\Big]_0^{40} = -e^{-(0.2)(40)} + 1.$

$$= -e^{-8} + 1 = 0.9997.$$

27. $\int_{12}^{\infty} 0.1e^{-0.1x}dx = \left(-e^{-0.1x}\right)\Big]_{12}^{\infty} = e^{-1.2} = 0.3012.$

28. $\int_{10}^{15} 0.05e^{-0.05x}dx = -e^{-0.05x}\Big]_{10}^{15} = e^{-(0.05)(15)} + e^{-(0.05)(10)}$

$$= -e^{-0.75} + e^{-0.5} = 0.1342.$$

29. Since $\alpha = \frac{1}{4}$,

$$Pr\{2 \leq X \leq 6\} = \int_{2}^{6} \frac{1}{4}e^{-\frac{1}{4}x}dx$$

$$= e^{-\frac{1}{2}} - e^{-\frac{3}{2}} = 0.6065 - 0.2231$$

$$= 0.3834.$$

30. $\int_{0}^{12} \frac{1}{20}e^{-\frac{1}{20}x}dx = \left(-e^{-\frac{1}{20}x}\right)\Big]_{0}^{12} = 1 - e^{-\frac{12}{20}} = 0.4512.$

31. $\int_{0}^{5} \alpha e^{-\alpha x}dx = 1 - e^{-5\alpha} = 0.221$

$\implies e^{-5\alpha} = 0.779 \implies -5\alpha = \ln 0.779.$

$\implies \alpha = -\frac{\ln 0.779}{5} \quad \therefore\ E(X) = \frac{1}{\alpha} = -\frac{5}{\ln 0.779} = 20.020482.$

32. $\int_{0}^{0.5} 0.5e^{-0.5x}dx = 1 - e^{-0.25} = 1 - 0.7788 = 0.2212.$